Microreaction Technology: Industrial Prospects

Springer

Berlin
Heidelberg
New York
Barcelona
Hong Kong
London
Milan
Paris
Singapore
Tokyo

W. Ehrfeld (Ed.)

Microreaction Technology: Industrial Prospects

IMRET 3:
Proceedings of the Third International Conference on Microreaction Technology

With 481 Figures

Springer

Professor Dr. Wolfgang Ehrfeld
Institut für Mikrotechnik Mainz GmbH
Carl-Zeiss-Straße 18-20
55129 Mainz
Germany

ISBN-13: 978-3-642-64104-6

CIP data applied for

Die Deutsche Bibliothek - CIP-Einheitsaufnahme
Microreaction technology: industrial prospects ; proceedings of the Third International Conference on Microreaction Technoloy / IMRET 3. W. Ehrfeld (ed.). - Berlin ; Heidelberg ; New York ; Barcelona ; Hong Kong ; London ; Milan ; Paris ; Singapore ; Tokyo : Springer, 2000
ISBN-13: 978-3-642-64104-6 e-ISBN-13: 978-3-642-59738-1
DOI: 10.1007/978-3-642-59738-1

Springer-Verlag is a company in the BertelsmannSpringer publishing group

Softcover reprint of the hardcover 1st edition 2000

Typesetting: Camera ready by authors
Cover: Medio, Berlin;
Printed on acid-free paper SPIN: 10751815 62/3020 hu - 5 4 3 2 1 0

Preface

Since the First International Conference on Microreaction Technology, held in Frankfurt / M., Germany in spring 1997, an extremely rapid development in fabrication and use of microreactors for a large variety of processes took place. Within only a few years time, the concepts of microreaction technology have found worldwide attention both in industry and basic R&D. Various applications in process engineering, screening purposes and basic research have been identified and investigated thoroughly. In many cases, microreaction systems already have left the state of lab-scale R&D, finding their way to industrial processing and commercialization.

Consequently, main focus of this years IMRET 3, again held in Frankfurt / M. on April 18 – 21, 1999, turned out to be on industrial prospects of this new technology and its neighboring fields. This symposium, like its 1998 predecessor, held in New Orleans, LA, was organized by Battelle Memorial Institute, the American Institute of Chemical Engineers (AIChE), the Society for Chemical Engineering and Biotechnology (DECHEMA) and the Institute of Microtechnology Mainz (IMM). The conference, featuring 84 contributions, was a successful forum for discussion of recent developments in design and fabrication of microreactors as well as applications in chemical production, combinatorial chemistry, catalysis, biotechnology and power generation.

The lectures and discussions once more showed a broad range of opinions regarding microreaction technology. These ranged from general scepticism concerning applicability of microreactors to recommendation of their use as tool for process optimization and convincing statements with respect to their utilization in industry. However, the first examples for successful industrial exploitation of microreaction systems seem to have a positive effect on the dissemination of microreactor concepts. These pioneering activities indicate the beginning of a new era in chemistry and reaction engineering characterized by increasing flexibility in production and R&D.

The book at hand comprises the full papers of the contributions to IMRET 3, giving a comprehensive overview on the present state-of-the-art in microreaction technology. We thank all contributors to this conference and hope, this book will prove to be a valuable and useful guide to scientists and researchers in this exciting new field in reaction engineering, chemistry and biotechnology.

September 1999

G. Kreysa (DECHEMA) W. Ehrfeld (IMM)

Table of Contents

Part 3
Microreactors in Combinatorial Chemistry

Lecture Session

Part 4
Microreactors in Production Processes

Lecture Session

Part 6
Microreactors for Energy Generation and Storage

Lecture Session

Poster Session

Part 7
Application of Microreactors in Drug Development and Biotechnology

Lecture Session

Author Index

Part 1

Opening Lectures

Process Intensification Through Miniaturization of Chemical and Thermal Systems in the 21st Century

Robert S. Wegeng, M. Kevin Drost, David L. Brenchley

Pacific Northwest National Laboratory, PO Box 999, Richland, WA 99352 USA

Abstract

The 21st century holds great promise for development of Micro Chemical and Thermal Systems (MICRO-CATS™). The quest for miniaturization will lead to greater process intensification. Miniaturization greatly reduces the resistances to heat and mass transfer. When heat and mass transfer rates are increased by orders of magnitude, revolutionary changes occur in technology.

Miniaturization of electronics over the past few decades has transformed the way we live. The invention of the transistor meant that small, lightweight portable radios could be carried anywhere. Not long ago the idea of chips was revolutionary; today chips make hand-held computers, cell phones, and many other hand-held electrical devices possible. These are examples of distributed technology made possible by miniaturization.

Miniaturization of chemical and thermal systems will take us on a similar technological journey. Already demands are driving the development of MICRO-CATS. The quest for miniature systems is leading to man-portable cooling, automotive fuel processing, and in situ resource utilization for space exploration. The 21st century will see great progress in the development and use of engineered nanosystems. In electronics, the concept of "molecular electronics" is seen as a way to further increase the power. The same will occur in MICRO-CATS. We envision using engineered nanosystems to create more powerful microscale devices with highly functional surfaces that incorporate enzymes and chemical catalysts. These surfaces will make the device or system more efficient and reliable; they will be able to repair, renew, or replace themselves; eliminate or heal themselves from surface fouling; and carry out step-by-step chemical reactions. Molecular coatings will create a molecular assembly line. By incorporating enzymes, reactions that are conventionally carried out at high temperatures and pressures can be achieved at ambient conditions. Finally, in the 21st century, the missions for outer space will create opportunities for MICRO-CATS to be major parts of the successful exploration of space. The future for MICRO-CATS is as wide open as outer space itself!

Keywords: miniaturization, microscale systems, reactors, heat pumps

1 Process Intensification

Micro Chemical and Thermal Systems (MICRO-CATS) use miniaturization to achieve process intensification. Process intensification is a measure of the amount of productivity that can be accomplished per unit volume. This increase in process intensification occurs because miniaturization significantly reduces the resistances to heat and mass transfer. Imagine heat and mass transfer rates increased by factors of 10 to 100 or more. Furthermore, when many components and functions are combined and integrated into one miniature system, good things happen. In fact, things happen that are almost too good to be true! That is where the age of miniaturization in MICRO-CATS is heading in the 21st century. Process intensification opens the way for revolutionary changes. Many things we can think of today and some things we haven't yet imagined will happen in the 21st century. For instance,

- Today, we adjust the room thermostat. Tomorrow, our clothing will be manufactured with microsystems that have the capability to automatically heat and cool us.
- Today, internal combustion engines move us around. Tomorrow, microsystems will allow different fuels and power systems to be used in our vehicles.
- Today, we launch almost everything that is needed on a space mission. Tomorrow, we "will live off the land" as we explore the universe. Microsystems will harvest raw materials and process them into what we need.
- Today, we are able to transplant some organs. Tomorrow, we will implant artificial organs composed of microsystems.

As this conference on process miniaturization demonstrates, work is under way at many research institutions and industries to fabricate microcomponents and systems that perform the same standard unit operations that are present in large chemical processing plants. Although miniaturization is in its early stages, it is driven by considerable incentive and the opportunity to revolutionize heat exchangers, heat pumps, reactors, gas absorbers, solvent extractors, fuel processors, pumps, valves, compressors, and other similar components and systems. In the foreseeable future, we anticipate the assembly of compact, chemical processing and energy conversion systems that range from smaller than a cubic centimeter to several cubic meters in size [1, 2, 3, 4, 5].

The quest for miniaturization is marching forward. The changes that have occurred in electronics in the 20th century enable us to envision what to expect from micro chemical and thermal systems in the 21st century. The advances in microelectronics have transformed computers, communication systems, and appliances, to name a few. The transistor is now about 50 years old. This invention, along with other innovations, provided for process intensification. The transistor radio changed things because it was small and light enough to carry anywhere. This created possibilities we didn't have before.

Miniaturization of components on computer chips has made it possible to have powerful hand-held computers, cellular telephones, global positioning systems, and many other electrical devices. In fact, the telephone booth may in one more generation be found only in museums. Everyone will have compact, low cost personal cell phones or something even better. A few years ago the idea of putting so many components and functions on one microchip was revolutionary. Today, chips are an integral part of almost every electrical device.

The advances in microelectronics spurred the development of microelectro mechanical systems (MEMS). Scientists working in this field have developed silicon actuators/levers, gears and gear reduction units, diaphragm pressure sensors, motors, transmissions, locks, mirrors, hinges, and more. Some of these devices have been combined to yield intricate mechanical systems-on-a-chip. Efforts are moving toward integrated microelectronic/micromechanical systems-on-a-chip; the acceleration sensor for deploying air bags is an example of the commercialization of micromachined systems-on-a-chip.

The strategy of miniaturized distributed systems is powerful; the potential impact worldwide is enormous. For example, this is probably one of the most significant strategies for the eradication of poverty. Until now, the countries with the resources, large-scale technologies and infrastructure, and skilled labor enjoyed the highest standards of living. But distributed processing can change that equation. For instance, some third-world countries will have communications without telephone poles and underground cables. Microtechnology presents new possibilities for poor countries to increase their standard of living using distributed micro chemical and thermal systems. With these technologies, all countries can have higher productivity, power, and communications anywhere, without needing to build expensive infrastructures. So miniaturization brings with it new hope for the 21st century.

2 Evidence for Success

The evidence for success of MICRO-CATS can be found in their demonstrated performance. They have higher heat transfer rates and faster reaction rates and can be operated with lower pressure drops than conventional systems. Some of the successful results for microreactors and solvent extraction are presented below.

Researchers at Pacific Northwest National Laboratory (PNNL) have investigated several microchannel reactor units. One, an integrated microchannel stainless steel combustor/evaporator, generates heat using the heat of reaction obtained from the gas-phase combustion of methane or hydrogen [6]. The technical objective is to demonstrate that a microchannel combustor can produce at least 25 watts of thermal energy per square centimeter of heat transfer profile area and transfer that energy to a cooling unit. This heat transfer rate is approximately 20 times higher than that of a conventional water heater. The most important problem for the microcombustors is maintaining the flame in a small volume. There are also problems maintaining the desired temperature in the combustion chamber and efficiently removing the heat and the combustion products.

The overall size of the system is 41 x 60 x 20 mm. Tests were conducted on micromachined combustor channels 300 microns wide, 500 microns deep, and 35 mm long; results showed that the system achieved heat fluxes greater than 30 W/cm^2 and combustion efficiencies greater than 85%. (Combustion efficiency is the ratio of the heat transferred to the water relative to the heating value of the fuel.) In this system, oxygen (or air) and fuel were introduced into the combustion chamber where an ignition wire initiated combustion. Microchannels were fabricated on the under side of the burner to provide cooling during the combustion process. The combustion products then entered a microchannel heat exchanger that enhances heat transfer between the combustion products and the combustor solid surfaces. Flame quenching problems increased as the combustor size decreased, indicating that flame quenching will set a limit on the degree of miniaturization possible in this type of combustor.

PNNL researchers have also investigated catalytic microchannel reactors made from high-temperature metal alloys for several applications. Engineered catalysts provide large surface areas, high heat-transfer rates, and low pressure drops. While slow kinetics is a current paradigm in conventional reactor systems, fast kinetics is possible in microchannel reactors. While multisecond reactions may be necessary in conventional reactors, millisecond reactions occur in microchannel reactors. This improvement is attributed to the diffusion lengths, which are reduced by at least two orders of magnitude, and to heat transfer, which prevents "hot spots" and "cold spots" and enables high throughputs.

A methane partial oxidation reactor was operated with fast kinetics at 900°C to produce carbon monoxide and hydrogen. Methane conversion efficiencies were more than 85% with 11 millisecond residence time and 100% with 25 millisecond residence time. The use of catalysts in microchannel reactors is so promising that PNNL is now developing engineered microstructures to find those best suited to microchemical reactors.

At PNNL, we are investigating microchannel reactors using the Sabatier reaction and the reverse water-gas shift reaction. The Sabatier reaction combines carbon dioxide with hydrogen to produce methane propellant and water vapor. In the reverse water-gas shift reaction, carbon dioxide combines with hydrogen to produce more fuel and water. An electrolysis unit dissociates the water into oxygen and hydrogen. Tests showed that the Sabatier reaction catalyst provides 85% conversion at 250°C with a reactor residence time of 0.1 second. The residence time for the reverse water gas shift catalyst was even shorter. These successful results are very promising for the development of MICRO-CATS for NASA's Mars missions.

PNNL researchers have designed and tested microchannel devices for efficient contacting of two liquids in solvent extraction [7]. Recent results are being presented by PNNL researchers at this conference [8]. Previous efforts have tested such devices as gas absorbers. This solvent-extraction work is a step in the development of compact efficient devices for chemical separations. The microchannel contactor results in substantially higher throughput per total system volume compared with conventional technologies.

Solvent extraction requires intimate contact of two immiscible liquids to facilitate mass transfer of one or more solutes from one fluid to another. The architecture in this work consists of two micromachined channels separated by a contact plate. The channels are separated by a micromachined contactor plate of 25-micron-thick Kapton substrate that has a matrix of uniform holes 25 microns in diameter. The solvent and feed streams can be operated co-currently or counter-currently. This arrangement allows for intimate contact of the two immiscible fluids as they flow through very thin channels smaller than the normal mass-transfer boundary layer.

The mass transfer resistance has contributions from each flow stream, from the contactor plate, and from the interface. The pressures inside the contactor must be carefully controlled to prevent breakthrough of the fluids through the holes in the contactor plate. If the interface is immobilized, the two liquid films can flow in counter-current directions, allowing for a more effective separation in the contactor. PNNL researchers are investigating reducing the overall size of the equipment and improve operating efficiency by reducing the thickness of the films and improving mass transfer efficiency in the contactor plate. Tests were conducted using microcontactor channels 1 cm wide, 10 cm long and 200 to 500 microns high. Results with cyclohexane-water feed and cyclohexane solvent indicate that the micromachined contact plates worked at least as well as commercial microporous membranes and have significant potential for improved performance. Also, with channel heights of less than 300 microns, diffusion through the contactor plate is the limiting factor in mass transfer.

Further evidence for success is our ability to model chemical and thermal processes at the microscale. PNNL researchers have developed a lattice-Boltzmann simulation capability for modeling the behavior of microscale fluidic systems. In fact, this work is being presented in a paper at this conference [9]. Complex fluid dynamics problems in this size range are not now amenable to being modeled by conventional simulation methods, which do not properly account for the importance of surface forces at fluid-gas, fluid-fluid, and fluid-solid interfaces. However, our lattice-Boltzmann modeling capability overcomes these limitations. This is critically important for supporting the design and testing of MICRO-CATS.

3 Progress Toward the Future

The quest for micro chemical and thermal systems is being driven by some important needs. One is the need for miniature analytical systems for DNA, toxic chemicals, and pathogens such as HIV. Such instruments are needed in the fields of medicine, forensics, agriculture, and environmental control. Other prominent needs driving development include 1) military requirements for human-portable heating, cooling, and power generating units; 2) transportation demands for hydrogen rich streams for fuel cells; and 3) chemical processing systems capable of processing the rocks, soil, and atmosphere on planets such as Mars. These are just examples. While we cannot be absolutely certain what the potential for microtechnology might be, we can speculate about areas where the demand caused by technical problems or social needs will draw us forward. At the present at PNNL, we are pursuing major opportunities for MICRO-CATS in buildings, carbon management, environmental restoration, industrial chemical processing, military, space exploration and transportation. More details on these applications are found on our web site: http://www.pnl.gov/microcats

At PNNL, we have several systems under development but just three are described in this paper: man-portable cooling, automotive fuel processing, and in situ resource utilization for space exploration. These three technologies demonstrate the diversity of potential applications.

3.1 Man-Portable Cooling

PNNL researchers are working on microsystems to provide man-portable cooling. The market is driving this work because there is a need to cool people who work in hazardous environments such as 1) soldiers in certain field operations, 2) emergency response workers for environmental cleanup, firefighters and other emergency workers, and 3) industrial workers in hot, unforgiving environments. A cooling device is needed for use with protective clothing, and it must be compact, lightweight, and able to operate for extended periods of time. We are developing absorption heat pumps using a lithium bromide and water solution as the working fluid. As this technology is successfully developed, we envision spin-off applications with slightly larger versions for cooling shipping containers and air conditioning vehicles and even commercial or residential structures.

In this conference, a poster session is being given by PNNL researchers describing the status of compact, heat-actuated heat pump development [10] (see Figure 1). Although the unit is for energy conversion and not for chemical processing and manufacturing, it is of interest because it includes micro chemical components that perform unit operations of interest to chemical engineers. It is a chemical heat pump that includes microchannel heat exchangers plus an assembly of components making up a thermochemical compressor. The components in this unit include a microchannel gas absorber and a microchannel gas desorber.

The high performance of this compact heat pump system is possible because of the rapid heat and mass transport available in microchannels. Current estimates suggest a size reduction of a factor of 60 compared with conventional heat pumps is possible. The weight of the system will be reduced by at least a factor of 2. The man-portable version of this system would operate by combustion of a liquid hydrocarbon fuel; the total system weight, including the fuel, for the man-portable system is estimated to be less than ten pounds.

3.2 Automotive Fuel Processing

Automotive fuel processing is another application where the potential exists for significant market demand for microtechnology. Some automotive manufacturers plan to introduce fuel-cell-powered automobiles into the market in the next few

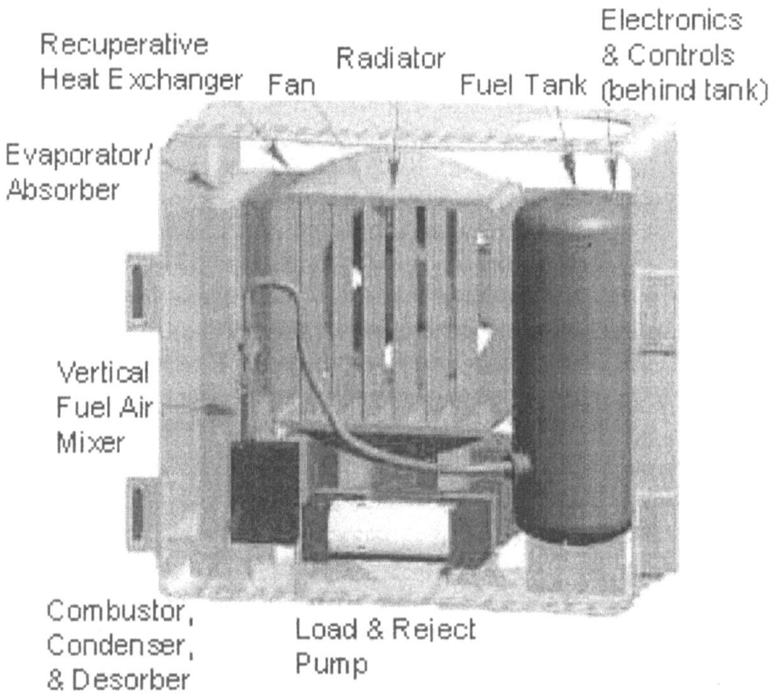

Figure 1. Mesoscopic Heat-Actuated Heat Pump

years. The fuel cells will generate electricity to drive the electric motors that move the vehicles. These vehicles will still carry liquid hydrocarbon fuels, plus a key additional item: an onboard fuel processing plant to produce hydrogen for the fuel cells.

The proton exchange membrane (PEM) fuel cell is the electricity generator for the baseline system being developed by the U.S. Department of Energy (with U.S. automotive manufacturers) in their Partnership for the Next Generation of Vehicles. The fuel processor is a critical reactor technology for the deployment of the PEM fuel cells, which operate using hydrogen. Figure 2 shows the fuel vaporizer portion of the automotive fuel processing system.

The fuel processor produces hydrogen rich streams from gasoline or methanol fuel using a heterogeneous, catalytic microchannel chemical reactor. This microreactor unit takes advantage of both the heat and mass transport effects in microchannels and is thus able to provide a compact system with high throughput. It is a multi-step process involving fuel vaporizer, primary conversion reactor to produce synthesis gas, water gas shift reactor, and CO cleanup reactor. The fuel processor has an occupied volume of less than 0.3 liters to support a 50 kWe output from a fuel cell. At this conference, PNNL researchers are presenting a paper that describes the development of this fuel processor (11).

Figure 2. Microchannel Gasoline Vaporizer for 50-Kwe Fuel Processing System

3.3 In Situ Resource Utilization

Microtechnology has attractive applications for supporting space exploration and orbit-based research. For such missions, payload size and weight is of paramount importance. Compact, lightweight microsystems cost less to launch and leave room on the spacecraft for other important equipment and supplies. Before the advent of microsystems, it was not possible to carry all fuels and other chemicals from Earth and it is not cost-effective to do so now. So the strategy is to "live off the land" in space exploration.

NASA wants to use compact chemical processing plants in its Mars missions. They are interested in reducing the costs associated with robotic and human missions to Mars and have determined that this can be accomplished by substituting process technology for propellants and oxygen for the return voyage. MICRO-CATS can produce fuel and oxygen from Martian atmospheric gases for NASA astronauts during their stay on Mars and for the return trip home.

So NASA's plan is to include lightweight, chemical systems as part of its Mission to Mars program to process indigenous space materials to produce propellants, oxygen, and other chemicals. The program uses a phased approach, with the following elements:

Robotic Missions		
Mars ISPP Precursor (MIP) mission	2001	Demonstrates small scale oxygen production
Mars PUMPP	2003	ISPP Utilization Demonstration
Mars Sample Return mission	2005	Utilizes ISPP propellants for propulsion
Mars Sample Return mission	2007	Produces and uses ISPP propellant and oxygen
Manned Mission		
Human Return Predeploy	2011	Propellants and O2 produced before crew departs Earth
Crew Launches	2013	

Extraterrestrial chemical processing plants will need to be compact, lightweight, efficient, and able to operate reliably for prolonged periods in reduced gravity environments. This presents unique challenges because conventional chemical process equipment relies heavily on gravity to operate effectively. If such systems are to be efficient and reliable, a fundamental understanding of operation in non-Earth environments is required.

We are working with the In Situ Resource Utilization (ISRU) program at NASA's Johnson Space Center to develop compact systems to produce fuel and oxygen from the Martian atmosphere. So far, the microchannel technologies have been conceived to convert carbon dioxide to methane and water by reacting it with hydrogen using the Sabatier process and reverse water gas shift reactions. These processes use carbon dioxide from the atmosphere to produce methane, water, and carbon monoxide. Hydrogen must be carried from earth for use in these reactions. Oxygen can be produced from the water in an electrolysis unit. The carbon monoxide can be further reacted with hydrogen to produce fuel.

4 MICRO-CATS in the 21st Century

The future is here for MICRO-CATS because many opportunities are now within our reach. We see three important fronts on the horizon of the 21st Century:

- The use of engineered nanosystems to produce "highly functional" surfaces
- The marriage of enzymes with highly functional surfaces
- Greater demands for MICRO-CATS to perform in the extreme and harsh conditions found on space missions.

4.1 Engineered Nanosystems

In the 21st century, engineered nanosystems will be used to develop MICRO-CATS. Why? Because engineered systems are often more powerful than natural ones. They present the possibility of further process intensification. Engineered nanosystems will be a platform for applying molecular sciences in the 21st century.

The ability to fabricate components is always one of the main limitations in the development of micro chemical and thermal systems. But overcoming these limitations presents opportunities for significant improvements. PNNL places major importance on developing the capability of microfabrication. Some of our progress is being presented in a paper at this conference [12]. Looking ahead, we believe that when engineered nanosystems (e.g., molecular systems) are included with microsystems, significant additional improvements in process technology may result.

Highly functional surfaces result from changes in surface properties that make the device or system more efficient, more effective, or more reliable. These surfaces may have the ability to repair, renew, or replace themselves. They could be surfaces that control or eliminate fouling—imagine a surface that can protect and heal itself from surface fouling. They could carry out a series of step-by-step chemical reactions. We envision using molecular coatings to create a molecular assembly line. In the 21st century, the use of highly functional surfaces will become routine.

Custom-designed catalysts will be developed. They will be designed and constructed to provide higher catalyst activity and greater specificity. When engineered and fabricated at the nanoscale, microsystems are more effective. This is what we can expect in the 21st century.

4.2 Enzymes and Functional Surfaces

All of life is full of microchemical systems. These have been working and developing for millions of years. Things like membranes, cells, lungs, kidneys, and hearts operate because they contain viable, effective, and reliable microtechnology. They involve chemical reactions, and virtually all of these chemical reactions require catalysts to arrange the reactants in energetically favorable orientations so that chemical products will be formed. Man-made chemical catalysts, such as platinum compounds, only crudely arrange reactants to form products, and, consequently, our chemical plants require high temperatures, pressures, and pH extremes to complete reactions. In contrast, cells require their catalysts to function under the mild conditions required for sustaining life. Nearly all reactions in cells are catalyzed by enzymes. Unlike chemical catalysts, enzymes are frequently catalytically perfect; i.e., once the enzyme binds its substrate, the reaction product is formed, sometimes at thousands of reactions per second. With the development of functional surfaces and the use of enzymes, many chemical processes that are not effected under high temperature and pressure conditions may be accomplished with enzymes and normal temperature and pressure conditions. Furthermore, the microdevices will then provide a molecular assembly line capability such that molecular changes occur in a sequenced step-by-step manner. These reactions will be dramatically different and probably dramatically better.

More than 5000 useful enzymatic reactions have been characterized to date, and many of these reactions are understood in molecular detail because the three-dimensional structures of the enzymes and reactants have been determined. DOE's genome sequencing programs and x-ray diffraction beam lines have been essential in identifying enzymes and determining their three-dimensional structures. Recombinant DNA technology makes it possible to produce virtually any enzyme in large quantities by isolating the gene, modifying the gene to enhance enzymatic properties, inserting the gene in an organism designed to express large quantities of protein, and then purifying the resulting protein. Recombinant enzymes provide attractive alternatives to constructing energy-intensive chemical plants that use crude catalysts.

A long-term goal is to someday manufacture artificial enzymes with highly functional surfaces. The structural and mechanistic insights elucidated from biological enzymes would provide the template for an advanced version of existing highly functional surfaces designed to mimic enzymes. Even if the artificial enzyme is orders of magnitude less efficient than the biological enzyme, the artificial enzyme might still be sufficient in most chemical applications and significantly less expensive than current crude catalysts.

4.3 Future in Space

The space age arrived more than 40 years ago, but it slowed when the costs became huge to carry out increasingly complex and ambitious space missions. Other priorities prevailed, and budgets for space exploration programs were not increased. But now, MICRO-CATS may help provide the way to accomplish missions at lower costs. In addition, and equally important is the fact that space exploration is one of the applications where microtechnology is better than conventional technology.

The extreme harsh environments coupled with the space mission requirements of compactness, light weight and high reliability present a quantum level change in requirements for MICRO-CATS. Can the system endure the rigors of temperature extremes? Does it operate under low-gravity or zero-gravity conditions? Can it be on stand-by for long periods and then operate reliably when needed? The rigors of developing technology for NASA and others may also spur the development for many other applications in the future.

5 Literature Cited

1. Wegeng R, C Call, and MK Drost. "Chemical System Miniaturization." February 1996. PNNL SA-27317, *AICHE 1996 Spring National Meeting*, New Orleans. Available on the world wide web: http://www.pnl.gov/microcats
2. Ehrfeld W, V Hessel, H Mobius, T Richter, V Hessel, and K Russow. February 1995. "Potentials and Realizations of Microreactors." Proceedings of the DECHEMA *Workshop on Microsystem Technology for Chemical and Biological Microreactors*.

3. Ehrfeld W, C Gartner, K Golbig, V Hessel, R Konrad, H Lowe, T Richter, and C Shulz. February 1997. "Fabrication of Components and Systems for Chemical and Biological Microreactors." *Microreaction Technology – Proceedings of the First International Conference on Microreaction Technology*.
4. Lerou J, M Harold, J Ryley, T Ashmead, C O'Brien, M Johnson, J Perrotto, C Blaisdell, T Rensi, and J Nyquist. February 1995. "Microfabricated Minichemical Systems: Technical Feasibility." Proceedings of DECHEMA *Workshop, Microsystem Technology for Chemical and Biological Microreactors.*
5. Wegeng R, MK Drost, T Ameel, and R Warrington. December 1995. "Energy Systems Miniaturization Technologies, Devices, and Systems." Proceedings of the International Symposium on Advanced Energy Conversion Systems and Related Technologies, RAN 95.
6. Drost MK, CJ Call, JM Cuta, and RS Wegeng. 1999. "Microchannel Integrated Evaporator/ Combustor Thermal Processes." *Journal of Microscale Thermophysics Engineering* (accepted for publication).
7. TeGrotenhuis WE, R Cameron, MG Butcher, PM Martin, and RS Wegeng. 1998. "Micro Channel Devices for Efficient Contacting of Liquids in Solvent Extraction." *Separation Science and Technology*. PNNL-SA-28743. Also available on the world wide web: http://www.pnl.gov/microcats
8. TeGrotenhuis WE, R Cameron, VV Viswanathan, and RS Wegeng. April 1999. "Solvent Extraction and Gas Absorption Using Microchannel Contactors." Third International Conference on Microreaction Technology. Also available on the world wide web: http://www.pnl.gov/microcats
9. Rector DR and BJ Palmer. April 1999. "Simulation of Chemical Separation Process Using the Lattice-Boltzmann Method." Third International Conference on Microreaction Technology. Also available on the world wide web: http://www.pnl.gov/microcats
10. Drost M, M Friedrich, C Martin, J Martin, and R Cameron. April 1999. "Recent Developments in Microtechnology-Based Chemical Heat Pumps." Third International Conference on Microreaction Technology. Also available on the world wide web at: http://www.pnl.gov/microcats
11. Tonkovich A, J Zilka, M LaMont, S Fitzgerald, D Vanderwiel, Y Wang, and RS Wegeng. April 1999. "Microchannel Reactors for Automotive Fuel Processors." Third International Conference on Microreaction Technology. Also available on the world wide web at: http://www.pnl.gov/microcats
12. Matson DW, PM Martin, DC Stewart, AY Tonkovich, GL Roberts, and M White. April 1999. "Fabrication of Microchannel Chemical Reactors Using a Metal Lamination Process." Third International Conference on Microreaction Technology. Available on the world wide web at: http://www.pnl.gov/microcats

Acknowledgment

Pacific Northwest National Laboratory is operated by Battelle for the U.S. Department of Energy under Contract DE-AC06-76RLO 1830.

Implementation of Microreaction Technology in Process Engineering

W. Ehrfeld, V. Hessel, S. Kiesewalter, H. Löwe, Th. Richter, J. Schiewe

Institut für Mikrotechnik Mainz GmbH, Carl-Zeiss-Str. 18-20, D-55129 Mainz, Germany, e-mail: wehrfeld@imm-mainz.de

Abstract

Microreaction technology offers several advantages for implementation in process engineering and biochemistry. Its application fields can be divided into two major areas, namely the conversion of matter and data acquisition. In the following, it will be shown that the advantages of microreaction systems, like increased yields and selectivities, process intensification as well as economical and ecological aspects, are unique properties and can be deduced to their basic characteristics. In this context, short pathways and response times, large surfaces, small volumes and defined flow characteristics have to be mentioned.

Today, a variety of microreactor systems as well as components already are available in small lot production. In this paper, focussing on recent developments in microreaction technology at IMM, the development of a variety of components for unit operations as well as their combination in integrated microreaction systems for special industrial purposes will be discussed.

1 Introduction

Meanwhile, it is widely accepted, that microreactors offer advantages, opening up the gate to a new chemistry including access to new process regimes as well as remarkable process improvements. First prototypes of complex integrated systems currently are tested in industry, not only as tools for process development but as instruments for production, process intensification and mobile as well as stationary energy production. Moreover, microreactor components of high versatility like micromixers and heat exchangers attract special attention. Some of these devices meanwhile are available in small scale production at low cost. Micromixers for example, have been applied successfully to a growing number of processes like emulsion and foam formation, multiphase reactions and polymerization processes, just to mention a few.

Furthermore, completely new aspects for chemical process engineering emerge, leading to new concepts in chemistry and reaction engineering. Without any doubt, an enormous application potential for microreaction technology can be predicted. This potential is known and accepted all over the world and competition concerning advantages in development and production using microreaction systems already has started. The dynamic of these technological developments and changes to be expected in chemistry and process engineering can be compared to the innovative developments that took place in biotechnology in recent years. Due to economic necessities, a change in paradigms will lead to a fast dissemination of microreaction technology in the near future.

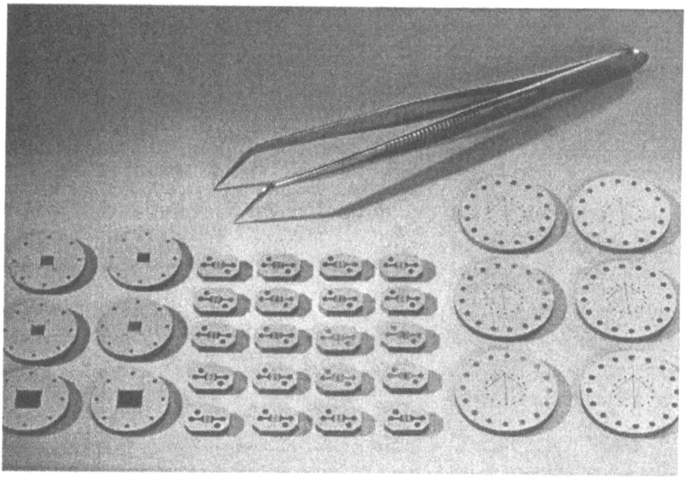

Fig. 1: Microreactor components, fabricated by means of LIGA-technology.

2 Basic characteristics of microreaction technology

Potentials and advantages of microreaction technology have been discussed intensively within the past few years. The unique properties of microreactors can be put down to their small characteristic dimensions and the ability to combine a large number of system components within small volumes. These characteristics affect the chemical process most effectively, e.g. by establishing strong gradients with respect to physical properties along the reaction pathway or enhanced heat management [1, 2].

Basically, microreactor applications can be divided into two distinct categories. The major field regarding reaction engineering can be seen in conversion of matter within miniaturized reaction systems and their components. Various processes in chemistry and pharmacy have been investigated so far with respect to production of chemicals in microreaction systems on a scale of grams, kilograms or even tons. In these fields, microreactors already are regarded as a competitive technology and open up economic alternatives to conventional process engineering.

In addition, microreaction systems are applicable for data acquisition utilizing the chemical process itself as a tool for identification of new qualities and properties of matter. Using small quantities of reactants, huge amounts of data can be generated for screening purposes by parallel processing. These systems already are used for many applications in pharmacy and biochemistry, e.g. in high throughput screening for lead discovery, medical technology and materials sciences. Consequent miniaturization increases the amount of information generated per unit volume permitting increased efficiency and fast screening efforts.

2.1 Fundamental properties of microreactors

Regardless of the categories of microreactor applications as described above, microreaction systems offer a number of inherent and fundamental advantages [2].

Short pathways and response times

With respect to dimensions along the reaction path, microstructurization yields strong gradients in temperature, concentration, density of the reactants or pressure allowing fast changes in reaction conditions regarding transport of heat or matter. Thus, microreactors enable new process regimes not accessible in conventional systems due to safety aspects as well as processes involving fast processing of unstable intermediates. Moreover, these properties of microreaction systems enable short response times regarding the chemical process.

Large surfaces and small volumes

The large surface to volume ratios of microreaction systems enable a significant enlargement in surface area available for transport of heat and matter. Thus, high transfer rates and efficient processes permit rapid changes in temperature, efficient heat management or fast mixing processes. At the same time, decreasing the system's volume significantly increases the information density obtainable in case of screening applications. Additionally, reduction of the amount of substances and chemicals employed in the process is advantageous regarding safety and ecological issues as well as aspects of sustainable development.

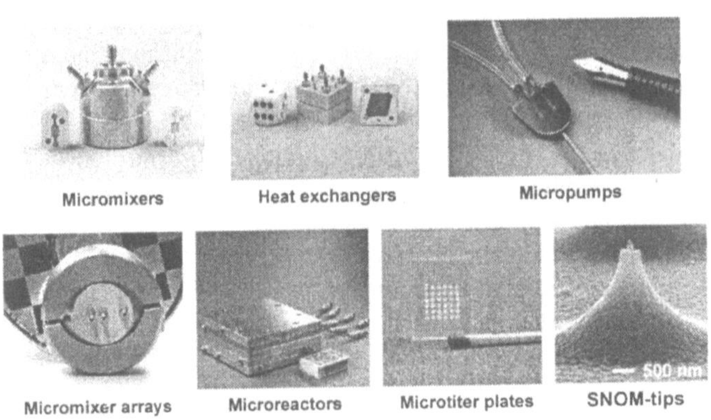

Fig. 2: Small series production of microreactor components and integrated systems.

Defined flow characteristics

In microchannels, laminar flow conditions are preferred compared to turbulent flow conditions, facilitating theoretical description of the flow process as well as an increase in reliability of simulations and theoretical predictions. Moreover, special flow characteristics like the highly ordered "hexagon flow" of gas / liquid dispersions only can be observed in microchannels [3]. Several advantages, like narrow residence time distributions and uniform mass transport kinetics are unique to microreaction systems.

2.2 Applications of microreaction systems

2.2.1 Process engineering aspects

Based on the inherent potentials of microreactors regarding process engineering aspects, the following advantages and improvements in chemical and biochemical engineering are accessible using microreaction technology.

Isothermal processing of exothermal reactions

The huge surface to volume ratios in microchannels have a strong impact on reaction conditions concerning heat transport phenomena. For example, efficient removal of excess heat of reaction gives access to new reaction regimes unique to microreactors and enables isothermal reaction conditions even for highly exothermal reactions. This was proven at BASF for a liquid multiphase process in an integrated microreaction system developed by IMM [4]. In this case, isothermal processing led to a 10 % increase in yield compared to the technical process [5].

Moreover, efficient heat removal is advantageous as well for processes employing highly reactive species, e.g. the direct fluorination of aromatics using elemental fluorine. This reaction, being hard to control on a larger scale for its high exothermicity and often explosive process conditions, can be performed safely and at considerable yields in microreaction systems [6].

Chemical processing within explosive regimes

Quench distances regarding explosive gas phase processes in most cases exceed the small dimensions and cross sections of microchannels. Therefore, gas phase reactions within the explosion limits of the reactants can be carried out safely in microreactors. This, inter alia, was proven for the ethylene oxide synthesis in a microreaction system [7] using pure ethylene / oxygen mixtures at an oxygen content up to 85 % as well as the explosive reaction of hydrogen and oxygen [8]. Thus, wide ranges of parameters become accessible for optimization of process conditions.

New process regimes in microreactors

Besides isothermal processing and explosive reaction regimes, other new process regimes in microreactors attract special attention regarding investigations of feasibility, specific process conditions and reaction engineering aspects as well. Here, a lot of new fields to explore can be found and the capability of microreactors for development of new processes has not been exploited thoroughly yet. Questions still open concern the fast processing of unstable intermediates as well as photochemical reactions or consecutive reaction sequences in microreactors, only to name a few.

2.2.2. Sustainable development

Besides engineering aspects as discussed above, there is a certain impact of microreaction technology regarding sustainable development and related economical and ecological aspects as well.

Waste reduction and saving of resources

Chemical processes in microreaction systems and their advantages and improvements e.g. on yields and selectivities lead to purer products. Therefore, less by-products have to be separated and fewer resources are needed for a certain amount of product. Moreover, for most screening applications, e.g. in drug discovery, catalyst or material development, reduction of the amount of reactants yield in dramatic cost saving.

Increased safety

Small volumes and hold-up as well as efficient heat management in microreactors significantly reduce the effort necessary concerning process safety. Without any doubt, today's industrial standards regarding safety are on a very high level. Nevertheless, microreactors can help to reduce costs in safety measures and allow the performance of reactions not applicable on a large scale for safety reasons.

Process optimization in microreactors

As already shown in several applications on a lab- as well as industrial scale, information on process optimization in microreactors allow predictions on optimal processing on a technical scale as well. This holds for parameters like residence time, transport of heat and matter and flow conditions. Often, these information can save years of development and costs regarding process engineering aspects. Thus, users in microreaction technology can gain a valuable lead over competitors on the world market.

Numbering-up instead of scaling-up

Traditionally, the development of production processes goes along with a time-consuming scaling-up process that often takes two to five years. Due to their small dimensions, microreactors are suitable for use on a lab-scale. Multiplication of the number of units easily gives access to higher production capacities while the specific reaction conditions of each microreactor unit is kept unchanged. Thus, years of development can be spared, saving time as well as money.

Smaller production units and distributed production

Conventional plants require a certain size for economical production. Microreactors do not obey the rule of decreasing costs with increasing plant size. In microreaction technology, the size of each production unit, available in large scale production at low costs, correlates to the absolute number of identical sub-units needed. Therefore, microreactors show a high flexibility in output capacity and can be used for decentralized small size production as well as production on-site and on-demand [9].

Fig. 3: Small series production of micromixers.

2.2.3. Microreactors for data acquisition

Applications of microreactors for data acquisition in pharmaceutical, medical and materials science mainly arise from the large number of independent units to be integrated within small volumes. Highly parallel screening processes in micro-systems are capable of generating enormous amounts of information at minimized input of substances, energy and time. Furthermore, screening processes in micro-reaction systems are quick due to the small volumes, short distances and response times. In the following, the main fields of application for microreactors in data acquisition as well as their main features are mentioned briefly.

High Throughput Screening (HTS)

- Fast identification of substances with specific properties.
- Fast processing of huge numbers of samples by parallel processing at a high degree of automation.
- Similar processing of each sample at a high degree of repeatability.

Combinatorial synthesis and materials science

- Identification of correlation e.g. between chemical structure and desired properties.
- Fast generation of huge numbers of samples, combined in substance libraries e.g. by combinatorial synthesis.
- Parallel examination of specific properties.

Lab-on-Chip

- Integration and combination of fluidic elements, sensor components and unit operations.
- Realization of complex processing steps at a high degree of integration within small volumes.
- Fast on-site monitoring of analytical parameters possible.
- Applications in pollution control, medical diagnostics, biotechnology and food stuff analysis.

3 Components for unit operations in microreactors

Since the early beginning of microreaction technology, several microreactor components have been reported on, mostly concerning mixing processes [10] and heat exchanger devices [1, 11]. Still there are a lot of activities in this field, now mainly dealing with industrial applications as well as questions of high throughput and cheap mass production of microreactor components.

3.1 Mixing devices

Recently, several micromixing devices, especially designed for industrial scale applications and enabling a high volume throughput, were developed. Compared to existing devices comprising an interdigital system of microchannels new design concepts were utilized.

Fig. 4: Caterpillar mixer suitable for high volume flows and multiphase systems. Material: stainless steel, process: precision engineering.

High throughput micromixers

The caterpillar mixer consists of two parts having an identical microstructure as shown in figure 4. Two fluids enter the mixer and are laminated vertically and split horizontally in two partial streams that are again laminated in the vertical plane. Consecutive mixing stages following this split and recombine principle double the number of fluid lamellae formed and simultaneously decrease their width by a factor of two. After passage of six to eight stages a typical width of each fluid lamella of approximately 2 μm is reached, enabling fast and efficient mixing by diffusion. The caterpillar mixer's design minimizes the pressure loss and enables volume flow rates up to 50 l/h.

Another type of mixer, the so-called "star laminator" (see figure 5) comprises a stack of metal foils. Here, a system of alternating lamellae are formed in a central boring by feeding both fluids radially through star-like formed structures. The width of the lamellae is determined by the width of the metal foils stacked, that is 20 or 50 μm. The star laminator was designed to enable flow rates up to 100 l/h at low pressure drops.

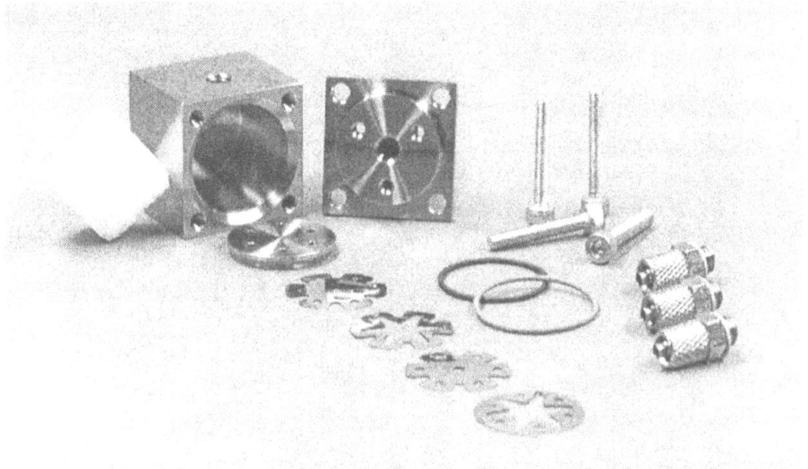

Fig. 5 Star laminator consisting of thin metal foils and suitable for high volume flows. Material: stainless steel, process: wire-EDM.

New processes in micromixers

A new approach for the generation of liquid / liquid emulsions or gas / liquid dispersions is given by the use of micromixers, characterized by a static operational principle. In micromixers comprising interdigital channel structures, the formation of dispersions with small droplet sizes and very narrow size distributions at low operating pressures can be observed, offering significant advantages as compared to standard dispersing techniques. Moreover, the use of micromixers enables an easy transfer of R&D results on a lab scale to the production scale. Increasing the number of mixing units working in parallel ("numbering up"), easily circumvents scale-up problems.

In the mixing elements used, dispersions are formed in a defined way by multiple splitting of two phases resulting in an alternating arrangement of small flow lamellae that decompose due to strong periodic velocity gradients. Main parameters with influence on the mean droplet size of the dispersions formed, are the total volume throughput, the ratio of flow rates with respect to the different phases as well as the microchannel dimensions [12]. Typical mean droplet diameters for oil-in-water emulsions are in the range of 4 to 50 μm. At large differences in the flow rates of both phases, the droplet size is smaller than the microchannel width because of fragmentation due to parabolic velocity profiles and the effect of hydrodynamic focussing.

In case of gas / liquid dispersions, the bubble size depends on the liquid flow rate and the ratio between the flow rates of liquid and gas flow. Mean bubble sizes were found in the range of 120 to 800 μm for the model system argon in glycerol / water with surfactant added. A continuous, well ordered flow behavior was found when using low to intermediate viscosities, low surface tensions and small liquid to gas flow ratios [3].

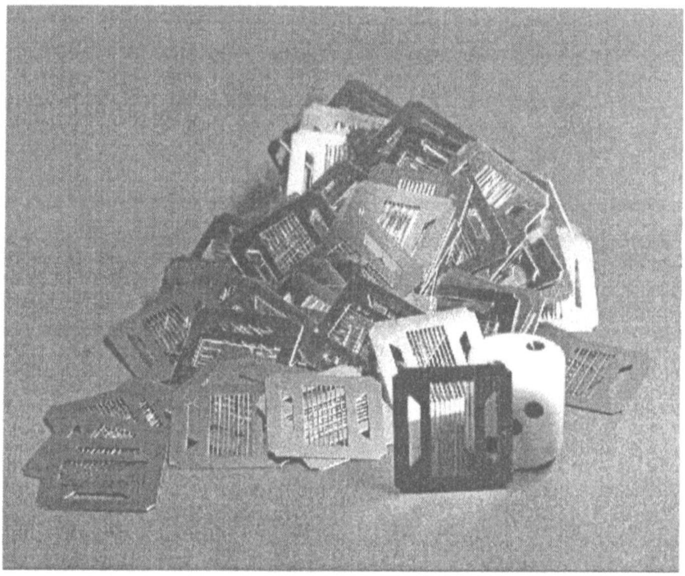

Fig. 6 Stainless steel heat exchanger structures, realized by a punching process.

3.2 Heat exchangers

Based on previous developments concerning countercurrent flow heat exchangers [1, 2], current activities focus on mass production techniques for cost-efficient realization of these important microreactor components. A five-step punching technique was employed for the production of microstructured stainless steel foils showing parallel fluid guiding channels 500 μm wide at a wall thickness of 300 μm. The foils are stacked upon each other using separation foils. A 90°-angle of the structured foils with respect to each other forms a cross-flow heat exchanger.

The number of layers to be stacked can be chosen almost freely, depending on the desired volume flows. Interconnection of the foil stack was assured by brazing techniques. So far, a maximum of 81 layers was tested, showing a heat transfer coefficient of 8,200 W / m^2 K at a volume flow of 4.8 l / h.

3.3 Analytical components

Miniaturized analytical components are crucial for the widespread and efficient use of microreaction technology in order to fully exploit the advantages of microreactors for chemical processing. In this context, integrated analytical components are needed for:

- Enhanced and fast process control using analytical data
- Fast and efficient determination of product distribution
- Quality control of products

Moreover, suitable miniaturized processes for product separation are of main interest. These processes and components are important features with respect to increased automation and process control in microreaction systems.

Fast and miniaturized GC-analysis

In order to monitor the reactant and product concentrations of a gas phase microreactor for ethylene oxide synthesis, a miniaturized device for fast GC-analysis was developed in a joint effort with University of Mainz, Germany. The introduction of short, packed capillary columns applied to the specific analytical problem of the separation of permanent gases, light hydrocarbons and their oxidation products significantly improved the separation efficiency compared to conventional, coated capillaries used in gas chromatography [13].

Additional developments dealt with the improvement of the heating and cooling process in GC-temperature programming in order to further increase analysis speed. Results achieved with a 2D-test module for fast heating of short packed GC-columns clearly show the capability of this device for speeding up analysis time. For example, using product mixtures of the ethylene oxide synthesis, retention time for the ethylene oxide peak was decreased to 80 seconds, a factor of four compared to conventional GC-analysis.

Permeation device

A miniaturized module for high pressure product enrichment using membrane principles was developed in co-operation with the Max-Planck-Institute for Polymer Research in Mainz, Germany [14]. Various microstructured supports, stacked upon each other and realized by photolithographie, electroforming and etching techniques assure a sufficient pressure stability of the system. Nickel structures with a thickness of 80 µm and containing highly ordered honeycomb-like openings (60 µm in diameter) support the membrane material. The module's layers are incorporated within a pressure-proof stainless steel containment. Currently, experimental characterization is under way.

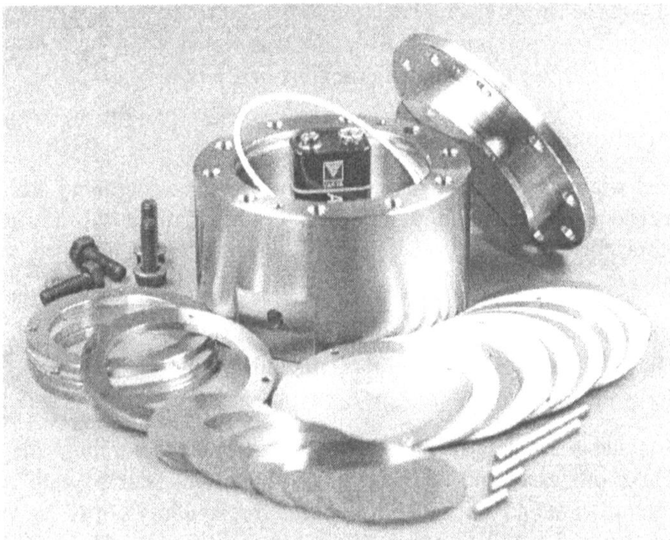

Fig. 7 Components of a membrane module for product enrichment in microreactors.

3.4 Standardized modular concepts

Standardization of microreactor components and their flexible arrangement to modular systems significantly enhances the availability of specific combinations of unit operations, e.g. for specific R&D-activities. This functional variation of microreaction systems is important for the widespread use of microreactors in R&D, education and small scale production, e.g. of fine chemicals. These standardized components can be divided into three major groups.

Basic modules

These basic units are the main part of any kind of reactor combination. They represent unit operations of reaction engineering, e.g. mixing, heat and mass transfer or separation processes as well as catalyst carriers and reaction units.

Accessory modules

In order to guarantee flexibility in the system's layout regarding variation in process conditions and reaction sequences, a number of different components for connection and distribution of fluids are needed. These assure interconnection between the system's components and to in- and outlet channels as well as fluid distribution to a certain number of microchannels and collection of individual reactant streams.

Individual modules

Depending on the desired degree of automation and control, modules including pumps, valves and devices for sensing and control are needed. These must be compatible to existing microdevices as well as to conventional components.

Fig. 8 Standardized modules for a modular microreactor concept.

So far, several components were realized using existing components like LIGA-mixing units [10], gas mixer and reaction zone used within a microreactor for ethylene oxide synthesis [15] and heat exchanger components. These were incorporated within cube-shaped housings as shown in figure 8. Special connectors allow free combination of each component with respect to the neighboring cubes. Applications of various sets of standardized microreactor components can be seen in screening purposes, realization of microplants or even decentralized production purposes.

4 Integrated microreaction systems

Microreactor components are combined to integrated systems in order to provide independent systems for defined purposes in microreaction technology. These highly versatile systems mainly are used for R&D activities and process optimization. Basic investigations focussed on product distributions [16], non-equilibrium chemistry [17] and reaction conditions, e.g. temperature control and feedback [18], in microreaction systems. More recently, prototypes are utilized for special investigations regarding industrial production purposes. The development of integrated microreactors is characterized by an increasing degree of integration.

4.1 Microreactors for gas phase processes

Ethylene oxide microreactor

In this microreactor, designed for the heterogeneously catalyzed direct oxidation of ethylene on silver surfaces, processes be performed in a pressure range between 1 and 25 bar. Typical reaction temperatures range below 300°C at total volume flows up to 5 l / h [15].

In general, the results of the experimental characterization at the Max-Planck-Institute for Coal Research in Mülheim, Germany, are similar to the industrial process. Selectivities are in the range of about 50 % while the industrial process shows selectivities up to 80 % at similar conversion levels. However, the industrial process is carried out in the presence of chloride compounds and promoters within the catalyst material. Here, no additives are used and pure silver surfaces were utilized as the catalytic component [7].

Moreover, feasibility regarding the use of explosive reactant mixtures was shown. In contrast to the industrial process, the oxidation of ethylene was carried out in pure oxygen. This way, space-time-yields exceeding those of the industrial process were obtained [7].

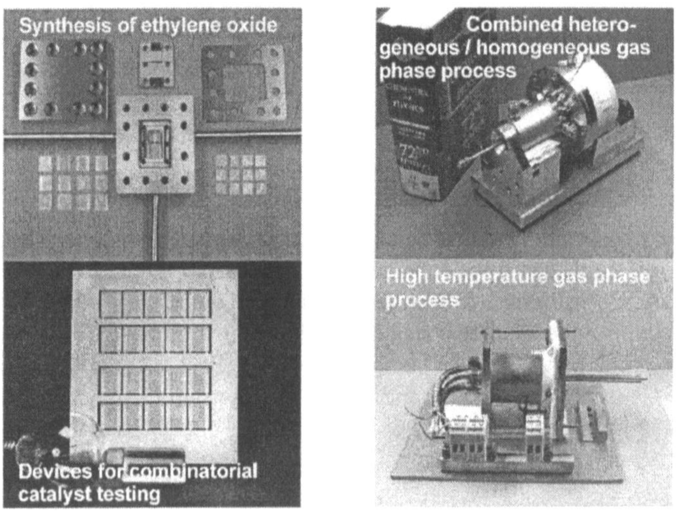

Fig. 9 Various microreaction systems for gas phase processes.

Propylene oxide microreactor

Short reaction paths, small amounts of reactants as well as fast and effective changes in reaction conditions are outstanding features of microreaction systems. These characteristics are of interest with respect to very fast consecutive reaction sequences utilizing different thermal conditions or reactants. Main advantage of microreaction technology for realization of these combined reaction processes in the gas phase is the effective control of temperature profiles and residence times in order to prevent undesired side reactions. Moreover, fast and efficient mixing processes are crucial for immediate changes in reactant composition.

A microreaction system, especially designed for the direct oxidation route of propylene to propylene oxide was developed and currently is under investigation at the Max-Planck-Institute for Coal Research in Mülheim, Germany [19]. Major aim of the experimental characterization is to show the feasibility of microreaction systems for combined reaction sequences employing unstable intermediates.

Microreactor for HCN generation

Within the frame of a multilateral industrial project including five industrial companies, a miniaturized reactor for high-temperature gas phase reactions was developed. The reaction chosen was the synthesis of HCN via the Andrussov route at temperatures above 1000°C. Aim was to investigate the influence of isothermal processing and extremely rapid cooling of a hot, reactive product gas for the example of a very fast and highly exothermic reaction. These conditions are expected to prevent undesired parallel as well as follow-up reactions, e.g. the hydrolysis of HCN to ammonia.

Central element of the microreactor is a platinum catalyst structure comprising an array of microchannels, 60 µm in diameter, which the gaseous reactants pass through. Within less than one millisecond, the reactants are heated up to a reaction temperature above 1000°C. Subsequently, the product gases enter a micro heat exchanger and are cooled instantaneously within a distance of some hundred micrometers [20].

The feasibility of this microreactor concept was proven within only a few experimental runs showing significant HCN formation. The selectivity based on ammonia of the industrial process almost was reached. For the microreactor, HCN yields up to 30 % were found [20].

Microreactor for periodic operation

A promising field for the application of microreactors is the periodic operation of heterogeneously catalyzed gas phase reactions. By switching e.g. between two reactant feeds, the conversion and selectivity of some reactions can be increased. Especially for short switching periods, small volumes and narrow residence time distributions are essential. These features are given by microreactors as an ideal tool to study the influence of sharp variations in the reactant composition on the performance of heterogeneously catalyzed gas phase reactions [21].

A modular reactor especially designed for this purpose was developed. The reactor consists of a stack of microstructured foils providing a variable number of catalyst supports and heat-exchanger layers for the effective control of the process temperature. First results of residence time distribution measurements performed at Technical University of Lausanne, Switzerland, and determining the response of the system to a rectangular pulse show, that switching periods regarding reactant composition down to one second are feasible.

4.2 Microreactors for gas / liquid processes

In order to allow the highly exothermal direct fluorination of aromatic compounds using elemental fluorine, a favorable process route due to various economic and environmental reasons, concepts for gas / liquid microreactors were developed. These base on the formation of thin fluid layers for gas / liquid contacting with high specific surface and enhanced heat and mass transfer between liquid and gaseous phase. These microreactors open up a possibility for a fast process and increased throughput at high yields. This work was conducted within a joint effort together with Institute of Applied Chemistry, Berlin, Germany, Karlsruhe Research Center, Germany as well as six industrial companies.

Micro bubble column

The micro bubble column contains an integrated multi-channel micromixer, reaction platelets as well as a miniaturized heat exchanger. A high-quality dispersion of fluorine into the liquid feed stream is achieved by multilamination within the micromixer by an interdigital channel structure. Thus, a segmented flow pattern with very small bubbles, separated by thin liquid layers was obtained. Their maximum size is restricted by the width of the reaction channels, being 50 μm [6].

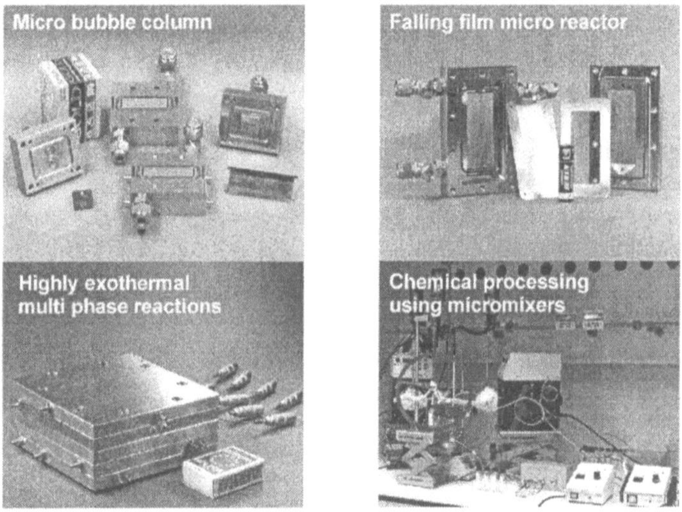

Fig. 10 Various microreaction systems for multiphase processes.

Falling film microreactor

The other microreactor used, a falling film microreactor, consists of a micro heat exchanger, a reaction plate and a gas chamber. The reaction plate comprises a large number of parallel channels, each 300 μm wide and 100 μm deep. Using this device, a stable, thin falling film of some ten micrometers thickness was observed. Flow patterns were visualized, e.g. by tracer experiments. Preliminary experiments show the feasibility of the direct fluorination route with this microreactor. Further investigations in order to specify product yield and selectivity as well as the substitution pattern of the side products were carried out [6].

Experimental results

Direct fluorination was feasible in both types of microreactors at yields of the monofluorinated products up to 20 %. Conversion in the falling film microreactor amounts to about 50 % while the micro bubble column obtains high selectivities up to 49 %. Yields and conversion in both cases were significantly larger compared to laboratory bubble columns, operated at similar process conditions [6].

4.3 Microreactors for liquid phase processes

Multi-Component microreaction system

The rising demand for microreaction systems suitable for various applications show the need for flexible components that can easily be adjusted to fulfill different tasks. A microreaction system was designed for highly exothermal liquid multiphase reactions with fast side reactions. The flat architecture of the system's components allows their easy exchange and additional incorporation providing high flexibility concerning reaction sequences and residence times. This system is available in small-lot production and especially suitable for basic R&D activities and process development [22].

Electrochemical microreaction system

Due to the fact that chemical reactions may be influenced by very sensitive regulation of the electric potential instead of heat management or the addition of further reactants that need to be removed later on, electrochemical methods offer many advantages for the performance of ionic and radical reactions.

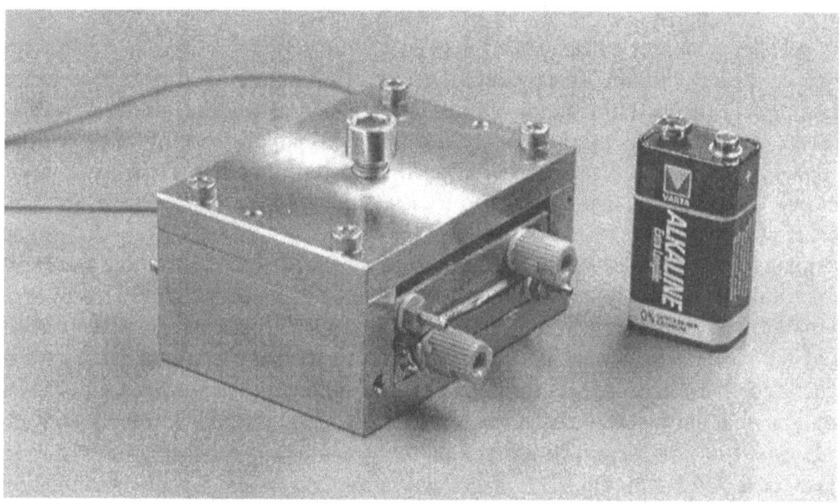

Fig. 11 Microreactor for electro-organic synthesis.

Utilizing miniaturized electrochemical cells, the advantages of thin layer cell technology and microstructuring techniques were utilized in a new microreactor concept for the electro-synthesis of organic compounds [23]. Using etching techniques and laser ablation, a miniaturized plate-to-plate electrode configuration within a non-conducting housing with several parallel reaction channels was realized (see figure 11). A homogeneous electric field at uniform current densities results from the reduction of the electrode distance. Uniform mass transport due to the homogeneous distribution of the electrolyte flow into the microchannels allows an effective heat management. Therefore, the heat generated by ohmic resistance of the electrolyte and the joule reaction heat respectively, may be removed easily due to the large surface-to-volume ratios even at high current densities. This rapid cooling with an integrated cooling device helps to avoid by-products.

In this reactor, a nearly quantitative oxidation of 4-Methoxytoluene to 4-Methoxybenzaldehyde was obtained at selectivities of 98% for low concentrations in methanol at room temperature. In comparison to the large-scale process, the concentration of the conducting salt was reduced significantly. The results evidently show very interesting perspectives in microreaction technology and new approaches to electrochemical organic synthesis [23].

5 Screening applications in microreaction technology

Recently, screening technologies play a major role not only in biotechnology but in materials science and catalyst testing as well. The requirements of these applications in microreaction technology are quite similar compared to reaction engineering. Therefore, there is no longer a sharp distinction between chemical and biotechnological applications in microreaction technology.

Microreactor for parallel catalyst screening

For parallel screening of catalysts, a special microreaction system was constructed in co-operation with the Technical University of Chemnitz, Germany. Due to the high number of parallel reactions, a fast replacement of reaction zones as well as the use of cheap mass production processes and a explicit marking of each inlay ("catalyst library") is crucial. Therefore, the concept employs a frame structure holding etched inlays that are easy to replace and can be produced in large numbers at low costs [24].

The reactor, shown in figure 12, consists of 35 frame structures with a thickness of 0.7 mm, arranged upon each other between a top and base plate. The number of frame structures is given by the maximum operation length of the probe nozzle used for product detection. Each frame holds an etched aluminum inlay micro-structured on both sides, thus forming a number of independent reaction zones for catalytic reactions. The inlay's microstructures show parallel channels 250 μm wide and 100 μm deep. The surface of these aluminum microstructures can be transferred into aluminum oxide, serving as a catalyst carrier, by anodic oxidation [24].

Fig. 12 Microreaction system for highly parallel catalyst screening.

Electrophoresis chip and nano titer plate

The development of a 96-channel disposable electrophoresis chip showed the applicability of microstructuring techniques for cheap mass production of analytical tools for applications in biotechnology. Here, a combined process employing advanced silicon etching (ASE), subsequent electroplating, yielding the complementary structure in nickel, and hot embossing in Polymethylmetacrylate (PMMA) was used.

The chip in credit-card size, allows a considerable reduction in analysis time at decreased sample consumption [25]. Sample transfer from a 384-well titer plate to the electrophoresis chip can be managed by means of a 96-pin array adapted to the system's size. Due to the special arrangement of the parallel separation channels, only two electrodes are needed. Thus, sample handling and process operation is simplified.

In order to enhance the trend of decreasing reaction volumes and assay formats, nano titer plates containing 384, 864 or 1536 wells were developed on a prototype basis using various materials [26]. These developments aim at the 3456 well format with a well-to-well spacing of 1.5 mm. Apart from miniaturization, further improvement of micro titer plates can be achieved by integration of functional elements, e.g. optical detection.

6 Outlook

Within various industrial applications, the enormous potential of microreaction technology as well as the decisive improvements already have been shown. In this context, process optimization of a metal organic reaction at Merck KgaA resulted in an 20 % increase of yield [27]. The use of micromixers within a polymerization process at Aventis Research & Technologies, led to a significant decrease in by-products [28]. These two examples vividly show the growing importance of microreaction technology for industrial processes. The success of these applications not only depends on the special properties of microreactors but mainly on their availability today and in the future. Although, microreactors already are commercially available, improvements in availability and production processes will play a major role in the future.

Momentarily, efforts focus on standardization of microreactor components for basic operations in process engineering. Major aim is the development of modular systems that allow easy variation of microreaction systems with respect to specific processes and needs.

In the near future, the development of standardized multi-functional systems including product separation and analytics will be the goal to reach. Therefore, the degree of integration in comparison to today's microreaction systems will increase further.

References

[1] Ehrfeld, W.; Golbig, K.; Hessel, V.; Konrad, R.; Löwe, H.; Richter, Th.: *Fabrication of components and systems for chemical and biological microreactors*, In: Microreaction Technology, Proceedings of the First International Conference on Microreaction Technology, Ed.: W. Ehrfeld, Springer, Berlin (1997) 72 – 90.

[2] Ehrfeld, W.; Hessel, V.; Haverkamp, V.: *Micoreactors*, In: Ullmann's Encyclopedia of Industrial Chemistry, Sixth Ed., 1999 Electronic Release, Wiley / VCH

[3] Hessel, V.; Ehrfeld, W.; Golbig, K; Haverkamp, V; Löwe, H.; Richter, Th.: *Gas / liquid dispersion processes in micromixers: The hexagon flow*, In: Process Miniaturization: 2nd International Conference on Microreaction Technology, March 9-12 1998, New Orleans, LA, Topical Conference Preprints, The American Institute of Chemical Engineers (1998) 259 - 266

[4] Richter, Th.; Ehrfeld, W.; Wolf, A.; Gruber, H.-P.; Wörz, O.: *Fabrication of microreactor components by electro discharge machining*, In: Microreaction Technology, Proceedings of the First International Conference on Microreaction Technology, Ed.: W. Ehrfeld, Springer Berlin (1997) 158 - 168

[5] Wörz, O.; Jäckel, K.P.; Richter, Th.; Wolf, A.: *Microreactors, a new efficient tool for optimum reactor design*, In: Process Miniaturization: 2nd International Conference on Microreaction Technology, March 9-12 1998, New Orleans, LA, Topical Conference Preprints, The American Institute of Chemical Engineers (1998) 183 - 185

[6] Hessel, V.; Ehrfeld, W.; Golbig, K.; Haverkamp, V.; Löwe, H.; Storz, M.; Wille, Ch.; Guber, A.; Jähnisch, K.; Baerns, M.: *Gas / liquid microreactors for direct fluorination of aromatic compounds using elemental fluorine*, published in this volume

[7] Kestenbaum, H.; Lange de Oliveira, A.; Schmidt, W.; Schüth, F.; Ehrfeld, W.; Gebauer, K.; Löwe, H.; Richter, Th.: *Synthesis of ethylene oxide in a microreaction system*, In: Microreaction Technology, published in this volume

[8] Hagendorf, U.; Janicke, M.; Schüth, F.; Schubert, K.; Fichtner, M.: *A Pt/Al_2O_3 coated microstructured reactor / heat exchanger for the controlled H_2/O_2-reaction in the explosion regime*, In: Process Miniaturization: 2nd International Conference on Microreaction Technology, March 9-12 1998, New Orleans, LA, Topical Conference Preprints, The American Institute of Chemical Engineers (1998) 81 - 87

[9] Rinard, I.: *Miniplant design methodology*, In: Process Miniaturization: 2nd International Conference on Microreaction Technology, March 9-12 1998, New Orleans, LA, Topical Conference Preprints, The American Institute of Chemical Engineers (1998) 299 – 312

[10] Ehrfeld, W.; Golbig, K.; Hessel, V.; Löwe, H.; Richter, Th.: *Characterization of mixing in micromixers by a test reaction: Single mixing units and mixer arrays*, Ind. Eng. Chem. Res. 38 (1999) 1075 - 1082

[11] Bier, W.; Keller, W.; Linder, G.; Seidel, D.; Schubert, K.: *Manufacturing and testing of compact micro heat exchangers with high volumetric heat transfer coefficients*, In: DCS – Vol. 19, Microstructures, Sensors and Actuators, Eds.: D. Cho, R. Warrington Jr., C. Blechinger, A. Pisano, W. Trimmer, L. Trefethan, Book No. G00527 (1990) 189 – 197, The American Society of Mechanical Engineers

[12] Haverkamp, V.; Ehrfeld, W.; Gebauer, K.; Hessel, V.; Löwe, H.; Richter, Th.; Wille, Ch.: *The potential of micromixers for contacting of disperse liquid phases*, Fres. J. Anal. Chem., 364 (1999) 617 - 624

[13] Schiewe, J.; Ehrfeld, W.; Hang, T.; Richter, Th.; Yan, X.L.; Kurganov, A.A.; Unger, K.K.: *Fast response heating module for temperature programmed GC-analysis in microreaction systems*, published in this volume

[14] Schiewe, B.; Vuin, A.; Günther, N.; Gebauer, K.; Richter, Th.; Wegner, G.: *Polymer membranes for product enrichment in microreaction technology*, published in this volume

[15] Richter, Th.; Ehrfeld, W.; Gebauer, K.; Golbig, K.; Hessel, V.; Löwe, H.; Wolf, A.: *Metallic microreactors: Components and integrated systems*, In: Process Miniaturization: 2nd International Conference on Microreaction Technology, March 9-12 1998, New Orleans, LA, Topical Conference Preprints, The American Institute of Chemical Engineers (1998) 146 - 151

[16] Wießmeier, G.; Hönicke, D.: *Strategy for the development of micro channel reactors for heterogeneously catalyzed reactions*, In: Process Miniaturization: 2nd International Conference on Microreaction Technology, March 9-12 1998, New Orleans, LA, Topical Conference Preprints, The American Institute of Chemical Engineers (1998) 24 - 32

[17] Tonkovich, A.L.Y.; Zilka, J.L.; Powell, M.R.; Call, C.J.: *The catalytic partial oxidation of methane in a microchannel chemical reactor*, In: Process Miniaturization: 2nd International Conference on Microreaction Technology, March 9-12 1998, New Orleans, LA, Topical Conference Preprints, The American Institute of Chemical Engineers (1998) 45 - 53

[18] Franz, A.J.; Quiram, D.; Srinivasan, R.; Hsing, I-M.; Firebaugh, S.L.; Jensen, K.F.; Schmidt, M.A.: *New operating regimes and applications feasible with microreactors*, In: Process Miniaturization: 2nd International Conference on Microreaction Technology, March 9-12 1998, New Orleans, LA, Topical Conference Preprints, The American Institute of Chemical Engineers (1998) 33 - 38

[19] Richter, Th.; Ehrfeld, W.; Erntner, D.; Gebauer, K.; Golbig, K.; Hausner, O.; Löwe, H.; Lange de Oliveira, A.; Schmidt, W.; Schüth, F.: *Microstructured reactor for consecutive heterogeneous / homogeneous gas phase reactions*, published in this volume

[20] Hessel, V.; Ehrfeld, W.; Golbig, K.; Hofmann, C.; Jungwirth, St.; Löwe, H.; Richter, Th.; Storz, M.; Wolf, A.: *High temperature HCN generation in an integrated microreaction system*, published in this volume

[21] Liauw, M.A.; Baerns, M.; Broucek, R.; Buyevskaya, O.V.; Commenge, J.-M.; Corriou, J.-P.; Falk, L.; Gebauer, K.; Hefter, H.J.; Langer, O.-U.; Löwe, H.; Matlosz, M.; Renken, A.; Rouge, A.; Schenk, R.; Steinfeldt, N.; Walter, St.: *Periodic operation in microchannel reactors*, published in this volume

[22] Richter, Th.; Ehrfeld, W.; Hessel, V., Löwe, H.; Storz, M.; Wolf, A.: *A flexible multi-component microreaction system for liquid phase reactions*, published in this volume

[23] Ziogas, A.; Löwe, H.; Küpper, M.; Ehrfeld, W.: *Electrochemical microreactors: A new approach for microreaction technology*, published in this volume

[24] Zech, T.; Hönicke, D.; Lohf, A.; Golbig, K.; Richter, T.: *Simultaneous screening of catalysts in microchannels: Methodology and experimental set-up*, published in this volume

[25] Konrad, R.; Ehrfeld, W.; Hartmann, H.-J.; Jacob, P.; Neumann, M.; Pommersheim, R.; Sommer, I.; Wolfrum, J.: *Disposable electrophoresis chip for high throughput analysis of biomolecules*, published in this volume

[26] Niggemann, M.; Ehrfeld, W.; Weber, L.: *Micro molding of fluidic devices for biochemical applications*, published in this volume

[27] Krummradt, H.; Kopp, U.; Stoldt, J.: *Experiences with the use of microreactors in organic synthesis*, published in this volume

[28] Bayer, T.; Pysall, D.; Wachsen, O.: *Micro mixing effects in continuous radical polymerization*, published in this volume

Part 2

Design and Production of Microreactor Systems

Fabrication and Application of Silicon-based Microchannels

J.G.E. Gardeniers, R.W. Tjerkstra and A. van den Berg

MESA Research Institute, University of Twente,
P.O.Box 217, NL-7500 AE, Enschede, The Netherlands

Abstract. Techniques for the fabrication of different types of microchannels, as well as their application in microfluidic devices and microreactors are described. The primary material in which the channels are fabricated is silicon, and the methods to be discussed include (an)isotropic and anodic etching in solutions, reactive ion etching in plasmas, and combinations of wet and dry methods. The fabrication of channels consisting of porous silicon will also be considered, and applications of such channels will be discussed. Finally, the use of silicon channels as moulds for the deposition of insulating materials like silicon nitride or glass will be described, which allow the fabrication of structures suitable for capillary electrophoresis and related separation methods.

Keywords. Silicon micromachining, microfluidics, microchannels, moulding

1. Introduction

Recently, progress in analytical sciences has gained an enormous impetus through the introduction of chip-based technologies and design strategies. An increasing number of new companies, mainly based in the USA, are penetrating the market with all kinds of pocket-size chemistry equipment, including HPLC assays, high-throughput drug-screening systems, portable environmental screening devices, miniature DNA analysers, etc. [1,2]. These chips are frequently being referred to as "microfluidics", in analogy with the well-known "microelectronics" chips. A similar revolutionary impact is expected of these so-called "labs-on-a-chip", with probably numerous as-yet-unforeseen applications.

Another field in which micromachined devices are of growing importance is that of microreactor development (see e.g. [3,4]). Microreactors have channels of a few hundred μm wide and a few mm long, and therefore exhibit extremely large surface-area-to-volume ratios, so that heat control can be optimised easily. The small amount of injected chemicals, combined with the high rate of heat removal, leads to inherently safe reactor systems, allowing highly exothermic reactions as well as reactions with highly toxic ingredients, which can be carried out at the location where they are needed, therewith avoiding the need for storage and transport. The small dimensions and the large surface-to-volume ratio allow a quick interruption of chemical processes, which may lead to higher selectivity and thus to higher-quality products.

Microchannels are basic entities of both microfluidic systems and microreactors. A large number of fabrication methods for channels in the size desired for such applications are available nowadays, like laser cutting, LIGA, and metal lamination (see other contributions to this conference). In this paper we restrict ourselves to the fabrication of photolithographically defined channels etched in silicon and the use of such channels as replicas for further fabrication. A number of examples of their application in micro chemical systems will be described.

2. Etching of channels in single-crystalline silicon

Methods to etch structures in single-crystalline silicon are well documented and are being applied in an increasing number of commercial products [5,6]. Here we focus on methods that are particularly suited for the fabrication of channels for micro chemical systems. For such devices, special requirements may hold, that may not be fulfilled by each of the etching methods available. Relevant issues are: the exact shape of the cross-section of the channels, the attainable number of channels per unit area, the freedom of channel design (e.g. sharp bends, tapered channels), and the roughness of the inner walls of the channels. We shall discuss these issues one by one for some relevant etching methods.

2.1. Anisotropic wet-chemical etching

This is the conventional method for the fabrication of silicon microstructures. The shape of channels attainable with this technique (most commonly performed in alkaline solutions) depends on the difference in etching rates for different crystallographic directions, but for the most commonly used substrates, i.e. {100}, channels with side-walls inclined ca. 55° to the substrate surface, corresponding to {111} planes, are obtained [5,6]. Such channels have been used with success in several micro chemical systems [7,8]. The maximum number of parallel, straight channels that can be achieved on a certain area depends on the desired depth of the channels, i.e. the width of one channel is always at least $\sqrt{2}$ times its depth, provided that the channels are carefully aligned along a <110> direction and that the {111} etch rate is negligible with respect to that of the {100} faces. The latter is, however, as was shown recently, rarely the case [9]. Choosing another type of substrate, like {110}, may lead to some improvement, but the freedom of design remains limited. The latter is especially true if one considers the fabrication of channels with sharp bends, where, because of the anisotropic etching rates, never the exact shape as defined by the mask is obtained, because convex corners tend to develop towards a shape limited by the fastest etching planes, and concave corners become enveloped by the slowest etching planes. Corner compensation structures in the masking layer will help, but this requires complex fine-tuning of the mask design and excellent knowledge of the orientation dependence of etching rates.

The surface roughness of anisotropically etched channels may be rather high, depending on the exact conditions of etching (temperature and composition of the etchant). Recent work has shown that this roughness is inherent to anisotropic etching: e.g. on {111} planes etching only proceeds when nucleation of etch pits is initiated by defects (e.g. oxygen-induced stacking faults) [9].

2.2. Isotropic wet-chemical etching

This method leads to channels with smooth sidewalls. Since etching is not limited by surface reactions, as is the case in the anisotropic method, but by mass transport in the solution, under optimised conditions any protrusion from a nominally flat surface will be removed at a higher rate than that flat surface. The channel shapes obtained with this method are tubular. Fig.1 shows a typical result.

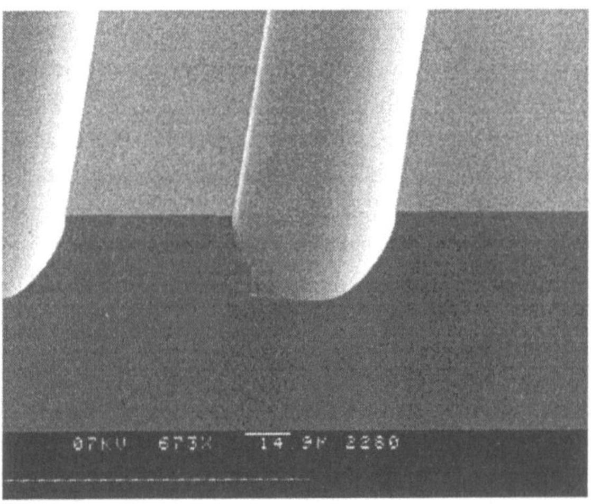

Fig.1. Cross-section of isotropically etched channels in silicon; note that the width of the channel is almost twice its depth.

Again the density of parallel lines will be limited, as in this case, due to underetching under the mask, the width of the channel will always be at least twice its depth. The underetching also limits the freedom of design considerably.

2.3. Anodic wet-chemical etching

This method, which exploits an anodic potential applied to the substrate to induce etching, gives essentially the same results as the previous method, except for the possibility to form porous silicon, which will be discussed below.

2.4. Reactive Ion Etching

Fig.2. gives a typical result of channels fabricated with Reactive Ion Etching (RIE), in this case in an SF_6-O_2-based inductively coupled plasma process. This method allows the largest degree of freedom of structural design: the pattern defined in the mask can directly be projected into the silicon substrate, allowing sharp corners and channels with in-plane tapers. The number of channels per unit area can be as high as the lithographic steps allow; state-of-the-art optical lithography achieves feature sizes of less than 0.25 μm, while with laser interference methods repetitive features smaller than 100 nm can be achieved [10].

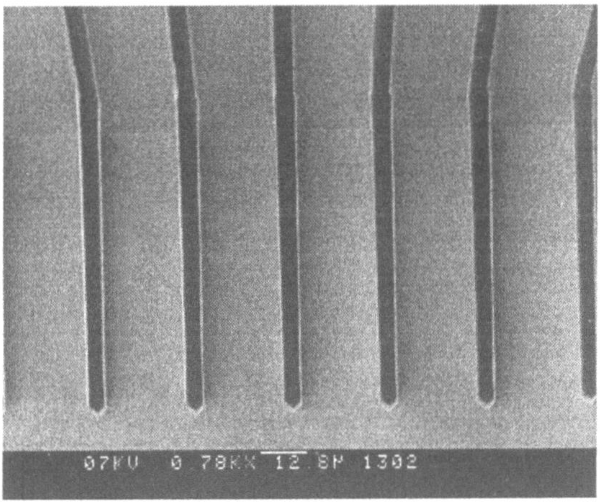

Fig.2. Deep trenches etched in an SF_6-O_2 plasma; the walls of the ca. 4 μm wide trenches are covered with a silicon nitride layer.

3. Porous silicon

If silicon is etched in HF-solutions at low anodic current densities, it transforms into a porous layer, of which the pore size distribution is well-defined and can be controlled accurately [11]. Recently we have demonstrated that, by altering the current density, freestanding structures of porous Si can be fabricated, as shown in Fig.3 [12], with a pore size of ca. 13 nm. Many applications of such a porous material in micro biochemical analysis and micro reactor devices can be imagined, and some interesting applications have already been realised. For example, it was reported that porous Si, used in an enzyme reactor, shows a 100-fold increased enzyme activity, compared to a non-porous reference [13]. Porous silicon can be oxidised to obtain porous silicon oxide [14], treated with ammonia to form silicon nitride [15] or transformed to other materials with high surface area either by gas phase reactions or by deposition of materials from the gas phase into its pores (like metals [16]). In this way many different materials with a high surface-to-volume ratio can be manufactured (if desired, in a tubular channel form like in Fig.3), which can be used in gas chromatography, as molecular sieves (microdialysis), as gas sensors [17] or as very effective catalysts in heterogeneous gas phase microreactors, to name a few. Structures like in Fig. 3 are very interesting for the latter application, since it is possible to micromachine a construction in which gases can be passed from the central to the next channel, while (although this requires quite complex processing) the outer channel in Fig.3. could be used to pass a cooling substance.

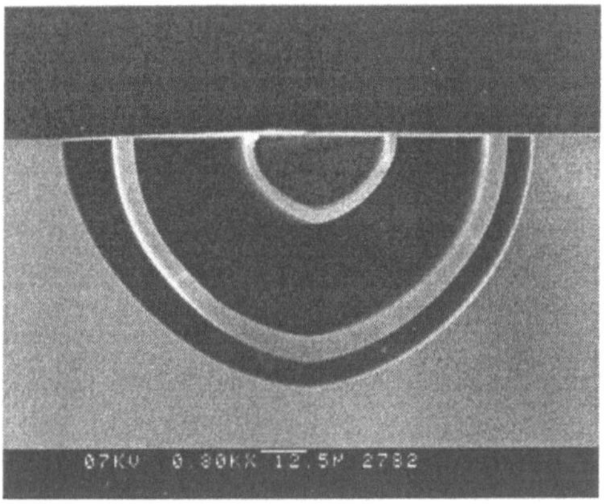

Fig.3. Several concentric channels composed of nano-porous silicon, attached to a silicon nitride mask

4. Microreactor applications

Microfabrication in silicon offers new opportunities for microreactors, mainly because of the possibilities of integration. Reactors containing catalytically active channel structures with integrated heating/cooling units and temperature sensors have already been demonstrated [8]. Future systems might also contain integrated sensors for flow and pressure, mixers, valves and flow controllers. For the time being, however, some important issues still have to be solved.

As demonstrated above, fabrication of micro channels in many different shapes and materials is possible, but for applications in microreactors, surfaces of sufficient catalytic activity are required. Silicon itself can not be considered a good catalyst for any reaction of industrial interest, and therefore has to be coated with such a material. This is a real challenge, as was found in the first attempts in this direction [8]. Catalysts of interest for high-temperature reactions are mostly based on noble metals like Pt, and such metals generally do not adhere well to silicon dioxide- or nitride-covered silicon (direct contact between the noble metals and silicon leads to alloying and should be avoided). To improve adhesion, a less noble metal is used, and this metal cannot withstand the severe oxidizing and high-temperature reaction conditions. An example is shown in Fig.4, which is the result of annealing sputtered Pt on different adhesion layers at temperatures up to 800 °C in an oxidizing atmosphere. The figure shows that Ta is a better choice than the Ti layer frequently reported in literature (see e.g. [18]). Once the problem of high-temperature stability of such metallic layers is solved, the integration of heaters and temperature sensors (e.g. based on Pt films) will be straightforward.

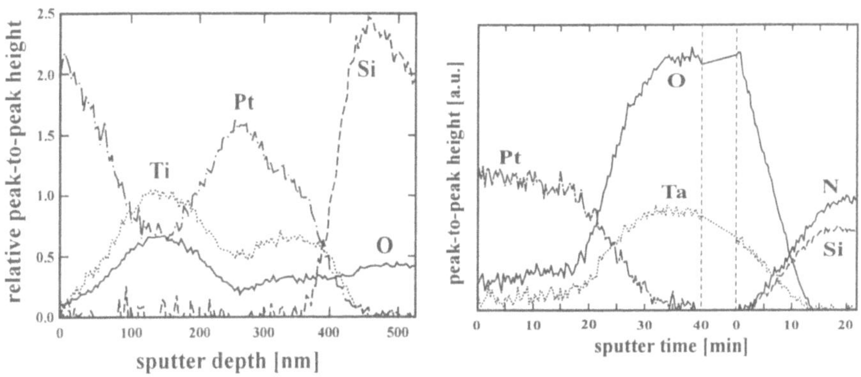

Fig.4. AES depth profiles of Pt films after annealing. Left: the original Pt-Ti-SiO_2 package shows severe interdiffusion, and evidence of TiO_2 formation; right: the original Pt-Ta-Si_3N_4 package shows no significant interdiffusion, although the Ta layer has completely oxidized to Ta_2O_5. Adhesion is preserved in the latter case.

5. Replication of silicon microchannels

Although the applications of silicon channels are numerous, their use is limited in some of the current fields of interest in chemical analysis. For example, in Capillary Electrophoresis (CE) high electrical fields along the length of a channel are required, which is not possible in silicon because of its high electrical conductivity. Another serious drawback of silicon is that it exhibits no optical transparency for the visible or UV wavelength regime used for detection. For such applications a more suitable material, like glass, is highly desirable.

One particularly interesting strategy is to use silicon channels as a mould for the deposition of insulating materials, like silicon nitride, silicon oxide or glass. This turns out to be very feasible [19,20]. Fig.5 shows some examples, in which a combination of silicon etching, silicon oxide deposition, and anodic bonding to a glass plate was applied, giving micro Transparent Insulating Channels (μTICs). Earlier results have shown that these channels support electrical field strengths of up to 4 kV/cm and are well-suited for CE [20].

Since these channels may have very thin walls and small diameters, completely new types of fluidic devices become feasible. An additional important advantage of the use of glass-like materials is that they are transparent for light in an interesting wavelength regime for optical detection of e.g. biochemical components. It is also possible to fabricate channels at different depths below the surface of a silicon substrate [19], thus leaving the silicon substrate surface free for the implementation of other microstructures or electronic circuitry. Such channel structures, fabricated by a combination of RIE, isotropic wet-chemical etching, and reactive sealing with a Low-Pressure Chemical Vapour Deposition silicon-rich silicon nitride coating, are shown in Fig.6.

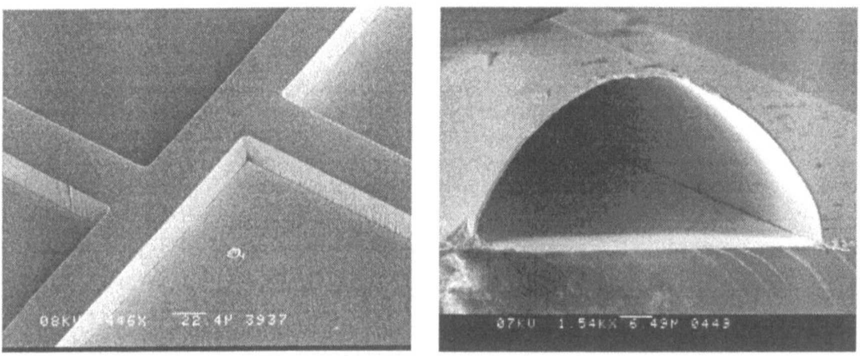

Fig.5. Examples of "µTICs". The left channel is based on a channel etched in Si by RIE, the right channel results is a replica of an isotropically etched Si pattern.

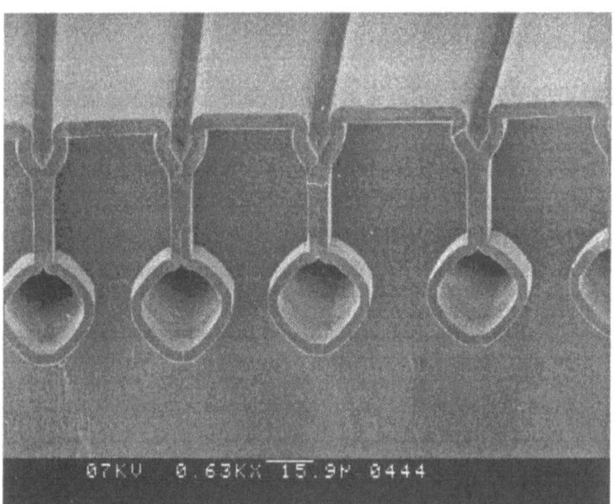

Fig.6. Silicon nitride tubes, buried underneath the surface of a silicon substrate

6. Summary

A number of micromachining techniques for the fabrication of different types of microchannels, including electrochemical and electroless (an)isotropic etching, and reactive ion etching, as well as their application in currently investigated microfluidic devices was described. Such channels, with dimensions of a few μm to a few hundred μm, are applied as e.g. fluid interconnects, mixing units and separation or reaction channels in integrated micro chemical analysis systems. A special topic is the fabrication of porous silicon channels, which, after surface treatment, may be used as sieves or as high-surface area catalysts, e.g. in microreactors for partial gas phase reactions or for enzyme reactions. The method with which the latter channels are made also allows the fabrication of multiple tube-in-tube channels, which become important as counterflow exchange channels for heat (e.g. for microcooling systems) or chemicals (e.g. for microdialysis). To avoid the use of a material with a high electrical conductivity and low optical transparency, like silicon, but still have access to the same flexibility of channel design, replication techniques to make channels in e.g. silicon dioxide were discussed, which use the etched silicon channels, described above, as a mould for the deposition of insulating and/or transparent materials. This allows the fabrication of channels to be used in e.g. capillary electrophoresis.

Acknowledgments

The authors thank Y. Fintschenko, V.L. Spiering (TMP), G.J. Burger (TMP), H.V. Jansen, J.W. Berenschot and M.J. de Boer for experimental contributions and discussions on microfabrication technology, and A.H.J. van den Berg for AES measurements. J.C. Schouten and M.H.J.M. de Croon of the Laboratory of Chemical Reactor Engineering at the Eindhoven University of Technology are gratefully acknowledged for discussions on microreactor developments. R.W. Tjerkstra acknowledges the financial support of the Dutch Technology Foundation, STW (project nr. TEL.3417).

References

1. R.F. Service, *Coming soon: The pocket DNA sequencer*, Science 282 (1998) 399.
2. M.A. Burns, B.N. Johnson, S.N. Brahmasandra, K. Handique, J.R. Webster, M. Krishnan, T.S. Sammarco, P.M. Man, D. Jones, D. Heldsinger, C.H. Mastrangelo and D.T. Burke, *An integrated nanoliter DNA analysis device*, Science 282 (1998) 484.
3. V. Hessel, W. Ehrfeld, H. Möbius, Th. Richter and K. Russow, *Potentials and realization of microreactors*, Proc. Int. Symp. Microsystems, Intelligent Mat. and Robots, sept. 27-29, 1995, Sendai, Japan, p. 45.
4. J.J. Lerou, M.P. Harold, J. Ryley, J. Ashmead, T.C. O'Brien, M. Johnson, J. Perotto, C.T. Blaisdell, T.A. Rensi and J. Nyquist, *Microfabricated minichemical systems: technical feasibility*, DECHEMA Monographs 132 (1996) 51.

5. M.J. Madou, *Fundamentals of Microfabrication,* 1997, CRC Press, Inc., ISBN 0849394511.
6. M. Elwenspoek and H.V. Jansen, *Silicon Micromachining,* 1999, Cambridge University Press, ISBN 052159054X.
7. Th. Laurell and L. Rosengren, *A micromachined enzyme reactor in <110> silicon,* Sensors & Act. B 18-19 (1994) 614.
8. R. Srinivasan, I-M. Hsing, P.E. Berger, K.F. Jensen, S.L. Firebaugh, M.A. Schmidt, M.P. Harold, J.J. Lerou and J.F. Ryley, *Micromachined reactors for catalytic partial oxidation reactions,* AIChE Journal 43 (1997) 3059.
9. A.J. Nijdam, J. van Suchtelen, J.W. Berenschot, J.G.E. Gardeniers and M. Elwenspoek, *Etching of silicon in alkaline solutions: a critical look at the {111} minimum,* J. Crystal Growth 198-199 (1999) 430.
10. S H Zaidi and S R J Brueck, *Multiple-exposure interferometric lithography,* J. Vac. Sci. Technol. B, 11 (1993) 658.
11. L.T. Canham and A.J. Groszek, *Characterization of microporous Si by flow calorimetry: Comparison with a hydrophobic* SiO_2 *molecular sieve,* J. Appl. Phys. 72 (1992) 1558.
12. R.W. Tjerkstra, J.G.E. Gardeniers, M.C. Elwenspoek and A. van den Berg, *Electrochemical fabrication of multi-walled microchannels,* Proc 3rd μTAS Workshop, Banff, Canada, Oct. 13-16, 1998, p. 133.
13. J. Drott, K. Lindstrm, L. Rosengren and T. LAurell, *Porous silicon as the carrier matrix in microstructured enzyme reactors yielding high enzyme activities,* J. Micromech. Microeng. 7 (1997) 14.
14. K. Imai and H. Unno, *FIPOS (Full Isolation by Porous Oxidized Silicon) technology and its application to LSI's,* IEEE Trans. Electron Dev. ED-31 (1984) 297.
15. R.C. Anderson, R.S. Muller and C.W. Tobias, *Chemical surface modification of porous silicon,* J. Electrochem. Soc. 140 (1993) 1393.14.
16. B.J. Aylett, L.S. Harding, L.G. Earwaker, K. Forcey and T. Giaddui, *Metallisation of porous silicon by chemical vapor infiltration and deposition,* Thin Solid Films 276 (1996) 253.
17. C. Cobianu, C. Savaniu, O. Buiu, D. Dascalu, M Zaharescu, C. Parlog, A. van den Berg and B. Pecz, *Tin dioxide sol-gel derived thin films deposited on porous silicon,* Sensors and Act. B 43 (1997) 114.
18. S.L. Firebaugh, K.F. Jensen and M.A. Schmidt, *Investigation of high-temperature degradation of platinum thin films with an in situ resistance measurement apparatus,* J. Micro Electro Mech. Syst. 7 (1998) 128.
19. R.W. Tjerkstra, M.J. de Boer, J.W. Berenschot, J.G.E. Gardeniers, M.C. Elwenspoek and A. van den Berg, *Etching technology for microchannels*, Proc. IEEE Workshop on Micro Electro Mechanical Systems, Nagoya, Japan, Jan. 26-30, 1997, p. 147.
20. Y. Fintschenko, P. Fowler, V.L. Spiering, G.J. Burger and A. van den Berg, *Characterization of silicon-based insulated channels for capillary electrophoresis*, Proc. 3rd μTAS Workshop, Banff, Canada, 13-16 Oct., 1998, p. 327.

Macroporous and Mesoporous Silicon: New Materials for Microfluidic and Microreaction Devices

V. Lehmann

Siemens AG, Dept. ZFE T ME 1
81730 München, Germany

Abstract. The formation of mesoporous films and macropore arrays with high aspect ratios by electrochemical etching of silicon electrodes in hydrofluoric acid is a well established technique, which has been applied to manufacture electronical and optical devices. Applications of this new materials in fluidic and microreaction devices, however, have not been realized so far. This work provides detailed information about pore dimensions, pore morphology, and surface conditions, which shall enable researchers from other fields to estimate the potential of porous silicon for their applications.

Keywords: porous silicon, microreactors, filters, microfluidic devices, channel plates

1 Introduction

That porous silicon is formed during anodization of silicon electrodes in hydrofluoric acid is known since roughly half a century [1]. A decade ago it was found that the growth of macropores on n-type Si substrates can be initiated by a small etchpit [2]. This led to applications in microelectronics or optics, like for example miniaturized capacitors [3] and photonic crystals [4]. But this technology offers also new possibilities for biotechnological, medical and chemical applications, due to its high freedom of design. In order to evaluate if porous silicon is an option for a certain application the feasibility of certain pore geometries, pore patterns and pore coatings has to be known. It is the aim of this report to give a conclusive picture of the possibilities offered by this new material. The mesopore and macropore fabrication process will be reviewed only briefly.

2 Size regimes of porous silicon and fabrication process

Since the details of the electrochemical pore formation process are published in detail elsewhere [2,5,6] only a brief description will be given here. Porous silicon is formed on silicon electrodes anodized in HF with current densities below a critical value Jps. The regime of pore formation in the I-V characteristic of the semiconductor electrolyte junction is shown in Fig. 1. The setup for pore formation is basically an o-ring cell with a window on the backside of the wafer for illumination, as shown in Fig. 2b. The structural dimensions and the morphology of a porous silicon layer, depend sensitively on anodization parameters such as HF concentration, doping type and density, current density, bias, illumination and temperature. Basically three size regimes can be distinguished:

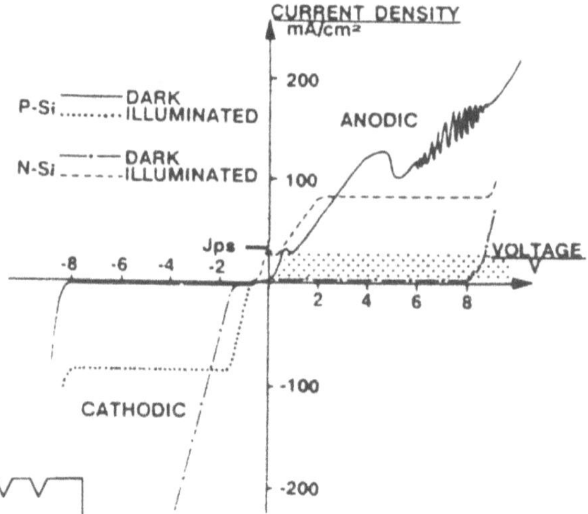

Fig. 1: The I-V-characteristic of a silicon electrode in hydrofluoric acid is shown in this figure. Electrochemical pore formation is observed for anodic current densities below a critical value Jps. This regime is indicated by a dot-pattern.

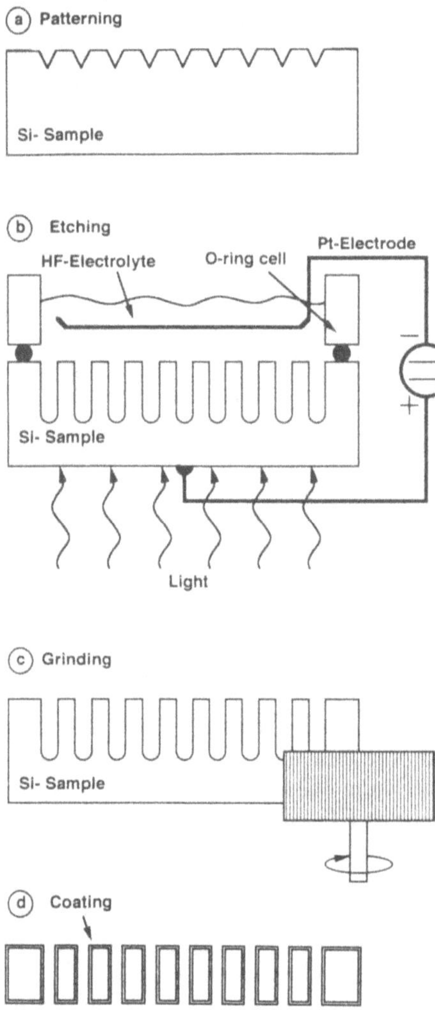

Fig. 2: The basic fabrication steps for a coated macropore array with thru-pores.
a) First the electrode is patterned by standard photolithograpy and subsequent alkaline etching.
b) Then electrochemical pores are formed by anodization.
c) Subsequently the pores are opened by grinding or etchback and
d) coated with a thin film.

A) Microporous silicon shows a random, sponge like pore structure, with pore diameters and pore wall thicknesses in the order of 2nm. Porosities range from 30% to 80% and the inner surface may be as large as $1000m^2/cm^3$. Such microporous films are very fragile, which limits the number of potential applications.

B) Mesoporous silicon is formed on highly doped substrates and shows random pore structures with pore diameters ranging from about 10nm to 50nm and pore wall thicknesses from 5nm to 200nm. Porosities from 10% to 80% and inner surfaces of about $100m^2/cm^3$ can be realized. The good mechanical stability of mesoporous silicon and its high inner surface makes it a potential candidate for applications such as filters or catalyst supports. Typical mesoporous samples are shown in Fig. 3a,b.

C) Macropores form in moderately doped n-type silicon if it is anodized in hydrofluoric acid while the backside of the wafer is illuminated. If this process is applied to a polished silicon substrate, the pores will grow in a random pattern. However if etch pits, for example generated by standard lithography and alkaline etching, are present, pore formation will start at these etch pits. By this procedure well ordered arrays of pores orthogonal to the electrode surface can be generated in the silicon substrate as shown in Fig.4. Diameters range from about 300nm to 10μm and pore wall thicknesses from 100nm to 50μm. Porosities from 1% to 90% can be realized. The upper limit for the surface enhancement is $1m^2/cm^3$. The high freedom in pore design and the superior mechanical stability of macroporous silicon offers a multitude of applications.

3 The pore diameter

The pore diameter of macropores in an array, as shown in Fig. 4 can be adjusted between 0.3μm and 10 μm by selecting the appropriate substrate doping (0.1 Ωcm to 100 Ωcm n-type single crystalline silicon) and etching current (usually in the range 1-100 mA/cm^2). The pore diameter can be well controlled and its deviation along a single pore and over the chip area is usually less than 10%. Pore diameters larger than 10 μm can be realized if the diameter of the pores is increased by subsequent chemical etching or oxidation. More complicated pore crossections like for example slits, can be realized in a similar manner by enlarging a row of pores, as shown in Fig. 6b. The pore crossection geometry after electrochemical etching is usually a square with rounded corners, as shown in Fig. 6a. This crossection can be changed to circle by subsequent oxidation or to a square or octagon by subsequent chemical etching.

Smaller pore diameters in the mesoporous regime can be realized using highly doped substrates for anodization (0.001 Ωcm to 0.05 Ωcm n-type and p-type single crystalline silicon). Then, however the pores distribution becomes random and branching of the pores is unevitable, as shown in Fig. 3a,b. The mesopore diameters show deviations in the order of 50%, which is still a quite narrow range if compared with other porous materials.

4 Pattern and porosity

The possibility to control macropore initiation and macropore geometry gives a high degree of freedom in design. Pores may be arranged in orthogonal, triangular, hexagonal, and many other patterns. However, there are limitations for the pore pattern inherent to the electrochemical etching process. Macropore array formation

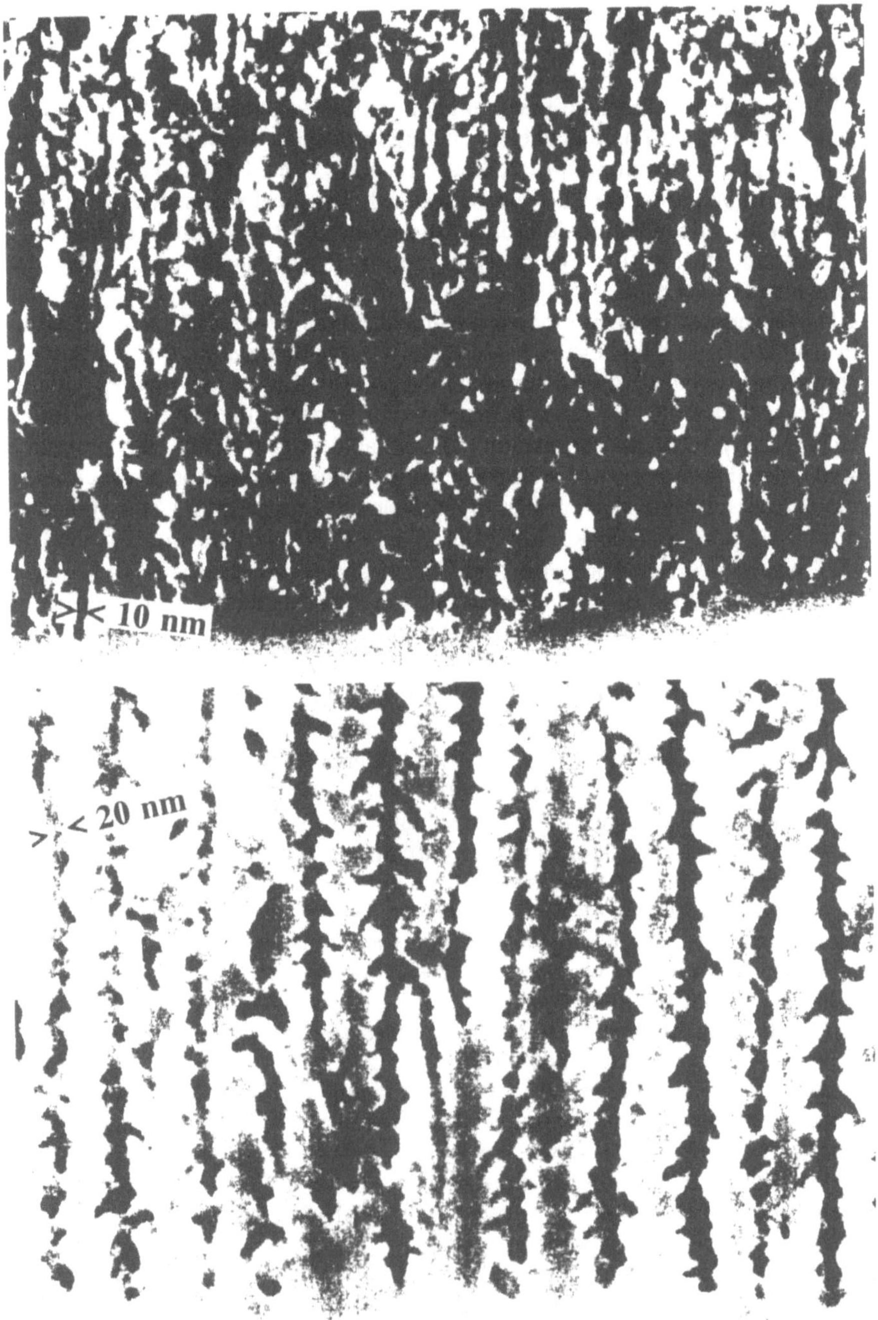

Fig. 3: SEM- micrographes of typical mesoporous structures formed by anodization. The porewalls are rough and show branching pores.
(top: p-type substrate, 50% porosity, bottom: n-type substrate, 20% porosity).

works best with patterns of a constant etchpit density. It does not matter if the pattern shows a symmetry (orthogonal, hexagonal, ect.) or if it is random. In the latter case the diameter of the pores shows variations depending on the distances to the neighboring pores. Stable pore formation is possible for ratios of pitch to pore diameter from 1 to about 10. This corresponds to porosities from close to 90% to about 1% or ratios of pore wall thickness to pore diameter of 0.1 to about 10.
So far only homogeneous pore patterns of constant etchpit density have been addressed. Small deviations from a perfect porosity like a missing pore or a missing row of pores are tolerable. A variation in porosity over larger areas would require a lateral variation of backside illumination intensity during electrochemical etching. This however, would generate a smoothed out current density distribution on the front side due to the random diffusion of the charge carriers in the bulk of the substrate. It is therefore difficult to realize closely spaced areas of porous and non-porous silicon.
A change of macropore array porosity over the pore depth can easily be realized by a change of the pore diameter as discussed in the next section.

5 The pore shape and pore depth

The remarkable straightness of macropores is a prominent feature of the etching process. It is caused by the <100> direction of the silicon single crystal being the prefered pore growth direction. The maximum macropore depth is limited to values below 1mm due to diffusional limitations. Under standard anodization conditions all pores in an array show the same depth. The required etching time shows a parabolic dependence on pore depth. Etching times of about a day, for example, are required for a pore depth of 0.5mm, while 0.1mm can be etched within one hour. By adjusting the etching current the pore diameter can be varied by about a factor of two. This enables us to realize different pore shapes like bottleneck, funnel or straight pores, as sketched in Fig. 6c. Even an array of sinewave shaped macropores, as shown in Fig. 5, can be realized.
Macropore arrays are usually etched to a certain depth, which is smaller than the total substrate thickness, thus realizing dead-end pores. If thru-pores are desirable the remaining substrate can be removed by grinding or etchback (fig. 2c). In the latter case only a certain fraction of the pores can be opened by defining a backside window for etchback, as shown in Fig. 6d.
The thickness of mesoporous silicon films is usually limited to values in the order of 0.1mm. Larger chip thicknesses which are required for good mechanical stability can be obtained by a combination of a mesoporous film with a macropore array.

6 Substrate dimensions

The maximum porous area is only limited by the wafer size. Macropore arrays have been manufactured on silicon wafers of 100 mm and 150 mm diameter. Etching of 200 mm and 300 mm would require larger electrolytic cells, however, there is no physical limit for the etchable wafer size. Smaller areas can be realized by cleaving the silicon wafer into chips.

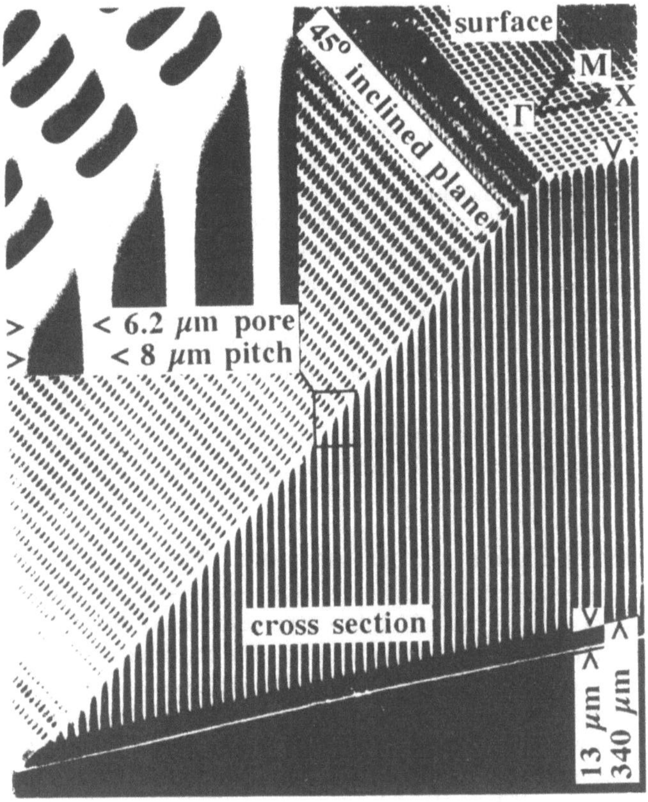

Fig. 4: Macropore array with a square pattern.
The remarkable straightness of the pores is due to the fact that macropores grow preferably along the <100> direction of the Si crystal.

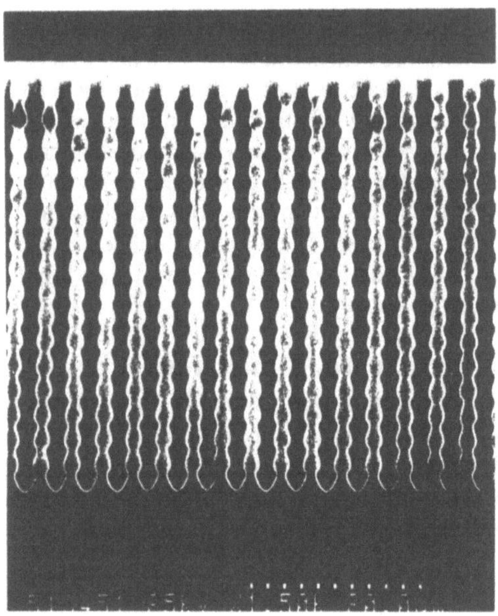

Fig. 5: A sinewave modulation of the etching current with etching time (+-50%) produces an array of pores with a modulation of the pore diameter. The pore depth is 40μm.

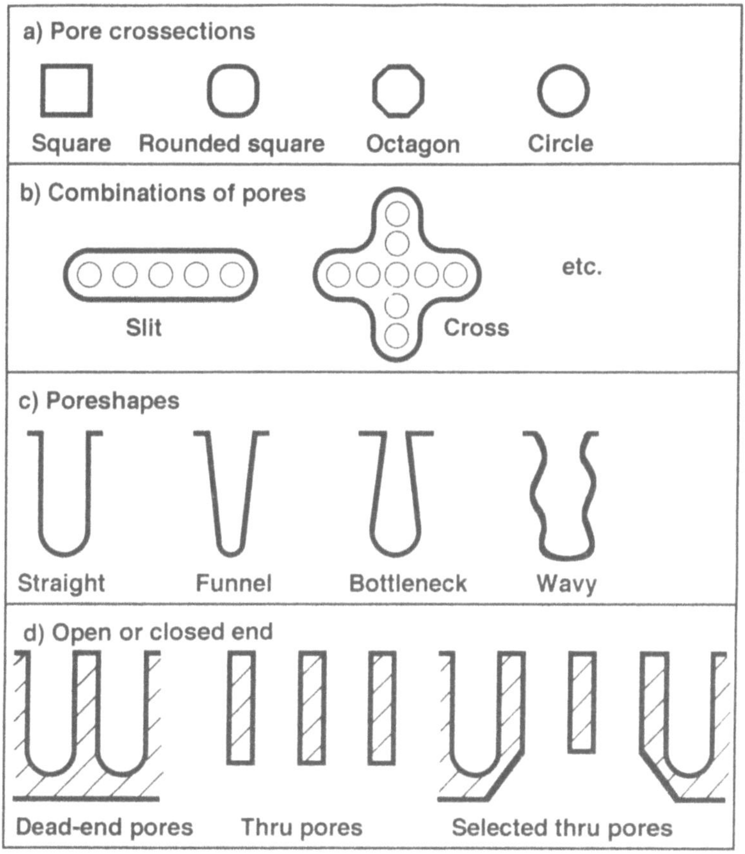

Fig. 6: This figure shows sketches of the feasible pore geometries: a) The pore crossection parallel to the electrode surface can be tuned between a square and a circle. b) Complex pore crossections can be realized by individual pores joint by subsequent chemical etching. c) The pore crossection orthogonal to the electrode surface (the pore shape) depends on the current-time profile of the electrochemical etching process and can therefore be precisely controlled. d) All pores or selected pores can be opened by etchback.

7 Coatings

The porous silicon surface is hydride covered immediately after etching. If stored under ambient conditions, however, a native oxide of about 1nm is formed within one day. By thermal oxidation a SiO_2 layer from 1 nm - 1000 nm can be formed on the Si surface. Silicon nitride (Si_3N_4) can be deposited in macropores using chemical vapour deposition (CVD). With such coatings the silicon surface becomes inert to most chemicals. Cleaning, for example, can easily be performed under harsh conditions, like in pure oxygen at 1200°C.
For catalyst applications coatings with metals, like for example Pt, are desirable. Thin metal films can in principle be deposited by CVD processes. Such processes have, however, not been applied to macropore arrays so far.
Mesoporous surfaces can as well be coated. Self-evidently the thickness of the coatings is limited by the pore dimensions to a few nanometer.

8 Conclusions

The high flexibility in pore size and pore geometry makes macroporous silicon a perfect material for the design of microfluidic and microreaction devices. The superior aspect ratios of macropore arrays and the option to combine them with mesoporous films increase the freedom in design. The inherent high temperature stability of crystal silicon and the possibility to apply various surface coatings enables us to realize custom-made solutions for aggressive chemicals. In contrast to microchannel glass or porous alumina macropore arrays show a long range order due to the photolithographic patterning. This is advantageous for integration with electronic devices or optical and electrical readout methods.

9 References

[1] D. R. Turner, *J. Electrochem. Soc., 105* (1958), 402
[2] V. Lehmann and H. Föll, *J. Electrochem. Soc., 137* (1990), 653.
[3] V. Lehmann,W. Hönlein, H. Reisinger, A. Spitzer, H. Wendt, and J. Willer, *Solid State Technol. 38*, (1995), 99.
[4] U. Grüning, V. Lehmann, S. Ottow, and K. Busch, *Appl. Phys. Lett. 68,* (1996) 747.
[5] V. Lehmann, *J. Electrochem. Soc.,140* (1993) 2836.
[6] V. Lehmann, *Thin Solid Films, 297* (1997) 13-17.

The Effect of Shape Variation in Microlamination on the Performance of High-Aspect-Ratio, Metal Microchannel Arrays

Brian K. Paul[1], Richard B. Peterson[2], and Wassanai Wattanutchariya[1]

[1] Department of Industrial and Manufacturing Engineering, Oregon State University, Corvallis, Oregon 97331-2407, USA

[2] Department of Mechanical Engineering, Oregon State University, Corvallis, Oregon 97331-2407, USA

Abstract. One means of enhancing heat and mass transfer rates is by pumping fluids through high-aspect-ratio microchannel (HARM) arrays. HARM arrays provide a high surface-area-to-volume ratio for both heat transfer and chemical reactions with minimal pressure drop. Past efforts to fabricate HARM arrays have included microlamination methods. Microlamination methods involve the forming, alignment, and bonding of thin laminates to produce composite microdevices. Past observations have been made of dimensional variation within microlaminated devices. It is well known that under certain fluidic conditions, the shape of a fluid conduit directly influences the heat and mass transfer rates within the conduit. Therefore, the emphasis of this paper is to investigate the effect of shape variation in microlamination on the performance of HARM arrays. Several metal HARM arrays are fabricated using a microlamination procedure. Prior microlamination methods have emphasized the use of diffusion bonding, soldering, or brazing techniques. This paper introduces an adhesive bonding technique for fabricating metal microchannel arrays. The performance of the microchannel arrays is evaluated by measuring the pressure drop across the array as a function of flow rate and comparing it with a theoretical model. Statistical results are analyzed to determine the effect on performance as well as the estimated dimensional variation within the arrays.

Keywords: High-Aspect-Ratio, Microchannels, Copper, Microlamination, Microfabrication, Micromanufacturing, Shape Variation

1. Introduction

Over the past three decades, sheet lamination techniques have been used to fabricate a variety of complex geometric structures which were difficult to manufacture with more conventional techniques.[1,2,3,4,5] Sheet lamination techniques involve the stacking and bonding of thin pre-formed laminates or sheets. Each laminate is pre-formed with a two-dimensional pattern so that when the laminates are bonded together, they form a functional solid geometry. Laminates may be either metal, ceramic, polymer or glass and can be machined, photoetched, or laser cut. The laminate stack may be diffusion bonded, fusion bonded, or even adhesive bonded.

More recently, these sheet lamination techniques have begun to be applied to the fabrication of microfluidic devices through microlamination processes. Microlamination processes involve the forming and bonding of thin (50 to 500 μm) laminates to produce a three-dimensional microdevices with internal features. One example of a microfluidic structure that has been fabricated successfully with microlamination is a high-aspect-ratio microchannel (HARM) array. A HARM array is a series of microchannels whose height-to-width ratio is greater than 20:1. High-aspect-ratio microchannels have been shown to improve lateral heat transfer properties of microchannels due largely to the increased surface area to volume ratio.[6,7] Additionally, in microscale chemical mixers and separators, HARM arrays have been found to improve material transport.[8]

Goldberg[9] fabricated copper, forced-air, HARM heat sinks using a microlamination procedure involving shearing and heli-arc welding of copper shims. Microchannel dimensions as small as 125 micrometers were produced. Krause et al.[10] microlaminated a water-cooled, HARM heat exchanger by registering five pre-formed copper laminates and diffusion bonding. Microchannel dimensions as small as 60 micrometers were produced. Kleiner et al.[11] assembled both copper and aluminum air-cooled, HARM heat exchangers with mechanical fasteners. Microchannel dimensions were on the order of 500 micrometers.

Several past research investigations have found that dimensional variation within microchannels result in degraded microchannel performance. Wu and Little[12] established that the size and substrate of microchannels both had significant effects on the friction factor, suggesting an influence of both surface roughness and dimensional accuracy on device performance. Pfahler *et al.*[13,14,15] conducted a series of experiments using both liquids and gases to measure the flow friction in microchannels. From these experiments, it was concluded that variation in the dimension of microchannels significantly increases friction. Cuta *et al.*[16] has confirmed these findings. Few studies have been conducted investigating the dimensional variation within microlaminated HARM arrays and the effect of this variation on device performance.

In this paper, a new adhesive bonding technique is introduced for microlamination of metal HARM arrays. To validate the new bonding technique, multiple HARM arrays are fabricated and pressure drop tests are performed. Because dimensional variation in microchannels affects the pressure drop across microchannels, the pressure drop tests are used to estimate the amount of dimensional variation within the microchannels. This is similar to the concept of air gaging used in industry to verify the size of small orifices. Results are compared with theoretical calculations.

2. HARM Array Description

A sketch of the HARM array and the test setup used for pressure drop tests is shown in Figure 1. Pressure drop tests were conducted across the array using water as the fluid medium. A constant pressure head was used to establish the required pressure drop and the resulting flow. The water flowed from the head to the HARM array through a tube with circular cross-section. After passing through the array, the water was collected in an open-air reservoir for positive displacement flow measurement. The array had a parallel plate-fin (PPF) architecture as shown

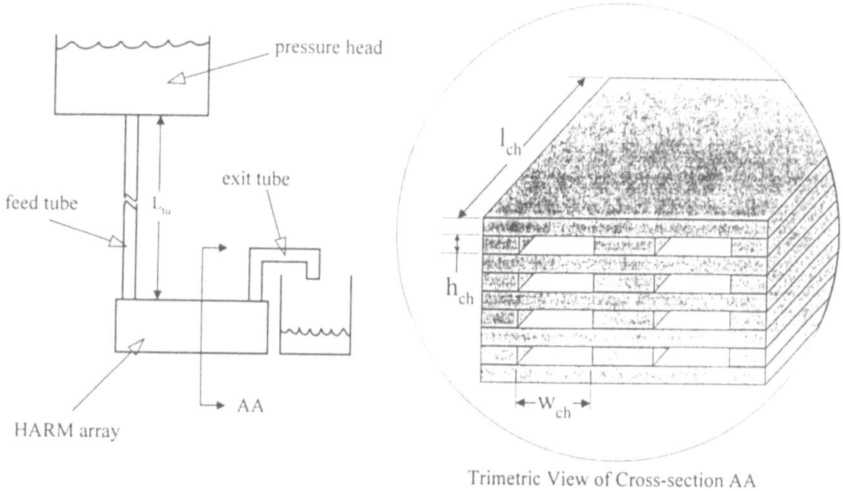

Figure 1. Schematic of HARM Array and Test Setup

in the cross-section view in Figure 1. Microchannel dimensions were 100 μm in height by 3 mm in width by 10 mm in length.

3. Theoretical

The total pressure supplied by the head, Δp_h, is equal to the sum of the total pressure drop across the HARM array, Δp_a, the pressure drop due to flow through the feed tube, Δp_{tu}, and the pressure drop across the entrance and exit of the microchannel array, Δp_{ee}:

$$\Delta p_h = \Delta p_a + \Delta p_{tu} + \Delta p_{ee} \tag{1}$$

Assuming that the fluid fully develops laminar flow, the pressure drop across the array, Δp_a, can be modeled as:

$$\Delta p_a = f_{ch} \cdot \frac{l_{ch}}{D_h}\left(\frac{\rho \cdot V_{ch}^2}{2}\right) \tag{2}$$

where f_{ch} is the friction factor within the microchannels, l_{ch} is the length of the microchannel, D_h is the hydraulic diameter for the microchannels, ρ is the fluid density, and V_{ch} is the mean fluid velocity within the microchannels. The friction factor is related to the Reynolds number for the microchannels, Re_{ch}, and, subsequently, to the kinematic viscosity of the fluid, v, as follows:

$$f = \frac{96}{\mathrm{Re}_{ch}} = \frac{96 \cdot v}{V_{ch} \cdot D_h} \tag{3}$$

The hydraulic diameter can be expressed as:

$$D_h = 2 \cdot \frac{w_{ch} \cdot h_{ch}}{w_{ch} + h_{ch}} \tag{4}$$

Total volumetric flow rate, $\dot{v}_{tot}$, through the array can be expressed as:

$$\dot{v}_{tot} = n \cdot V_{ch} \cdot A \tag{5}$$

where n is the number of microchannels in the HARM array and A is the cross-sectional area of a single microchannel:

$$A = w_{ch} \cdot h_{ch} \tag{6}$$

Rearranging Eq. 5 and substituting equations 3, 4, 5 and 6 into equation 2 yields:

$$\Delta p_a = \left(\frac{12 \cdot \nu \cdot \rho}{n} \right) \cdot \frac{l_{ch} \cdot (w_{ch} + h_{ch})^2}{(w_{ch} \cdot h_{ch})^3} \cdot \dot{v}_{tot} \tag{7}$$

The pressure drop due to flow through the feed tube, Δp_{tu}, may be predicted by:

$$\Delta p_{tu} = f_{tu} \cdot \frac{L_{tu}}{D_{tu}} \left(\frac{\rho \cdot V_{tu}^2}{2} \right) \tag{8}$$

where f_{tu} is the friction factor within the feed tube, l_{tu} is the length of the feed tube, D_{tu} is the diameter of the tube, and V_{tu} is the mean fluid velocity within the microchannels. Assuming that the flow in the tube is turbulent, the friction factor is calculated from the Blasius formula for fully developed turbulent flow through smooth pipes[17]:

$$f_{tu} = \frac{0.3164}{\mathrm{Re}_{tu}^{.25}} \tag{9}$$

which works well for Reynolds numbers between 3,000 and 100,000. The Reynolds number is given by:

$$\mathrm{Re}_{tu} = \frac{V_{tu} \cdot D_{tu}}{\nu} \tag{10}$$

and the mean velocity of the water flowing through the tube is:

$$V_{tu} = \frac{4 \cdot \dot{v}_{tot}}{\pi \cdot D_{tu}^2} \tag{11}$$

Neglecting the hydrodynamic entrance and exit effects within the microchannels and substituting Eqs. 7-11 into Eq. 1, the theoretical pressure supplied by the head is:

$$\Delta p_h = \left[\left(\frac{12 \cdot \rho \cdot \nu}{n} \right) \cdot \left(\frac{l_{ch} \cdot (w_{ch} + h_{ch})^2}{(w_{ch} \cdot h_{ch})^3} \right) \cdot \dot{v}_{tot} \right] + \left[\left(0.2415 \cdot \rho \cdot \nu^{1/4} \right) \cdot \left(\frac{L_{tu}}{D_{tu}^{19/4}} \right) \cdot \dot{v}_{tot}^{7/4} \right] \tag{12}$$

4. Experimental

A microlamination process was used to fabricate the HARM arrays. A diagram of the process is shown in Figure 1. The array consists of two sets of four microchannels with each set between two sets of header regions. Inlet and outlet ports into the header regions were machined into the top end cap. In addition to making external interfacing simpler, endcaps provided stiffness within the array. The array was produced by adhesive bonding alternating laminates of pre-formed, 100 µm thick copper shimstock and Kapton KJ polyimide. Microchannel-patterned laminates were fabricated from polyimide while microfin-patterned laminates were fabricated from copper. In addition to providing structure, polyimide laminates provided the mechanism for bonding. Laminates were produced using an ESI 8000c Laser Micromachining System with a 532 nm Nd:YAG laser. To consolidate the stack, the laminates were registered one to another, compressed, and heated within a precision jig. An observed tendency of the polyimide adhesive was for it to wick into the microchannels resulting in clogging. A response surface methodology was carried out on the parameters of time and pressure to minimize fouling while maximizing bond strength. Final bonding conditions were 265°C and 200 kPa for 1 minute. Pressure was estimated based upon the screw force provided by the jig. Immediately prior to the bonding of

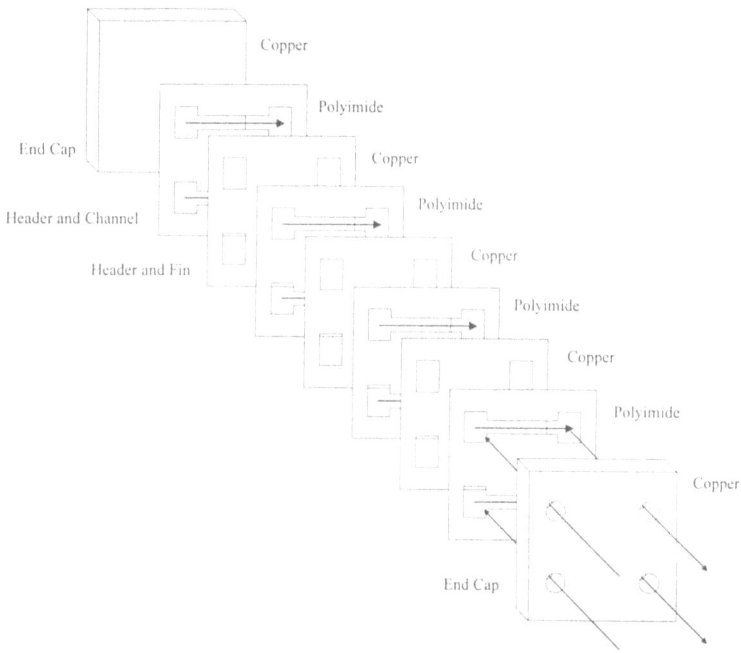

Figure 2. The microlamination scheme used to fabricate the HARM arrays. Arrows show the flow of fluid through the device.

the laminate stack, each laminate was acid etched ($HCl:HNO_3$) to remove contaminants that could hinder the bonding process. All preformed laminate dimensions were measured before bonding.

All arrays were tested for airtightness by pressurizing the devices to approximately 120 psi and submerging them in water. To determine the effect of the microlamination procedure on device performance, a pressure drop experiment was conducted on each HARM array and compared with theoretical calculations. The HARM array was tested by flowing water through the device at different head pressures and measuring the resultant volumetric flow rate through the device. A schematic of the test setup was shown in Figure 1.

5. Results

Figure 3 shows a picture and a cross-section of one of the HARM arrays fabricated using the procedure described above. Channel width was measured before and after bonding and found to be 3.005 and 2.992 mm, respectively, indicating a 0.27% change in width. The total height of the array was measured before and after bonding and found to be 7.446 and 7.428 mm, respectively. Assuming that this change in width was due to compression of the polyimide, this would indicate a 4.1 % (or 4.45 μm) change in the height of the microchannels. Cross-sectioning revealed that the microchannels are open and unclogged. In fact, no flash was found in any of the microchannels examined.

Airtightness tests of the HARM arrays indicated that the devices were hermetically sealed. Pressure drop tests were conducted across the arrays to investigate the effect of dimensional variation on device fluidics. Results of these pressure drop tests are shown in Figure 4. On average, actual volumetric flow rate results were 16.2% lower than the theoretical calculations which is in good agreement with theory. It is expected this deviation is due to entrance and exit effects as well as the development of laminar flow. This finding suggests that there is no clogging within the HARM arrays.

A statistical analysis was performed on the experimental flow curves from the four different arrays. It was found that the average coefficient of variance was consistent across the head

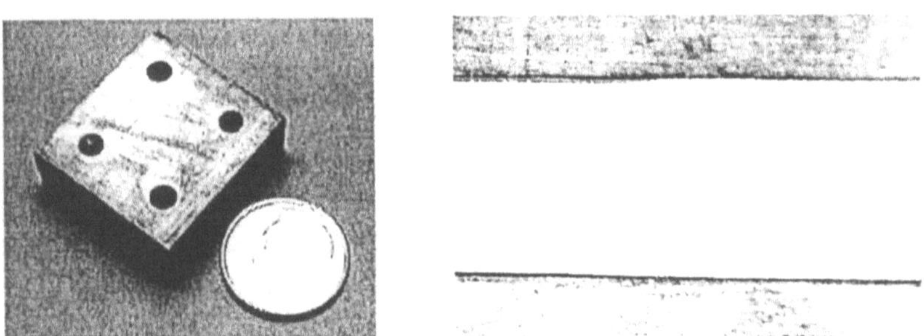

a) b)

Figure 3. Photographs of the HARM array fabricated by laser micromachining and adhesive bonding: a) next to a dime; and b) cross-section of width at 10X.

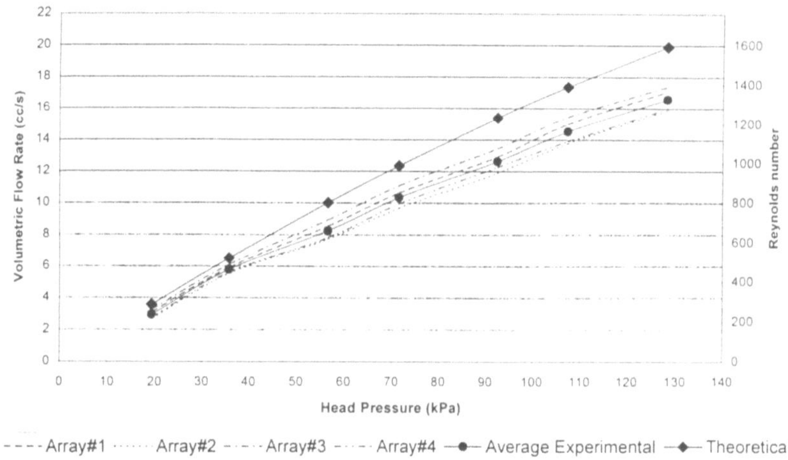

Figure 4. Comparison of actual pressure drop results with theoretical calculations.

pressure range at about 6.1%. This suggests that at a 95% confidence interval, the volumetric flow rate would vary less than 18.3% from the average shown in Figure 4. Assuming all of the pressure drop variation is due to dimensional variation within the HARM array and simplifying the hydraulic diameter as two times the microchannel height (since high-aspect-ratio microchannels), Eq. 12 simplifies to:

$$\Delta p_{ch} = \left[\left(\frac{12 \cdot \rho \cdot v}{n}\right) \cdot \left(\frac{l_{ch}}{w_{ch} \cdot h_{ch}^{3}}\right) \cdot \dot{v}_{tot}\right] \tag{13}$$

Assuming that all of the pressure drop variation is due to dimensional variation in the channel height and is proportional to the variation in volumetric flow rate, Eq. 13 can be rearranged:

$$\Delta h_{\max} = \left[\left(\frac{12 \cdot \rho \cdot v}{n}\right) \cdot \left(\frac{l_{ch}}{w_{ch}}\right) \cdot \frac{\Delta \dot{v}}{\Delta p_{h}}\right]^{1/3} \tag{14}$$

where Δh_{max} is the maximum channel height variation and $\Delta \dot{v}$ is the change in volumetric flow rate at a 95% confidence interval. Based on this equation, the maximum channel height variation with the HARM arrays would be approximately 12.6 μm per laminate or an 11.6% change in dimension. This is in good agreement with the actual channel height variation measured in the experiment and suggests that this is an overestimate of the actual dimensional variation within the microchannel array.

6. Conclusions

Four HARM arrays were fabricated with an average channel width of 109 μm at 30:1 aspect ratio using a microlamination process involving a new adhesive bonding technique. All arrays were found to provide hermetic sealing. Pressure drop tests as well as visual inspection indicate that all arrays were unclogged without evidence of flash. Internal channel dimensions were found to vary on average about 0.27% in the lateral direction and about 4.1% in the clamping direction after bonding. These dimensional variations resulted in a coefficient of variance for the volumetric flow rate of about 6.1%. Further studies are required to investigate the effect of geometric variation due to various methods of laminate forming, registration, and bonding on microfluidic device performance.

7. Acknowledgment

This work was funded by the Office of Naval Research. Also, the authors would like to express their thanks to Dr. Edward McDowell for his assistance.

8. References

[1] Young, M. "Platelet-Cooled Plasma Arc Torch", DOE Contract No. DE-AR21-93MC30361, Technology Development Data Sheet, (Morgantown, WV: Federal Energy Technology Center, 1997), pp5-6.

[2] Esposito, A. *Fluid Power with Applications*, (Englewood Cliffs, NJ: Prentice-Hall, 1988), pp380-381.

[3] Haas, R.W., D. Brest, H. Mueggenburg, L. Lang, and D. Heimlich, "Fabrication and performance of MMW and SMMW platelet horn arrays," *Intl. J. Infrared and Millimeter Waves*, (v14 , n11, 1993), pp2289-93.

[4] Haas, R.W., "Further development of MMW and SMMW platelet feed horn arrays," *Astron. Soc. Pac. Conf. Ser.*, (v75, 1995), pp99-105.

[5] Kennedy Space Center, *National Space Transportation System (NSTS) Shuttle Reference Manual*, (NASA, 1988).

[6] Pfahler, J.N., J. Harley, H.H. Bau and J. Zemel, "Liquid and Gas Transport in Small Channels," *Proc. ASME Winter Meeting*, (vDSC-19, 1990), pp149-157.

[7] Pfahler, J., J. Harley, H.H. Bau and J. Zemel, "Gas and Liquid Flow in Small Channels," in D. Cho *et al.* (eds.), *Micromechanical Sensors, Actuators, and Systems*, (ASME, vDSC-32, 1991), pp49-60.

[8] Matson, D.W., P.M. Martin, W.D. Bennett, D.C. Stewart and J.W. Johnston, "Laser Micromachined Microchannel Solvent Separator," *Proc. SPIE*, (v3223, 1997), pp253-9.

[9] Goldberg, N., "Narrow channel forced air heat sink," *IEEE Transactions on Components, Hybrids, and Manufacturing Technology*, (v7, n1, 1984), pp154-9.

[10] Krause, V., H.-G. Treusch, P. Loosen, T. Kimpel, J. Biesenbach, A. Kosters, F. Robert, H. Oestreicher, M. Marchiano, B. DeOdorico. "Microchannel Coolers for High Power Laser Diodes in Copper Technology." *SPIE Proceedings*, (v2148, 1995), pp351-358.

[11] Kleiner, M. B., S. A. Kuhn, K. Haberger, "High Performance Forced Air Cooling Scheme Employing Microchannel Heat Exchangers," *IEEE Transactions on Components, Packaging, and Manufacturing-Part A*, (v18, n4, 1995), pp795-804.

[12] Wu, P.Y. and W.A. Little, "Measurement of Friction Factor for the Flow of Gases in Very Fine Channels Used for Microminiature Joule-Thompson Refrigerators," *Cryogenics*, (v24, n8, 1983), pp273-7.

[13] Pfahler, J.N., J. Harley, H.H. Bau and J. Zemel, "Liquid Transport in Micron and Submicron Channels," *J. Sensors Actuators*, (vA21-23, 1990), pp431-434.

[14] Pfahler, J.N., J. Harley, H.H. Bau and J. Zemel, "Liquid and Gas Transport in Small Channels," *Proc. ASME Winter Meeting*, (DSC-19, 1990), pp149-57.

[15] Pfahler, J., J. Harley, H.H. Bau and J. Zemel, "Gas and Liquid Flow in Small Channels," in D. Cho *et al.* (Eds.), *Micromechanical Sensors, Actuators, and Systems*, ASME DSC-32, pp49-60.

[16] Cuta, J.M., A. Shekarriz and C.E. McDonald, "Forced Convection Heat Transfer in Microchannel Heat Exchangers," *Proc. ASME Winter Annual*, (November, Atlanta, GA, 1996).

[17] Massey, B.S., *Mechanics of Fluids*, (London: Van Nostrand Reinhold, 1989).

Fabrication of Microchannel Chemical Reactors Using a Metal Lamination Process

Dean W. Matson, Peter M. Martin, Donald C. Stewart,
Anna Lee Y. Tonkovich, Michael White, Jennifer L. Zilka, and Gary L. Roberts

Pacific Northwest National Laboratory, PO Box 999, Richland, WA 99352 USA

Abstract

Incorporation of microchannel arrays into small-scale chemical processing devices offers significant advantages in improving thermal control in the reaction region. Development of methods to produce high aspect ratio microchannels within solid metallic structures is an important step in the evolution of man-portable and other small-scale chemical processing units requiring high temperature and/or high pressure capabilities. Microchannel arrays can be used to facilitate heat removal or addition at the reaction zone or can be used to pre-heat gases prior to reaction. Staff at the Pacific Northwest National Laboratory (PNNL) have developed a method for producing solid metal components incorporating complex microchannel arrays by using a lamination and diffusion bonding process. This method uses metal shims that have been machined such that, when properly stacked, alternating microchannels and fins are produced in the laminated structure. The widths of the microchannels are determined by the thickness of the shim material, and heights and lengths of the channels are determined by the machined areas on the shims forming the channels. Machining of the shims can be accomplished in large quantities at low cost per shim by using photochemical machining or stamping processes. Consolidation of the laminated stack into a solid, leak-tight metal device is accomplished at elevated temperature and pressure by diffusion bonding. Microchannel arrays formed using this process can be produced either on the interior of the finished device or on an exterior surface. Typical microchannel dimensions in chemical processing devices produced at PNNL are 250 microns wide by 5000 microns deep. Application of the lamination process combined with a sheet-flow architecture can be used to produce highly compact chemical processing units. Among the devices produced at PNNL using this method are catalytic fuel processors and fuel vaporizers. Examples of all-metal stainless steel microchannel chemical processing devices produced using the lamination/diffusion bonding process will be presented and discussed.

Keywords: microchannel, chemical reactor, lamination, combustor

1 Introduction

Small-scale chemical processing devices are receiving increasing interest for military, automotive, and remote applications where size and weight limitations are an important consideration. Typical applications include heating and/or cooling devices, fuel processors, and chemical synthesis units [1,2]. Such small-scale devices provide unique design challenges that are inherent to the scale to which they are limited. For example, heat transfer mechanisms must be an integral part of the design of any chemical processing device in which heat is either released or absorbed by the reactions involved. In conventional-scale devices, this is generally accomplished through some form of heat exchange mechanism involving a heat transfer fluid confined within secondary containment and that flows either into or around the reaction zone. Thermal management in chemical processing devices intended for man-portable or other similar-scale applications can be particularly challenging as a result of the need to maintain highly compact, lightweight designs while simultaneously accommodating the materials requirements for high temperature and/or high pressure operation.

Staff at the Pacific Northwest National Laboratory (PNNL) have developed a method for fabricating all-metal small-scale chemical processing units using a lamination process [3,4]. These devices are produced from patterned metal shims that, when assembled, form complex features embedded in the internal structure of the final product. The stacked shims are diffusion bonded under high temperature and pressure, yielding a monolithic metal block. This method allows the production of solid, leaktight metal devices containing integral internal microchannel flowpaths for reactants, products, and heat exchange fluids. Thus, issues related to the use of gaskets or other such sealing devices used in high temperature/high pressure devices can be essentially eliminated. Because all liquids and gases involved in the process can be confined within the as-built device, fluid interconnects are also minimized.

An additional important benefit of the laminate fabrication method can be found in the fact that it allows the formation and incorporation of unique microchannel arrays and headers within the completed device without the need for additional machining. For the purposes of this discussion, we define a microchannel as having at least one dimension (typically the width) <250 μm. Arrays of parallel microchannels separated by thermally conductive fins provides an efficient mechanism by which to transfer heat either to or from fluids moving through the microchannels. Thus, microchannels can be used to promote enhanced heat transfer within the device at locations where heating or cooling is necessary or desirable. We believe that the fabrication method described provides a unique approach to the production of solid metal chemical processing units containing built-in thermal control and heat transfer mechanisms.

We describe the general procedures used at PNNL for fabricating laminated stainless steel devices for small-scale chemical processing applications. Specific considerations related to the use of this fabrication method are discussed, and examples of stainless steel units fabricated to date are presented.

2 Microchannel Device Fabrication

In the simplest configuration, parallel arrays of microchannels can be formed by alternately stacking thin shims containing channel features with solid shims acting as fins separating the channels. The elongated channels in the channel shims are machined or etched through the shim so that the channel area is typically open. The pattern for the fin shims contains holes aligned with either end of the channel ends of the channel shims. When the channel and fin shims are alternately stacked, a series of parallel channels are formed that are bounded on either side by solid fins. A header is formed at either end of the channels by the holes in the fin shims that are aligned with the ends of the open channels in the channel shims. Widths of microchannels produced by this method are determined by the thickness of the shim material used to produce the channel shims. Shim thicknesses used for producing the types of devices discussed here typically range from 25 to 250 μm. The microchannel heights and lengths are determined by the patterned area of the channel shims. Using this assembly method, microchannels having a wide range of aspect ratios (height/width) can be produced. Typically, endplates having sufficient thickness to accommodate fluid connections are added to either end of the stack of patterned shims to complete the laminated device.

The design for a simple heat exchanger device consisting of two adjacent series of parallel microchannels is shown in Figure 1a. In this illustration, a hot fluid is directed through the lower series of channels. Heat is conducted through the fins between adjacent channels and into the thin layer of solid material separating the two levels of channels. It is then similarly conducted into the upper set of fins and into a cooler fluid flowing through the upper channels. Typically the intervening layer of solid material is kept as thin as possible to facilitate heat transfer. In an alternative design (Fig. 1b), two fluids can be made to transfer heat directly through the fins by flowing the hot and cold fluids through alternating channels in the microchannel array. In this design the headers for the fluids are located in the corners of the microchannels and alternating channel shims are flipped 180° to allow access of the respective fluids only to alternating channels. Using this type of design, heat transfer need only occur across the thickness of the fin shim, reducing heat transfer resistance by minimizing the conduction pathway. Flows of hot and cold fluids in this type of microchannel device can be configured in either a co-flow or counter-flow mode, as required for a specific heat transfer application.

Production of solid metal microchannel devices such as those described here can be broken up into several distinct steps, each with its own considerations. In summary, those steps can be grouped into the following categories:

1. device design
2. shim patterning
3. shim cleaning and assembly
4. bonding
5. post-bonding processes (optional)

We discuss each of these steps in more detail below.

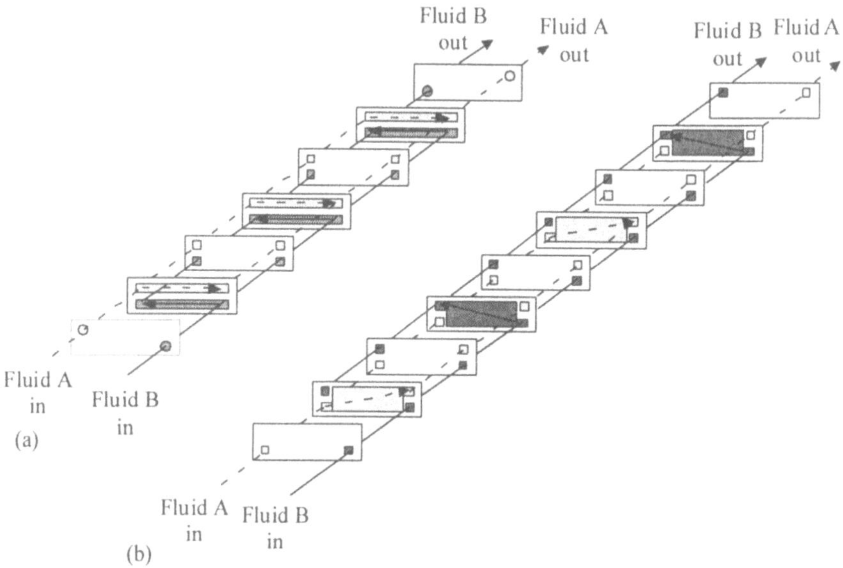

Figure 1. a) Design for a simple heat exchanger device; b) alternative design

2.1 Device Design

Generally, microchannel devices to be fabricated by the lamination process contain a series of parallel fluid flowpaths at different levels within the device. The sets of microchannels are vertically aligned and heat is transferred between fluids flowing in adjacent levels. The designs of the individual shims making up the device must be considered carefully to allow ingress and egress of fluids to the levels desired. At the same time, care must be taken to ensure that, when stacked into the final configuration, the shims will allow uniform transfer of pressure from one side of the stack to the other on all surfaces that require bonding.

For some applications, laminated devices may need to be built as separate components (e.g., to facilitate catalyst loading or replacement). Sealing issues between the components must be addressed using o-rings or other sealing mechanisms. Sealing surfaces generally need to be machined after the bonding process to achieve an adequate finish.

2.2 Shim Patterning

For proof-of-principle and development work, we have found that the individual shims used to fabricate stainless steel laminated devices can be produced at relatively low cost (<$1 US/shim) using a photochemical etching process. In larger quantities, the cost/shim could be reduced by at least a factor of two using this process. Photochemical etching produces shims having acceptable tolerances in the thicknesses of material (50 to 500 μm) and given the flowpath dimensions commonly employed (0.1 – 1.0 cm). This patterning process has the capability to produce shims having highly complex patterns with no surface burring. Other patterning processes such as laser machining or stamping may also prove viable for producing shims for specialized applications or in mass production.

The endblocks used to sandwich the stacked shims and provide fluid interconnects are machined on a per-piece basis. This process could be automated for producing a laminated device in large quantities.

2.3 Shim Cleaning and Assembly

Patterned shims are vapor degreased prior to assembly to remove residual photoresist from the patterning process and any other organic contaminants. Prototype devices produced by this fabrication method have been hand-stacked, but this process could be automated large-scale production.

2.4 Bonding

Stacked shim/endblock assemblies are bonded into a single solid piece under vacuum using a high temperature/high pressure diffusion bonding process. Laminated parts produced to this point have been bonded by stacking the endblocks and shims into a pre-oxidized high temperature alloy clamping device to provide alignment and side support. Bonding is accomplished in a vacuum hot press. An alloy endplate and ram extension are used to transmit pressure from the hot press ram to the stacked laminate. A set of conditions that has been used to bond laminated stainless steel devices such as these described here is 920°C and 4000 psi for 4 hr.

Alternative diffusion bonding processes are currently being explored that would allow bonding to be accomplished in a vacuum furnace without the need for an external ram. The modified process uses the difference in thermal expansion coefficients between the clamping device and the laminated material to produce the pressure required for bonding at elevated temperature. Ultrasonic bonding processes are also being evaluated for producing laminated devices.

2.5 Post-Bonding Processes

Because the bonded laminate acts as a solid piece of stainless steel, additional machining may be performed on the laminated parts after bonding. Such machining may be necessary to achieve smooth surfaces for o-ring or gasket seals, to provide additional fluid inlet or outlet ports, to reduce the overall mass of the resulting part, or for aesthetics.

Stainless steel tubing is commonly brazed or welded onto the fluid inlet and outlet ports of the laminated part after bonding. This connection method takes up much less surface area on the finished part and requires less material depth than other connection methods such as machined fittings.

3 Laminated Microchannel Devices

3.1 Microchannel Catalytic Combustor/Reactor

A small-scale combustor/reactor was designed to provide man-portable production of hydrogen for proton exchange membrane (PEM) fuel cells. The device was designed with a catalytic reactor section to process hydrocarbon feedstocks (methane, alcohols, gasoline, etc) into partial oxidation products, and a combustor section to pre-heat reactant gases. An all-metal design was stipulated to allow operation at temperatures approaching 1000°C in the combustion zone. Although not necessary from a fabrication standpoint, the device was produced in three sections to allow loading and replacement of foam metal catalyst supports. Two of the three parts comprising the device were fabricated using the lamination method to incorporate microchannels for enhanced heat transfer. The third component was a solid stainless steel plate required to seal the top of the device.

Patterned shims used to produce the laminated combustor/reactor are shown in Figure 2. All shims were photochemically etched in 250 μm-thick type 316 stainless steel. Eight different shim patterns (Fig. 2a) were required to produce the reactor body in order to accommodate the various coolant, fuel, and gas product flow paths required in this section of the device. The assembly order for the reactor part of the device was:

1. solid end block
2. 12 each shim #3 and 13 each shim #4, alternating
3. 26 each, shims #1 and #2, alternating
4. 8 each, shims #5 and #6, alternating
5. 26 each, shims #1 and #2, alternating
6. 12 each shim #7 and 13 each shim #8, alternating
7. endblock with coolant outlet

The combustor section of the device consisted of only two shim designs (Fig. 2b) and had no internal flow channels when assembled. The assembly order for the combustor was:

1. solid end block
2. 85 each shims #9 and #10, alternating
3. solid end block

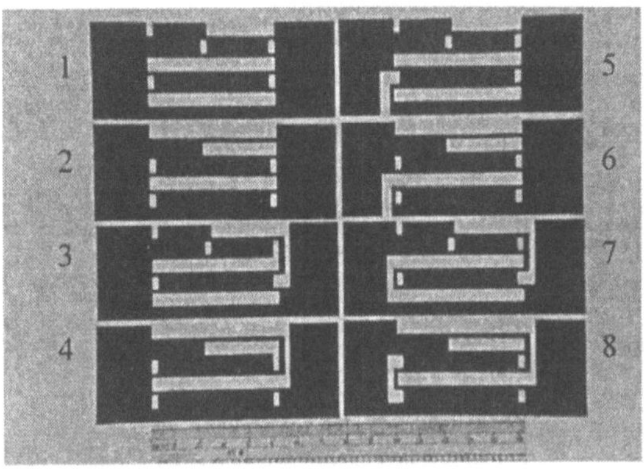

a)

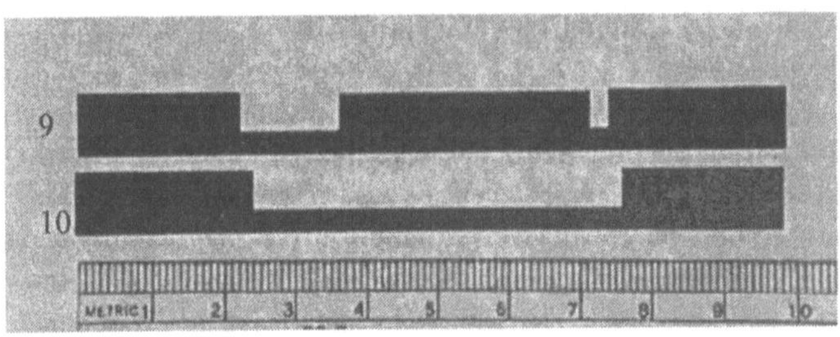

b)

Figure 2. Shim designs used for fabrication of microchannel combustor/reactor. a) Shim designs for reactor sections. b) Shim designs for combustor section.

After diffusion bonding at the conditions outlined in the previous section, the components were machined to add grooves for "C"-cross section metal seal rings between the components and a series of bolt holes around the perimeter of the device. Excess metal was also machined from the corners to reduce the weight and thermal mass of the finished device. Stainless steel tubing was welded onto the device at all inlet and outlet ports to facilitate fluid connections. A photograph of the three parts of the completed combustor/reactor is shown in Figure 3.

Figure 3. Three components of the laminated microchannel combustor/reactor

3.2 Microchannel Gasoline Vaporizer

Another device fabricated using the laminate process was the microchannel gasoline vaporizer, the first stage within a fuel processor to be used for automotive applications. As with the combustor/reactor described above, the gasoline vaporizer was designed with a catalytic combustion zone using an auxiliary fuel to provide the heat required to drive the primary function of the device. Microchannel heat exchangers were used to effectively transfer combustion heat to its area of use.

The gasoline vaporizer was produced using only two shim designs (Fig. 4) and two solid end plates. The shims were designed such that, when alternately stacked, two parallel series of microchannels were formed, separated by 1.7 mm of solid metal. Similarly, all fluid headers were produced by the design of the stacked shims. A total of 268 shims were used to produce the device. When bonded, the device formed open cavities designed to hold metal foam-supported combustion catalyst. After bonding, excess metal was removed to achieve a final shape. A groove was machined to accommodate a metal "C"-section seal ring, bolt holes were added around the perimeter, and ports were added for the liquid gasoline inlet, vaporized gasoline outlet, and the reaction gas exhaust. A solid top plate containing the inlet port for combustion gas was also fabricated. The flow diagram for the completed gasoline vaporizer is shown in Figure 5. Hot product gas from the reaction zone was directed into a series of microchannels on the level below the combustion zone before being removed through exhaust ports in the sides of the device. Heat was transmitted through fins of the microchannels and into the gasoline being pumped through the series of channels below. The microchannel gasoline vaporizer was demonstrated at full-scale for an automotive application, vaporizing 265 cc/min of gasoline.

4 Summary

A lamination/bonding procedure was developed for producing novel all metal small-scale chemical processing devices capable of operating at high

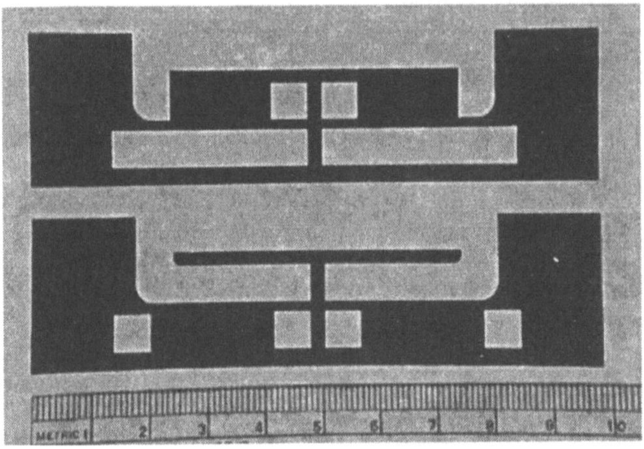

Figure 4. Shim designs used to fabricate the laminated microchannel gasoline vaporizer.

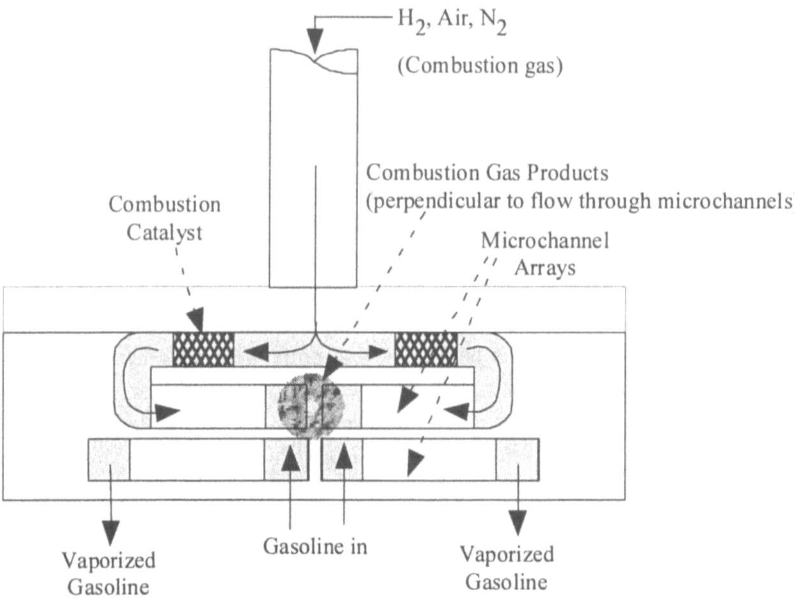

Figure 5. Flow diagram for laminated microchannel fuel vaporizer.

temperature/high pressure conditions. Such devices are suitable for portable or small-scale chemical processing applications. This fabrication method allows the construction of unique, highly compact, leak tight, devices containing integral microchannel arrays for thermal control and heat exchange processes. Microchannels and other fluidic features can be completely contained within the internal structure of the solid laminated component. The microchannel dimensions produced using the lamination process are determined by the machined pattern on the shims and on the thickness of the shims used in the construction of the device. Costs of fabricating the solid laminated devices can be reduced through volume production, the use of automated assembly methods, and improved bonding procedures.

Examples of laminated metal devices produced using the lamination/bonding method include a man-portable catalytic combustor/PO_x reactor and a fuel vaporizer for automotive applications. A number of additional laminated microchannel devices are currently in various stages of planning and development.

Acknowledgements

The Pacific Northwest National Laboratory is operated for the U. S. Department of Energy by Battelle Memorial Institute under contract DE-AC06-76RLO 1830. Support for this work was provided by the U. S. DOE and by PNNL through the Microscience and Engineering Initiative, a Laboratory Directed Research and Development program.

References

1. R. S. Wegeng, C. J. Call, and M. K. Drost, "Chemical system miniaturization" Proceedings of the 1996 Spring National Meeting of AIChE, 1996.
2. A. Y. Tonkovich, D. M. Jimenez, J. L. Zilka, M. LaMont, Y. Wang, and R. S. Wegeng, "Microchannel chemical reactors for fuel processing" Proceedings of the Second AIChE International Conference on Microreaction Technology, pp. 186-195. 1998.
3. P. M. Martin, W. D. Bennett, D. J. Hammerstrom, J. W. Johnston, and D. W. Matson, "Laser micromachined and laminated microchannel components for chemical sensors and heat transfer applications" *SPIE Conference Proceedings Vol. 3224*, pp. 258-265. 1997.
4. D. W. Matson, P. M. Martin, A. Y. Tonkovich, and G. L. Roberts, "Fabrication of a stainless steel microchannel microcombustor using a lamination process" *SPIE Conference Proceedings Vol. 3514*, pp. 386-392. 1998.

Microreactor Technology for Biological Applications

Gillian M. Greenway and Tom McCreedy

Analytical Science Group, Department of Chemistry, University of Hull, Hull HU6 7RX, UK

Abstract. The immobilisation of enzymes on a porous silica structure fabricated within a capillary is demonstrated for the determination of glucose. Glucose oxidase and peroxidase were immobilised onto the support by physical adsorption, and the extent of *o*-dianisidine oxidation was used as the glucose concentration indicator. The covalent immobilisation of urease is also reported, but the selected detection method proved to lack sensitivity. The porous silica structures used in this work were fabricated *in-situ* by the reaction of potassium silicate and formamide solutions. Enhanced conversion rates were observed due to the large surface area of the reactor, but it was very susceptible to blocking due to the fine pore structure. The fabrication of a more porous structure is presented, based on the use of diluted potassium silicate and formamide solutions.

Keywords. Enzyme immobilisation, bioreactor, glucose determination

1. Introduction

The specificity of enzymes has long been employed in analytical chemistry for the selective determination of a wide range of compounds. In early systems, a significant disadvantage with the use of enzymatically catalysed reactions was the large quantity of pure enzyme required for analytical measurement, and the associated high cost of purchase. By immobilising enzymes onto a solid surface, it has been possible to drastically reduce enzyme usage. Devices employing immobilised enzymes form an important class of biosensor; indeed this can be confirmed by the large number of publications on the topic. There are two primary types of biosensor; those that employ a layer of enzyme directly immobilised onto a transducer element and those that use a flow through reactor. Only the latter will be considered further in this paper. However, the former represents an important class of biosensor.

The major choices that influence the design of a flow through bioreactor are the method of immobilisation, the reactor material, and the type of reactor i.e. open tubular or packed reactor. The choice of enzyme is dictated by the analyte of interest, with enzymes targeted to specific analytes e.g. glucose oxidase for glucose. The selection of open tubular or packed reactor depends to some extent on the requirements of the application. For example, in a recent paper, Jurkiewicz and co-workers [1] compared the two types of reactors with three different enzymes (urease, creatinine iminohydrolase, and creatinine aminohydrolase). The packed reactors were prepared from covalently immobilised enzymes on controlled pore glass packed into tubes, whereas the enzymes were covalently immobilised onto the inner walls of nylon tubing for the open tubular reactors. Under optimised conditions, it was evidence that the packed reactors offered a greater conversion ratio, shorter response times, superior sensitivity, and a lower limit of detection than the open tubular reactors. However, the overall analysis time was shorter with the open tubular reactors, and they also offered higher reproducibility than the packed reactors. Controlled pore glass offers a reliable and reproducible support for immobilised enzymes. However, other support materials have also been used, such as polyamide and collagen membranes [2] for immobilised glucose oxidase and lactate oxidase. Open tubular reactors have been prepared from a range of materials including nylon tubing [1], PTFE tubing [3], and fused silica capillary [4]. A range of immobilisation reactions has been employed with these materials.

There are a number of immobilisation techniques that can be employed with enzymes; these include adsorption, microencapsulation, entrapment, cross linking, and covalent bonding. The simplest of these techniques is adsorption, but is usually considered only suitable for short-term investigations due to the weak nature of the bond between enzyme and support. However, as demonstrated by Li and Townshend [3], it permits the rapid desorption and re-immobilisation of the enzyme when it is denatured. In that work, the physically adsorbed enzyme reactor was stable for at least a month if kept in a refrigerator, and did not show appreciable loss of activity when substrate solution was pumped through it at 1.5 mL min^{-1} for 40 hours. This ease of enzyme replacement is attractive in the present study.

For systems involving flow through a packed reactor, perhaps the most useful alternative to physical or chemical adsorption is covalent bonding. Here the enzyme is attached via a non-essential functional group to a solid support by the formation of a covalent bond. The most significant advantage of this approach is that the enzyme will not be released into the flowing solution. The most widespread use of this technique has been for the attachment of enzymes to controlled pore glass. One widely studied enzyme was urease, due to its importance in the rapid determination of urea in body fluids. Johansson and Ögren [5] reported the immobilisation of urease onto controlled pore glass in 1976. In that paper, the glass was first boiled in 5% nitric acid, then reacted with 3-aminopropyltriethoxysilane to form alkylamino glass. A solution of glutaraldehyde was used to activate the glass prior to the covalent attachment of urease at +4°C. This general approach has been extensively used since, such as for the immobilisation of acetylcholinesterase for the determination of inhibitors neostigmine and galanthamine [6].

For packed reactors, the enzyme is normally attached to a particulate support prior to packing the reactor. This approach lends itself well to multistep chemical reactions, where the intermediate products are easily manipulated and washed. For conventional scale flow injection analysis, packing the reactor with particulate matter presents few problems, and regular packing is easily achieved. Difficulties become apparent in miniaturised systems where the regular packing of a channel, perhaps 200 μm across, is more problematic. In addition, the process of physical packing would not lend itself well to automated processes due to the difficulties in obtaining a reproducible and regular packing without damaging the enzyme. The obvious alternative would be to use open tubular reactors, but at the expense of the detection limit, conversion ratio and sensitivity. Since only minute sample volumes would be used, these parameters would be critical; thus packed reactors would be preferable.

It is evident that the ideal production method for a micro-bioreactor would be the *in-situ* fabrication of the enzyme support. The support would need to be sufficiently porous to allow flow through the reactor, and provide a surface suitable for enzyme attachment. Christensen *et al.* [7] reported a method for the *in-situ* fabrication of a porous silica plug within capillaries with the dual purpose of retaining catalysts and providing a surface capable of supporting electroosmotic flow (EOF). The use of EOF within micro-systems offers an attractive method for pumping small volumes of solution in channels and small diameter capillaries. However, this surface should also be suitable for the attachment of enzymes either by physical adsorption or covalent bonding.

In this paper, we report the initial findings of the evaluation of this support for such an application and show the viability of this material. During the course of the research several difficulties were encountered; these are also reported. Reagent delivery during the immobilisation process was perhaps the paramount challenge, particularly at elevated temperatures.

2. Experimental

2.1 Reactor fabrication

The microporous silica structure was fabricated as previously reported [7]. 140 μL of potassium silicate solution (Prolabo, Manchester) was mixed with 18 μL of formamide (Avocado Research Chemicals, Heysham, Lancashire) for 30 seconds before drawing the liquid into the capillary channel. A comparative study was performed with the above solution diluted by 50% before drawing it into the capillary channels. While such devices are easily fabricated on-chip, the development work reported in this paper has taken place with discrete bio-reactors. These were fabricated from glass tubing, 500 μm internal radius and 6 mm in length (LIP Equipment and Services, Shipley). For the immobilisation reactions, the various reagents could be drawn through the reactors using a peristaltic pump.

2.2 Physical adsorption of enzyme

This was the simplest method for enzyme immobilisation. The enzymes selected for this study were glucose oxidase (Sigma) and peroxidase (Sigma). All flow rates were 0.1 mL min^{-1} unless otherwise stated. Boiling 10% nitric acid was first drawn through the bioreactor in order to activate the surface. This step was continued for 1 hour, then purified water was passed for 1 hour. 100 mg of glucose oxidase and peroxidase were dissolved in 10 mL cold 0.1 M phosphate buffer. This solution was drawn through the bioreactor for 30 minutes, followed by cold phosphate buffer for a further hour. The reactors were stored in a refrigerator at 4°C.

2.3 Covalent immobilisation of enzyme

The procedure followed for the covalent immobilisation was similar to that reported by Johansson and Ögren [5], but modified to work with such porous supports. The immobilisation of urease was undertaken in the present work. All flow rates were 0.1 mL min^{-1} unless otherwise stated. Boiling 10% nitric acid was drawn through the porous silica reactor for 1 hour; then washed with purified water for 1 hour before drying at 95°C overnight. 5 g of 3-aminopropyltriethoxy-silane (Sigma) was dissolved in 45 mL of water, then adjusted to pH 3.45 using 6 M HCl. This solution was drawn through the bioreactor for 4 hours at 75°C. Once this step was complete, the reactor was washed with purified water for 1 hour and then dried at 95°C overnight. A solution of 2.5% glutaraldehyde was prepared in 0.1M phosphate buffer at pH 7. This solution was drawn through the bioreactor for two hours; initially the solution was degassed with helium. Later studies used a 1% solution of glutaraldehyde. Prior to introducing the enzyme, the reactor was washed with water for 1 hour. 100 mg of urease enzyme (Sigma) was dissolved in 5 mL of 4°C phosphate buffer and then drawn through the reactor for 2½ hours, then the bioreactor was washed with cold buffer solution for a further 2½ hours.

2.4 Detection methods

The method of detection for the glucose/peroxidase bioreactor was spetrophotometric. An Ocean Optics model S2000 fibre optic spectrophotometer will be used for on chip detection; however, a Lambda 10 spectrophotometer (Perkin Elmer) was used with the discrete bioreactor. The reaction employed for the detection of glucose was the formation of hydrogen peroxide by the action of glucose oxidase on glucose, followed by the oxidation of *o*-dianisidine, in the presence of peroxidase, by the hydrogen peroxide from the glucose oxidase reaction. This was monitored at 436 nm.

The determination of urea by its conversion to ammonia by urease was followed with a pH electrode. In order to work with the small volumes produced by such systems, a pH electrode with a 100 μm sensing tip was used. The Beetrode (World Precision Instruments, Stevenage) could operate in minute volumes of solution; however, a separate reference electrode was required. This was employed for the study, since only the sensing tip needed to be located immediately adjacent to the reactor.

3.0 Results and discussion

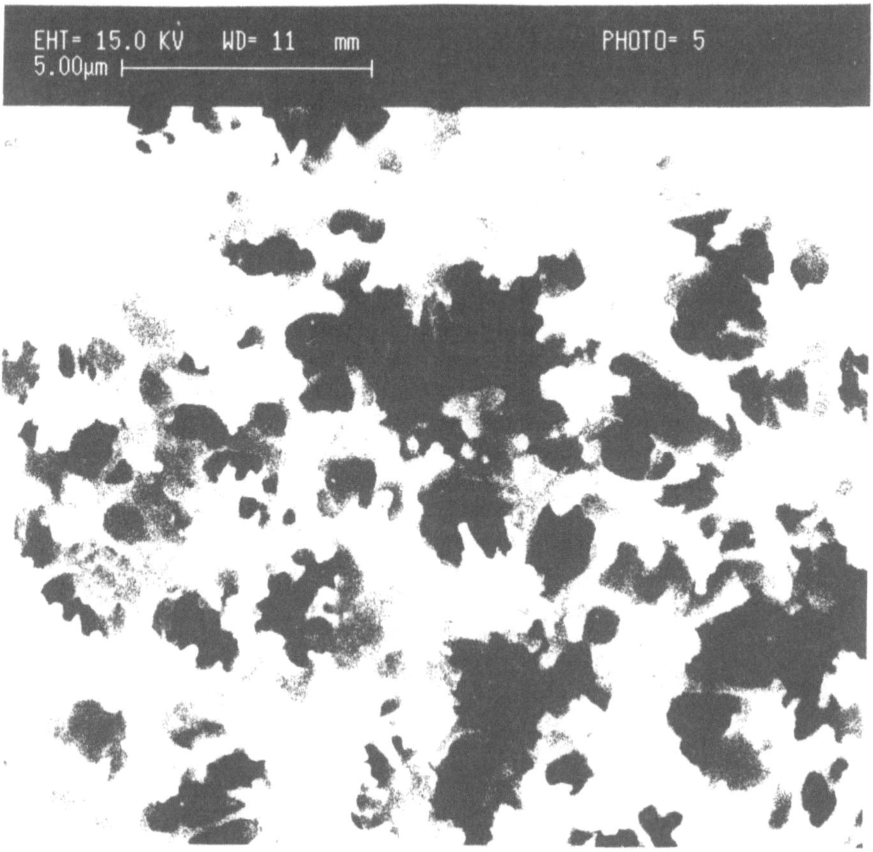

Figure 1: A scanning electron micrograph of the silica structure formed from the undiluted reaction mixture. The presence of the regular pore structure is evident.

3.1 The porous support material

The reaction to form the silica support liberates ammonia and the reaction can be speeded up by heating the mixture in the capillary channel. A temperature of 85-100°C should be used for the first hour, followed by stronger heating (up to 250°C) for 3 hours. This is particularly important for the 50% diluted solution, which has a higher water content. A scanning electron micrograph of the support material can be seen in Figure 1, and clearly shows a highly porous structure that resembles a natural sponge. The pores are approximately 3-5 μm in width. The support formed from the 50% diluted reagents offers less resistance to flow, and this is thought to be due to larger pores. The structure of this support is currently under investigation using scanning electron microscopy.

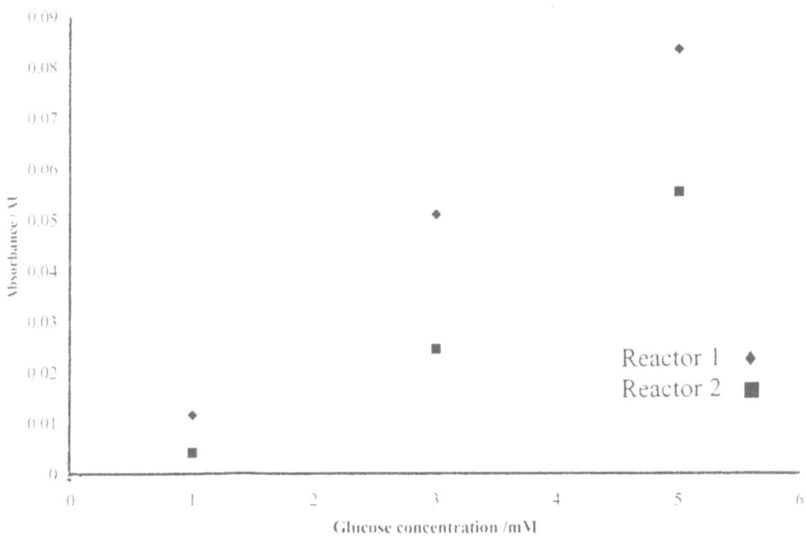

Figure 2: A plot showing the change in oxidised *o*-dianisidine absorbance as a function of glucose concentration. Flow rates through reactors 1 and 2 were 0.1 and 0.3 mL min^{-1}, respectively. Reactor 2 was prepared from 50% diluted reagents while reactor 1 was prepared from non-diluted reagents.

3.2 Physically adsorbed glucose oxidase/peroxidase bioreactor

This reactor was found to be stable for at least one week if stored in the refrigerator at +4°C, and the reactors were simple to produce. Typically four reactors could be produced simultaneously; however, this was theoretically only limited by the number of channels on the pump. Glucose solutions in the range 0-

5 mM were prepared from phosphate buffer containing 6.6 μg ml^{-1} *o*-dianisidine. The bioreactor and glucose solutions were kept thermostatted at 37°C during the experiments. Detection of the oxidised *o*-dianisidine was achieved by measuring the absorbance of the solutions after having passed through the bioreactor compared to the stock solution. All collected solutions were stored at 0°C to prevent any further reaction. Between each experiment, the reactor was washed with 0.1 M phosphate buffer solution.

Typical results for two bioreactors can be seen in Figure 2, and show the increase in oxidised *o*-dianisidine with respect to glucose concentration. It is interesting to note the different slopes (i.e. sensitivity) obtained for the two plots. These are inversely related to the flow rate, a result to be expected since the extent of reaction will be proportional to the residence time in the reactor. The measured flow rates for reactors 1 and 2 were 0.1 and 0.3 mL min^{-1}, respectively. The flow rate through reactor 2 is greater than that through reactor 1, and since both were connected to the same pumping system it appears that reactor 2 produced less resistance to flow. This is not unexpected, since the reagents that produced the support were diluted 50%, and therefore it is likely that the material produced will be more porous.

3.3 Covalently immobilised urease bioreactor

A significant problem was encountered with the production of this bioreactor. During the step where 2.5% glutaraldehyde was drawn through, the resistance to flow exhibited by this reactor increased, sufficient to block the reactor. The exact reason for this was not known, but the problem was significantly reduced if 1% glutaraldehyde was used. This problem was also reduced if the support was formed from 50% diluted reagents; this was thought to result in a more porous structure. During the glutaraldehyde step, the reactor colour was seen to change from white to pale orange/brown. When the enzyme solution was drawn through, the colour of the reactor changed to a darker red. This was to be expected and showed the enzyme had been immobilised. There was a problem with the selected method of detection in that it lacked sufficient sensitivity, only giving 7.5 mV per decade urea concentration. With this system, the variation in background noise due the pH electrode was sufficient to mask the pH change. Future work will need to refine the detection method, and either employ an ammonium ion selective electrode or spectrophotometric detection employing a modified reaction process yielding a light absorbing compound.

4.0 Conclusions

This work has demonstrated the potential of microfabricated bioreactors, which are far smaller than current systems. The small size of the device is more than compensated for in terms of the large internal surface area. The problems of detection and fluid transport across the reactor have proved to be the most significant. The use of electroosmotic flow is currently under investigation as a pumping method, but its interaction with the enzymes will need investigation. The next phase of the study is to further reduce the size of the reactors to less than 250 μm channel width and incorporate them onto microfabricated structures.

5.0 References

[1] M. Jurkiewicz, S. Alegret, E. Fabregas, Anal. Chim. Acta, 370, 1998, 47-58.
[2] C.A. Margette, L.J. Blum, Anal. Chim. Acta., 381, 1999, 1-10.
[3] Y.Z. Li, A. Townshend, Anal. Chim. Acta., 359, 1998, 149-156.
[4] Y. Shi, S.R. Crouch, Anal. Chim. Acta., 381, 1999, 165-173.
[5] G. Johansson, L. Ögren, Anal. Chim. Acta., 84, 1976, 23-29.
[6] T. Ghous, A. Townshend, Anal. Chim. Acta., 372, 1998, 379-386.
[7] P.D. Christensen, S.W.P. Johnson, T. McCreedy, V. Skelton, N.G. Wilson, Anal. Comm., 35, 1998, 341-343.

Laserprocessing for manufacturing microfluidic devices

E. Bremus*[1], A. Gillner[1], D. Hellrung[1], H. Höcker[2], F. Legewie[1], R. Poprawe[1], M. Wehner[1], M. Wild[1]

[1] Fraunhofer Institute for Lasertechnology, 52074 Aachen, Germany
[2] Dept. of Macromolecular and Textile Chemistry, 52062 Aachen, Germany

Summary:
Microreactors which combine several synthetic operations with online analysis of reaction products are of increasing interest for combinatorial chemistry.

Laser technique is a very versatile tool for flexible, small scale production of microstructures like channels, cavities, bores etc. in a number of materials (polymers, ceramics, metals). Excimer lasers are ideally suited for planar processing while structuring with solid state lasers is useful for machining of complex geometries. If 3-dimensional shapes should be structured with high resolution a laser writing process should be preferred. In this case the high pulse rates of solid state lasers are imperative to obtain high writing speeds.

The described technologies of microstructuration are ideally accompanied by laser joining processes which can weld various materials together. These two laser technologies enable us to design and manufacture virtually any microreactor or microfluidic device. To exemplify these technologies a microreactor concept and corresponding prototypes are shown.

Keywords: excimer laser, solid state laser, ablation, polymers, ceramics, micro tools, microreactor, combinatorial chemistry

1 Introduction

In combinatorial chemistry as a modern methodology in drug discovery there is an ongoing trend towards smaller reaction vessels due to the fact that the needed amounts for chemical characterization and HTS (high throughput screening) are decreasing dramatically. This leads from standard laboratory equipment and todays methods of automated synthesis towards microreactors having volumes less than 100 µl. These microreactors are ideally suited for parallelization in matrices having dimensions like microtiter plates.

---

* For further author information:
email: bremus@ilt.fhg.de; WWW: http://www.ilt.fhg.de/; Telephone: ++49-241/8906-202; Fax: ++49-241/8906-121

Microfluidic devices like microchannels e.g. for capillary electrophoresis, microreactors and microtiterplates are produced in a wide range of materials. For the manufacturing of microfluidic devices from polymer materials methods like injection molding, hot embossing and direct ablation are used, depending on the lot size and the materials choice. Silicon and glass are also common materials for producing microfluidic devices. For manufacturing them methods like wet and dry etching are proven. Ceramics and metals are applied in the microreactor field especially for high temperature approach. Such hard and/or brittle materials show remarkable features (e.g., the high corrosion resistance) which are highly appreciated by application but pose serious problems in respect to machining because of high wear of the tools. This difficulty can be overcome by laser machining. Mentioned materials as well as temperature sensitive materials like polymers can be laser structured with precision in the µm-range. So the laser processing offers appropriate solutions for tool manufacturing and rapid prototyping as well as packaging and interconnection technologies.

Laser technology has been applied for manufacturing of micro-scale components in electronics and micro systems technology. Laser-assisted etching of read/ write heads for hard discs[1], drilling of vias in printed circuit boards[2] , or drilling of ink jet nozzles for printer heads[3] has been proven as reliable techniques. The experience in production technology gained there could be easily transferred to the production of microfluidic devices. Depending on requirements, nearly all materials could be structured by laser radiation. In the following, examples for laser machining will be presented.

2 Laserprocessing

2.1 Ablation of Polymers

Excimer-lasers are well known as excellent tools for the ablation of biological or macromolecular materials[4]. Due to the high photon energy of the UV-radiation most polymers exhibit strong absorption of the laser radiation. Depending on the absorption coefficient of the materials, the optical penetration depth is typically 1 µm or less. This results in very clean cuts or bores with a negligible heat affected zone. Patterning of grids with 100 nm spatial resolution has been demonstrated[5].

Structuring of polymers using excimer laser radiation is preferentially done by a mask projection scheme, where the laser beam is shaped and homogenised to illuminate a mask. The mask is imaged onto the substrate by a projection lens in reduced size. Therefore the pattern on the substrate is defined by the pattern of the mask which, e.g., could be made by laser cutting from thin sheets of steel. In our set-up (Fig. 1) we use a KrF (λ = 248 nm) or ArF (λ = 193) excimer laser with reduction ratios from 4 : 1 to 20 : 1.

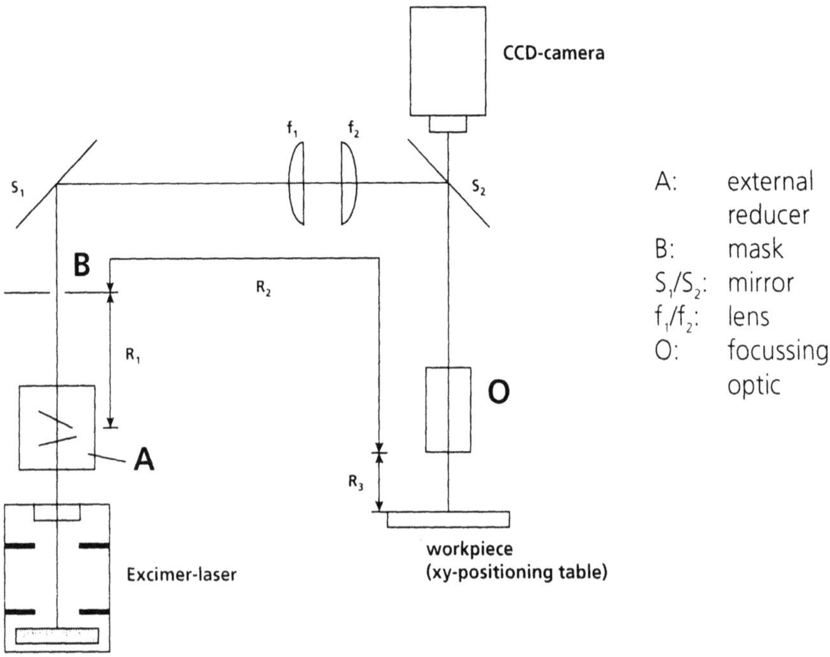

Fig. 1: Schematic drawing of an Excimer-laser system.

Polymer materials ranging from poly(methylmethacrylat) (PMMA) and poly(bisphenol-A-carbonate) (PC) to the high performance polymers polyimide (PI) and poly(etherether)ketone (PEEK) as well as polymers used for biomaterials like poly(dimethylsiloxane) (PDMS), polyurethanes (TPU) and polylacid are microstructured by our group using the described mask projection scheme. This way the production of structures < 20 µm with steep walls and clearly defined edges can be achieved. By using cnc-controlled processes complex structures of microchannels and holes for capillary electrophoresis cells or microreactors as shown in chapter 3 can be realized.

Further examples are those for biotechnological applications where micro- and nanotiterplates are preferentially made from optically transparent materials which facilitate optical screening techniques. PC is a well suited polymer for precision machining by excimer lasers which exhibits high transparency and high optical quality. The typical ablation rate for a KrF-excimer laser is about 0.2 µm/ pulse. Therefore, many pulses have to be applied to form even small dips. In order to prevent cumulative heating of the substrate, the pulse repetition rate is limited to 20 pps. An array of 225 compartments each of 0.14 nl volume is shown in Fig. 2 each hole could be machined in less than 3 s.

Different patterns could be formed by changing the mask. Even individual structures which are applied to the design of microreactors (cf. Fig. 3) can be achieved by mask projection or scanning of contact masks.

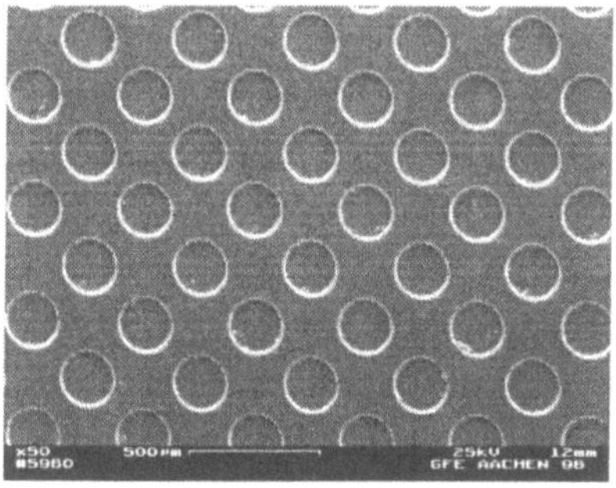

Fig. 2: Nanotiterplate in PC.

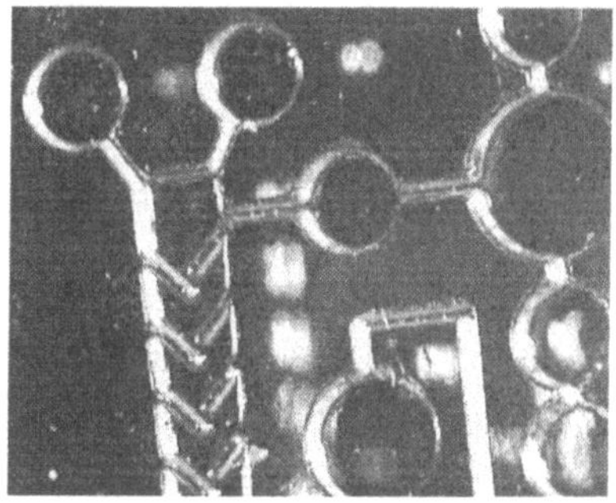

Fig. 3: Microreactor structure in PMMA[6].

2.2 Machining of ceramics and hard metals

Hard metals and high performance ceramics like Si_3N_4, SiC, Al_2O_3, ZrO_2 and also diamonds can be hardly machined by conventional methods. Short pulse lasers, especially frequency-tripled, diode pumped Nd:YAG-lasers (λ= 355 nm) with a high beam quality offer the possibility to ablate these materials with high quality. With a spot size of about 10 µm, high fluences (> 100 J/cm^2) can be achieved, so that the materials are vaporized without or with only a small amount of molten material. This technique is applicable for drilling small holes with diameters ≥ 5 µm (aspect ratio up to 60) and cutting of thin ceramic substrates (thickness < 0.5 mm). The edges are sharp and the face of the cut is very smooth. Furthermore it is possible to ablate three dimensional microstructures. Therefore, the laser beam is scanned over the surface and the material is ablated pulse beside pulse. The surface roughness depends on the overlap of the laser pulses, the pulse energy, and the scanning strategies. With optimized machining parameters the surface roughness can be reduced to Ra ≤ 0.1 µm. Due to the low ablation rate of around 0.05 µg / pulse the ablation depth of a single slice can be controlled very precisely. Depending on the material and the machining parameters the depth is in the range of 1 to 10 µm.

Similar to the ablation or removal process of an Excimer-laser, the removal depth per pulse has a maximum value at a material specific energy density. After that the amount of removed material decreases. The best results however are achieved at lower intensities, because higher intensities cause a damage of the surrounding material. In the range of investigation it is shown that the machining quality is dependent on the maximum pressure of the vaporized material while the laser pulse length influences only the amount of molten material.

As already mentioned using an overlap of 30-50 % a good surface roughness can be achieved. Out of this range the roughness increases steeply[7]. Furthermore, the energy density influences the depth of a single layer and the roughness. The surface roughness increases at higher aspect ratios of a single pulse mould (removal depth of a single laser pulse in comparison to the effective working diameter).

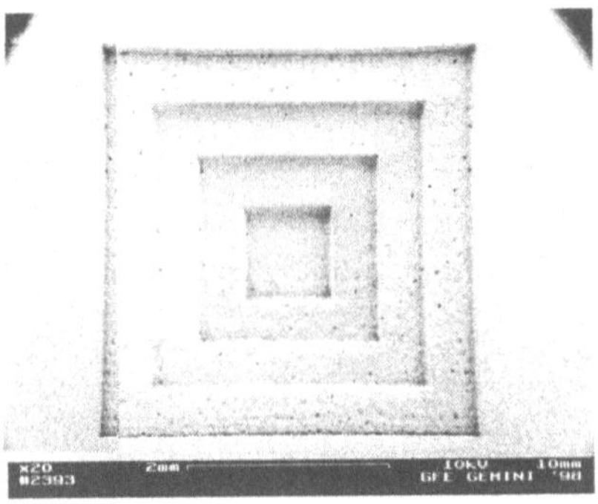

Fig. 4: Pyramid in ceramics (Si_3N_4).

An optimized surface roughness can be achieved with flat removed areas, that means a low aspect ratio of the single pulse mould. Especially the pyramid structure made of Si_3N_4 (Fig. 4) has a very good surface roughness, which is in the range of the raw material. The height of the steps is around 100 µm. The quality of the edges is very good. Due to the low energy densities[8] damage caused by removed grains could not be detected. The debris is very low and the removed material has only very weak adhesion to the basic material, so that the structures can be cleaned in an ultrasonic bath.

Compared to the combined mask and direct writing process with an Excimer-laser, the direct writing process with a Q-Switch Nd:YAG-laser offers a higher flexibility concerning the possible shapes[9], because curved lines can be generated with the CAD/CAM-system . The main restrictions are caused by the accuracy of the moving system. With this technique real three-dimensional structures can be produced. Since the main advantages of ceramics are high electrical resistance, good heat conductivity, high hardness, high chemical resistance, the final structures can be used as tools for the production of embossed metal micro parts. They are also applicable as microforming tools for polymers by injection molding or even directly for high temperature reaction cells or microscopic heat exchangers.

3 Microreactor concept

For applications in the field of combinatorial chemistry a microreactor prototype with 20 µl volume of the reaction chamber was made. It is suitable for the synthesis of <1-2 mg product (Fig. 5). The solutions of starting materials are supplied through a static mixer where they are homogeneously distributed by diffusion. In the reaction chamber the chemical reaction is carried out as a batch process induced by laser radiation through a transmitting quartz glass cover.

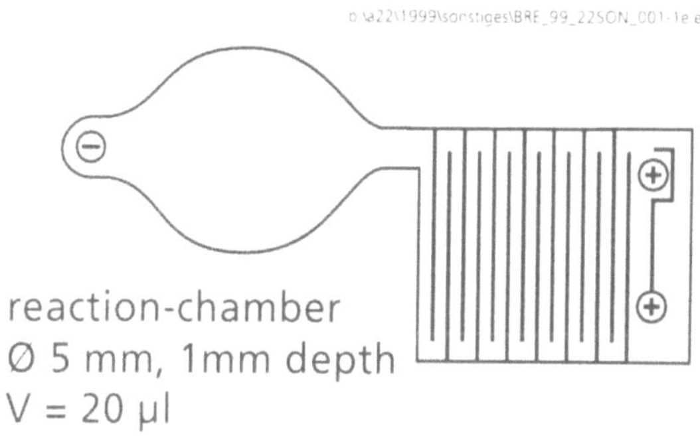

Fig. 5: Scheme of the microreactor with static mixer and reaction chamber.

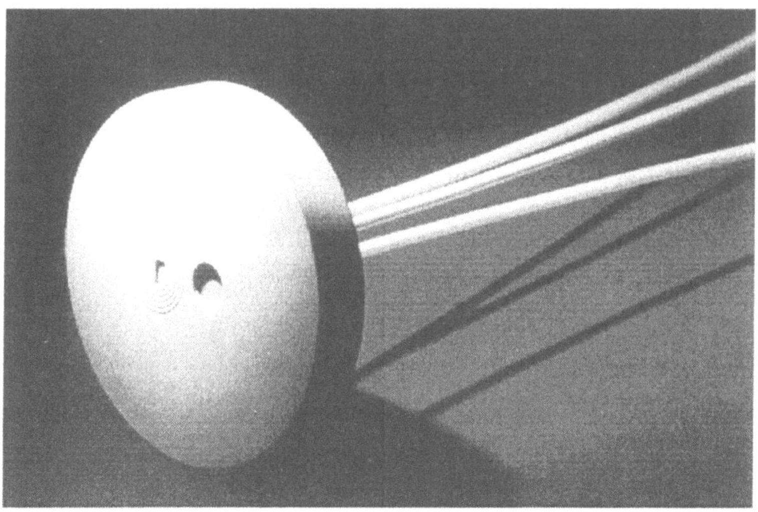

Fig. 6: Photograph of microreactor prototype.

An array of microreactors arranged as a 8 x 12 reaction matrix is beeing manufactured from PEEK by laser ablation with excimer laser radiation by mask projection and direct writing technique (patent pending).

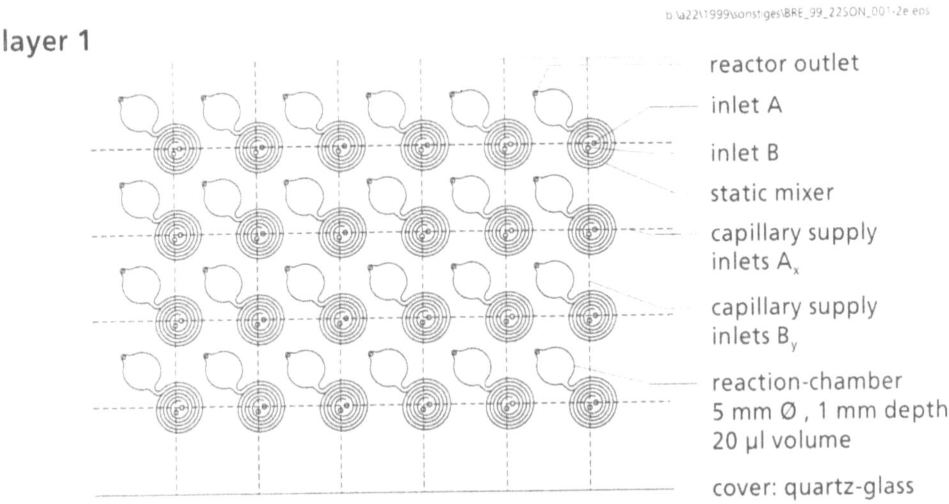

Fig. 7: Microreactor array designed for chemical applications.

The following possibilities for reaction conditions are involved in our system:

- addition of 8 / 12 starting materials through capillaries in sandwich-like polymer layers to the 8 x 12 matrix of reaction chambers
- simultaneous production of 96 new products (standard microtiter plate)
- energy supply through IR-radiation (laser heating)
- supply of laser energy in the UV and visible range for photoreactions
- supply of microwaves for direct stimulation of chemical bonds which results in accelerated reactions
- generation of pressure by laser evaporation or by laser induced extension of the reaction mixture in the filled chamber
- acceleration of the reactions by ultrasonic stimulation
- immobilisation of catalytically active molecules on the polymer surface in the reaction chamber

The sandwich-like structure of our microreactor system is shown in the following cross section.

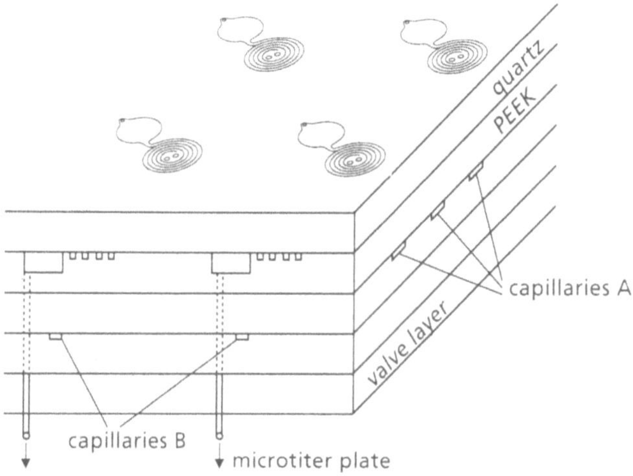

Fig. 8: Cross-section of the microreactor matrix.

The reaction can directly be observed by photometric methods; especially laser fluorescence is an promising tool due to the high sensitivity. The reaction products can be released into a standard microtiter plate for immediate use in high throughput screening of pharmacological activity. Otherwise online analytic methods can be applied by serializing the products by a special valve technique. The product fractions are arranged in lines each separated from the following by a gas bubble and a droplet of pure solvent. This way it is possible to pump the whole line through the 1 µl FTIR-flow-cell shown in Fig. 9 which was developed by our group in co-operation with *Resultec analytic equipment*[10]. Afterwards there can be further characterization, e.g., by GC/HPLC-MS or CE-MS.

Fig. 9: Flow-cell (PEEK) with an ATR-element for FTIR-spectroscopy, available at *Resultec*[10].

4 Laserjoining

The described technology of microstructuration is ideally accompanied by laser joining processes which offer appropriate solutions for packaging and interconnection of microfluidic devices such as µ-tas and microreactors using processes with high precision and long term stability of the joint. With new developments laserjoining can be used for joining and assembly of systems made from silicon, glass and polymers in monomaterial devices as well as dissimilar devices.

With an adaptation of the laser wavelength with respect to the absorption properties of the material a large number of different materials can be joined. The specific advantages of the joining technologies are minimum of energy input without heating the whole device, possibility of selective joining, no additional materials like adhesives required, and joining geometries < 20 µm.

The laser bonding technology has been used for bonding pyrex and silicon wafers as well as polymers. First attempts have been made for joining dissimilar materials like silicon and polymers with any additional materials.

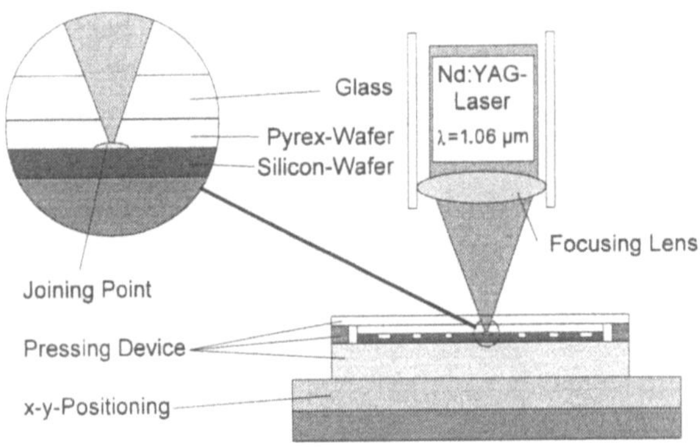

Fig. 10: Principle set up for laser joining of dissimilar materials.

5 Literature

1. M. Takai "Laser Chemical Processing of Magnetic Materials for Microelectronic Applications", LASERION '93 June 21-23 Munich.
2. F. Bachmann, P. Holzer, "Bohren von Leiterplatten mit Excimerlasern", in: Industrielle Laseranwendung, Carl Hanser Verlag, Juni 1989 S. L31.
3. Ch. Rowan, "Excimer lasers drill precise holes with higher yield", Laser Focus World, August 1995, pp. 81.
4. R. Srinivasan "Ablation of Polymers and Biological Tissue by Ultraviolet Lasers", Science 234 (1986), pp. 559.
5. H.M. Phillips, R.A. Sauerbrey, "Excimer-laser-produced nanostructures in polymers", Opt. Eng. Vol. 32 No. 10, Okt. 1993.
6. Bartels Mikrotechnik GmbH, Dortmund.
7. D. Hellrung, A. Gillner, R. Poprawe, „Laser beam removal of micro-structures with Nd:YAG-lasers", Proc. Lasers in Material Processing Laser'97, SPIE 3097, pp. 267-273, Munich, 1997.
8. J. Jandeleit, „Materialabtrag mit intensiven, ultrakurzen Laserpulsen für Anwendungen in der Mikrosystemtechnik", Dissertation RWTH Aachen, 1997.
9. H. K. Tönshoff, J. Mommsen, „Process of generating three-dimensional microstructures with excimer lasers", Proc. ECLAT'92, Göttingen, 1992.
10. Contact: Dr. M. Feustel, Resultec analytic equipment, Garbsen (email: resultec@t-online.de).

Functional Coatings for Microstructure Reactors and Heat Exchangers

M. Fichtner, W. Benzinger, K. Haas-Santo*, R. Wunsch and K. Schubert

Forschungszentrum Karlsruhe, 76021 Karlsruhe, Germany
*KiKi Ingenieursgesellschaft, 76725 Ettlingen, Germany

Abstract

Functional coatings for microchannel devices have been developed for protection and catalytic activation of microstructured metallic supports. Micron-thick coatings of vapor-deposited PTFE or fluorine-doped nanomers have surface characteristics similar to PTFE bulk material. Porous ceramic sol/gel coatings of varying composition show micro- and mesoporosity and have surface enhancement factors of up to 430. Other coatings based on sintered nanoparticles, CVD and anodic oxidation have been tested successfully.

Keywords: microchannel, microstructure, microchannel reactor, catalytic coating, sol/gel coating, polymer coating

1 Introduction

For applications of microstructure reactors and heat exchangers it can be necessary to have functionalized inner surfaces of the microchannel system instead of the pure metal surfaces of the microstructured body itself. For this purpose, microstructure-compatible coatings have been developed and investigated in order to achieve tailored properties of the device.

In this contribution, examples shall be presented for microstructure-compatible ceramic and polymer coatings which have surface energy-reducing and protective properties and can be applied to reduce fouling and corrosion in the microchannels. Furthermore, ceramic coatings will be presented, which can be used as carrier catalysts for microstructure reactors.

2 Protective Coatings

Due to their small dimensions, microchannels are endangered by fouling and corrosion. As in other process technologies, the use of coatings may be advantageous to protect the devices.

2.1 Coatings for Corrosion Prevention

Corrosion is expected to be enhanced in microchannels. Due to the high mass transfer rates between the fluid and the wall surface, the concentration of oxidized and diluted wall material at the phase boundary is low compared to the concentration at the surface of a macroscopic fluidic system, for example a stirred tank or a tube reactor. According to the Nernst equation, lower ion activities at the phase boundary lead to a reduction of the surface potential of the wall, to a higher corrosion potential and therefore to a higher corrosion susceptibility.

Corrosion of a microstructure is likely when a microstructured base material which is normally susceptible to corrosive attack is used as support for a protective coating. If such a material is equipped with a corrosion-resistant coating which has to be a thin layer because of the microchannel dimensions, even small imperfections in the layer will lead to corrosion pits. It should therefore be a good approach first to choose a "right" base material with anticorrosion properties, which is then microstructured and later coated for *additional* protection. As metallic base materials, Ni-based alloys, precious or refractory metals may be a good choice. However, not all microstructuring methods are able to process these materials. Nevertheless, microstructure reactor prototypes made of micromachined stainless steel and Hastelloy have already been presented [1].

Many materials with anticorrosion properties have a tight passivating layer of oxides on their surface. This self-constituting layer grows by diffusion or segregation of elements such as Ni, Cr, Al, Zr which form stable oxides at the surface. The segregation process is induced when the surface is in contact with air or the medium and/or the material is heated. This is a thermodynamically driven process and responsible for the self-sealing properties of such materials.

2.1.1 Coating for High-temperature Applications

To investigate a metallic system with surface segregation particularities, a Fe-base alloy with a content of 72.6% Fe, 22% Cr, 4.8% Al, 0.3% Si and 0.3% Y (Fecralloy®) has been microstructured mechanically and heated for 15 hours in air at 1323 K. During the tempering process, a 2 – 3 μm thick oxide layer formed, which consisted mainly of Al_2O_3. The layer thickness did not increase substantially when the sample was left in the oven for a total of 30 hours, which indicates that the reaction with the oxygen did not go further. Figure 2.1 shows that the layer has a grainy fine structure which prevents the inward diffusion of detrimental elements, O_2 in this case. Hence, it is likely that this surface is more corrosion-resistant against hot gases than the non-treated surface.

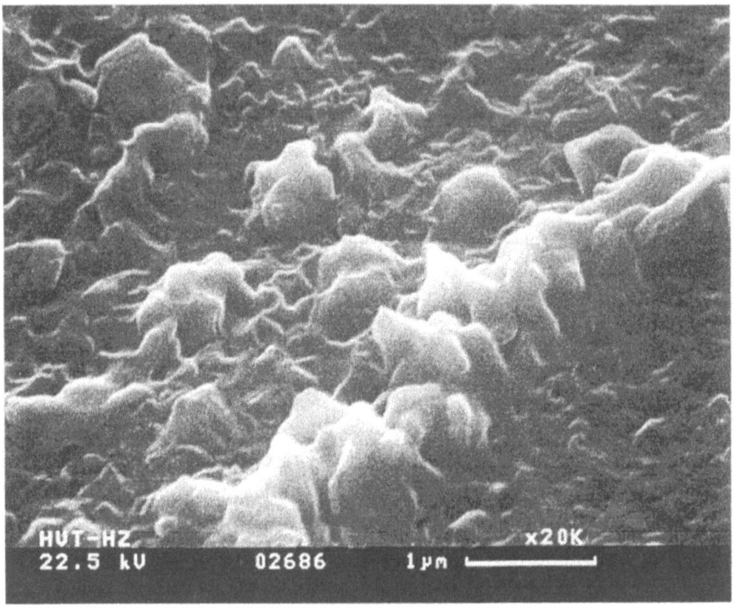

Figure 2.1: SEM of a tempered Al-containing Fe alloy surface, coated with a protective Al-oxide layer which was produced by a tempering process.

2.2 Coatings for Fouling Prevention

Fouling may occur when the adhesive forces between solid material in the fluid and the wall exceed the shear forces caused by the fluid stream. If there was an independently forced fluid flow in every microchannel, shear forces would be high and the risk of fouling would correspondingly be low.

In practice, this may not be the case because a microchannel passage usually consists of many microchannels which are connected in parallel [1][2]. When a channel cross section becomes smaller by fouling, the fluid can flow through the neighboring channels and the only effect would be an increase in the common pressure drop. This increase may be too small for an additional shear force to remove the solid deposit from the wall again. Hence, the only approach left is to attain a decrease of the adhesive forces. A reduction of these can principally be achieved by smoothing the surface - which shall not be regarded here - and by a reduction of its surface energy.

Materials with fluorine-containing functional groups are well known for their low surface energies and should therefore be suitable coating materials to reduce the risk of fouling in microchannel systems. For example, polytetrafluoroethene (PTFE) is known for its low wettability, its high chemical resistance and its permanent temperature resistance up to 540 K [3].

As already mentioned, coatings for microchannel systems should be thin, which limits the methods that can be applied to coat the structures. Among these, vapor deposition methods and sol/gel processes are favorable. In the following sections, two types for coating of metallic foils with protective layers will be presented:

- Vapor deposition of a PTFE layer on a metallic support by pulsed **EL**ectron **B**eam **A**blation (ELBA process) [4].
- Sol/gel method with deposition of a flourine-doped nanocomposite layer.

For the characterization of the coatings, analytical techniques such as SEM / EDX and Secondary Neutral Mass Spectrometry (SNMS) have been used. The latter is described in detail elsewhere [5] and allows to detect the in-depth composition of a layer with a high depth resolution in the range of 10 nanometers.

For a further characterization of the coatings the contact angles of the surfaces were measured with a conventional contact angle measurement system. Water and n-propanole have been used as test liquids. For comparison the contact angles of the uncoated substrate, PTFE bulk material, PTFE layer and the nanocomposite layer were determined.

2.2.1 Fluoropolymer Coating

For the development of tight and uniform layers of PTFE on microstructured substrates, microstructured metal samples have been coated with evaporated PTFE bulk material.

To evaporate PTFE, surface excitation of a bulk PTFE target may be realized with a pulsed ion beam, UV light via a pulsed excimer laser or a pulsed electron beam. An economically favorable ablation method, the so-called ELBA process, has recently been developed at the Karlsruhe Research Center [4]. The method is based on channel spark ablation where a pulsed, self-pinched high-power electron beam hits the surface of a solid target, heats it up and evaporates it in a small explosion. The ablated material has approximately stoichiometric composition and the velocity of the front of the ablation cloud is about 10 km/s. Hence, the kinetic energy of the ablated material is much higher compared to the PVD process, and good adhesion properties of the deposited material are generally observed.

Microstructured foils of stainless steel have been coated with PTFE ablated by the ELBA process. To examine the uniformity of coating along the microstructure walls, cross sections of the samples have been prepared and investigated by SEM / EDX. The experiments show that a uniform layer can be achieved by tilting the sample in two different positions during the coating process. Fig. 2.2 shows a SEM of the edge of a PTFE layer which is partially removed from the support. The thickness of the fluoropolymer coating is in the range of 700 – 800 nm.

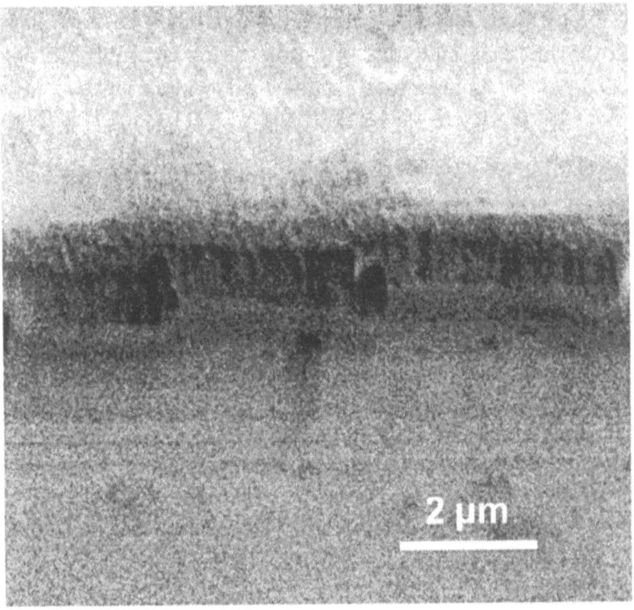

Figure 2.2 : SEM of the edge of a tapered PTFE, film-deposited on a stainless steel substrate by the ELBA process.

Top views of the sample show a tight layer with a grainy fine structure. The fine structure dimensions are typically below 100 nm. It is likely that the grains originate from tiny PTFE droplets which are formed during an ablation event and deposited on the sample. Moreover, the deposition process is influenced by irregularities at the sample surface.

2.2.2 Fluorine-doped Ceramic Sol/Gel Coating

Based on a preparation method for nanocomposite coatings [6], a mixture of 80 mol % methyltriethoxysilane (MTEOS) and 20 mol % tetraethoxysilane (TEOS) was stirred at room temperature for 2 h with hydrochloric acid and silica sol. After hydrolysis a small amount (2-3 mol %) of a fluoroalkylsilane was added and the mixture was stirred for 1 h. The obtained solution was used for coating immediately after its preparation.

Coatings were prepared by dipping of the microstructured stainless steel foils which were used as substrates. Small parts of the metallic foils (5 cm in length, 1 cm in width) were dipped into the solution and subsequently dried at 50°C in a drying oven. For thermal treatment the samples were baked at 150 °C for 10 h. Then they were removed from the hot oven and cooled in air.

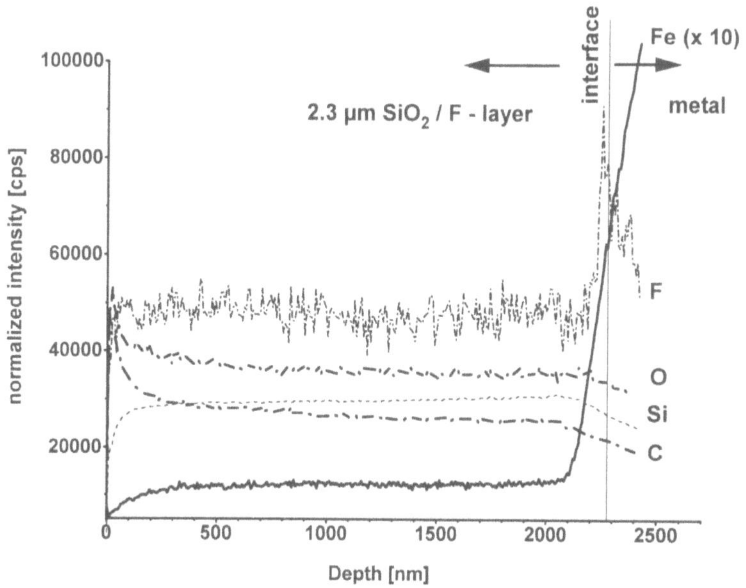

Figure 2.3: SNMS depth profile of the flourine-doped ceramic layer. The lines in the diagram represent the measured intensities, normalized to their isotopic ratio and detection factor, respectively.

The coatings are scratch-resistant and stay attached to the surface when the metal foil is bent. Top views of the sample with SEM show a uniform and tight layer free of defects. From a SNMS depth profile (see Figure 2.3) the layer thickness was determined to be 2.3 µm. The thickness was determined by the position of the point of inflection in the Fe intensity curve.

The SNMS depth profile reveals a fine structure of the layer. Fluorine was found to be enriched at the second boundary, which was probably due to its high affinity to the metal and probably responsible for the very good sticking properties of the ceramic coating to the metal support. An enrichment of carbon and oxygen was found in the outer 100 nm, which may be caused by the baking process. Corresponding to the decrease in the O and C intensity in the first 200 nm, the Si intensity increased until they reached constant values of the undisturbed layer.

2.2.3 Properties of the Coatings

The results of the contact angle measurement for the PTFE- and the ceramic-coated samples are shown in Table 1. An increase in the contact angle means a decrease in the wettability. As can be seen, the increase of the contact angles from the uncoated metallic foil to the coated samples is significant, independent of the test liquid. Only little differences in the properties of the modified surfaces and the PTFE bulk material were observed: The variation between the contact angles of the flourinedoped ceramic layer and the PTFE layer to the PTFE bulk material is less than 1%. For n-propanole the variation is less than 9%.

	water	n-propanole
Uncoated metallic foil	74°	(0°)*
PTFE bulk material	110°	31.3°
PTFE layer	109.7°	28.7°
Flourine-doped ceramic layer	108°	28.5°

*Table 2.1: Contact angles of PTFE bulk material, coated metallic foils and an uncoated metallic foil, determined with different liquids; *measurement impossible, drop disappears immediately.*

Hence, both preparation methods seem to be convenient for the coating of metallic microstructures with a thin protective layer. Both types of coatings have surface energy-reducing properties and should therefore be appropriate to reduce fouling in a microchannel system. However, further tests are necessary and a field study of the behavior of surface-modified microstructures in a microchannel system shall be performed.

3 Catalytic Coatings

Due to their exceptional mass and heat transfer capabilities, catalytic microstructure reactors have the potential to attain high space-time yields of the desired chemical product. First examples of laboratory reactors have already been presented (e.g. [7][8][9]).

In all of these applications, an appropriate amount of catalytically active centers has to be provided along a reaction channel. Their number depends on the turnover frequency and the mass transport to and from the catalytically active centers. Whereas fast reactions such as partial oxidations [8][10] may get along with smooth catalytically active metal surfaces which can be deposited e.g. by sputter coating or fabricated by micromachining of noble metal foils, a variety of applications need many more active centers [9][11][12][13]. Therefore, the task is to provide a fine dispersion of the catalytically active material in a microstructure-compatible ceramic layer with a high porosity. In the following sections, examples of such coatings will be presented.

3.1 Sol/Gel Coatings

Porous oxide coatings on the basis of alumina, silica and titanium have been developed as support for catalytically active components on a Fe-based alloy. The sols have been prepared according to procedures from the literature [15], which have been optimized in our laboratory. For the preparations, alkoxides have been used as precursors, acetylacetone as stabilizer, nitric acid as catalyst and an alcohol as solvent. As alkoxides, tetraethylorthosilane, aluminum-sec.-butylate, aluminum triisopropylate and titanium tetraisopropylate have been used.

The sols were prepared by adding a mixture of demineralized water and acid to a solution of the alkoxide in the solvent. The preferred solvent was isopropanol or 2-butoxyethanol. In case of binary metal oxides, solutions of the alkoxides were mixed before hydrolysis. The sol was stirred at ambient temperature for 1 - 6 h. Further condensation took place in an oven at 60 °C. The prepared sols were then ready for use. Before dip-coating, a Fe-based alloy used as metallic support was pretreated by thermal treatment. The dip-coated samples were dried first at room temperature and then in an oven at 60 °C for 12 hours. As a final preparation step, temperature-programmed drying in air was used. With a heating rate of 4 K/min the samples were heated to the final calcination temperature with a holding time of 5 hours.

After the coating process was finished, the samples were inspected by light-optical microscopy and scanning electron microscopy. First experiments led to a non-continuous surface layer with small particles which could easily be wiped off. Good adhesion was obtained after modification of the sol by adding acetylacetone as stabilizing chelating agent. The use of a chelating agent as stabilizer was found to be very important for a good adhesion of the coating to the support. The whole hydrolysis-condensation process is modified promoting the formation of polymeric gels.

The specific surface area and pore structure were determined by physisorption with N_2 and Kr as adsorbates. The specific surface area of the coatings was determined by the BET method from the krypton adsorption isotherms at 77 K. The resulting surface area of the coated samples is expressed in terms of an enhancement factor F defined by the ratio of the measured BET surface area to the geometric surface area of the support. The factor is calculated for a coating thickness of about 1- 2 μm.

Influence of Preparation Parameters on the Surface Area

The various preparation methods led to a variation in the specific surface area caused by influences of the type of alkoxide used, the solvent and the temperature treatment.

Thermal annealing of the coatings was performed in the range between 400 and 800 °C. The resulting surface area was found to be dependent on the temperature. The highest specific surface area was obtained at a calcination temperature of 500 °C for all samples. Table 3.1 shows the values obtained for an Al_2O_3 coating made of aluminumtriisopropylate.

T_{calc} / °C	400	500	600	700	800
$F_{Surface}$ [m²/m²]	3	152	97	34	30

Table 3.1: Surface enhancement factor of Al_2O_3 coatings made of aluminum tri-isopropylate for various calcinating temperatures.

The influence of the alkoxide used in the preparation of the sol can be seen when comparing the Al_2O_3 coatings made of aluminum-sec.-butylate and aluminum isopropylate. The surface enhancement factor F rises from 152 (Table 3.1) to 430 (Table 3.2) when aluminum-sec.-butylate is used.

Coating	F [m²/m²]	V_{Micro} [cm³/m²] HK Method	V_{Meso} [cm³/m²] BJH-Method
Al_2O_3 (al.sec.butyl.)	430	0.15	0.52
TiO_2	74	0.02	0.18
$Al_2O_3 - TiO_2$	100	0.03	0.18
$Al_2O_3 - SiO_2$	193	0.06	0.42

Table 3.2: Surface enhancement factor F and pore volumes for sol/gel-coatings on a metallic support, calcinated at 500°C. The thickness of the coatings is about 1-2 μm.

Pore Structure

Information about the pore structure of the coatings was obtained from the analysis of the nitrogen adsorption isotherms. The shape of the measured isotherms was the same as IUPAC type IV which is typical for mesoporous solids. In the case of Al_2O_3 a significant contribution of microporosity reflected by enhanced sorption at low relative pressure was observed. Furthermore, different shapes of the hysteresis loops were observed, which indicates different shapes of the mesopores.

Mean pore size distribution was calculated from the nitrogen adsorption isotherms applying the HK (Horvath-Kavazoe) method for micropores and the BJH (Barrett, Joyner and Halenda) method for mesopores. The calculated micro- and mesopore volumes are shown in Table 3.2, the plots of the pore size distributions are shown in Figure 4. The alumina samples have a higher porosity than the binary oxides of aluminum with either silica or titanium.

The position of the maximums of the pore size distribution is the same as obtained from bulk samples of the gel. The maximum of the pore size distribution is in the range of 5 – 6 Å for micropores and 40 Å for mesopores. The plots in

Fig. 4 show that the pore size distributions of the samples are similar while the adsorbed volume varies with the sample.

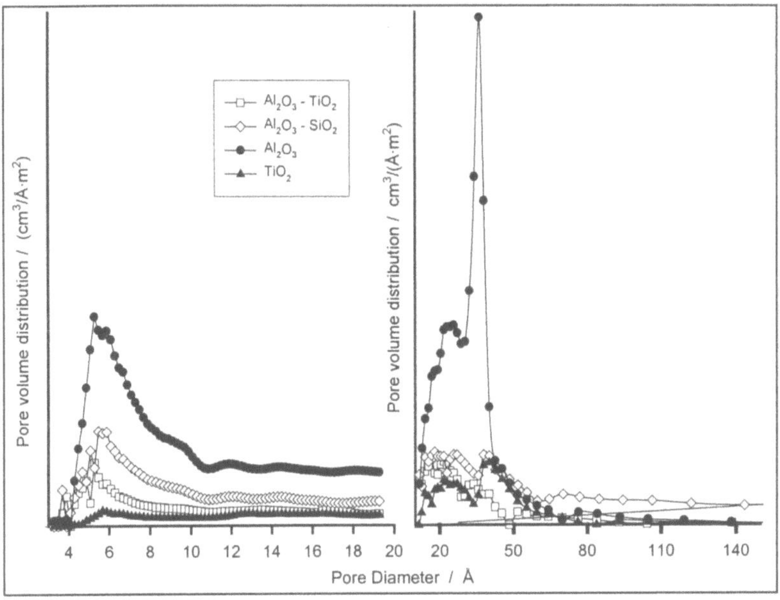

Figure 4: Pore size distribution according to the HK (for micropores) and BJH model (for mesopores) for various sol/gel coatings on a metallic support.

3.2 Nanoparticle-based Coatings

Another approach which has been pursued is the development of coatings based on immobilized and sintered nanoparticles. The dispersion of catalytically active material in a carrier structure is achieved by an intense mixture of commercially available nanoparticles of the catalyst and the promoter or carrier material. The mixture is immobilized along the microchannel walls and sintered. The sintering process exerts influence on the pore structure and the formation of intermetallic phases which, in some cases, are responsible for the catalytic behavior of the coating. The advantage of this method is that, depending on the procedure, no additional impregnation step is necessary after immobilization of the nanoparticles. The pore structure of the coating is dominated by macropores and depends on the type of the nanoparticles and the thermal treatment procedure.

First coatings of this type have been developed and utilized for the methanol-steam reforming reaction to H_2 and CO_2. Within a microchannel device, high turnovers at low residence times have been observed. A more detailed description of the coatings is given elsewhere in these proceedings [14].

3.3 Other Coatings

CVD Process

The CVD process is a valuable tool to obtain porous ceramic coatings at the inner walls of a microchannel system. A catalytic microreactor equipped with an alumina layer which was deposited by atmospheric pressure CVD and with platinum as active component was developed in collaboration with the university of Frankfurt. The coating consists of γ-alumina and has a surface enhancement factor of about 100 with a layer thickness of about 10 μm. The microreactor is able to convert explosive H_2 / O_2 mixtures safely to water. It is equipped with an internal heat transfer structure and described elsewhere [11].

Anodic Oxidation of Aluminum

Porous alumina coatings have a regular and mesoporous fine structure with a surface enhancement factor up to about 100 per μm thickness of the layer [7][9]. Pore width and length can be adjusted by variation of the electrolyte solution, bath temperature and duration of the anodic oxidation.

Microchannel reactors based on such coatings have been developed in collaboration with the Technical University of Chemnitz [9][13]. With these applications, the combination of the regular pore system together with the defined residence time distribution of the microchannels was proven to be advantageous to attain high selectivities for a desired product.

Another microchannel reactor device has been developed by FZK to attain a complete turnover for the combustion of small amounts of hydrocarbons in air. The reactor is described in detail elsewhere in these proceedings [12].

4 Manufacturing of Coated Microchannel Devices

The choice of a suitable coating and manufacturing process of an internally coated microchannel device depends on the boundary conditions which are imposed by the later application of the device. Temperature, pressure and permitted leak rate are important factors which define the bonding method and not every coating can be combined with every bonding method. For example, pressures over several 100 kPa and temperatures above 500 K require joints with a high stability, e.g. by welding or soldering whereas under mild conditions, glued joints may be sufficient. Some of the coatings presented above can only be prepared *before* the device is joined (Al_2O_3 and PTFE coating) and arrangements have to be made for the parts to be joined being covered by an appropriate mask during the coating process. With all other coatings, a post-coating of the completely manufactured microstructure is possible.

Acknowledgements

The authors want to thank Dr. Schultheiß for the support rendered with regard to the PTFE coatings by the ELBA process and Dr. Carsten Natzeck for the SNMS measurements.

References

[1] K. Schubert, W. Bier, J. Brandner, M. Fichtner, C. Franz, G. Linder, Proc. 2nd Int. Conf. on Microreaction Technology, AIChE Spring Meeting, New Orleans (1998) 88 - 95

[2] A.L.Y. Tonkovich et al., Proc. 2nd Int. Conf. on Microreaction Technology, AIChE Spring Meeting, New Orleans (1998) 186 - 195

[3] D.P. Carlson, W. Schmiegel, Organic Fluoropolymers, Ullmann Encyclopedia of Industrial Chemistry, 6th Edition, Electronic Release (1998) Chapter 2.2.

[4] G. Müller, M. Kronijnenberg, G. Krafft, C. Schultheiss, in: Science and Technology of Thin Films, F.C. Matacotta, G. Ottaviani (ed.), World Scient. Publ. Co. Pte. Ltd. Singapore (1995) 89 - 119

[5] M. Fichtner, M. Lipp, J. Goschnick, H.J. Ache, Surface and Interface Analysis, Vol. 17 (1991) 151 - 157

[6] M. Menning, G. Jonschker, H. Schmidt, SPIE Vol. 1758, Sol-Gel Optics II (1992) 125 - 134

[7] M. Fichtner, T. Ludwig, R. Wunsch, K. Schubert, FZKA 6080 (1998) 121 - 128

[8] O. Wörz, K.P. Jäckel, Th. Richter, A. Wolf, Proceedings of the 2nd Int. Conf. on Microreaction Technology, AIChE Spring Meeting, New Orleans (1998) 183 - 185

[9] G. Wießmeier, K. Schubert, D. Hönicke, Proc. 1st Int. Conf. on Microreaction Technology, Springer (1997), 20 - 26

[10] J. Mayer, M. Fichtner, D. Wolf, K. Schubert, this issue

[11] U. Hagendorf, M. Janicke, F. Schueth, K. Schubert, M. Fichtner, Proc. 2nd Intern. Conf. on Microreaction Technology, AIChE Spring Meeting, New Orleans (1998) 81 - 87

[12] R. Wunsch, M. Fichtner, K. Schubert, this issue

[13] A. Kursawe, E. Dietzsch, S. Kah, D. Hönicke, M. Fichtner, K. Schubert, G. Wießmeier, this issue

[14] P. Pfeifer, M. Fichtner, M. Liauw, G. Emig, K. Schubert, this issue

[15] M. Galán, J. Llorens, J.M. Gutiérrez, C. González and C. Mans, Journal of Non-Crystalline Solids 147&148 (1992) 518-522

Polymer nanowell plates with variable well slope angles

Holger Becker[1], Thomas Klotzbücher[2]

[1]Jenoptik Mikrotechnik GmbH, Prüssingstr. 41, D-07745 Jena, Germany
[2]Institut für Mikrotechnik Mainz GmbH, Carl-Zeiss-Str. 18-20, D-55129 Mainz, Germany

Abstract: Microwell plates are standard elements in any biotechnological system dealing with high throughput screening or drug discovery. Progressing from the current standards, we have investigated the microfabrication of higher density well plates with respect to an optimized well geometry. This included the defined variation of the wall slope allowing particularly bead based dispensers to minimize distribution errors as well as the fabrication of wells with a very small surface area but high depth to minimize evaporation losses, simultaneously increasing the optical path length for detection. The well-plates with variable slope were fabricated using UV-laser ablation of a spin-coated PMMA layer. Other geometries were realized using advanced silicon etching for the fabrication of the embossing masters.

Keywords: UV-laser ablation, polymers, nanowell plates, biotechnology

1 Introduction

Over the last years, the speed of generating data in life sciences has tremendously increased. The driving forces of this development can mainly be found in the field of drug discovery, where the introduction of combinatorial chemistry generated very large component libraries, which need to be analyzed. Another driving force are the ultra-high throughput screening applications, which span the gap between diagnostic data generation and lead identification, and the shotgun-approach in genomic sequencing, generating genetic information with a rate of several completed genomes a year. The technological breakthroughs enforcing these developments, are the introduction of microsystem technologies into the life sciences [1] and the laboratory automation. Most robotic laboratory automation systems are based on microwell plates. The parameters of the currently commercially available microwell plates are listed in Table 1. In this paper we describe the microfabrication of nanowell plates as a further development of existing microwell plate standards.

Format	Volume [µl]	Well diameter [mm]	Well spacing [mm]	Price [approx USD]
96	300	6.7	9.0	0.5
384	75	3.5	4.5	2.5
864	20	2.2	3.0	8
1536	4-10	1.8	2.25	20

Table 1: Overview on existing microwell plate formats.

We have concentrated our research onto the following topics:

- Wells with variable well angles

  Particularly in the case of bead-based combinatorial synthesis or analysis or screening applications, miniaturization is often limited by the position accuracy of the bead dispensing apparatus, a dispenser head or bead handling system. In combination with conventional well plate design with vertical well walls and a comparatively large flat area on top of the well plate, a bead which is dispensed not directly over the well opening (due to e.g. a misaligned dispenser head) remains on top of the well plate and is lost for any purpose. In a sloped well format instead, the bead automatically rolls down the slope into the well. In addition, the well walls can have an additional function due to their geometry. If the walls are not evenly sloped, but have the surface profile of an optical mirror, in case of fluorescence detection of reaction products the light gathering properties of a well are greatly enhanced and the sensitivity of the setup is increased.

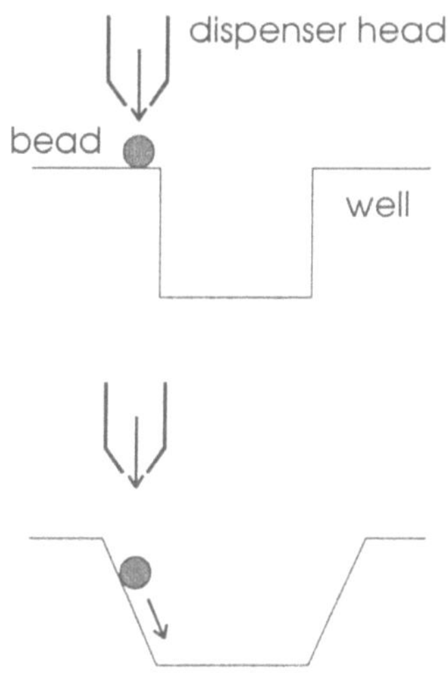

Fig. 1: Schematics of the effect of a displaced dispense head in straight well plates and in sloped well plates.

- Nanowell plates with high density arrays

  Currently, the 1536-well plate marks the end of the commercially available standardized well formats, with some custom-specific designs having 3456 and 9600 wells. In terms of microfabriation by polymer replication technologies, a much higher well density can easily be achieved. We have therefore fabricated high density well formats, on the one hand square wells with a high aspect ratio, on the other hand circular wells with a large variation of well diameters (2µm up to 450 µm) and densities of up to 16900 wells in an area of 380 mm^2 (array of 130 by 130 wells with a spacing of 150 µm). This is equivalent to about 350.000 wells on the standard footprint of a 96 well plate.

- Wells with a high aspect ratio

  A critical parameter in the reduction of well volumes is the evaporation rate of the sample in the wells. Already in 1536 well plates, under normal atmospheric conditions the contents of the well can be evaporated in several minutes. Therefore, in well plates with even smaller well volumes, the design of the wells has to be adjusted to the increasing surface-to-volume ratio of the wells. One way of reducing the evaporation rate is the minimization of the open well surface, which in case of a given well volume, determines the depth and therefore the aspect ratio of the well. We have fabricated wells with aspect ratios of up to 10.

2 Microfabrication

The commercialization of microsystem technologies requires low-cost microfabrication methods, which are suitable for high volume production. Particularly in the field of microsystems in the life sciences, disposable devices on a biocompatible substrate are in great demand. In terms of microfabrication, an ideal microfluidic biochemical device should have the following properties:

- biocompatible surfaces without additional treatment (no sticking or molecular decomposition)
- ease of manufacturing
- suitable microfabrication technologies for a large variety of geometries (rectangular, circular, high aspect ratio, etc.)
- low conductivity for electrokinetic pumping or electro-osmotic separation
- low autofluorescence in case of fluorescence detection
- low cost for high-volume fabrication (disposable devices)

These requirements are best fulfilled using polymers as substrate materials for the fabrication of well structures together with galvanoforming and replication methods like hot embossing for the component fabrication, as enabled by the LIGA process [2, 3, 4]. The polymers used, mostly polymethylmethacrylate (PMMA) and polycarbonate (PC) offer suitable surface properties without modification, low conductivity, an acceptable fluorescence behaviour and availability at very low material cost.

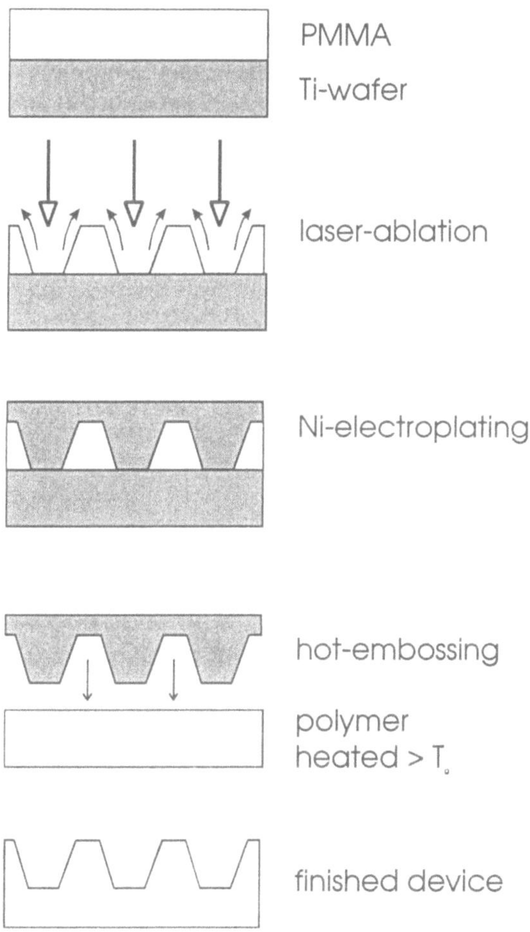

Fig. 2: Schematics of the Laser-LIGA process.

The fabrication of the well plates is done by the Laser-LIGA process, schematically shown in Fig. 2 [5, 6]. A layer of PMMA with a thickness of 170 and 320 μm, respectively, was spin coated on a Ti-wafer, which, in the ablated regions serves as a starting layer during galvanoforming. The PMMA layers were then exposed to pulsed excimer laser radiation of wavelength λ=193 nm via a motorized projection mask. Choosing a well defined starting width of the mask and matching the opening speed of the projected mask to the ablation speed, allows controling the well angle in the range of α=45° to α=90° (Fig. 3). The ablation of the wells is a sequential process, one well after the other is formed, where the material ablation rate is about 0.5 μm per pulse and 10μm per second if the laser is operated at a pulse repetition rate of ν=20 Hz.

Alternatively, the fluence of laser radiation on the polymer surface was varied in the range of 150 to 320 mJ/cm². In this region the ablation rate of PMMA shows a strong dependence on the fluence [7]. Because heat conduction takes place at the edges of the wells during laser ablation, the slope of the well can be varied by changing the fluence on the sample surface. The patterned PMMA surface can be electroplated in a nickel galvanic bath, until a solid nickel shim is created. After separation of the titanium wafer the nickel structure can be used as a master for a subsequent polymer replication step. The replication can be done by hot embossing or injection moulding [8, 9]. During hot embossing the master and the substrate are both heated to just above the glass transition temperature T_g of the polymer substrate. The master is then brought in contact with the substrate and embossed with a force of typically several kN. After application of this embossing force, the complete master-substrate sandwich is cooled down to just below T_g and the master is separated form the substrate, which now contains the replicated geometrical features. The nanowell plates are geometrically well suited for the replication process, as due to their well slope of <90°, the contact between master and substrate in the demoulding process is immediately broken and only very little frictional forces are built up during this process. Using a replication method has the further advantage that the comparatively slow and expensive microfabrication step has to be completed only once for the master fabrication and the subsequent component fabrication can be done in the replication step only [6].

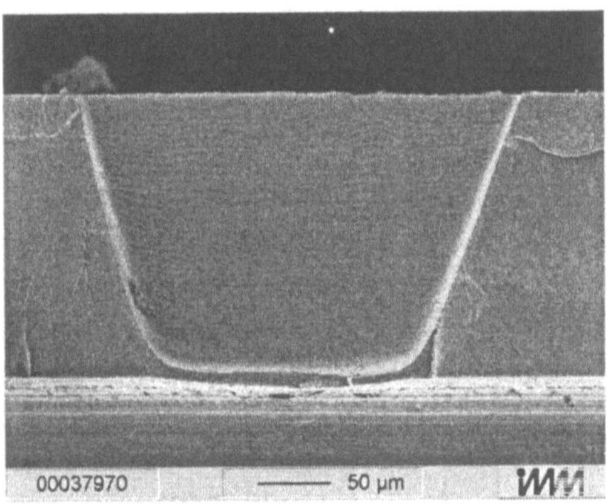

Fig. 3: Well with a slope of 71.6 degrees (side view).

3 Results

3.1 Wells with variable slope angles

Figures 3 and 4 show the side view and the top view of a well with a wall angle of 71.6°, as created by laser ablation following the description of Sect. 2. The cross-section pictures were obtained by SEM analysis after treating the PMMA

Fig. 4: Well with a slope of 68.5 degrees (top view with 15° declination).

samples with liquid nitrogen and subsequent cracking along the cross sectional area. For a quantitative analysis of the cross sectional areas the corresponding SEM pictures were measured with geometrical parameters as indicated in Fig. 5. The data as a function of the projected mask start width are given in Table 2, where the wall angle was calculated as the mean value of the angles on the left and on the right side of the wall, respectively. The well volume was calculated from: $V=[(le+bo+ri)^2+bo^2]/2*he$ (Fig. 5).

Plotting the measured wall angles versus the projected mask start width, as indicated in Fig. 6, gives a nearly linear behaviour. For projected mask start width's smaller than about 200 µm, corresponding to small wall angles, a deviation from the theoretical curve to larger wall angle values takes place, whereas for start width's larger than 200 µm a deviation to smaller wall angle values occurs (Fig. 6). The reason for that behaviour is suggested to be caused by different heat conduction mechanisms. For large wall angles the heat flow in the center of the ablation region is only downwards (onedimensional heat flow), whereas near the edges the thermal energy can

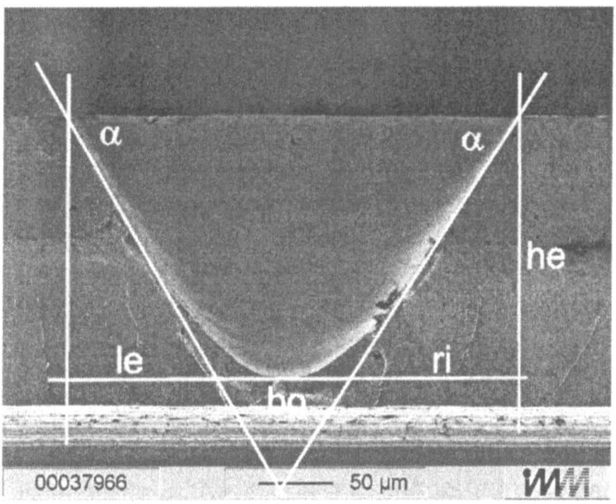

Fig. 5: Cross-section of a laser ablated well with parameter as used in Table 2: le: left width of the well; ri: right width, bo: bottom width, he: height of the well, α: wall angle.

travel downwards as well as sidewards into the bulk (twodimensional heat flow), resulting in less material removal and therefore smaller wall angles than calculated. For small wall angles, that means small projected mask start widths, the ablation progresses from sloped walls, corresponding to twodimensional heat flow in the center of the ablation region.

start width [μm]	le [μm]	ri [μm]	Bo [μm]	He [μm]	mean wall angle α [°]	Well volume [nl]
300	5.2	22.4	284.5	174.1	85.6	15.5
270	20.0	35.0	250.0	171.7	80.9	13.4
240	21.7	46.7	233.3	175.0	79.0	12.7
210	40.0	58.3	205.0	170.0	73.9	11.4
177	48.3	68.3	183.3	175.0	71.6	10.8
130	65.0	80.0	158.3	173.3	67.3	10.1
104	80.0	93.3	126.7	175.0	63.7	9.3
60	95.0	108.3	101.7	176.7	60.1	9.1
14	111.7	115.0	78.3	163.3	55.2	8.1

Table 2: Geometrical data for the laser ablated nanowells (in all cases the end width of the projected mask is 300 μm, corresponding to the upper opening width of the wells).

This results in less material removal in the center of the well, causing a rounding of the edges at the bottom of the well, as can be seen in Fig. 5, and therefore larger wall angels as theoretically predicted. The reduced material removal in the center of the ablation region in the case of small projected mask start widths also results in less deep wells, especially for the smallest width used (Tab. 2). The slight differences in the wall angle at the left and right hand side of the well resulting in different values of li and re (Tab. 2) may be due to a slight deviation of the beam propagation direction to the surface normal of the sample. Producing wells with a maximum opening of 300 μm at the top and a depth of 170 μm by the method of a variable mask opening during laser ablation allowed the variation of the well volume in the range of 8.1 to 15.5 nl (Tab. 2).

Because the maximum width of the wells produced by laser ablation via a motorized mask is limited to 300 μm, due to a limited width of the mask and focal length of the projection optics, wells with a maximum diameter of 600 μm and variable well slope angles were produced by varying the fluence on the sample surface.

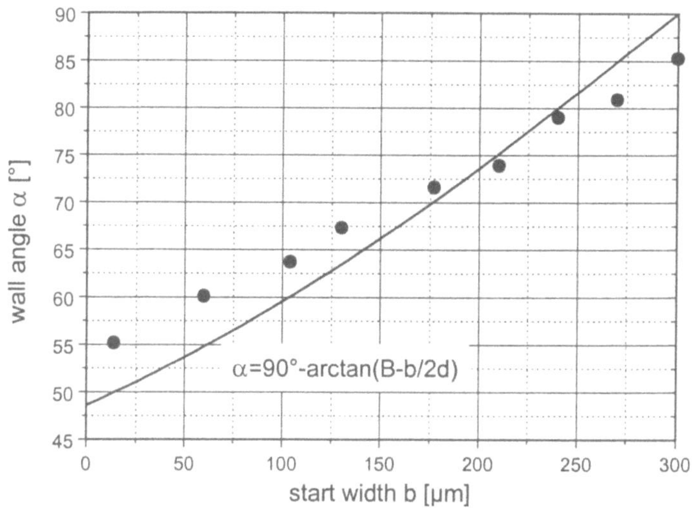

Fig. 6: Wall angle as a function of the projected mask start width (end width in all cases 300 µm, solid line: theory with mask start width b, mask end width B and depth d, dots: measured values).

Fig. 7 and Fig. 8 show two examples of wells fabricated with fixed mask width but different laser fluences on the sample surface (lowest and highest fluence used in the fluence variation are shown). The projection mask width was set to give a maximum well width of 600 µm and ablation depth was set to 320 µm. The well volumes were determined in a similar way as described above. The well volumes as a function of fluence are listed in Tab. 3. A variation between 67.6 and 91.3 nl could be achieved. For lower fluences smaller wall angles form during laser ablation, because the relative loss of laser energy

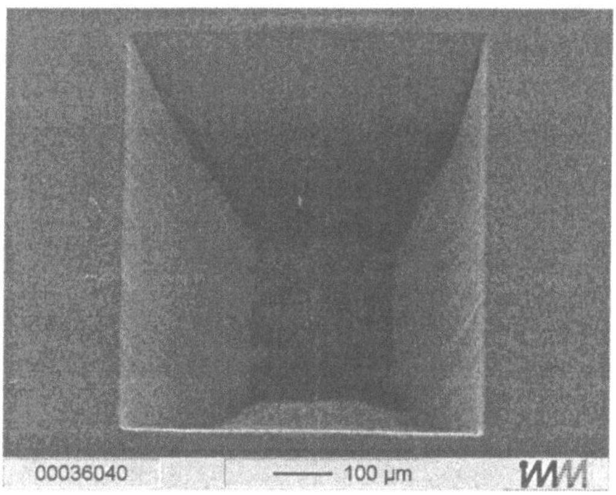

Fig. 7: Well produced by laser ablation with a fluence of 0.15 J/cm² and fixed projection mask (top view with 15° declination)

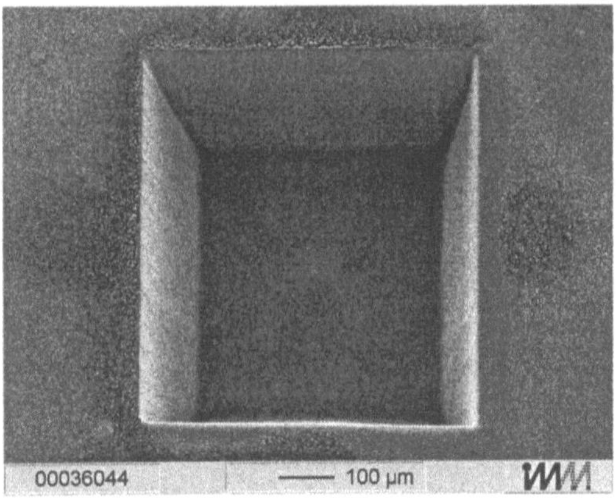

Fig. 8: Well produced by laser ablation with a fluence of 0.32 J/cm² and fixed projection mask (top view with 15° declination)

due to heat cunduction at the edges is higher (see discussion above).

An example of wells with optical sidewall profiles is given in Fig. 9. This picture shows an array of optical elements, fabricated with laser ablation. The period of these optical elements is 25 µm, their height is about 20 µm. Using this structure as an embossing tool, an array of nanowells with a well volume of 6.25 pL and a well density of 160.000 wells /cm² can be fabricated, where each single well has an optical well profile and collects efficiently all the fluorescence generated within each well.

Fluence [J/cm²]	Volume [nl]
0.15	67.6
0.20	79.3
0.24	82.1
0.29	87.0
0.32	91.3

Table 3: Well volume as a function of laser fluence

3.2 High density and high aspect ratio wells

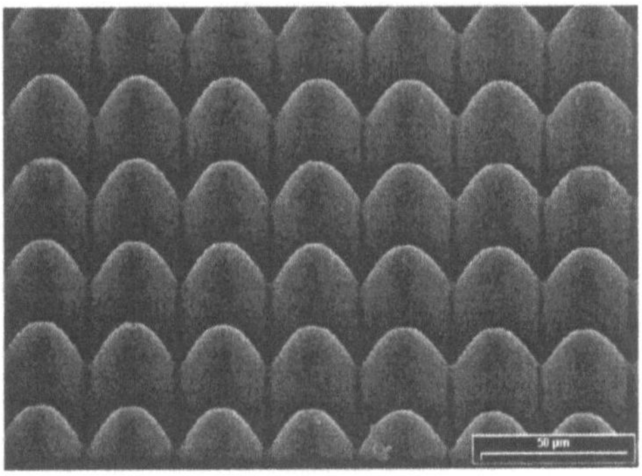

Fig. 9: Array of microoptical elements, which form wells with optical surfaces.

Arrays of high density wells were produced with a different technology for the master fabrication. In this case, the master substrate was a 5“ silicon wafer. The silicon was etched by an advanced silicon etch process (ASE) with an ICP plasma source. A thermal oxide was used as masking material during the etching process. The oxide was structured by standard photolithography and subsequent wet etching. Table 3 shows the various well dimensions

Well diameter [micron]	Well spacing [micron]
1800	2250
900	2250
450	2250
200	2250
100	2250
100	150
50	500
20	500
5	500
2	200

Table 3: Parameters of wells fabricated with ASE.

fabricated in this manner. In Fig. 9 a high density array of circular wells (diameter 100 μm) as produced by ASE can be seen. For comparison in Fig. 10 is shown an array of square wells (side length 250 μm).

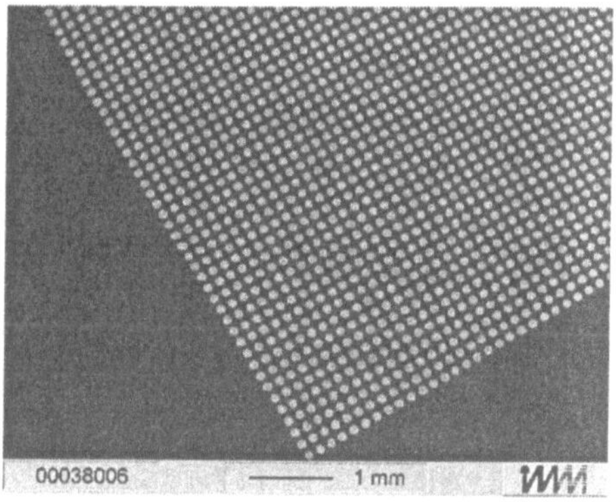

Fig. 9: High density array with a well spacing of 150 μm, fabricated with ASE.

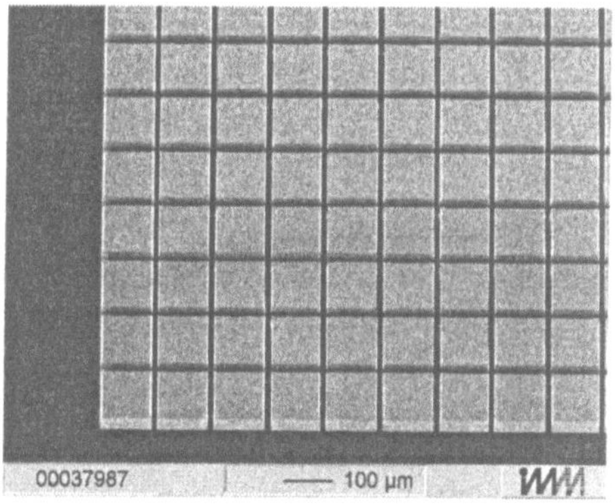

Fig. 10: High density square well array (side length 250 μm) fabricated with ASE.

4 Conclusions

We have fabricated a variety of nanowell plates using different techniques for the fabriation of the replication master. Nanowell plates with a variable slope angle were fabricated using laser ablation which allowed wall angles in the range between 57 and 90 degrees to be adjusted with high accuracy. High density and high aspect ratio well arrays with square and round well shapes have been fabricated using the ASE process. With these fabrication processes, customer-specific well-plates can easily be realized at a low cost and with a high degree of freedom in the well geometry in comparison to the fabrication of nanowell arrays either in silicon [10], glass [11] or silicone rubber (PDMS) [12]. The results show a development of nanowell plates beyond the 1536 well plate, with densities increased up to two orders of magnitude.

Acknowledgements

The experimental work was carried out at IMM Mainz under EU contract CHGE-CT93-0052. We would like to thank Klaus Lowack and Thomas Zetterer for the performance of the photolithography and the ASE etch.

References

[1] A. Manz, H. Becker, Microsystem Technology in Chemistry and Life Science, Topics in Current Chemistry 194, Springer Heidelberg 1998

[2] E. W. Becker, W. Ehrfeld, P. Hagmann, A. Maner, D. Münchmeyer, Microelectron. Eng. 4 (1986) 35

[3] W. Ehrfeld, D. Münchmeyer, Nucl. Instrum. Methods A 303 (1991) 523

[4] W. Ehrfeld, H. Lehr, Radiad. Phys. Chem. 45 (1995) 349

[5] J. Arnold, U. Dasbach, W. Ehrfeld, K. Hesch, H. Löwe, Appl. Surf. Sci. 86 (1995) 251

[6] J. Arnold, W. Ehrfeld, K. Hesch, H. Möbius, F&M, Feinwerktechnik, Mikrotechnik, Messtechnik 103 (1995) 48

[7] M. S. Kitai, V. L. Popkov, V. A. Semchishen, Macromol. Chem., Mocromol. Symp. 37 (1990) 257

[8] H. Becker, W. Dietz, P. Dannberg, Proc. Micro-TAS '98, Banff, Canada, 1998, 253

[9] M. Heckele, W. Bacher, K. D. Müller, Microsystem Technologies 4 (1998) 122

[10] J. M. Köhler, R. Pechmann, A. Schaper, A. Schober, T. M. Jovin, M. Thürk, A. Schwienhorst, Microsystem Technologies 1 (1995) 202

[11] P. Pantano, D. R. Walt, Chemistry of Materials 8 (1996) 8832

[12] R. J. Jackman, D. C. Duffy, E. Ostuni, N. D. Wilmore, G. M. Whitesides, Anal. Chem. 70 (1998) 2280

Micro molding of fluidic devices for biochemical applications

Monika Niggemann, Wolfgang Ehrfeld, Lutz Weber

Institut für Mikrotechnik Mainz GmbH (IMM)
Carl-Zeiss-Straße 18-20, D-55129 Mainz, Germany
e-mail: niggemann@imm-mainz.de

Abstract

Micro molding techniques are gaining more and more importance as low cost alternatives to the established silicon based microfabrication methods. This paper is focused on the development of micro molding processes for the production of microstructured plastic devices for biochemical applications. This is illustrated by the realization of nano titer plates with µl well volumes and transparent well bottoms for applications in screening systems with integrated confocal microscopes as optical detection devices.

Keywords: Micro molding, Biotechnology, Polymers, Nano titer plates

1 Introduction

The idea to use miniaturized fluidic devices as an alternative to conventional analytical chemistry laboratories has been pursued for some time now. The great advantage of this approach is that it allows parallel processing, integration, cost reduction and transportability. Many microstructured chemical and biological components have been fabricated, e.g. various microfabricated capillary electrophoresis chips for DNA analysis[1-3] and miniaturized micro plates for high throughput screening (HTS)[4]. Up to now most micro fluidic devices have been manufactured in silicon or glass[1-4]. This is due to the fact that microfabrication technologies for these materials are well established in the microelectronics industry.

However, for many applications silicon is not the best material choice. The fragility as well as the chemical/biological compatibility sometimes cause problems. In addition, optical detection devices (cameras or microscopes) are often integrated in screening systems and as silicon is not transparent detection is limited to take place from the top. If a large number of disposable devices is needed, the manufacturing costs of silicon devices are too high. Concerning costs polymers are very interesting materials to realize these products. Another point in favor of polymers is the fact that characteristic material properties like mechanical strength, optical transparency, chemical stability, biocompatibility etc. can be chosen from a wide variety of polymers to meet the specific requirements of a certain application.

Recently increasing developments in the field of plastic microfabrication can be observed due to the growing demand for single use consumables in diagnostic applications[5-7]. This paper first gives an overview of the state-of-the-art of micro molding technology for the production of microstructured plastic devices at IMM. Then, as an example, the development of nano titer plates is described that have very thin transparent well bottoms enabling optical detection from underneath.

2 Molding of plastic microstructures

Molding techniques for the production of plastic microstructures available at IMM include compression molding, injection molding, injection compression molding and vacuum casting. However, the molding of micro components requires a number of modifications to these commonly well known processes as well as specialized techniques for the fabrication of microstructured molds.

2.1 Molding processes

At the beginning of the micro molding development reaction molding (or casting) was predominantly used for microstructure replication[8]. Today injection molding, a fully automated and from conventional applications well known short cycle time process, and compression molding (or hot embossing), a simple and very flexible process, are the main replication technologies for the production of three dimensional plastic microproducts. These processes and the necessary modifications for the production of micro parts have been described previously[9,10].

Injection compression molding is a special injection molding process combining some advantages of injection molding and compression molding. The polymer melt is injected (pressure < 1000 bars) into a half-open positive mold. By closing the mold the melt is shaped. As a result only low shear rates occur and the molded part has less orientations but all the same the cycle times are short. Disadvantageous is the high technical effort concerning the mold.

2.2 Mold inserts

Micro molding's central task is to transfer the micron or even sub-micron precision of microstructured metallic molds to molded plastic products. Therefore, besides controlling the micro molding process, the fabrication of the mold inserts has a key position in micro molding technology. Depending on product requirements like structure dimensions, accuracy requirements or target fabrication costs microfabrication technologies like LIGA technique (which is the German acronym for the main process steps lithography (**Li**thographie), electroforming (**G**alvanoformung) and replication (**A**bformung)), LIGA-EDM, Laser-LIGA, silicon micromachining or micro milling are applied for mold insert fabrication[9-11]. In many cases complex molds are build up in a modular way by combining components made by different techniques[12]. Table 1 shows different processes that are available today with typical structure dimensions and aspect ratios that can be achieved.

Process	Typical Structure Dimensions [μm]	Typical Aspect Ratios	Mold Insert Material
Ultraprecision milling	100 – 1000	10 – 50	Steel, nickel
Electric discharge machining	10 – 1000	10 – 100	Steel
Laser ablation and electroplating	5 – 500	1 – 10	Nickel and –alloys
Micromachining of silicon and electroplating	2 – 500	1 – 10	Nickel and –alloys
UV-lithography and electroplating	2 – 500	1 – 10	Nickel and –alloys
X-ray-lithography and electroplating	0.5 – 1000	10 – 100	Nickel and –alloys
E-beam-lithography and electroplating	0.1 – 0.5	1 - 2	Nickel and –alloys

Table 1: Mold insert fabrication methods

2.3 Materials

At IMM commercially available standard thermoplastic materials are used for micro molding. Due to the fact that micro molding of polymers has not yet reached a state where material amounts are consumed that are of commercial interest for the material producers there has been only a small number of developments of polymers that are especially tailored to micro molding. As a result, in the last years a lot of commercially available materials with properties interesting for the production of microparts have been tested for micro molding. Regarding the reproduction quality, filling and mold separation properties polyoxymethylene (POM) is one of the best materials for micro molding. Therefore this material is often used in basic investigations or for first test structures. Otherwise, today not only standard polymers, e.g. PMMA, PC, PP, PE, PBT, PSU and PPE are established but also engineering plastics like LCP, PEEK and PEI as well as biodegradable materials like polylactide. In addition, even ceramics and metals can be micro molded. The result of molding strongly depends on the geometry of the mold cavities, the runner system and on the conditions of the molding process.

Apart from the molding behavior the material for the fabrication of a micropart has to be chosen to meet the specific requirements of the application. As far as biotechnology is concerned chemical and thermal stability play an important role. For micro optical applications amorphous polymers like PC, PMMA and COC (a new material developed by Ticona GmbH, Germany[12]) are used. Optical transparency is usually of importance for biochemical devices when optical analysis is performed. Table 2 and 3 show typical values of basic properties of polymers that are often used in biochemical applications.

	PMMA	PC	PS	COC	PP
Glass transition temperature / crystalline melting point [°C]	105	140	100	130	160
Density [g/cm^3]	1,19	1,2	1,05	1,02	0,9
Refractive index	1,42	1,58	1,59	1,53	opaque

Table 2: Characteristic properties of different polymers

	Resistant to	Not resistant to
PMMA	Aliphatic hydrocarbons, aqueous acids and alkaline solutions, alcohol up to 30 %	Polar hydrocarbons, esp. chlorinated hydrocarbons, conc. acids, alcohol > 30 %
PC	Aliphatic hydrocarbons, grease, oil, alcohol	Organic solvents, alkaline solutions, ammonia, aromatic hydrocarbons, amines
PS	Aliphatic hydrocarbons, conc. acids, alkaline solutions, alcohol	Organic solvents (benzine, acetone, chlorin. hydrocarbons)
COC	Acids & diluted alkaline solutions, alcohol < 30 %, acetone, acetonitrile	Chlorin. hydrocarbons, aliphatic hydrocarbons, aromatic hydrocarbons
PP	Aqueous inorganic salt solutions, slightly inorganic acids and alkaline solutions, alcohol	Aliphatic and aromatic hydrocarbons (benzine, benzole), halogenated hydrocarbons

Table 3: Chemical resistance of some polymers

3 Nano titer plates

Over the last few years high throughput screening and methods of combinatorial chemistry have developed to be important tools in search for new pharmacologically active compounds. In order to break the trend of continuously increasing costs reaction volumes and assay formats are being miniaturized. Since the implementation of the 96-well micro titer plate technology considerable developments have taken place to increase micro titer plate density. New formats are available now: 384 wells, 864 wells and 1536 wells. On a prototype basis even higher density formats have been fabricated.

Apart from miniaturization integration of functional elements (e.g. lenses or filter membranes) is a possibility to improve micro plates. A first step in this direction is made by enabling optical detection from underneath through thin, transparent well bottoms. Optical detection devices like CCD-cameras or confocal microscopes are often integrated in screening systems.

In cooperation with Evotec BioSystems GmbH, Germany, test structures with thin well bottoms were developed at IMM for use in an ultra high throughput screening system with confocal fluorescence detection. The optical image formation is influenced by parameters like bottom thickness, roughness, planarity etc and deviations may lead to incorrect results. Therefore, the quality of the bottom membrane, tight tolerances concerning thickness and planarity, is essential for the confocal fluorimetric method (see table 4). The necessary well bottom thickness depends on the refractive index of the polymer material. For PC the necessary bottom thickness amounts to 120 μm, for COC to 160 μm and PP to 200 μm. Intrinsic fluorescence of the sample carrier is also an important issue, but as the focal plane of the detection system is located approx. 100 μm above the bottom this effect turned out to be not critical at least for polymers like PC, COC or PP.

Issue		Tolerance
Bottom thickness (PC)	120 μm	10 μm
Transparency	90 %	5 %
Surface roughness (RMS)	50 nm	10 nm
Flatness (diameter 1mm)		5 μm
Intrinsic flourescence		

Table 4: Tolerances of well bottoms of PC nano titer plates suitable for confocal optical detection

The developments at IMM aim to a 3456 well format with a well-to-well-spacing of 1.5 mm. Simple test structures with different well shapes (well volumes 1.4 μl and 0.9 μl) have been designed and compared in different fabrication processes:

- Injection molding of the wells including thin well bottoms,
- Injection molding using inserted foils,
- Injection compression molding of the wells including thin well bottoms and
- Injection molding open wells without bottoms and subsequent sealing of the well bottoms with glass.

3.1 Injection molded nano titer plates

PC and COC chips with 120 wells with a well-to-well-spacing of 1.5mm were produced by injection molding as described previously[10]. The well diameter and the overall chip thickness are approximately 1 mm, the well volume is 0.9 μl. The mold insert was fabricated by precision engineering techniques due to the relatively large structure dimensions in the millimeter range. It consists of precisely fabricated pins which have been fixed in the holes of a base plate.

Due to the fact that the flow behavior of polymer melts in injection molding strongly depends on the thickness of the cavity to be filled, first the spacing between the mold pins is filled. Only then the melt flows into the well bottom. This results in so called flow marks which could be seen under a microscope (see fig. 1). Hence, conventional injection molding seems not to be suitable for the fabrication of nano titer plates with such thin well bottoms.

Figure 1: Wells with flow marks

3.2 Injection molding using inserted foils

In order to avoid the problem described above we placed commercially available PC foils with a thickness of 125 μm in the molding tool. The melt was injected on top which yielded a permanent and tight joint. In fact, the interface can hardly be seen. Fig. 2 shows a photo of the chip. The tolerances of the bottom thickness depend mainly on the quality of the inserted foil. As far as the planarity is concerned our quality control measurements showed that depending on the position of the well with respect to the overall chip size the well bottoms are curved between 0.5 μm (in the middle of the chip) and 10 μm (on the edge).

Figure 2: PC chip with 120 wells (well volume: 0.9 μl, bottom thickness: 125 μm)

The process with inserted foils works fine as long as foils of the desired material, thickness and transparency are available. And even then a drawback is the fact that we do not know exactly what happened to the foil material during the production. Therefore it is interesting to produce the micro plate and the bottom membrane from the same material, i.e. the same granulate, in one process and of course if possible in one step.

3.3 Injection compression molded nano titer plates

We fabricated a new mold insert that fits into a compression injection molding tool. The advantages of this process compared to injection molding are the following:

- realization of the required bottom thickness without flow marks,
- process parameters (pressure, temperature) on a lower level,
- lower shear rates and as a result less orientations in the molded parts (significant for the optical detection).

At the same time the important advantages of injection molding remain, short cycle times and full automation.

The chip design contains 48 wells. 24 wells are similar to those on the 120 well chip, but here the well volume is 1.4 µl as the chip thickness is 1.5 mm. In addition, 24 wells with a larger volume of 11.1 µl and a spacing of 4.5 mm were integrated on the chip for testing purposes. The chip was molded in PP due its better chemical resistance and a photo is shown in fig. 3. It turned out that although PP is an opaque material the well bottoms were transparent because of the low thickness.

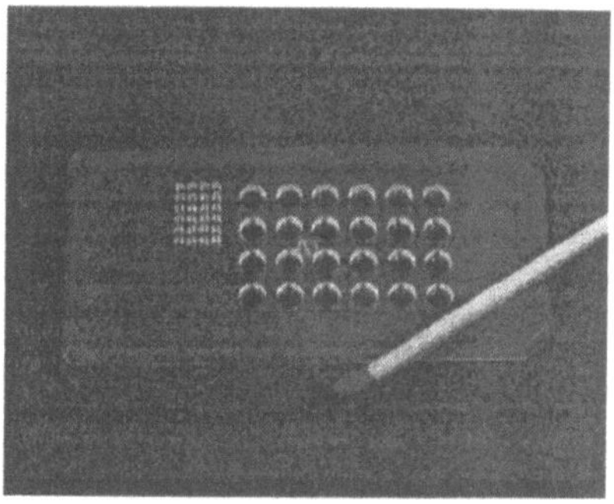

Figure 3: PP chip with 48 wells (bottom thickness: 200 µm)

Our measurements revealed that depending on the process parameters (temperature, compression pressure etc.) the bottom thickness could be varied between 190 ± 10 µm and 210 ± 10 µm. The well bottoms are curved between 0.5 µm and 10 µm. The well to well tolerances mainly result from the tolerances of the mold insert. Tighter tolerances can be achieved with an ultra precise mold insert.

3.4 Two-component nano titer plates

The next alternative for manufacturing the nano titer plates is the composition of sieve-like plates (plates with through holes) with transparent bottom membranes. In this area the combination of plastic sieve plates with glass bottoms is interesting. An obvious advantage are the excellent optical properties and the high planarity of glass. Of course for the production this brings along greater costs and workload.

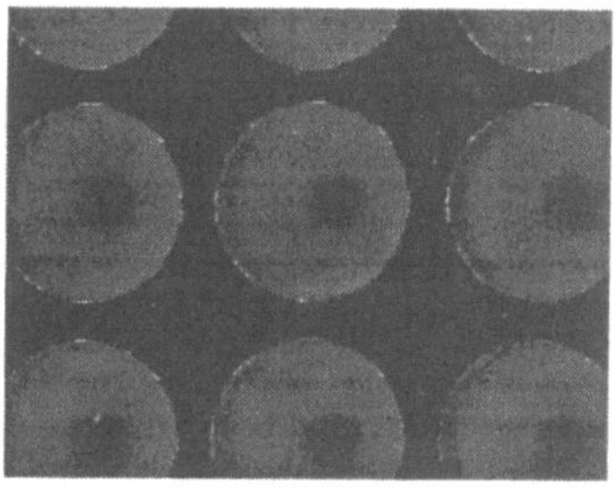

Figure 4: Glass bottom on PP plate

We attached glass bottoms (commercially available cover glasses with a thickness of 120 µm) to PP plates by gluing, a microscope picture can be seen in fig. 4. We have developed an adhesive printing process for the sealing of plastic microparts suitable for mass production. The adhesive is transferred in a structured way to the micropart as indicated in fig. 5. Here the PP plates were pre-treated in an oxygen plasma. As adhesive a UV-curing epoxy resin was used. The adhesive layer has a thickness of 13 µm.

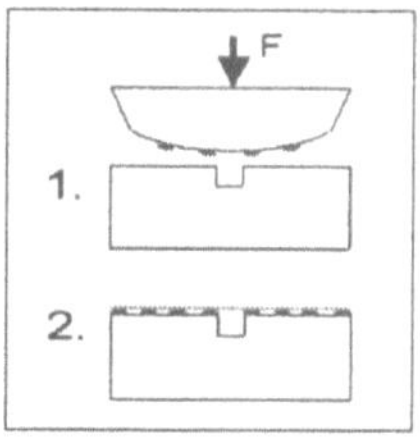

Figure 5: Adhesive printing

4 Conclusions

With injection molding the required optical properties of the well bottoms could not be obtained. Injection molding on inserted polymer foils yielded reasonable optical quality of the well bottoms which is however strongly dependant on the tolerances and material properties of the foils. Similar results were obtained with injection compression molding, which allows the fabrication of the micro plates in one step. Here, the tolerances of the well bottoms are dominated by the precision of the mold insert. Best optical quality was achieved by molding of open, bottomless wells and subsequent sealing of the well bottoms with thin glass plates.

Further work has to be done in order to increase the quality of molded nano titer plates, i.e. achieve tighter tolerances concerning thickness and planarity, and to realize the 3456 well format.

5 Acknowledgements

This work was performed within various research projects at IMM. The authors want to thank Evotec BioSystems GmbH for the confidence as well as the support and the technical staff of IMM for the preparation of the mold inserts, the molding experiments and the quality control measurements.

6 References

1. A. Manz, *Analytical Methods and Instrumentation, Special Issue μTAS '96*, pp. 28-30
2. J. M. Ramsey, S.C. Jacobsen, R.S. Foote, *Proc. of 1st Int. Conf. On Microreaction Technology* (1997), pp. 204-218
3. R. Mathies, P. Simpson, A. Wooley, *Proc. μTAS '98*, pp. 1-6
4. A. Schober, G. Schlingloff, A. Thamm, H.J. Kiel, D. Tomandl, M. Gebinoga, M. Döring, J.M. Köhler, G. Mayer, *Microsystem Technologies* 4 (1997), pp. 35-39
5. C.S. Effenhauser, G. Bruin, A. Paulus, M. Ehrat, *Anal. Chem.* 1997, 69, pp. 3451-3457
6. H. Becker, W. Dietz, *Proc. SPIE Symp. Microfluidic Devices and Systems*, Vol. 3515 (1998), pp. 177-182
7. R. McCormick , R. Nelson, M. G. Alonso-Amigo, D. Benvegnu, H. Hooper, *Anal. Chem.* 1997, 69, pp. 2626-2630
8. P. Hagmann, W. Ehrfeld, *International Polymer Processing*, Vol. IV, No. 3 (1989), pp. 188-195
9. L. Weber, W. Ehrfeld, H. Freimuth. M. Lacher, H. Lehr, B. Pech, *Proc. SPIE Symp. Micromachining and Microfabrication*, Vol. 2879 (1996), pp. 156-167
10. M. Niggemann, W. Ehrfeld, L. Weber, *Proc. SPIE Symp. Micromachining and Microfabrication*, Vol. 3511 (1998), pp. 204-213

11. L. Weber, *Proc. Darmstädter Erfahrungsforum Werkzeug- und Formenbau -Schneller zu präzisen Werkzeugen*, Darmstadt, 06.12.1995, pp. 2-9
12. L. Weber, *Course booklet LIGA Technique*, UETP Series on Micro Electro Mechanical Systems, edited by H.-D. Bauer, Neuchatel (1994), pp. 90-99
13. N.N., "Topas (COC)", company paper of Hoechst AG, Germany

Part 3

Microreactors in Combinatorial Chemistry

Combinatorial Organic Compound Libraries on Continuous Surfaces: Towards Chemical Chips

L. Germeroth[a*], U. Reineke[a], K. Dietmeier[a], C. Piossek[a], N. Heine[b], D. Scharn[b], T. Ast[b], M. Schulz[c], H. Matuschewski[c], A. Kramer[b], J. Schneider-Mergener[a,b] and H. Wenschuh[a]; [a]Jerini Bio Tools GmbH, Rudower Chaussee 29, D-12489 Berlin; [b]Institut für Medizinische Immunologie, Universitätsklinikum Charitè, Humboldt-Universität zu Berlin; Schumannstr. 20-21, D-10098 Berlin; [c]Poly-An GmbH, Rudower Chaussee 6a, D-12489 Berlin; Germany.

Keywords: combinatorial chemistry, membranes, screening, SPOT-synthesis

1. Introduction

Combinatorial chemistry is well suited for the rapid and systematic optimization of molecular properties. The approach is commonly used for the generation of large numbers of compounds which can be applied for the examination of novel materials, catalysts or receptors. However, most frequently assembly of compound libaries is directed to the identification and optimization of novel therapeutic leads to accelerate the drug discovery process. The majority of combinatorial libraries assembled so far was synthesized on solid supports. The advantages of this method are the opportunity to use excesses of reagents driving reactions to completion and the fact that several technologies were invented to automatize and miniaturize these reactions. Beside high throughput parallel synthesis on resin beads, polymeric pins and chips [1], SPOT-synthesis using continuous membranes has been described to be an efficient synthetic approach [2]. The major feature of the latter method is the positionally addressed delivery of small volumes of liquids to a membrane surface. The droplets dispensed by a robot form separate spots which can be considered as micro-reactors provided that a non-volatile solvent system is used (Fig. 1).

Figure 1.) SPOT-robot for spatially addressed dispension of reagents.

Thus, solid phase synthesis of biooligomeric compounds and small organic molecules accessible via non-inert and robust operations can be performed in analogy to conventional multiple synthesis systems. The compounds assembled can either be applied covalently linked for solid phase assays or upon cleavage for high throughput screening techniques in solution.

2. Surfaces

The SPOT-synthesis technology was originally developed for the parallel generation of high numbers of peptides. Cellulose membranes were found to be of sufficient stability towards chemicals routinely used for peptide assembly. In addition, cellulose was reported to display a high compatibility for application of various assay systems [3].
Despite the advantageous properties of cellulose membranes for some applications, this material has a number of limitations such as low chemical and mechanical stability and high concentration of reactive, hydroxyl groups causing side reactions. Therefore, a novel polymeric membrane for solid phase SPOT-synthesis was developed based on the grafting of flexible polymer chains carrying reactive groups for synthesis [4]. A microporous polypropylene membrane was modified via photo-induced graft copolymerisation with functional acrylates (Fig. 2) preserving the membranes' porous structure and stability and allowing further derivatisation. The capacity of the membrane could be adjusted from 0.1- 1 $\mu mol/cm^2$.

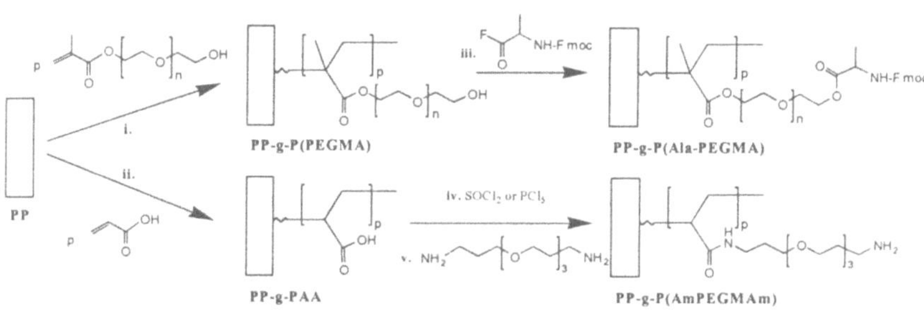

Figure 2.) Overview on the preparations of novel, stable polymeric membranes for solid-phase syntheses (i. & ii. coating of polypropylene, PP, with photoinitiator benzophenone, BP, then UV induced graft copolymerization of monomer from water onto PP; iii. "spot-wise" modification of hydroxyl functions (10 eq. Fmoc-AA-F/ NMI; 0.6 M in NMP); iv. activation with $SOCl_2$ or PCl_5 in DCM; v. reaction with 4,7,10-trioxa-1,13-tridecanediamine.

3. Chemical Synthesis on Polymeric Membranes

A variety of chemistries applying repetitive reaction cycles for assembly of biooligomeric compounds as well as small organic molecules can be adapted for applications using the SPOT-technology on polymeric surfaces. An impressive characteristic of the technology is the freedom to choose the arrangement and size of spots on the membrane. The volumes dispensed create specific spot sizes determining both the scale of reaction and the absolute number of compounds which can be arranged on a membrane area. SPOT-synthesis of large arrays of compounds is feasible by application of automated SPOT-robots for distribution of activated monomers and reagents (Fig. 1).

3.1 Peptides

Peptides were the first compounds extensively prepared by SPOT-synthesis because a long time of experience in peptide chemistry could be exploited. Furthermore, in some cases, peptides derived from a protein sequence retain their function by resembling the structual requirements of a binding or active site. Over the years different screening assays for peptides on cellulose membranes were established (Fig. 3) and successfully applied for the elucidation of molecular recognition events occurring in nature or used in biotechnology:

(1) Binding assays to study peptide-protein interactions directly on the peptide bearing membranes were the initial application. The detection of peptide-bound protein is carried out radioactively, by coupling a reporter enzyme directly to the protein or immunologically using a labelled antibody (Fig. 3A) [3]. These techniques served for the identification of linear as well as discontinuous protein-protein binding sites, which can subsequently mimicked synthetically [5].
(2) The investigation of peptide phosphorylation by kinases is an important tool for the identification and optimization of kinase substrates. Vice versa, known peptidic substrates synthesized by SPOT-synthesis can be used to determine kinase activity in cell lysates or enzyme preparations. The kinase activity is quantified by incubation in the presence of [γ-^{32}P]ATP and measurement of the incorporated activity (Fig. 3B) [6].
(3) Analogous to the kinase assays peptides on membranes are applied for protease substrate identification and the measurement of protease activities. Therefore, the peptide which is tested for stability or instability against proteolytic degradation is synthesized with a fluorescence marker on the N-terminus and punched out into a microtiter plate. Upon cleavage the marker with the N-terminal part of the peptide is released into the surrounding solution and quantified after transfer into another microtiter plate (Fig. 3C). Both enzymatic assays can also be used for the identification of inhibitors by adding other compounds to the incubation solution.
(4) The assay for the identification of metal binding peptides is predominantly used in biotechnology and for the analysis of environmental pollution. Visualization of metal binding is achieved using radionuclids in combination with a „phosphoimager“ or by formation of insoluble chromogenic metal complexes (Fig. 3D) [7].

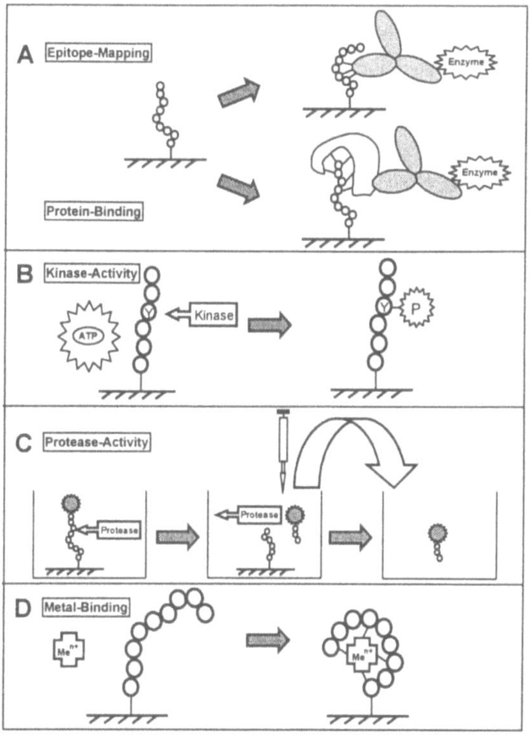

Figure 3.) Binding and enzyme assays for peptides assembled by SPOT-synthesis.

All of these assays can be applied for different types of peptide collections. (i) Given the case that a binding or active site within the primary structure of a protein has to be identified, the protein sequence is scanned with overlapping peptides which are typically 13-mers shifted by three amino acids (Fig. 4A). Linear binding sites are characteristically identified by a few consecutive peptides which bind the respective protein, whereas discontinuous sites have two or more binding regions separated in the primary structure [3]. (ii) Once having identified a stretch of the protein sequence which is important for binding or serves as an enzyme substrate, the key residues mediating the respective function are identified with a so-called substitutional analysis (Fig. 4B). Each residue of a peptide is substituted (rows) by all other L-amino acids and analyzed for binding. The sequences corresponding to the left column are identical and represent the starting peptide. Other spots are single substitution analogs [8]. (iii) For the *de novo* generation of peptides with a predefined binding specificity combinatorial libraries were described [3,8]. The first libraries had the general structure XXB_1B_2XX (X = randomized position with statistical incorporation of all proteinogenic amino acids, B = defined positions). These libraries (Fig. 4D) were arrayed in a 20 x 20 matrix with a mixture of 20^4 = 160.000 different sequences with one defined dipeptide combination on each spot. The increase of synthesis capacity and miniaturization of the SPOT-synthesis formats made the synthesis of more complex libraries of the type $XXXB_1B_2B_3XXX$ with $20^3 = 8000$ peptide mixtures possible. For the distribution of the defined positions over a longer peptide a $XXXX[B_1B_2B_3XXX]XXXX$ library consisting of 68.590 peptide mixtures was designed. The library comprising 10 sublibraries with the different hexapeptide cores $[XXB_1B_2B_3X]$, $[XB_1XB_2B_3X]$, $[XB_1B_2XB_3X]$, $[XB_1XXB_2B_3]$, $[XB_1XB_2XB_3]$, $[XB_1B_2XXB_3]$, $[B_1XXXB_2B_3]$, $[B_1XXB_2XB_3]$, $[B_1XB_2XXB_3]$, $[B_1B_2XXXB_3]$ corresponds to all possible distance patterns of three defined positions (cysteine omitted) [8].

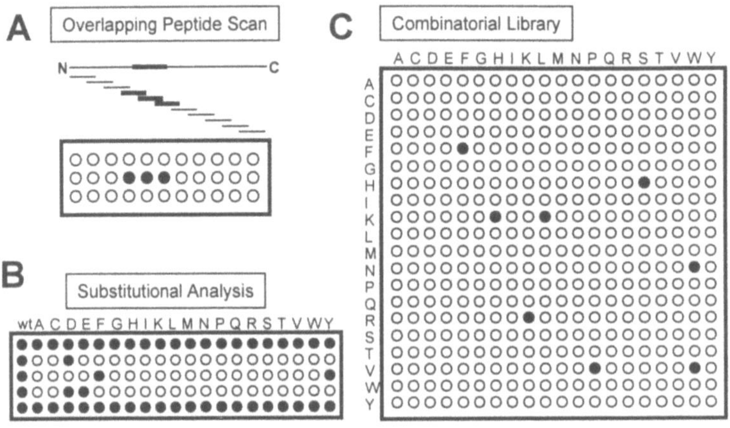

Figure 4.) Library formats of protein-sequence-based and *de novo* approaches for the identification, characterization and optimization of peptides or other oligomeric compounds.

3.2 Peptidomimetics

Although SPOT-synthesis of peptides on cellulose membranes has been shown to be a powerfool tool to screen for binding and bioactivity (see section 3.1 and references cited therein), the applicabilty of ordinary peptides as potential therapeutics is limited due to metabolic degradation and poor oral availability. Several promising approaches have been described to circumvent these drawbacks such as incorporation of D-amino acids or design of various oligomeric peptidomimetic analogs [9]. Peptoids (oligomeric *N*-alkylated glycines) were among the first described and most widely examined class of highly diverse peptidomimetics which are stable against proteolytic degradation [10]. Diversity can be further increased via hybridisation of the peptoids with peptides yielding so-called peptomers [11]. In addition, peptomer synthesis provides a promising tool for the stepwise and systematic transformation of biologically active peptides into non-peptidic analogs by means of high-throughput synthetic methodologies. Success with the SPOT-synthesis of peptides inspired a study on extension of the

method to the generation of spatially addressable peptomer libraries and related peptidomimetics synthesized directly on continuous membranes. In order to develop a reliable SPOT-synthesis protocol for cellulose membranes the following prerequisites had to be fulfilled: i) a linkage system allowing efficient attachment of peptomers as well as mild and orthogonal final cleavage from the support and ii) an efficient synthetic pathway for the *N*/*O*-selective and parallel incorporation of peptoidic and peptidic building blocks.
In order to apply a suitable linker system cellulose membranes were amino functionalized by treatment with epibromohydrin and 4,7,10-trioxa-1,13-tridecanediamine (Fig. 5), to provide a homogeneous amino-functionalisation of the cellulose (0.1-1.0 µmol/cm^2). Subsequently, a photo-labile linker system was chosen (Fig. 5) allowing dry-state cleavage by UV-irradiation at 365 nm (90 min) of the assembled compounds yielding the products adhesively bound to the membrane.

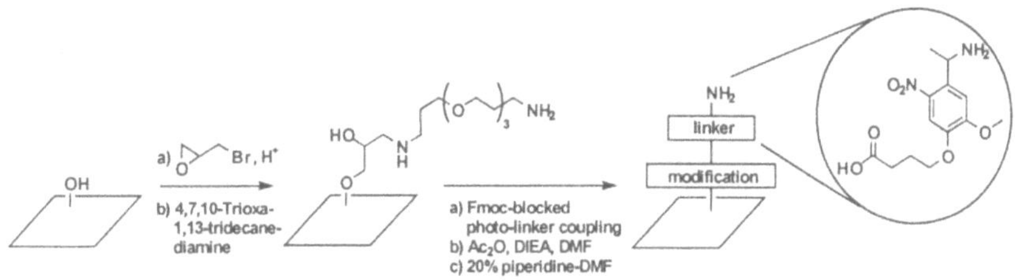

Figure 5.) Modification of a cellulose sheet and coupling of a linker to the membrane.

Peptomer assembly via the SPOT-technique on planar surfaces requires the parallel incorporation of peptoidic and peptidic building blocks (Fig. 5). Synthesis of peptoidic building blocks is usually performed by a two step procedure using bromoacetic acid for acylation and primary amines for nucleophilic substitution at the bromomethyl group (submonomer method) [10]. In order to adapt standard protocols of peptide- and peptoid-chemistry for SPOT-synthesis on cellulose surfaces a number of modifications and optimizations had to be accomplished. Bromoacetic acid 2,4-dinitrophenylester was found to yield complete amino-acylation of peptidic or peptoidic building blocks with low ester formation at remaining hydroxyl functions of the cellulose and at peptoid units carrying unprotected hydroxyl functions at the side chain (Fig. 6). The primary amines were used as 50% solutions in NMP or, depending on their solubility, as saturated solutions in NMP or water.
Fmoc deprotection following the introduction of an Fmoc amino acid needs to be carried out in parallel with treatment of an adjacent spot with an amine for the generation of a peptoid (Fig. 6). Piperidine, the standard Fmoc deprotection reagent, was shown to be too volatile using SPOT-synthesis conditions (displacement of bromine at adjacent bromoacetylated peptoid spots). Therefore, the less volatile DBU was substituted for piperidine as deblocking agent. A solution of 4% DBU in NMP resulted in complete Fmoc deprotection. Finally,

effective acylation of the sterically hindered secondary amino function of a peptoidic building block with an Fmoc amino acid in the presence of free hydroxyl functions was required. The combination DIC/HOAt, reported to be well suited for peptomer synthesis on resins [11], gave only mediocre coupling yields using SPOT-synthesis conditions on cellulose membranes. Therefore, several additional activation methods were examined. It was found that highest acylation values were obtained via symmetric anhydrides and isolated amino acid fluorides.

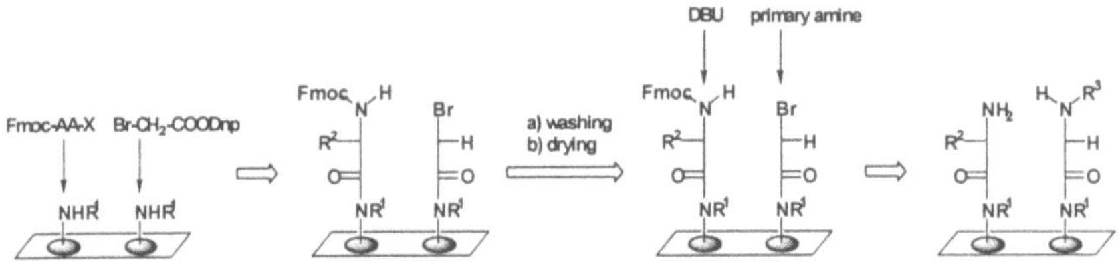

Figure 6.) Reaction scheme for the parallel incorporation of peptidic and peptoidic building blocks in peptomer synthesis on continuous surfaces.

The optimized SPOT-synthesis protocol was applied to a systematic peptide-peptoid-transformation of the peptide PPPPY, the consensus motif of the hYAP-WW-protein domain [12]. More than 350 peptomers were synthesized in sufficient yields and purities. Figure 7 shows typical HPLC-profiles of crude peptomers obtained by photolytic cleavage from dry membranes.

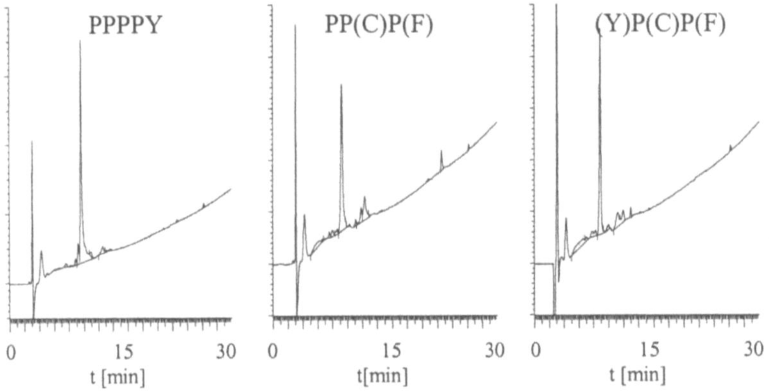

Figure 7.) HPLC-profiles of peptomers synthesized via SPOT-technique on continuous membranes. (C): *N*(2-aminoethyl)glycine, (F): *N*[2-(4-hydroxy-phenyl)ethyl]glycine, (Y): *N*(2,3-dihydroxypropyl)glycine.

3.3 Small Organic Molecules

Currently, the major effort of combinatorial chemistry is focused on combinatorial libraries of non-peptide like small organic molecules [13]. Adaption of SPOT-synthesis technique for efficient assembly of such compounds on membranes and their screening directly on the surfaces would tremendously facilitate identification of novel leads.

Compounds derived from 1,3,5-triazines where chosen as suited targets since they play an important role as modulator of multidrug resitance during cancer therapy, in the treatment of depression and as aromatase inhibitors. Earlier reports were aimed on the synthesis of triazine-based libraries starting from cyanuric chloride and using use of the temperature depending successive displacement of the chlorines. However, these syntheses either yielded in compound mixtures or needed tedious solution phase methods using extensive instrumentation and harsh reaction conditions [14,15].

In contrast, the present SPOT-synthesis strategy allows the incorporation of various amines and phenols into the libraries yielding spatially adressed single compounds directly applicable for solid or solution phase screening techniques. In order to perform efficient synthesis on continuous surfaces several requirements needed to be accomplished: i.) selection of an appropriate membrane, ii.) development of a suitable linker, iii.) establishment of reaction conditions allowing mild and complete assembly of target compounds.

Polypropylene membranes modified with acrylic acid (Fig. 2, pathway ii.) and amino functionalized (Fig. 2, pathway iv.) were applied to overcome problems related to cellulose surfaces (see section 2). After spotwise introduction of a Rink-linker via its OPfp-ester several peptides and peptoids where synthesized according to section 3.2 and deprotected amino-functions immersed with an excess of cyanuric chloride (Fig. 8).

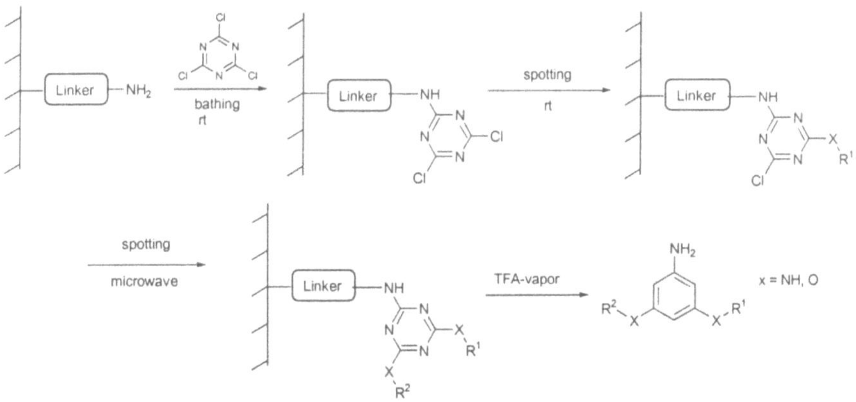

Figure 8.) General reaction scheme of assembly of 1,3,5-triazine librairies on continuous surfaces.

Attachment of the cyanuric chloride to the membrane is followed by successive replacement of the two remaining chlorines. The second chlorine atom can be substituted at room-temperatur by spotting a 3 M solution of primary or secondary amines in NMP directly onto the membrane. Moreover a 0.5 - 1 M solution of cesiumsalts from phenol-derivatives in DMSO can be employed as well to attach phenolates efficiently.
In order to achieve complete substitution of the last chlorine atom of the mono-chlorotriazines harsh conditons such as 80°C for five hours are required [15]. As the SPOT-technique does not need any reactor vessels these conditions are not suited for reaction at cellulose- or polypropylene- membranes due to the fast evaporation of reactants. Thus, a hot plate was not useful for the membranes since the slow energy transfer from the surface of the plate to the membrane resulted in a evaporation of the 5 M amine solution before the appropriate temperature was reached. Interestingly, complete substitution could be achieved at all membranes within minutes by means of a microwave oven as a heating source (600 W, 3 min) while employing both amines and cesiumsalts as nucleophiles.
The release of the 1,3,5-triazines from solid support was performed by TFA-vapor, which is advantageous since the compound libraries remain adsorbed at the same spatial areas as the substances were covalently linked before. Thus, completely cleaved products can subsequently be used conveniently for various screening techniques. The spots can be punched out for product identity- and purity-control or placed into a microtiter plate for high throughput screening assays. One spot (0.23 cm^2) gave approximately 50-200 nmol appropriate for HPLC-MS analysis (Fig. 9). Work on the biological screening of the assembled libraries is in progress and will be published in due course.

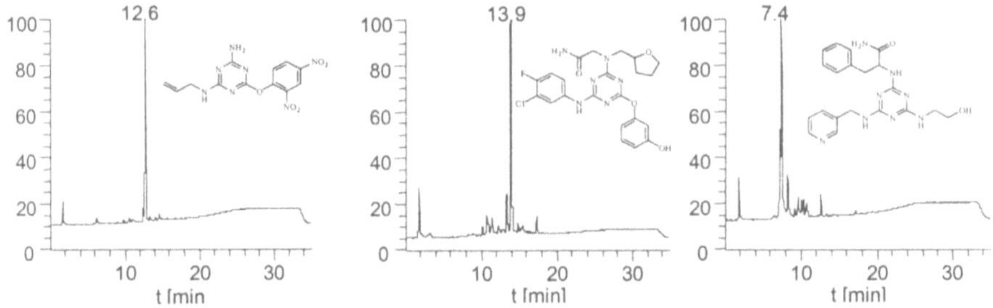

Figure 9.) HPLC-traces of several 1,3,5-triazines assembled via the SPOT-synthesis technique on novel polypropylene membranes.

4. Abbreviations & References

Abbreviations: Ac_2O: acetic anhydride; DBF: dibenzofulvene; DBU: 1,8-diazabicyclo[5,4,0]undec-7-ene; DCM: dichloromethane; DIC: diisopropyl-carbodiimide; DIEA: diisopropylethylamine; DMF: *N,N*-dimethylformamide;

Dnp: 2,4-dinitrophenyl; Fmoc: 9-fluorenyl-methyloxycarbonyl; Fmoc-AA-F: Fmoc amino acid fluoride; HATU: *N*-[(dimethylamino)-1H-1,2,3-triazolo[4,5-*b*]pyridin-1-ylmethylene]-*N*-methylmethanaminiumhexafluorophosphate *N*-oxide; HOAt: 1-hydroxy-7-azabenzotriazole; NMI: *N*-methylimidazole; NMP: *N*-methylpyrrolidone; OPfp-: pentafluorophenyl-; PyAOP: 7-azabenzotriazol-1-yl-oxy-tris-(pyrrolidino)-phosphoniumhexafluorophosphate; PyBOP: benzotriazol-1-yl-oxy-tris(pyrrolidino)phosphonium hexafluorophosphate; PyBroP: bromo-tris(pyrrolidino)-phosphoniumhexafluorophosphate; TBTU: *N*-[(1H-benzo-triazol-1-yl)-(dimethyl-amino)methylene]-N-methylmethanaminium tetrafluoroborate *N*-oxide.

[1] Pirrung, M. *Chem. Rev.* **1997**, *97*, 473 and references cited therein.

[2] Frank, R. *Tetrahedron* **1992**, *48*, 9217.

[3] Reineke, U., Kramer, A., Schneider-Mergener, J. *Curr. Top. Microbiol. Immunol.* **1999**, in press.

[4] Volkmer-Engert, R., Germeroth, L., Haupt, T., Heine, N., Scharn, D., Ast, T., Matuschewski, H., Schedler, U., Ulbricht, M., Schneider-Mergener, J., Wenschuh, H. in *Peptides 1998* (S. Bajusz, Ed.), Mayflower Scientific Ltd., Kingswinford, UK, **1999** in press.

[5] Reineke, U., Sabat, R., Misselwitz, R., Welfle, H., Volk, H.-D., Schneider-Mergener *Nature Biotechnol.* **1999**, *17*, 271.

[6] Tegge, W., Frank, R., Hofmann, F., Dostmann, W.R. *Biochemistry* **1995**, *22*, 10569.

[7] Malin, R., Steinbrecher, A., Semmler, W., Noll, B., Johannsen, B., Frömmel, C., Höhne, W., Schneider-Mergener, J. *J. Am. Chem. Soc.* **1995**, 117, 11821.

[8] Kramer, A., Keitel, T., Winkler, K., Stöcklein, W., Höhne, W., Schneider-Mergener J. *Cell* **1997**, *93*, 799.

[9] Gante, J. *Angew. Chem.* **1994**, *106*, 1780 and references cited therein.

[10] Zuckermann, R.N., Kerr, J.M., Kent, S.B.H., Moos, W.H. *J. Am. Chem. Soc.* **1992**, *114*, 10646.

[11] Ostergard, S., Holm, A., *Mol. Div.* **1997**, *3*, 17.

[12] Chen, H.I., Einbond, A., Kwak, S.J., Linn, H., Koepf, E., Peterson, S., Kelly, J.W., Sudol, M. *J. Biol. Chem.* **1997**, *27*, 17070.

[13] Thompson, L.A., Ellman, J.A. *Chem. Rev.* **1996**, *96*, 555.

[14] Stankova, M., Lebl, M., *Mol. Div.* **1996**, *2*, 75.

[15] Gustafson, G.R., Baldino, C.M., O'Donnell, M.M.E., Sheldon, A., Tarsa, R.J., Verni, C.J., Coffen, D. L. *Tetrahedron* **1998**, *54*, 4051.

Part 4

Microreactors in Production Processes

Electrochemical Microreactors: A New Approach for Microreaction Technology

A. Ziogas, H. Löwe, M. Küpper, W. Ehrfeld

Institut für Mikrotechnik Mainz GmbH, Carl-Zeiss-Str. 18-20, D-55129 Mainz, Germany

Abstract

Combining the advantages of thin layer cell technology and microstructuring techniques, a new microreactor concept for the electro-synthesis of organic compounds was developed. A miniaturized plate-to-plate electrode configuration with a non-conducting housing comprising several parallel reaction channels was realized by etching and laser techniques. For investigating the features of this electrochemical microreactor, the oxidation of 4-methoxytoluene to 4-methoxybenzaldehyde was chosen as a benchmark and carried out at various reaction conditions. The results were monitored by HPLC-analysis. At room temperature, a nearly quantitative conversion of 4-methoxytoluene was found at a selectivity of 98% for low concentrations in methanol. The concentration of conducting salt was significantly reduced compared to the large-scale process. Further experiments using higher reactant concentrations as well as various concentrations of conducting salt led to very interesting perspectives in electrochemical organic synthesis and new approaches to microreaction technology.

Key words: electrochemical microreactor, mass transfer, heat transfer, current density distribution, anodic oxidation, toluene oxidation, benzaldehyde.

Introduction

Electrochemical methods offer many advantages for the performance of electroactive organic compounds. Chemical reactions may be influenced very sensitively by regulation of the electric potential rather than by heat management or the addition of further reactants that need to be removed later on. Furthermore, electrochemical reactions in most cases are carried out at room temperature. In competition with other process technologies, organic electrochemistry has made large progress over the past few years. However, there is only a small choice of electrochemical reactions showing high selectivities or particularly advantageous reaction conditions in large scale production. Examples are anodic substitution reactions for the functionalization of olefins or aromatic compounds or the cathodic hydrodimerization of activated olefins and carbonyl compounds, providing a simple method for C-C coupling [1-4].

From the point of view of process engineering, electrochemical reactions offer a number of advantages. Conversion and reaction rates, as well as, within certain limits, selectivity can be influenced and controlled easily by variation of current density and potential. Electrochemical reactions in general do not entail any waste problems. However, these advantages are accompanied by a number of limitations when using conventional cells, such as inhomogeneities of the electric field, differences in flow velocity or deficiency in heat transfer. Therefore, the benefits of miniaturization were investigated by a feasibility study based on miniaturized reaction systems, so-called microreactors.

Microreactors are characterized by small dimensions and an extremely large surface to volume ratio. Therefore, they offer completely novel reaction conditions compared to large-scale reactors. Besides the potential to increase process safety, microreactors enable, the use of potentially explosive process routes. Furthermore, microreactors reduce the impact of chemical processing on environment by introducing safer production processes and the ability to produce on demand and on-site [5].

The Electrochemical Microreactor: Basics and data

Combining the advantages of thin layer cell technology [6] (until now mainly a concept in analytical technologies) and microstructuring techniques, a new microreactor concept for the electro-synthesis of organic compounds was developed. In the following, the benefits of miniaturization will be outlined in terms of the mass transfer coefficient and space-time yield achievable.

A miniaturized plate-to-plate electrode configuration with a non-conducting housing with several parallel reaction channels was realized by wet chemical etching of metallic foils and laser cutting techniques (Fig.1).

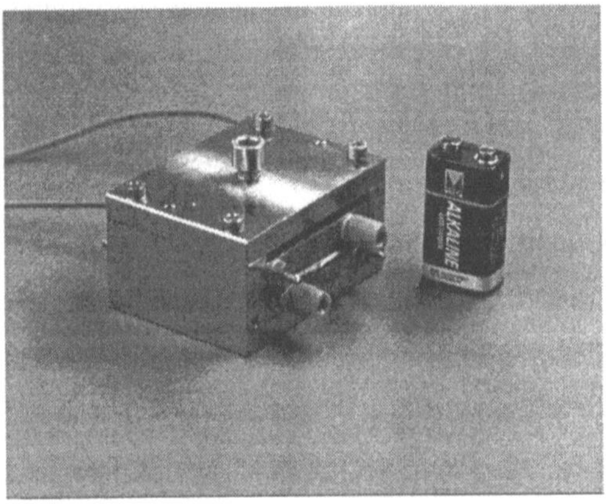

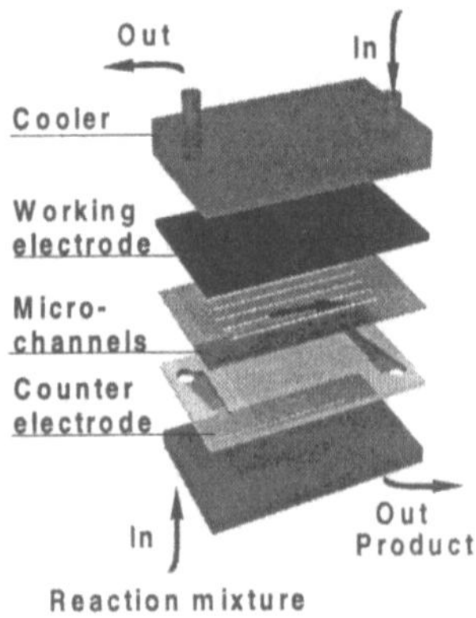

a b

Fig.1a,b. The electrochemical microreactor with integrated cooler. **b** Schematic presentation of assembly

The ionic conductivity, κ , of the electrolyte together with the cell geometry and the current density, $\mathbf{i}$, determines the ohmic voltage-drop in the interelectrolytic gap [7].

For a parallel plate electrolyzer with electrode distance $\mathbf{d}$ and electrode surface $\mathbf{A_e}$ the ohmic-voltage drop in electrolyte amounts to

$$\Delta U = \mathbf{IR}_{\mathbf{drop}} = \mathbf{I}\frac{1}{\mathbf{A_e}}\frac{\mathbf{d}}{\kappa} = \mathbf{i}\frac{\mathbf{d}}{\kappa} \quad \text{(Eq.1)}$$

The reduction of the electrode distance $\mathbf{d}$ to magnitudes shorter than 50 µm decreases the ohmic-voltage-drop. Compared to today's mono- and bipolar cells [7] the ohmic resistance is decreased for at least one order of magnitude.

The distribution of current density along the electrode is of great industrial importance, since, in case of electroplating processes, it may cause a local variation of the thickness of the deposited metal, a non-uniform corrosion of electrodes in industrial electrolysis cells, a poor energy efficiency in primary and secondary batteries, or a decreased space time yield with three-dimensional electrodes [8].

Owing to the design and construction of the microreactor, the current density distribution should be uniform [9] in order to achieve a homogeneous current distribution and enhancement of selectivity and current efficiency. The uniform distribution of the electrolyte flow into the micro channels without dead volumes leads to uniform mass transport and allows an effective heat management. The characteristic Reynolds number for a volumetric flow rate of $\dot{\mathbf{V}}$ = 0.1 ml/min is **Re**= 0.2, implying a laminar flow conditions.

Table1. Dimensions and process values of the microreactor

Characteristics	*Symbol*	*Value*
Channel surface	A	0.8 x 64 mm^2
Channel length	L	64 mm
Channel volume:	$V_{channel}$	1.3 μl
Number of channels	N	27
Reactor volume	V	35 μl
Surface to volume ratio	Av	400 cm^{-1}
Flow rate	$\dot{\mathbf{V}}$	0.1 ml/min
Residence time	τ	21 s
Inlet concentration,	c_0	2 mol/l
Nernst diffusion layer	δ_N	$25 \cdot 10^{-4}$ cm
Diffusion constant	D	$\approx 1 \cdot 10^{-5}$ cm^2 s^{-1}

The laminar flow and the constant channel length allow a reaction control with steady residence time (Tab.1). Due to the very small space between electrodes in the plug-flow system (25 μm), the diffusion distances are very short and the Nernst diffusion layer thickness, $\delta_{N,}$ can reach **d**= 25 μm at the most. For this reason, the mass transfer vertical to the flow is high.

The maximum velocity is $\mathbf{w}_{max}$= 4.6 mm/s, whereas the flow velocity $\mathbf{w} = \mathbf{w}_{max}/2$ amounts to 2.3 mm/s. The space-time yield ρ for any electrochemical synthetic process should, for costs reasons, be as high as possible. The ratio of the total area of the electrode surface to the cell volume $\mathbf{A}_v$ (in cm^{-1}) is $\mathbf{A}_v = 400\ cm^{-1} = \mathbf{d}^{-1}$. This value is very high in contrast of classical reactors like monopolar cells (approx. 0.1 cm^{-1}) and bipolar cells (approx. 1 cm^{-1}) and even chemical fixed-bed reactors (approx. 20 cm^{-1}).

We obtain a mass transfer coefficient

$$k_m = \frac{D}{\delta_N} \approx 4 \cdot 10^{-3}\ \text{cm s}^{-1}, \qquad \text{(Eq.2)}$$

and a space-time yield,

$$\rho = \frac{c_0}{\tau} \cdot \left(1 - e^{-k_m A_V \tau}\right) \approx 62000 \cdot \text{kg} \cdot \text{dm}^{-3} \cdot \text{h}^{-1} \qquad \text{(Eq.3)}$$

which is an extremely high value.

The rapid cooling using a heat exchanger, directly integrated in the electrodes, allows not only a very fast heat transfer, but also a very easy heat management as well as a thermostatic temperature control and helps to avoid by-products, since no hot spots appears [10]. Even at high current densities, the heat generated by the ohmic resistance of the electrolyte and the joule reaction heat respectively may be removed easily. Fig.1 demonstrates the a) electrochemical microreactor and b) the schematic view of assembly. If necessary, bipolar or divided cell systems can be built up.

Reaction process

We chose 4-methoxytoluene as a model substance for the application in the microreactor. The results of direct anodic oxidation can be compared to those of electrosynthesis in industrial plants [11].

The technical synthesis of benzaldehyde typically is performed by catalytic high temperature oxidation of toluene with air. Substituted benzaldehydes, however, are more often produced by other processes like side chain chlorination (with subsequent hydrolysis of benzalchlorides) or partial side chain oxidation. Examples are the chromate oxidation of anethol to anisaldehyde or the formylation of anisol [12]. As many of these reactions are not very selective, several attempts to use the potentially higher selectivity of anodic toluene and its oxidation derivates mainly dealt with direct anodic oxidation processes [13]. H. Wendt and et al. [14,16] reported on the analytical and the batchwise preparative anodic oxidation using three different toluenes. They investigated p-methoxytoluene as an activated (with respect to anodic oxidation), and

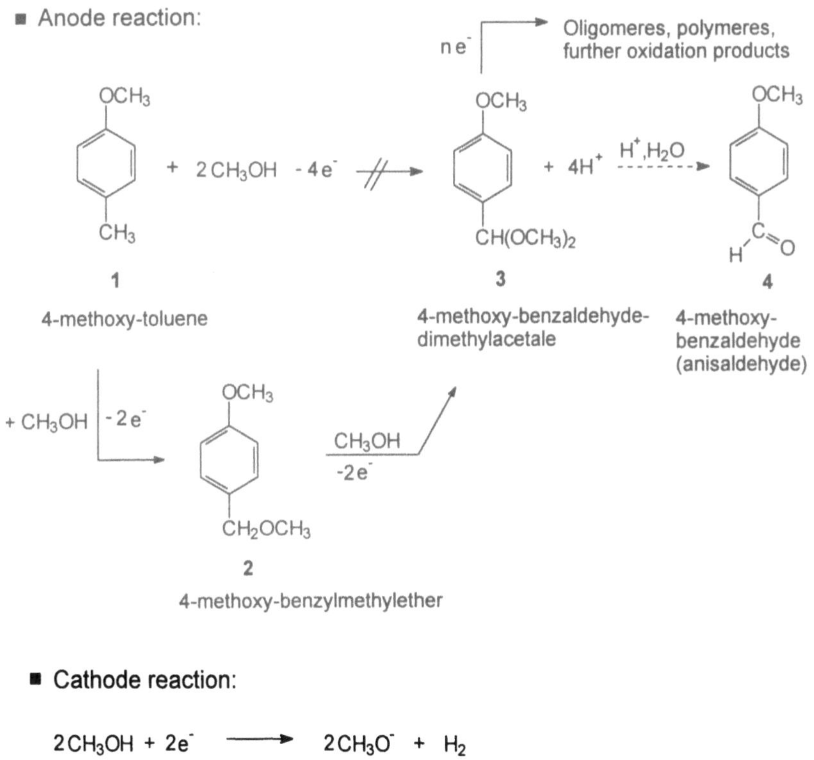

Fig.2. Reaction route of the anodic oxidation of 4-methoxytoluene

p-chlortoluene as a deactivated toluene. Summarizing the results of these investigations, three well defined oxidation steps (Fig.2) can be distinguished, namely:

1. The $2e^-$ oxidation of the toluenes to benzylalkylethers
2. The $2e^-$ oxidation of benzylalkylethers to benzaldehyde dialkylacetals and
3. The multi-electron oxidation of benzaldehyde dialkylacetals to oxidized and poorly defined products.

Thus, simultaneous oxidation processes of substrate, intermediate products and solvent are expected. A reaction pathway via the intermediately formed methoxy radical is probable too [15].

Experimental

Preparative electrolyses were continuously performed in an undivided micro cell with constant current density of **i**= 79 mA cm^{-2}. The p-methoxytoluene concentration in methanol was 20 vol% (2.0 mol/l). The electrolyte was pumped with a volumetric flow of 0.1 ml/min, yielding a maximum linear velocity of 0.46 cm s^{-1} per channel and a pressure drop of $4 \cdot 10^{-2}$ bar/channel or 1.1 bar in the complete microreactor.

Solvent (methanol, Merck) and substrate (4-methoxytoluene, Aldrich) were of p.a. grade and used as obtained. The anode material used was glassy carbon (Sigratur K/Sigri) whereas stainless steel formed the cathode. A potentiostat/galvanostat (HEKA PG 284/ IEEE) and galvanostat (Hewlett Packard 663A) was used in combination with a charge counter (Neukum Elektronik GmbH AZ 4000). The microcell was thermostatized using a conventional thermostat (Lauda).

The reaction products were determined by HPLC (Shimadzu *VP* series with UV-Vis detector) with acetonitrile (Aldrich) as eluent and the following chromatographic conditions: isocratic flow 0.9 ml/min, (65/35) CH_3CN/H_2O, 250 x 4 mm Inertsil ODS-3, 5µm column (GL Sciences Inc. Japan) and UV-detection at 254 nm.

Results

For finding the best process conditions, we first investigated the dependence of conversion and selectivity on potential in case of low substrate concentration. Subsequently, the dependence of conversion, yield, selectivity, current efficiency and product distribution on temperature was measured for high substrate concentration.

- **Potential dependence**

Regarding the oxidation products **2**, **3** and the acetal hydrolyzed **4** (see fig.2), potentiostatic electrolysis showed a high conversion and selectivity at an initial concentration of 0.1 M and a flow rate of 0.1 ml/min.

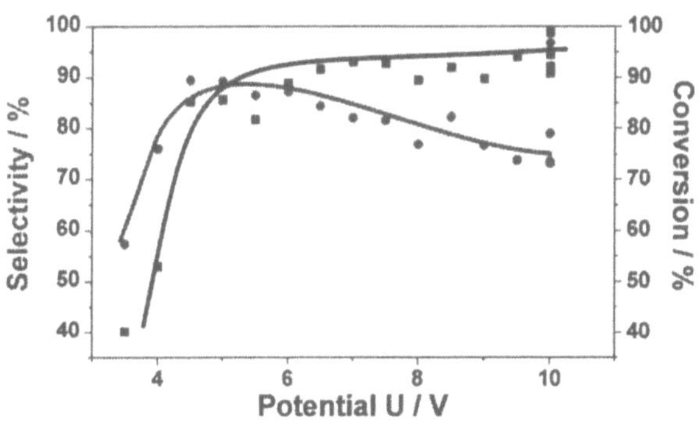

Fig.3. Dependence of ■ conversion and ● selectivity on potential with 0.1 M KF and 0.1 M 4-methoxytoluene

Fig.3 shows the dependence of the conversion and the selectivity on the potential. Between 3.5 and 4 V, conversion and selectivity increase rapidly while selectivity exceeds conversion. Between 4 and 4.5 V, selectivity amounts to 98% at a conversion of 96%. Thus, with increasing potential, the conversion rises to 100% at simultaneously decreasing selectivity due to an increasing amount of side reactions.

With respect to substrate concentrations commonly applied in industrial plants, galvanostatic conditions were tested depending on concentration of conducting salt and temperature. The concentration of the conducting salt KF was reduced significantly to 0.01 M or no KF was added at all, compared to the large-scale process (0.1-0.2 M).

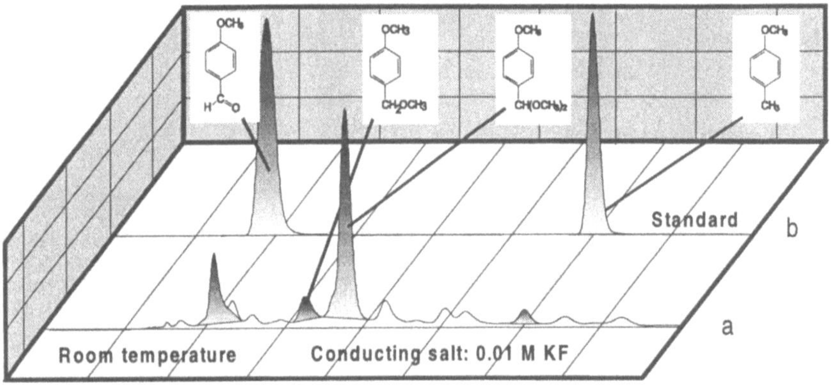

Fig.4. Analysis with reversed phase HPLC of **a**) anodic oxidation products of 4-methoxy-toluene **b**) standards. Chromatographic conditions: isocratic flow 0.9 ml/min (65/35) MeCN/H_2O, 250 x 4 mm Inertsil ODS-3, 5μm Column (GL Sciences Inc. Japan) and UV-detection at 254 nm.

Besides the expected products, the chromatogram (Fig.4) shows additional peaks (in the order of 8 to 25% peak area), which were identified by means of mass spectroscopy. Most of them were recognized as by-products, namely the dimer of the 4-methoxybenzyl radical formed in the first oxidation (242M^+), as well as secondary acetal oxidation product 4-methoxybenzoic acid orthoester (212M^+). Additionally two signals at 184 M^+ and 169 M^+, respectively, were identified as 1,1,4-Tri-methoxy-4-methylcyclohexa-2,5-dien and semi-acetal or fragments. Similar products were observed by Wendt et al [16]. Oligomeres, polymers or methanol solvent oxidation products, like the formaldehyde, formic acid or CO_2 were not detected.

- **Conversion and Yield**

Fig.5 compares conversion and yield from oxidation products **2**, **3**, and **4** as a function of the temperature and different salt concentrations. With KF for temperatures from 5 to 30 °C, a constant conversion in the order of 95-98% was observed, slightly decreasing with increasing temperature. The yield, on the contrary, increases slightly with increasing temperature in a range between 5 and 30 °C and remains constant at 73% at temperatures exceeding 30 °C. This proves that temperature has a minor influence on conversion and yield. Without KF,

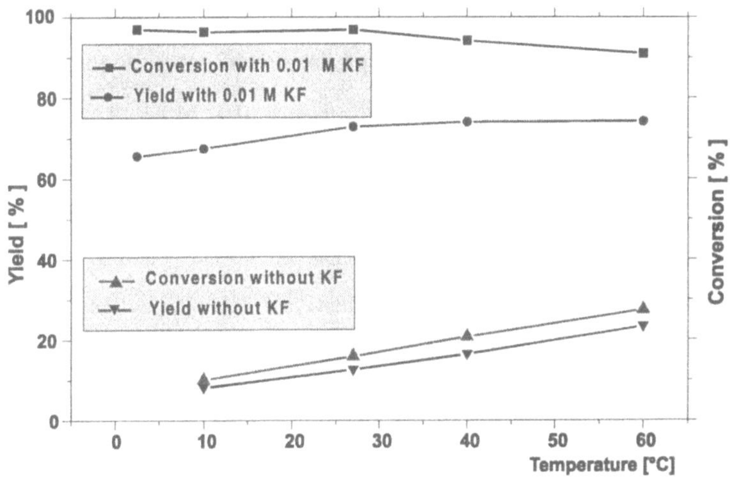

Fig.5. Electrolysis of 4-methoxy-benzaldehyde: Conversion and yield in dependence of the temperature

conversion increases continuously from 10% to 27%, whereas the yield increases from 8% to 24%.

The reason for this gain in conversion and yield at increasing temperature is due to increasing diffusion velocity in case of electrolyte without conducting salt.

Yield and selectivity

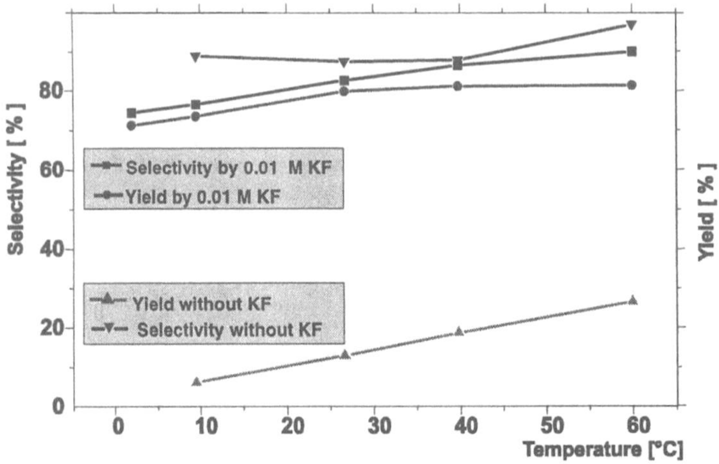

Fig.6. Electrochemical synthesis of 4-methoxy-benzaldehyde: Selectivity and yield in dependence of the temperature

Fig.6 compares yield and selectivity of products with respect to various conducting salt concentrations and temperature. Using 0.01 M KF at increasing temperature, a slight increase in selectivity (from 73% to 89%) can be observed. Without conducting salt, selectivity reaches over 90%. This is evidenced by chromatograms, showing, apart from the oxidation products **2, 3** and **4,** only one peak corresponding to the dimer bis-4-methoxybenzyl.

- **Product distribution**

The bar diagrams (Fig.7) shows the product distribution a) without, and b) with

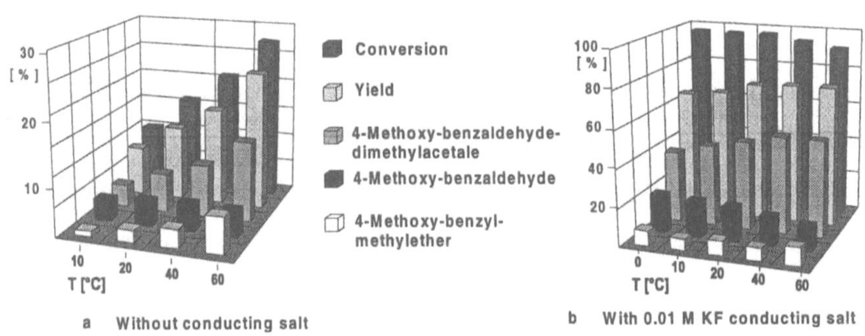

Fig.7.a,b. Conversion, yield and product distribution for the anodic oxidation of 4-methoxy-toluene.

0.01 M KF.

In case a), the conversion and yield, for all species except aldehyde, increases with increasing temperature in a linear way. A small part of the acetal **3** is catalytically hydrolyzed by acid enriched in the surrounding of the anode. The obtained hydrolysis rates depend on the water content of the electrolyte as well as catalytic speed of reaction. The formation of the aldehyde **4** does not depend on the temperature within the range investigated (10 - 60 °C).

In case b), a slight decrease in the aldehyde yield is observed at increasing temperature due to further aldehyde oxidation [16].

- **Current efficiency**

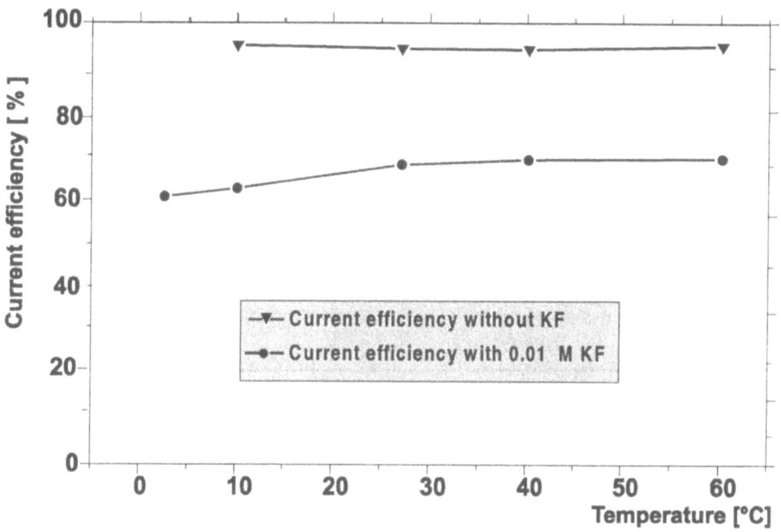

Fig.8. Current efficiency with 0.01 M KF and without conducting salt.

The current efficiency (Fig.8) within the electrolyte without conducting salt amounts to 96-98%, while with a concentration of 0.01 M of conducting salt, a linear value of 60-65% is found (Ref. [11]: 49 - 54%). This indicates the absence of side reactions in the conducting salt free electrolyte.

Conclusions

Comparing the results reported here, to those obtained in industrial processes, a large number of advantages of the microreactor compared to conventional reactors have been identified, e.g. short residence times, increased conversion, selectivity, yield, and lower conducting salt concentrations. Even without conducting salt, high selectivities (>90%) and almost quantitative current efficiencies were obtained. Leaving out the conducting salt, significantly cuts the operating costs and eliminates the problem of its separation. Besides, continuous operation at 90% conversion reduces the energy consumption with regard to conventional production lines, where the electrolyte has to be circulated. The ratio of the total area of the electrode surface to the cell volume $\mathbf{A_v}$ (in cm^{-1}) is about 400 times higher than for bipolar cells and about 20 times higher compared to chemical fixed-bed reactors, leading to extraordinary space-time yields.

List of Symbols and Abbreviations

A_e	electrode surface
A_v	ratio of surface area to reactor volume
c	concentration
D	diffusion coefficient
d	electrode distance
I	current
i	current density
k_m	mass transfer coefficient
L	characteristic length
M	molarity
M	molecular weight
R	resistance
Re	Reynolds number
T	temperature
U	potential
V	volume
$\dot{\mathbf{V}}$	volume flow
w	flow velocity
w_{max}	maximum velocity
δ_N	Nernst diffusion layer
κ	specific conductivity
ρ	space-time yield
τ	residence time

References

[1] Schäfer, H. J.: *CC-Verknüpfung und Umfunktionalisierung an der Elektrode* (Teil 1 und 2), Kontakte (Darmstadt) (1987).

[2] Beck, F.: *Organic Electrosynthesis*, In: Ullmann's Encyclopedia of Industrial Chemistry, Vol. A9, VCH Verlagsgesellschaft mbH, Weinheim (1987).

[3] Weinberg, N.L.: *Introduction to Industrial electrosynthesis*, In: Electrosynthesis from Laboratory to pilot plant production, Genders, J.D.; Pletcher, D.; eds., Published by the Electrosynthesis Company Inc., P.O.Box 430, E. Amherst, New York (1990).

[4] Danly, D.E.; King, C.J.H.: *Industrial Electroorganic Chemistry*, In: Organic Electrochemistry, thd. Eds., Edited by Lund, H.;. Baizer, M.M.; Published by the Marcel Dekker Inc. New York (1991).

[5] Ehrfeld, W.; Golbig, K.; Hessel, V., Konrad, R.; Löwe, H.; Richter, Th.: *Fabrication of components and systems for chemical and biological microreactors*, In: Microreaction Technology, Proceedings of the First International Conference on Microreaction Technology, Ed.: Ehrfeld, W.; Springer-Verlag, p.72 - 90, Berlin (1997).

[6] Woodard, F.E.; Reilley, C.N.: *Thin Layer Cell Technigues*, Compr. Treatise of Elect. vol. 9, Edited by Yeager, Bochris, Conway and Sarangapani, Published by the Plenum Press, New York (1984).

[7] Beck, F.; Guthke, H.: *Entwicklung neuer Zellen für elektro-organische Synthesen*, Chem. Ing. Techn. 41, p 943 - 990 (1969).

[8] Wendt, H.; Kreysa, G.: *Electrochemical Engineering*, Science and Technology in Chemical and Other Industries, Springer Verlag, Heidelberg (1999).

[9] Heitz, E.; Kreysa, G.: *Grundlagen der Technischen Elektrochemie*, Verlag Chemie, Weinheim (1978).

[10] Ehrfeld, W.; Hessel, V.; Haverkamp, V.: *Microreactors*, Ullmann's Encyclopedia of Industrial Chemistry, Sixth Ed., 1999 Electronic Release, Verlag Wiley/ VCH (1999).

[11] DE 28 48 397; BASF; D. Degner, M. Barl, H. Siegel; CA 93 : 56963 d (1980).

[12] Neumüller, O.-A.: Römpps Chemie-Lexikon 7th Edn. Vol. 1, p. 187. Frankh'sche Verlagshandlung, Stuttgart (1972).

[13] Degner, D.: *Organic Electrosinthesis in Industry*, Topics in current Chemistry Vol. 148, Springer-Verlag, Berlin, Heidelberg (1988).

[14] Wendt, H.; Bitterlich, S.: *Anodic Synthesis of Benzaldehyds–I. Voltammetry of the Anodic Oxidations of Toluenes in Non-Aqueous Solutions*, Electrochimica Acta, Vol. 37, p.1951 - 1958, (1992).

[15] Krüger, R.: *Benzaldehydsynthese durch mittelbare anodische Oxidation*, Dissertation, TH Darmstadt, (1990).

[16] Wendt, H.; Bitterlich, S.; Lodowicks, E.: *Anodic Synthesis of Benzaldehyds–II. Optimation of the Direct Anodic Oxidation of Toluenes in Methanol and Ethanol*; Electrochimica Acta, Vol. 37, p.1959 - 1969, (1992).

High Temperature HCN Generation in an Integrated Microreaction System

V. Hessel, W. Ehrfeld, K. Golbig, C. Hofmann, St. Jungwirth, H. Löwe, Th.Richter, M. Storz, A. Wolf
Institut für Mikrotechnik Mainz GmbH, Carl-Zeiss-Str. 18 - 20, D-55129 Mainz
Tel.: 49-6131-990450; Fax: 49-6131-990475; e-mail: hessel@imm-mainz.de

O. Wörz
BASF AG, 67056 Ludwigshafen, Germany
Tel.: 49-621-6055227; Fax: 49-621-6079020; e-mail: otto.wörz@zhf.x400.basf-ag.de

J. Breysse
Rhône-Poulenc, 24 Av. Jean Jaures, 69153 Decines-Charpieu Cedex, France
Tel.: 33-4-72935721; Fax: 33-4-72935850; e-mail: jacques.breysse@rp.fr

Abstract

A high-temperature microreactor was designed for the performance of the Andrussov process. The microreactor enables isothermal processing and rapid cooling times of this ultrafast, highly exothermal reaction. An array of 20 channels, each 70 μm in diameter, forms the reaction zone in the centre of a catalyst structure which was heated to a reaction temperature of about 1050°C. Electrical connectors were realized from a metallic solder by micromilling allowing a reversible exchange of the catalyst structure. Gas tightness of the reactor was achieved using a new developed metal-ceramic joining technique and standard flat sealing by graphite foils.

First experiments prove the feasibility of the concept. They showed that the selectivity with respect to ammonia corresponds to that of the industrial process, whereas the methane related selectivity reached nearly half of the reference value. Since methane conversion was nearly complete high contents of totally oxidized products are formed in the product mixture. These findings were compared to laboratory monolithic reactors with larger channel diameters. These monoliths showed lower conversions and, in consequence, lower yields compared to the microreactor.

1. A Microreactor for the Performance of the Andrussov Process

A number of microreactors recently have demonstrated the feasibility of performing gas phase reactions. In some cases even process intensification, e.g. an increase in selectivity, was achieved by means of isothermal processing. To mention just a few, a selective hydrogenation was performed in a palladium coated cross-flow catalyst carrier [1], the ethylene oxide formation in a reactor with bulk silver catalytic platelets [2], the selective ammonia oxidation in a platinum coated tee-type reactor [3], and the partial oxidation of methane in a reactor with rhodium coated sheets [4]. These microreactors benefited from their uniform temperature distribution due to the small channel dimensions and the high thermal conductivity of the catalyst carrier. The operating conditions of the microreactors covered a temperature range from 120 to about 700°C and a variation of residence times in the order of 50 milliseconds to some seconds.

In this article, a high-temperature microreactor is described for performance at temperature exceeding 1000°C and with sub-millisecond reaction times followed by rapid cooling. This microreactor was designed to demonstrate process improvements achievable by isothermal processing and rapid cooling for an extremely fast, highly exothermic reaction. The Andrussov process, the formation of hydrogen cyanide from methane, ammonia and air, was chosen because it fulfills these characteristics and, in addition, is a well-known industrial process which has been performed since decades [5-7].

$$CH_4 + NH_3 + 3/2\ O_2 \dashrightarrow HCN + 3\ H_2O \qquad \Delta H = -\ 474\ kJ/mol$$

The selectivity of this reaction is decreased by thermally sensitive parallel and consecutive reactions, e.g. total oxidation of methane to carbon dioxide and hydrolysis of the product hydrogen cyanide.

The conventional Andrussov process is performed using a platinum catalyst comprising several nets, the so-called gauzes, made from wires of platinum [6,7]. Gauzes of various mesh numbers and wire diameters are assembled as a stack to flexibly adjust the overall net thickness and porosity. The educt gases are preheated to about 600°C, pass the gauze catalyst and enter a heat exchanger. Typical process parameters are reaction temperatures around 1050°C and residence times of 0.1 milliseconds. The methane is converted almost completely with selectivities of about 60%. The ammonia selectivities are still higher, in the order of 60 to 70%, but considerable amounts of ammonia remain unreacted in the product gas.

2. Components of the High-Temperature Microreactor

The high-temperature microreactor was designed to operate using similar process parameters as chosen in the traditional Andrussov process. In a preheater the three educt gases are guided in sepearate feed lines and are heated to temperatures ranging from 250 to 600°C. The gases enter the mixing zone through three outlet holes of 60 μm diameter. Complete mixing is achieved within a distance of 4 mm. The educt mixture passes a 250 μm thick platinum catalyst structure inserted on a 5 mm thick ceramic support, is heated to more than 1000°C and is catalytically converted to the product. The reactor is designed that by varying the volume flow, residence times between 0.1 and 1 milliseconds were achieved. After passing a short postreaction zone of 300 μm length, the hot gas mixture is cooled down to 120°C in a micro heat exchanger. In Figure 1 a schematic of the assembly of the microreactor is shown.

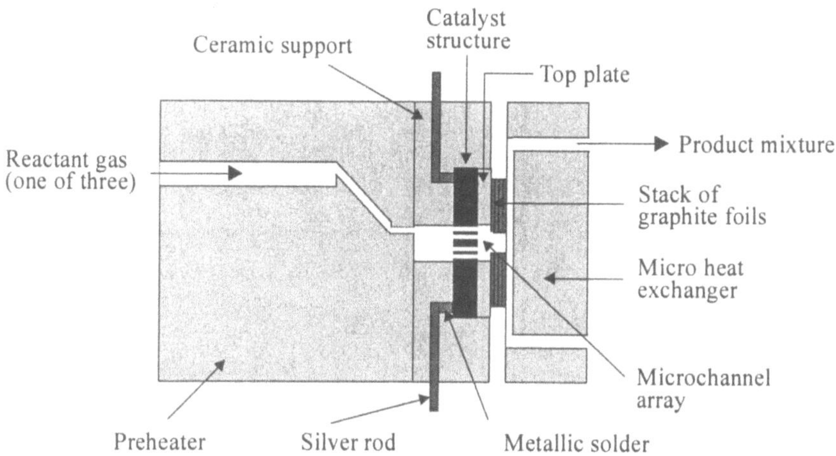

Fig. 1 Schematic of the microreactor assembly. The design is simiplified, in particular, conerning the dimensions of the components.

The components of the disassembled microreactor including heating cartridges, insulation jackets and other supplementary equipment are shown in figure 2. Apart from the ceramic disk and platinum catalyst structure, all other components were realized in a high-alloy stainless steel.

Two distinct differences between this microreactor and the large-scale reactor for the Andrussov process have to be mentioned. First, by using an electrical heating of the catalyst zone an isothermal operation can be achieved. This is realized by passing the reactant gas through a special channel configuration of high symmetry in the catalyst material. In the case of the large-scale process, the reaction is self-sustaining, i.e. once the reaction has started, the reaction heat released is sufficient to maintain processing. Second, a very close connection of the hot reaction zone and the cold heat exchanger within the microreactor enables to realize extremely short cooling rates.

Apart from these arguments concerning process optimization [8], the proof of feasibility of the performance of a high-temperature reaction yielding hazardous gases is an interesting aspect on its own [9,10]. A small microreactor capable of synthesizing quantitative amounts of hydrogen cyanide can principally be used for a hydrogen cyanide production and conversion on-site instead of transporting this dangerous compound in storage vessels [9,10]. However, the installation of the infrastructure for feeding of fluids and supplying current may pose a severe problem.

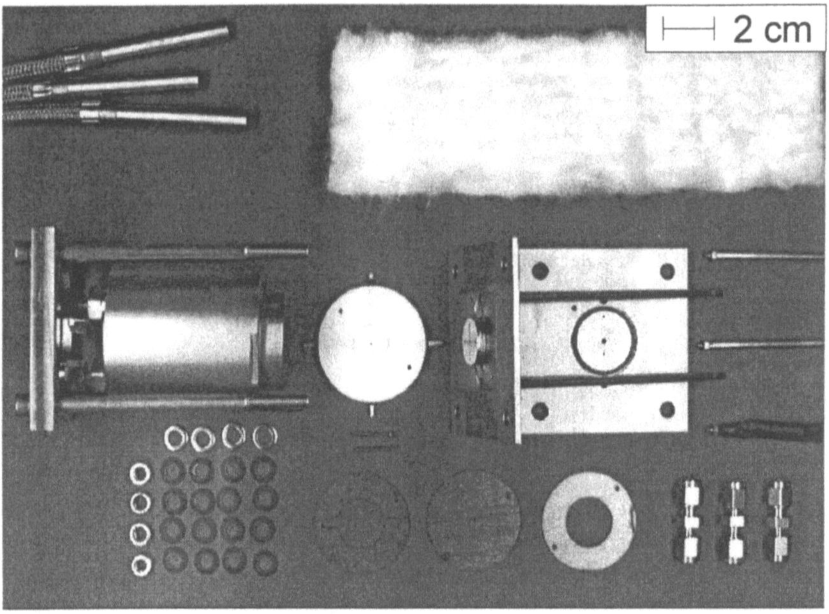

Fig. 2 Photograph of the components of the microreactor and supplementary equipment of the microreactor.

The microreactor is a flexible system concerning throughput and represents a safe alternative due to the small holdup in the reactor volume. It is transportable, may be rapidly installed and the reaction product can be directly converted at the site of generation. Moreover, this type of microreactor is an interesting measuring tool for studying the process performance of high-temperature reactions. Although it was designed with focus on the Andrussov process, it may be easily adapted to other reactions with similar process characteristics.

2.1 Platinum Catalyst Structure with Microchannel Array

The central element of the microreactor is a platinum catalyst structure inserted on a ceramic support shown in figure 3. The catalyst structure has two basic functions concerning process performance. First, it comprises the reaction channels made of the catalytic active material. Secondly, it controls the process temperature by using resistance heating with variable electrical power input. Thereby, the temperature of the reaction channels can be flexibly adjusted.

The reaction zone, an array of 20 microchannels with circular cross-section, is located in the central part of the catalyst structure on a 0.54 mm wide and 4 mm long strip (see figure 3).

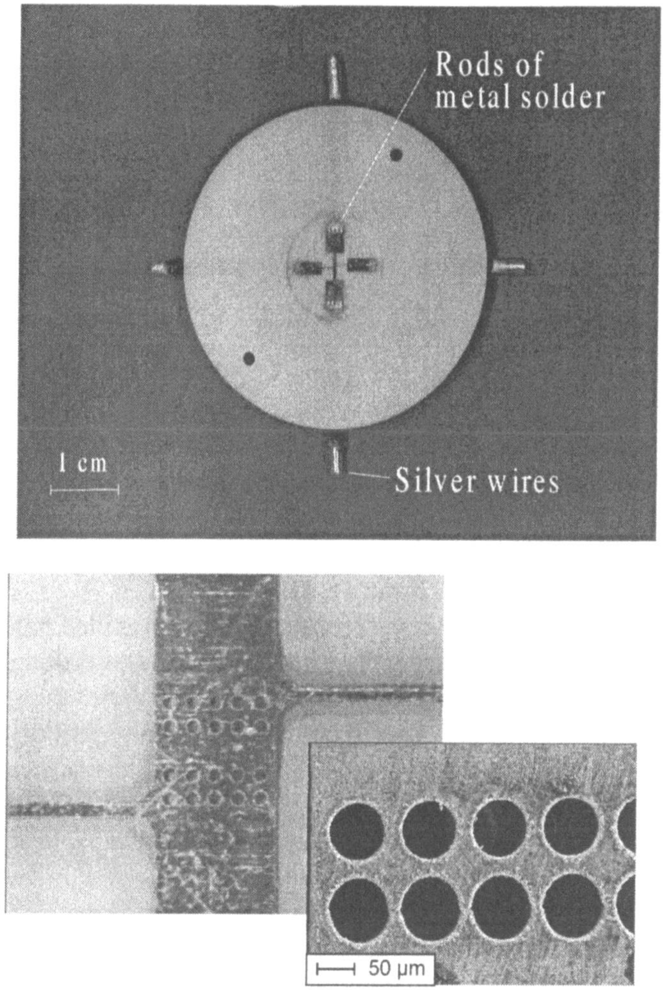

Fig. 3 Photograph of the catalyst structure inserted on the ceramic support.

Heating of this array is performed by guiding electrical current through thin remaining material bridges separating the channels. In order to achieve a uniform temperature profile, a particular channel configuration was chosen according to heat flux calculations. The array was splitted into two groups of channels, each consisting of two rows with five channels. In this highly symmetrical arrangement each channel is surrounded by three thin bridges which act as heat sources. Thus, the equidistribution of heat generation is guaranteed by the uniform border conditions of each reaction channel. In order to reach the desired process temperatures of more than 1000°C, the thickness of the channel walls had to be restricted to 40 μm.

For the microchannels a diameter and length of 70 μm and 250 μm, respectively, were chosen according to mass transfer considerations. The contours of the catalyst structure were fabricated by means of micro thin-wire erosion using a thin platinum foil of 250 μm thickness. The channels were drilled into the strip using a die sinking process with cylindrical electrodes of about 30 μm. Details of the fabrication were already reported elsewhere [11].

2.2 Ceramic Support Enabling Reversible Catalyst Insertion

The main function of the ceramic support is to carry the catalyst structure and to insulate it from the surrounding reactor parts. By this it forms a passage from the preheater to the micro heat exchanger via the mixing and postreaction zones, respectively. Moreover, it contains an electrical line connecting the catalyst structure to an external power source. This source has to be connected to two silver rods which stand out of the edges of the ceramic support (see figures 3 and 4).

The rods penetrate the ceramic material through drilled holes and are contacted to a metallic solder [12] close to the centre of the support. This solder was filled in drilled holes oriented perpendicular to the rod's axes. The holes were deliberately overfilled in order to leave part of the solder material on the support's surface. After melting the solder at high temperature, the surface part was shaped by micromilling to yield small plugs. The plugs contain cog-like structures which enable a reversible mechanical connection to the connectors of the catalyst structure which have cogs as well (see figure 4).

The material of the ceramic support is a mixture of various oxides, silicon oxide being the major component. Details about the geometry of an older version of the ceramic support and the microfabrication based on micromilling were reported elsewhere [11]. The geometry of the support used in this study had to be slightly adapted to the needs of the new connecting concept.

This introduction of a reversible concept for catalyst exchange is a significant improvement compared to the use of irreversible techniques as reported in [11]. In that study, laser welding and pressing techniques were applied for first feasibility tests and proved their suitability for catalyst heating as well, but led to an irreversible connection of the catalyst structure to wires attached to an electrical power source. Experiments showed that, using the new concept, an exchange of the catalyst structure is even possible after long-term usage of about one month at temperatures of more than 1000°C.

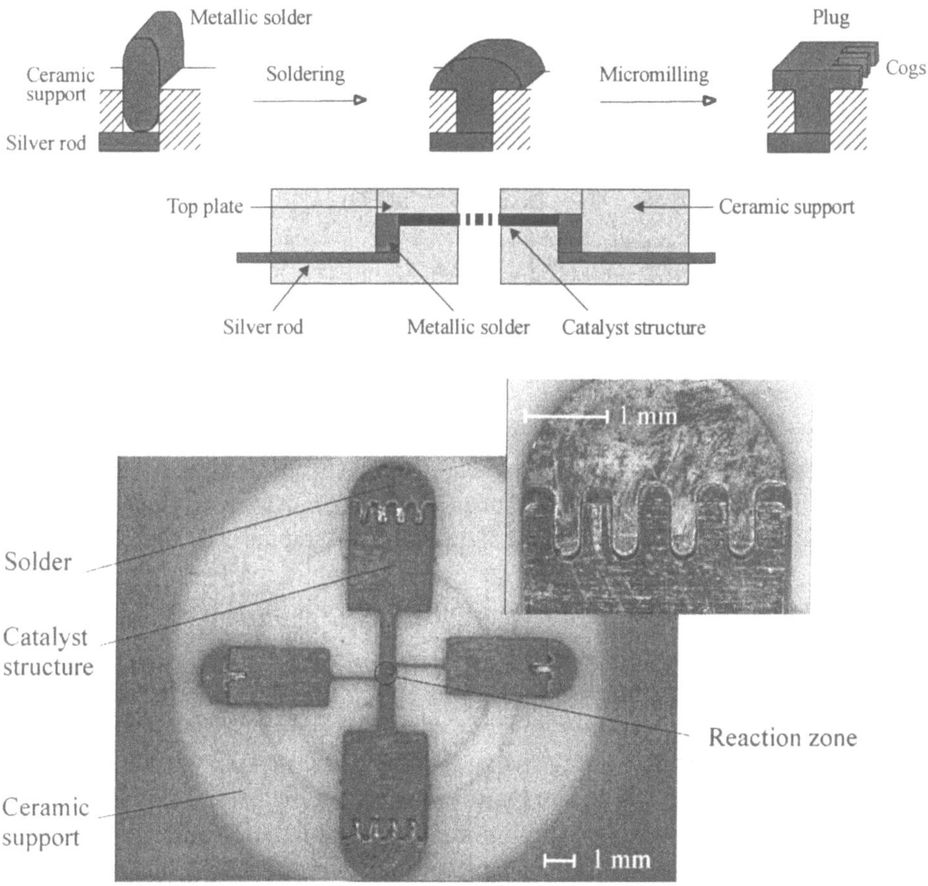

Fig. 4 Schematic of the joining process applied and cross-section of the ceramic support (top) and photograph of a plug with cogs for connection of the catalyst structure to an external power source (bottom).

3. Cross-type Micro Heat Exchanger

After passing the reaction zone, the product gas was guided through a postreaction zone and enters a micro heat exchanger (see figure 5). From an entrance area, which is located at the centre of the heat exchanger, the product gas was fed to four microchannels which form a cross-shaped structure (see figure 5). In order to efficiently cool the gas, channels of small width and large depth were chosen to provide large contact surfaces

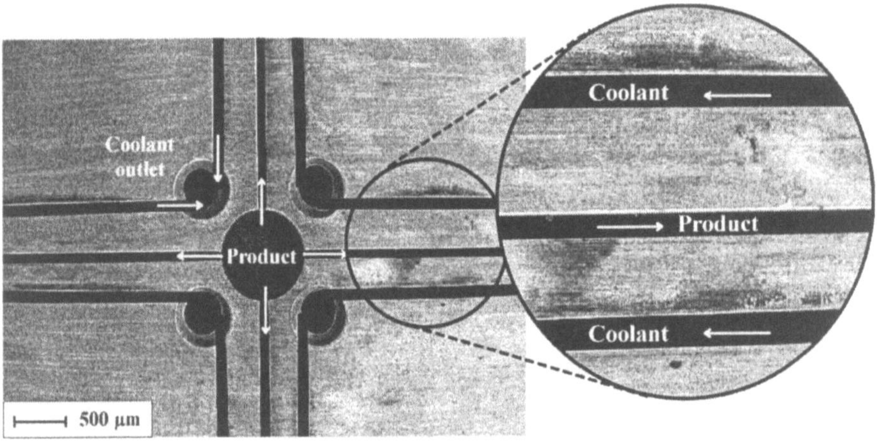

Fig. 5 Cross-type micro heat exchanger operating in counter-current flow mode.

to the cooling channels carrying the heat exchange medium air. Each product channel is surrounded by two cooling channels operating in a counter-current mode. After passing the four product channels, the product gas is recollected on a feed line and guided through a drilling into an outlet tube.

4. Sealing Techniques for High Temperature Applications

The main issues of the sealing concept were to prevent gas leakage along the electrical line and the various interfaces of the reactor components. The first issue was achieved by the development of an irreversible metal-ceramic joining technique using the metallic solder mentioned in chapter 2.2. The second issue was realized by using standard flat sealing techniques based on graphite foils. Some of these foils had to be microstructured to be specially adapted to the geometries of the reactor components. In particular, cross-over of fluid flows of the heat exchanger medium and reaction mixture had to be prevented in the micro heat exchanger.

The mechanical flexibility of the graphite foils allowed, to a certain extent, to compensate the effect of thermal expansion of the hot reactor components and, thus, to prevent damage of the ceramic support by applying a too high fitting pressure. During experiments at high temperatures the foils proved thermal stability, since a direct contact of the graphite material to the 1000°C hot reaction zone was avoided.

The graphite foils were realized by laser cutting. Due to this relatively inexpensive process the foils could be used as disposable parts and were exchanged after each experimental run.

Apart from flat sealing, the major achievement concerning sealing was the development of a metal-ceramic joining technique (see first issue). This technique was based on the formation of a metallic-ceramic interphase by sintering the metallic solder and the ceramic material above a certain process temperature in an inert atmosphere. By formation of this phase, a gas tight interface was achieved between the connecting plugs and the ceramic support. In addition, the catalyst structure was covered by a ceramic top plate with a central hole. A gas tight interconnection was achieved for these two parts by sealing with a graphite foil.

This new sealing concept for the catalyst structure has considerable advantages over an older procedure using a ceramic glue as discussed in [11]. The former concept did not allow a reversible exchange of the catalyst structure and the microreactor was not totally gas tight. Measurements of the gas leakage of the microreactor using this new sealing concept showed that even for several hours the amount of leakage streams was close to the detection limit.

5. First Performance Tests

A complete experimental set-up was realized at IMM integrating the microreactor, several fluid connecting lines and boxes for electrical contact on a support plate with a size half of that of a standard laboratory table. Two set-ups were parallely tested by two industrial partners, BASF and Rhône-Poulenc. The set-ups were connected in the laboratories to a high-energy source for resistance heating and feed lines for the reactant gases and cooling media. The reactant gases were preheated to 250°C before entering the catalyst zone. Analysis of the product spectra was performed using gas chromatography (GC).

A gas flow of 640 ml/h was chosen in order to realize the same contact time as in the technical process. The gas flow was controlled by flow meters. Gas sampling was performed using a gas mouse. A direct measurement of the temperature was not possible, instead a calibration curve had to be used. The error of temperature detection using this method was estimated to be about 50°C.

The operating conditions and results of all experiments at BASF are listed in table 1. The volume contents of the various gases in the product mixture were derived from GC analysis. These data were used to calculate the conversion of methane and ammonia as well as the selectivity based on ammonia. The volume contents of the carbon-containing product gases were normalized to the volume content of the reactant methane. The dependence of this quantity on the temperature is shown in figure 7.

The results of table 1 clearly show that with rising temperature the methane conversion increased (see also figure 7).

Feed			Analysis					Temperature	Conversion		Selectivity
Air [ml/h]	CH_4 [ml/h]	NH_3 [ml/h]	Vol% CH_4	Vol% NH_3	Vol% HCN	Vol% CO_2	Vol% CO	T [°C]	X CH_4	X NH_3	S NH_3
500	70	70	7.57	0.84	0.03	2.98	0.36	865	31	92	0.2
500	70	70	2.48	2.03	0.05	3.22	3.63	900	77	81	1
500	70	70	3.73	4.26	0.60	3.02	2.41	912	66	61	9
500	70	70	2.18	4.52	1.98	2.36	3.49	923	80	59	31
500	70	70	1.40	7.35	2.19	2.04	2.13	939	87	33	61
500	70	70	0.88	3.74	1.92	4.34	4.32	950	92	66	27
500	70	70	0.86	4.59	0.77	1.81	4.61	963	92	58	12
500	70	70	0.24	2.76	1.07	1.76	5.22	977	98	75	13
500 *	70	70	0.38	15.30	3.33	1.54	4.77	963	96	***	***
500 *	70	70	0.17	5.97	3.08	1.44	5.11	977	98	45	62
500**	70	70	0.16	0.79	1.36	2.00	4.73	963	99	93	14

Tab. 1 Operating conditions and results of the measurements of BASF.
* Gas mouse was heated before the experiment
** Gas mouse was evacuated before the experiment
*** Values were not calculated (see discussion below)
T: temperature; X: conversion; S: selectivity

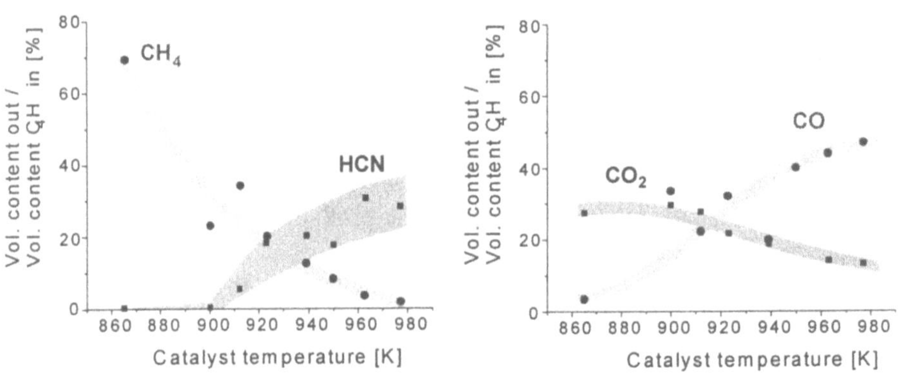

Fig. 7 Dependency of the volume contents of methane, hydrogen cyanide, carbon monoxide and carbon dioxide, normalized to the methane content, on the reaction temperature (BASF measurement).

From 900°C upwards hydrogen cyanide was formed and two byproducts, carbon monoxide and dioxide, were observed. With increasing temperature the carbon dioxide content decreased and the carbon monoxide content increased according to the Boudouard equilibrium (see figure 7).

During these experiments it became obvious that the values of the ammonia content were not correct, e.g. the value of 92% at 865°C appears to be unbelievably high. In addition, the data of table 1 indicate a decrease of ammonia conversion with rising temperature, in contrast to methane conversion, and finally a fluctuation. One explanation for this unusual behaviour are the formation of cold spots inside the microreactor. This leads to a condensation of water and decreases the volume contents of water-soluble species, in particular ammonia and to a minor extent hydrogen cyanide, by dissolution.

The correctness of this assumption was proven by careful heating of all reactor parts. Thereafter, the ammonia content increased to 15%, i.e. this value was beyond the ammonia content of 11% in the feedstock. The values of all other gases were not affected similarly. The following experiment showed a yield of 31% (see table 1). The ammonia related selectivity was 62% which is in the range of the technical process [6,7,13]. But with respect to methane, a selectivity of about 30% was achieved which is only half of the value of the industrial process. Therefore, the conversion of methane was twice as large as that of ammonia.

At this point, the experiments had to be stopped at BASF due to leakage of the microreactor. At present, the measurements are continued at Rhodia, formerly Rhône-Poulenc, using a second microreactor prototype. First experiments showed yields of hydrogen cyanide similar to the results of BASF. At a temperature of 1200°C complete ammonia conversion was achieved, but at a low selectivity.

A comparison of the results of BASF, Rhodia and the industrial process is given in figure 8. The diagrams reveal the volume contents of relevant components of the product gas. The remaining components, e.g. nitrogen or hydrogen, are not listed in this figure for reasons of clarity.

In order to explain these results, it was tried to model the micro-channel structure with regard to contact time, gas velocity and geometry. It was calculated that in the microchannels a concentration equilibration is achieved, i.e. every ammonia molecule at least has one contact with the catalyst. However, experiments show that this did not lead to complete conversion. One explanation are thermal side reactions, yielding carbon monoxide and dioxide, which compete with the catalyzed hydrogen cyanide formation. This competition may be suppressed by increasing the contact time with the catalyst. Thus, an improved design of the microreactor should reconsider longer microchannels.

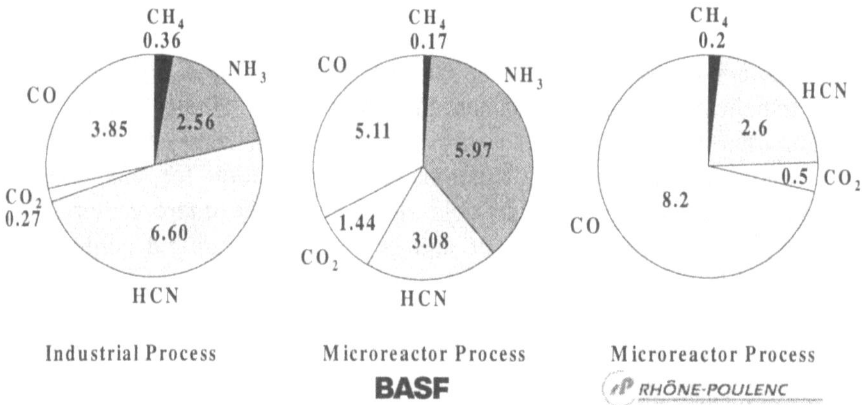

Fig. 8 Typical product spectra obtained by measurements of BASF and Rhone-Poulenc and comparison to results of the industrial Andrussov process [13].

In addition to the reference data from the industrial process, laboratory reactors have been chosen for comparison. Regularly structured monolithic reactors are geometrically analogous to the reaction zone of the microreactor. Figure 9 shows that the best yield of the microreactor is higher compared to other monolithic reactors [14].

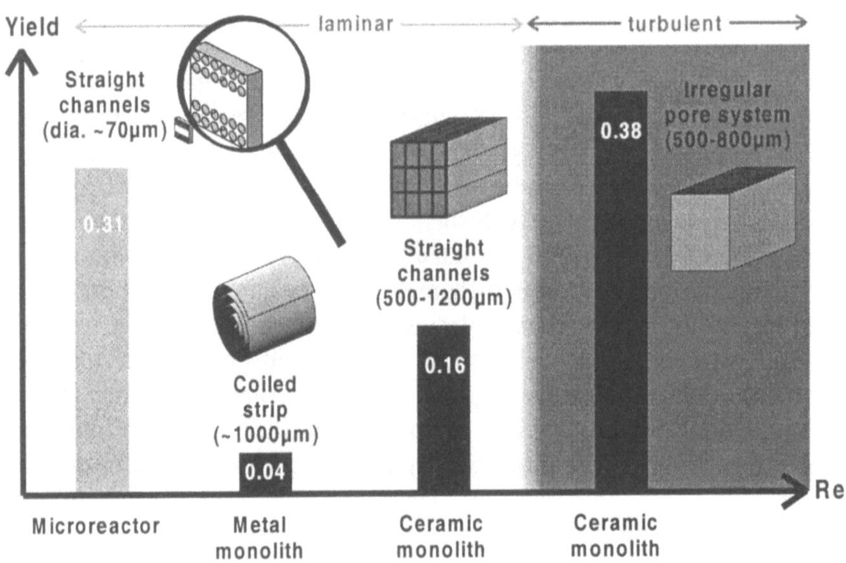

Fig. 9 Comparison of the yields of the microreactor to monolithic reactors operating in laminar and turbulent regimes [14].

The low yields of these monoliths, as well as their low overall conversion, was attributed to insufficient mass transport due to laminar flow in channels of a relatively large diameter (500 µm to 1 mm). This assumption was supported by the finding of higher conversions and yields in another reactor operating in a turbulent regime due to its irregular porous structure (foam monolith) [14]. The results of this article show that, in addition, a similar increase of mass transport can be achieved by using small microchannels (see also figure 7), even when operating in the laminar regime.

6. Conclusions

It has to be stated that an optimization of the Andrussov process [6,7,13] could not be achieved as was shown by the measurements of BASF and Rhodia so far. Since the real aim of this study was to construct a simple microreactor operating at temperatures exceeding 1000°C, the results prove the feasibility of the concept, i.e. a mobile reactor was developed capable of generating quantitative amounts of hydrogen cyanide by a high-temperature process.

Moreover, it could be shown that microchannels of a length as short as 250 µm and a contact time of about 0.1 ms, respectively, are sufficient to increase the temperature of a reactant mixture to more than 1000°C. In consequence, nearly quantitative conversion was achieved, although the selectivity to the product hydrogen cyanide has to be improved. Based on this experience, IMM has already specified an improved microreactor design. In combination with the usage of other catalyst materials, this may enable a more reliable process control and extend the range of applications of the reactor. These conceptual improvements should allow to use this microreactor for the mobile production of small amounts of hazardous gases and as a measuring tool to study the influence of process regimes on the reaction performance.

7. Acknowledgments

The results presented in this article have been made in the frame of a multilateral industrial project. The authors wish to thank the following industrial companies for their financial and technical support during this project: Aventis, BASF, Degussa, DuPont, Hüls, and Rhône-Poulenc. Furthermore, the technical assistance of A. Wolf for microfabrication based on µ-EDM techniques and N. Zimmerschitt for micromilling is greatly acknowledged.

8. References

[1] G. Wießmeier, D. Hönicke, Strategy *for the development of micro channel reactors for heterogenously catalyzed reactions,* in: Proc. of 2nd International Conference on Microreaction Technology, W. Ehrfeld, I.H. Rinard, R.S. Wegeng, eds., Topical Conference Preprints, March 9th - 12th 1998, New Orleans, U.S.A., pp. 24 - 32 (1998)

[2] Th. Richter, W. Ehrfeld, A. Wolf, H.P. Gruber, O. Wörz: *Fabrication of microreactor components by electro discharge machining,* in: Microreaction Technology, Proc. of the 1st Int. Conf. on Microreaction Technology, ed.: W. Ehrfeld, Springer-Verlag, Berlin (1997), pp. 158 - 168

[3] K.F. Jensen, I-Ming-Hsing, R. Srinivasan, M.A. Schmidt, M.P. Harold, J.J. Lerou, F. Ryley: *Reaction engineering for microreactor systems,* in: Microreaction Technology, Proc. of the 1st Int. Conf. on Microreaction Technology, Ed.: W. Ehrfeld, Springer-Verlag, Berlin (1997), pp. 2 - 9

[4] A.L.Y. Tonkovich, J.L. Zilka, M.R. Powell, C.J. Call: *The catalytic partial oxidation of methane in a microchannel chemical reactor,* in: Proc. of 2nd Int. Conf. on Microreaction Technology, eds.: W. Ehrfeld, I.H. Rinard, R.S. Wegeng, Topical Conference Preprints, New Orleans (1998), pg. 54 - 62

[5] L. Andrussow: *Über die katalytische Oxydation von Ammoniak-Methan-Gemischen zu Blausäure*, Angew.Chem. **48 (37)** (1935) 593 - 604

[6] L. Andrussow: *Blausäuresynthese und die schnell verlaufenden katalytischen Prozesse in strömenden Gasen*, Chem. Ing. Techn. **27 (8/9)** (1955) 469 - 472

[7] C.T. Kautter, W. Leitenberger: *Großtechnische Herstellung von Cyanwasserstoff nach Andrussow*, Chem.-Ing-Tech. **25** (12) (1953) 697 - 768

[8] O. Wörz, K.P. Jäckel, Th. Richter, A. Wolf: *Microreactors, a new efficient tool for optimum reactor design,* in: Proc. of 2nd Int. Conf. on Microreaction Technology, eds.: W. Ehrfeld, I.H. Rinard, R.S. Wegeng, Topical Conference Preprints, New Orleans (1998), pp. 183 - 185

[9] J.W. Ponton: *Observations on hypothetical miniaturised, disposable chemical plant,* In: Microreaction Technology, Proc. of the 1st Int. Conf. on Microreaction Technology, ed.: W. Ehrfeld, Springer-Verlag, Berlin (1997), pp. 10 - 19

[10] I.H. Rinard: *Miniplant design methodology,* in: Proc. of 2nd Int. Conf. on Microreaction Technology, eds.: W. Ehrfeld, I.H. Rinard, R.S. Wegeng, Topical Conference Preprints, New Orleans (1998), pp. 299 - 312

[11] H. Löwe, W. Ehrfeld, K. Gebauer, K. Golbig, O. Hausner, V. Haverkamp, V. Hessel, Th. Richter; *Microreactor Concepts for Heterogenous Gas Phase Reactions; in:* Proc. of 2nd International Conference on Microreaction Technology, W. Ehrfeld, I.H. Rinard, R.S. Wegeng, eds., Topical Conference Preprints, March 9th - 12th 1998, New Orleans, U.S.A., pp. 63 - 74 (1998)

[12] Commercial product of Degussa AG, Hanau, Germany

[13] F. Endter: *Die technische Synthese von Cyanwasserstoff aus Methan und Ammoniak ohne Zusatz von Sauerstoff*, Chem. Ing. Techn. **30 (5)** (1958) 305 - 310

[14] D.A. Hickman, M. Huff, L.D. Schmidt: *Alternative catalyst supports for HCN synthesis and* NH_3 *oxidation*, Ind. Eng. Chem. Res. **32** (1993) 809 - 817

Micro mixing effects in continuous radical polymerization

Thomas Bayer, Detlev Pysall, Olaf Wachsen

Aventis Research & Technologies GmbH & Co KG
Production Technologies, Industriepark Höchst
D-65926 Frankfurt am Main, Germany

Summary

In this study the development of a continuous radical solution polymerization process for acrylates in a lab scale tube reactor was performed. Fouling in the reactor was always observed if there were bad mixing conditions at the feed point of the reactor. In order to increase the homogenization of monomer and initiator a micro mixer was installed at the reactor inlet. The micro mixer ensured a very fast mixing at the micro scale of the preheated monomer and initiator solutions, and fouling was reduced tremendously.

The reactor employed has a theoretical production capacity of 50 metric tons per year. A first prebasic design has shown that a production capacity of some thousand tons per year can be achieved by numbering up instead of conventional scale up.

Introduction

Radical polymerization is an important and well known method for the synthesis of different polymers and copolymers, e.g. the production of acrylic resins. Up to now batch and semi batch processes are widely used for production /1/. There is a growing interest in realizing processes with higher reproducibility, quality and minimized hazard potential.

These requirements are met by a continuous process. Additionally it requires less operation steps than batch processes and allows higher space-time-yields.

Experimental setup and conditions

The general technology of continuous processes in tube reactors is well known, but hardly used for polymerization processes /1,2/.

A setup with static mixers (Sulzer SMX and Kenics) was installed. The lab scale equipment with vessels for educt and initiator and a vessel for the product is shown in Figure 1. To achieve optimum mixing conditions at the feeding point of the reactor a Sulzer SMX with 5 mm diameter was implemented as a premixer. The characteristics of the tube reactor are given in Table 1.

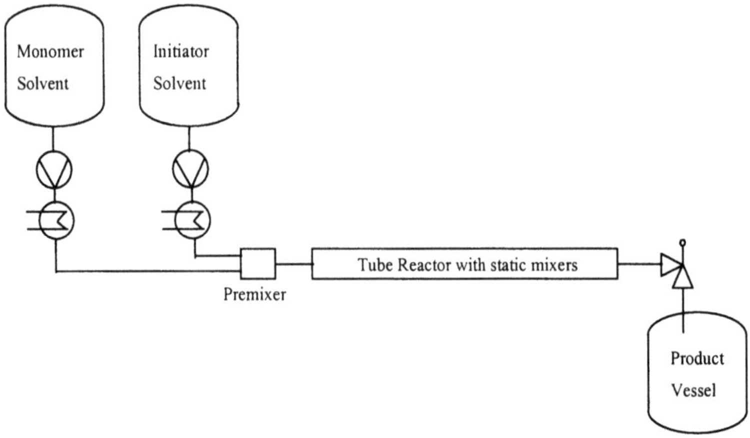

Figure 1: Experimental setup of the lab scale reactor

Table 1: Characteristics of the lab scale tube reactor

Reactor diameter	5...20 mm
Volume	4.7 liter
Reactor length	23...60 m
Residence time	40 min
Temperature	max. 150 °C
Pressure	max. 16 bar
Throughput	6...8 kg/h
$\text{Re-number}_{\text{Premixer}}$	400
$\text{Velocity}_{\text{Premixer}}$	0.11 m/s
$\text{Bo-number}_{\text{Reactor}}$	40

Polymerizations were carried out using different recipes for acrylic copolymers. The reaction temperature was varied from 120 °C to 145 °C. Characterizations of the products were done by determination of conversion of each monomer by NMR-Spectroscopy, solid content and the molecular weight distribution by Size Exclusion Chromatography.

More than a dozen experiments with both static and micro premixers were performed. Steady state was achieved after about one hour and kept constant for several hours.

Results (static premixer)

First experiments showed a molecular weight distribution with a second peak at higher molecular weights (Figure 2).

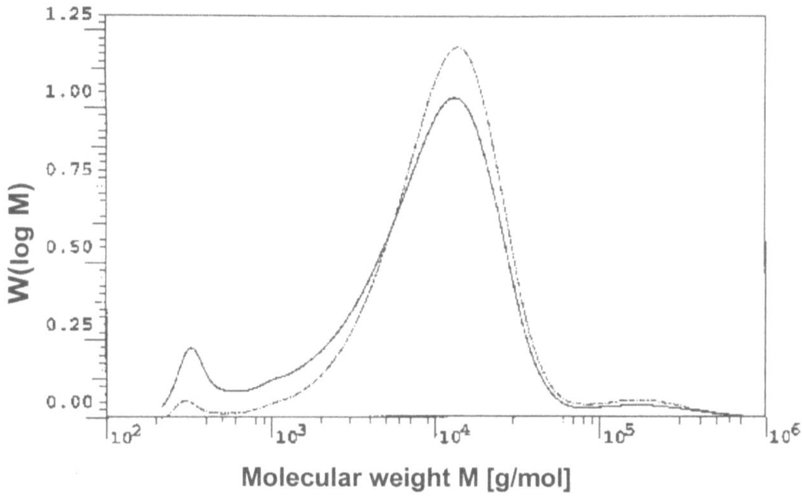

Figure 2: Molecular weight distribution (static mixer at the feeding point) measured by SEC (size exclusion chromatography) using UV and RI detectors

This effect can be explained by well known micro mixing phenomena /1/, because the mixing time of a few seconds is longer than the reaction time for the formation of polymer chains. Poor micro mixing conditions can result in undesired decrease in yield and selectivity or in changes in molecular weight distributions in case of polymerization reactions. During radical polymerization the formation of a particular polymer chain is terminated after a time less than one second /3/. Therefore bad mixing conditions (segregation) can lead to broader molecular weight distributions than expected in the ideal case, because there may occur zones in the mixture with a higher or lower local initiator concentration than desired. In solution polymerization of acrylic monomer, molecules with a high molecular weight can become insoluble in the solvent and fouling in the reactor may occur (Figure 3). This means that a static mixer as premixer is insufficient for fast polymerization reactions. An other type of premixer has to be used.

Figure 3: Fouling in the reactor with a static mixer at the feeding point (photo taken at 2/3 of the tube length)

Results (micro premixer)

In the study presented fouling in the tube reactor was always observed if there were bad mixing conditions at the feeding point of the reactor. Different tools are known to achieve enhanced micro mixing. For turbulent mixing a tangential jet mixer for the very fast ionic polymerization of methyl methacrylate is published /4/. For laminar mixing different types of micro mixers with lamella structures are available /5-7/. This kind of mixers consists of lamella channels in the micro scale. The components to be mixed were fed into this channels. This kind of mixers leads directly to a micro scale mixture.

A micro mixer device from IMM (Institut für Mikrotechnik), Mainz, Germany (Figure 4) was installed at the reactor inlet. A very fast mixing in the micro scale of the preheated monomer and initiator solutions is ensured.

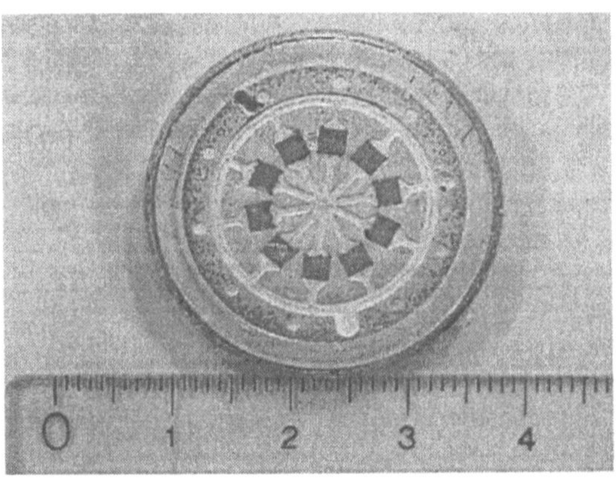

Figure 4: Micro mixing device used as premixer, scale in cm

In contrast to first suspicions the increased surface of the micro mixer didn't increase fouling. It was tremendously reduced (Figure 5) compared to the formerly used apparatus with a static mixer. The experiments showed that a acrylic resin with a better quality compared with the existing production process could be achieved. Especially a more uniform molecular weight distribution was obtained (Figure 6).

Figure 5: Fouling in the reactor with a micro mixer at the feeding point (photo taken at 2/3 of the tube length)

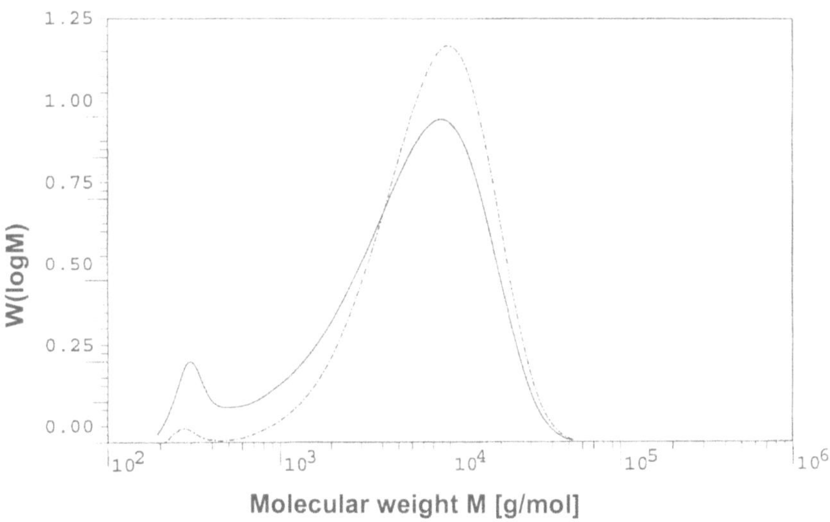

Figure 6: Molecular weight distribution (micro mixer at the feeding point) measured by SEC (size exclusion chromatography) using UV and RI detectors

Scale up

While in case of conventional static mixers scale up is possible, a numbering up has to be done in case of laminar micro mixers. The lab scale reactor employed has a theoretical production capacity of about 50 metric tons per year (running 8,000 hours). Thus, a numbering up of the micro mixing device is possible and a scale up was performed for the tube reactor.

A first prebasic design has shown that some twenty micromixing devices are sufficient to produce a capacity of thousand metric tons per year.

References

(1) Echte, A., Handbuch der Technischen Polymerchemie, Verlag Chemie, 1993

(2) Gerrens, H., Polymerisationstechnik, Ullmanns Enzyklopädie der technischen Chemie, 4. Ed., Verlag Chemie, Weinheim, 1980, p. 107ff

(3) Elias, H.-G., Makromoleküle, Band 1, 5. Aufl., Hüthig & Wepf Verlag, Basel, 1990, p. 441ff

(4) Luzlau, Nelly et al., Procédé de polymérisation anionique en continu d`au moins un monomère(méth)acrylique pour l`obtention de polymères à haut taux de soliede, EP 0 749 987

(5) Hessel, V., Ehrfeld, W., Möbius, H., Richter, Th., Russow, K., Potentials and Realization of Microreactors, Proc. Int. Sym. On Microsystems, Intelligent Materials and Robots, Sept. 27-29, 1995, Sendai, Japan

(6) Schubert, K., Bier, W., Linder, G., Seidel, D., Statischer Mikro-Vermischer, DE 4 416 343

(7) Häligren, K., Mischvorrichtung, DE 3 114 195

Novel Liquid Phase Microreactors for Safe Production of Hazardous Specialty Chemicals

Tamara M. Floyd, Matthew W. Losey, Samara L. Firebaugh‡, Klavs F. Jensen, and Martin A. Schmidt‡

Department of Chemical Engineering, ‡Microsystems Technology Laboratories
Massachusetts Institute of Technology, Cambridge, MA, 02139, USA

Abstract. A novel liquid-phase microreactor with thin film temperature sensing, good thermal management, and fast mixing has been fabricated for the production of hazardous specialty chemicals, specifically for organic peroxides. The reactor has been characterized and shown to be suitable for these types of applications. Moreover, fast mixing at 10ms and good heat transfer based on an overall heat transfer coefficient of 1445 $W/m^{2}{}^{o}C$ were demonstrated. The established linear resistive response of the temperature sensors was confirmed experimentally, and the reactor was shown to operate at 11 psi at the 1.0 ml/min design flow rate.

Keywords. microreactor, microfluidics, mixing, heat exchangers, thin-film temperature sensors

1 Advantages of micro-reactors for chemical production

Several researchers have studied microreactors for applications in liquid phase chemical production [1],[2]. In particular, these systems would be appropriate for safe production of hazardous specialty chemicals with fast, exothermic reaction rates. Fast, exothermic reactions may be mixing limited. Consequently, the small channels that can be etched using micro-fabrication techniques reduce characteristic mixing times so that maximum conversion of the reactants can be achieved. In macroscopic reactors, exothermic reactions may be subject to thermal runaway [3]. For these systems, it is important to monitor and control the temperature to avoid unsafe operating regimes. Microsystems offer the advantage of low thermal mass, fast control response and advanced temperature sensing using thin-film methods. Furthermore, on-demand, point-of-use production methods would become important for processes with hazardous intermediates due to the safety concerns associated with the storage and transport of these chemicals[4],[5].

2 Microreactors for Safe Production of Hazardous Specialty Chemicals

Organic peroxides, used industrially for polymerization initiation and other applications, are candidates for chemical production in the microreactor. Explosions of peroxides have caused millions of dollars in property damage and loss of life [6]. Industries that produce and consume peroxides are more sensitive to the hazards of organic peroxides. Nevertheless, studies are still conducted to minimize the hazards associated with producing, storing and handling these materials [7]. Dilution of the peroxide is essential for heat management and good mixing assures a uniform composition and temperature distribution throughout the reactor. Also, temperature sensing is important for detecting and preventing dangerous peroxide decomposition during production and storage. A standard molar ratio of initiator to monomer is 0.001 [8]. Based on this estimate and an assumed polymer production capacity of 1Mlbs/yr, the minimum product flow rate for the micro-chemical system is ~1000 lbs/yr. These production volumes are within the capabilities of a single microreactor, and the production capacity can be expanded by scaling out with multiple micro-chemical reactors. Thus, the volumes of organic peroxides are potentially small enough for production in micro-chemical systems. The production recipes for diacyl peroxides are relatively simple. Essentially, acid chlorides react with alkaline hydrogen peroxide or alkali metal peroxides [9],[10]. The lack of process complexity makes on-site, on-demand production of organic peroxides in micro-chemical systems more feasible.

For this work a novel, liquid phase microreactor was fabricated to address temperature sensing, thermal management and mixing issues identified as important parameters for organic peroxide production and handling. The reactor design is simple, yet highly integrated with a heat exchanger and thin film temperature sensors. Temperature sensing using thin film methods, good heat exchange characteristics and fast mixing were demonstrated.

3 Microreactor Design

The reactor design incorporates mixing by lamination, hydrodynamic focussing and high aspect ratio channels for fast mixing and higher throughput. A first generation device, shown in Fig. 1, was fabricated to study mixing in the system. In this system, there are two sets of inlet ports. Fluids enter a reservoir and are distributed among 5 inlet ports per reactant. After the flows have developed, they are laminated upon contact [11]. After lamination, the fluids are hydrodynamically focussed to reduce the diffusive mixing length. Then, the fluids mix and react in a straight, high-aspect-ratio channel.

After conducting a mixing study to understand the species transport properties for the system, a second generation reactor, shown in Fig. 2, was designed and fabricated. The second generation reactor incorporates a heat exchanger and thin-film temperature sensors.

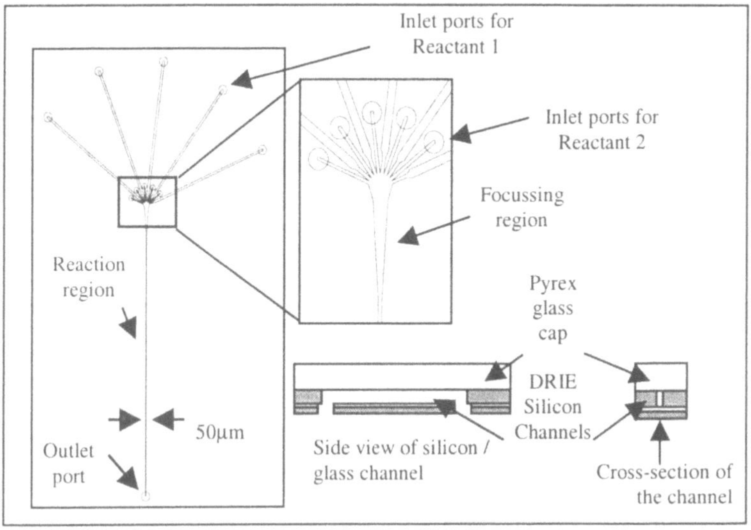

Fig. 1. First generation reactor design

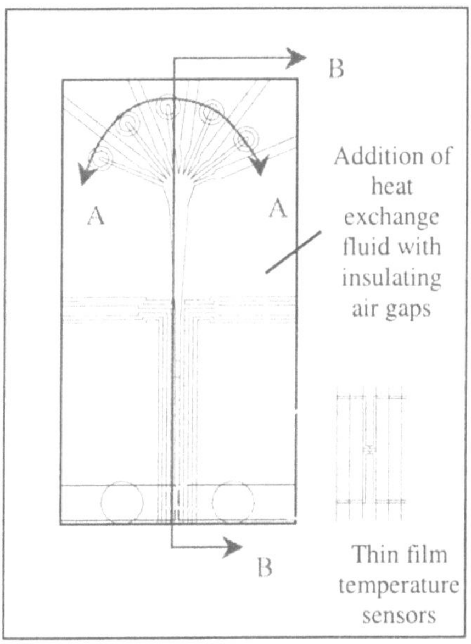

Fig 2. Second generation reactor design

4. Microreactor Fabrication

Fig. 3 is a combined schematic of the first and second generation fabrication processes. As shown in Fig. 3, the first generation device was fabricated using a silicon-on-insulator (SOI) wafer with a 40 μm silicon film and a 1 μm layer of silicon dioxide . The inlet ports were defined and etched using standard photolithographic techniques and a deep reactive ion etch (DRIE) [12]. After defining the inlet ports, the wafer was flipped and patterned using a thick photoresist layer. Next, a DRIE was performed to define the mixing and reacting channels. Finally, the wafer was capped by anodically bonding to pyrex [13],[14]. Each reactor was 2.4 cm X 2.9 cm.

The second generation device was fabricated using an SOI wafer with a 2.2 μm silicon film and a 1 μm layer of silicon. First, the SOI wafers were doped with phosphorus using a $POCl_3$ source. Then, the temperature sensors were defined, and the wafers were re-patterned to define the inlet ports. Next, the wafer was flipped, and the channels for the reaction and the heat exchange fluid were etched. Subsequently, a patterned glass wafer was anodically bonded to the silicon thin film. Then, a blank glass wafer was anodically bonded to the silicon substrate to cap the device. After dicing, blanket layers of titanium and platinum were deposited to make contact to the temperature sensors. Then, the reactors were annealed at 550°C to stabilize the temperature sensors and the deposited metal. The reactor dimensions for the second generation design were reduced to 2.1 X 2.7 cm.

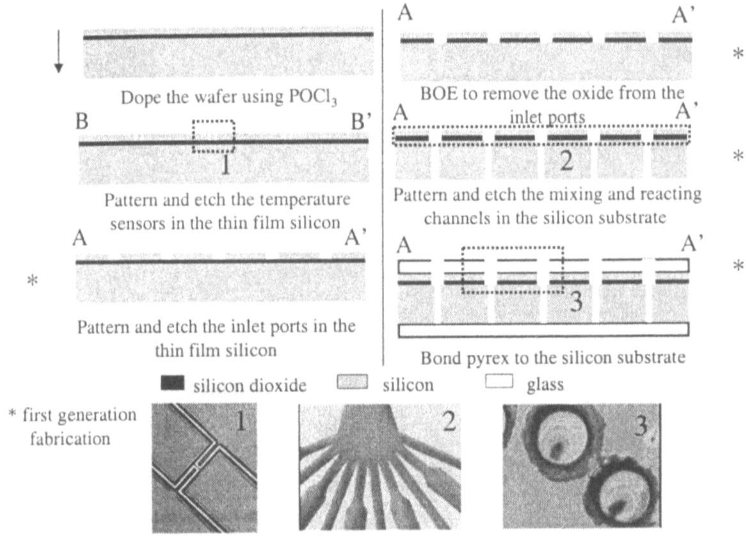

Fig. 3. Fabrication of the liquid phase microreactor

5. Experimental Test Station

Fig. 4 is a schematic of the test station for the liquid phase micro-reactor. The reactants are filtered and delivered to a plexy-glass chuck using a syringe pump. Polyether-etherketone (PEEK) tubing connects the syringe pump and the chuck, and the chuck interfaces the pump to the microreactor. The micro-reactor is compressively sealed between a poly-dimethyl siloxane (PDMS) gasket with a crescent-shaped reservoir and a compressive force plate.

The system was observed visually using a charge-coupled device (CCD) camera with a long-working distance magnifying lens. Visible spectrometry was used for detection. For this system, light in the visible range (400–700 nm) was passed through the reaction channel. The light was collected using an optical fiber and sent to the spectrometer. The reactor effluent was then collected for further analysis or disposal.

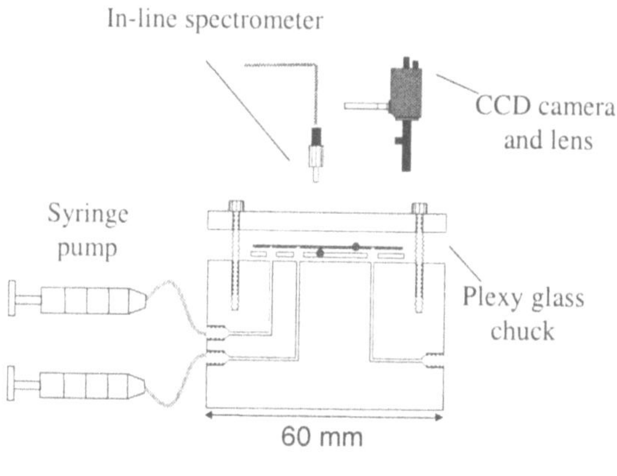

Fig. 4. Liquid phase micro-reactor test station

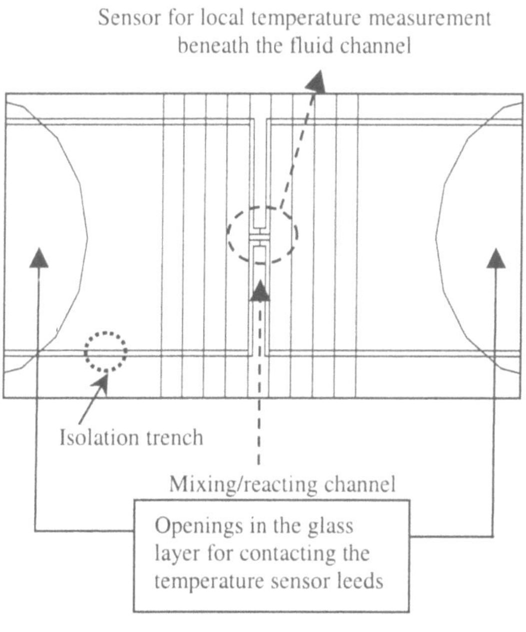

Fig. 5. Thin film temperature sensors

6. Temperature Sensors

Thin-film temperature sensors were etched in the 2.2 μm silicon film. The temperature sensor leads are 400 μm wide and 2mm long whereas the sensing resistor is nominally 10 μm wide and 50 μm long to assure that the resistance is high in the area of interest. Fig. 5 shows how the sensors are located in the silicon film directly beneath the reacting channel to monitor the temperature of the reacting fluid.

The temperature sensors were contacted using 0.01" diameter wires bonded with electrically-conductive epoxy. The temperature coefficient of resistance (TCR) was measured to demonstrate that the sensors are stable and operational. The established temperature coefficient of resistance of silicon is 0.0024/°C. Fig. 6 compares the experimental TCR, calculated as 0.0024/°C, and the literature value.

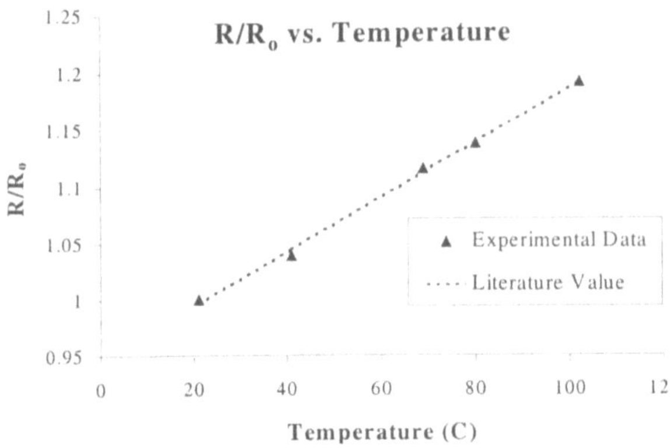

Fig. 6. Temperature sensor characterization

7. Heat Exchangers

Fig. 7 shows the heat exchangers that were defined for the liquid phase micro-reactor by DRIE. The heat exchange fluid enters the chip and flows beside the mixing and reacting channel in a double channel heat exchanger configuration. The heat exchanger channels and walls are 50 μm wide and 350-500μm deep. The heat exchangers were tested using hot and cold water in the inner and outer shells respectively. The average experimental heat transfer coefficient was 1445 W/m^2°C, which is comparable to macroscale overall heat transfer coefficients for water-to-water heat exchangers [15].

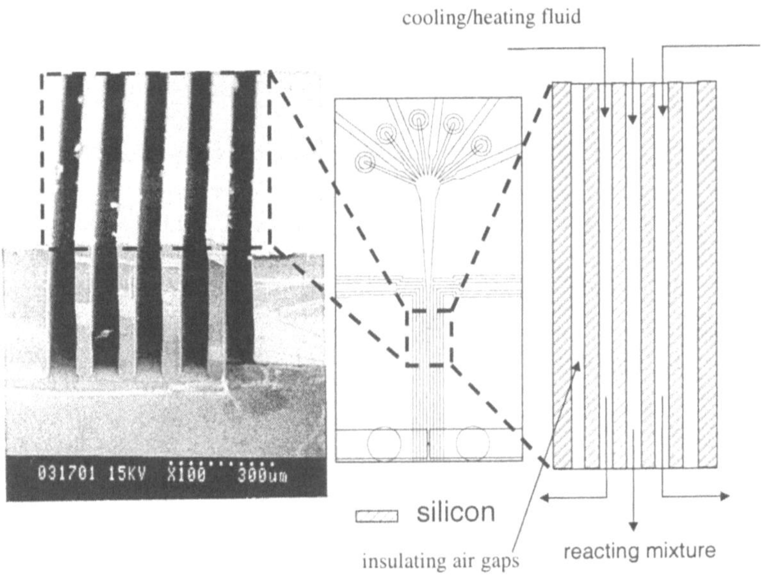

Fig. 7. Liquid phase heat exchanger

8. Throughput and Power Requirements

Throughput and power requirements are an issue for liquid phase micro-chemical systems to be used for chemical production. The throughput of the liquid phase micro-reactor was maximized by using a deep reactive ion etch to produce 350-

500 μm deep channels. In-line piezo-resistive pressure sensors were used to determine the pressure drop as a function of the liquid flow rate. The pressure drop for the design flow rate of 1 ml/min is 11.0 psi. The maximum flow rate achieved for this system was 8 ml/min. The primary failure mode was not the device, but the back-pressure on the syringe pump. Thus, flow rates in excess of 8 ml/min could be attainable with different feed schemes.

9. Mixing Characteristics

A mixing study characterization was performed using the acid-base reaction of phosphate buffers [16]. Phenol red, a color indicator, was used to investigate the mixing behavior of the system. Typically, spectrometry is used to determine the concentration of a component. For this study, we wish to differentiate between elements of a fluid within a vessel. The laminated steam made by contacting two unreacting components will be heterogeneous when monitoring at a specific position while varying contact times. But, under Beer-Lambert conditions, the result gives an average concentration and that is the same whether the fluids are completely mixed, completely unmixed or have achieved some intermediate state of mixing. Thus, to probe mixing, the desired property should be non-linear. One such property exists for most color indicators. Absorbance as a function of pH is non-linear, so even though the concentration of the absorbant is constant over the channel width, the absorbance captures the pH variation from point to point. For the phenol red indicator, at pH 8.0, the absorbance is high, and it drops non-linearly to zero at about pH 6.0. Thus, mixing was determined by charging pH 6.4 and pH 8.0 solutions to the two sets of inlet ports. The concentration of phenol red was the same for both solutions so only the non-linear pH absorbance was detected using visible spectroscopy.

Fig. 8 was captured from a video-taped experiment. It shows the 10 streams contacting and being hydro-dynamically focused. Two phenol red streams at pH 6.4 and pH 8.0 are laminated. Fig. 8 also shows a plot of the experimental results. Absorbance is plotted versus the wavelength. The reaction was monitored at 560 nm, and the data was extracted to determine the mixing time for the system. The absorbance of premixed solutions was measured as 100% mixed. Then, the absorbances were normalized to estimate the degree of mixing based on a linear interpolation between 0% mixed and 100% mixed. At high flow rates, corresponding to low residence times, the absorbance was higher. This was expected because the contact time is shorter and more of the fluid is close to pH 8.0 where the absorbance is highest. At lower flow rates and correspondingly longer residence times, mixing has progressed and the absorbance drops because the solution is closer to the fully mixed value of pH 7.2. Based on the absorbance data, the mixing time for this system is ~10 ms.

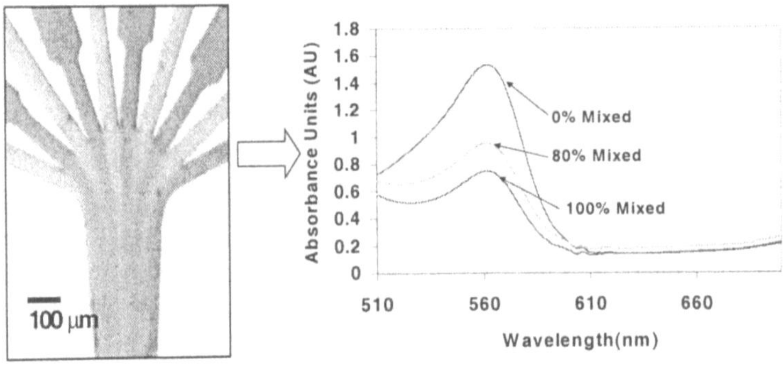

Fig. 8. Visualization of laminating layers and spectroscopic measurements for the mixing determination

CFD-ACE^{TM} software was used to model the system behavior in order to gain further insight into the system behavior. Fig. 9 shows the simplified, 2-D model of the system. Only a segment of the inlet channels was simulated. The concentrations are dilute so the momentum and species transport equations are uncoupled. The model assumes mixing by diffusion of a solute. As shown in Fig. 6, the model and the experiments are in good agreement. Hence, the model was used to probe the system and extract concentration profiles to gain a better understanding of the species transport.

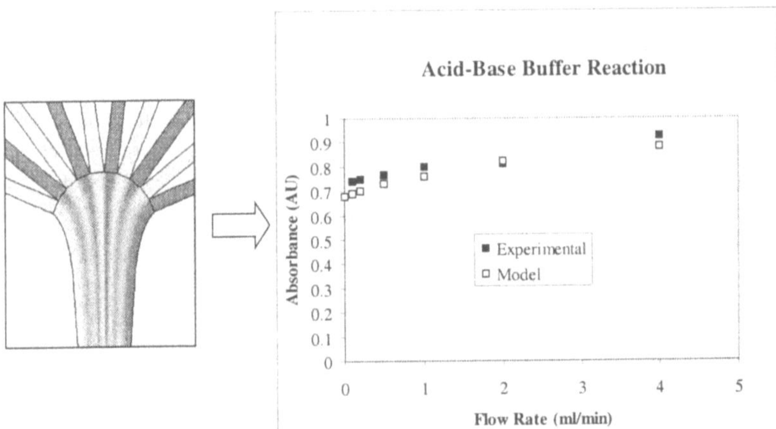

Fig. 9. Laminating layers using CFD-ACE^{TM} and comparison of the mixing model and experiments

10. Reactions

Several reactions have been demonstrated for the liquid phase reactor. Fig. 10 shows spectroscopic data for the Dushman reaction. The Dushman reaction is the reverse hydrolysis of iodine according to Equation 1.

$$6H^+ + IO_3^- + 5I^- \rightarrow 3\, I_2 + 3H_2O \quad (1)$$

$$I_2 + I^- \rightarrow I_3^- \quad (2)$$

Iodine and iodide react further to produce triodide according to Equation 2. Both, iodine and triodide absorb in the visible region. Thus, as demonstrated in Fig. 10, conversion trends can be identified by monitoring the reaction mixture at 450 nm using visible spectrometry. As the residence time decreases for flow rates ranging from 0.05 ml/min to 0.5 ml/min, the extent of the reaction drops as indicated by decreasing absorbances due to lower product concentrations. The hydrolysis of proprionyl chloride has also been demonstrated.

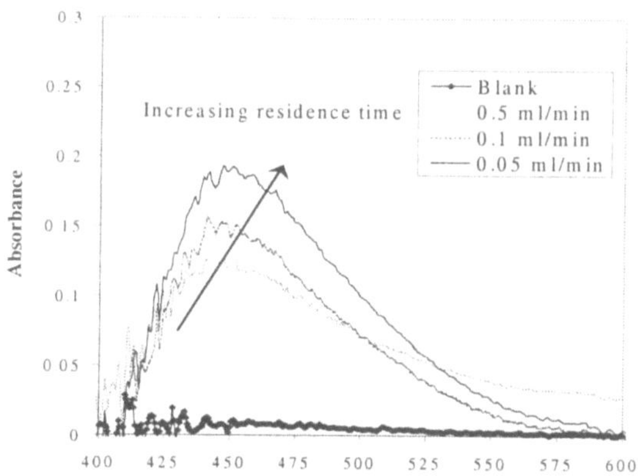

Fig. 10: Preliminary spectroscopic data for the Dushman Reaction

11. Conclusions

Temperature sensing capabilities, thermal management and fast mixing have been identified as key issues that need to be addressed when designing a micro-chemical system for production of hazardous specialty chemicals. A novel liquid phase reactor with a heat exchanger and thin film temperature sensors has been fabricated for the production of hazardous specialty chemicals. Thin-film sensing capabilities and a heat transfer coefficient of 1445 $W/m^{2\circ}C$ were demonstrated. Furthermore, the microreactor was shown to operate at 11 psi for flow rates of 1 ml/min. A mixing time of ~10 ms was determined.

Acknowledgments

The authors would like to acknowledge the National Science Foundation (NSF) and the Defense Advanced Research Projects Agency (DARPA) for funding. Devices were fabricated at Microsystems Technology Laboratories, and glass wafers were drilled by Jim Ashmead at the DuPont Company .

Citations

[1] Worz, O., Jackel, K.P., Richter, T., Wolf, A., "Microreactors, a New Efficient Tool for Optimum Reactor Design," 2nd *International Conference on Microreaction Technology*, 183-185, (1998)

[2] Burns, J. R., Ramshaw, C., "Development of a Microreactor for Chemical Production," 2nd *International Conference on Microreaction Technology,39-44, (1998)*

[3] Fogler, H.S, *Elements of Chemical Reaction Engineering,* 2nd *ed.*, New Jersey: Prentice Hall, 447-463, (1992)

[4] Benson, R.S., and Ponton, J.W., "Process Miniaturization – A Route to Total Environmental Acceptability?," *Trans. Chem. Eng., **71A**, 160 (1993)*

[5] Lerou, J.J., Harold, M.P., Ryley, J., Ashmead, J., O'Brien, T.C., Johnson, M., *et.al*, "Microfabricated Minichemical Systems: Technical Feasibility," *Microsystem Technology for Chemical and Biological Microreactors,* Mainz, Germany, DECHEMA, **132**, 51 (1996)

[6] Fowler, E.W., "Fire, Explosion and Health Hazards of Organic Peroxides," Research Report No. 11, American Insurance Association, (1966).

[7] Steensma, M., Westerterp, K.R., "Safe Operation of a Batch Reactor: Safe Storage of Organic Peroxides in a Supply Vessel," *Chem. Eng. Technol. **14***, 147-161 (1991).

[8] Rosen, S.L., *Fundamental Principles of Polymeric Materials*, New York: John Wiley & Sons, 115-151, (1982)

[9] Lower, E., "Synthesis and Application of Organic Fatty Peroxides," *Specialty Chemicals*, **18**, 2 (1998)

[10] Martin, L.F., *Organic Peroxide Technology*, Park Ridge, NJ: Noyes Data Corp., 220-239, (1973)

[11] Braneberg, J., Gravensen, P., Krog, J.P., Nielsen, C.R., "Fast mixing by lamination." *Proc. IEEE Micro Electro Mechanical Systems,* 441-446, (San Diego, Ca, 1996)

[12] Ayon, A.A., Braff, R., Lin, C.C., Sawin, H.H., Schmidt, M.A., "Characterization of a Time Multiplexed Inductively Coupled Plasma Etcher," *Journal of the Electrochemical Society, **146**, 339-349, (1999)*

[13] Mizra, A.R., Ayon, A.A., "Silicon Wafer Bonding: Key to MEMS High-Volume Manufacturing," Sensors, (Dec. 1998)

[14] Schmidt, M.A., "Wafer to Wafer Bonding for Microstructure Formation," *Proceedings of the IEEE*, **86**, 1575-1585, (1998)

[15] Holman, J.P., *Heat Transfer, 7th ed.,* New York: McGraw-Hill, Inc., 543 (1990)

[16] Scampavia L.D., Balnkenstein G., Ruzicka J., Christain G.D., "A Coaxial Jet Mixer for Rapid Kinetic Analysis in Flow Injection and Flow Injection Cytometry." *Anal. Chem.*, **67**, 2743-2749, (1995)

Experiences with the use of microreactors in organic synthesis

H. Krummradt, U. Koop, J. Stoldt, Merck KGaA, Abt. ZVE-I, 64271 Darmstadt, Germany

1. General

For many chemical reactions the temperature, the concentrations and the residence time should be kept in a narrow range in order to achieve the best results concerning yield and selectivity.
Meanwhile it is well known, that microreaction technology offers several advantages to meet these prerequisites. The characteristics of microreactors comprise beside others

- surface to volume ratios much higher than in conventional reactors,
- very high heat transfer coefficients and
- a reasonable variety of possible materials.

From these characteristics some technical advantages can be derived such as

- improved mass- and heat transfer,
- improved control of residence time distribution,
- inherently safe operation due to small hold up,
- possibility of higher selectivity, yield and quality and
- "numbering up" instead of "scaling up" by multiplying microsystems (parallel operation).

On top of this there are some economical advantages such as

- "on-demand" & "at point of use" production (flexible output, no transport and storage),
- rapid development of new processes or new substances ("time to market"),
- chance of less investment costs and
- minimal environmental impact.

2. Example for use of micro- and minireactors in process development and production

In central process development of Merck KGaA we searched for processes, where the above mentioned facts would lead to a significant benefit, if microreaction technology will be applied. In order to find a suitable candidate the reaction should be

- highly exothermic or endothermic, respectively,
- sensitive to temperature,
- fast and

- difficult to handle in a conventional stirred vessel.

After some discussion we found a reaction according to the scheme indicated in Figure 1. It is a step in an established multi-stage production process that leads to a fine chemical as final product.

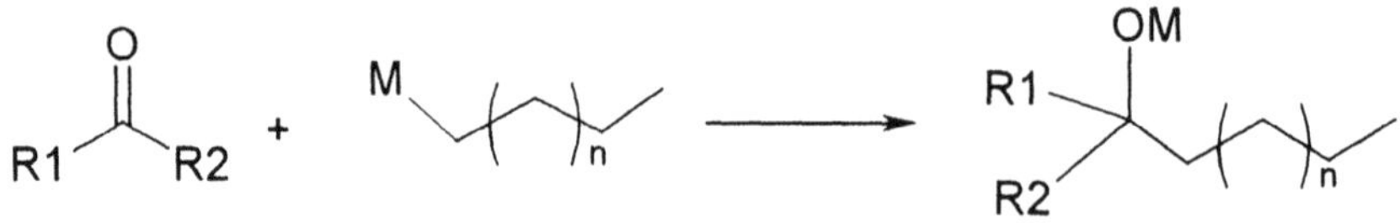

Figure 1: A carbonyl compound and a metallo-organic reagent form a product

A carbonyl compound and a metallo-organic reagent form a product.
The reaction is characterised by the following facts. It is

- a liquid phase reaction,
- exothermic with $\Delta_R H^0$ of ca. -300 kJ/mol,
- sensitive to temperature,
- parallel- and consecutive reactions can occur,
- the main reaction and most side reactions are comparatively fast (residence time < 10 s).

We built up a lab scale reaction system with three pumps according to figure 2. Two pumps were used to feed the two reactants into the microreactor, one pump was used for the solvent that rinses the system. The microreactor was the array of ten mixing devices of IMM [1] with 2 x 16 channels of w x h = 40 x 220 µm each. The material was silver. The micromixer/microreactor was immersed into a thermostatic bath.

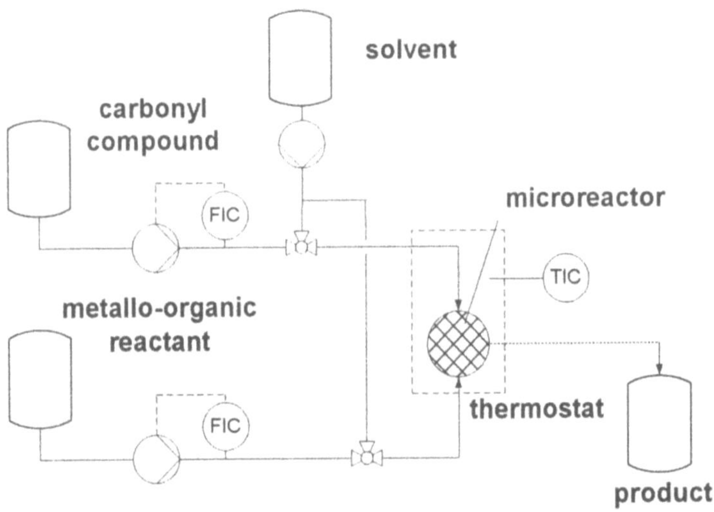

Figure 2: The equipment for lab-scale trials with the microreactor

The experiments with the microreactor followed a statistical design. In this design the parameters as temperature, ratio of metallo-organic reagent and carbonyl component and residence time were varied.

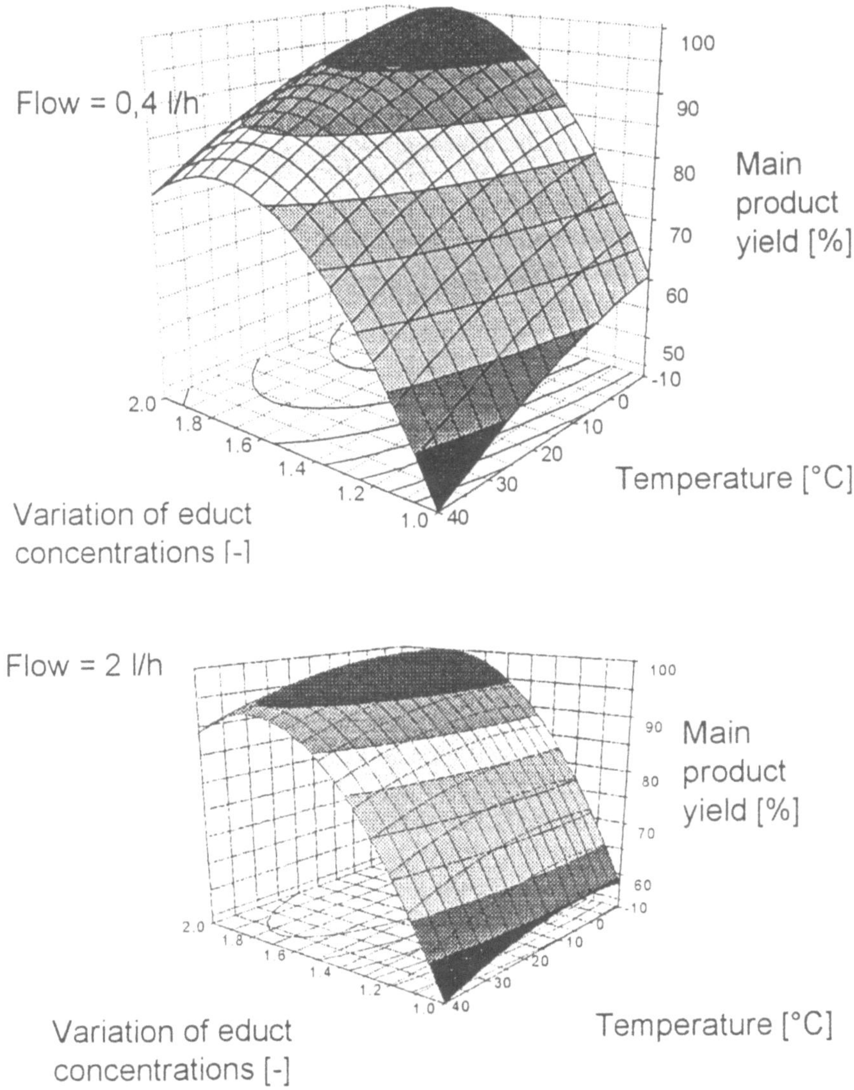

Figure 3: Results of trials according to the statistical design of experiments.

The graphic shows the dependence of the yield from temperature and the variation of the educt concentrations for two different flows. The curve is similar for both flows, but for the higher flow yields above 90% can be reached, even at higher temperatures. This correlates well with the hydrodynamic behaviour of the used IMM-micromixer array. According to this, a certain flow is necessary for optimum distribution of the feed streams to the ten chambers of the mixer array. With the statistical design of experiments we were able to determine a parameter set very quickly, that led to a yield of 95%. The best result in a conventional 0,5l stirred flask, where one component was in the flask and the second dropped into it, was a yield of 88% at the significantly lower temperature of –40° C. In the 6000 l production vessel however the yield dropped to 72%. This is due to technical limitations of the equipment. The cooling system used allowed only a temperature of –20° C. To stay at this temperature, the second component has to be dropped into the vessel over 5 hours giving enough time for side and consecutive reactions to take place. Because of the worse surface to volume ratio of the 6000l vessel of around 4 m^2/m^3 we cannot establish isothermal conditions. To the contrary we have a temperature distribution via the vessel, which results in the formation of unwanted components too . For comparison the surface to volume ratio of the lab flask was about 80 m^2/m^3, for the microreactor about 10.000 m^2/m^3.
Eventually the microreactor was blocked due to deposition of solids. This fact is not a problem in the lab because the microreactor could be cleaned easily and quickly before the following experiment. By doing this we

- carried out the statistical design quickly,
- needed only small amounts of chemicals,
- and were able to use a continuous process in the course of the lab scale investigation.

Of course the blocking is not acceptable for a stable production process. Therefore a new mixing element with wider channels was designed, that we called "minireactor". It was small enough to keep the beneficial features discussed before, and big enough to omit the system from blocking. It had a surface to volume ratio of around 4000 m^2/m^3. We used the minireactor in pilot-scale experiments with a more sophisticated set-up, where we reinvestigated the favourable conditions from the statistical design and practised for the production. An acceptable yield of 92% was obtained under the same conditions as before. With the experience of the pilot-scale trials we designed the production unit following the concept of numbering-up instead of scaling-up. This implicated the automation of the entire process and the by no means trivial splitting of the feed streams to the individual minireactors As indicated in table 1 we were able to maintain the yield of 92% as expected. The capacity of one minireactor is about $3.10^{-5} m^3/s$. As the preceding step is batch, the process with the minireactors which itself is a continuous process, works on intervals.

Reactor type	Temperature	Residence time	Yield
Flask, 0,5 Liter	-40°C	0,5 h	88 %
Production (stirred vessel, 6m³)	-20°C	5 h	72 %
Microreactor (lab)	-10°C	< 10 s	95 %
Minireactor (pilot-scale)	-10°C	< 10 s	92 %
5 Minireactors (production)	-10°C	< 10 s	92 %

Table 1: Results for various reactor types and selected parameters

In figure 4 the dosage stations for carbonyl component and organometallic reagent are seen.
The development can be summed up in the following way. A difficult organometallic reaction could be improved significantly by means of microreaction technology. A batch reaction was successfully transferred into a continuous process in a short period of time. The results from lab led to the development of a production plant with 5 parallel minireactors that is operating since August 1998.

Figure 4: Dosage station of the carbonyl compound and of the metallo-organic reactant

[1] W. Ehrfeld et al., Fabrication of Components and Systems for Chemical and Biological Microreactor, in W. Ehrfeld (Ed.), Microreation Technology, Proceedings of the First International Conference on Microreation Technology, 72-90, Springer, 1998.

A Microstructured Reactor for the Catalytic Partial Oxidation of Methane to Syngas

J. Mayer, M. Fichtner, D. Wolf* and K. Schubert
Forschungszentrum Karlsruhe,
Hauptabteilung Versuchstechnik, Postfach 3640, D-76021 Karlsruhe
*Institut für Angewandte Chemie Berlin-Adlershof, D-12484 Berlin

Abstract

The partial oxidation of methane to syngas as a fast, slightly exothermic reaction has been chosen as example for studying the properties and applicability of microstructure reactors. In conventional fixed-bed reactors problems may be caused by the large heat of reaction which occurs mainly at the reactor inlet. There, the high temperature may cause a fast degradation of ceramic-based catalysts. By the use of microstructured honeycomb-type catalysts made of rhodium, inflammable methane/oxygen mixtures could safely be converted into syngas even at higher operating pressures. A microstructured honeycomb catalyst with channel cross sections of about 120 x 130 μm^2 and channel lengths of 5 mm was investigated with regard to its suitability for and operating behavior during partial methane oxidation. At residence times of a few milliseconds and varying operating pressures, methane and oxygen conversion rates and the product distribution were determined experimentally. Furthermore, aspects of structuring rhodium foils and their assembly shall be discussed.

Keywords: partial oxidation, methane, syngas, hydrogen, rhodium catalyst, microreactor

1 Introduction

As an alternative to steam reforming, partial oxidation of methane to syngas is gaining importance in industrial research. Advantages of partial methane oxidation are the higher energy efficiency and less cost-intensive facilities compared to other processes [1]. Syngas production by heterogeneously catalyzed partial oxidation of methane has been investigated by several working groups [2-5], so that this reaction should be a good example for studying the properties and applicability of catalytic microstructure reactors with a fast, slightly exothermic high-temperature process.

The partial oxidation of methane premixed with oxygen at a ratio of 2:1 produces syngas which consists of hydrogen and carbon monoxide.

$$CH_4 + \frac{1}{2} O_2 \rightleftarrows CO + 2 H_2 \qquad \Delta_r H_{298} = -35.7 \text{ kJ/mol}$$

With a heat of reaction of -22 kJ/mol at 1000 °C an autothermal process should be possible. Depending on the catalyst system, contact times of a few up to several hundred milliseconds can be found in literature. In addition to syngas, carbon dioxide, water and hydrocarbons such as ethane, ethene etc. are obtained as by-products. The formation of this product distribution is explained by exothermic oxidation and endothermic reforming reaction steps along the catalyst bed.

combustion step	$CH_4 + 2\,O_2 \longrightarrow CO_2 + 2\,H_2O$	$\Delta_r H_{298} = -802.3$ kJ/mol
reforming steps	$CH_4 + H_2O \rightleftarrows CO + 3\,H_2$	$\Delta_r H_{298} = 206.3$ kJ/mol
	$CH_4 + CO_2 \rightleftarrows 2\,CO + 2\,H_2$	$\Delta_r H_{298} = 247.4$ kJ/mol

In conventional systems the catalyst usually consists of a noble metal-loaded ceramic support of low thermal conductivity. Extreme hot-spot temperatures have been expected for elevated operating pressures. For example, for a pressure of 2.5 MPa, a hot-spot temperature up to 2600 K has been predicted [4]. Under these conditions, a fast degradation of the catalyst is likely to occur. However, a metallic microstructured catalyst should help to avoid this problem. Due to the high thermal conductivity of the metal structure, heat transfer is supported in the direction of the flow axis, where the endothermic reaction steps take place.

To realize a reactor with high axial heat dissipation and mass transfer rates and to learn about principal facts related with heterogeneously catalyzed chemical reactions in microchannels, a catalytic microstructure reactor has been developed and tested at the Karlsruhe Research Center (FZK) in cooperation with the Institute of Applied Chemistry, Berlin-Adlershof (ACA).

2 Experimental

2.1 Experimental set-up

The experiments were performed in an experimental set-up which was constructed for total feed flows between 1 and 10 l/min (STP) with a CH_4/O_2 ratio of 2:1. Pressures in the range of 0.1 and 2.5 MPa can be set by a pneumatically controlled valve. The test rig for the investigation of microstructured rhodium catalyst honeycombs is equipped with a micromixer [6] for the mixing of methane and oxygen. For preheating of the educt gas mixture an electrically heated micro heat exchanger is applied [7]. These devices also serve as flame traps.

To minimize heat losses, the microstructured rhodium catalyst honeycomb is installed in a refractory holder as depicted in Figure 2.1. This catalyst holder is made of a shrinkproof ceramic material that can be processed in the green state and has been developed at the FZK [8]. The catalyst holder is provided with a helical groove for accommodation of the heating wire. With this heater of about 150 W electrical power, ignition temperature of the catalyst can be reached. Furthermore isothermal operation is allowed by additional heating to investigate the operation behavior at catalyst temperatures of up to 1200 °C. The catalyst with the ceramic holder is installed gastight into a pressure vessel by a screw connection. For pyrometric measurement of the catalyst surface temperatures, inspection glass flanges (quartz glass) are fixed to both ends of the pressure vessel, which is externally heated to 300 °C. A double-pipe heat exchanger, installed directly at the outlet of the pressure vessel, serves as a quencher for cooling down the gas to 140 °C.

Figure 2.1: Photography of the ceramic holder for the rhodium catalyst honeycomb consisting of individual microstructured foils and a cover plate welded to the face of the honeycomb.

Plant control and data acquisition is done by means of the measurement and control software LABVIEW®. For inline chemical analysis, a quadrupole mass spectrometer is employed. This allows measurement of all species involved in the reaction. Sampling is accomplished using a heated quartz capillary of 1 m length. The signal intensities are then evaluated by means of a multivariate calibration model for the separation of the superposed CO, CO_2 and C_2H_6 signal contributions from the intensity at mass 28. The following signals were used for the evaluation:

species	H_2	CH_4	H_2O	C_2H_6	CO,etc.	O_2	Ar	CO_2
m/e	2	15	18	26	28	32	40	44

The superposition of the methane signal at the mass 15 by the CH_3^+ fragment of ethane could be neglected. 10 vol.% argon were added to the reaction gas mixtures as internal standard for evaluation of the mass spectra and calculation of the mass balance.

2.2 Fabrication of the microstructured rhodium catalyst

For the development of the honeycomb catalyst mechanical micromachining and electric discharge machining by wire erosion were applied for microstructuring hard-rolled and thermally treated rhodium foils of 220 μm thickness. Two light microscopies of wire-eroded and micromachined rhodium foils are represented in Figure 2.2.

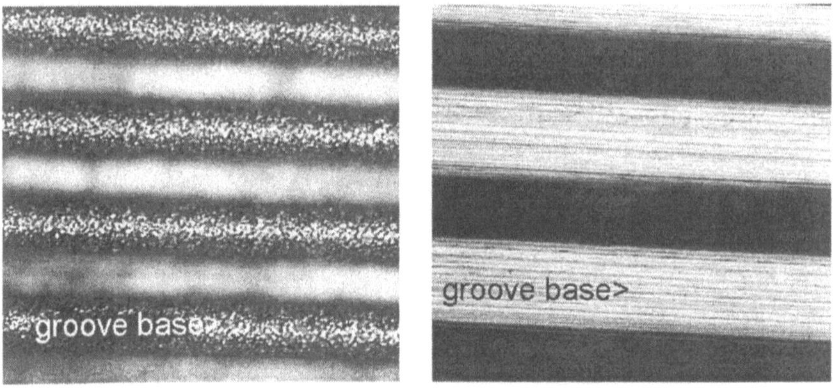

Figure 2.2: Light microscopies of microstructured rhodium foils, focussed on the groove base. Left: Wire-eroded foil with 100 µm wide grooves and 100 µm thick walls. Right: Micromachined foil with 200 µm wide grooves and 100 µm thick walls.

Generally, both structurization techniques can be applied to rhodium foils. By mechanical micromachining groove widths of 200 µm in hard-rolled foils have been realized. Groove widths of 60 µm have been achieved by wire erosion.

Depending on the method, the structures exhibit varying surface properties. Mechanical micromachining yields e. g. rectangular groove cross sections and comparably smooth surfaces. In contrast to this, wire-eroded microstructures have a U-shaped cross section with a rough surface. Wire erosion was applied for the fabrication of the rhodium catalyst honeycomb described in this paper.

The catalyst consisted of 23 microstructured rhodium foils, each having 28 microchannels arranged in parallel. The cross-sectional dimensions were about 120 µm wide and 130 µm deep, separated by 80 µm thick walls. The channel lengths were 5 mm.

To integrate the microstructured rhodium foils in the ceramic holder with a high fitting accuracy, the foils are welded to a compact catalyst honeycomb, as shown in Figure 2.1. A cover plate welded to the honeycomb face ensures tight fitting in the ceramic holder and prevents gas from passing by the honeycomb. Due to the high melting point of rhodium of 1966 °C and the external dimensions of the honeycomb of a few millimeters, electron-beam and laser welding were applied. The catalyst honeycomb obtained by electron beam (EB) welding of the individual microstructured foils and the ceramic holder is represented in Figure 2.1. For EB and laser welding the foils had to be provided with an unstructured margin of 1 mm. Five unstructured foils each form the unstructured bottom and top of the foil stack.

As another assembly technique, diffusion welding was studied for its applicability to rhodium foils. According to metallurgical examinations, the weld is dense and has a minor microporosity. The material is coarse-grained. As can be seen from Figure 2.3, the large-angled grain boundaries have grown beyond the welded surfaces.

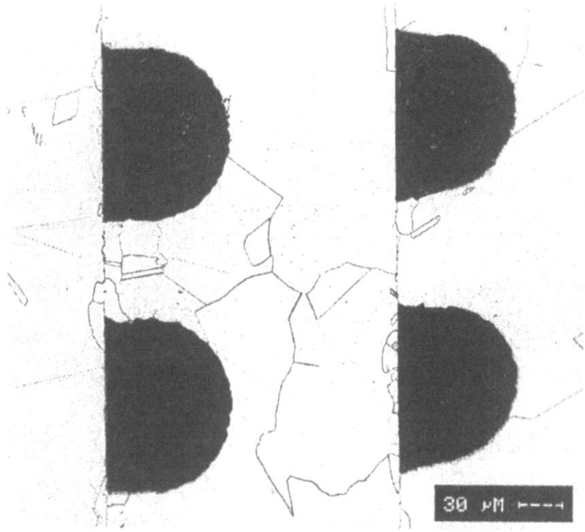

Figure 2.3: Cut section of a diffusion-welded stack of rhodium foils (the specimen was subjected to ion etching)

2.3 Experimental procedure and calculations

Prior to the reactor ignition, the operating pressure was adjusted with argon. Subsequently the signal background of the mass spectrometer was determined. The composition of the reaction gas mixture consisting of 60 vol.%. methane, 30 vol.% oxygen and 10 vol.% argon was controlled via a bypass line. To start the reaction, the temperature of the catalyst was increased in a stepwise manner. Under ignited conditions the heater power was subjected to a stepwise reduction when reaching a heating temperature of 1000 °C. Then the product yields were determined. If the reaction did not reach autothermal conditions it was stopped at 800 °C to prevent the formation of soot.

The volume changes as a result of partial methane oxidation. For the calculation of the methane and oxygen conversion rates, the product volume flow $\dot{V}$ is therefore calculated from the decrease in the partial pressure of argon $p(Ar)$.

(1) $$\dot{V} = \dot{V}_0 \cdot \frac{p(Ar)_0}{p(Ar)}$$

$\dot{V}_0$ is the total flow of educt gas into the reactor with the initial argon partial pressure $p(Ar)_0$.This allowed to obtain the element balance with a mean relative error of 3%. Due to a measurement error of up to 0.5 vol.% for several species, the mass balance could not be totally closed, especially at small methane conversion rates. The conversion rates *X(i)* and selectivities *S(i)* are calculated as follows:

(2)

$$X(i) = 100 \cdot \frac{\dot{V}(i)_0 - \dot{V}(i)}{\dot{V}(i)_0},$$

where $i = CH_4, O_2$

(3)

$$S(H_2) = \frac{100 \cdot p(H_2)}{p(H_2) + p(H_2O) + 3p(C_2H_6)}$$

(4)

$$S(CO) = \frac{100 \cdot p(CO)}{p(CO) + p(CO_2) + 2p(C_2H_6)}$$

(5)

$$S(H_2O) = \frac{100 \cdot p(H_2O)}{p(H_2) + p(H_2O) + 3p(C_2H_6)}$$

(6)

$$S(CO_2) = \frac{100 \cdot p(CO_2)}{p(CO) + p(CO_2) + 2p(C_2H_6)}$$

(7)

$$S(C_2H_6) = \frac{100 \cdot 3 \cdot p(C_2H_6)}{p(H_2) + p(H_2O) + 3p(C_2H_6)}$$

3 Results and discussion

The experiments with the freshly manufactured rhodium honeycomb showed an increasing activity for methane and oxygen conversion with each experiment. After a formation time of approximately 20 hours, no further activity changes were observed and a maximum conversion of about 62% for methane and 98% for oxygen was found.

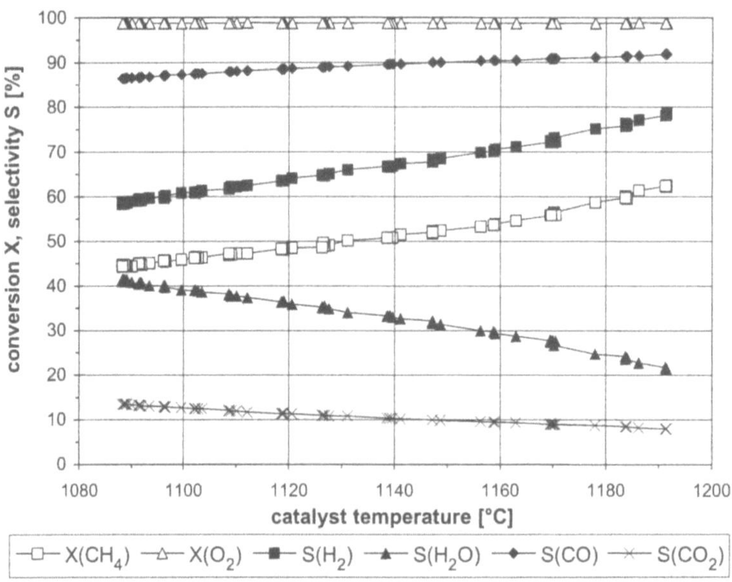

Figure 3.1: Conversions and product selectivities for partial oxidation of methane at a pressure of 0.1 MPa and a flow rate of 2 l/min (STP) as a function of the pyrometrically measured catalyst outlet temperature. The residence time was 1.4 ms (STP).

At atmospheric pressure the reaction started at surface temperatures of 650 °C up to 750 °C, depending on feed flow. This is explained by cooling caused by the relatively small gas inlet temperature into the microstructured catalyst of 300 °C.

Before ignition methane and oxygen conversion rates of less than 10% were observed. At these temperatures mainly steam ($S(H_2O) \approx 80\%$) and carbon dioxide ($S(CO_2) \approx 60$-70%) are detected. After ignition, the catalyst temperature rises to more than 1000 °C within one minute. As depicted in Figure 3.1, carbon monoxide and hydrogen were formed as main products under ignited conditions. The measured maximum selectivity for hydrogen was 78% and for carbon monoxide 92%. Within the experimental tolerance oxygen was totally converted. An increasing methane conversion was found when temperature was increased. Simultaneously, the hydrogen and carbon monoxide selectivities were increasing. The thermodynamic equilibrium was not reached (see Table 3.1). Probably, the residence times in the order of 1 ms at the respective operating pressure and 900 °C were too short. The gas passes the catalyst too rapidly for significant reforming processes to take place after oxidation. This explains the generally high water and carbon dioxide selectivity. On the other hand, a distortion of the gas composition by homogeneous reactions has to be taken into consideration before quenching. In the present set-up the gas needs more than 50 times the duration for reaching the quench compared to the residence time in the honeycomb.

Table 3.1: Comparison between conversions *X* and product selectivities *S* at different operating pressures and feed flows with the calculated equilibrium data[#].

Feed [l/min], STP	$X(CH_4)$ [%]	$X(O_2)$ [%]	$S(H_2)$ [%]	$S(CO)$ [%]	$S(H_2O)$ [%]	$S(CO_2)$ [%]
p = 0.1 MPa, T(catalyst) = 1150 °C						
Equil.[#]	99.3	100	99.7	99.8	0.3	0.16
1	56.7	98.6	73.7	89.2	26.2	10.7
2	52.1	98.8	68.4	90.0	31.6	10.0
3	48.0	98.6	70.0	90.8	30.0	9.1
p = 0.6 MPa, T(catalyst) = 1200 °C						
Equil.[#]	98.5	100	99.4	99.7	0.6	0.3
2	48.0	98.8	66.0	85.5	33.9	14.3
3	45.5	98.8	64.0	87.0	35.8	12.7
6	53.8	98.2	64.2	89.6	35.8	10.4
p = 1.2 MPa, T(catalyst) = 1200 °C						
equil.[#]	97.1	100	98.8	99.5	1.2	0.5
3	44.8	98.4	58.5	71.6	37.0	19.5.4

When the residence time was reduced by increasing the feed gas flow at a constant pressure no significant influence on methane conversion and product selectivities was found. According to this result, a mass transfer limitation of the heterogeneously catalyzed reactions is likely. For a more detailed investigation of mass transfer limitation additional microstructured honeycombs with appropriate channel dimensions are in preparation.

At feed flows above 3 l/min (STP) the heat of reaction was enough for reaching an autothermal state of the microreactor.

With a constant feed flow, the rhodium honeycomb was examined in further experiments at different operating pressures up to 1.2 MPa. The influence of the pressure on conversions and selectivities is shown in Figure 3.2. With increased pressures methane conversion, hydrogen and carbon monoxide selectivity are decreasing. For pressures up to 0.6 MPa, no soot was found in the product filter after the quench. Hence, gaseous products are assumed exclusively for calculating the mass balance. The formation of ethane and ethene was observed up to a selectivity of 3% at a pressure of 1.2 MPa. Probably as a result of the Boudouard equilibrium, the formation of soot between catalyst outlet and quench prevents the application of the present laboratory set-up at higher pressures. Without the measurement of the catalyst temperature by a pyrometer it is possible to place the quench immediately downstream of the catalyst. Pressures above 1.2 MPa should be achievable with this improvement.

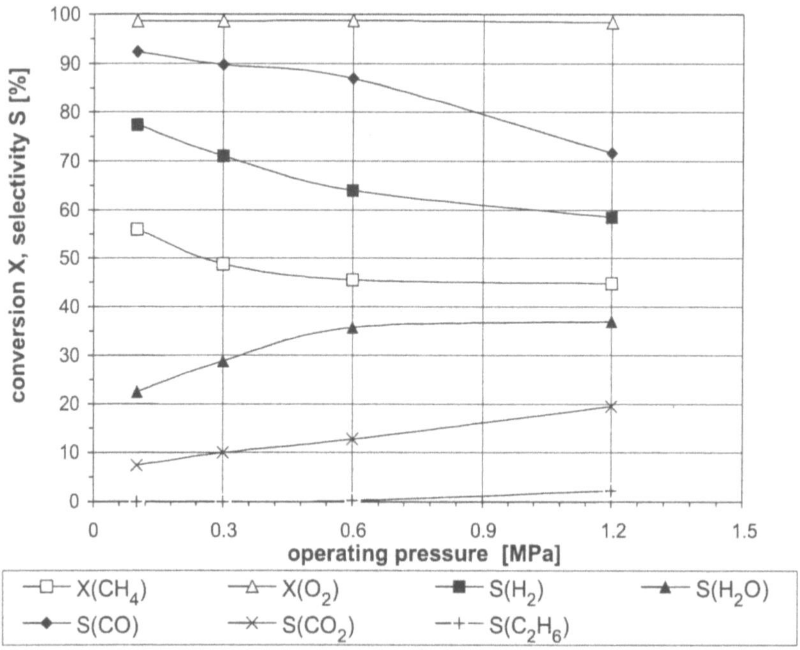

Figure 3.2: Conversions and product selectivities for partial oxidation of methane at different operating pressures and a constant feed flow of 3 l/min (STP) and a catalyst outlet temperature of 1200 °C.

Calculations in a computer model where fluid dynamics, chemical kinetics as well as mass and heat transfer are coupled, have shown the possibility to control the heat of reaction. An overheating of the catalyst due to the high partial pressure of oxygen at elevated operating pressures and the deposition of carbon at low oxygen pressures should be prevented by a distributed and continuous addition of methane. This reactor concept was studied by computational fluid dynamics (CFD) simulations considering a porous plate or a nozzle for methane addition. The simulation results showed that an undesired back diffusion of products (CO and H_2) through the porous plate into the methane feed has to be expected. Due to jet formation such back diffusion should be avoided using nozzles.

4 Summary

A microstructured honeycomb made of rhodium for partial oxidation of methane has been developed and tested. Fabrication techniques for the honeycomb have been discussed. Both electric discharge machining and mechanical micromachining can be used for structuring the rhodium foils. For welding the foil stack to a compact honeycomb, electron beam, laser or diffusion welding have been applied. In the laboratory set-up, a micromixer and a micro heat exchanger for preheating of the gas were used. To keep the required temperature of 800 – 1200 °C, the rhodium catalyst was inserted into a heatable ceramic insulation which was fixed in a suitable pressure vessel. The catalyst performance was examined at residence times of 0.2 – 2.5 milliseconds and pressures up to 1.2 MPa. Depending on feed flow an autothermal state is achieved. Catalyst temperatures measured by a pyrometer were observed to be in the range of 1000 – 1250 °C. A non-equilibrium product distribution for the partial oxidation of methane was observed under these conditions. The experimental results obtained so far demonstrate the usability of the microstructured rhodium catalyst for syngas production at increased operating pressures. A mass transfer limitation of heterogeneously catalyzed partial oxidation in the microchannels is likely.

Acknowledgements

We thank Mr. H. Schlenker for his valuable contributions to the construction of the experimental set-up and the manufacture of the microreactors.
Partial funding of this work by the German Ministry of Education, Science, Research and Technology (BMBF) within the joint project of "Mikreak" is gratefully acknowledged.

References

[1] S. S. Bharadwaj, L. D. Schmidt. Catalytic partial oxidation of natural gas to syngas. *Fuel Processing Technology*, 42(1995)109-127
[2] D. A. Hickman, L. D. Schmidt. Production of Syngas by Direct Catalytic Oxidation of Methane. *Science*, 259(1993)343-346
[3] J. Frauhammer, G. Veser. Elementarkinetische Modellierung der katalytischen Oxidation von Methan zu Synthesegas in einem Monolithreaktor. *Chem. Ing. Tech.* 70(1998)1020-1027

[4] D. Wolf, M. Höhenberger, M. Baerns. External Mass and Heat Transfer Limitations of the Partial Oxidation of Methane over a Pt/MgO Catalyst. *Ind. Eng. Chem. Res.* 36(1997)3345-3353

[5] A. Jess. Synthesegaserzeugung durch katalytische partielle Oxidation von Methan mit Luft. *Dissertation, Universität Karlsruhe*, 1991

[6] K. Schubert, W. Bier, J. Brandner, M. Fichtner, C. Franz, G. Linder. Realization and Testing of Microstructure Reactor, Micro Heat Exchangers and Micromixers for Industrial Applications in Chemical Engineering. 2rd International Conference on Microreaction Technology, March 09-12, 1998 New Orleans, Louisiana, pages 88-95

[7] J. Brandner, M. Fichtner, K. Schubert. Electrically Heated Microstructure Heat Exchangers and Reactors. 3rd International Conference on Microreaction Technology, April 18-21, 1999 Frankfurt, Germany

[8] V. D. Hennige, H.-J. Ritzhaupt-Kleissl, J. H. Haußelt. Verfahren zur Herstellung von schrumpfungsfreier $ZrSiO_4$-Keramiken. *Keramische Zeitschrift.* 50(1998)262-265

Expansion of Microreactor Capabilities through Improved Thermal Management and Catalyst Deposition

Aleksander J. Franz, Sameer K. Ajmera, S. L. Firebaugh‡, Klavs F. Jensen, and Martin A. Schmidt‡

Dept. of Chemical Engineering, ‡Microsystems Technology Laboratories
Massachusetts Institute of Technology, Cambridge, MA 02139, USA

Abstract. Capabilities of a gas phase membrane-based microreactor have been expanded, enabling new chemistries and applications. The expanded capabilities include improved control of temperature and selectivity for highly exothermic partial oxidation reactions, a larger thermal and pressure operating window, and the development of a catalyst library.

Keywords. Microreactors, partial oxidation, catalyst deposition, kinetics.

1 Introduction

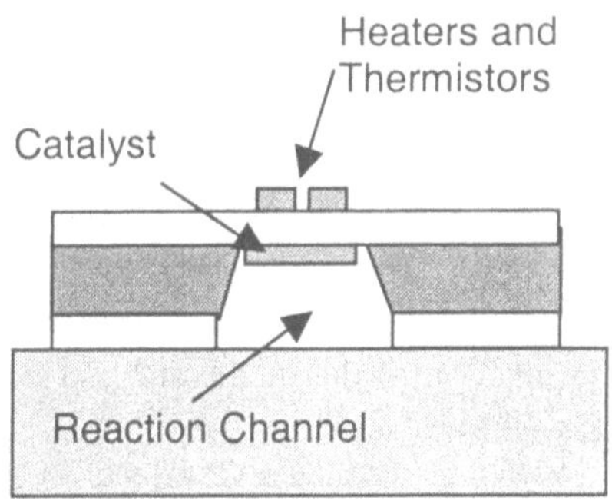

Fig. 1. MIT Gas phase catalytic microreactor

Microreactors offer potential advantages in the areas of safety, control, and time of scale-up [1]. The MIT gas phase microreactors, shown in Fig. 1, consist of a channel etched in silicon, capped by a thin membrane coated with catalyst. Patterned temperature sensors, heaters, and flow sensors are integrated into the design on the topside of the membrane. The fabrication and testing of a prototype microreactor for partial oxidation reactions has been previously described [2,3]. Safe ignition of several highly explosive mixtures in the flammable regime has been demonstrated in this microreactor. Here, we report progress in expanding the number of catalysts usable with this type of microreactor design, in controlling the catalyst temperature for highly exothermic reactions, and in expanding the operating range of the microreactors.

2 Catalyst Deposition

Catalyst deposition in microreactors often presents a unique set of challenges. In the original microfabrication scheme, catalysts were deposited in the microreactor using physical vapor deposition (PVD). The catalyst was sputtered onto the capping membrane either in a uniform coat, or in areas defined using a shadow mask. This approach to catalyst deposition had the advantages of control over deposition area and utilization of standard microfabrication technology. However, only a very limited number of catalysts could be deposited using PVD, and PVD catalyst films are smooth, resulting in low catalyst surface areas.

To increase the numbers of utilizable catalysts and catalyst surface area, we have adapted wet catalyst deposition techniques to the constraints imposed by the microreactor design. Wet catalyst deposition methods involve exposing the catalyst support to an aqueous or an organic solution of a metal salt. As the solvent evaporates, the catalyst precursor is deposited on the support. The precursor can then be further treated through reduction and calcination steps to yield a high surface area catalyst.

In the case of the MIT gas phase reactor, the membrane acts as the catalyst support. However, we have discovered that simple wetting of the microreactor channel with the metal salt solution does not yield satisfactory results. As shown in Fig. 2, surface tension causes the catalyst to coat the channel corners, resulting in very limited catalyst deposition in the desired area in the center of the membrane. Addition of surfactant to the solution improves the deposition process, but majority of the catalyst continues to deposit near the junction of the membrane and the channel wall.

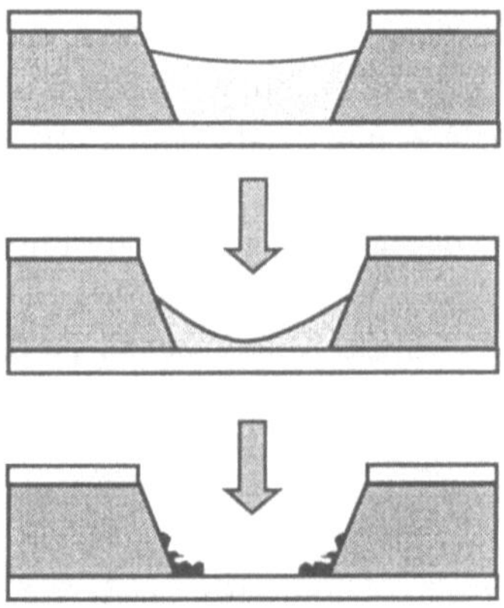

Fig. 2. Catalyst deposition in microreactor using a metal salt solution: surface tension causes

To overcome the negative effects of surface tension, we have developed a method for deposition of high surface area catalysts using aerosols of metal salt solutions. The aerosol is applied to the reactor membrane, which can either be heated or at room temperature during the deposition process. As long as the aerosol drops do not coalesce to form a meniscus across the channel, the solution evaporates and deposits the catalyst precursor at the point of droplet impact. As shown in Fig. 3, this approach is very effective in depositing high surface area catalyst in the central region of the membrane. In addition, similar to the PVD approach, a shadow mask can be used to control the area of catalyst deposition. Using the aerosol approach we have successfully deposited platinum, silver, and rhodium catalysts.

Other catalyst preparation techniques have also been investigated. Platinum black was electroplated in the channel using a PVD blanket Pt seeding layer and an aqueous Pt salt solution. In addition, mixed oxide catalysts have been deposited on the reactor membrane from a slurry of precursors. Finally, two different approaches were used to produce activated carbon catalyst. One method

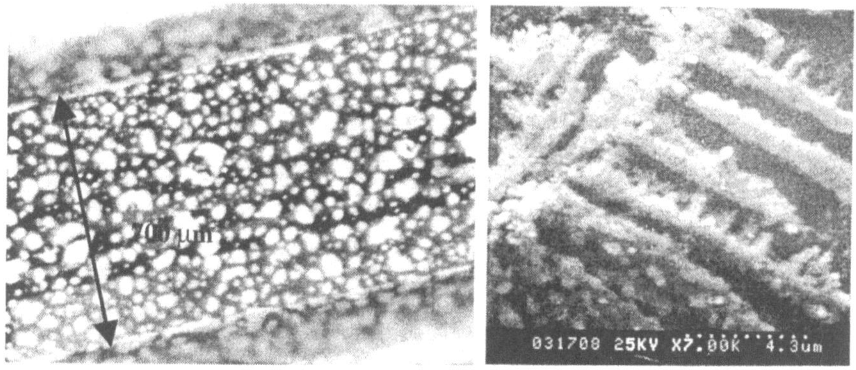

Fig. 3. Aerosol deposited platinum catalyst: reactor membrane with uniform catalyst coverage and a SEM close up of a catalyst particle showing high surface area.

involved polymer CVD, followed by carbonization. The second approach used a sooting flame to deposit the carbon. The increased ability to deposit catalysts opens up the doors to a variety of industrially relevant chemical reactions, previously inaccessible using physical vapor deposited noble metals.

3 Catalyst Temperature Control

The unique design of the MIT gas phase microreactor results in novel chemical reactor behavior, which presents new opportunities. As described previously, the catalytic membrane based design results in thermal isolation of the catalyst region from the neighboring silicon walls [2,4]. The thermal isolation combined with the low membrane thermal mass result in unusually fast reactor thermal time constant of about 10 ms, enabling excellent feedback temperature control [3]. The reactor design lends itself readily to simulations, enhancing the understanding of reactor performance.

One potential area of application for these microreactors is for safely carrying out fast, highly exothermic reactions. Oxidation and partial oxidation reactions of hydrogen, ammonia, carbon monoxide, methane, ethane, and ethylene have been successfully carried out in the flammable regime [2]. The ability to control the catalyst temperature, and therefore the selectivity, of partial oxidation reactions is the key factor determining reactor performance. We have been exploring the ability to control catalyst temperature for partial oxidation reactions through small design modifications to the current microreactor platform.

The thickness and thermal conductivity of the catalytic membrane, along with the heater design, determine the catalyst temperature profile. Fig. 4 shows the heat fluxes during microreactor operation. During an exothermic reaction, the heat is supplied to the membrane by both the heaters and the reaction. The primary heat loss mechanism is through conduction in the membrane to the

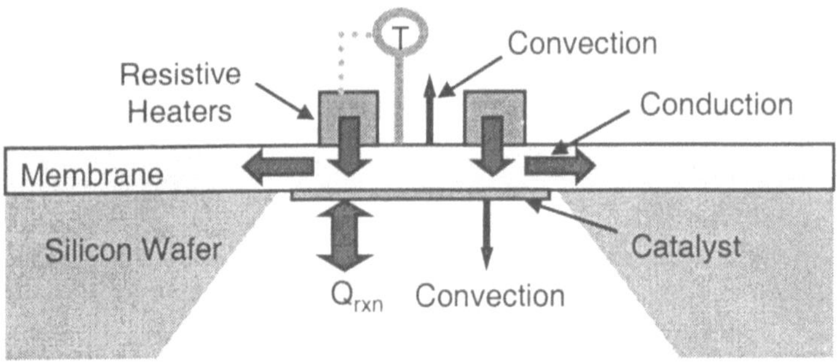

Fig. 4. Heat streams in the membrane based microreactor

reactor walls [4]. Therefore, by varying the thickness and thermal conductivity of the membrane, the ability of the reactor to dissipate the heat of the reaction can be controlled.

In the prototype microreactor with a silicon-rich SiN membrane (1 μm thick), fast, exothermic oxidation reactions exhibit typical ignition/extinction behavior. That is, when the heat of the reaction exceeds the ability of the membrane to dissipate it, the temperature increases rapidly to a new steady state. Under many conditions these reactors can be operated autothermally, that is no power needs to be supplied following ignition. This type of operation is non-ideal from the standpoint of catalyst temperature control, since the reactor cannot be operated at intermediate temperatures between the ignition point and the upper steady state. We have previously shown that if the reactor membrane is instead made out of 2.6

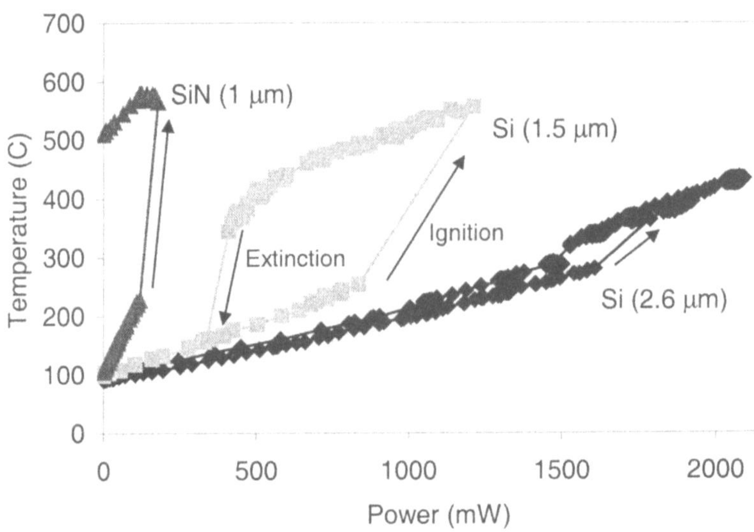

Fig. 5. Ignition/extinction curves for ammonia oxidation over platinum in different membrane based microreactors.

μm thick silicon, the ability of the reactor membrane to dissipate heat increases by about an order of magnitude [3].

The increased ability of the silicon membrane microreactors to dissipate heat has a dramatic effect on the reactor performance. In Fig. 5 we show the ignition/extinction curves for ammonia oxidation over aerosol deposited platinum in three different types of reactors. The silicon rich SiN membrane exhibits ignition, followed by autothermal operation. In contrast, for identical reactor feed conditions, silicon membrane reactors exhibit ignition/extinction behavior. The thicker silicon membrane reactor (2.6 μm) ignites at higher temperatures and reaches lower steady state temperatures than the 1.5 μm thick silicon membrane reactor. Increasing the ability of the membrane to dissipate heat expands the microreactor thermal operating range for highly exothermic reactions.

The ability to control the reactor temperature over a broader temperature range due to improved heat dissipation enables greater control over the reactor selectivity. In Fig. 6, the NO/N_2 selectivity for ammonia oxidation is shown for the three reactors in Fig. 5. The SiN membrane reactor operated autothermally at temperatures of about 550 °C; lower operating temperatures could not be accessed. The NO/N_2 selectivity has been shown to increase with temperature in this temperature range [5]. Therefore, the SiN reactor has a relatively high selectivity to NO. In contrast, the silicon membrane reactors are capable of carrying out the ammonia oxidation at much lower temperatures, resulting in

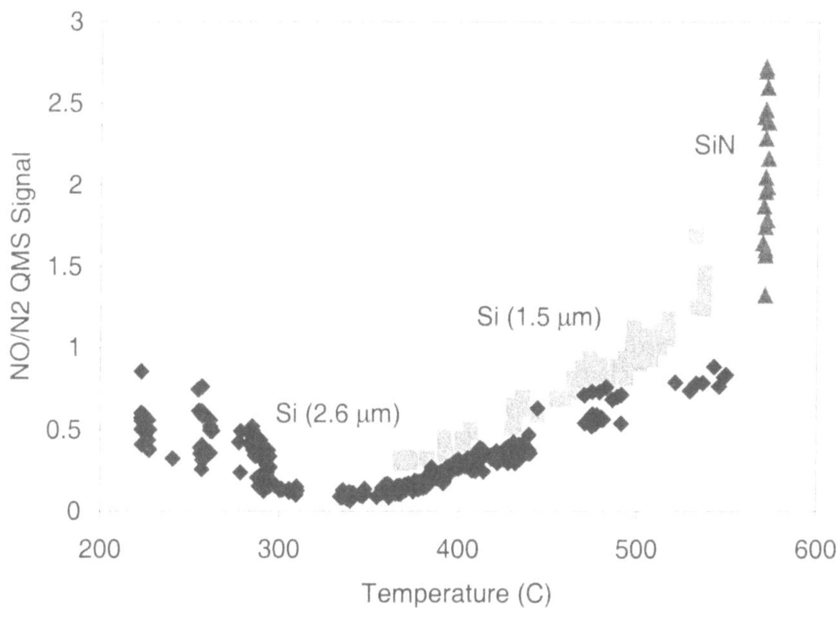

Fig. 6. Dependence of NO/N_2 selectivity for ammonia oxidation on reactor membrane characteristics

much lower NO selectivities. The thicker silicon membrane reactor expands the thermal operating window enough to apparently access a different kinetic operating region, where NO selectivity begins increasing with decreasing temperature.

Typical platinum wire experimental reactors exhibit steady state reaction temperatures no lower than 650 °C for comparable NH_3/O_2 ratios, but with air, rather than pure oxygen, fed into the reactor [5]. The nitrogen in the air stream acts as a diluent and a heat sink, lowering the minimum reactor operating temperature. In contrast, the microreactors use pure oxygen feeds and are capable of carrying out the oxidation reaction at temperatures of 300 °C or lower. Although in this case the NO selectivity decreases at lower operating temperatures, for most partial oxidation reactions it is desirable to avoid high operating temperatures, which lead to complete oxidation products. Therefore, the microreactors expand the range of operating conditions and selectivities for oxidation reactions, and could potentially outperform conventional scale reactors. In addition, the ability to control the temperature over an expanded range is useful for kinetic studies, which are used to determine rate laws for design of large-scale equipment. Increasing the temperature range over which the rate law is determined increases its accuracy and applicability.

In addition to improving catalyst temperature control, it is also desirable to maintain the catalyst at a uniform reaction temperature, both for optimal selectivity control, as well as for kinetic studies. The membrane based

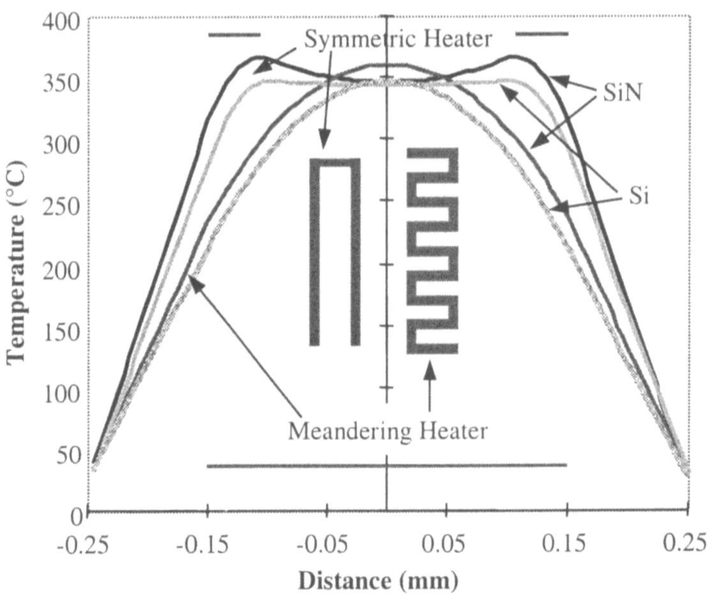

Fig. 7. Effect of heater design and membrane material on the catalyst temperature profile: FEM simulations.

microreactor design possesses great flexibility in terms of layout of heaters and temperature sensors. We have explored modifications to heater design to achieve thermal uniformity across the catalyst region.

The prototype microreactor employed a meandering heater design. It can be shown that this type of design results in a non-uniform temperature profile across the channel, with a maximum in the center and the temperature sharply dropping off towards the sides (Fig. 7). Since conduction is the primary mechanism of heat loss from the membrane, the heat delivered to the center of the membrane by the meandering heater must be conducted to the sides, causing a temperature gradient and a maximum in the center. The heater design can be modified to improve the thermal uniformity. In a symmetric heater design, the heat is delivered only at the edges of the catalyst region of the membrane, with no heat supplied in the middle of the membrane. For non-reactive flow, simulations and simple analysis show that this type of design results in a flatter temperature profile, with a temperature minimum in the membrane center. The minimum is caused by convective heat losses. Increasing the membrane thermal conductivity by using silicon instead of silicon nitride further improves the thermal uniformity for the symmetric heater design.

Microreactors with the symmetric heater design have been microfabricated and tested. The thermal uniformity of the design can be experimentally evaluated by placing temperature sensing resistors and different axial locations inside the heater loop. As shown in Fig. 8 for non-reactive flow, the symmetric heater results in good thermal uniformity for the SiN membrane reactor, and the silicon membrane further improves the thermal uniformity.

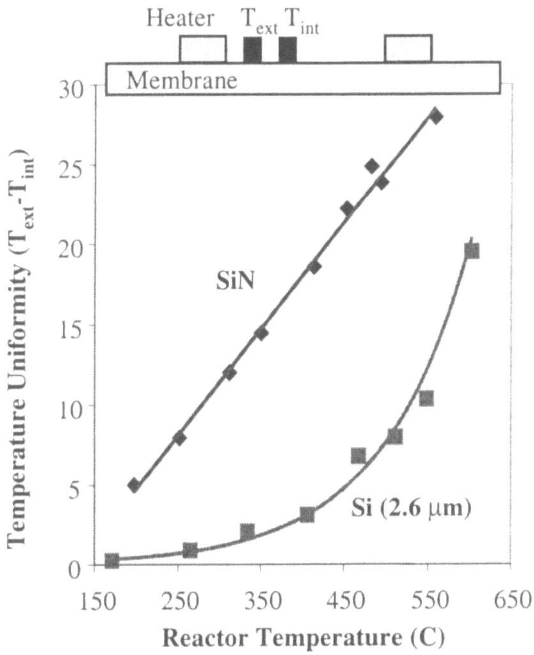

Fig. 8. Temperature uniformity across the catalyst region for the symmetric heater design in non-reactive flow.

The thermal uniformity over the catalyst region can be maintained even for highly exothermic reactions. Fig. 9 shows the temperature difference between the external and internal temperature sensors plotted against the natural log of the heater power, for ammonia oxidation carried out using the symmetric heater design. For the SiN membrane reactor, the temperature difference exhibits a clear ignition/extinction loop. The temperature difference between the two sensors changes sign as ignition takes place. The change in sign is due to the heat of the reaction. Prior to ignition, convective losses in the center cause the temperature in the middle of the reactor to be lower, as in Fig. 8. After ignition, the reaction adds heat to the center of the membrane, resulting in a temperature profile similar to that determined for the meandering heater design in Fig. 7, with a temperature maximum in the center. In the case of the silicon membrane the ignition/extinction loop is less pronounced. The addition of the heat of the reaction seems to offset the convective heat losses almost exactly, resulting in a virtually flat temperature profile following ignition, with a near zero temperature difference between the two temperature sensors. Uniform catalyst temperatures can therefore be achieved, even for highly exothermic oxidation reactions.

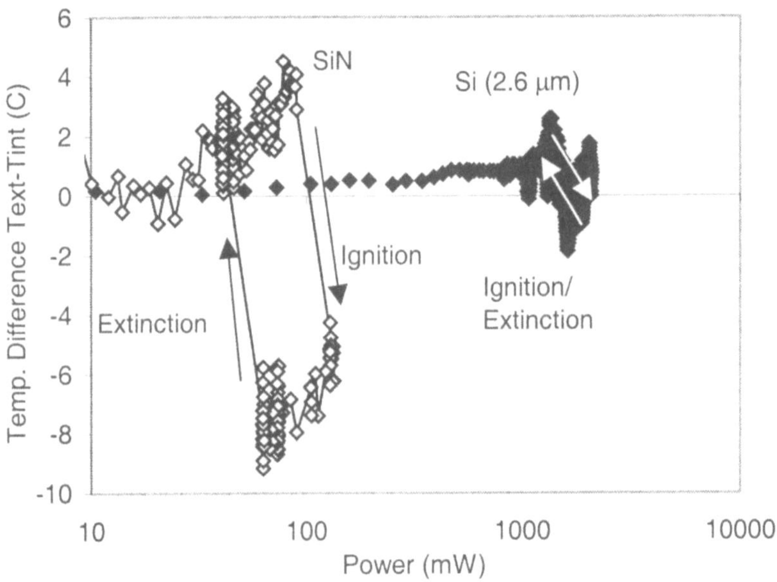

Fig. 9. Temperature uniformity across the catalyst region in reactive flow: ammonia oxidation over Pt.

As described above, understanding of the heat balance in the membrane based microreactor design can be used to modify the membrane and heater design to improve catalyst temperature control. Excellent temperature control over highly exothermic oxidation reactions has been demonstrated. The ability to extend the controllable temperature operating range for these types of reactions provides advantages in terms of controlling the reactor selectivity. Control of selectivity is further enhanced by maintaining a uniform reaction temperature over the catalyst region using a redesigned resistive heater. The catalyst temperature uniformity and expanded operating range suggest kinetic studies as one application for this type of micoreactor. The flexibility of heater design and membrane construction makes reaction specific reactor design feasible.

4 Expansion of Operating Conditions Range

Although the membrane based microreactor design for gas phase reactions has significant advantages in terms of catalyst temperature control, the very thin membranes (~1 μm) can be fragile. In addition, the silicon rich SiN membranes buckle at elevated catalyst temperatures, eventually leading to reactor failure. The buckling is due to the thermal isolation of the membrane, which expands as it heats up, relative to the room temperature reactor chip. Since the SiN membrane is initially at only a very low tensile stress, the expansion leads to a compressively stressed membrane and buckling. The previously reported mixing-T reactors had maximum operating pressures of about 0.5 atm, and ~650 °C.

The thermal and pressure operation windows of the gas phase microreactors were expanded by improving membrane robustness. First, the reactor geometry was modified to increase operating pressures. The T microreactor design was fabricated using a KOH etch, which resulted in a large unsupported membrane area at the T junction, as shown in Fig. 10. Two approaches were used redesign the reactor. First, the KOH fabrication approach was used to produce slot reactors with off chip gas mixing. The second design maintains on-chip mixing, but requires a deep reactive ion etch to define the channel geometry in silicon. These modifications in the membrane reactor geometry resulted in improving the pressure operating range from vacuum up to 2.7 atm.

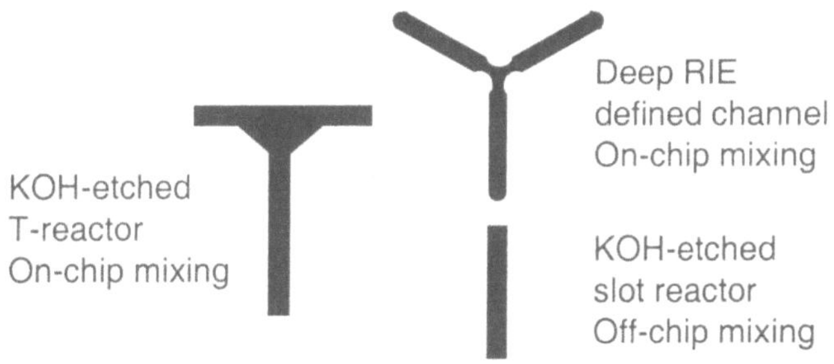

Fig. 10. Microreactor geometry modifications in increase membrane robustness and expand pressure operating range.

The buckling membrane failure at elevated temperatures can be addressed by controlling the membrane residual stress. We have microfabricated reactors with composite membranes of 0.25 μm SiO_2 and 0.3 μm LPCVD Si_3N_4. This membrane is stress compensated, due to the high residual tensile stress of the stoichiometric nitride. As the membrane heats up, the expansion results in reducing the tensile stress, without crossing over into the compressive stress regime. Similar stress compensated membranes were previously used in a microfabricated hot stage for chemical gas sensing applications [6]. The composite membranes were tested under both non-reactive and reactive flow conditions. The membranes show no signs of buckling at operating temperatures as high as 800 °C. At these temperatures the previously described thermal degradation of the Pt/Ti heaters and temperature sensors begins limiting the microreactor thermal operating range [7].

5 Summary

The capabilities of the previously reported membrane based gas phase reactors have been expanded by improving the ability to control catalyst temperature, expanding the thermal and pressure operating ranges for the microreactor, and adapting traditional wet catalyst preparation methods. Increasing the ability of the membrane to dissipate heat and redesigning the heater allowed the reactor to carry out highly exothermic reactions over a much broader temperature range than previously reported, while maintaining good catalyst thermal uniformity. Higher operating temperatures and pressures were made possible by redesigning the reactor membranes to improve strength and compensate for thermal stress.

Acknowledgements

The authors thank the DARPA Micro Flumes Program (F30602-97-2-0100) for financial support.

Citations

[1] J. J. Lerou, M.P. Harold, et al. "Microfabricated Minichemical Systems: Technical Feasibility," *Microsystem Technology for Chemical and Biological Microreactors*, DECHEMA Monographs **132**, 1996, p. 51.

[2] R. Srinivasan, I-M. Hsing, et al. "Micromachined Reactors for Partial Oxidation Reactions," *AICHE Journal*, **43**, 1997, p. 3059.

[3] A.J. Franz, D. Quiram, et al. "New Operating Regimes and Applications Feasible with Microreactors" *2nd International Conference on Microreaction Technology Proceedings*, 1998, pp. 33-39.

[4] I-M Hsing, PhD Thesis, MIT, 1997.

[5] T. Pignet and L.D. Schmidt, "Selectivity of ammonia oxidation on Pt," *Chemical Engineering Science* **29**, 1974, pp.1123-1131.

[6] S. Majoo, J. L Gland, et al. "A silicon micromachined conductometric gas sensor with maskless Pt sensing film deposited by selected-area CVD" *Sensor. Actuat. B-Chem.* **36**(1-3), 1996, pp. 312-319.

[7] S.L. Firebaugh, K.F. Jensen, M.A. Schmidt MA "Investigation of high-temperature degradation of platinum thin films with an in situ resistance measurement apparatus" *Journal of Microelectromechanical Systems* **7**(1), 1998, pp.128-35.

Synthesis of ethylene oxide in a microreaction system

H. Kestenbaum[1], A. Lange de Oliveira[1], W. Schmidt[1], F. Schüth[1*],
W. Ehrfeld[2], K. Gebauer[2], H. Löwe[2], Th. Richter[2]

[1] Max-Planck-Institut für Kohlenforschung, Kaiser-Wilhelm-Platz 1, D-45470 Mülheim / Ruhr, Germany

[2] Institut für Mikrotechnik Mainz GmbH, Carl-Zeiss-Str. 18-20, D-55129 Mainz, Germany

Abstract. Since 1937 ethylene oxide has been produced by direct oxidation of ethylene over a supported silver catalyst and it has become a very important industrial product. Consequently, the synthesis of ethylene oxide has been the subject of a large number of scientific and industrial studies. These examinations of the ethylene oxidation have suggested that this process can be used for the comparison of potential microreactor systems to current industrial processes. Therefore, this reaction was chosen as a benchmark to evaluate the potential of a microstructured reactor in the synthesis of ethylene oxide. Here, a microreactor, designed and constructed by IMM, was used for the investigations. Silver foils (126 channels: length 9 mm, height 50 µm, width 500 µm) served as the catalytically active component. Most of the experiments worked at pressures between 5 and 15 bar in a temperature range of 500-570 K. In order to optimize the selectivity while still having a satisfying yield, the reactant gas composition was varied over a wide range. In this study it was found that the selectivities and conversions reached were comparable to those achieved in industrial processes and the selectivity was not strongly dependent on the oxygen concentration. Additionally, the possibility to run the reactor with gas compositions within the explosion limits for ethylene was demonstrated.

Keywords. ethylene oxide, partial oxidation, silver, catalysis, microreactor

1 Introduction

Since T. E. Lefort discovered the oxidation of ethylene to ethylene oxide by silver catalysts, the industrial production of ethylene oxide by direct oxidation with air or oxygen over Al_2O_3 supported silver catalysts has become a widely recognized process[1]. In 1995, the world capacity for ethylene oxide was approx. 11.2×10^6 tons per year. For this reason much research has been focused on this reaction from a fundamental point of view, as well as with respect to prospective applications. This was the reason to choose the synthesis of ethylene oxide as an

* to whom correspondence should be sent

ideal benchmark to test the performance of catalytic microstructured reactors and to compare it with existing industrial processes.

Ethylene may react in a selective oxidation process with oxygen to form either ethylene oxide or carbon dioxide and water by total oxidation. Furthermore, ethylene oxide may be oxidized further to carbon dioxide and water as illustrated in Fig. 1. The selective partial oxidation of ethylene is a moderately exothermal reaction (-105 kJ/mol), whereas the total oxidation of ethylene and ethylene oxide is highly exothermal (-1327 kJ/mol and -1223 kJ/mol). The total heat of reaction in a conventional industrial reactor reaches –350 to-550 kJ/mol, causing 'hot spots' and heat removal problems[2].

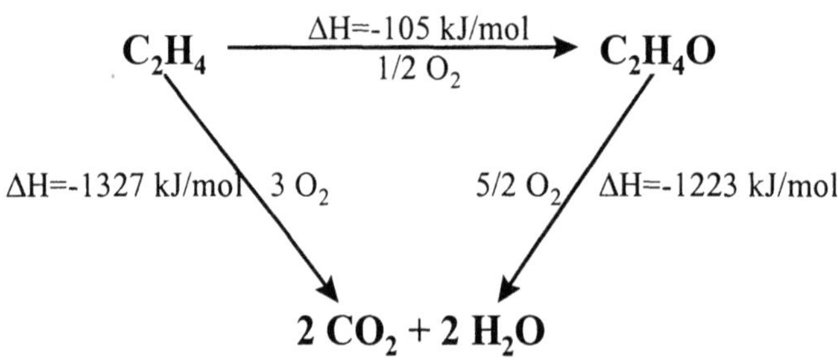

Fig. 1: Reaction network for ethylene partial oxidation and competing combustion reactions.

The microreactor is a useful tool to investigate the 'hot spot' problem. Calculated temperature profiles from IMM and Dr. O.-U. Langer and T. Stief at the DECHEMA Frankfurt show a maximum temperature deviation of 1K, even with high ethylene conversions. On the other hand, it is possible to use gas compositions within the explosive limits of the reactants since microstructured reactors show an inherent explosion safety. Hazardous gas mixtures can be handled safely since the quench distance for ethylene in O_2 is larger than the characteristic dimension of the microstructured channels. Thus, microreactors enable the use of reaction conditions not feasible in conventional reactors, while avoiding troubling hot spots.

2 Reactor and Preparation

In this study, a microreactor, designed and constructed by IMM was used. It contains a mixing unit in which reactants are injected from opposite sides. The specific channel geometry directs the gases in a ninety degree angle in order to achieve exact arrangement of the single gas streams above each other. The thickness of each layer is about 200 µm while the width of the channels increases from 148 µm up to 469 µm in order to establish a constant pressure drop[3] As shown in Fig. 2, a diffusion zone of 1 mm length is located between the mixing unit and the catalyst zone for establishing a homogeneous gas mixture before entering the catalytic area. Silver foils in parallel flow geometry where used as catalytically active compound. The reactor section consists of 126 channels, each of them 9 mm long, 50 µm high and 500 µm wide[4].

Since the catalyst is made of polycrystalline silver, an attempt was made to increase the catalytically active surface by using a modified OAOR process[5]. During this procedure the catalyst was oxidized at 530 K in oxygen atmosphere at a volume flow of 0.5 l/h oxygen and followed by reduction at 630 K in hydrogen atmosphere at the same flow rate. Because a precise measurement of the surface area could not be achieved, an approximate value of the geometrical surface area has been used for all calculations. However, chemisorption data suggest an increase in surface area by a factor of two to three. Additionally, this preparation step was helpful in removing organic impurities resulting from the microstructuring process from the catalyst foils surface.

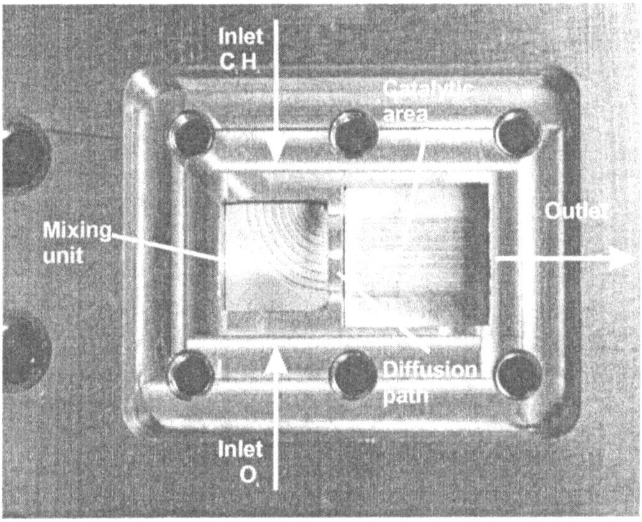

Fig. 2: Overview on mixing unit[3] (left side), diffusion zone (in between) and silver catalyst[4] (right side).

3 Reactor Operation

A forced convection flow oven was used to generate the necessary temperatures ranging from 480–630 K. The reactant gas tubes were guided through the oven in order to preheat the gases before entering the microreactor. Gas concentrations were regulated using mass flow controllers, adjusting the gas concentration of ethylene from 1-6 vol-% and oxygen from 5-50 vol-% respectively. Nitrogen has been used as carrier gas. The total gas flow varied from 1-5 l/h at pressures between 1-20 bar.

In order to avoid condensation of synthesized ethylene oxide, the exhaust gas tube between microreactor and gas chromatograph was heated. Product gas mixture analysis was carried out via on-line GC with a FID/TCD combination. Thus, concentrations of ethylene oxide and carbon dioxide could be obtained simultaneously, a crucial measure for determination of mass balance and selectivity.

4 Results and Discussion

Upon first use, the new microstructured silver foils showed no catalytic activity. After treating the silver foils with the OAOR process, a slight increase in catalytic activity was observed. During the first 100 h of operation at reaction conditions the previously known catalyst formation of silver was observed. Further on in the reaction process, the catalyst produced a sufficient amount of ethylene oxide for analysis. The catalyst activity observed corresponds to the range of values given in literature. At a temperature of 545 K, a pressure of 5 bar and a feed composition of 3 vol-% ethylene and 16.5 vol-% oxygen, a reaction rate of 1.5×10^{-4} $mol^{-1}m^{-2}$ has been observed, which is fairly close to the value reported by Tsybulya *et al.* for silver powder[6].

It was found that the selectivity in the microreactor depends on the partial pressure of oxygen. Thus, the yield of ethylene oxide increases with increasing oxygen pressure even at a constant conversion rate. As shown in Fig. 3, the conversion rate of ethylene does not increase with the concentration of oxygen added. However, an increase in selectivity was noted unexpectedly up to an oxygen concentration of approx. 40 %. This experiment was conducted at a constant ethylene concentration of 3 %, a total pressure of 5 bar, a gas flow of 4 l/h at 550 K. Even though these concentrations place the reaction gas mixture within the explosion limits for ethylene in oxygen, the inherent safety of this microreactor design enables reaction processes at these conditions at minimal risks.

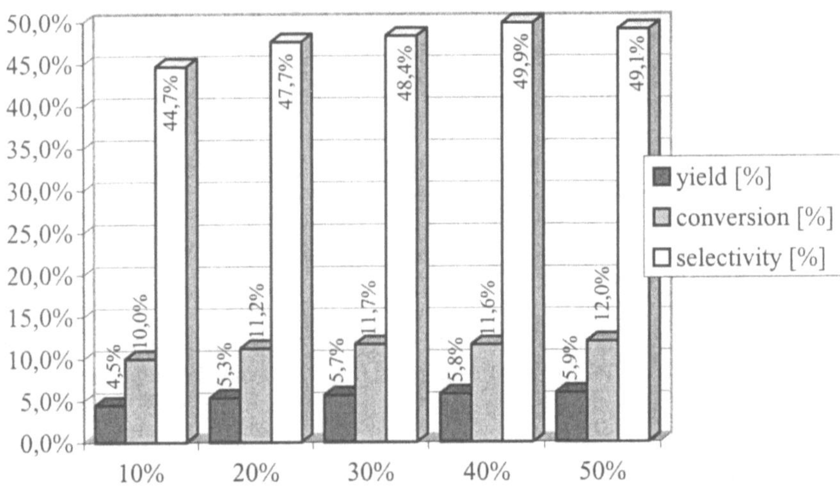

Fig. 3: Effect of the oxygen concentration on conversion, yield and selectivity.

In comparison, a selectivity of about 50 % is lower than the values established in industrial processes. However, industrial conditions include the presence of chlorine compounds such as 1,2-dichlorethane. Decreasing the temperature to moderate values, e.g. 510 K, a selectivity of about 65 % can be reached at the same conditions. These values exceed those being observed for polycrystalline silver catalysts[7]. Variation of reaction temperature between 510 K and 600 K, enabled calculation of the activation energy at the process conditions already mentioned. The observed value of 48 kJ/mol corresponds to the data given in the literature[7].

Variation of the gas composition over a wide range leads to higher conversion rates. Using 0.75 % ethylene and 33 % oxygen at a flow rate of 2 l/h, a pressure of 5 bar and 550 K, a conversion higher than 24 % at a selectivity of 47 % was observed. In this case, the reaction rate decreases to $4.7 \times 10^{-5}\,mol^{-1}m^{-2}$. On the other hand, the reaction rate increases to $2.8 \times 10^{-4}\,mol^{-1}m^{-2}$ while conversion decreases to 7 % when a gas composition of 6 % ethylene and 33 % oxygen, flow rate of 5 l/h, 5 bar and 540 K is used.

Initial experiments showed that an increase in pressure from 5 to 15 bar does not lead to an increase in conversion as expected. Instead, constant conversion and yield was observed. Further experiments led to a maximum conversion at 5 bar and 540 K. Dependence of conversion rates on pressure only was observed between 1 and 5 bar.

At this point in time, the catalyst has been in operation for over 1000 h under reaction conditions without any deactivation. Fig. 4 shows two SEM-images of parts of a reaction channel as well as of the "wall" between two channels, which is the darker portion in the images. It can be clearly seen, that the catalyst surface in the channels has been roughened during the reaction process. Moreover, the surface is pitted substantially. This interaction between silver and oxygen has already been described by Ertl *et al.*[8].

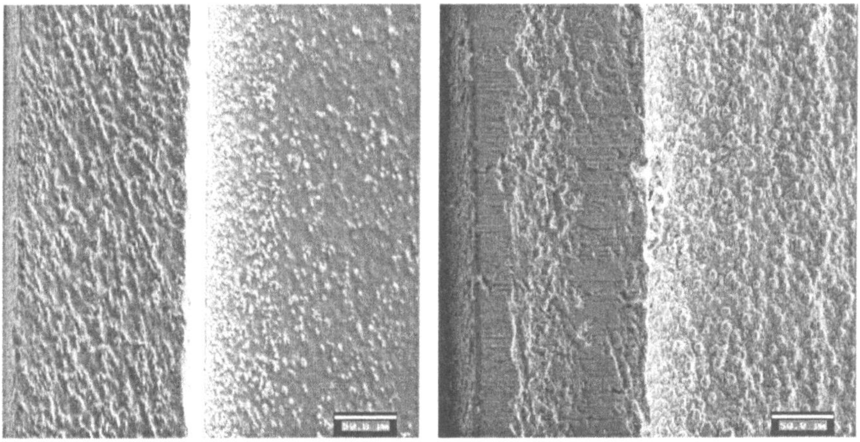

Fig. 4: Silver surface before (left) and after 1000 h time-on-stream (right).

5 Conclusions

The results obtained using a microreactor for ethylene oxide synthesis in general are comparable to industrial processes known. Selectivities of about 50 % were found for the microreactor. The industrial process is operating with a somewhat higher selectivity of up to 80 % if pure oxygen is used as the oxidant at similar conversion levels, albeit these results were obtained in the presence of chloride and usually other promoters on the catalyst, in contrast to the pure silver investigated here. Thus, microreactors seem to be able to reach a performance similar to industrial processes without using chlorine promotors in the feed. The performance in presence of chlorine on the surface will be investigated in the near future to further increase the selectivity. Moreover, experiments in the presence of hydrogen for mechanistic investigations are still under way, and no conclusive data can be presented yet. However, it can already be stated with confidence, that catalytic microreactors are in principle suitable to carry out demanding catalytic reactions with at least a comparable performance to industrial processes. Whether such reactors will ever be used in industrial practice certainly will depend on economic factors, the possible hazards of the products and the amounts of product necessary at certain points of use. In favorable cases on-demand production in microstructured reactors will be useful.

6 Acknowledgements

The authors would like to thank Thomas Scholl at IMM for valuable help in microreactor development and D. Wolf (ACA) for valuable discussions. Furthermore the authors would like to thank the BMBF (German Ministry of Education, Science, Research and Technology) for funding this work and all the partners within this project for their cooperation and discussions.

References

[1] *Ullmanns Encyclopedia of Industrial Chemistry*, 5th Ed., Vol. A 10, 117ff

[2] K. Weissermel, H.-J. Arpe, *Industrial Organic Chemistry*, 3rd Ed., VCH, 153ff (1994)

[3] Th. Richter, W. Ehrfeld, K. Gebauer, K. Golbig, V. Hessel, H. Löwe, A. Wolf: *"Metallic Microreactors: Components and Integrated Systems"*, in: Process Miniaturization: 2nd International Conference on Microreaction Technology, AIChE, 146-151 (1998)

[4] H. Löwe, W. Ehrfeld, K. Gebauer, K. Golbig, O. Hausner, V. Haverkamp, V. Hessel, Th. Richter: *"Microreactor Concepts for Heterogeneous Gas Phase Reactions"*, in: Process Miniaturization: 2nd International Conference on Microreaction Technology, AIChE, 63-74 (1998)

[5] A. W. Czandera, *J. Phys. Chem.* **70**, 2120 (1966)

[6] S. V. Tsybulya, G. N. Kryukova, S. N. Goncharova, A. N. Shmakov, B.S. Bal'zhinimaev, *J. Catal.* **154**, 194 (1995)

[7] R. A. van Santen, P. C. E. Kuipers, *Advances in Catalysis* **35**, 265 (1987)

[8] C. Rehren, M. Muhler, X. Bao, R. Schlögl, G. Ertl, *Zeitschrift für Physikalische Chemie* **174**, 11 (1991)

Selective Reactions in Microchannel Reactors

Ansgar Kursawe[1], Enrico Dietzsch[1], Stefan Kah[1], Dieter Hönicke[1], Maximilian Fichtner[2], Klaus Schubert[2], Georg Wießmeier[3]

[1] Technische Universität Chemnitz, Lehrstuhl für Technische Chemie, D-09107 Chemnitz

[2] Forschungszentrum Karlsruhe, Hauptabteilung Versuchstechnik, D-76021 Karlsruhe

[3] Bayer AG, Reaktionstechnik, D-51368 Leverkusen

Abstract. Microchannel reactors made of packed wafer stacks were used for selective hydrogenation of benzene to cyclohexene as well as for partial oxidation of 1-butene to maleic anhydride and ethene to ethene oxide. The microstructured wafers were catalytically activated by anodic oxidation of aluminum wafers followed by an impregnation process with ruthenium for benzene hydrogenation and V_2O_5/TiO_2 for 1-butene oxidation or by physical vapor deposition of silver for ethene epoxidation.

1. Introduction

The aim of our current investigations is to determine the applicability of microchannel reactors for selective and exothermic reactions such as partial hydrogenations and oxidations of unsaturated hydrocarbons. The partial gas phase hydrogenation of benzene to cyclohexene in microchannel reactors has been investigated. Since the partial hydrogenation of benzene, which has been previously introduced in industry, is performed as a liquid phase process [1], an alternative gas phase process could be of industrial interest. In addition, the highly exothermic partial oxidation of 1-butene to maleic anhydride and ethene to ethene oxide have been carried out in a microchannel reactor. These investigations are focused on the performance of the reactors using high feed partial pressures and high total pressures. Both reactions are normally carried out in cooled multitube fixed bed reactors. Thus, microchannel reactors with their superior heat transfer ability [2] are a suitable, novel, and interesting reactor type for those kind of reactions.

2. Microchannel Reactor Construction and Catalyst Preparation

The reactions have been investigated in microchannel reactors (Fig. 1a), containing a stack of microstructured aluminum wafers (Fig. 1b) with rectangular microchannels of 200 x 200 μm, manufactured by means of mechanical micromachining, a technology developed by Forschungszentrum Karlsruhe GmbH [2].

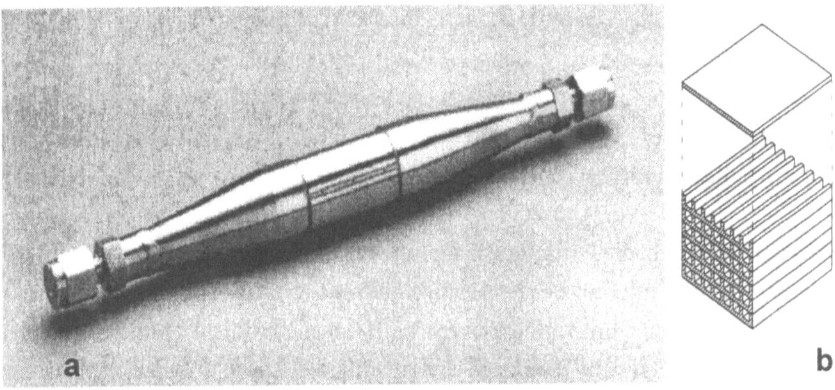

Fig.1: Example of a sealed microchannel reactor (a), with attached diffusors containing a stack of packed wafers (b) [3]; 24 wafers (300 μm x 8.5 mm x 30 mm), channel geometry 200 μm x 200 μm, 28 channels / wafer (672 channels total).

The microstructured wafers were modified in order to yield catalytic active species at the walls of the microchannels. There are several methods to activate microstructured Al-wafers because there is no catalytic activity by the wafer material itself. One method is the immobilization of catalytic active species within a suitable porous catalyst support material, e.g. porous alumina. In order to form such alumina at the walls of the microchannels, anodic oxidation of the Al-wafers was carried out (Fig.2) [4]. The resulting alumina oxide layer was used as catalyst support. Since this layer is highly porous, the BET surface area can be increased by a factor up to 10^3 compared to the geometric surface area. Texture parameters like pore diameter, pore length, and pore density depend highly upon electrolysis parameters like voltage, duration, and electrolyte [5]. In this well defined pore system the required catalytic active species were immobilized by impregnation. This technique was used for the preparation of catalysts suitable for the selective hydrogenation of benzene and partial oxidation of 1-butene.

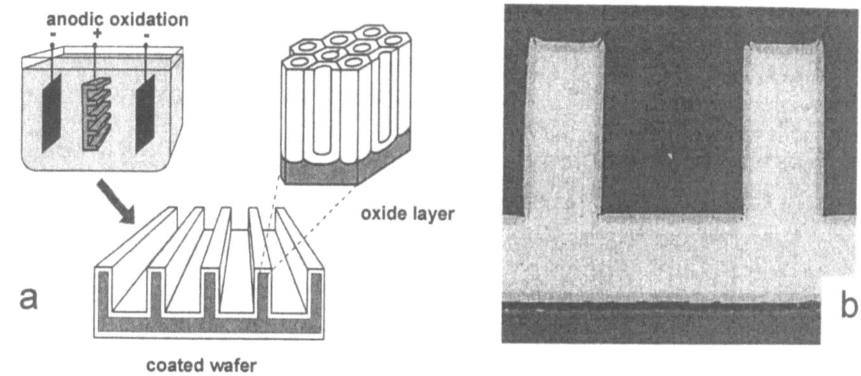

Figure 2: Anodic oxidation of the microstructured wafers, model of the resulting pore system and coated microstructured wafers (a); REM micrograph of a cross section of an anodic oxidized microstructured wafer (b), channel size 200μm x 200μm

Other ways to obtain catalytically active wafers are coating processes like PVD, CVD, or sputtering techniques. Silver is known to be a catalytically active component for partial oxidation of ethene to ethene oxide and thus microstructured aluminum wafers were coated with a thin silver layer by PVD in order to activate the wafers for ethene epoxidation.

3. Results

3.1 Selective Hydrogenation of Benzene to Cyclohexene

The hydrogenation of benzene (B) is characterized by a strong thermodynamic preference of the fully hydrogenated cyclohexane over the more desired cyclohexene [6]:

$$C_6H_6 \quad \overset{H_2}{\rightleftharpoons} \quad C_6H_{10} \quad \overset{H_2}{\not\rightleftharpoons} \quad C_6H_{12}$$

The reactivity of the intermediate cyclohexene is much higher than that of the aromatic nucleus [7]. Therefore, satisfactory selectivities to cyclohexene are difficult to achieve. The partial gas phase hydrogenation is commonly performed over non-supported or supported ruthenium (Ru) catalysts [8, 9]. For the formation of the desired cyclohexene, the addition of a gaseous reaction modifier (M) to the feed is necessary. The modifier influences the adsorption and desorption as well as mass transfer processes and the kinetics of the consecutive reaction [8,10]. Therefore, microchannel reactors with their superior mass and heat transfer properties were used, which sould be a suitable reactor type to attain a cyclohexene formation in the present study.

A ruthenium catalyst for the benzene hydrogenation was made by the following procedure: microstructured aluminum wafers were anodically oxidized, calcined in air and afterwards soaked with toluene after a vacuum treatment. Impregnation of the porous alumina was performed with $Ru(acac)_3$ dissolved in toluene for several hours at slightly elevated temperature followed by calcination in air. 15 wafers with a geometry of 10mm x 50mm x 300μm having 450 channels were assembled to a $Ru/Al_2O_3/Al$ microchannel reactor by a mechanical packing process.

Fig. 3 shows the time on stream behavior of conversion degree and selectivity obtained using the $Ru/Al_2O_3/Al$-microchannel reactor. A second order deactivation of the catalytic active ruthenium was observed. The selectivity to cyclohexene was about 20% at 13% conversion degree after a time on stream of 15 h. The selectivity to cyclohexene depends highly upon the partial pressure of the reaction modifier, which was methanol. This dependence is shown in Fig. 4.

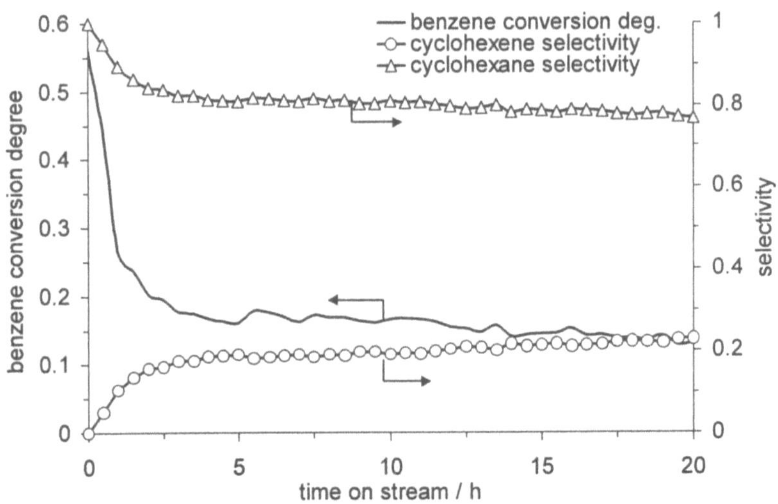

Fig. 3: Time on stream behavior of the conversion degree and selectivities in the gas phase hydrogenation of benzene using a $Ru/Al_2O_3/Al$-microchannel reactor. Reaction conditions: $p = 110$ kPa, $T = 353$ K, $p_B = 0.11$ kPa, $p_M = 0.33$ kPa, $p_{H2} = 44$ kPa, residence time = 530 ms.

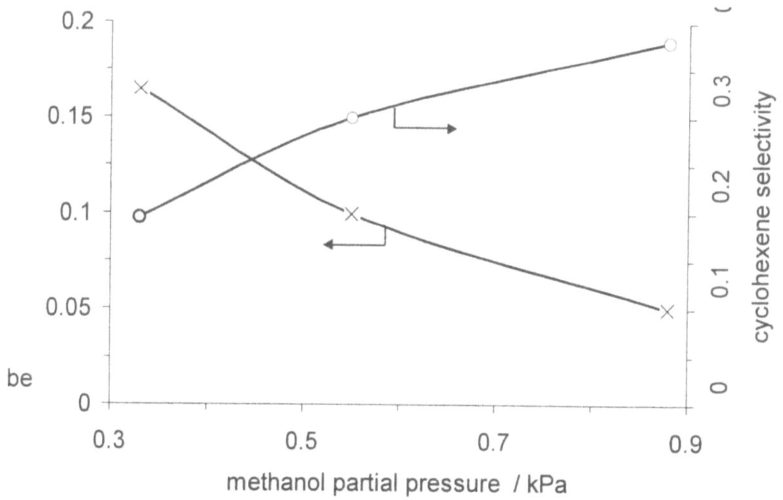

Fig.4: Conversion degree and cyclohexene selectivity in the partial gas phase benzene hydrogenation using a $Ru/Al_2O_3/Al$-microchannel reactor as a function of the methanol partial pressure. Reaction conditions: p = 110 kPa, T = 353 K, p_B = 0.11 kPa, p_{H2} = 44 kPa, residence time = 530 ms, TOS = 10 h.

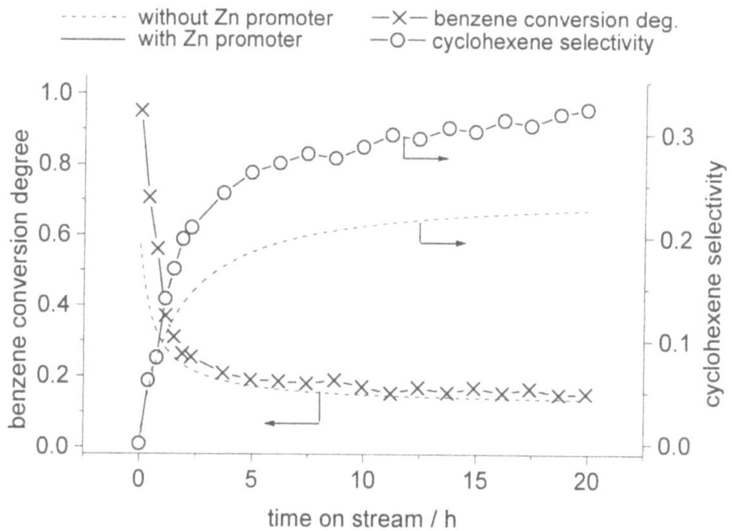

Fig. 5: Time on stream behavior of conversion degree and selectivity to cyclohexene in the gas phase hydrogenation of benzene using a $Ru/Al_2O_3/Al$-microchannel reactor. Reaction conditions: p = 110 kPa, T = 353 K, p_B = 0.11 kPa, p_M = 0.33 kPa, p_{H2} = 44 kPa, residence time = 530 ms.

An increase of the methanol partial pressure from 0.33 kPa to 0.88 kPa improved the selectivity up to 38% at conversion degrees close to 5%.

Zinc as promoter of the $Ru/Al_2O_3/Al$-catalyst is also able to improve the selectivity to cyclohexene. As shown in Fig. 5, Zinc improved the selectivity by 10% to 32% in contrast to 22% without any promoter at nearly the same degree of conversion (TOS = 15 h). In further investigations, the influence of other promoters than Zn on cyclohexene selectivity will be a matter of concern.

3.2 Selective Oxidation of 1-Butene to Maleic Anhydride

The partial gas phase oxidation of 1-butene to maleic anhydride is an extremely exothermic reaction:

$$C_4H_8 + 3\ O_2 \rightleftharpoons C_4H_2O_3 + 3\ H_2O \qquad \Delta_R H = -\ 1315\ \text{kJ/mol}$$

This reaction is normally carried out in fixed bed reactors using 1.5% hydrocarbon in the feed. V_2O_5 and TiO_2 are known to be catalytically active species in this partial oxidation [11] and thus catalysts active for the selective oxidation of 1-butene to maleic anhydride were prepared by anodic oxidation of aluminum followed by impregnation of the resulting porous alumina layer with suitable precursors.

Preliminary investigations and catalytic runs were carried out in conventional fixed bed reactors using V_2O_5/TiO_2 impregnated anodically oxidized aluminum wires. Therefore, time consuming microstructuring of the wafers and reactor construction could be omitted.

Catalytic runs were performed at a reaction temperature of 693 K using 0.41% 1-butene in air as feed. Several organic products besides maleic anhydride were observed, i.e. mainly acetic acid, acetaldehyde, and phthalic anhydride as well as butadiene, crotonaldehyde and furane in smaller amounts. The selectivity towards maleic anhydride shows a maximum of nearly 30 % at a conversion degree of 87% (Fig. 6). Currently $V_2O_5/TiO_2/Al_2O_3/Al$-microchannel reactors are under investigation.

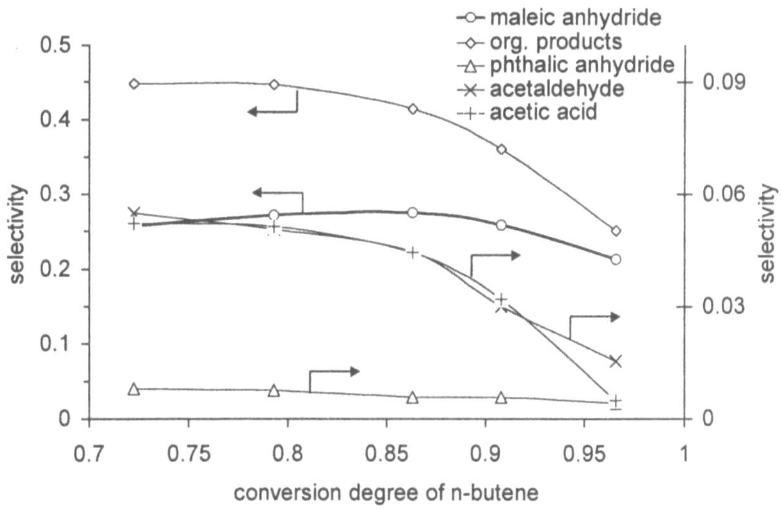

Fig.6: Selectivity as a function of conversion degree in the partial oxidation of 1-butene with air using a fixed bed reactor; Reaction conditions: p = 110 kPa, T = 693 K, $p_{1\text{-butene}}$ = 4.51 kPa.

3.3 Selective Oxidation of Ethene to Ethene Oxide

The partial oxidation of ethene (E) is an exothermic reaction, normally carried out using Ag/α-Al_2O_3 catalysts [12] at low conversion degrees and reaction temperatures of 500-550 K at high total pressures up to 3 MPa.

$$C_2H_4 + \tfrac{1}{2}\,O_2 \rightleftharpoons C_2H_4O \qquad \Delta_R H = -105 \text{ kJ/mol}$$

As mentioned before, physical vapor deposition is another way to activate microstructured wafers catalytically. Thus, microstructured aluminum wafers were coated with silver in order to achieve catalytic activity.

The deposition process was performed in vacuum by evaporation of metallic silver, resulting in a 200 nm thin silver layer on the microstructured wafers. Known promoters like Cs-salts [12] were not used in the preparation process. 20 wafers with a geometry of each 10 mm x 50 mm x 300 µm were packed in order to form an Ag/Al-microchannel reactor (600 channels).

Preliminary experiments showed that the selectivity to ethene oxide increased with increasing oxygen partial pressure. Thus, no inert gas was used as diluent in the following experiments. The microchannel reactor was tested using two different ethene concentrations of 4% and 20%, respectively. Using 4% ethene in oxygen as feed, selectivities close to 45% at conversion degrees from 40 to 66% were obtained (Fig. 7). It is noteworthy that the selectivity to ethene oxide using an Ag/Al-microchannel reactor is nearly independent from the degree of conversion in this experiment. Yields of 19 to 29% were calculated without recycling unconverted ethene. The estimated adiabatic temperature rise was about 630K, however, at a conversion degree of 66%. An increase in ethene partial pressure leads to higher selectivities of about 55% at lower conversion degrees of 15 to 39% (Fig. 8) and yields of 7.5% to 20.5%. Again, the selectivity changes only slightly with increasing degree of conversion. The estimated adiabatic temperature rise in this case was close to 1500 K at 39% conversion degree. Currently the influence of alkaline metals and thickness of the silver layer on ethene oxide selectivity are a matter of concern.

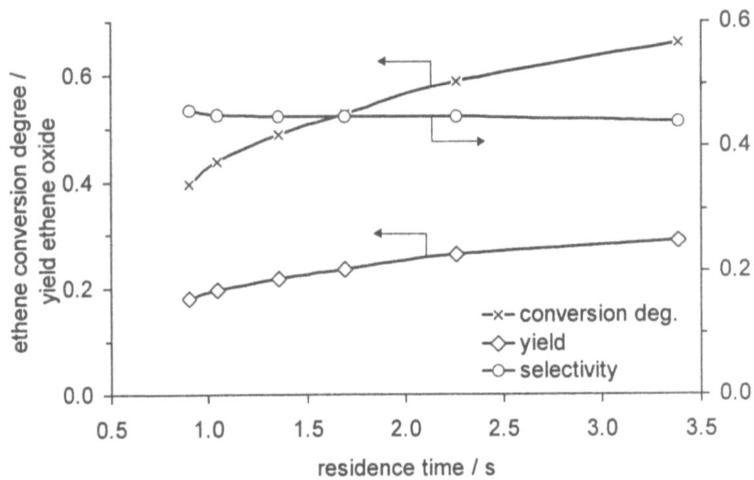

Fig. 7: Degree of ethene conversion, yield and selectivity to ethene oxide as a function of residence time using an Ag/Al-microchannel reactor (600 channels) for partial oxidation of ethene; Reaction conditions: T = 523 K, p = 0.3 MPa, p_E = 12 kPa, bal. O_2.

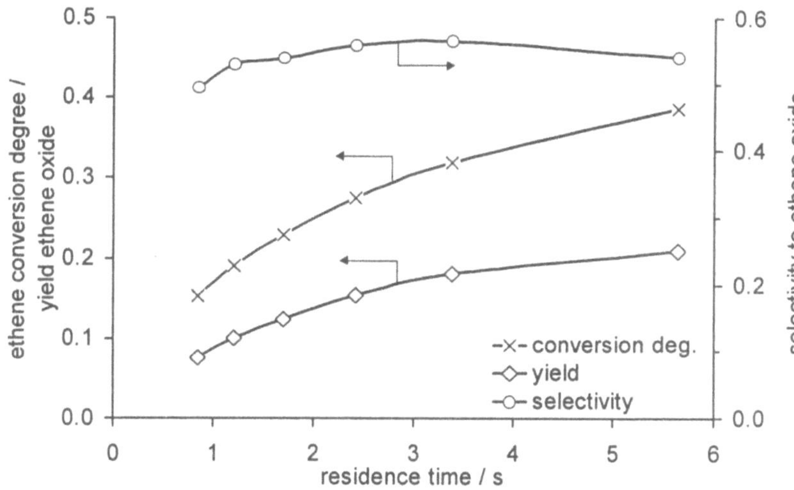

Fig. 8: Degree of ethene conversion, yield and selectivity to ethene oxide as a function of residence time using an Ag/Al-microchannel reactor (600 channels) for partial oxidation of ethene; Reaction conditions: T = 523 K, p = 0.3 MPa, p_E = 60 kPa, bal. O_2.

4. Conclusions and Outlook

It is possible to perform selective hydrogenation and exothermic partial oxidation reactions in microchannel reactors. Anodic oxidation followed by impregation techniques and PVD coating had been successfully used for catalytic activation of microstuctured aluminum wafers. Microchannel reactors are suitable for hydrogenations with reactive intermediate compounds such as cyclohexene from benzene. Even exothermic partial oxidation reactions, such as the partial oxidation of 1-butene and ethene can be performed at high degrees of conversion and high hydrocarbon partial pressures in the feed.

In future investigations crossflow microchannel reactors will be tried out in the reactions in order to improve the heat transfer. A prototype of such a reactor made by Forschungszentrum Karlsruhe is depicted in Fig. 9.

An alternative process for catalytic activation of microstructured wafers is the sol-gel method. Those wafers are catalytically activated by dip-coating with the sol followed by a controlled drying and calcination process of the resulting gel. A microchannel reactor made of those wafers is currently under investigation.

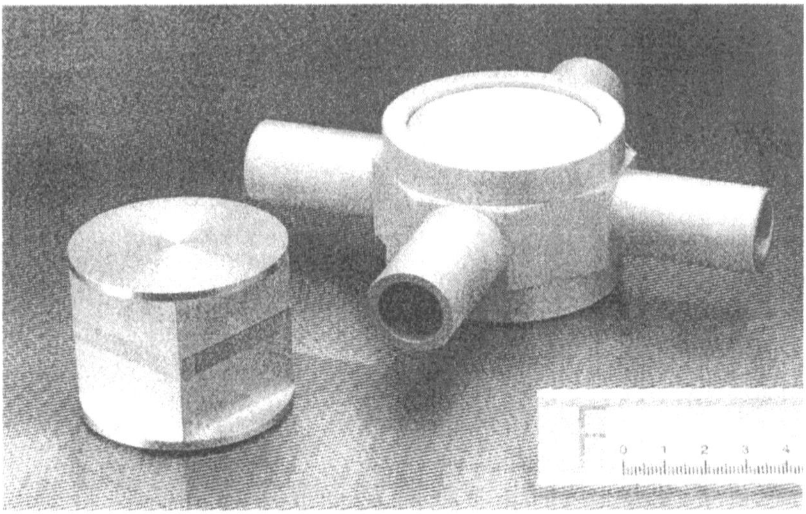

Fig.9: Prototype of a crossflow microchannel reactor made of aluminum (developed by Forschungszentrum Karlsruhe GmbH)

Acknowledgments

This study was supported by the "Arbeitsgemeinschaft industrieller Forschungsvereinigungen" (AiF), project number 11146B/1. We also like to thank the "Max-Buchner-Forschungsstiftung" and the "Fond der chemischen Industrie" (FCI) for financial support.

References

[1] H. Nagahara, M. Ono, M. Konishi and Y. Fukuoka, Appl. Surf. Sci. 121/122 (1997) p. 448

[2] W. Bier, W. Keller, G. Linder, D. Seidel, K. Schubert, ASME, DSC Vol. 19 (1990) p. 189

[3] G. Wießmeier, K. Schubert, D. Hönicke, Proc. 1st Int. Conf. on Microreaction Technology, Springer (1997), p. 20

[4] G.E. Thompson and G.C. Wood in Treatise Material Science and Technology, Vol. 23, ed. J.C. Scully (Academic Press, New York, 1983) p. 205

[5] D. Hönicke, Appl. Catal. 5 (1983) p. 179

[6] I. Bahrin, *Thermochemical Data of Pure Substances*, 3rd ed. (Verlag Chemie, Weinheim, 1995)

[7] H.A. Smith, H.T. Meriwether, J. Am. Chem. Soc. 71 (1949) p. 413

[8] J.A. Don, J.J.F. Scholten, Faraday Discuss. Chem. Soc. 72 (1982) p. 145

[9] J. Patzlaff, J. Gaube, Chem.-Ing.-Tech. 69 (1997) 10, p. 1462

[10] J. Struyk, J.J.F. Scholten, Appl. Catal. 62 (1990) p. 151

[11] K.Mori, A.Miyamoto, Y.Muramaki, J. Chem. Soc. Faraday Trans. I 82 (1986), p. 13

[12] R.A.Van Santen, H.P.C.E. Kuipers, Adv. Catal. 35 (1987), p. 265

Periodic Operation in Microchannel Reactors

M. A. Liauw,[1,*] M. Baerns,[2] R. Broucek,[3] O. V. Buyevskaya,[2] J.-M. Commenge,[4] J.-P. Corriou,[4] L. Falk,[4] K. Gebauer,[5] H. J. Hefter,[6] O.-U. Langer,[7] H. Löwe,[5] M. Matlosz,[4] A. Renken,[8] A. Rouge,[8] R. Schenk,[5] N. Steinfeldt,[2] St. Walter[1]

[1] Lehrstuhl für Technische Chemie I, Universität Erlangen, D-91058 Erlangen, Germany
[2] Institut für Angewandte Chemie ACA Berlin, D-12484 Berlin-Adlershof, Germany
[3] Akzo Nobel, D-63784 Obernburg, Germany
[4] LSGC-CNRS, F-54001 Nancy, France
[5] Institut für Mikrotechnik Mainz GmbH, D-55129 Mainz-Hechtsheim, Germany
[6] DuPont de Nemours (Deutschland) GmbH, D-61352 Bad Homburg, Germany
[7] Karl-Winnacker-Institut der Dechema e.V., D-60486 Frankfurt, Germany
[8] LGRC-EPFL, CH-1015 Lausanne, Switzerland

Abstract. Microchannel reactors promise to display unusually fast response behavior to sharp changes of feed composition in heterogeneous catalytic systems. In an attempt to use this feature for the elucidation of the transient kinetics and possibly for performance enhancement, three model reaction systems are selected. CFD assisted reactor design and catalyst preparation studies are presented.

Keywords. Microchannel Reactor, Catalyst Application, Periodic Operation, Propane Dehydrogenation, Isoprene Oxidation, Stop Effect, CFD

1 Introduction

With the advent of microchannel reactors, a quest has begun for areas where the new tools have significant advantages over conventional (macroscopic) reactors. Enormous heat transfer properties and inherent safety are features that hint to certain applications. Low inventory, fast dynamics and intrinsic explosion safety of microchannel reactors for heterogeneous catalytic reactions are the properties that should give them a competitive edge: safe periodic operation over a wide frequency range comes within reach.

It has been shown both theoretically and experimentally that switching between two reactants may lead to an increase in conversion or even selectivity over the co-feed mode of operation [1]. The optimal switching period may be less than a second (automotive catalysts) up to hours.

The slow response of conventional (macroscopic) reactors significantly limits the attainable range of frequencies. Also, their large holdup may lead to the formation of dangerous explosive mixtures. Microchannel reactors are much better suited for this type of operation. Model discrimination and process optimization may be prime applications [2]. Industrial-scale production with this approach, e.g. of fine chemicals, is conceivable once basic issues like corrosion, clogging and process control have been studied.

Three basic problems are dealt with in this paper: how to select suitable model systems (Section 2), how to ensure well-defined concentration modulation (Section 3), and how to apply catalyst to the microchannel walls (Section 4).

2 Selection of Model Systems

For a discussion of generic aspects of periodic operation in microchannel reactors, it is desirable to study more than just one reaction system. Two criteria are that a marked effect of concentration cycling should be anticipated and that the systems should not display excessive product condensation or polymerisation.

2.1 Isoprene Oxidation

It has been found that total oxidation may be suppressed during butane oxidation to maleic anhydride over vanadium-based catalysts if the catalyst is exposed to butane and oxygen in an alternating manner [3]. The catalyst acts as a storage for a selective oxygen species while the butane is never exposed to the nonselective gas phase oxygen. In the present study, a similar system will be studied: the partial oxidation of isoprene to citraconic anhydride.

In a first step towards comparing steady-state and unsteady-state performance, preliminary experiments over a titania-supported vanadia catalyst in a lab-scale reactor were carried out with a 20% yield to citraconic anhydride at steady-state [4]. In the next step the oxygen storage capacity will be studied as this is a prerequisite for selectivity enhancement by unsteady-state operation.

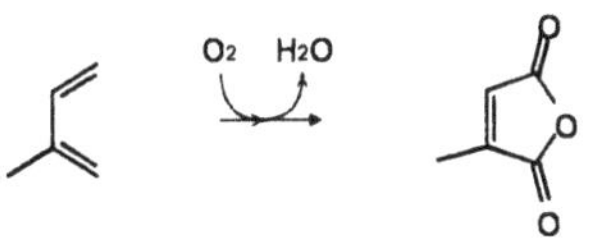

2.2 Alcohol Dehydration

For the catalytic dehydration of alcohols (and de-amination of amines) to olefins on alumina or other amphoteric metal oxides, a drastic increase of the reaction rate is observed when the reactant feed is switched to zero [5]. This phenomenon has been named *stop effect*. Different models were proposed to describe it. Two basic models were discussed by Thullie and Renken [2]:

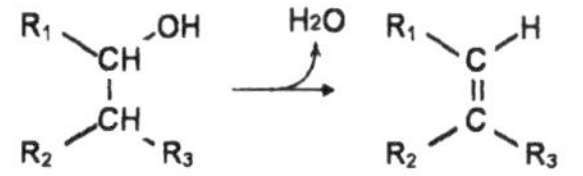

- Model I: Adsorption of the reactant A on two different sites S1 and S2, with a consecutive reaction involving an adsorbed intermediate (on S1) and a free site S2. A on S2 acts as an inhibitor.
- Model II: Adsorption of the reactant A with a consecutive reaction/product desorption, or an inhibiting adsorption of the A on the first adsorbed layer.

When the reactant feed is stopped, the inhibitor desorbs rapidly, and the reaction rate increases. Both models are consistent with the measured ethanol dehydration rates, but the surface concentrations are better described by model I [6].

Model calculations show that in a reactor operating under periodic variation of the inlet concentration of the reactant, the average productivity can be much higher than the maximal steady-state value. The higher productivity is expected at short cycling times (see Fig. 1a). As the predicted reaction rate at high frequency is very sensitive to the model used, rapid concentration cycling may serve as a useful tool for model discrimination. However, the minimum attainable cycling time in conventional lab-scale reactors is limited to a minute or more. While such experiments display rate enhancement (Fig. 1b), they do not allow for model discrimination.

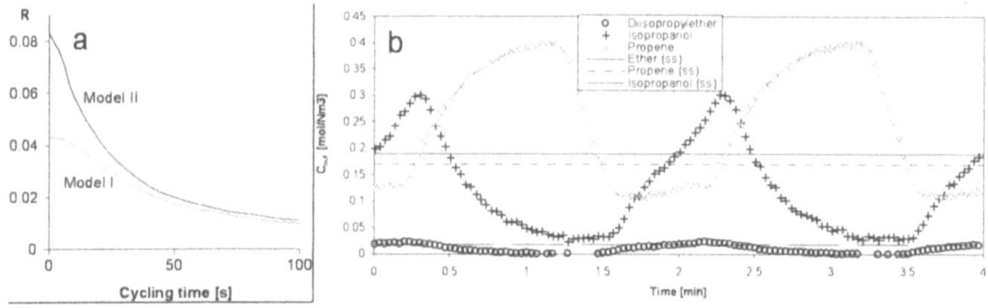

Figure 1. a) Average reaction rate as function of cycling time assuming instantaneous equilibrium for the reaction. Rate enhancement and model discrimination are possible at low cycling times. b) Experimental observation of rate enhancement. Periodic versus stationary: Gas phase measurements at the reactor outlet. Experimental conditions 0.26 g γ-alumina, T 200°C, p 1.3 bar, inlet flow 100 Sml/min. Steady state: $C_{0,iPrOH}$ = 0.45 mol/Sm³. Periodic operation: $C_{0,iPrOH}$ = alternatively 0 and 0.9 mol/Sm³, period 2 min, split 0.5.

2.3 Propane dehydrogenation

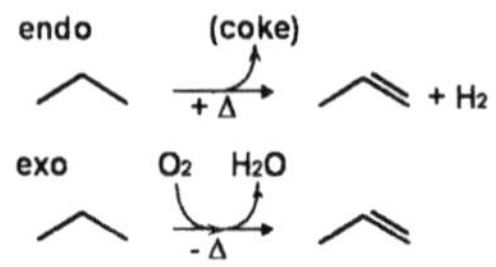

The use of microchannel reactors for processes with a coupling of endothermic and exothermic catalytic reactions is a new and attractive field (see e.g. [7]). In particular, the advantages regarding an efficient heat transfer are of interest. The suitability of microchannel reactors for the heat management during catalytic reactions has yet to be proven. In the present work, a combination of non-oxidative and oxidative dehydrogenation of propane will be examined as a model reaction. The former process takes place in the absence of oxygen, consuming heat and producing coke, while the oxidative process is accompanied by the release of heat:

		ΔH at 550°C
Non-oxidative Dehydrogenation	$C_3H_8 \rightarrow C_3H_6 + H_2$	**+ 130 kJ/mol**
Side reaction: Coke formation	$C_3H_8 \rightarrow 3\ C_s + 4\ H_2$	+ 129 kJ/mol
Oxidative Dehydrogenation	$C_3H_8 + {}^1/_2\ O_2 \rightarrow C_3H_6 + H_2O$	**- 126 kJ/mol**
Side reactions: Total oxidation	$C_3H_8 + {}^7/_2\ O_2 \rightarrow 3\ CO + 4\ H_2O$	- 1193 kJ/mol
	$C_3H_8 + 5\ O_2 \rightarrow 3\ CO_2 + 4\ H_2O$	- 2043 kJ/mol

By running the two processes in alternating layers of a microchannel reactor, heat may be efficiently transferred from the exothermic to the endothermic process, hence permitting autothermal operation (Fig. 2). The propane flow (phase

I) should be periodically interrupted by a flow of air (phase II) to burn off the coke in the non-oxidative zone.

For the oxidative dehydrogenation, catalysts based on vanadia were most intensively studied. While unselective combustion severely limits the yield, a transient kinetic analysis on a V-Mg-O catalyst showed that propene selectivity in the absence of gas phase oxygen is superior to co-feed steady state selectivity at the same propane conversion [8]. Thus, the selectivity may be improved by periodically switching between propane (phase I) and oxygen feed (phase II). The oxygen storage capacity has been observed in a test run with V_2O_5/TiO_2.

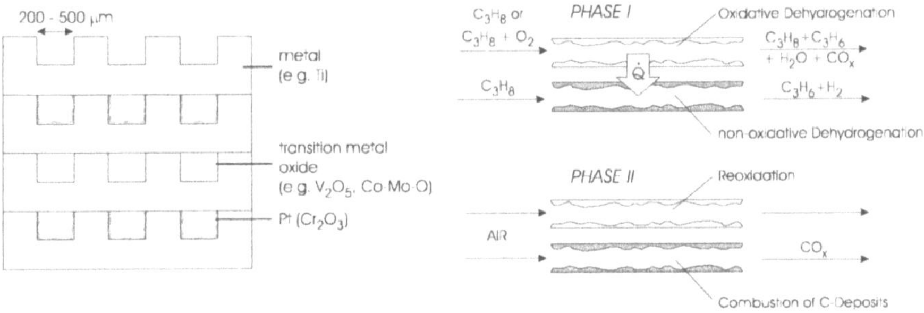

Figure 2. Oxidative propane dehydrogenation in one layer of a catalytically activated microchannel reactor serves as a heat source for the non-oxidative endothermic dehydrogenation in the adjacent layer.

3 Reactor Design and Fluid Dynamics

3.1 Basic considerations

In this study, periodic operation means to switch between two or more feed flows into a microchannel reactor. Two four-port valves allow for feeding three different gases (e.g. reducing, inert, oxidizing) in an alternating manner. The experimental setup is in some respects different from a conventional laboratory setup. In order to generate and maintain very sharp concentration fronts, it is crucial to have

- fast switching valves (e.g. Valco) with a zero dead volume fitting system;
- no sudden changes of the cross section of the tubing; and
- pressure regulators at the outlets of all branches for adjusting the pressure drops and hence to ensure a constant flow during switching.

Special care must be taken for the design of the microchannel reactor as the high-frequency concentration modulation requires a narrow residence time distribution.

3.2 Conical inlets

For conical inlet geometries it has been shown that a stationary vortex may form (Fig. 2a) [9]. This vortex acts as a dead volume and is detrimental to the residence time distribution, smearing out any step changes in concentration applied [10]. One possible solution is to shape the inlet such that the vortex is replaced by wall material, thereby minimizing the dead volume (Fig. 3b). This approach may be extended to an extreme by coupling an extremely flat conical inlet with a frit with high pressure drop, thus ensuring both minimal vortex volume and equal flow distribution (Fig. 3c) [11]. However, other geometries may avoid the high Δp frit.

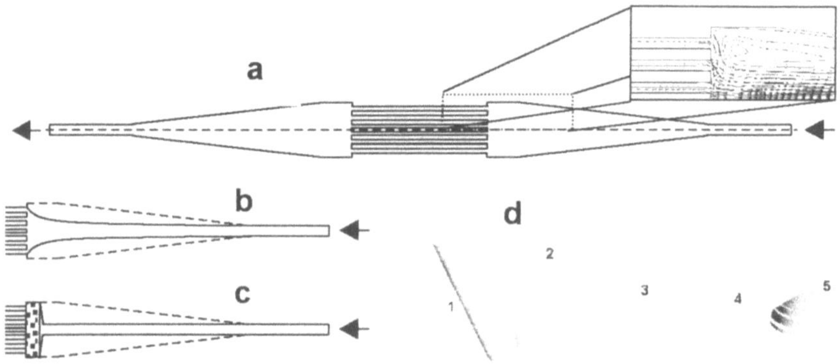

Figure 3. a) Conical inlet: vortices are formed that are detrimental to concentration changes. b) Modified conical inlet: vortex volume is minimized. c) Very shallow conical inlet with frit: near-zero dead volume and even flow distribution. d) Micro plate heat exchanger (schematic). Inlet tube (1), inlet chamber (2), channels (3), outlet chamber (4) and outlet tube (5).

3.3 Micro plate heat exchanger

In the attempt to use standard components, the countercurrent heat exchanger from IMM should also be considered. The reactant mixture enters through an inlet tube, is distributed over plates at various heights where it is distributed over numerous parallel channels and then led into an outlet tube (Fig. 3d). For a narrow residence time distribution it is necessary that the flow in all plates be similar, as should be the flow in all the channels. These velocity distributions of an incompressible fluid have been studied using a simplified pressure drop model as well as finite volume calculations with the software code Fluent®. Good agreement has been observed between the two methods.

If both inlet tube and outlet tube are located on the top plate of the assembled reactor ("U-configuration"), the distribution between plates indicates a maximum flow rate in the upper plate of the reactor and a minimum flow rate in the lowest plate. This problem may be solved by positioning the inlet and outlet on opposite sides of the stack of plates ("S-configuration").

Within a given plate, the velocity distribution in the channels is symmetric: the maximum and minimum flow rate are obtained for the channels on the edges of the plate and the middle channel, respectively (Fig. 4). The maldistribution may be reduced to about 1% by using a higher number of channels with smaller size.

The response to a step function is estimated from these results (Fig. 4). It appears possible to attain sufficiently fast response even at cycling periods of as low as 1 s.

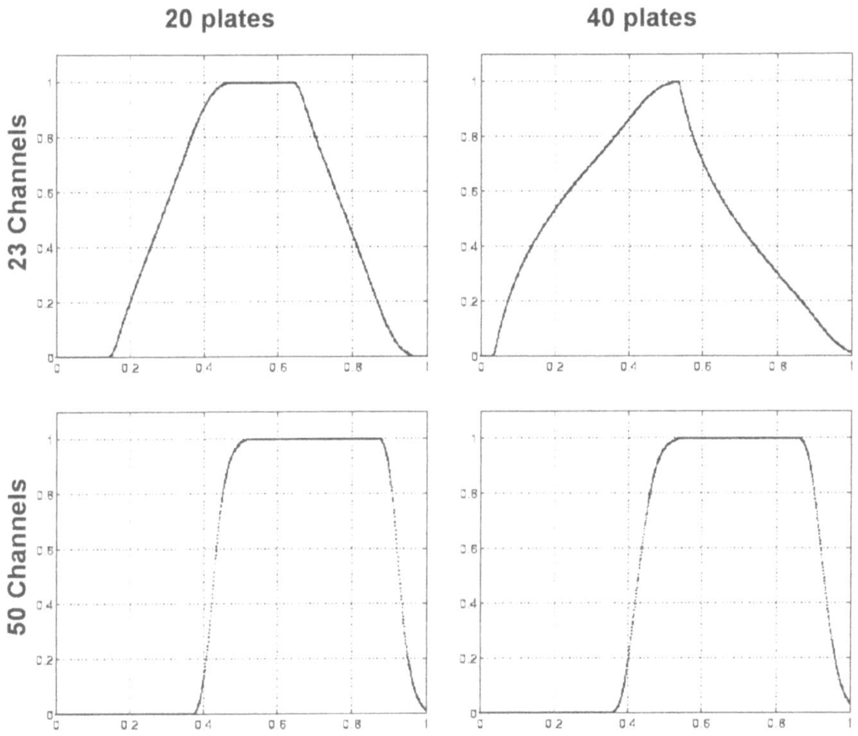

Figure 4. Response to a rectangular concentration pulse ($C_{inlet} = 1$ for $0 < t < 0.5$ s; $C_{inlet} = 0$ otherwise) with inlet and outlet on the same side. Stack of 20 (left) or 40 (right) plates with 23 channels (Ø = 500 µm, top) or 50 channels (Ø = 200 µm, bottom), respectively.

3.4 Reactor construction

In this project the specific advantages of microreaction technology as minimizing the hot-spots should be combined with exchangeable plates which contain either the reaction channels which are coated with catalyst or the heat exchanger structures.

The microsystem should be suitable for different chemical reactions and different materials are necessary for the plates (aluminum, stainless steel, titanium). So a modular concept is developed: the same stainless steel housing is used for all three reactions and the geometry of the reaction channels as well as the plates are individually adapted to the demands of the different chemical reactions (Fig. 5).

As the catalyst activity and the kinetics of the different reactions are unknown, the housing of the microreaction system should be flexible so that different numbers of catalyst plates can be used. For efficient removal of the reaction heat, plates with reaction channels and heat exchanger structures have to be arranged alternatively. A sufficiently gas-tight connection for different materials in a temperature range of 20-700°C may be reached with diffusion welding, gluing or even laser welding.

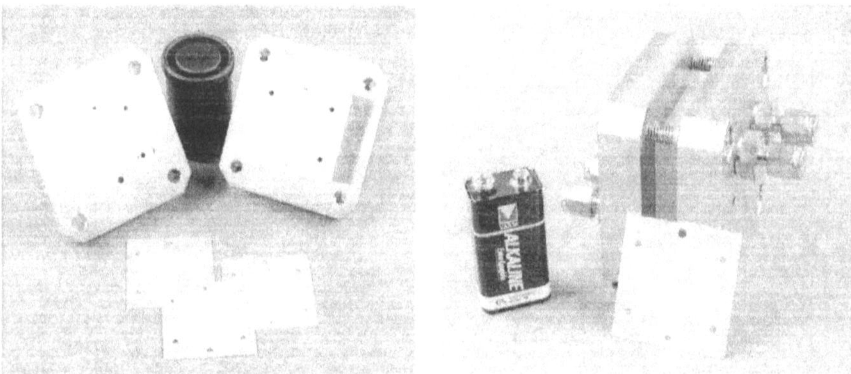

Figure 5. The modular microchannel reactor for periodic operation. The inlays may be made of various materials, depending on the reaction conditions. Inlet and outlet are arranged on opposite sides of the housing (S-configuration), as suggested by the simulations.

4 Catalyst preparation

4.1 Anodic Oxidation of Aluminum

The selection of a catalyst for the partial oxidation of isoprene is guided by the patent literature as well as by analogy to the butadiene partial oxidation to maleic anhydride. Mixed oxides of metals like V, Mo and Sb on supports like TiO_2, SiO_2, Al_2O_3, ZrO_2 appear to be promising (see Section 2.1). Reaction temperatures below 450°C allow for using aluminum as reactor material. Wießmeier and Hönicke introduced the anodic oxidation of microstructured aluminum plates to prepare a porous alumina support layer [12].

Preliminary experiments were performed on an aluminum wire (Ø 1 mm) in oxalic acid of 1.5 wt.-% (50 V, 12°C, up to 8 h). The pore length increases with increasing oxidation time. A BET surface area of 4 m^2/g was obtained for the alumina layer after subtraction of the mass of the aluminum core. When the anodic oxidation was applied to microstructured aluminum plates, the sealing surfaces were protected by the resist already used for the microstructuring process.

For the immobilization of the active metal oxide component on the support, a method suggested by van Hengstum et al. [13] has been selected. Batch adsorption of vanadyl acetylacetonate solved in toluene as a precursor will be performed at 65°C for several days with the anodically oxidized aluminum wire. V_2O_5 is obtained after subsequent calcination. It may be interesting to study batch adsorption with mixtures of different metal acetylacetonate complexes (e.g. vanadyl and molybdenum) as a preparation method for multicomponent catalysts.

4.2 Washcoating method

A pronounced stop-effect has been found on γ-alumina, so this oxide will be used as catalyst. Considering the known kinetic data for the dehydration and deamination, the catalyst should be 10-20 μm thick (~25-50 g/m^2) and the reaction temperature will be kept between 140 and 350°C.

γ-alumina is not accessible by anodic oxidation. Layers of porous γ-alumina that are stable up to 900°C may be obtained by washcoating with either a colloidal suspension of boehmite in water or a solution of aluminum alcoxide and subsequent heating to 450-500°C [14]. However, thick films (> 2 μm) will crack during drying, detaching themselves from the substrate [15,16].

Thicker films may be obtained by adding γ-alumina powder to the aqueous suspension of boehmite. Rather than homogeneous layers, the films are agglomerates of particles with boehmite acting as a binder for the particles [17]. In our case, an aqueous suspension containing 10% boehmite, 10% γ-Al_2O_3, 1% polyvinyl alcohol and 1.3% acetic acid is used. In order to ensure a homogeneous deposition in the channels and to maintain the entrance and exit chambers clean, a six-step procedure is proposed (Fig. 6). This procedure was applied to microstructured stainless steel plates (Fig. 7). Steps 2-4 were repeated twice for a thick layer of 22 - 24 g/m^2.

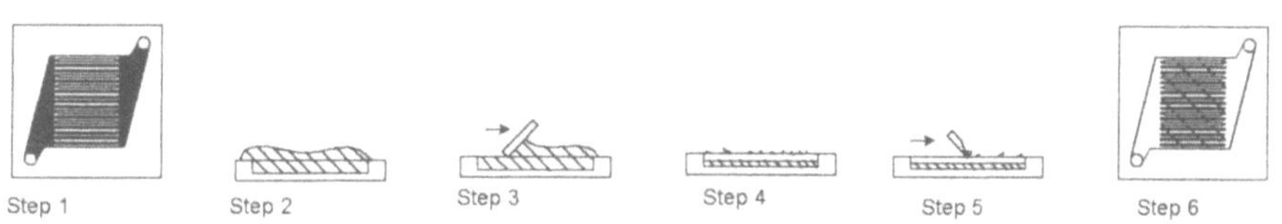

Figure 6. Different steps during the suspension-based catalyst deposition. Steps 2 through 4 may be repeated for thicker layers. 1) Protecting the inlet and outlet chambers with a thin polymer film. 2) Depositing the suspension on the plate. 3) Wiping off any excess suspension. 4) Drying at room temperature. Layer will shrink. 5) Cleaning mechanically the upper

surface of the plate. 6) Heating up to 500°C within 3 hours in order to burn the polyvinyl alcohol and the protecting polymer film and dehydrate the boehmite to γ-alumina.

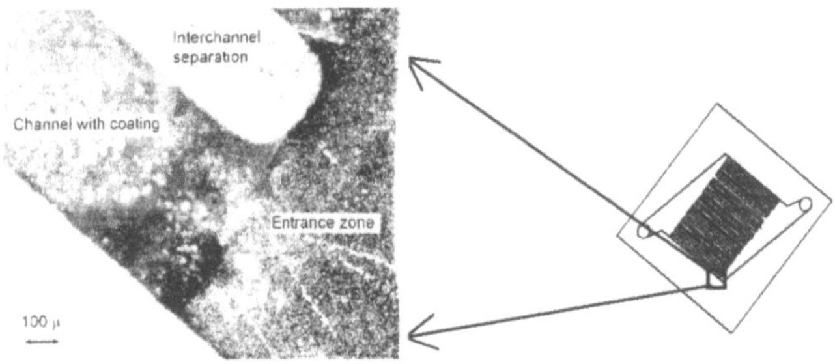

Figure 7. Channel coated with γ-alumina. Substrate is stainless steel, channel width 500 μm, height 100 μm. The photograph is a blow-up of the transition zone from entrance to channel.

4.3 Plasma-assisted sputtering and chemical vapor deposition (CVD)

For the oxidative autothermal dehydrogenation of propane, the use of aluminum (melting point too low) or stainless steel (catalytically active) was prohibitive. Titanium was chosen as the reactor material as it proved rather inert under oxidative conditions at high temperatures.

In the oxidative zone of the microchannel reactor, thin layers of deposited vanadia will be first examined for their suitability as a catalyst. In the non-oxidative zone, platinum is of interest due to its high activity for C-H bond rupture but poor ability to cleave C-C bonds (main component of most dehydrogenation catalysts). Plasma assisted sputtering is used as deposition method in both cases. Preliminary experiments yielded uniform vanadia layers with a thickness of 10 - 50 nm (Fig. 8). However, the catalytic activity was extremely low. Further optimization of the sputtering approach is in progress.

Figure 8. Microchannel reactor plate coated with a 10 nm V_2O_5 layer using magnetron sputtering (P = 0.039 mbar, 100 W (261 V, 389 mA), 20 ml/min Ar flow, 30 s).

As an alternative, chemical vapor deposition (CVD) has been performed with aluminum isopropylate at 370°C. XPS revealed carbon compounds that could be removed at 550°C in air. Both sputtering and CVD will be further studied.

5 Summary

The various motivations for periodic concentration modulation in microchannel reactors are reflected by the selected model reaction systems: reversible storage of selective oxygen during partial oxidation reactions in a wide frequency range (isoprene oxidation), periodic removal of inhibitory reactant for rate enhancement and model discrimination (alcohol dehydration), and periodic removal of coke in microchannels in a combination of an exothermic and an endothermic process (propane dehydrogenation). An adapted micro heat exchanger promises to allow for sharp concentration fronts. A variety of catalyst application procedures like anodic oxidation, suspension method, sputtering or CVD may be used.

Acknowledgments

This project was funded by the Bundesministerium für Forschung und Technologie (grant N° 03C0282), by the CNRS (France), and the Swiss Federal Science Foundation (grant N° 510-155). One of us (M.L.) acknowledges financial support through a Liebig grant from the BMBF/FCI. Thanks to Dr. J.-P. Müller for carrying out sputtering experiments, H. Redlingshöfer for CVD experiments and B. Spötzl for support with the anodic oxidation.

References

[1] P. Silveston, R.R. Hudgins, A. Renken, "Periodic operation of catalytic reactors - introduction and overview", *Catal. Today* **25** (1995), 91-112; F.J.R. van Neer, A.J. Kodde, H. Den Uil, A. Bliek, "Understanding of Resonance Phenomena on a Catalyst

under Forced Concentration and Temperature Oscillations," *Canad. J. Chem. Eng.* **74** (1996), 664-673.

[2] J. Thullie, A. Renken, "Model discrimination for reactions with a stop-effect", *Chem. Eng. Sci.* **48** (1993), 3921-3925.

[3] R.M. Contractor, "Improved vapor phase catalyic oxidation of butane to maleic anhydride", U.S. Patent 4,668,802 (May 26, 1987).

[4] R. Philipps, M. Liauw, "Selektive Oxidation von Kohlenwasserstoffen zur Erzeugung chemischer Wertprodukte", FORKAT Statusbericht, March 1999.

[5] J. Koubek, J. Pasek, V. Ruzicka, "Exploitation of a Nonstationary Kinetic Phenomenon for the Elucidation of Surface Processes in a Catalytic Reaction", in: *New Horizons in Catalysis*. 1980, Elsevier-Kodansha: Amsterdam-Tokyo. p. 853-862.

[6] S. Golay, O. Wolfrath, R. Doepper, A. Renken, "Model Discrimination for Reactions with Stop-effect", in: *Dynamics of surface and reaction kinetics in heterogeneous catalysis*, G.F.Froment and K.C. Waugh, Eds. Amsterdam: Elsevier, 1997, 295-304.

[7] A. Gavriilidis, S. Konstantinidis, D. Gobby, "Interaction of Endothermic and Exothermic Reactions in Microreactors", Proceedings of the 2nd Int. Conf. On Microreaction Technology, New Orleans, USA, March 8-12, 1998, 211-219.

[8] D. Creaser, B. Andersson, R.R. Hudgins, P.L. Silveston, "Transient Kinetic Analysis of the Oxidative Dehydrogenation of Propane", *J. Catal.* **182** (1999), 264-269.

[9] G. Wießmeier, *Monolithische Mikrostrukturreaktoren mit Mikroströmungskanälen und regelmäßigen Mesoporensystemen für selektive, heterogen katalysierte Gasphasenreaktionen*. Aachen: Shaker, 1997.

[10] St. Walter, G. Frischmann, R. Broucek, M. Bergfeld, M. Liauw, "Fluiddynamische Aspekte in Mikrostrukturreaktoren", *Chem. Ing. Tech.* 171 (1999), 447-455.

[11] St. Heß, M. Liauw, G. Emig, "Partial Oxidation of n-Butane to Maleic Anhydride over a Vanadiumpyrophosphate Catalyst in the Riser Regenerator System", DGMK-Conference "Selective Oxidations in Petrochemistry", Hamburg-Reinbek, 8./9. 10. 1998.

[12] G. Wießmeier, D. Hönicke, "Strategy for the Development of Micro Channel Reactors for Heterogeneously Catalyzed Reactions", Proceedings of the 2nd Int. Conf. On Microreaction Technology, New Orleans, USA, March 8-12, 1998, 24-32.

[13] A.J. Van Hengstum, J.G. Van Ommen, H. Bosch, P.J. Gellings, "Preparation of Supported Vanadium and Molybdenum Oxide Catalysts Using Metal Acetylacetonate Complexes", *Appl. Catal.* **5** (1983), 207-217.

[14] J. Brinker, G.W. Scherer, *Sol-Gel Science*. Boston: Academic Press, 1990, 908.

[15] R. Nass, H. Schmidt, "Synthesis of alumina coating from chelated aluminum alkoxydes", *J. Non-crystalline Solids* **121** (1990), 329-333.

[16] O.A. Harizanov, M. Surtchev, "Glycerol Treated Aluminum Trihydroxide Sol for Coatings", *Materials Letters* **32** (1997), 25-28.

[17] T. Dupin, "Alumina Coating Compositions for Catalyst Supports and Process for their Formulation". June 11, 1984, Rhône Poulenc Spécialités Chimiques, Courbevoie, France, US-Patent 4,529,718.

Micro-reactor synthesis: synthesis of cyanobiphenyls using a modified Suzuki coupling of an aryl halide and aryl boronic acid.

Victoria Skelton, Gillian M. Greenway*, Stephen J. Haswell, Peter Styring and David O. Morgan[1]

School of Chemistry, Faculty of Science and the Environment, University of Hull, Cottingham Road, Hull, HU6 7RX, UK

[1] Smithkline Beecham Pharmaceuticals, Old Powder Mills, Nr Leigh, Tonbridge, TN11 9AN, UK

Abstract.

The synthesis of cyanobiphenyls using a modified Suzuki coupling was performed in a micro-reactor system (channel geometries of 300 μm wide and 115 μm deep) etched in borosilicate glass, using a photolithographic patterning technique [1]. The organic starting reagents were moved by electro-osmotic flow (EOF). A micro-porous silica structure [2] was used within the channels, to act as a micro-pump and also assist in the immobilisation process for the heterogeneous catalyst bed.

The synthesis of cyanobiphenyl was performed at room temperature, within 30 seconds, using a heterogeneous catalyst. The yield obtained was *62 ± 7%* without the addition of a base and this was achieved using an injection profile whereby 4-bromobenzonitrile was injected for 5 seconds, with a 25 second injection interval, into a continuous flow of phenylboronic acid. Palladium contamination levels from the supported catalyst were determined in the product by ICP-MS and were found to be in the range of *1.2 - 1.6 ppb*.

Conventional laboratory methodology was adopted using a heterogeneous catalyst, with and without the addition of a base, at both room and elevated temperatures (75 – 80°C) for 8 hours. The maximum yield obtained was significantly lower (10 % yield) than with the micro-reactor illustrating the potential for clean efficient synthesis.

Keywords: Micro-reactor, Suzuki synthesis, electro-osmotic flow, catalysis, micro-porous silica structures.

1 Introduction.

A common synthetic method for the formation of a carbon to carbon bonds is the Suzuki-Miyaura reaction [3] due to the selectivity in product formation. With this type of reaction however, the homogeneous catalyst, commonly $[Pd(PPh_3)_4]$ undergoes degradation, leaving undesirable by-products such as colloidal palladium requiring expensive product purification. Typically, homogeneous catalyst yields are in the range of *44 – 78%* in dimethoxyethane (DME), benzene and tetrahydrofuran (THF) with reactions requiring between 2 to 6 hours [4].

To date, analytical publications of literature have described the Micro-Total analytical system (μ-TAS) [5-7], which have mainly been directed towards separation and characterisation of DNA [8-13]. Ponton [14] however identified the considerable potential micro-reactors could have in the field of chemical synthesis potentially offering high energy efficiencies, clean, selective and controllable synthesis with versatile distribution and scale-out methodology leading to parallel processing.

This paper uses similar technology to μ-TAS but focuses on the development of a liquid based catalytic synthetic micro-reactor system enabling the synthesis of organic species under highly controlled conditions using EOF. The micro-reactor has unique features (spatial and temporal control) enabling the rapid synthesis of high purity products with effective reaction kinetics. The potential yield and speed of analysis with minimal product purification could be developed for an industrial process by parallel scale out, producing bulk quantities, which could be an ideal development platform for the pharmaceutical and chemical refining industries. The characteristics of the micro-reactor have been highlighted in this paper by the development of a modified Suzuki coupling reaction, Figure 1, which is an ideal reaction candidate for study due to the catalytic nature of the synthesis.

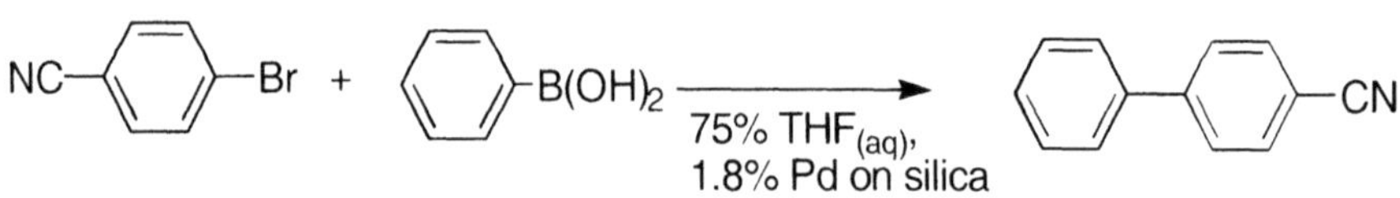

Fig. 1. Reaction scheme for the aryl-aryl coupling of 4-bromobenzonitile and phenylboronic acid.

2 Experimental.

The micro-reactor was etched in borosilicate glass using a photolithographic technique (channel geometries 300 μm wide and 115 μm deep). The plate was etched using 1 % hydrofluoric acid with 5% ammonium and exposed to strong UV light before being dipped in photoresist remover.

The catalyst bed was manually positioned in the channel indicated in figure 2, at room temperature, using a10% v/v of formamide and potassium silicate [2]. The micro-reactor bottom plate was annealed using a microwave furnace to the top plate (17 mm thick) sealing the channels. The top plate also included 2 mm id predrilled holes aligned at the ends of each channel to act as reservoirs.

A modified synthetic method by Styring [15] was used in the micro-reactor system. Equal molar concentrations and volumes of phenylboronic acid and 4-bromobenzonitrile, in 75% $THF_{(aq)}$ were measured into reservoir B and A respectively. Into reservoir C, 30 µl of the THF solution was measured and the electrodes were then placed in each reservoir (A & B positive, and C negative), which were then sealed with laboratory film. An external applied voltage (in the range 100 - 400V) was applied to channel B. However, channel A was switched on for various injection times by applying an external applied voltage of 200V. There was continuous flow from B to C for 25 minutes during which ranges of injection times were investigated. The product was analysed by GC-MS.

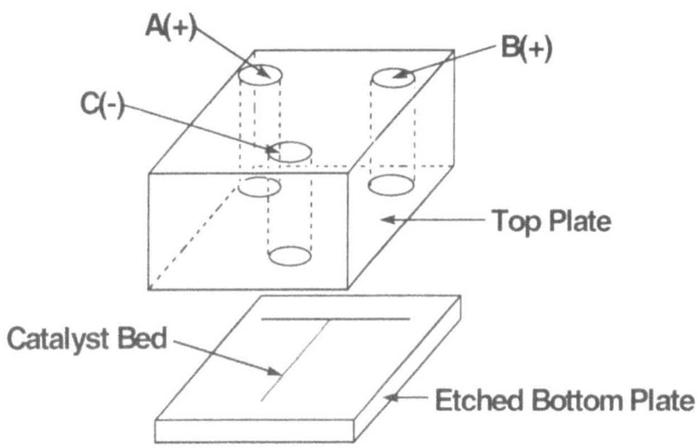

Fig 2. Schematic of 'T' shaped manifold used for the synthesis of cyanobiphenyl.

3 **Results & Discussion**.

Study 1 centred on the effect the injection time (sample plug volume) of 4-bromobenzonitrile would have on the yield of cyanobiphenyl. The longer the injection time used; the effective concentration of the aryl halide in the micro-channels increased i.e. the longer the injection time produced a decrease in the yield of cyanobiphenyl. By increasing the injection length, the effective concentration of aryl halide in the micro-reactor channels reduced the effective concentration of the phenyl boronic acid due to dilution. The optimum injection length was found to be 5 seconds injection time giving a yield of 14%. The injection time of 5 seconds reflected the volume time for diffusive mixing to occur in the micro-channels between the two organic reagent species. This injection time was used for the remainder of the study.

Study 2 investigated the maximum number of 4-bromobenzonitrile injections per minute that could be tolerated prior to catalyst saturation and the concentration of phenyl boronic acid effected the reaction rate. In order to achieve this, the time interval between injections was varied, consequently affecting the number of injections in a 25 minute reaction period.

In figure 3, a low yield is initially seen for the following injection times intervals 5 - 20 seconds and this occurs due to catalyst saturation. The shorter time intervals between the injections increase the concentration of 4-bromobenzonitrile (effective dilution of phenylboronic acid) in the micro-reactor channel therefore active competition for catalyst sites occurs. The effective concentration of the phenylboronic acid is decreased in the micro-reactor, also effecting the reaction kinetics high lighting the yield.

Using a 5 second injection with a 25 second time interval between injections there is a high enough concentration of phenylboronic acid for the transmetallation step to occur. This gave a mean yield (n = 4) of **62 ± 7%** (GC retention time = 10.22 minutes, $[M^+] = 179$).

The final section of the graph (30 - 55 seconds) indicated that the effective concentration of 4-bromobenzonitrile was too low. As the interval between injection time increased the effective concentration of the aryl bromide decreases in the micro-channel. As the reaction proceeds, the 4-bromobenzonitrile cannot reoccupy the catalytic sites effectively or fully after the reductive elimination step due to the high concentration of the phenylboronic acid in turn reduing catalyst turnover.

Study 3 compared the yield of cyanobiphenyl obtained for various applied voltages or flow rates using an injection profile of 5 seconds injection every 25 seconds.

Figure 4 indicates that at 200V (0.8 μl min^{-1}), the flow rate of the phenylboronic acid gave a reported yield of cyanobiphenyl between 49 % and 68 % whilst the other voltages investigated produced negligible yields. At 1.0 and 1.3 μl min^{-1} (300 and 400 V) the flow rate is too fast, reducing the residency time of the 4-bromobenzonitrile on the palladium catalyst. At lower voltages (0.65 μl min^{-1}), the flow rate is too small for effective transfer of the 4-bromobenzonitrile to occur efficiently.

Conventionally, the Suzuki synthesis requires the addition of a base, commonly aqueous sodium carbonate [16], which activates the boronic acid group. The synthesis of cyanobiphenyl in this micro-reactor system is novel as the addition of a base is not required, shown in figure 5.

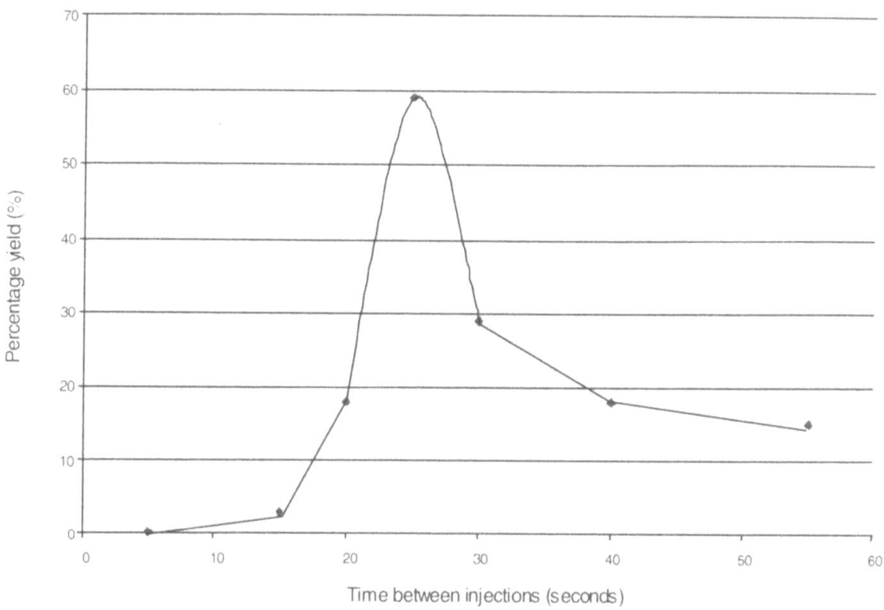

Fig. 3. Graph represents the increase in phenylboronic acid concentration with increasing interval time.

In the micro-reactor the water present under the influence of an external applied voltage could undergo solvation to form hydroxide. The additional surface wall area, due to the incorporation of the silica frit, probably accelerates this solvation rate. Clearly ions free in solution will be at a negligible concentration but due to the presence of the palladium catalyst, it is postulated that hydroxide ions are localised on the palladium(II) surface. This forms a high concentration near the catalytically active sites, which in turn produces a high enough concentrations of hydroxide ions for the transmetallation step to proceed.

The Suzuki reaction was performed on a bulk scale with and without a base both at room and elevated temperature (75 - 80°C) using conventional laboratory methods based on the sample solvent and reagents outlined for the micro-reactor system. As expected, the reaction did not yield any cyanobiphenyl product, which confirmed that the micro-reactors characteristic features and the immobilisation of the catalyst bed are highly efficient at converting the starting reagents to the cyanobiphenyl product at room temperature within 25 minutes. The same reaction was also performed at elevated temperatures with the addition of sodium carbonate (0.2 M, 20 ml water). The conversion of the reagents to cyanobiphenyl was found to be only 10 %.

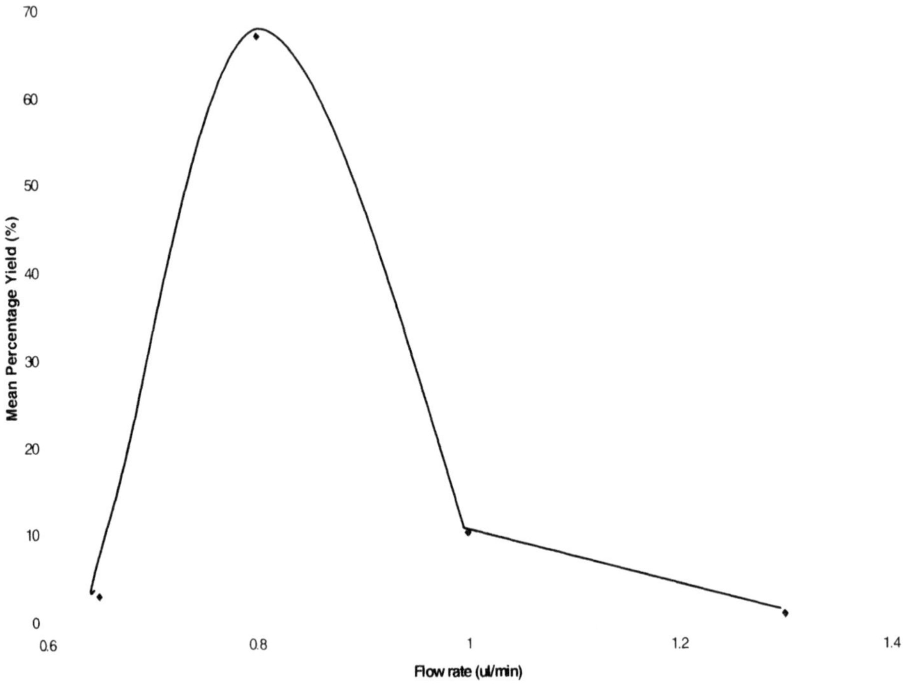

Fig. 4. Plot to show the yield for various applied voltages for a 5 second in injection time.

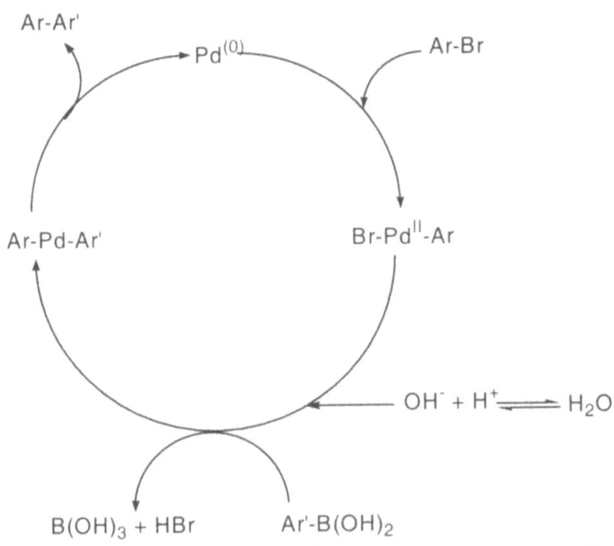

Fig. 5. Proposed catalytic mechanism for the catalyst palladium on silica.

Analysis of the reaction products by ICP-MS was performed establishing the levels of palladium contamination. The analysis indicated that the level of palladium was in the range of 1.2 to 1.6 ppb, confirming that the current immobilisation method using a micro-porous silica structure was environmentally robust. The low levels of product contamination from the heterogeneous catalyst including the removal of colloidal palladium allows the removal of costly sample 'clean' up, producing a pure compound, that is ideal for the chemical or pharmaceutical industries.

4 Summary of Micro-reactor features.

Summarised below are the key features of a liquid based catalytic system, which have been highlighted in this paper.

a) Yield of cyanobiphenyl reported to be *62 ± 7%*, producing a six-fold increase compared to the macro-scale reaction using the same catalyst.

b) The Suzuki reaction is unique due to the fact that the reaction *uses a heterogeneous catalyst without the addition of a base (Na_2CO_3), at room temperature.*

c) VG Elemental PlasmaQuadII+ ICP-MS determined the levels of contamination for palladium to be in the range *1.2 - 1.6 ppb.*

d) The use of a supported catalyst compared to the traditional homogeneous catalysts *reduces* the normally observed high levels of palladium impurities observed in batch reactions.

e) Conventional heterogeneous catalysts require regeneration due to the formation of colloidal palladium, which is difficult to isolate and recycle. Using a supported immobilised palladium catalyst not only allows immobilisation on a micro-porous silica structure within the micro-reactor but also prevents the diffusion of the catalyst into the product solution *removing the costly process of separating catalyst impurities.*

f) The current reaction has been used for a large number of reactions *(operating for over 35 hours)* with no apparent loss of catalytic activity.

5 Further Work

With the successful synthesis of cyanobiphenyls, a range of starting materials will be investigated broadening the variety of Suzuki reaction products. Also enhancement of the product yield will be achieved by the incorporation of a separation system, enabling the recycling of the starting materials in turn reducing waste. Using the natural fluorescence observed by many liquid crystals that can be generated by the Suzuki method will also require the development of in-situ fluorescence detection.

6 References

[1] Daykin, R.N.C., Haswell, S.J., *Anal. Chim. Acta*, 1995, **313**, 155.

[2] Christensen, P.D., Johnson, S.W.P., McCreedy, T., Skelton, V., Wilson, N.G., *Anal. Commun.*, 1998, **35**, 341.

[3] Moreno-Manas, M., Perez, M., Pleixats, R., *J. Org. Chem*, 1996, **61**, 2346.

[4] Miyaura. N, and Suzuki, A., *Chem. Rev.* 1995, **95**, 2457.

[5] Manz, A., Effenhauser, C.S., Barggraf, N., Verpoorte, E.J.M, Raymond, D.E., Widmer, H.M., *Analysis Mag.*, 1994, **22**, M25.

[6] Manz, A., Harrison, D.J., Verpoorte, E., Fettinger, J.C., Ludi, H., Widmer, H.M, 1991, **45**, 103.

[7] Manz, A., Harrison, D.J., Verpoorte, E., Widmer, H.M., *Adv. Chromatogr.*, 1993, **33**, 1.

[8] Woolley, A.T., Mathies, R.A., *Anal Chem*, 1995, **67**, 3676.

[9] Woolley, A.T., Hadley, D., Ladre, P., deMello, A.J.; Mathies, R.A., Northrup, M.A., *Anal. Chem.*, 1996, **68**, 4081.

[10] Woolley, A.T., Sensabaugh, G.F., Mathies, R.A., *Anal. Chem.*, 1997, **69**, 2181.

[11] Woolley, A.T., Mathies, R.A., *Proc. Natl. Acad. Sci. U.S.A.*, 1994, **91**, 11348.

[12] Jacobson, S.C., Ramsey, J.M., *Anal. Chem.*, 1996, **68**, 720.

[13] Simpson, P.C., Roach, D., Woolley, A.T., Thorsen, T., Sensabaugh, G.F., Mathies, R.F., *Science*, **1998**, 282, 399.

[14] Ponton, J.W., Microreaction Technology, Proceedings of the First International Conference on Microreaction Technology, Ehrfield, W. (Ed.), Springer-Verlag Berlin Heidelberg, New York, Part 1, 10.

[15] Styring, P., Vuijk, J.D., Wright, S.A., Takatoh, K., Dong, C., *J. Mater.Chem.*, 1994, **4**, 1365.

[16] Smith, G.B., Dezeny, G.C., Hughes, D.L., King, A.O., Verhoeven, T.R., *J. Org. Chem.*, 1994, **59**, 8151.

Optimization of Reaction-Separation Networks via Mass Integration on the μ-Scale

František Štěpánek and Miloš Marek

Department of Chemical Engineering
Prague Institute of Chemical Technology
Technická 5, 166 28 Praha 6, Czech Republic

Abstract. Simultaneous reaction and product separation in a single multifunctional microreactor – consisting of continuously operated microchannels with a wall-immobilized enzyme, and periodically (adsorption/regeneration cycles) operated separating channels using an ion-exchange resin – is investigated. The production of nicotinamide by immobilized nitrile hydratase is considered as an example of a reaction whose yield, otherwise limited by product inhibition, can be substantially increased by *in-situ* product separation. Optimum design (sizing and arrangement of functional blocks) and operating (flowrates, switching periods) parameters are calculated, the objective being maximum product throughput per unit volume of monolith.

Keywords. Process design, mixed periodic/continuous operation, simultaneous reaction-separation, product inhibition, nicotinamide.

1 Background

Techniques for the fabrication of sub-milimeter size components for microreactors have advanced rapidly in recent years, nowadays allowing the implementation of micro-scale analogues of many unit operations used in the chemical process industries. Mechanisms for active transport of liquids have successfully been applied (Gallardo et al., 1999, Gau et al., 1999, or Sammarco and Burns, 1999), micro heat exchangers with large heat-transfer rates were demonstrated (Schubert et al., 1998), and the fabrication of micro-channels and vessels of arbitrary shapes, which can serve as reactors or liquid-liquid contactors, has also been mastered (Jackman et al., 1998, or Martynova et al., 1997).

When momentum-, heat- and mass-transfer operations (which form a basis of the chemical engineering toolbox) can be performed at the micro-scale, it is also desirable to apply (or newly develop) methodologies for the design, "scale-down", and feasibility evaluation of microreactor-based manufacturing processes (cf. Rinard, 1998). It is important to select processes where the main comparative advantages of microreactors – i.e., large surface/volume

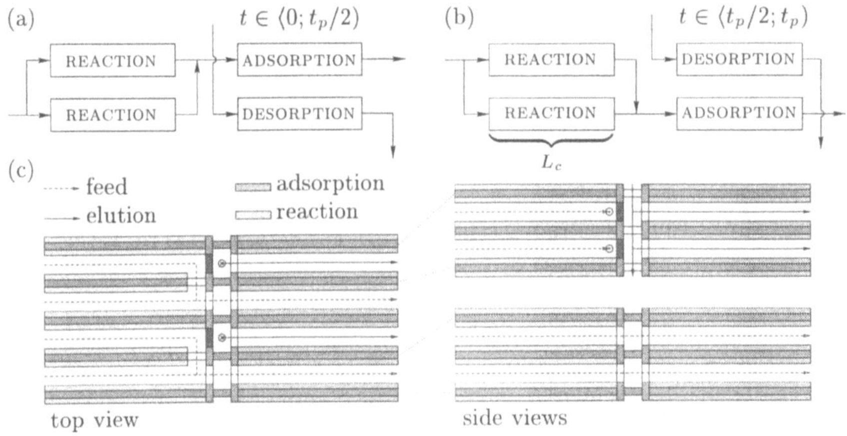

Figure 1: Process flow diagram and possible physical arrangement of microchannels inside a structured monolith. (a) and (b) PFDs corresponding to two half-periods; (c) close-up view on the junction of reaction, cross-flow, and adsorption channels, corresponding to the second half-period.

ratios and the possibility to accurately control the spatial and temporal distribution of small quantities of reactants and products – can bring significant improvements (in terms of conversion, selectivity, operation safety, etc.). Particularly appealing is the possibility to fabricate multifunctional microreactors, and thus integrate the function of several unit operations into a single device.

Some thermodynamic (low equilibrium constant) or kinetic (product inhibition) limitations, which can render otherwise desirable reaction pathways economically infeasible, can be overcome by the coupling of a reaction with product separation. For example, by combining solid mass-separating agents (adsorption, ion exchange) with an immobilized catalyst in a single microchannel, true *in-situ* reaction-separation coupling can be achieved (Štěpánek et al., 1999), providing higher efficiency than mixed catalyst/adsorbent packings used at the large scale.

2 Design Problem

Recently, design methodologies for co- and countercurrent multistage extractive reaction processes were developed (Minotti et al., 1998, Samant and Ng, 1998). The present work deals with the design of a multistage, cross-flow adsorptive reaction process, which differs mainly in that the reaction and elution streams do not come into contact directly, but the product is transferred from one stream to the other via an immobile solid phase (adsorbent), thus requiring a mixed continuous/periodic operation regime (Barton and Pantelides, 1994).

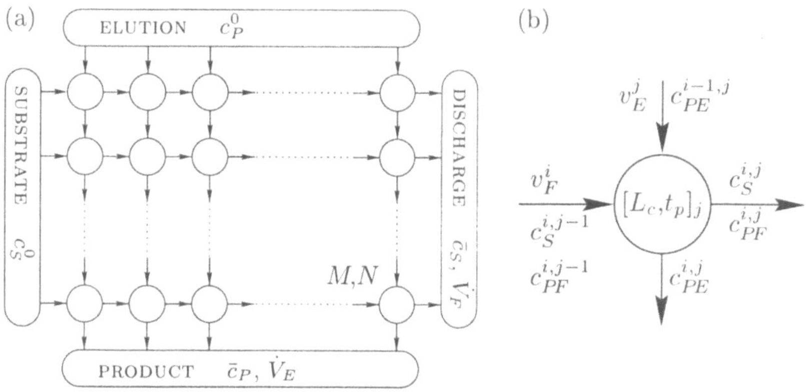

Figure 2: (a) Flowsheet of a cross-flow arrangement. (b) Detail of a balance node with design parameters and properties of inlet and exit streams.

Consider a heterogeneously catalysed liquid-phase reaction $S \to P$, whose kinetics is limited by product inhibition, carried out in a bifunctional structured monolith, consisting of a system of micro-channels divided into segments of length L_c whose walls contain either an immobilised enzyme (catalyst) or an ion-exchange resin (for selective product binding). Catalytic channel segments are operated continuously; separation channels must be operated periodically (in a sequence of pH-induced adsorption/elution cycles with period t_p) in order to regenerate the ion-exchange resin and recover the product accumulated therein. Figure 1 shows the flow sheets and physical arrangement of channel segments in the microreactor. To maintain a continuous operation of the production unit as a whole (i.e., constant flow rate per cross section of monolith), the flow is split into two parallel reaction channels for each pair of separation channels, because only one of them is available for product separation at a time (the other one being regenerated) – Fig. 1b.

The system of two reaction and two separation channel segments is considered as a single continuous-flow reaction-separation node (Fig. 2b) in the overall material balance, with outlet stream properties corresponding to time-averaged values. The design parameters specifying one vertical layer of channels in the microreactor (Fig. 2a) are the number, N, and lengths, L_c [m], of reaction and separation segments in the axial direction, and the position of longitudal "service" channels for distributing the regeneration solution and collecting the recovered product, given by the number, M, of interleaved "working" channels. A parameter α^j assigned to each vertical layer of nodes, $\alpha^j = 1$ meaning reaction-separation, $\alpha^j = 0$ meaning reaction only nodes. In the latter case, the elution flowrate in the j-th cross-flow layer is zero and each node contains four reaction-only segments. The operating parameters are the periods t_p^j [s] of adsorption/elution switching, and the flow velocities of feed ($v_F^i = \dot{V}_F^i / 2\epsilon A$, $i = 1 \dots M$) and regeneration ($v_E^j = \dot{V}_E^j / \epsilon A$,

$j = 1 \ldots N$) streams, where $\dot{V}$ [m^3/s] is the volumetric flowrate, A [m^2] the cross-section area belonging to one channel, and ϵ [-] the void fraction.

Eyal and Charles (1990) studied the kinetics of cyanopyridine hydration to nicotinamide over immobilized nitrile hydratase, and proposed the relationship

$$r_s = \frac{k_r c_S}{c_S + K_M(1 + c_P/K_P) + c_S^2/K_S} \tag{1}$$

where r_s [mol/s·kg of enzyme] is the specific reaction rate, c_S and c_P [mol/m^3] are the concentrations of substrate and product, respectively, $k_r = 0.027$ mol/s·kg is the reaction rate constant (at optimum pH = 7.5), $K_M = 0.122$ mol/m^3 is Michaelis, $K_P = 0.126$ mol/m^3 product inhibition, and $K_S =$ 0.185 mol/m^3 substrate inhibition constant. The mass balance in the reaction channels (assuming ideal plug flow) leads to

$$\frac{dc_k}{dz'} = \nu_k \frac{\tilde{\rho}}{v_F'} r_s \qquad k = S, P \tag{2}$$

where $z' \equiv z/L_c$ is dimensionless axial coordinate, $v_F' \equiv v_F/L_c$ [s^{-1}] is the space velocity, $\tilde{\rho} \equiv \rho w(1-\epsilon)/\epsilon = 6.60$ kg/m^3 is apparent enzyme density (ϵ being the void fraction, ρ the true density of solid and w the mass fraction of enzyme therein); ν_k is a stoichiometric coefficient, equal to -1 for the substrate ($k = S$) and $+1$ for the product ($k = P$). Assuming plug flow and instant fluid-solid equilibrium, the material balance of P (S is not adsorbed) in a separation channel gives

$$\left(1 + \frac{1-\epsilon}{\epsilon} K_{a/d}\right) \frac{\partial c_P}{\partial t} = -v_{a/d} \frac{\partial c_P}{\partial z} \tag{3}$$

where $K_{a/d}$ is an adsorption equilibrium constant, equal to $K_a = 8.20$ during the feed pH (i.e., (t mod t_p) $< t_p/2$) and $K_d = 0.75$ at elution (for details about the pH-dependence of the ion-exchange equilibrium, see Štěpánek et al., 1999); $v_a = 2v_F$ during the feed sub-period and $v_d = v_E$ during elution. By introducing a transformed coordinate $y \equiv z + u_{a/d}t$ into Eq. 3, where $u_{a/d}$ is the shock wave velocity, the switching period can then be calculated from

$$t_p/2 = L_c/u_a = \left(1 + \tfrac{1-\epsilon}{\epsilon} K_a\right)/(2v_F') \tag{4}$$

The space velocity of the elution stream, $v_E' \equiv v_E/L_c$, follows from the requirement of a complete regeneration of the separation channel in $t_p/2$.

$$v_E' = 2v_F'\left(1 + \tfrac{1-\epsilon}{\epsilon} K_d\right)/\left(1 + \tfrac{1-\epsilon}{\epsilon} K_a\right) = 2v_F' r_K \tag{5}$$

where the ratio $r_K = 0.214$ (given the values of the adsorption equilibrium constants, specified above) was introduced for convenience. A requirement on the total achieved substrate conversion, $\bar{x}_S = (c_S^0 - c_S^N)/c_S^0$, where

$c_S^N = \sum_{i=1}^M c_S^{i,N} v_F^i / \sum_{i=1}^M v_F^i$, follows from the costs of recovery and recycle of unreacted substrate, and is assumed to be an exogenous variable, set independently to the design sub-problem. The substrate concentration in feed, c_S^0, is also specified externally. The initial product concentrations in the feed and elution streams, c_P^0, are assumed to be zero.

The process design objective is the maximization of the operating profit (i.e., revenues minus costs) created by an incremental investment into the fixed capital being designed (the microreactor in this case). Let π_P and π_S be the internal prices (in \$/mol) of the product and substrate, respectively, and π_r [\$/m^3 reactor space] be the asset reproduction price; let the cost of fluid transport be π_f [\$/m^3]. An objective function $O_P = (\pi_P \dot{n}_P - \pi_S \dot{n}_S - \pi_f \dot{V}_f)/(\pi_r V_r / t_l)$ can then be defined, where t_l is the asset lifetime (assuming linear depreciation), $V_r = 4(M+2)NL_c\epsilon A$ is the reactor space, and $\dot{V}_f = 2v'_F L_c \epsilon A(M + r_K \sum_{j=1}^N \alpha^j)$ is the total fluid flow rate (recall Eq. 5). From the reaction stoichiometry it follows that $\dot{n}_P = \bar{x}_S \dot{n}_S$, where $\dot{n}_S = 2Mc_S^0 v'_F L_c \epsilon A$. Inserting the above expressions into the definition of O_P leads, after some rearrangement, to

$$O_P = v'_F \frac{\pi_1 c_S^0 M - \pi_2 (M + r_K \sum_{j=1}^N \alpha^j)}{(M+2)N} \tag{6}$$

where $\pi_1 = (\pi_P \bar{x}_S - \pi_S)/(\pi_r/t_l)$ is a cost factor accounting for the product throughput and $\pi_2 = \pi_F/(\pi_r/t_l)$ can be thought of as a penalty for the increased total flow rate. The decision variables are M, v'_F, and the vector of α^j; N follows from the requirement on the overall substrate conversion $\bar{x}_S$.

3 Solution Methods

Let us represent the model of one reaction-separation node (Fig. 2b) by a vector function, denoted $U_{i,j}$, which maps the properties of the inlet streams, $[c_S^{i,j-1}, c_{PF}^{i,j-1}, c_{PE}^{i-1,j}]$, and the parameter $v_F'^i$ to the unknown operation parameters $[t_p^j, v_E'^j]$ and the composition of the outlet streams $[c_S^{i,j}, c_{PF}^{i,j}, c_{PE}^{i,j}]$. As a result of the equilibrium assumption in Eq. 3, it holds that $c_{PF}^{i,j} = c_{PE}^{i-1,j}$. Since the adsorption equilibrium is described by a linear relationship, and the shock-wave velocity is thus concentration-independent, t_p^j and $v_E'^j$ can be calculated directly from Eqs. 4 and 5, respectively. The outlet concentrations $c_S^{i,j}$ and $c_{PE}^{i,j}$ are obtained by a short-cut method, which makes use of the fact that the substrate conversion per one reaction segment is relatively small, thus the two ODEs (Eq. 2) can be decoupled by assuming a constant mean product concentration $\tilde{c}_P^{i,j} = (c_{PF}^{i,j-1} + c_{PE}^{i,j})/2$ and integrating the remaining equation for c_S analytically with $\tilde{c}_P^{i,j}$ inserted into the reaction rate expression, Eq. 1. The resulting nonlinear algebraic equation for $c_{SE}^{i,j}$ is then solved by the Newton method. From the reaction stoichiometry it follows that $c_{PE}^{i,j} = c_{PF}^{i,j-1} + (c_{SE}^{i,j-1} - c_{SE}^{i,j})$. For the reaction-only nodes ($\alpha^j = 0$),

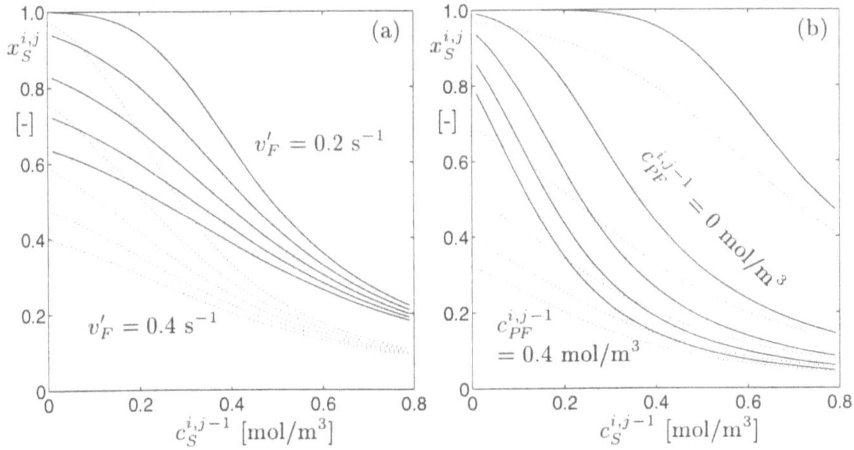

Figure 3: Dependence of the substrate conversion, $x_S^{i,j}$, on the inlet substrate concentration, $c_S^{i,j-1}$, the space velocity, v'_F, and the product concentration in the feed stream, $c_{PF}^{i,j-1}$, for a single reaction segment. (a) Curves are parametrized by $c_{PF}^{i,j-1} = 0.0, 0.2, 0.4, 0.6, 0.8$ mol/m^3, top to bottom. (b) Curves are parametrized by $v'_F = 0.1, 0.3, 0.5, 0.7, 0.9$ s^{-1}, top to bottom.

no elution stream is present ($\Rightarrow$ no c_{PE}), the exit stream composition $c_{PF}^{i,j}$ is obtained by applying twice the procedure described above (two pairs of reaction segments in series form a reaction-only node). The dependence of the substrate conversion in a single reaction segment on the inlet substrate and product concentrations and the space velocity, is shown in Fig 3.

From the continuity condition and the the cross-flow relationship between the flow rates v'_E and v'_F (Eq. 5) it follows that $v_E'^j = v'_E \forall j$ and $v_F'^i = v'_F \forall i$, which substantially reduces the number of degrees of freedom. For specified M, v'_F, and $\boldsymbol{\alpha}$, N is calculated by a "stage-to-stage" method starting at the feed side. The maximization of O_P is a problem of mixed integer nonlinear programming (MINLP), because M and α^j are integer variables, while v'_F can take any real positive value. The problem is further complicated by the fact that the dimension N of the vector $\boldsymbol{\alpha}$ is not fixed but depends on M and v'_F, as well as on the elements of the vector itself.

4 Results

The decision variables are M, v'_F, and $\boldsymbol{\alpha}$, N is a dependent variable. The objective function O_P (Eq. 6) is directly proportional to v'_F, and inversely proportional to N. The functional dependence of N on M and $\boldsymbol{\alpha}$ at different v'_F will thus be crucial for determining the optimum O_P. For larger M, the additive term $+2$ in the denominator of Eq. 6 can be neglected and the whole fraction simplified by $1/M$; then the only term which remains dependent on

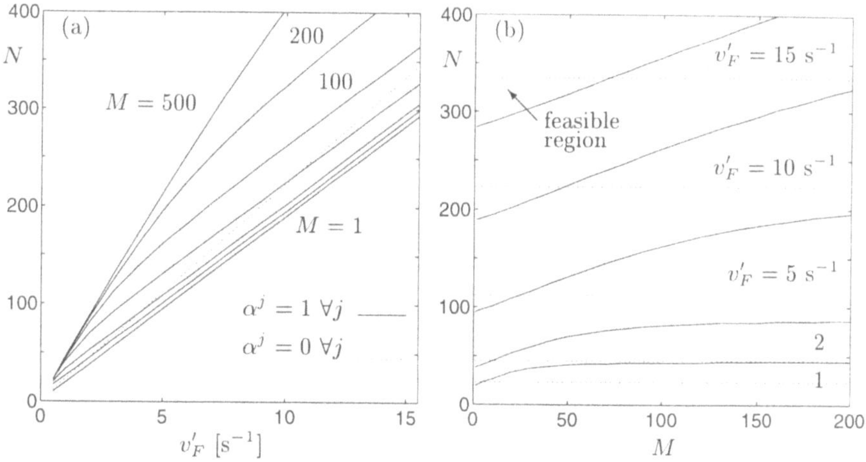

Figure 4: (a) Dependence of the number of reaction-separation nodes on the space velocity and the number of horizontal layers, with elution at every cross-flow channel. (b) Comparison between complete elution (full lines) and no elution (dotted lines).

M is $-\frac{1}{M}\pi_2 r_K \sum_{j=1}^{N} \alpha^j$ in the numerator. Thus, higher values of M will tend to improve the objective function, other things being constant. The more layers a given cross-flow elution stream transverses, the higher is its contribution to the final concentration of the combined product-rich stream, $\bar{c}_P$ (recall from Fig. 2a). O_P thus favors a highly concentrated product stream over a diluted one. The total number of cross-flow elution streams, $\sum_{j=1}^{N} \alpha^j$, contributes to the same term in the numerator as M, and determines the total flowrate and concentration of the final product stream. The number of nodes, N, necessary in order to achieve the required substrate conversion $\bar{x}_S$, effectively scales the whole objective function. Its dependence on the decision variables: v'_F, M, and $\boldsymbol{\alpha}$, is studied next.

The trade-off between an increase in the flowrate and an increase in the number of nodes is plotted in Fig. 4a for several values of M and a case where every node is separating ($\alpha^j = 1\ \forall j$). The ratio $\Delta N/\Delta v'_F$ attains an asymptotic value already for moderate v'_F and varies only slightly with M. Therefore, it will be the influence of v'_F on the N-vs.-M and N-vs.-$\boldsymbol{\alpha}$ relationships, what eventually determines the optimum feed flowrate. Besides, v'_F is limited from above by two constraints: first, by the switching period via Eq. 4; v'_F corresponding to $1/t_p = 1$ Hz (which can be considered as a technological limit) is about 9.0 s^{-1}. Second, the chosen mechanism for liquid transport (pressure gradient, electroosmotic flow, thermocapillary pumping) sets an upper limit for v'_F.

Figure 4b shows the dependence of N on M, the number of horizontal layers insterted between each pair of elution-distributing and product-collecting

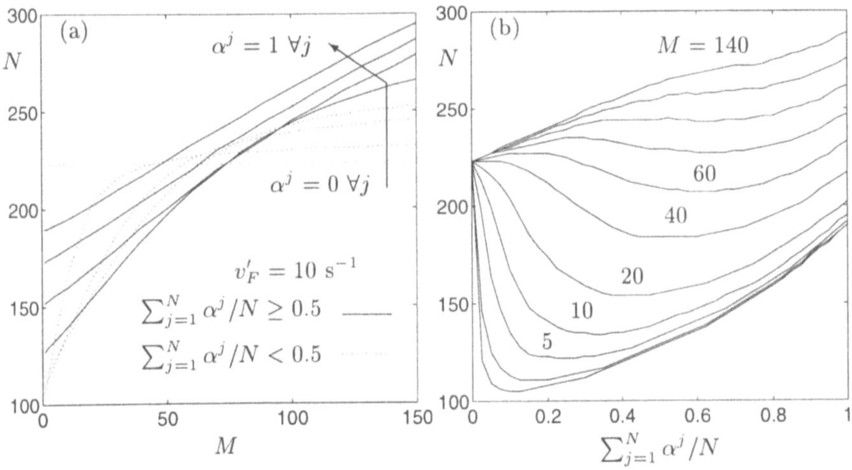

Figure 5: (a) A detail of the situation $v'_F = 10\ s^{-1}$ from Fig. 4b, for varying ratios of reaction-only and reaction-separation nodes. (b) The dependence of the required reactor size on the fraction of reaction-separation nodes to the total number of nodes, for $v'_F = 10\ s^{-1}$ and a range of M values.

channels (cf. Fig. 2a). Two cases are compared for every feed space velocity: one where all nodes are bifunctional (i.e., $\alpha^j = 1\ \forall j$), and a situation where unifunctional, reaction-only nodes are used throughout the microreactor ($\alpha^j = 0\ \forall j$). In the latter case, N is a function of v'_F only, as no cross-flow coupling exists and the convesion profiles in all M layers are identical. Depending on M, either case can result into a lower total number of nodes. The ranges of M for which the product separation by cross-flow streams is feasible (i.e., leads to lower $N \Rightarrow$ smaller reactor) can be seen directly for every v'_F. The reason why for larger M is the cross-flow arrangement with product elution actually worse than the reaction-only arrangement, is that the product accumulates in the cross-flow stream as more M layers are passed, and its concentration eventually reaches such levels that product inhibition in subsequent M layers offsets the benefits of product removal in the upper layers.

The adverse effects of product build-up in the cross-flow streams can be partially eliminated and the feasible region extended by changing the structure of the vector $\boldsymbol{\alpha}$. Let us further consider that $\boldsymbol{\alpha}$ consists of regularly distributed 0's and 1's with certain frequency and is thus characterized by a single quantity, the fraction of reaction-separation nodes, $f_E \equiv \sum_{j=1}^N \alpha^j/N \in \langle 0; 1\rangle$. Figure. 5a shows a detailed view on the N-vs.-M curves from Fig. 4b for $v'_F = 10\ s^{-1}$. By reducing the fraction of elution streams below one, the feasible region can be expanded towards larger M, and the minimum necessary number of nodes can also be reduced dramatically. For $M = 1$, it drops from $N = 223$ for $f_E = 0$ and $N = 190$ for $f_E = 1$ to only $N = 105$ for $f_E = 0.1$, being a reduction of remarkable 53 % of reactor volume compared with the reaction-only case, or 43 % compared with the separation-only case, for the same absolute product throughput.

The results of a parametric study of the dependence of N on f_E and M for a constant v'_F are plotted in Fig. 5b, showing clearly the existence of a minimum with respect to f_E, at which a smaller reactor volume is achieved than in either the elution-only or reaction-only case. The minimum N increases with increasing M, thus revealing a trade-off between two objectives – increasing M increases the product throughput (as well as the concentration of the product stream), but requires a larger N, and also a higher proportion of elution streams, f_E. The shift of the optimum f_E towards higher values for increasing M is caused by the fact that higher absolute amounts of product have to be transported by the cross-flow streams. Two threshold values of M can be identified – the first (around $M \sim 75$) occurs when the local minimum of N with respect to f_E becomes larger than N corresponding to $f_E = 0$ (reaction-only). Beyond this point, it does not pay off to introduce any elution stream, since a simple PFR (or a bundle of them) result into a smaller reactor volume for given throughput. Above the second threshold value ($M \sim 100$ in this particular case), even the local minimum disappears and N becomes a monotonously increasing function of f_E.

Once the analysis of the influence of the decision variables on N has been completed, the actual optimization of O_P with respect to v'_F, M, and f_E for specified values of the cost parameters (π_1 and π_2) would be a straightforward task for which any method of nonlinear scalar optimization can be used. For small M (say, up to 5) and f_E (i.e., only a few elution streams along N), not only the total fraction $\sum_{j=1}^{N} \alpha^j/N$, but also the positions of the individual streams will determine N. The knowledge of the local strength of product-inhibition (which is not uniform along the substrate conversion trajectory, cf. Štěpánek et al., 1999) can be used when searching for the optimal distribution of the elution streams. This analysis, however, exceeds the scope of this article.

5 Conclusions

The purpose of this contribution was to demonstrate on an example of a product-inhibited reaction, that reaction-separation coupling in a structured microreactor can significantly increase the product throughput per unit reactor volume, compared with a uni-functional reactor. It was also shown that by a proper analysis of the relationships between decision and dependent variables, even a relatively complex mixed design-operation problem with an initially large number of degrees of freedom can be reduced to manageable dimensions after decomposition. This work is an example of a quantitative (i.e., not conceptual) systems-engineering analysis of a microreactor-based manufacturing process.

References

Barton, P. I., and Pantelides, C. C. (1994) Modelling of combined discrete/continuous processes. *AIChE J.*, **40**, 966.

Eyal, J., and Charles, M. (1990) Hydration of cyanopyridine to nicotinamide by immobilized nitrile hydratase. *Biochemical Engineering VI*, p. 678, Annals of the New York Academy of Sciences, New York.

Gallardo, B. S., Gupta, V. K., Eagerton, F. D., Jong, L. I., Craig, V. S., Shah, R. R., and Abbott, N. L. (1999) Electrochemical priciples for active control of liquids on submilimeter scales. *Science*, **283**, 57.

Gau, H., Herminghaus, S., Lenz, P., and Lipowsky, R. (1999) Liquid morphologies on structured surfaces: From microchannels to microchips. *Science*, **283**, 46.

Jackman, R. J., Duffy, D. C., Ostuni, E., Willmore, N. D., and Whitesides, G. M. (1998) Fabricating large arrays of microwells with arbitrary dimensions and filling them using discontinuous dewetting. *Anal. Chem.*, **70**, 2280.

Martynova, L., Locascio, L. E., Gaitan, M., Kramer, G. W., Christensen, R. G., and MacCrehan, W. A. (1997) Fabrication of plastic microfluid channels by imprinting methods. *Anal. Chem.*, **69**, 4783.

Minotti, M., Doherty, M. F., and Malone, M. F. (1998) Design for simultaneous reaction and liquid-liquid extraction. *Ind. Eng. Chem. Res.*, **37**, 4748.

Rinard, I. (1998) Miniplant design methodology. *IMRET-2*, p. 299, March 9–12, 1998, New Orleans, USA.

Samant, K. D., and Ng, K. M. (1998) Design of multistage extractive reaction processes. *AIChE J.*, **44**, 2689.

Sammarco, T. S., and Burns, M. A. (1999) Thermocapillary pumping of discrete drops in microfabricated analysis devices. *AIChE J.*, **45**, 350.

Schubert, K., Bier, W., Brandner, J., Fichtner, M., Franz, C., and Linder, G. (1998) Realization and testing of microstructure reactors, micro heat exchangers and micromixers for industrial applications in chemical engineering. *IMRET-2*, p. 88, March 9–12, 1998, New Orleans, USA.

Štěpánek, F., Kubíček, M., Marek, M., and Adler, P. M. (1999) Optimal design and operation of a separating microreactor. *ISMR-1*, April 25–28, 1999, Amsterdam, NL, *Chem. Eng. Sci.*, in press.

Acknowledgements. Financial support from project VS 96073 (Czech Ministry of Education) and grant 104/99/1408 (Czech Grant Agency) is gratefully acknowledged.

Modelling of Gas-Liquid Catalytic Reactions in Microchannels

P. Angeli, D. Gobby, A. Gavriilidis

Department of Chemical Engineering, University College London, Torrington Place, London WC1E 7JE.

Abstract. Gas-liquid catalytic reactions in flat plate reactors with separated feed were investigated with the use of Computational Fluid Dynamics. Gas and liquid reactants were brought into contact through a diffusing wall, while catalyst was deposited on the opposite wall. Hydrogenations were used as model reactions and parametric studies were performed in order to investigate the effect of channel height, reaction mixture velocity, diffusion coefficient and reaction rate constant on the concentration profiles and reactant conversion. From the parameters studied, the channel height and the fluid velocity affected conversion most.

Keywords. Microreactor, gas-liquid, catalytic, CFD, hydrogenations.

1 Introduction

Microchemical reactors have been hailed as the future of chemical industry. Operation in these microscales leads to higher heat and mass transfer rates than are possible on the macro-scale (Wegeng et al., 1996, Shanley, 1997). Together with the confinement of reaction in small volumes, they can enable exothermic reactions to be carried out safely at higher temperatures and pressures, while they also allow safe operation within flammability limits due to flame-trap dimensions of micro-channels (Lerou, 1997, Srinivasan et al., 1997, Burns et al. 1998). Furthermore, uniform velocity fields within the microstructures, short reaction pathways and residence times and fast quench rates all provide the potential for greater product selectivity through the avoidance of side reactions. These characteristics also open up possibilities for new reaction regimes and novel reaction routes and can lead to increased environmental protection through better control and greater reaction specificity which greatly reduce the formation of waste products (European Chemist, 1999). Another major advantage is that microreactors enable the production of chemicals to be carried out at any scale by simply varying the number of reactor units i.e. "number-up" instead of "scale-up" (Lerou et al, 1996). This increases flexibility and allows production of chemicals on demand and at the point of application, thereby reducing transportation hazards and storage costs (Chopey et al, 1997).

The above advantages make microreactors suitable for the fine chemical and pharmaceutical industry which deals with high value/small volume products. The viability of microreactors has already been demonstrated for gas phase reactions (Shrinivasan et al., 1997). A large number of industrial applications though involves multiphase systems, i.e. gas-liquid and liquid-liquid. Fluid flow in microchannels has been found to diverge from what is expected from classical theory (Peng et al., 1994). In multiphase systems there is the added difficulty of how the two phases are brought into contact. Research in large scale channels has shown that a variety of flow regimes can be established depending on the properties of the two fluids, the geometry of the channel and the flowrates. In microchannels the different flow patterns which can form have not been established yet, but initial findings in gas-liquid systems suggest that regimes such as bubbly and stratified flow can exist (Stanley et al., 1997, Shaw et al., 1998). Liquid-liquid extractors have been used with the two phases in direct contact (Robins et al., 1997, Shaw et al., 1998).

Three-phase hydrogenations are of interest to fine chemicals synthesis. They are also fast reactions, and thus suitable for microreactors. For gas-liquid reactions one way to bring the two phases in contact is through a diffusion barrier such as a porous wall, where the gas diffuses to the liquid phase without forming bubbles in it. The advantage of this method is that the two phases do not need separation after reaction. Also due to the well defined area of contact between the phases this type of microreactors is suitable for kinetic measurements.

2 Modelling of gas-liquid catalytic reactor

For the modelling of the two-phase gas/liquid catalytic reactor a commercial CFD code (CFDRC) was used. This is a structured finite-volume based code where the fundamental equations for momentum, heat and mass transfer are solved with the application of the appropriate boundary and initial conditions.

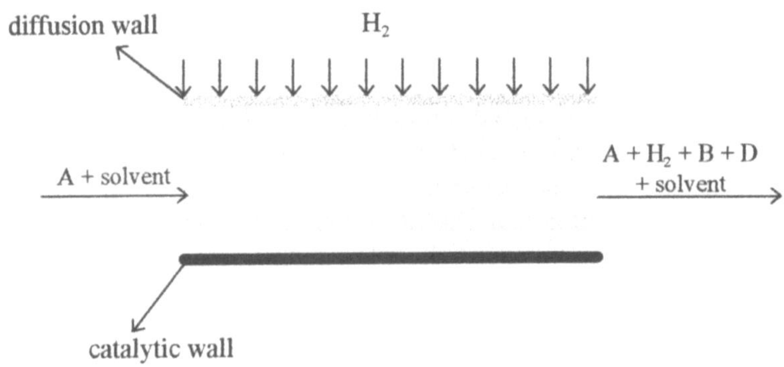

Fig. 1. Schematic diagram of the flat plate reactor with separated feed.

A model reaction of was used which was 1st order with respect to H_2, which is typical for this type of reactions (cf. Lu et al., 1996). In the model, hydrogen was diffusing through the top of the reactor and the catalyst was deposited on the opposite wall (see Figure 1). Hydrogen was assumed to be at its solubility concentration in the reaction solution at the top wall surface. The concentration of reactant A in the inlet solution was 5wt%. Density and viscosity of the reaction mixture were kept constant during the simulations. Assuming that efficient heat transfer would be favoured in microchannels, the reaction was modelled as isothermal. A zero-thickness catalyst was considered, while the code had to be modified through FORTRAN subroutines, to account for reaction on the catalytic wall and hydrogen addition through the upper diffusion barrier.

The base case as well as the range of conditions, characteristic for hydrogenations (Lu et al., 1996, Reid et al., 1987), used for the parametric studies are shown in Table 1. Channel length was taken equal to 2 cm. Under these conditions the Reynolds number varied from 0.04 to 0.78, and the laminar flow model was used for the calculations.

Table 1. Range of parameters and dimensionless groups used in the CFD simulation.

	Base case	Range of parameters
channel height, d	100 μm	10 - 100 μm
reaction mixture velocity, u	10^{-2} m/s	10^{-3}-$2x10^{-2}$ m/s
diffusion coefficient, D	$2.8x10^{-9}m^2/s$	10^{-8}-10^{-10} m^2/s
reaction rate constant, k	$2.5x10^{-3}$ m/s	10^{-3}-10^{-2} m/s

3 Results and discussion

Investigators have found that flow in small sized channels can divert from classical fluid flow theory (Bailey et al., 1995) but at the moment there are no conclusive theories to explain this behaviour and furthermore no models are available which take this behaviour into account. Hence, the simulations were based on the solution of the Navier-Stokes equations. Pressure drop was found to vary from 70 to 7900 Pa, apart from the case of the very small channel diameter where the pressure drop was 0.7 bar. Concentration profiles for all species were obtained and those of product B can be seen for two cases in Figures 2 and 3. Figure 2 is for the base case while Figure 3 is for the same case but with decreased reaction mixture velocity (u = 10^{-3} m/s). It can be observed that in the base case (Figure 2) reaction starts further down the channel, and a large part of it is occupied by unreacted solution. For lower velocity (Figure 3) the reaction starts almost next to the inlet of the reaction channel and the radial concentration profiles of B are less steep, indicating effective utilisation of the reactor volume.

Fig.2. Concentration profile for product B for the base case simulation.

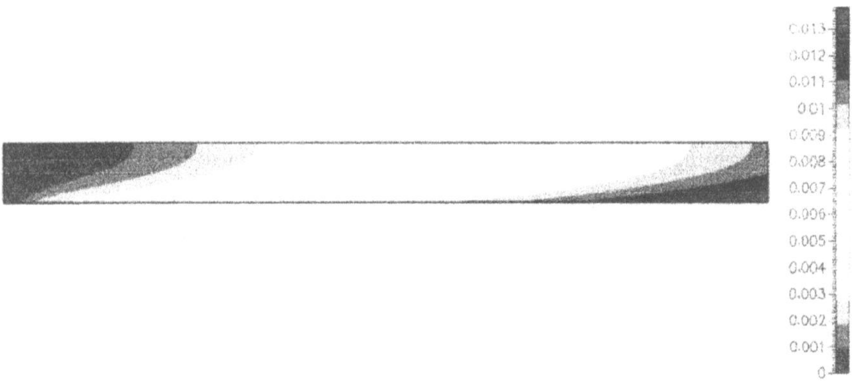

Fig. 3. Concentration profile for product B for reaction mixture velocity u = 10^{-3} m/s.

In Figures 4 to 7 the effect of the different parameters on the conversion of reactant A are shown. As can be seen the decrease in channel height plays an important role in reactant conversion. In the 100 μm channel the conversion is very small; as the height increases the conversion increases and at 30 μm it has already reached 25% (Figure 4). Decrease of channel height further to 10 μm results in 100% conversion but at the expense of high pressure drop. From axial concentration profiles it was found that the average concentration of the products was increasing approximately linearly along the channel, and this increase was more significant as the channel height decreased.

Decrease of the reaction mixture velocity, or equivalently increase of residence time, increases the conversion. Under the conditions investigated, residence times above 10 s are required to achieve conversion above 20% (Figure 5). It is worth noting that higher residence times are also accompanied by lower pressure drops. For example for 4 s residence time (4.5% conversion) the pressure drop was 356 Pa while for 20 s residence time (30% conversion) the pressure drop was 71 Pa.

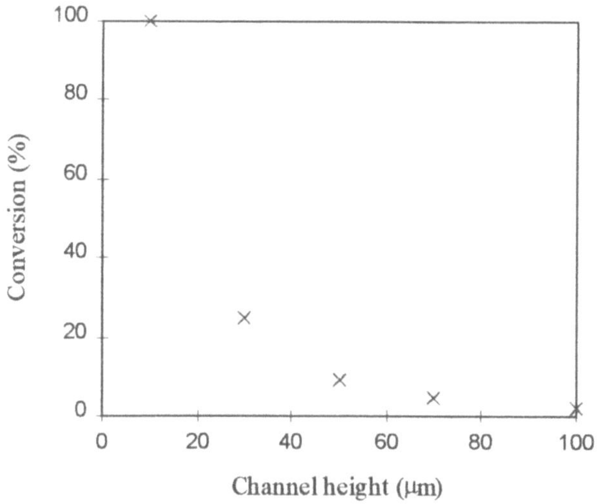

Fig. 4. Effect of channel height on conversion.

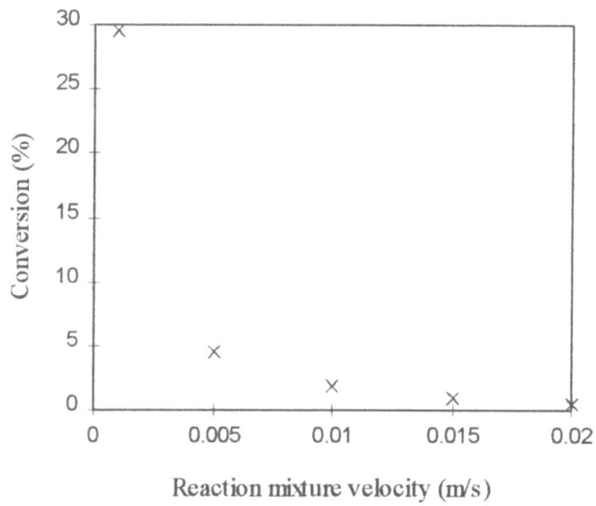

Fig. 5. Effect of reaction mixture velocity on conversion.

Conversion increases almost linearly with the diffusion coefficient, indicating that mass transfer of H_2 to the catalyst is the controlling mechanism (Figure 6). This is corroborated by Figure 7 which shows that higher reaction rate constant, which can be achieved for example by higher catalyst loading, does not have a significant effect on conversion.

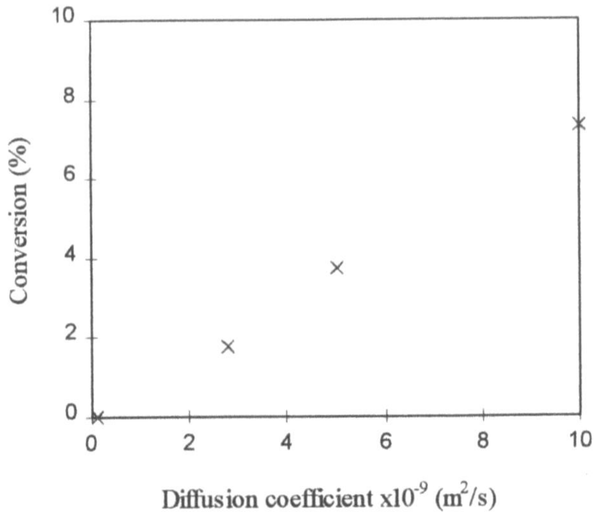

Fig. 6. Effect of the diffusion coefficient on conversion.

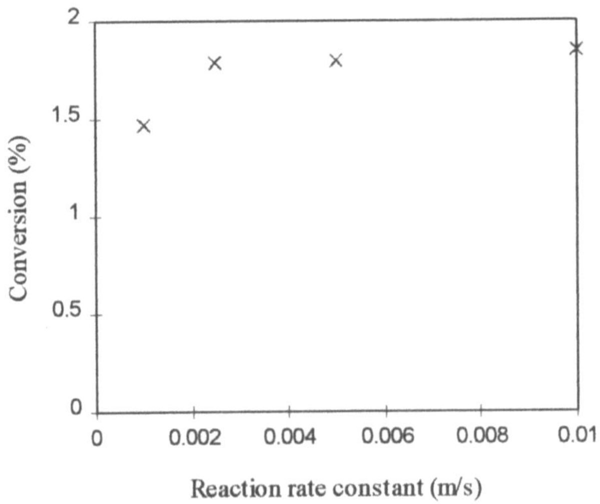

Fig. 7. Effect of reaction rate constant on conversion.

4 Conclusions

The results demonstrated that the separated feed microreactor can be utilised for gas-liquid hydrogenations provided that resistance to mass transfer is kept small by decreasing the channel height. A limiting factor for achieving high conversions is the amount of hydrogen that can dissolve in the reaction mixture. The type of reactor used in these simulations allows hydrogen to diffuse only from the top of the reaction channel and thus limits the total hydrogen amount available for the reaction. For reactions with higher intrinsic reaction rates than the ones investigated, other reactor designs which allow hydrogen to be at its solubility concentration in a larger portion of the reactor may be preferable. Such reactors can include for example those where the hydrogen is in the form of bubbles within the reaction mixture. This is going to form part of future work in gas-liquid microreactors.

Nomenclature

d	channel height, m
D	diffusion coefficient, m/s
k	reaction rate, m^2/s
u	reaction mixture velocity, m/s

References

Bailey D.K., Ameel T.A., Warrington R.O. and Savoie T.I., 1995, IECEC Paper no. ES-396, ASME, p. 301.

Burns J.R., Ramshaw C. and Harston P., 1998, Proc. 2nd Int. Conf. on Microreaction Technology, AIChE, p. 39.

Chopey N.P., Ondrey G. and Parkinson G., 1998, Chemical Engineering, March, p. 30.

European Chemist, 1999, January, p. 17.

Lerou J.J., Harold M.P., Ryley J., Ashmead J., O'Brien T.C., Johnson M., Perrotto J., Blaisdell C.T., Rensi T.A. and Nyquist J., 1996, Dechema Monographs, vol. 132, p. 51

Lu X.X., Boyes A.P. and Winterbottom J.M., 1996, Chem. Eng. Sci., vol. 51, no. 11, p. 2715

Peng X.F., Peterson G.P. and Wang B.X., 1994, Exp. Heat Transfer, vol. 7, p. 249.

Reid R.C., Prausnitz J.M. and Poling B.E., 1987, The properties of gases and liquids, 4th Edition, McGraw-Hill, Inc..

Shaw J., Turner C., Miller B. and Harper M., 1998, Proc. 2nd Int. Conf. on Microreaction Technology, AIChE, p. 176.

Srinivasan R., Hsing I.M., Berger P.E., Jensen K.F., Firebaugh S.L., Schmidt M.A., Harold M.P., Lerou J.J. and Ryley J.F., 1997, AIChE J., vol. 43, no. 11, p. 3059.

Robins I., Shaw J., Miller B., Turner C. and Harper M., 1997, Proc. 1st Int. Conf. on Microreaction Technology, Dechema e.v., p. 35.

Shanley A., 1997, Chemical Engineering, March, p. 30.

Stanley R.S., Barron R.F. and Ameel T.A., 1997, ASME, Micromechanical Systems, DSC-Vol. 62/HTD-Vol. 354, p. 143.

Wegeng R.S., Call C.J. and Drost M.K., 1996, Chemical System Miniaturisation, AIChE Spring National Meeting, New Orleans.

Simultaneous Screening of Catalysts in Microchannels: Methodology and Experimental Setup

Torsten Zech[1], Dieter Hönicke[1],
Astrid Lohf[2], Klaus Golbig[2], Thomas Richter[3]

[1] Technische Universität Chemnitz, Lehrstuhl für Technische Chemie, D-09107 Chemnitz
[2] Institut für Mikrotechnik Mainz GmbH, Carl-Zeiss-Straße 18-20, D-55129 Mainz
[3] AMTEC GmbH, Annaberger Straße 240, D-09125 Chemnitz

Abstract. Given the excellent properties of microfluidic reactors our project aims to develop a fast and parallelized catalyst screening unit. In a first step the channel-walls of microstructured wafers are coated with catalytic active species. For the screening of catalytic properties a stack of 35 catalyst wafers is operated under steady-state conditions and continuous flow. Each wafer is exposed to the same amount of reactant gases at the same time, temperature and pressure. The product analysis is carried out using mass spectrometry for each catalyst wafer in order to gain full chemical information of the product distribution, conversion degrees and selectivities. A fast, sequentially sampling device is used to sample each catalyst wafer at least every 60 minutes.

Keywords. combinatorial catalysis, heterogeneous catalysis, microchannel, screening, sampling, methane oxidation

1 Introduction

The development of catalysts has traditionally been an empirical and therefore time consuming process. Nowadays, efforts towards the speeding up of the preparation of catalysts as well as of testing technologies have been made by industry [1, 2] and in the field of academic research [3, 4, 5]. Obviously, a systematic and integrated approach is needed to succeed in developing efficient catalysts as it is a very complex task. Diversity-based or combinatorial strategies as used in the pharmaceutical industry seem applicable to new methods for the development of efficient catalysts. The miniaturization of the equipment used is essential to the success of such combinatorial approaches.

Our project aims to use the excellent properties of microfluidic reactors within a fast, parallelized catalyst screening unit, whereby the catalysts are coated onto microstructured wafers. The screening unit should be flexible, robust and designed for the use in heterogeneously catalyzed gas-phase-reactions under harsh chemical conditions.

2 Experimental setup

2.1 Reactor module, microstructured catalyst wafers and catalyst preparation

The catalyst screening unit consists of four main parts: a newly developed reactor module, appropriate microstructured catalyst wafers and a newly developed sampling device which is attached to a quadrupole mass spectrometer for product analysis.

The first and simplest generation of the reactor module consists of a stack of 35 metallic frames having compartments which can hold one catalyst wafer each, as shown in Fig. 1. The wafers are mounted into their specific compartment before the screening. They are easily to remove when the screening has finished. The module is heated externally to achieve reaction temperatures up to 450 °C. The product outlet at the end of each element has a cross section of 2 mm to 0.5 mm.

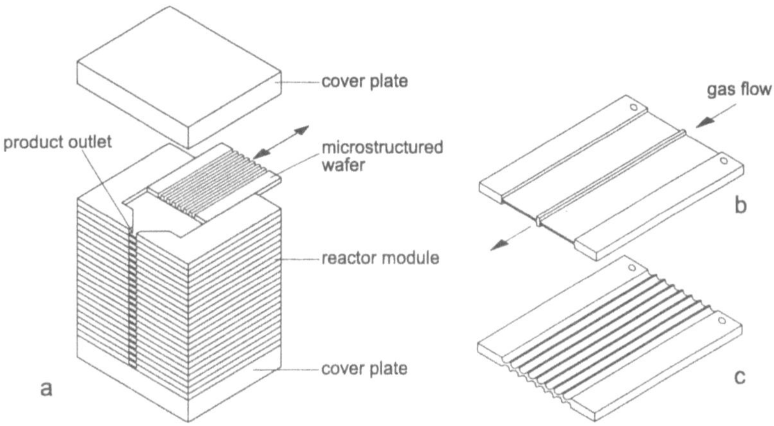

Fig. 1: a) Reactor module consisting of 35 stacked frames. The catalyst wafers are mounted and removed in the directions of the arrow. b) Microstructured catalyst wafers made of aluminum either by mechanical micromilling or c) by wet etching (channel radius: 130 μm).

The module design allows the use of microstructured catalyst wafers made of different materials such as metals, silicon, ceramics, and glass. That is an important strength which increases the flexibility of the system greatly. Thus, it is very easy to test different catalyst supports and hence catalyst preparation techniques as well as microchannel geometries. In the first run, microstructured catalyst wafers made of aluminum by wet etching and mechanical micromilling have been used, as illustrated in Fig. 1. With anodic oxidation of aluminum, it is possible to generate a regular pore system onto the channel walls of such a microstructured catalyst wafer [6]. Pore depth and pore diameter can easily be adjusted through variation of the process conditions in anodic oxidation. Having prepared the catalytic support, one or more catalytic active species can be immobilized in different concentrations within the pores of the Al_2O_3-layer by wet impregnation. Finally, the wafer is calcined for several hours at a defined temperature and atmosphere. The catalyst synthesis procedure, although not yet automated, is carried out in parallel for 35 catalysts and can be further speeded-up, e.g. by conventional automated pipetting systems. Certainly it is also possible to use other methods for coating the channel walls with a catalytic active layer, e.g. through use of sol-gel-, CVD-, or PVD-technologies.

Thus, the design of the reactor module and the catalyst wafers allows many different catalyst synthesis procedures that are essential to the generation of suitable catalyst libraries for different heterogenously catalyzed reactions. It is obvious that the design of the reactor module can be further improved, e.g. the module can be designed as cross-flow heat exchanger [7] to ensure a constant temperature along the reaction channels and the height of the reactor module.

2.2 Automated sampling

The product sampling is performed by using a newly developed sampling device. The product outlets of the reactor module lie very close together which implies that the reactor module can almost be considered as a static micromixer. Thus, products from an individual wafer are mixed immediately with the products from the adjacent wafer when leaving the corresponding frame. As the products from each catalyst wafer should be analyzed with low background noise from the other ones the sample must be taken *inside* the product outlet of the corresponding frame.

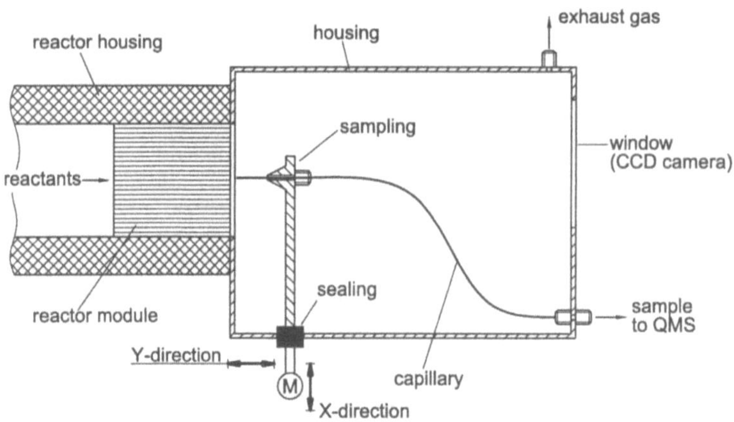

Fig. 2: Principle functioning of the developed sampling device. The sample is taken by the X-Y-positioned capillary.

The principle functioning of the sampling device is illustrated in Fig. 2. The sample is taken by a replaceable 300 μm wide silica capillary in a gas-tight housing. Fig. 3 shows the capillary in front of the product outlets of the reactor module. The capillary can be X-Y-positioned with an accuracy of about 5 μm under hazardous chemical conditions and high thermal demands. During the sampling the capillary moves first in X-direction to the corresponding frame and afterwards in Y-direction up to 5 mm *into* the product outlet of that frame. The moving parameters and the

Fig. 3: The capillary in front of the reactor module.

position of the capillary which is adjusted and supervised by a CCD camera are freely programmable. The materials of the housing and the sealing were carefully chosen to withstand temperatures up to 450 °C.

Obviously, it would be possible to extend the X- and Y-moving ranges of the sampling device and to add a Z-direction to allow the simultaneous screening of more than 35 catalysts.

2.3 Flowsheet of the catalyst screening unit

The catalyst screening unit consists of several mass flow controllers (Brooks) supplying the reactant gases and the described reactor module in a housing that is heated externally. Furthermore, there is the sampling device, a quadrupole mass spectrometer (Hiden Analytical) and a PC controlling the reaction, the sampling and the data acquisition (Fig. 4).

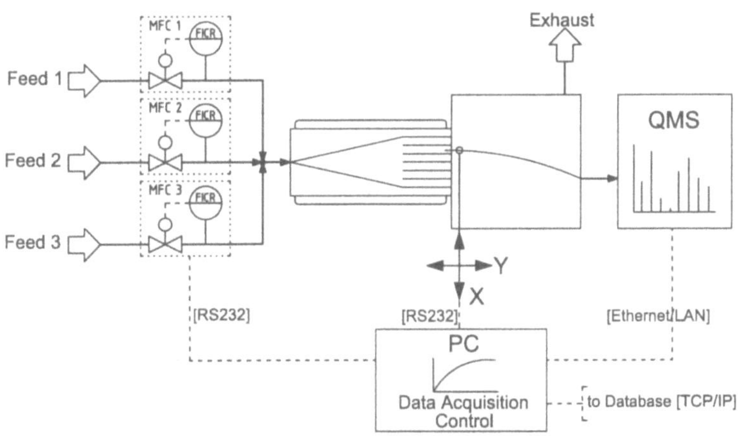

Fig. 4: Simplified flowsheet and communication links of the catalyst screening unit consisting of: gas supply, reactor, sampling device, mass spectrometer and PC.

Software tools have been developed in order to automate the screening completely. Thus, it is possible to run a predefined program varying, for instance, the feed gas composition, temperature, pressure and the sampling frequency.

3 Results

3.1 General system behaviour

In order to determine the general behaviour of the system, some basic investigation were carried out. The response function of the complete system including mass flow control, tubing, reactor and analysis system is shown in Fig. 5. As can be seen, it takes about 15 seconds until the system reaches a new steady state following the jump of the tracer concentration initiated by a mass flow controller. That means it is possible to analyze the products of a catalyst wafer at least within 60 seconds having finished the analysis of the products of the adjacent wafer and having moved the capillary to the new wafer. One sample cycle, i.e. sampling each of the 35 catalyst wafers once, can therefore be carried out within about 35 minutes. Statistic calculations of the mass spectra show that a variation coefficient of the corresponding M/Z-peaks of less than 1 % can be achieved within that short period of time under reaction conditions.

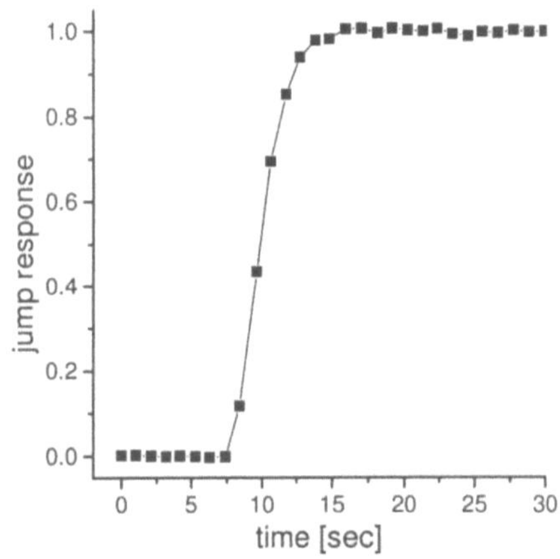

Fig. 5: Response function of the complete screening unit to the jump of the tracer concentration from 1.3 vol-% to 6.3 vol-% O_2 at 450 °C (capillary position: 4.5 mm inside the frame).

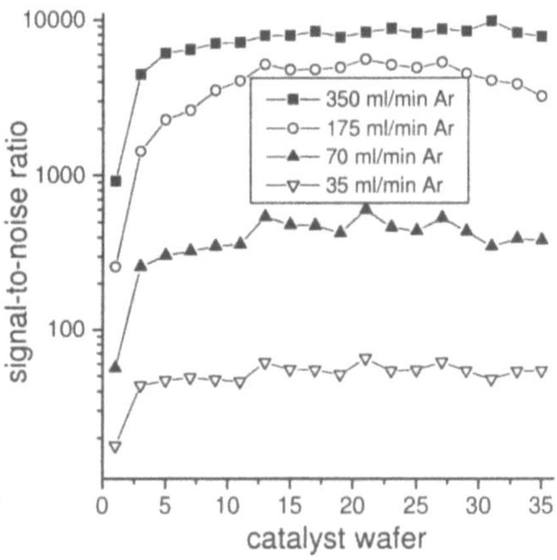

Fig. 6: Signal-to-noise ratio over the individual wafer at different total gas flows (capillary position: 4.5 mm inside the corresponding frame).

For a high accuracy of the results the signal-to-noise ratio of the analysis should be as high as possible. A simple test determining this ratio was carried out: a constant argon flow through the reactor module was maintained while at the same time the region of the product outlets in the sampling chamber was cross-flowed with air. The ratio of argon to air within one frame was used as a measure for the background noise. If the ratio is high, only a small amount of gas diffuses back into the reactor module. As shown in Fig. 6, the signal-to-noise ratio increases with a higher argon throughput. Even at a very low argon throughput of 35 ml/min, i.e. 1 ml/min for each wafer, the signal-to-noise ratio is sufficiently high. Another test has shown a strong dependence of the length that the capillary moves into the frame and the signal-to-noise ratio. The deeper the capillary moves into the frame the purer a sample will be.

3.2 Oxidation of methane: preliminary examinations

In a first preliminary examination, a set of 35 catalysts wafers prepared by anodic oxidation of aluminum followed by wet impregnation with Pt-, V- and Zr-precursors was tested in the oxidation of methane at 430 °C and 1.1 bar. The methane-to-oxygen ratio in the feed was 2 in order to find out the potential of the catalysts to form intermediate products. Fig. 7 illustrates the results for the first 5 hours of screening. The graph shows the conversion degree of oxygen over the individual catalyst wafers and the cycle number. 60 minutes were chosen for each cycle. As can be seen, there are some catalyst wafers that show a very high conversion degree while others are much less active under the reaction conditions used. The active catalysts are high surface area Pt/Al_2O_3-catalysts while Pt-catalysts with a lower surface area are much less active.

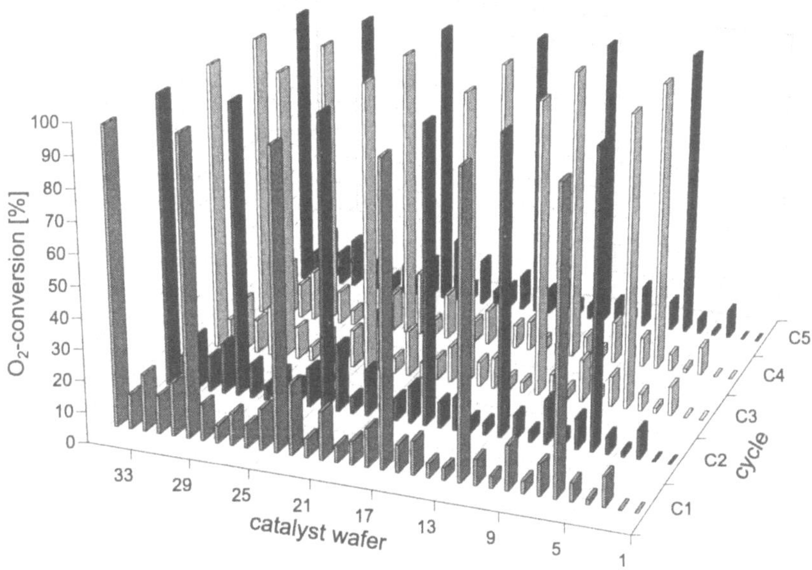

Fig. 7: Screening of a set of 35 catalyst wafers: conversion of oxygen in methane oxidation at 430 °C and 1.1 bar.

The products CO, CO_2 and H_2O have always been observable, whereas organic intermediate products have not been found. The screening was left running overnight and finished after 25 hours on stream. The activities of almost all catalyst wafers decreased with time on stream.

4 Conclusion and outlook

Within only nine month, a catalyst screening unit for 35 catalyst wafers has been constructed and tested. It consists of a gas supply, a novel reactor module, a novel sampling device and a mass spectrometer. The system is flexible, robust and designed for heterogeneously catalyzed gas-phase-reactions under harsh chemical conditions. First preliminary investigations were carried out and emphasize the potentials of the system. Thus, it is possible to test the activity and selectivity of a catalyst wafer within only 60 seconds as well as under steady-state conditions. The screening can be automated completely varying the feed gas concentrations, temperature and pressure.

In future investigations, the system will also be used for the partial hydrogenation of 1,3-butadiene. Some improvements of the system will be focused on increasing the temperature and pressure during the screening.

Acknowledgements

The authors would like to thank the German Ministry of Education and Research (Bundesministerium für Bildung und Forschung) which supports this project under contract number 03D0068C8.

References

[1] X.-D. Xiang, X. Sun, G. Briceno, Y. Lou, K.-A. Wang, H. Chang, W. G. Wallace-Freedman, S.-W. Chen, P. G. Schultz, *Science* **1995**, 268, 1738-1740

[2] P. Cong, R. D. Doolen, Q. Fan, D. M. Giaquinta, S. Guan, E. W. McFarland, D. M. Poojary, K. Self, H. W. Turner, W. H. Weinberg, *Angew. Chem. Int. Ed.* **1999**, 38, 484-488

[3] F. C. Moates, M. Somani, J. Annamalai, J. T. Richardson, D. Luss, R. C. Wilson, *Ind. Eng. Chem. Res.* **1996**, 35, 4801-4803

[4] S. M. Senkan, *Nature* **1998**, 394, 350-352

[5] A. Holzwarth, H.-W. Schmidt, W. F. Maier, *Angew. Chem. Int. Ed.* **1998**, 37, 2644-2647

[6] G. Wießmeier, D. Hönicke, *Ind. Eng. Chem. Res.* **1996**, 35, 4412-4416

[7] K. Schubert, W. Bier, G. Linder, D. Seidel, *Chem.-Ing.-Tech.* **1989**, 61 (2), 172

Palladium Membrane Microreactors

Aleksander J. Franz, Klavs F. Jensen, Martin A. Schmidt[‡]

Dept. of Chemical Engineering, [‡]Microsystems Technology Laboratories
Massachusetts Institute of Technology, Cambridge, MA 02139, USA

Abstract. Novel palladium membrane microreactors have been designed and microfabricated. Selective hydrogen flux has been demonstrated through the palladium membrane in the microreactor. The membrane microreactors can be used for a variety of hydrogenation and dehydrogenation reactions, as well as in hydrogen purification applications. The microfabrication process allows for integration of heaters and temperature sensors into the device. The e-beam deposited thin Pd film achieves excellent hydrogen selectivity at high permeation rates. The membrane has good mechanical strength and has been shown to withstand pressure gradients of over 5 atm. A model hydrogenation reaction has been demonstrated in the palladium membrane reactor. Microfabricated palladium membrane reactors enable new applications because of their small size, fast thermal response times, and high efficiency.

Keywords. Palladium membranes, microreactors, hydrogen, separations, hydrogenation

1 Introduction

Microchemical systems for chemical synthesis and product purification functions are under development. Microchemical systems could have applications in manufacturing, by providing an option for point of use generation, and improving on efficiency and safety of large scale systems [1]. As with conventional scale chemical operations, the microchemical reactors are a key component of the microchemical systems.

Palladium membrane reactors represent an interesting technology for miniaturization. The palladium membrane reactors possess many advantages over other types of reactors [2]. First, they are capable of beyond equilibrium conversions, as the products and reactants are continuously separated. Second, palladium membrane reactors enable use of impure hydrogen feed streams, such as syngas, since only hydrogen diffusing through the membrane contacts the other reactant. Finally, there is some evidence that improved selectivities can be achieved using palladium membrane reactors. Unfortunately, more extensive use of palladium membrane reactors is limited by their high cost, due to the thick palladium films employed.

The thick palladium films are necessitated by the method of fabrication that entails coating a porous ceramic support with the palladium film [2]. Thus, the support pore size dictates palladium film thickness (10-25 μm). In addition to adding expense, the thick palladium films greatly reduce efficiency, since the hydrogen flux through the film is inversely proportional to its thickness. Microfabrication methods offer an opportunity to produce palladium membrane devices with much lower thickness palladium films, potentially resulting in large efficiency increase and lower reactor cost.

In this paper, a novel, microfabricated palladium membrane reactor is presented. The palladium membrane achieves a selective flux of hydrogen and can be used for highly selective hydrogenation/dehydrogenation reactions. The device can also be used for hydrogen separation or purification from gas mixtures [3]. A fabrication process, which allows for integration of heaters and temperature sensors into the device, has been developed. The membrane's performance was characterized with respect to mechanical strength and hydrogen flux. A model hydrogenation reaction has been demonstrated.

2 Palladium Membrane Microreactor Design

The main principal of operation for palladium membrane reactors is to achieve a selective hydrogen flux through the membrane. Diffusion limited hydrogen flux through a palladium film depends mainly on film thickness, hydrogen pressure gradient across the membrane, and temperature. Therefore, the palladium membrane microreactor design must include the ability to control the temperature, and must be robust enough to handle pressure gradients while utilizing a very thin palladium film.

A schematic of the MIT palladium membrane microreactor design is shown in Fig. 1. The design is based on the previously described microreactor for catalytic gas phase reactions [4]. The device consists of two flow channels, separated by a thin membrane. The membrane itself is a composite of three layers: silicon nitride, silicon oxide, and palladium. The silicon nitride and oxide layers are perforated and serve as a structural support for the active element of the membrane, the palladium film. The nitride and oxide layers also electrically

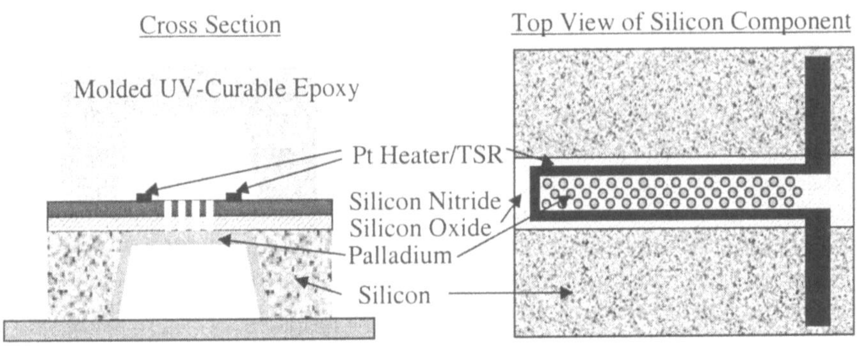

Fig.1. Schematic of the integrated palladium membrane microreactor design

insulate the palladium film from the integrated temperature sensing and heater elements. One channel flows through the silicon wafer. The second channel is formed by attaching an epoxy cap to the topside of the device.

3 Reactor Fabrication

The palladium membrane microreactor fabrication process is shown in Fig. 2. A three-mask process is used to produce the bottom half of the device. A silicon wafer, preferably double side polished, is coated with 0.2 μm of SiO_2, and 0.3 μm of LPCVD silicon nitride. The backside of the wafer is patterned using standard photolithography, followed by a dry SF_6 nitride etch and removal of the oxide layer, using wet buffered oxide etch (BOE). The patterned nitride film will serve as an etch mask to form the membrane and one of the channels of the device. If the wafer is not double-side polished, a brief KOH etch (~50 μm) is necessary to better define the backside pattern, and allow for front to back alignment.

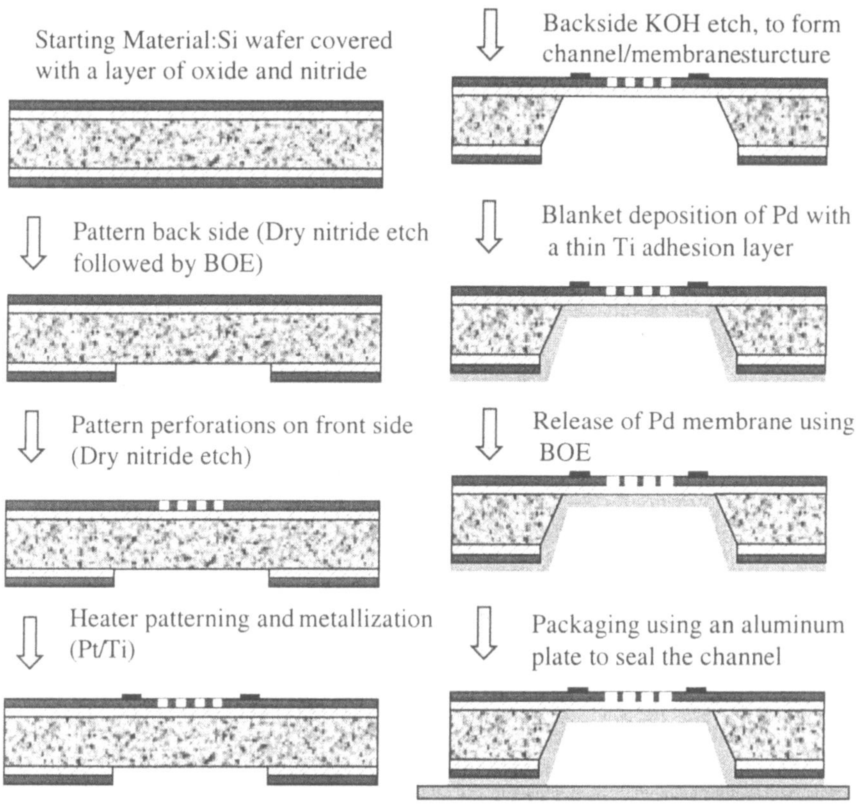

Fig. 2. Microfabrication sequence for the silicon component of the palladium membrane microreactor

The nitride layer on the front side of the wafer is patterned to define the perforations in the membrane, using photolithography with an IR front to back alignment, followed by a dry nitride etch. An e-beam deposited Pt/Ti film and a standard lift-off process are subsequently used to pattern temperature sensing resistors (TSR's) and heaters on the wafer front side. The membrane is defined and the channel in the silicon is formed simultaneously, using a KOH etch, with the oxide layer serving as an etch stop. A special chuck is used to protect the front side of the wafer during the KOH etch. Next, a blanket of Pd is deposited on the wafer backside, with a thin Ti film serving as an adhesion layer. The final step in fabricating this half of the device is to release the Pd film by removing the oxide layer in the perforations with BOE. The microfabrication process described above enables use of very thin Pd films as a membrane, since the support is made porous after the deposition of the Pd film.

A top view of the bottom half of the device is shown in Fig. 3. The channel in the device is 1.2 cm long, and the membrane width is slightly less than 700 μm. Use of a symmetric heater design is a key feature. The symmetric design allows for heating of a large area of the membrane [5], while maximizing the area in the center available for perforations and hydrogen flux. A thin TSR is patterned inside the heater loop, and provides a measurement of the average temperature for a given heater segment. There are two independent heater segments for each device. A single device chip measures 0.8x1.6 cm.

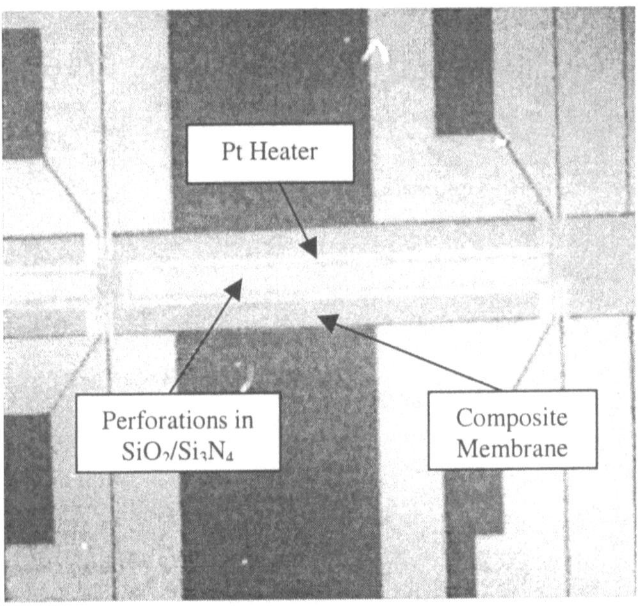

Fig. 3. Top view of the microfabricated silicon component of the Pd membrane microreactor.

The second channel in the device is fabricated using a molding process [6]. First, channels identical to the channel in the bottom half of the device are etched

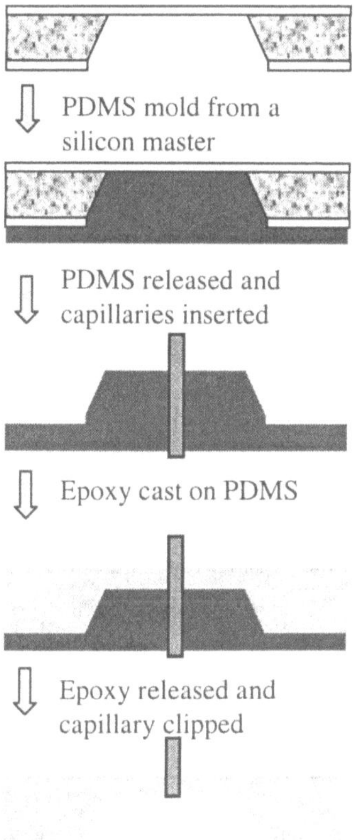

Fig. 4. PDMS molding process for fabrication of the capping channel.

in a silicon wafer. A negative mold is made using polydimethylsiloxane (PDMS) polymer. The PDMS mold is then used to reproduce the channels in a UV curable epoxy. Prior to applying the epoxy, small capillary tubes are inserted into the soft PDMS mold at opposite ends of each channel. As the epoxy hardens, the capillary tubes are sealed in, providing an inlet and outlet for the channel. When the hard epoxy cast is separated from the soft PDMS mold, the capillaries are also released from the PDMS, and can be clipped even with the bottom, on the channel side. The capillaries protrude from the topside of the cast, resulting in relatively easy fluidic connections. Individual epoxy caps are narrower than the silicon half of the device, to allow for probes to contact the sensing and control elements. The cap is attached to the silicon chip using epoxy. The tapered shape results in a capping channel that is wider than the membrane, leaving room for error in alignment. The bottom channel is completed by fixing the device onto an aluminum mounting plate.

During operation, the membrane itself is thermally isolated from the surrounding silicon, as demonstrated using computational fluid dynamic (CFD) simulations [4]. This thermal isolation enables use of low temperature materials, such as plastics and epoxies, in construction of the top device channel. Although polymers may not be used in the final design because of their gas permeability, they do offer attractive features. These features include low cost, ease of manufacture, and optical transparency, enabling visual inspection of the membrane even after the device is assembled.

4 Device Characterization

Palladium membrane microreactors must be structurally sound and achieve a selective hydrogen flux. Both aspects of the microfabricated devices were tested prior to carrying out a model hydrogenation reaction.

4.1 Mechanical Properties

The structural strength of the fabricated microreactors was tested. The mechanical membrane strength is an important aspect of the device, since hydrogen flux is driven by a pressure gradient across the membrane, usually associated with a net pressure gradient. The rupture strength of the membranes at room temperature was measured. We found the rupture pressure to strongly depend on direction of the pressurization, i.e. which side of the membrane was exposed to the high pressure. With the membrane pressurized from the silicon channel side, an average rupture pressure of 1.4±0.3 atm was measured. However, the membrane survived average pressures of 5.2±0.8 atm when the pressure was applied in the opposite direction. These results suggest that the microreactors should be operated with hydrogen flowing from the epoxy to the silicon channel. The membranes also survived operation at elevated temperatures (~500 °C) with the silicon channel under vacuum, and the top channel at 1 atm.

The structural membrane strength results are very positive. Although the maximum pressure drop at operational membrane temperature has not been determined at this time, room temperature results suggest that this unoptimized design is capable of operation under desirable pressure gradients. Several design modifications should result in even higher rupture strengths. First, the tested devices were microfabricated using single-side polished wafers. The roughness on the backside of the wafers results in rough KOH-etched channel walls, and therefore a rough interface between the membrane and the supporting silicon. This roughness causes localized stress points and a lower rupture strength. Use of double-side polished wafers should therefore improve the membrane structural properties, although the magnitude of this improvement is uncertain. Isotropic rounding of the fillet radius would also increase membrane strength. In addition, the silicon nitride and oxide layers, which determine the structural properties of the membrane, have not been optimized. Variations of residual stress and film thickness should allow considerable room for optimization of the first-pass design presented in this paper. Finally, the design of the device itself can be modified if very high operational pressure gradients are required. Design parameters with the largest impact on membrane strength include membrane width and length, as well as size and distribution of the perforations in the nitride and

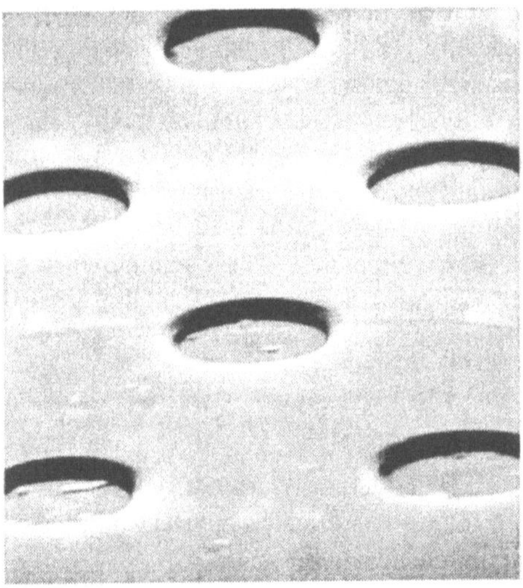

Fig. 5. SEM image of free standing palladium inside perforations in Si_3N_4 and SiO_2 membrane layers.

oxide layers.

Several design aspects contribute to the membrane strength results presented above. The membranes are stress compensated, with a net tensile stress. These stress properties are important for several reasons. The LPCVD nitride film is under high residual tensile stress. The stress compensation by the oxide film likely enables formation of stable perforations in the nitride layer, shown in Fig. 5. The perforations are stable despite the slight undercut of the nitride layer during the BOE etch, which removes the underlying oxide. The membrane net tensile stress is also important because of the operational properties of the membrane. The membrane is thermally isolated from the surrounding silicon. Therefore, when the heater is on, only the membrane is at an elevated temperature (~500 °C). The thermal expansion of the membrane causes a decrease in the tensile stress, and could lead to a net compressive stress on the membrane. This behavior is observed in the microreactor devices studied by our group, where low tensile stress silicon-rich nitride reactor windows buckle at elevated temperatures, eventually leading to device failure [4]. Clearly buckling in our perforated membranes is not desirable. Because of the large residual tensile stress of the stoichiometric nitride film used, the separation micromembranes do not buckle at elevated temperatures.

4.2 Hydrogen Flux

The microreactor membrane permeability and selectivity for hydrogen were characterized as a function of the hydrogen pressure gradient and average membrane temperature. The devices, fixed onto an aluminum plate, were mounted on an XYZ stage, which contained the gas manifold and allowed for alignment with the probe card. Either pure hydrogen or a 1:9 mixture of hydrogen and nitrogen was fed into one device channel. The other channel was either under

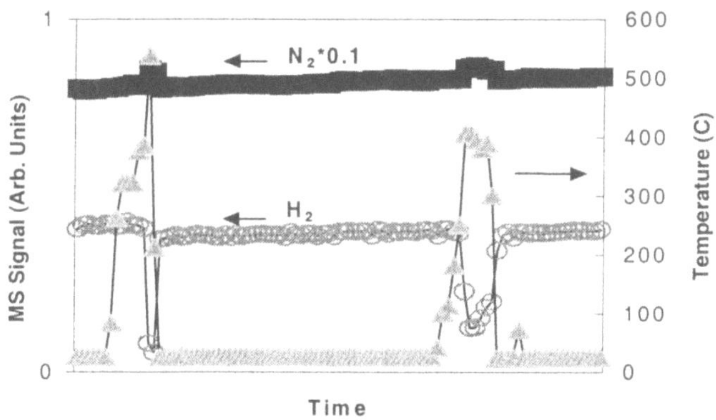

Fig. 6. Selective flux of hydrogen from a mixture of 90/10 N_2/H_2 at elevated membrane temperatures. Selectivity indicated by simultaneous drop in H_2 and rise in N_2 concentrations.

vacuum, or pure argon gas was flown at a known flow rate to sweep out the hydrogen, and maintain a driving hydrogen pressure gradient across the membrane. A quadrupole mass spectrometer (QMS) was used to measure gas partial pressures.

The microreactor achieves a selective hydrogen flux when a nitrogen/hydrogen mixture is fed into the device. Fig. 6 shows hydrogen and nitrogen QMS signals at the exit of the microreactor, and the membrane temperature, as a function of time. When the membrane temperature is elevated, the exit hydrogen concentration drops and the nitrogen concentration increases. The temperature dependant change in the mixture concentration clearly indicates selective hydrogen flux out of the channel through the membrane. Membrane separation factor of over 1800 was determined. The final determination of selectivity was limited by the sensitivity of our experimental setup.

Dependence of hydrogen flux on temperature and driving hydrogen pressure gradient was also studied [3]. In good agreement with theory, hydrogen flux increases for increasing temperature. Hydrogen flux also has a strong dependence on the hydrogen pressure gradient. When pure hydrogen is flown through the device at atmospheric pressure, much lower membrane temperatures are needed to achieve the same flux as when a mixture with a hydrogen partial pressure of 0.1 atm is used.

Hydrogen flux across the membrane was quantified by measuring the hydrogen concentration change in the device for a known hydrogen/nitrogen mixture flow rate, with argon sweeping the hydrogen on the permeate side. At an average membrane temperature of 500 °C, hydrogen flow rates of ~0.5-1 sccm were observed for a single active heater segment. The maximum hydrogen driving pressure gradient to achieve this flow rate was 0.1 atm. Using the total area of Pd exposed in the membrane perforations, this flow rate corresponds to a flux of about 600 sccm/cm^2. These results suggest that the microfabricated palladium membranes are potentially much more efficient than large-scale devices, due to incorporation of an extremely thin Pd layer (0.2 μm).

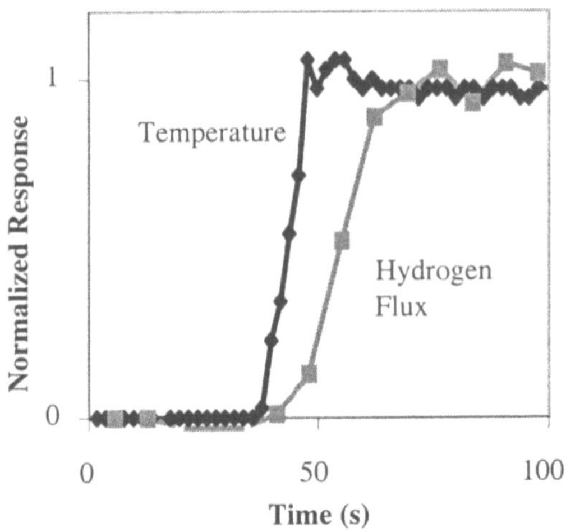

Fig. 7. A response of hydrogen flux through the palladium membrane microreactor to a step change in temperature

Extremely low thermal mass and high thermal isolation of the membranes result in open-loop thermal response time of the membranes of about 10 ms. The membrane thermal response time is not rate limiting in achieving hydrogen flux. A delay of several seconds between changes in membrane temperature and corresponding changes in hydrogen flux, shown in Fig. 7, indicates diffusion-limited hydrogen flux through the membrane.

5 Hydrogenation Reactions

We have demonstrated the ability to carry out hydrogenation reactions in the membrane microreactor. A hydrogen/nitrogen mixture was flown through the silicon channel of a device, with the topside of the membrane uncapped, and exposed to air. The hydrogen permeating through the membrane reacted with oxygen in the air, forming water. Due to elevated membrane temperatures, the water formed in the reaction desorbs from the membrane into the gas phase. The reaction could be visually observed when the produced water vapor condensed on the cold top surface of the chip, adjacent to the heated membrane. Fig. 8 shows the micromembrane exposed to air prior to heat up and after the reaction is initiated. The device performance is currently limited by the rate of hydrogen flux across the membrane, and design modifications to improve that flux are being implemented.

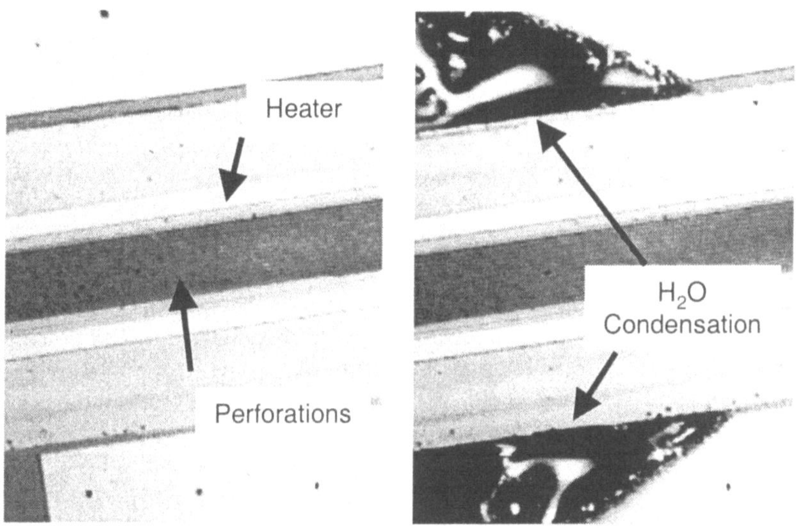

Fig. 8. Water condensation on the cold reactor walls during reaction of hydrogen diffusing through the palladium membrane with oxygen in air.

6 Conclusions

Integrated, palladium membrane microreactors, which achieve a highly selective flux of hydrogen, have been microfabricated and tested. This technology represents a novel application of MEMS microfluidics. The selective flux of hydrogen through the membranes allows for use of these devices in several applications. The devices can be used as reactors, to carry out hydrogenation-dehydrogenation reactions. The devices can also be used for hydrogen separation or purification from gas mixtures. The device performance was characterized with respect to mechanical strength and hydrogen flux. The membranes exhibit good mechanical properties, and are capable of withstanding pressure gradients of up to 5 atm. Hydrogen flux through the membranes increases with increasing temperature and increasing hydrogen pressure gradients, in good agreement with theory. Excellent selectivity is observed for high membrane fluxes. A simple hydrogenation reaction has been demonstrated in the microreactors. Microfabricated palladium membrane reactors enable new applications because of their small size, fast thermal response times, and higher efficiencies.

Acknowledgements

The authors thank the DARPA Micro Flumes Program (F30602-97-2-0100) for financial support. Discussions with Samara Firebaugh helped define the microfabrication process. Matt Losey was instrumental in adapting the epoxy casting process.

Citations

[1] J. J. Lerou, M.P. Harold, et al. "Microfabricated Minichemical Systems: Technical Feasibility," *Microsystem Technology for Chemical and Biological Microreactors*, DECHEMA Monographs **132**, 1996, p. 51.

[2] J. Shu, B. P. A. Grandjean, et al. "Catalytic Palladium-based Membrane Reactors ..." *The Canadian Journal of Chemical Engineering* **69**, 1991, p. 1036.

[3] A. J. Franz, M. A. Schmidt, K. F. Jensen, "Palladium Based Micromembranes For Hydrogen Separation and Hydrogenation/Dehydrogenation Reactions" Proceedings of the IEEE International MEMS '99 Conference, 1999.

[4] R. Srinivasan, I-M. Hsing, et al. "Micromachined Reactors for Partial Oxidation Reactions," *AIChE Journal*, **43**, 1997, p. 3059.

[5] A.J. Franz, D. Quiram, et al. "New Operating Regimes and Applications Feasible with Microreactors" *2nd International Conference on Microreaction Technology Proceedings*, 1998, pp. 33-39.

[6] D. Qin, Y. N. Xia, et al., "Microfabrication, microstructures and microsystems" *Microsystem Technology in Chemistry and Life Science* **194**, 1998, pp. 1-20.

A Micro Packed-Bed Reactor for Chemical Synthesis

M.W. Losey, M.A. Schmidt[‡] and K.F. Jensen

Department of Chemical Engineering, [‡]Microsystems Technology Laboratories,
Massachusetts Institute of Technology, Cambridge, MA, 02139

Abstract. A microchemical device has been fabricated using deep-reactive-ion etch technology for the purpose of gas-liquid contacting and for use as a micro packed-bed reactor. A 4 μL volume reaction chamber holds the packing as reactants are fed continuously in a co-current fashion and distributed at the inlet of the bed. The gas and liquid reactant streams are brought into contact by a series of interleaved, high-aspect ratio (25μm X 300μm) inlet channels. The inlet channels meet at the inlet of the reaction chamber which is sized to accommodate a variety of packing materials. These packing materials are fed into the reaction chamber through separate inlets and are held in place by a grating of 40μm wide posts with 25μm gaps. Several types of materials have been successfully loaded into the reaction chamber, such as glass spheres 50μm diameter, polystyrene beads, and granular catalyst particles.

Fluid flow regimes and flow behavior have been characterized with and without packing in the reaction chamber. Co-current flow of the gas and liquid has been visualized in-situ to ascertain the nature of the two-phase flow over the packing and to determine the efficiency of the fluid-solid contacting. Granular catalysts of various sizes have been loaded into the micro packed-bed reactor and several classical heterogeneous gas-liquid reactions have been performed, including hydrogenation reactions. Mass transfer effects have been characterized to gauge the efficiency of gas-liquid contacting and to establish the process capabilities of the microchemical device as compared to conventional large-scale multiphase reactors.

Keywords. Packed-bed, gas-liquid, heterogeneous catalysis, hydrogenation

1 Introduction

Microfabrication technology offers the opportunity to manufacture novel chemical reactor designs which could offer improvements in process capability and control and allow the more efficient elucidation of chemical kinetics. Particularly for multiphase reactions, the opportunity to manipulate at the micron scale the means and conditions by which two or more phases contact, mix, react, and separate afford the reaction engineer an unprecedented degree of control over the chemical reaction. Reactor dimensions can now be scaled with high-aspect ratio features smaller than 100 microns of any given planar complexity. Batch fabrication allows these designs to be manufactured with a high level of control on tolerance. Operating a multitude of microreactor units in parallel could achieve desired production capacities and offer distinct advantages over traditional modes of processing. In addition, by constructing reactors using silicon micromachining methods, the opportunity to integrate a wide variety of sensing elements could allow more efficient control and rapid probing of chemical kinetics.

Multiphase reactions present a unique opportunity for microfabricated reactors. In addition to the traditional reactor design paradigms, i.e. efficient thermal control, there is the added complexity of forcing a reactant of one phase to mix, diffuse, and react with that of another. For gas-liquid-solid reactions, the chemical kinetics are often limited by the mass transfer rate of the gaseous species through the liquid to the catalyst surface. Hydrogenations represent a typical and ubiquitous gas-liquid-solid reaction. Hydrogen is reacted with an organic substrate over a supported noble metal catalyst in either a slurry reactor or in a packed bed arrangement. Its been estimated that nearly 20% of all the reaction steps in a typical fine chemical synthesis are catalytic hydrogenations.

Owing to the limited solubility of hydrogen in organic substrates and solvents, mixing is a primary concern. The high energy requirements for vaporization precludes gas-phase hydrogenation for a wide range of compounds. Continuous industrial reactors are usually classified according to their flow dynamics.[1] In trickle bed reactors, for example, the volume of the reactor is predominantly the gas phase and the liquid forms a thin film around the catalyst pellet. Reaction engineers are concerned with the uniform distribution of the fluids within the reactor. Uneven flows can lead to incomplete utilization or local zones of varying reaction rate and heat transfer. Poor distribution of the fluids can thus lead to local 'hot spots' which can decrease the selectivity, or lead to side reactions which may cause reactor runaway[2]. Hydrogenations are exothermic and consequently thermal uniformity and adequate temperature control is of primary concern. In addition to losses in selectivity, poor thermal management can decrease the life of a catalyst and further reduce the efficiency of a reactor.

The high surface-volume ratios attainable in microfabricated structures can make gains in terms of thermal management and in improving mass transfer rates. Many liquid phase micromixer designs have emerged in recent years, most of which rely upon lamination and rapid diffusion at short length scales. By increasing the interfacial contact area between two mixing fluids and by reducing the characteristic length scale, the rate of transfer can be accelerated. For rapid multiphase reactions which are limited by mass transfer effects, a reduction in scale commensurate with increased surface area may ultimately improve process efficiency.

In this work, we describe the design and initial characterization of a microfabricated reactor intended for multiphase reactions. Gas and liquid reactants are mixed by maximizing the interfacial contact area: the inlet streams are split into multiple channels which are interleaved and meet at the start of a main reaction channel. Standard catalyst powders are used to take advantage of their high surface areas and known chemistry. In order to retain the catalyst particles within the reaction channel, a filter was formed near the outlet using a series of posts. The classical heterogeneous hydrogenation of α–methylstyrene(AMS) to cumene has been used as a model reaction.

1.2 Mass transfer analysis

For multiphase reactions, the overall reaction rate can be expressed in terms of a sum of resistances. The mass transfer resistances consist of gas absorption into the liquid, diffusion of the dissolved gas from the bulk liquid to the catalyst surface, and diffusion of the species inside the porous catalyst. Since diffusion times scale with the square of the characteristic length, appreciable gains in diffusion can be made at the small length scales achieved in microfabricated structures.

The established method for assessing diffusion within the porous catalyst is to use an effectiveness factor, η, which is a function of the Thiele modulus, ϕ. For a spherical catalyst particle and first order reaction, this takes the form,

$$\phi = \frac{D_p}{2}\sqrt{\frac{k\rho_p}{D_e}}$$

The square of the Thiele modulus represents the ratio of the characteristic time for diffusion to the characteristic time for the reaction. Values have been determined for the AMS hydrogenation reaction for palladium impregnated alumina catalysts.[5] Extending calculations for the effectiveness factor to scales on the order of 50μm indicate that diffusion in the pores is unlikely to pose a significant resistance. For a 50μm diameter spherical catalyst particle, $\phi^2 = 0.03$, indicating that the surface reaction is the dominant time scale. For the catalyst dimensions (as small as 550μm) employed in the trickle bed reactor study, $\phi^2 = 3.7$.

The other diffusional resistance involves the external transfer of the dissolved gas within the liquid film surrounding the catalyst. The Frössling correlation can be used to correlate the mass transfer from a liquid to a spherical particle.

$$Sh = \frac{k_c d_p}{D} = 2 + 0.6\,\mathrm{Re}^{1/2}\,Sc^{1/3}$$

The Schmidt number for hydrogen in AMS at 40°C is 56.7. The Reynolds number is 0.05 based on a particle diameter of 50μm and the low liquid flow rates (0.01mL/min) used in the microreactor. Consequently, the Sherwood number is nearly approximated by a stagnant thin film model. The rate for external mass transfer, R_E depends on the mass transfer coefficient k_c and the surface to volume ratio of the catalyst particles in the reactor. Ultimately, the mass transfer rate scales as

$$R_E \propto \frac{1}{d_p^{\ 2}}$$

so that reducing the particle diameter or thickness of the liquid film would have increasingly larger impact on the mass transfer rate for external diffusion.

The third resistance to mass transfer involves the absorption of the gas into the liquid. The rate of absorption can be expressed as

$$R_A = k_L a(C_i - C_b)$$

Where k_L is the gas absorption mass transfer coefficient, a is the interfacial gas-liquid contact area per volume, C_i is the interfacial gas concentration, and C_b is the bulk concentration of dissolved gas. Provided that the mixing phases are evenly dispersed throughout the packed bed, the interfacial contact area per volume can be made to be large within microchemical systems. Through novel microfabrication designs, it is possible to create high surface area structures which can force the gas and liquid reactants to mix more effectively by promoting dispersion and preventing macro-scale coalescence of the two phases.

2 Reactor Design

The motivation for the particular reactor design of this work is based on the nature of the chemistry. For heterogeneous reactions, the high surface areas of traditional porous supports allow for a greater dispersion of the active surface sites and consequently, a higher reaction rate per volume of support. For reactions of moderate rate, such as some hydrogenations, processing in small volumes demands the highest attainable active surface area in order to have a practical reactor throughput. Planar metal films, while easy to integrate using thin film methods, can not attain the same active area through geometry alone. So while thin film catalysts are sufficient for fast gas-phase reactions such as ammonia oxidation, higher surface area catalysts can extend the range of applicable reactions in the microenvironment. The advantage of using standard catalysts, such as noble metals supported on inert porous materials is the readily available information on chemical kinetics. Catalyst preparation becomes a matter of separating discrete particle size ranges and loading the reaction channels. The difficulty lies in choosing the proper scaling of reactor and catalyst dimensions in order to maintain an acceptable pressure drop.

Consequently, a critical issue for these micro-packed channels is the pressure drop. As the surface to volume ratio increases, the resistance to fluid flow also increases. For microreaction systems in general, the engineering design challenge is to balance the gains made in heat and mass transfer with the losses in pressure drop. For packed channels, the length scales which determine the pressure drop are the diameter of the particle D_p, the cross-sectional area of the flow A_s, and the channel length, L. The flow rate is dictated by the kinetics of the reaction, and fortunately for multiphase reactions, the kinetics are such that the space velocity at maximum conversion is not too high.

The functional dependence of the pressure drop for a packed bed based on a number of correlations(i.e. the Ergun equation) is given by,

$$\frac{\Delta P}{L} \propto \frac{\mu \cdot Q}{\rho \cdot D_p^{\,2} \cdot A_s} \cdot \frac{(1-\varepsilon)^2}{\varepsilon^3}$$

where Q is the flow rate and ε is the void fraction. For a given volume of packing with constant residence time, this functional dependence would suggest a short, large diameter reaction channel as having the optimum configuration for minimizing the pressure drop. If the reactions are not mass transfer limited, then the capacity for conversion should depend on the amount of catalyst and not on the specific geometry. The problem with operating such a design is 1) obtaining an even distribution of reactants over the inlet diameter and 2) controlling the temperature profile. Since the temperature of most tubular reactors is controlled through the exterior radial walls, a large diameter reactor will be increasingly hindered by radial gradients in temperature. Therefore, splitting the flow into multiple channels so that the effective cross-sectional area is large will ultimately

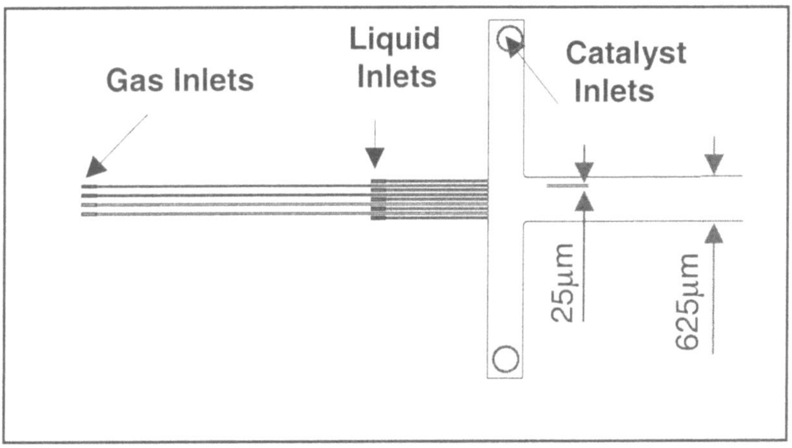

Figure 1. Micro packed bed reactor design

reduce the pressure drop while maintaining the same reactor throughput and high surface-to-volume ratio. Microfabrication methods have the potential for efficiently constructing such a reactor design.

Figure 1 illustrates the inlet portion of the reactor. Gas and liquid inlet flows are split among several 25µm wide channels which meet at the main reaction chamber which is 625µm wide. Perpendicular to these inlet channels are two 400µm wide channels used for delivering the catalyst slurry. Figure 2 is a scanning electron micrograph of a cross-section of the inlet channels where they meet at the main reaction chamber. At the outlet of the 2 cm long reaction chamber is a series of posts etched in the silicon which are used to retain the packing material. Figure 3 contains their design and the corresponding scanning electron micrograph.

3 Fabrication

The fabrication process involves two etch steps and an anodic bond. The topside channels are first formed in the silicon substrate by using a time-mulitplexed inductively coupled plasma etch process.[3]

The depth is controlled by timing the etch to approximately 300μm. As seen Figure 2, a loading effect in the etch process causes the 25μm wide channels to be approximately 20% more shallow than the main reaction channel. The wafer is then patterned from the bottom side to

Figure 2. SEM of inlet channels

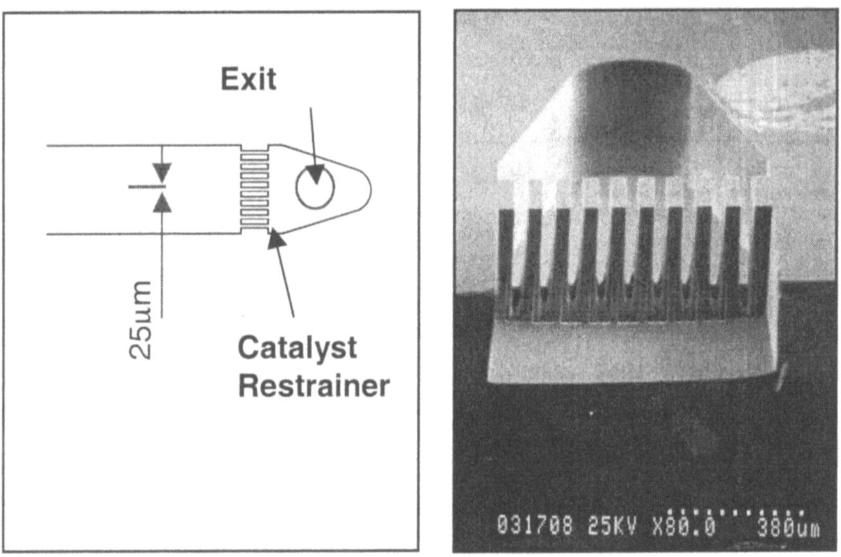

Figure 3. Reaction channel outlet with structures to hold packing material

allow ports to access the various channels. The final step in the process is to cap the channels with Pyrex glass using an anodic bond. The wafer is then diced into twelve reactors 1.5cm X 3.5cm. Figure 4 shows a cross-section of the silicon/glass reaction channel.

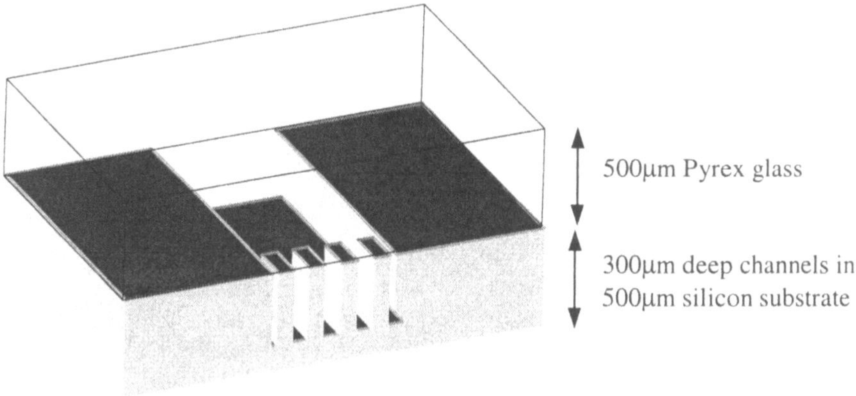

Figure 4. Illustration of the silicon/glass device

4 Experimental Results

4.1 Packaging

Figure 5 illustrates the system and packaging for the micro packed-bed reactor. The silicon-glass reactor is compressed against a thin Viton sheet in order to form

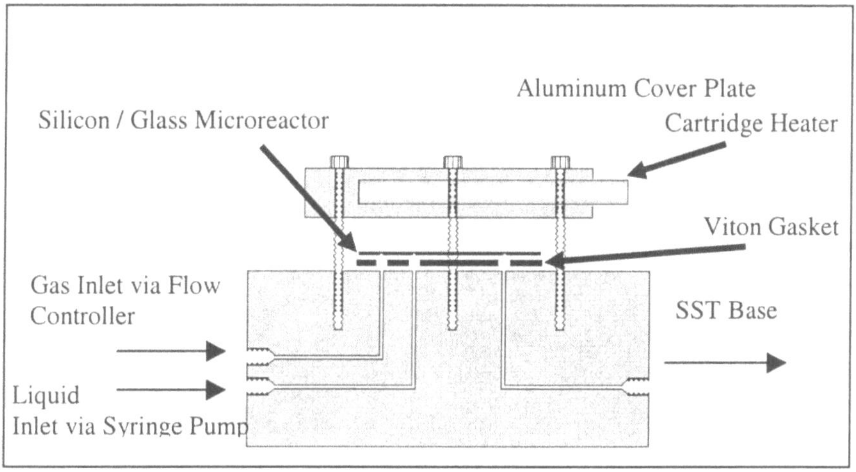

Figure 5. System configuration and reactor packaging

the fluidic connection to the stainless steel base which has been machined for standard high-pressure fittings. The aluminum cover plate contains a single cartridge heater. A mass flow controller delivers the hydrogen while a syringe pump delivers the liquid reactant.

4.2 Reactor Loading

A variety of materials have been loaded into the microreactor such as glass microspheres, polystyrene beads, and catalyst powders. The carbon and alumina supported metal catalysts were obtained from Aldrich Corporation and various particle sizes were separated using standard sieving equipment. The 50-75μm

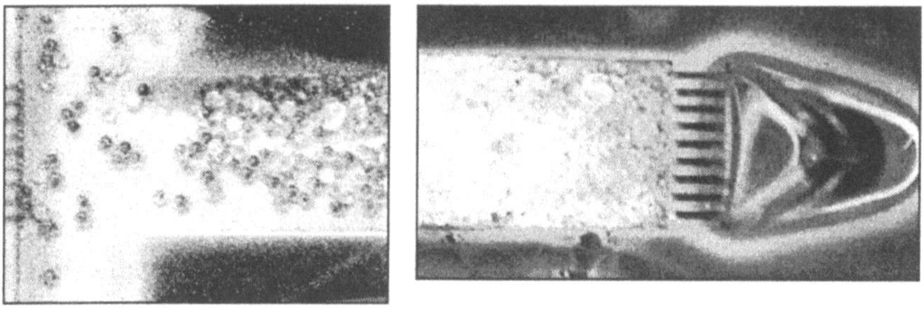

Figure 6. Photomicrograph of 50μm glass spheres

fraction was retained and a slurry formed using ethanol. The slurry was delivered to the microreactor using the larger 400μm wide side channels (Figure 1). A plexiglass cover plate was used in order to visualize the process. One channel was used as an inlet and the other as an outlet in order to maintain a steady flow perpendicular to the main reaction channel. Flow from the main liquid inlets would then redirect a portion of the slurry flow towards the main reaction channel.

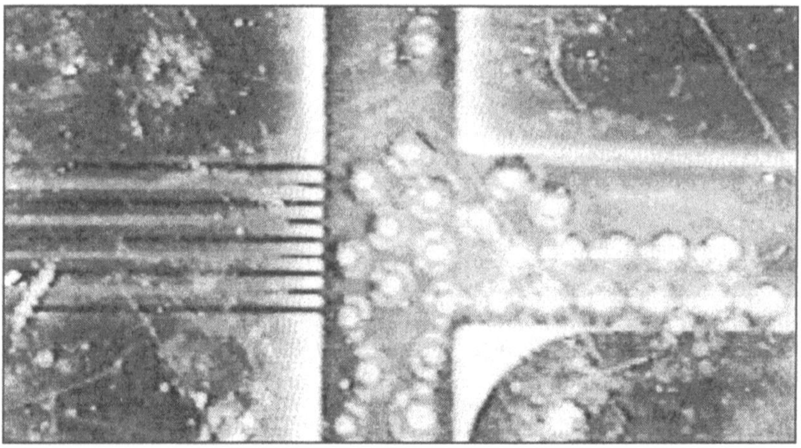

Figure 7. Co-current flow of water and air

For the case of the 50μm glass spheres, packing was obtained without a solvent: the dry spheres could be loaded by creating a well above the inlet ports and continuously agitating the reactor. An image of the glass spheres in the microreactor can be found in Figure 6. Once the reactor is loaded with material, the catalyst inlets are closed off by either substituting a different gasket or capping the external fittings.

4.3 Fluid flow visualization

Some representative images of dual gas-liquid flow in the microreactor are presented in Figures 7and 8. Figure 7 demonstrates gas-liquid flow in the absence of packing in the channel. The fluids are water and air and the bubble size is approximately 150μm. Figure 8 shows the mixing of hydrogen with heptane over a bed of catalyst. Here, the flow illustrates rapid mixing of the reactants before distribution at the catalyst bed.

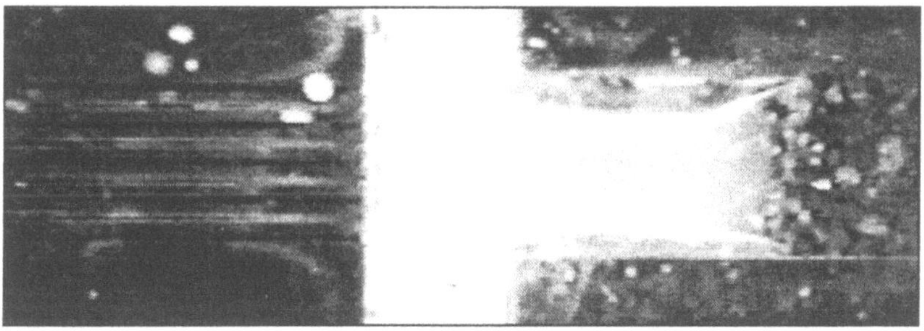

Figure 8. Co-current flow of H_2 and heptane over packed particles

4.4 Reaction Studies

The hydrogenation of α-methylstyrene to cumene is a standard reaction that has been used to gauge mass-transfer effects in catalyst pellets and in fixed bed reactors.[4] Various physical properties, such as the solubility and diffusivity of hydrogen in α-methylstyrene, are readily available in the literature. The reaction is exothermic (ΔH = -100kJ/mol) and the intrinsic kinetics are known for a variety of catalysts. The kinetic behavior is often first order in hydrogen and zero order in substrate. The reaction is moderately fast at room temperature and 1 atm pressure hydrogen.

Carbon supported palladium catalysts were used in several reactions at 50°C. The effluent from the reactor was collected in fractions and analyzed using GC/MS. Conversions from 20% to 100% have been observed depending on the feed composition and flow rates. Initial reaction rates for the hydrogenation of α-methylstyrene were near 0.01 mmol/min per reaction channel without prior activation of the catalyst and compared favorably with the intrinsic kinetics reported in the literature.[5]

5 Conclusions

Using silicon microfabrication technology, a microchemical reactor has been constructed for the purposes of performing multiphase reactions. The facilitation of gas-liquid dispersion and mixing has been observed in the microreactor, standard catalysts have been integrated, and conversions measured for a model hydrogenation reaction. For many of the exothermic multiphase reactions, the improved thermal control offered by microchemical reactors could also improve reactor safety and efficiency. Operation of multiple reactor units in parallel should give an optimum configuration for pressure drop, provided that even flow distribution among the multiple channels can be achieved. The result is a reactor configuration that could potentially operate with greater efficiency and extend process capability over its macro-scale counterpart.

Acknowledgements
The authors wish to thank the DARPA Micro Flumes Program (F30602-97-2-0100) for partial finanical support.

References

1. Mills, P.L. and R.V. Chaudhari, *Multiphase catalytic reactor engineering and design for pharmaceuticals and fine chemicals.* Catalysis Today, 1997. **37**(4): p. 367-404.
2. Eigenberger, G. and U. Wegerle. *Runaway in an Industrial Hydrogenation Reactor.* in *Chemical Engineering Research Symposium.* 1982. Boston.
3. Ayon, A.A., *et al.*, *Characterization of a time multiplexed inductively coupled plasma etcher.* Journal of the Electrochemical Society, 1999. **146**(1): p. 339-349.
4. Satterfield, C.N., *Mass Transfer in Heterogeneous Catalysis.* 1981: Krieger Publishing.
5. Herskowitz, M., R.G. Carbonell, and J.M. Smith, *Effectiveness Factors and Mass Transfer in Trickle-Bed Reactors.* AIChE Journal, 1979. **25**(2): p. 272-282.

Parallel Synthesis and Testing of Heterogeneous Catalysts

U. Rodemerck, P. Ignaszewski, M. Lucas, P. Claus, M. Baerns

Institut für Angewandte Chemie Berlin-Adlershof e.V., Rudower Chaussee 5, 12489 Berlin, Germany, e-mail: rodemerk@aca-berlin.de, fax: +49-30-63924350

Abstract

We are presenting an effective method to prepare and test heterogeneous catalysts much faster than on the conventional way. The catalysts were prepared by incipient wetness method by means of an automatic liquid handler. For catalytic testing two systems were developed: (i) 15 quartz glass reactor tubes were simultaneously heated in a stainless steel block to reaction temperature and the reaction gases passed through the reactor tubes were sequentially analysed by a QMS. (ii) Up to 250 catalyst compositions can be prepared and tested in parallel in a ceramic monolithic reactor module the channels of which containing different catalyst compositions.

Keywords

high throughput screening, combinatorial chemistry, heterogeneous catalysts

1 Introduction

The development of new heterogeneous catalysts is commonly based on the chemical knowledge and intelligent intuition as well as on trial and error. This involves time consuming preparation and testing of a great number of catalysts needing appreciable manpower and material resources. In the last few years combinatorial methods which are well established in drug synthesis have been used to synthesise new inorganic magnetoresistance [1], luminescent [2] and superconducting [3] materials, too. Such materials can be tested very fast automatically owing to their characteristic physical properties.

Properly speaking, heterogeneous catalysts are inorganic materials, too. To find out a good catalyst for a chemical process, however, is much more complicated because the concentration of the educt and of all reaction products in the effluent gas must be measured to calculate conversion degree and selectivities. For that reason the development of heterogeneous catalysts by combinatorial methods started a few years ago [4-12]. Extensive work in the field of heterogeneous catalysis have been done by H.W. Weinberg et al. at Symyx in Santa Clara, Calif. who prepared catalyst arrays in the form of thin films on small wafers and analysed reaction products in the ppm region by a high speed mass spectrometer [4, 5]. Such thin material films, however, have different structural, morphological and maybe chemical and catalytic properties in comparison with conventionally prepared heterogeneous catalysts.

The aim of the present work was to develop a method for fast parallel synthesis and catalytic tests of heterogeneous catalysts, which give results similar to conventional procedures. Oxidative coupling of methane was chosen as a test reaction.

2 Parallel Synthesis of Heterogeneous Catalysts

The synthesis of heterogeneous catalysts which includes preparation of the catalyst precursor, drying and calcination is a time consuming process when each catalyst is prepared separately. Parallel synthesis of arrays of catalysts different in chemical composition can help to save a lot of time because all catalysts are dried and calcined in the same period.

For parallel preparation of arrays of heterogeneous catalysts a automated liquid handler (GILSON 215) was used. Catalysts were prepared by incipient wetness method, that means a porous carrier material is impregnated by such a volume of a solution of precursor components what is equal to the pore volume of the carrier.

We developed a computer program to prepare catalyst arrays automatically: The desired concentrations of the components on the carrier material were entered into the program. After this, the liquid handler mixes and dilutes the stock solutions of the components and transfers the calculated volume to the portions of carrier material which are in small glass containers or inside the channels of a monolithic reactor. All prepared catalyst precursors can be dried and calcined together at the same time.

As an example an array of 30 Mn/Na_2WO_4 catalysts on SiO_2 (Davisil 646, ALDRICH) was prepared by using 5 different loadings of Mn and 6 loadings of Na_2WO_4. By this parallel technique the preparation time for the 30 catalysts was decreased to 8 hours. It is possible to synthesise even much greater arrays in one day because the time consuming step is the drying of the array and not the mixing of solutions and impregnation of the carrier.

3 Systems for High Throughput Catalyst Screening

For the high throughput testing of heterogeneous catalysts two automatic reactor systems were developed. The first system combines 15 conventionally quartz reactors whereas the second system is a monolithic reactor which contains the different catalysts in each individual channel.

3.1 Multitube Reactor System

One possibility to realize a faster testing of catalysts is to combine conventional tubular reactors in one apparatus with a common temperature control and analytic system. We used 15 quartz reactors (3.5 mm i.d.) which are heated in an ingot of stainless steel (Fig. 1). The gases supplied by mass flow controllers are fed to one of the reactors by a 16-port valve and the effluent gases of this reactor flow through the second 16-port valve to the analytic system, a quadrupole mass spectrometer (QMS). To the other reactors an inert gas is fed.

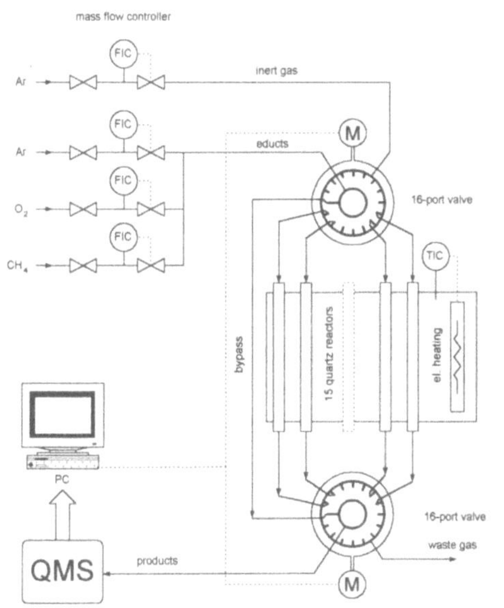

Fig. 1. Set-up of a computer controlled apparatus for simultaneous testing of up to 15 heterogeneous catalysts

The catalytic test of 15 catalysts is automatically done as follows: First, the reaction gases flow through the bypass and a QMS analysis is done. Then, the flow was switched to the first reactor. By QMS analysis the concentration of one or more of the reaction products (e.g. CO_2) is measured. If no more concentration changes are observed the stationary state is reached and the concentration of all educts and products is determined to calculate the conversion degree, the yields and product selectivities. After that, the valves switch to the next position and the catalysts in the second reactor is tested in the same way.
In this apparatus one catalytic test takes ca. 5-10 minutes so that 15 catalysts can be tested in one day at different reaction conditions (temperature, GHSV).

3.2 Monolithic Reactor

Up to 250 catalyst compositions can be prepared and tested in parallel in a ceramic monolithic reactor module the channels of which contained different catalyst compositions (Fig.2). The ceramic monolith is heated in a stainless steel reactor and fed by the reaction gas. Gas samples from each channel of the monolith were analysed by QMS. The fused silica QMS inlet capillary is moved

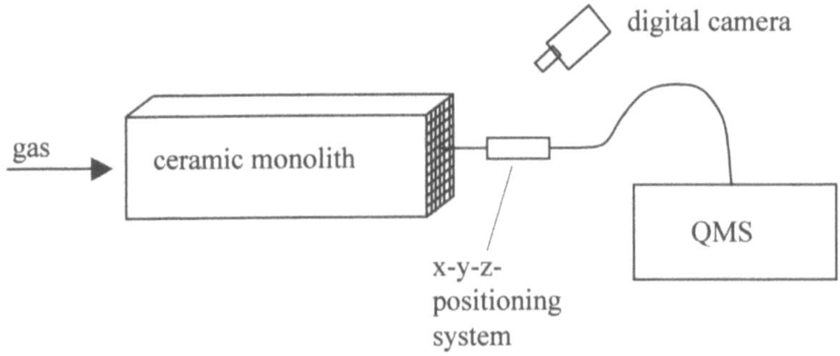

Fig. 2. Set-up of the monolithic reactor system for high throughput screening of heterogeneous catalysts

into the channels with inner diameters of 2 mm one after another by a xyz positioning system (3 stepping motors) which can be used at high temperatures, too (Fig. 3). The actual position of the capillary was watched by a digital video camera. The system is fully computerised so that up to 250 catalysts can be tested automatically. Moving the capillary from one channel into the next, waiting for stationary state and recording a mass spectrum can be carried out in one minute so that 250 catalysts can be tested in one day.

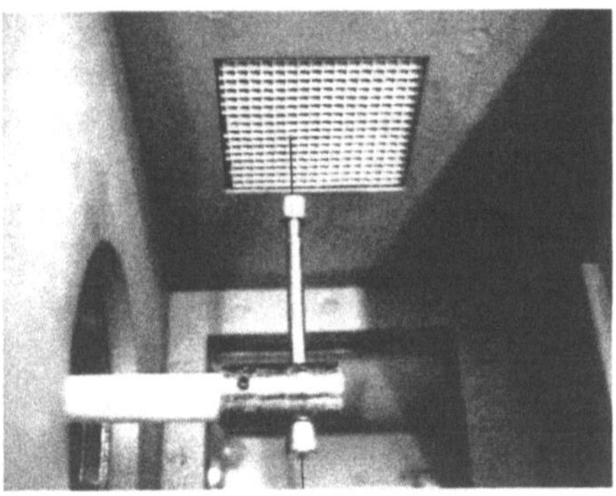

Fig. 3. Positioning system scanning the channels of the monolith

The accuracy of the testing results, however, is not as good as in the quartz reactors because it is difficult to obtain equal gas flow in each channel.

4 Test of an Catalyst Array for Oxidative Coupling of Methane

$Mn/Na_2WO_4/SiO_2$ catalysts are effective catalysts for oxidative coupling of methane. The catalytic efficiency of these catalysts are influenced by the concentration of both components on the carrier. It is known that activity and selectivity will be increased even on small Mn concentration and that Na is necessary for a suitable Mn distribution on the catalyst surface. Usually, a catalyst composition of 1.9 wt% Mn and 4-5 wt% Na_2WO_4 on SiO_2 is used which gives C_{2+} selectivities (sum of the selectivities of ethane, ethylene, propane and propylene) of 80 % at a methane conversion degree of ca. 20 % [13].

We used the above catalytic system to test the automatic preparation and fast testing of catalysts. An array of 30 $Mn/Na_2WO_4/SiO_2$ catalysts which was prepared as described above was tested in the multitube reactor system. Fifteen catalyst precursors were simultaneous put into the quartz reactors and were calcined 5 hours in a stream of oxygen at 850 °C and, after this, tested for the oxidative coupling of methane. For quantitative MS analysis the following fragments were used: m/e=16 (CH_4), m/e=25 (Ethylene), m/e=28 (CO), m/e=30 (Ethane), m/e=32 (O_2), m/e=40 (Ar, internal standard), m/e=41 (Propene), m/e=43 (Propane) und m/e=44 (CO_2).

Table 1. CH_4 conversion and C_{2+} selectivities for oxidative coupling of methane on $Mn/Na_2WO_4/SiO_2$ catalysts, T = 800 °C, τ = 0.1 g s ml^{-1}, feed: 76% CH_4, 14% O_2, 10% Ar

	Conversion of Methane / %					Selectivity C_{2+} / %				
	0 % Mn	0.1 % Mn	1.0 % Mn	1.9 % Mn	5.0 % Mn	0 % Mn	0.1 % Mn	1 % Mn	1.9 % Mn	5 % Mn
0 % Na_2WO_4	9	14	12	10	11	35	32	31	31	29
0.1 % Na_2WO_4	13	15	15	15	16	31	35	37	36	37
1.0 % Na_2WO_4	4	17	17	17	17	34	39	40	39	45
2.5 % Na_2WO_4	5	19	11	16	17	41	51	79	47	46
4.0 % Na_2WO_4	3	17	19	11	9	75	79	77	78	81
5.0 % Na_2WO_4	2	15	12	10	9	79	82	82	80	82

In Table 1 and Fig. 3 the catalytic results are summarised in dependence on the loading of the carrier with the both components Mn and Na_2WO_4: Catalysts not containing Mn have only low activity which can be attributed to the carrier material itself as can be seen for the catalyst which contain neither Mn nor Na_2WO_4. High activities were found if the catalysts contain 2.5-4 wt% Na_2WO_4 and at the very least small amounts of Mn. The selectivity of the catalysts is determined by the Na_2WO_4 loading; on this occasion, the Mn content has no influence. The maximum yield of C_{2+} hydrocarbons is obtained on catalysts witch contain 0.1-1

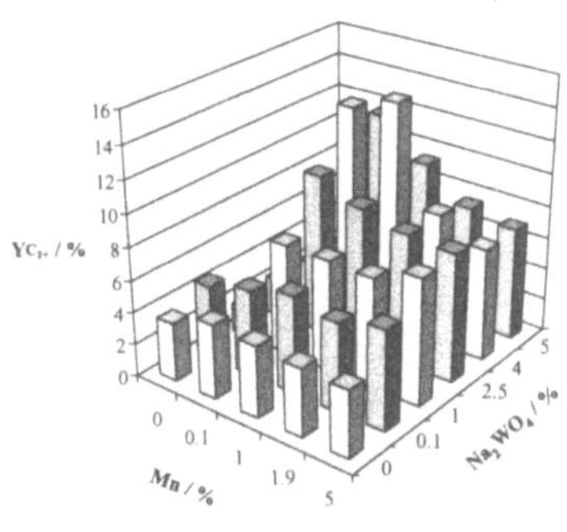

Fig. 3. C_{2+} yields for oxidative coupling of methane on $Mn/Na_2WO_4/SiO_2$ catalysts, experimental conditions see Table 1

wt% Mn and 2.5-5 wt% Na_2WO_4. The results obtained are in agreement with the published literature [13].

5 Conclusions

Parallel preparation of heterogeneous catalysts and fast catalytic testing is a valuable tool to select good catalysts from a great catalyst matrix.
The catalytic test of an array of $Mn/Na_2WO_4/SiO_2$ in the 15-reactor system followed by MS analysis gave results which were similar compared to a conventional tubular flow reactor system equipped with a gas chromatographic system.
One QMS analysis could be done in ca. 5-10 minutes with good accuracy so that methane conversions and product selectivities for 15 catalysts could be determined at different within one day.
With the monolithic reactor the time for catalytic tests can be decreased by a factor of 10 in comparison with the 15-reactor-system or by a factor of 100 compared to conventional catalytic test equipment, respectively. The accuracy of the results, however, will be lower due to shorter times for collecting MS data for each catalyst. Moreover, it is difficult to generate equal gas flows in all channels of the monolith. This problem will be solved in the near future.

6 Acknowledgement

The authors thank the Bundesministerium für Bildung und Forschung (BMBF, contract number 03D0068D0) for financial support.

7 References

[1] G. Briceño, H. Chang, X. Sun, P.G. Schulz, X.-D. Xiang, Science 270 (1995) 273.
[2] E. Danielson, J.H. Golden, E.W. McFarland, C.M. Reaves, W.H. Weinberg, X.D. Wu, Nature 389 (1997) 944.
[3] X.-D. Xiang, X. Sun, G. Briceño, Y. Lou, K.-A. Wang, H. Chang, W.G. Wallace-Freedman, S.-W. Chen, P.G. Schultz, Science 268 (1995) 1738.
[4] W.H. Weinberg, B. Jandeleit, K. Self, H. Turner, Curr. Opinion Sol. State Mat. Sci. 1998, 3, 104.
[5] P. Cong, R.D. Doolen, Q. Fan, D.M. Giaquinta, S. Guan, E.W. McFarland, D.M. Poojary, K. Self, H.W. Turner, W.H. Weinberg, Angew. Chem. 111 (1999) 507.
[6] S.M. Senkan, S. Ozturk, Angew. Chem. 111 (1999) 867
S.M. Senkan, Nature 1998, 394, 350.

[7] J. Klein, C.W. Lehmann, H.-W. Schmidt, W.F. Maier, Angew. Chem. 110 (1998) 3557.
[8] D.E. Akporiaye, I.M. Dahl, A. Karlsson, R. Wendelbo, Angew. Chem. 110 (1998) 629.
[9] C.L. Hill, R.D. Gall, J. Mol. Catal. A: Chemical 1996, 114, 103
[10] F.C. Moates, M. Somani, J. Annamalai, J.T. Richardson, D. Luss, R.C. Willson, Ind. Eng. Chem. Res. 1996, 35, 4801.
[11] A. Holzwarth, H.-W. Schmidt, W.F. Maier, Angew. Chem. 110 (1998) 2788.
[12] M. Baerns, O. Buyevskaya, U. Rodemerck, D. Wolf, MRS Fall Meeting, Boston, USA, 30. 11. – 4. 12. 1998.
[13] A. Palermo, J.P.Holgado Vazquez, A.F. Lee, M.S. Tikhov, R.M. Lambert, J. Catal. 177 (1998) 259.

SINGLE FLOW ELECTROCHEMICAL MICROREACTOR APPLICATION TO FURAN METHOXYLATION

R. Ferrigno[(1)], V. Reid[(2)] and H. H. Girault[(1)]

(1) Laboratoire d'Électrochimie, École Polytechnique Fédérale de Lausanne
CH - 1015 Lausanne, Switzerland
(2) Direction des Études et Recherches, Électricité de France.

Abstract

Interdigitated band electrodes are fabricated using screen printing technology and laser ablation methods in order to pattern electrodes featuring inter-electrode gaps down to 30 μm. The electrochemical microcell is tested with a scheme reaction : the furan methoxylation. Various parameters are investigated including cell voltage, supporting electrolyte composition and bromide concentration. The results are presented in terms of conversion and current efficiencies as well as energy consumption. For instance, cells with inter-electrode gap of 100 μm allow 70% conversion with a current efficiency of 80% and an energy consumption of 1.5 kWh kg^{-1}, for inlet methanol solutions containing 50 mM of furan and 5 mM of bromide salt at a flow rate of 85 μl min^{-1}. It is worth noticing that the improvement brought about by this configuration relates to energy as well as selectivity. When compared to others electrochemical cells (bipolar packed bed cells or pump cells) in batch configuration, the microreactors feature 50% decrease of the energy consumption and considerable improvement in selectivity as only one by-product is detected in the outlet solution compared to the numerous by-products usually obtained. Moreover, the reduced amount of salt, the high selectivity and the quasi-total conversion of the furan reduce the successive purification processes of the outlet solution.

Keywords

Interdigitated band electrodes, electrochemical microreactor, furan methoxylation

1 - Introduction

Miniaturisation techniques are currently well established in a wide range of analytical applications and preparative syntheses. Microvalves, micropumps[1-3] and micro-heat exchangers[1,4] have been designed in fully integrated systems allowing complete and controlled chemical processing. For instance, microchip systems (μTAS) are available for capillary electrophoresis applications with integrated injector, separation column and detector in one single chip. The small dimensions involved in microreactors in conjunction to the high surface to volume ratio provide an effective enhancement of mass and heat transfer within the device. In the case of an electrochemical microreactor, the potential distribution also plays an essential role in selectivity and energy consumption. The control of the current and potential distribution mainly depend on the reduction of the ohmic losses which can be achieved using for instance microelectrode geometries as interdigitated band electrodes. In this specific configuration, it has been demonstrated that the decrease of the characteristic dimension of the cell (inter-electrode gap) to the micrometer range provides some advantages[5,6]. Moreover, the integration of this geometry in restricted reactive cell volume should allow new preparative reactions which cannot be investigated in classical reactors. Amatore et al. [7] have described the use of interdigitated band electrodes for reactions between two unstable intermediates such as a strong electrophile and a strong nucleophile.

The test reaction here is the furan methoxylation where the final product 2,5-dihydro 2,5-dimethoxy furan (DHDMF) is obtained through the indirect oxidation of the furan via electrogenerated bromine molecules. This reaction is already characterised in classical reactors as pump cells[8], packed bipolar cells[9] and Solid Polymer Electrolyte cell (SPE)[10]. In the SPE reactor, the electrolyses are carried out in absence of supporting electrolyte and bromide salt, the furan being directly oxidised at the anode. However, this strategy is energy consuming as a current efficiency of 75% is reached with an energy consumption of 5 to 9 kWh kg^{-1} depending on the cell voltage (from 8 to 20 V).

In the present work, we will attempt to show that an interdigitated band electrode microreactor is able to undertake organic electrosyntheses in absence of supporting electrolyte with no detrimental increase of the energy consumption.

2 - Experimental details

The electrodes are fabricated using both classical screen printing technique and laser ablation method. The screen printing is performed as already published using platinum ink (Heraeus DC 430 79, 63% of platinum) on alumina substrate[6]. For inter-electrode gap width larger than 250 μm, the interdigitated geometry is directly printed on the substrate through the use of a patterned mask. For narrower gap (till 30 μm), the geometry is drawn with a laser beam directly on a printed platinum layer. The electrodes are then cured at 850°C for one hour. Two different geometries are investigated : an interdigitated band electrodes featuring a gap of 500 μm and band width of 1 mm and an interdigitated band electrode featuring a 100 μm wide gap and 500 μm band width. The 100 μm electrode geometry is characterised by Scanning Electron Microscopy (Figure 1).

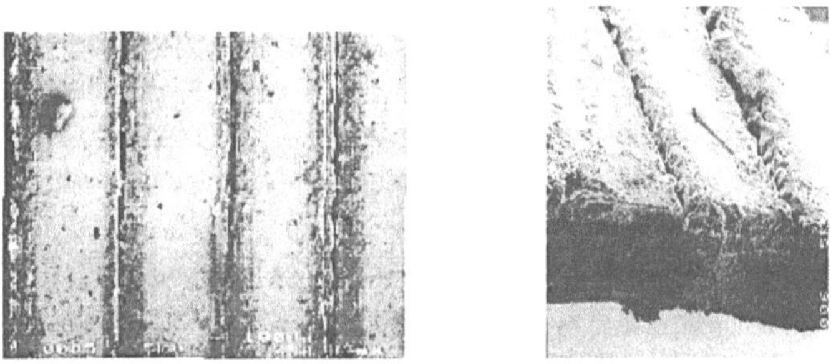

a b

Figure 1 : SEM picture of 100 μm interdigitated band electrodes. (a) Top view, magnification 50, (b) Side view, magnification 65.

During the ablation process, the platinum ink made of SiO_2 and platinum particles is smelt by the laser beam at high temperature and the smelt material settled on both sides of the gap. From the SEM pictures, debris are noticed and seem to create a short-cut between two adjacent bands. Nevertheless, a high open circuit resistance (10 MΩ) is measured between the two electrodes which is enough to polarise anodically or cathodically each electrode.

The interdigitated band electrodes are tested in a 800 μm deep perspex microreactor. During the electrolyses, samples are collected to be analysed by gas chromatography. The detection and quantification of the furan and DHDMF compounds allow to calculate figures of merit as conversion (τ), current efficiency (η) and energy consumption (γ).

3 - Results

Various parameters are investigated in single flow configuration : the cell voltage, the supporting electrolyte composition, the inter-electrode gap and finally the bromide concentration.

3.1 - Cell voltage influence:

The evolution of the furan and DHDMF concentrations with the cell voltage are presented in Figure 2 for a flow rate of 150 μl min^{-1} and an inlet solution containing 0.1

M $NaClO_4$, 100 mM furan and 5 mM NaBr. The DHDMF concentration follows the sigmoidal shape of the bromide oxidation current.

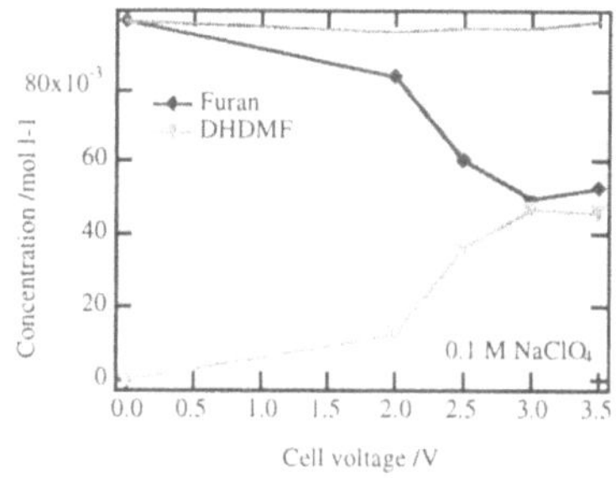

Figure 2 : Furan and DHDMF concentration evolution with cell voltage in a 800 μm deep cell configuration. Solution 0.1 M $NaClO_4$, 100 mM furan, 5 mM NaBr and flow rate 150 μl min^{-1}.

A conversion of nearly 50% is obtained after one pass above the electrodes for a cell voltage above 3 V and a current of 30 mA. DHDMF production is limited by bromine production which is a mass transfer controlled reaction. Therefore, the DHDMF concentration depends on both the flow rate and the bromide concentration.

3.2 - Supporting electrolyte and inter-electrode gap influence:

When the supporting electrolyte is removed from the inlet solution, the conversion behaviour depends on the inter-electrode gap width as represented in Figure 3.

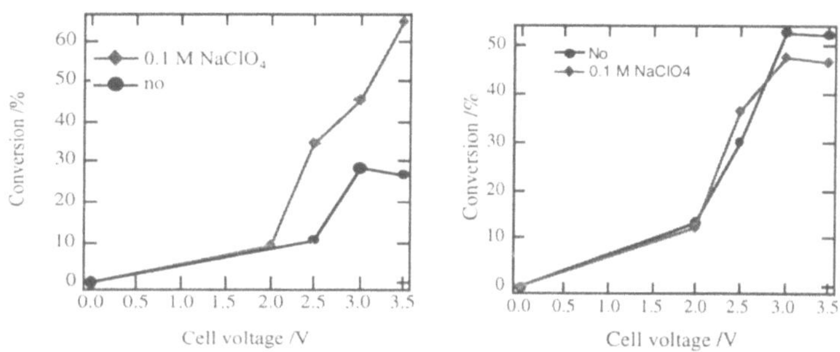

a b

Figure 3 : Supporting electrolyte influence on the conversion for a 800 μm deep perspex cell. (a) 500 μm wide interdigitated band electrode, 150 μl min^{-1}, (b) 100 μm wide interdigitated band electrode, 75 μl min^{-1}. Solution : 100 mM furan, 5 mM Nabr.

For a 500 μm wide gap, the conversion reached in absence of supporting electrolyte is lower than the conversion attained with a perchlorate salt concentration of 0.1 M (Fig. 3a). In contrast for a 100 μm wide gap, no significant differences are noticed in term of conversion (Fig. 3b). When the inter-electrode gap is too wide, the cell voltage distribution is affected by the inter-electrode ohmic losses and is not sufficient anymore for the anodic polarisation. The minimisation of the gap to 100 μm allow preparative electrolyses in absence of supporting electrolyte.

3.3 - Bromide concentration influence:

The optimisation of the solution composition is investigated by varying the bromide concentration. In Figure 4, the evolution of the conversion with the bromide concentration (from 5 to 200 mM) is represented for a flow rate of 150 μl min^{-1}. At 5 mM, half of the furan is converted into DHDMF for a cell voltage above 3 V. The conversion increases above 50% when the concentration is 20 mM and then stabilised at 70% when the concentration is increased to 100 and 200 mM.

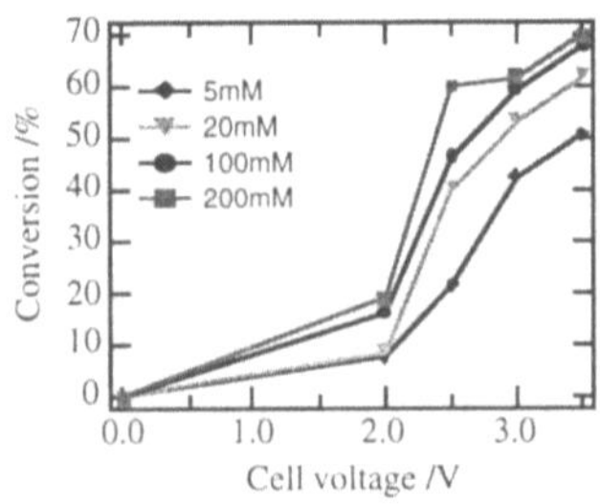

Figure 4 : Conversion evolution with the bromide concentration.

For concentrations higher than 20 mM, the increase of the measured current has no effect on the conversion. In the case of single flow electrolysis, a high bromide concentration is useless from the point of view of the conversion. Moreover, this large concentration is responsible of worse current efficiency and energy consumption as described by the figures of merit derived from preparative electrolyses at 3 V and presented in Table I. The experiments were carried out with an inlet solution of 50 mM furan at a flow rate of 85 μl min^{-1}.

	100 mM NaBr	5 mM NaBr
Current efficiency /%	30	80
Energy consumption /kWh kg^{-1}	4.4	1.5
Conversion /%	70	70

Table I : Bromide concentration influence on the conversion. the current efficiency and energy consumption in a 800 μm deep perspex cell.

Concentrated bromide solutions are not always convenient in electrolysis as a large concentration of bromide induces the recombination of bromine and bromide ions as $Br_2 + Br^- \rightarrow Br_3^-$. This phenomena takes place in single flow experiments as the initial furan concentration is low.

4 - Conclusion

When compared to classical reactors (Table II), the interdigitated band electrode configuration presents obvious advantages for the energy saving standpoint demonstrated by the 50% energy saving. Moreover, for sufficiently thin gap (100 μm) the reduction of the total amount of salt (bromide and perchlorate) slightly affects the energy consumption and current efficiency of the furan methoxylation reaction. During the electrolyses, only one by-product is detected demonstrating the control of the process. Moreover, the low salt concentration, the high selectivity and the quasi-total conversion of the furan reduce the successive purification processes of the outlet solution.

	Current efficiency /%	Energy consumption /kWh kg^{-1}
100 μm interdigitated band electrode cell (no supporting electrolyte)	80	1.5
Packed bed bipolar cell	90	3

Table II : Comparison of the figures of merit between the interdigitated band electrode cell and the packed bed bipolar cell.

Acknowledgments

The authors thank Electricité de France, Direction des Etudes et Recherches who financed the project and also S. Favre (Institut d'Optique Appliqué – EPFL) for the manufacture of the electrodes.

Bibliography

1 - R. W. Wegeng, C. J. Call, M. K. Drost Proc. 1996 Spring National Meeting AIChE, Feb. 25 - 29 New Orleans, Seiten 1 - 13 (1996).

2 - J. Döpper, M. Clemens, W. Ehrfeld, K. P. Kamper and H.Lehr, (1996) Proceedings Actuator'96, Bremen.

3 - M. Stehr, S. Messner, H. Sandmaier and R. Zengerle, (1996) IEEE MEMS Conference proceedings.

4 - V. Hessel, W. Ehrfeld, H. Möbius, Th. Richter and K. Russow, (1995) Proc. Int. Symp. Microsystems, Intelligent Materials Robots, Sendai, Japan 45.

5 - C. Belmont and H. H. Girault, J. Appl. Electrochem. (1994) **24** 475.

6 - C. Belmont, H. H. Girault, J. Appl. Electrochem. (1994) **24** 719.

7 - C. Amatore and A. R. Brown, J. Am. Chem. Soc., (1996) **118** 1482.

8 - N. R. Tomov and R. E. W. Jansson, J. Chem Techn. Biotechnol. (1980) **30** 110.

9 - K. Kusakabe, S. Morooka and Y. Kato, J. Chem. Eng. Jpn. (1986) **19** 43.

10 - E. Raoult, J. Sarrazin and A. Tallec, J. Appl. Electrochem. (1985) **15** 85.

Part 5

Microfluidics

Microfluidic Devices for Sample Preparation (Liquid) in the Nanoliter Range

H. Fiehn, March 1999 IMRET3, Frankfurt

1. Core manufacturing technologies
2. Microfluidic basic components
3. Customized microfluidic systems

Our company has developed a modular set of microfluidic devices manufactured by the silicon / glass micromachining technology. These devices have become commercially available now and can support a wide range of new analysis methods dependent on handling and distributing of smallest amounts of liquid. Essential target fields are drug screening, combinatorial chemistry and genetics (life science). Latest developments focus on the adaptation of multilayer ceramics technology (up to 80 layers) to design high complex microfluidic devices.

The presentation will outline the capabilities of our piezoelectrical multipurpose micropipette. It fits to the needs of micro dosage tasks requiring a precision of a less than single nanoliter. The pipette can be used as standalone element attached to robotic systems (e.g. to fill micro well plates and nano well plates, to plot dot arrays on biochips) or as integrated part of a (micro-)system (e.g. to suck liquid through miniaturized manifolds).

The second part of our contribution deals with microfluidic flow through systems manufactured by assembling single components onto the surface of planar substrates . Small and complex fluid containing structures like capillaries and reaction chambers can be integrated on these substrates - considering customer specified design. A variety of microfluidic components like pumps, mixers and the patented GeSiM microinjector is available.

Our microfluidic products operate both with aqueous solutions and several solvents which might contain small particles like large biomolecules or cells.

Main applications are micro array technology, biochip preparation and combinatorial biochemistry.

Core Technologies and Know How

Preparation of silicon and glass wafers:
- Deposition of thin layers by PVD and CVD
- Bulk micromachining (dry and wet)
- Double side lithography
- Anodic bonding
- Diebonding, wirebonding
- Micro adhesive bindings
- QA, e.g. by surface profilometer

Micromachining of LTCC- components (ceramics) and polymer devices
- Design of highly complex microfluidic systems
- Manufacturing of stacks containing more than 50 layers

Lithography:
- Manufacturing of glass masks for MEMS (> 10 µm) using LCD technology (very cost effective)

Electrical and mechanical engineering:
- Hard- and Softwaredesign
- Rapid Prototyping (Fused Deposition Modeling)
- Development of robotic systems for micro liquid handling

Assembling technology:
- Bridging the gap between microsystems and conventional lab environments
- Reversible mounting of microfluidic devices
- Combination of microparts made by different manufacturing technologies

Computer Aided Development:
- Simulation: FEA and EDA
- Design: CAD (2d and 3d)

Customized Micromachined Manifolds -

Silicon / Glass versus LTCC ?

- Both technologies have their source in electronics - manufacturing.
- Both technologies are suited for prototyping and small quantities.

Silicon- Glass micromachining

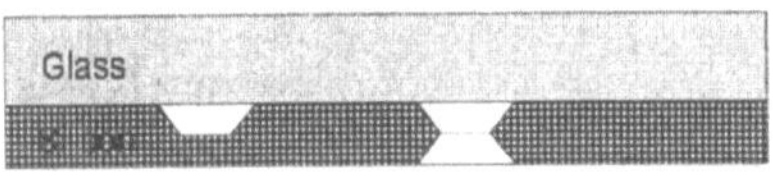

- Two layer technology
- Silicon structuring with high reproducibility and precision

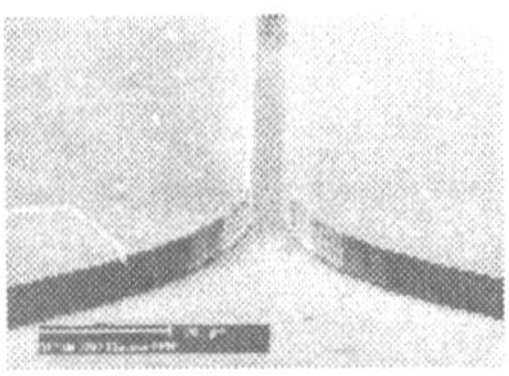

Channel width: 50 μm

- Good chemical resistance
- Good surface quality
- Optically transparent glass layer
- Planar design

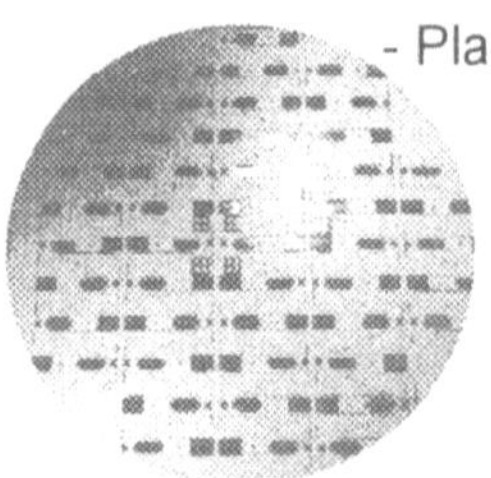

Low Temperature Cofired Ceramics (LTCC)

- Multi layer technology (max. 50)
- Structuring of LTCC by laser ablation

LTCC:
- SiO_2
- Al_2O_3,
- Organic Binder

Channel width: 300 μm

- Rough material
- Buried channels and conductors
- Channels can be separated by aranging them in different layers

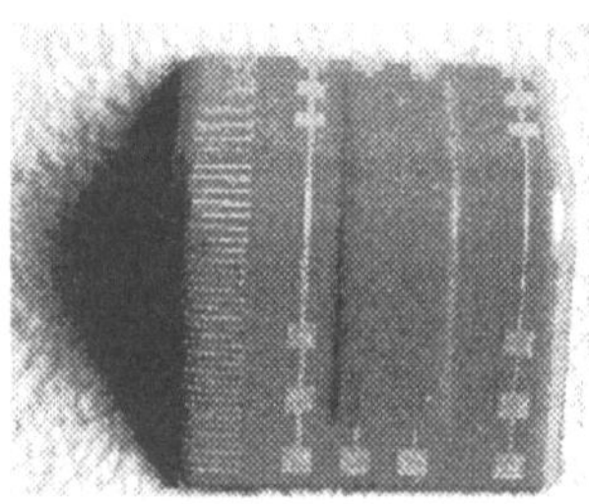

LTCC can be shaper before burning

Piezoelectrically powered Micropipettes

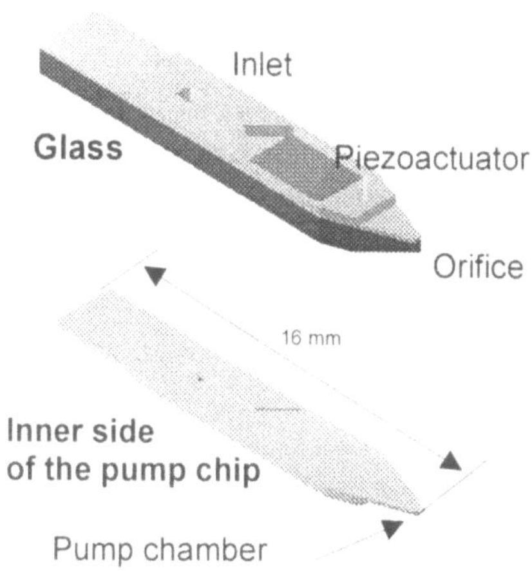

- Valveless
- Drop on demand function

Applications:
Loading of micro well plates
Printing of dot arrays
Dispensing of rare and expensive substances

Features ***to dispense*** liquid:

- Minimum dispensing volume: (50 .... 2000) pl
- Maximum transporting capacity: approx. 700 µl/min
- Nozzle tip and piezoactuator take up only a few square millimeters
- Extremely compact design
- Sophisticated hard- and softwareinterface

Features ***to aspirate*** liquid:

- External support is necessary (diluter, syringe) Suitable robotic systems are available
- Minimized bottom surface
- Optional: Liquid level detection and heatable nozzle

Droplet emitting pumps in micro analysis systems

- Similar function, optimized design
- Supported by the user software of GeSiM micropipettes
- Adapter for silicon-glass manifolds available
- Functions as part of a microsystem: Valve ("Fluidic Diode"), Mixer, Injector

Active Micro Mixer based on a Microdroplet Pump

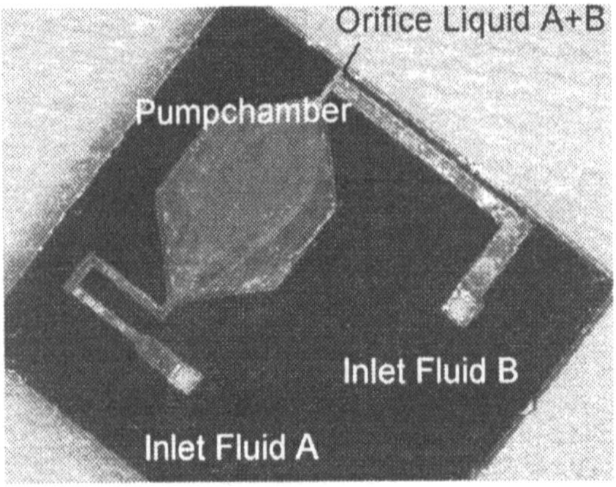

Silicon micromachining allows design modifications of micro-fluidic devices

The picture shows a microdrop-let pump mixing two different liquids immediately before passing the outlet nozzle

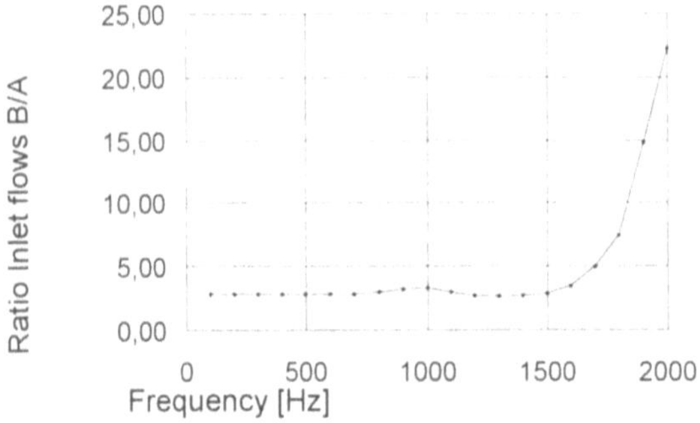

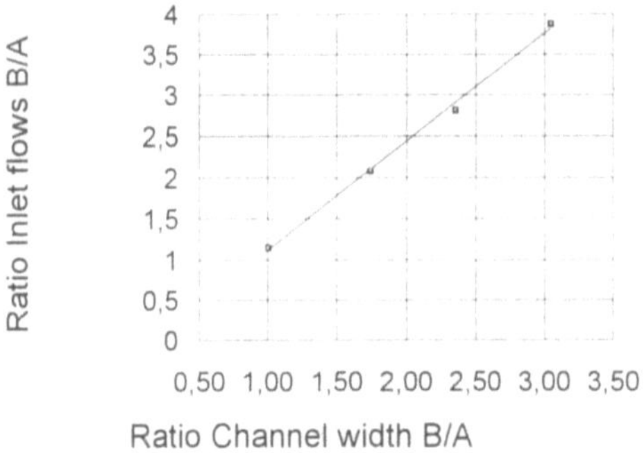

The performance of the mixer pump depends both of the operating parameters and the design

Thermal flow sensor for smallest volumes of liquid

Schematic cross section

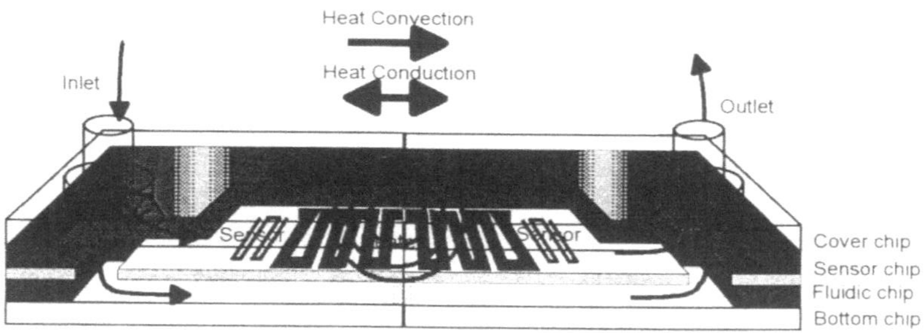

Each impulse can be assigned to a single droplet leaving the orifice

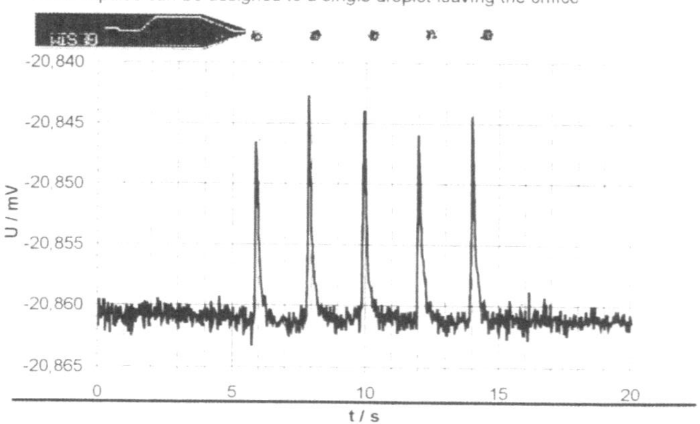

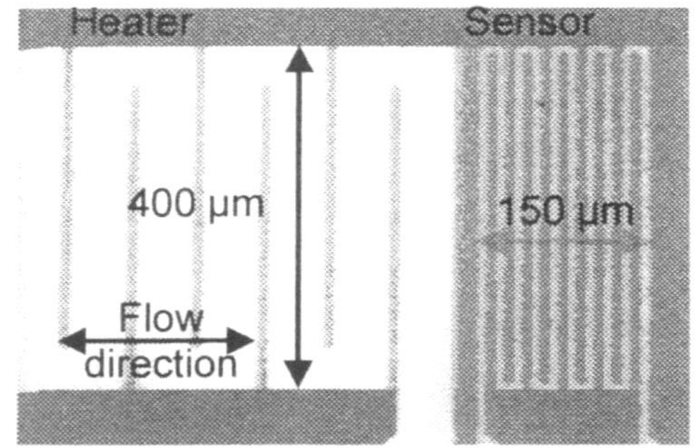

SEM - view of the sensor layout

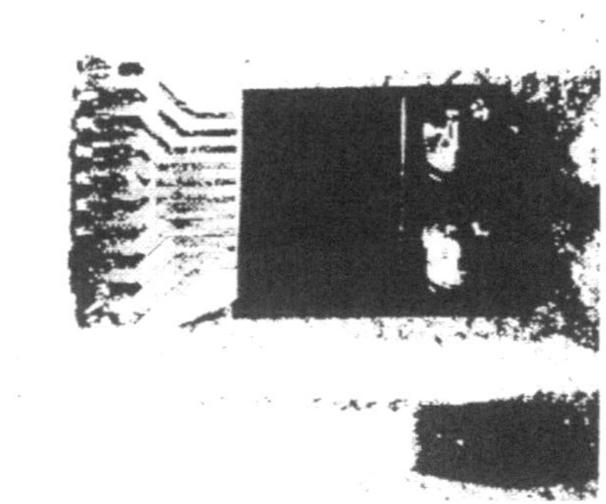

The encapsulated sensor chip

Microscope Visualization Chamber
Manufacturing Steps

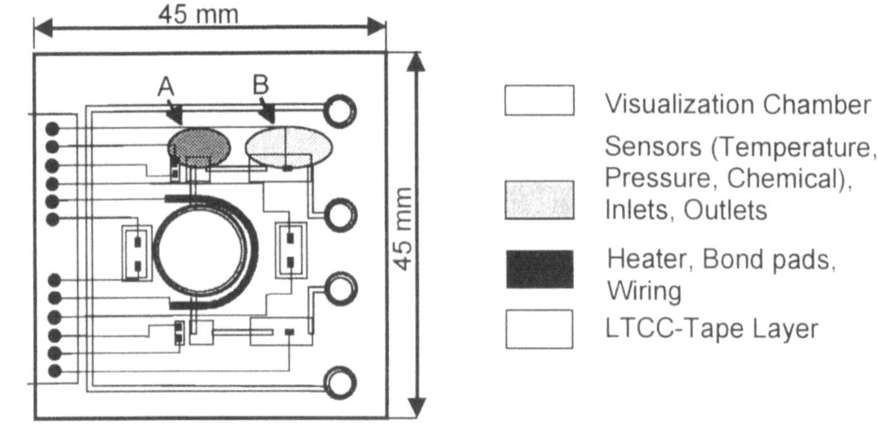

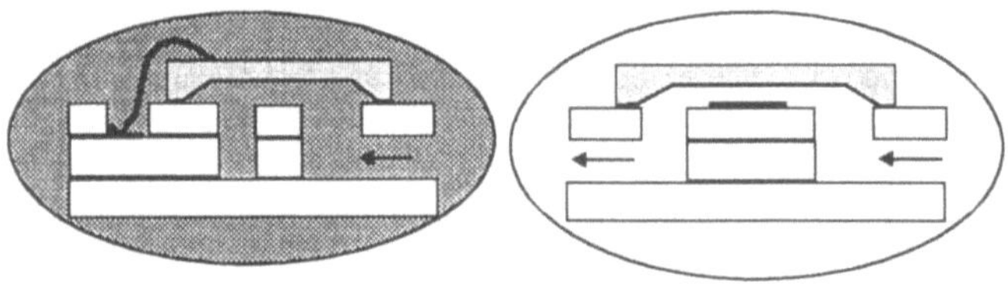

Detail A (Pressure Sensor)   Detail B (Chemical Sensor)

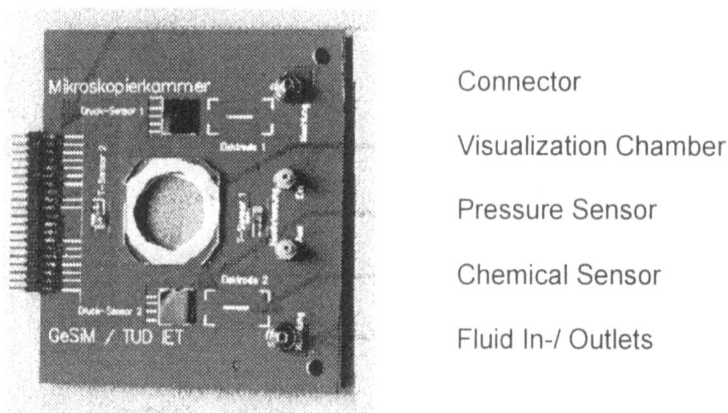

Microscope Visualization Chamber
Application

1. Immobilization of test particles (cells, biomolecules) on glasses

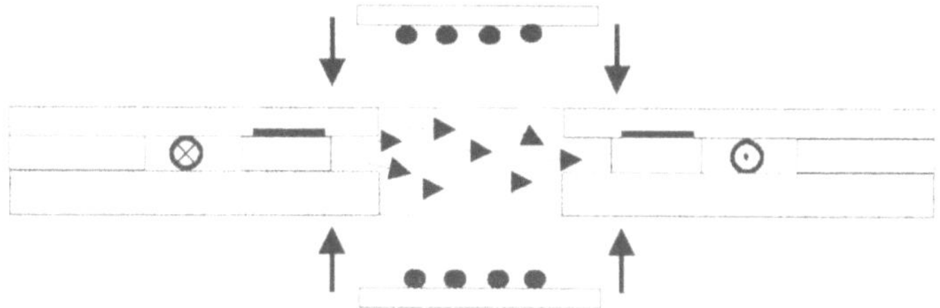

2. Priming of the microscope chamber with appropriated substances (Toxic substances .... nutrition solutions)

3. Applying of temperature cycles

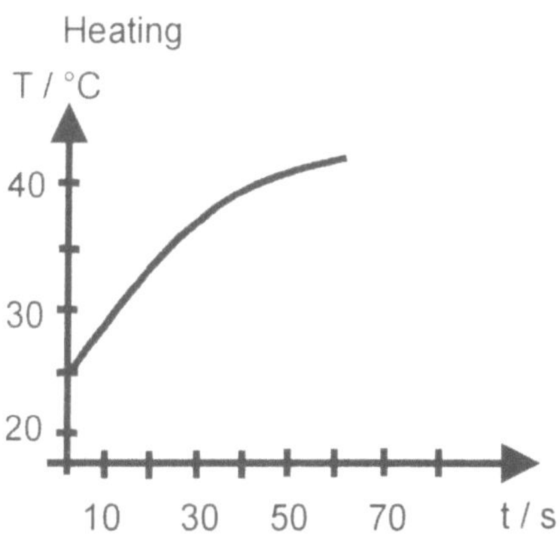

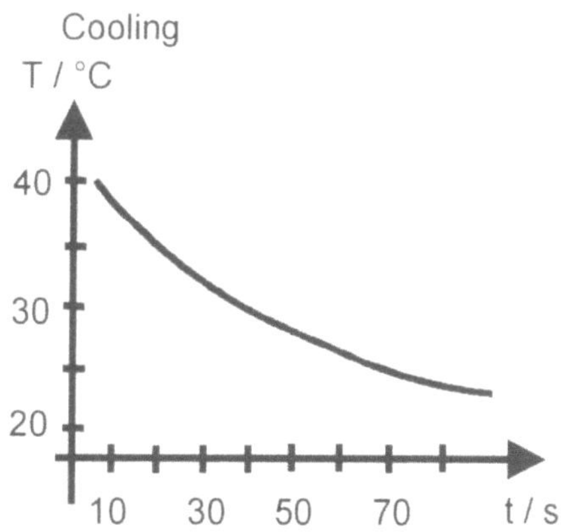

Microfluidic Device for Single Cell Manipulation

(Collaboration work with Humboldt-Universität zu Berlin, Dept. of Biology)

Funded by the German Government, BMBF, 16 SV 662/0

Scope: Handling of single cells in a microchannel by electrostatic actuation,
Cells are moving independently from the surrounding fluid flow

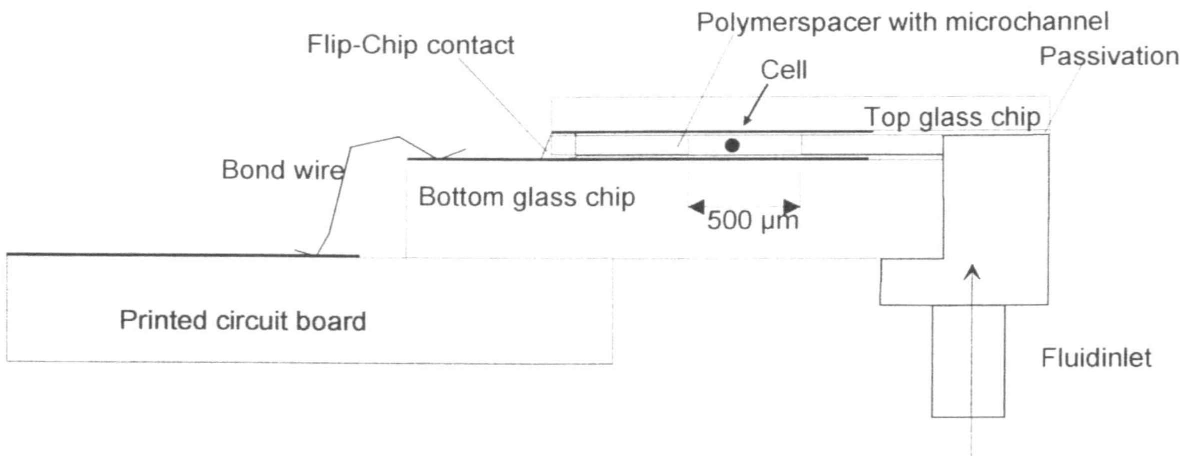

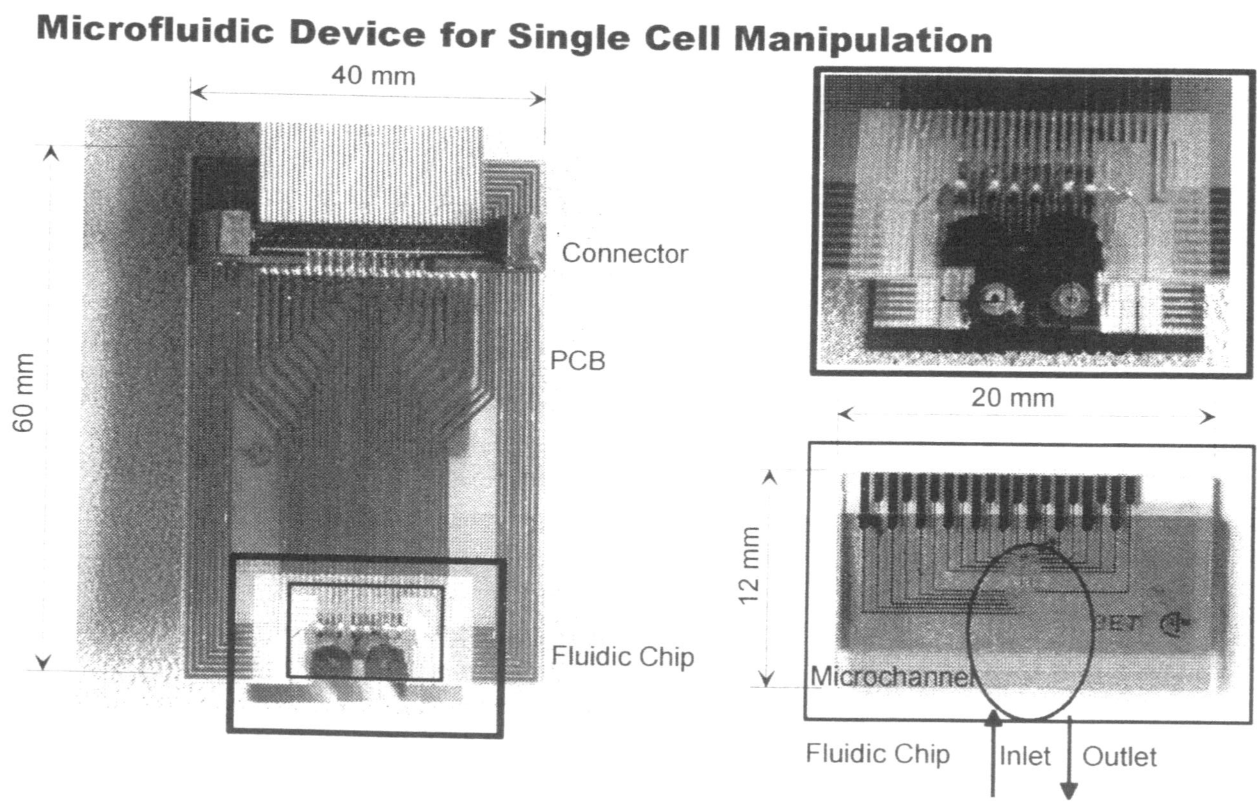
Microfluidic Device for Single Cell Manipulation
40 mm
60 mm
Connector
PCB
Fluidic Chip
20 mm
12 mm
Microchannel
Fluidic Chip
Inlet
Outlet

Liquid Handling in the Pico- and Nanoliter Range

Wilhelm Meyer, Michael Döring

microdrop Gesellschaft für Mikrodosiersysteme mbH, Mühlenweg 143, 22844 Norderstedt, Germany

Abstract. The fast development of Biochemistry as well as that of Microtechnology more and more requires an advanced technique for liquid handling of minimal volumes. Not only small volumes are of interest, but also a precise and fast handling is needed in drug screening and production lines entirely. The MicroDrop technology provides a tool covering many of the requirements for micro dosing in scientific and industrial applications. The microdrop dispensers as a spin-off from the ink jet technology are based on a piezo electric actor allowing to dispense droplets with volumes in the pl-range very precisely without contact to the substrates or work pieces. The „simple" design and geometry makes the handling with agressive materials possible, also cleaning the devices is very easy. The wide range of viscosity that can be handled with the dispensers makes the devices useful in very different applications as Biochemistry for very different liquids with very different solvents like water and DMSO, as well as in the field of microelectronics or optics for dispensing of solder flux and adhesives.

In this paper the MicroDrop dispenser technology is described. The potential is reflected by the presentation of different applications where this technology plays a key role or is the base for new products. Also the enhanced effects and actual limitations of this technique are described: Due to the small volumes in the pl - range that can be generated, a quantum leap for liquid handling, physical and chemical behavior occurs. The small volumes require careful handling and preparation of the used materials. Surface effects negligible in large volume applications now are the determining factors in the physics and chemistry of the liquids. Their interaction with substrates and environment compared with the bulk applications may change entirely. New developments at microdrop are presented. Especially developments as feedback control systems, monitoring devices and miniaturization for higher automatisation and integration of the technology are of interest and under work at microdrop.

Keywords. Microdispenser, small volume, microdosing, drop, droplet, feedback control, volume measurement

1 Introduction

The MicroDrop technology is based on the principle of the ink jet technology which was developed in the early sixties. The devices made by microdrop look completely different compared to recent ink jet printheads. The goal is to optimize the dispenser design for the demands of different kinds of liquids to be dispensed. The core device consists of a glass capillary which is surrounded by the tubular piezo actuator. The one end the capillary is formed to a nozzle (diameter: 30 - 100µm). Applying a voltage pulse on the piezo actuator forces it to contract. This creates a pressure wave which propagates through the glass into the liquid. At the end of the nozzle - the interface between liquid and air - the pressure wave is transformed into a motion of the liquid which is accelerated with up to 100,000g. A small liquid column leaves the nozzle, breaks off and forms a droplet which freely flies through the air. In this way volumes of 30 to 500 picoliters can be dispensed. The variation of volume is a few percent and better. The droplet size is mainly determined by the nozzle dimensions. The diameter can be produced with high accuracy (±1µm) depending on the desired drop volume and the liquid. The viscosity range where the system is operating is from 0.4 to 20mPas for unheated systems and from 20 up to 10000 mPas if the liquid is heatable and the viscosity is caused to decrease.

Due to numerous applications the demands on a microdosing system are very different. Therefore Microdrop provides mainly two different dispenser types. One is designed for pipetting, where the change of liquid is very often needed; i. e. in Biochemistry. The other type is with a larger storage bin, here always the same liquid is dispensed for production; glueing of fiber optics is one example. The micropipette type is entirely designed for the needs is Biochemistry. The last 7mm of the glass capillary at the outlet are without housing. So it can easily merged into the liquid and sucked up through the nozzle by applying a negative pressure at the inlet side of the dispenser. The Micropipette AD-K-501 has a storage volume of only 25µl. It's small outer diameter of 7mm allows to integrate several devices to rows for use with microtiterplates with a 9mm well spacing. Since droplet generator and storage volume are one glass capillary there is no incontinuity of surface, and cleaning is very effective. The pipette can be plugged into a holder, connecting electric and pressure lines at the same time.

Because of the fact that the dispenser is in principle an acoustic driven system, it has to be made sure that damping effects of the acoustic coupling are minimized. Very high damping is caused by large changes in density what is established at interfaces of gas bubbles. This is the reason why some efforts have to be made to avoid air bubbles within the dispenser device. As a prevention it is recommended to use degassed liquids only. On the other hands air bubbles are preferentially trapped by gaps or edges within the fluidic channel. Due to the design of the MicroDrop dispensers such gaps and edges are avoided; the glass capillary has a smooth surface and is made of one piece and its cross section is circular.

Due to wide range of viscosity that can be handled with the MicroDrop dispensers the applications for these devices are very manifold. The heated version - for lowering the viscosity - provides the use for adhesives and fluxes. For example the glueing of micromechanical parts like optical fibers or loud speakers for cellular phones are fields where the demands for small and precise volumes do exist. Also in the field of life science the demands of liquid handling of volumes in the pl- to nl-range are increasing. Namely the development and production of biochips and the high throughput screening are growing markets. In both cases the demands on the dispensing results are mainly the same; a very precise and small volume independent from the surface properties of the substrates is needed with a large repetition rate. This is achievable by the contactless MicroDrop dispenser technique. To control the precision and reliability of such systems, some feedback mechanisms under development.

2 Feedback Mechanisms

In most cases of microdispensing, small volumes with a high accuracy are needed. The drop volume is effected by the dispenser performance and the properties of the liquid. The dispenser performance is controlled by the nozzle diameter and the driving parameters voltage, pulse length and - in case of a heated dispenser - also temperature. The conditions are shown in fig. 1. An increase in voltage causes an increase in drop speed; means an increase of the kinetic energy of the system leading to a larger drop volume. The behavior is almost linear.

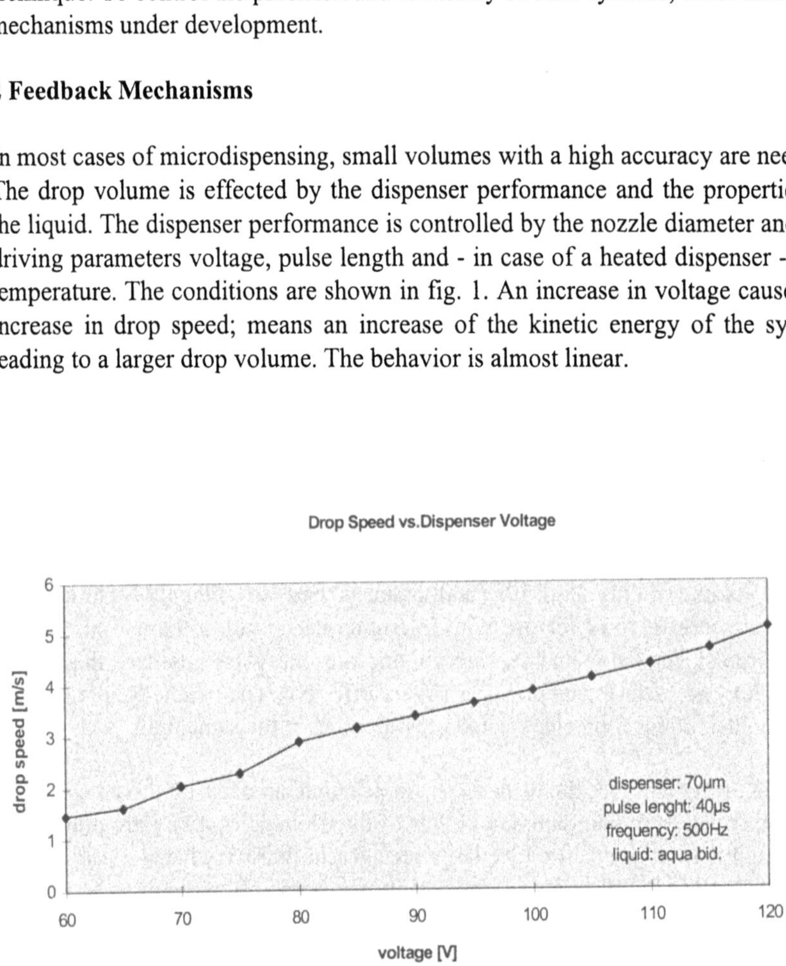

fig. 1: drop speed vs. dispenser voltage

The liquid properties mainly effecting the dispense result are viscosity and surface tension. The larger the viscosity of the liquid is, the larger are the energy losses of the dispenser due to the inner friction of the liquid. The limits of the dispenser device described above are at about 20mPas at room temperature. A decrease of drop speed and a smaller drop volume is obtained as a result the more viscous the dispensed liquid is. Due to the relatively large ratio of surface to volume in the relevant regime - the drop diameter ranges from 30-100μm - the surface tension has a very large effect. Almost the same amount of the acoustic energy transformed into kinetic energy is in addition put into the surface energy of the droplet. This means changes of the surface tension do have a dramatic effect on the drop speed and therefore on the drop volume.

The dispensing result - as stated above the drop volume is of most interest - is effected by a multiple set of parameters. Due to the fact that the properties of the liquid in most cases are fixed by the application, the parameters of the dispensers have to be controlled for an optimal result. This can de done by the calibration of the system for a certain liquid; one gets a set of parameters - voltage and pulse length - for a certain drop volume. Our measurements have shown, that the accuracy (standard deviation) of the drop volume can be kept within 1% for a very high number of dispenses. These measurements have been made by weighing an ensemble of several thousand drops and computing the mean volume per drop. An other method is to add a known amount of fluorescent dye to the liquid and determine the volume of dispensed drop by photon counting. In both cases the measurement of absolute values is difficult. In the first case the evaporation my play a role and the result is a mean drop volume of a statistical ensemble. The other method is also very sophisticated because of it's calibration; a lot of efforts are needed to keep a high accuracy. This means both methods are not very useful for „every day“ measurement.

To fulfill this demand, we looked for other methods. One is an image processing of individual drops. Due to the imaging with a strobe LED available in the microdrop electronics, such drop pictures are easily made and digitized with commercially available hardware. The setup (fig. 2) consists of the Autodrop Electronics for the dispenser and strobe LED control and a PC-System with a frame grabber board. A CCD-camera and a strobe LED are placed on opposite sides underneath the dispenser outlet. The CCD-camera optics (magnification of 5) is focused on the drop trajectory. The Autodrop Electronics is controlled by the PC, that way that pictures of the ejected drop can be made at different positions below the nozzle by controlling the strobe delay. The digitized pictures are loaded into the PC and prepared with several software tools for the three different demands on the system: drop detection, drop speed measurement and drop volume measurement. Therefore three software utilities are developed.

The drop detection utility gives a feedback whether the dispenser is operating or not. It inspects the appearance of a droplet in an image three times a second. For each detection a current image of the scene is grabbed into an image buffer. The detection criterion is the dark appearance in front of the bright background. The detection of artifacts based on dirty optics etc. is avoided by the algorithm.

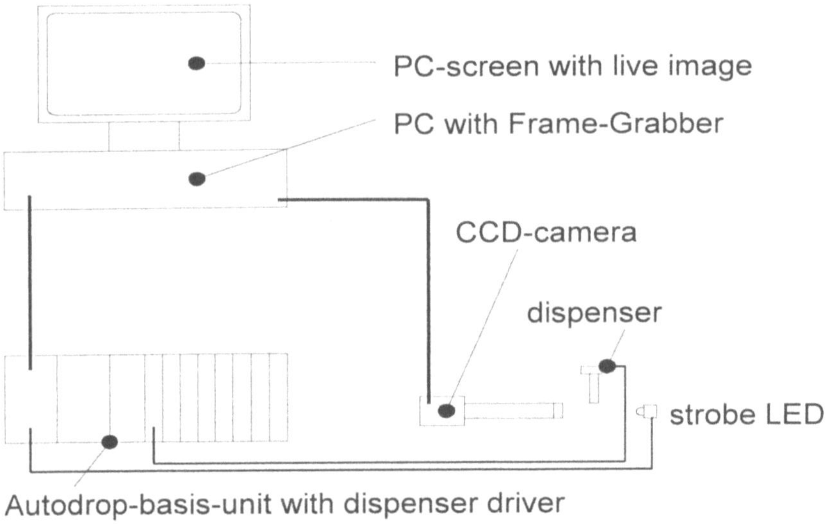

fig. 2: setup - drop image processing

The determination of the drop speed is based on a two-point-measurement. The calculation is based on the difference of the two drop locations (top and bottom) and their appropriate strobe delays. The positions at the top and the bottom of a current image are determined to calculate the averaged speed in the visible area of the scene. To minimize the methodical error the longest distance is found out by a successive approximation; therefore the image software changes the strobe delay.

In addition the 2D-pictures are used for drop volume measurement. Based on the assumption that the droplet is symmetrical about it's trajectory the volume is calculated by integration of thought cylindrical drop slices. These are received from cuts perpendicular through the plane of drop image. The diameter of the cylinder results from the actual cross section of each cut, the height is one pixel.

In the following the results of drop volume measurements are shown. Measurements by balance were chosen for evaluation of the software measurements. Testfluid was DMSO to minimize evaporation. A dispenser with a 50μm nozzle was chosen. Within each series of software volume measurement and reference measurement by balance the dispenser parameters were kept constant. For reference measurement five weights of 99990 drops were taken for calculation of the average drop volume. For each series 100 individual software measurements were made. The frequency distribution is shown below (fig. 3). The standard deviation was calculated to get a value for the error of this method. The errors that effect the measurements randomly are the blur of the drop due to its motion during illumination and the pixel noise due to the optical and electrical image processing. The results show a CV of 2% for the mean drop volume of 200pl. Considering the error sources mentioned above and that a 3D-value is computed from 2D-datas, this is a promising good result.

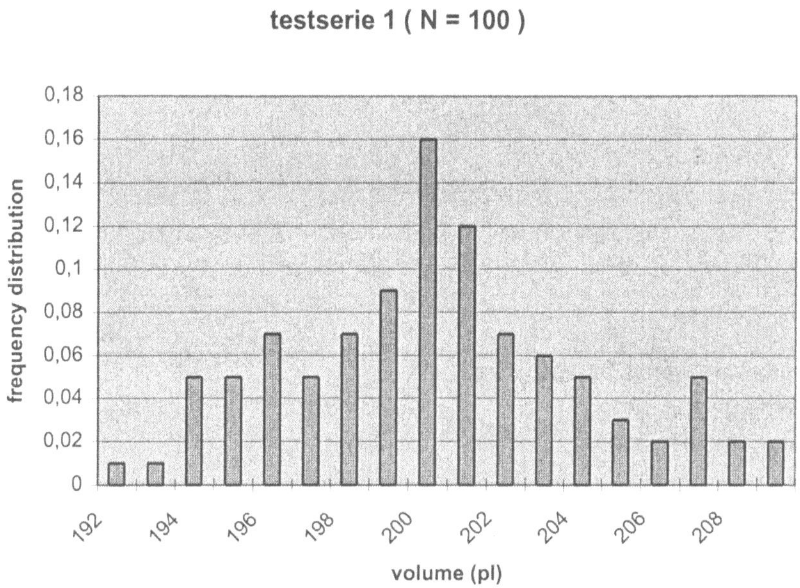

serie 1:	**testfluid: DMSO**
measurements by software:	
average volume:	200,24 pl
standard deviation:	3,8 pl
CV:	1,9%
measurements by balance:	**5 scales á 99990 droplets**
average volume:	201,9 pl
comparison balance to software:	
relative mean error:	0,8%
max. error of single measurement	8,8 pl (4,3%)

fig. 3: results drop volume measurement

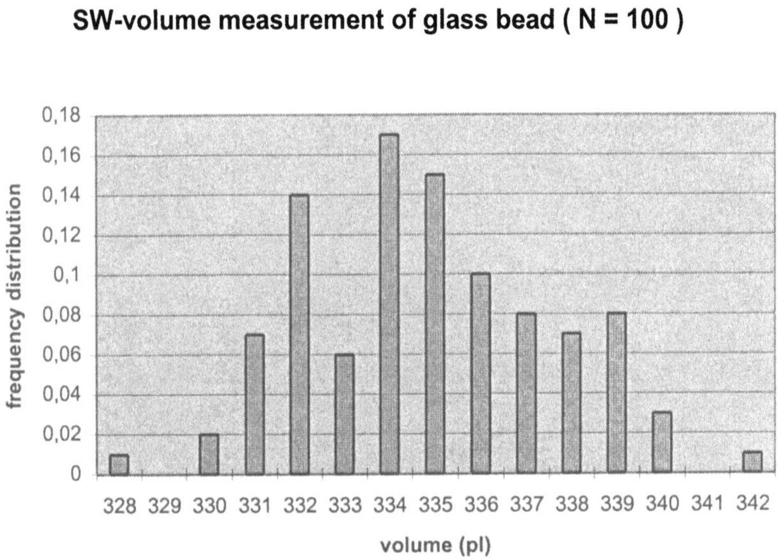

	glass bead
measurements by software:	
average volume:	334,83 pl
standard deviation:	2,7pl
CV:	0,80%
measurement by microscope:	
mean diameter:	86µm
calculated volume:	333pl

fig. 4: results glass bead volume measurement

For evaluation of the drop motion (blur) effect and asymmetries of the circular drop shape in about the axis of propagation on the software volume measurement, the volume of a fixed glass bead was measured with the same method. For this purpose a glass bead (diameter 86µm) was put into the focus of the CCD-camera optics and illuminated under same conditions as above. As a reference for the absolute volume the diameter of the glass bead was measured by microscope and the volume, was calculated on the assumption that the bead is a sphere. The results of 100 measurements are shown above (fig. 4). The CV is obviously smaller than the results shown with the DMSO-drops in motion. This gives the hint that the method described above might not at it's limit and can be improved by decreasing the exposure time of the LED (currently at 10 - 12µs), the use of other SW-processes and improved optics. But it has also to be considered that the assumption of a circular symmetry of the drop about the propagation axis is not entirely valid.

Another method for drop detection and feedback control is also under development. Here a light barrier which detects each drop disturbing the light path between an LED and a light detector is detected. Because of the low signal to noise ratio a special discriminator electronics was developed to detect each drop reliable out of the signal. Drops with a diameter from 30µm to 100µm and with a drop speed from 1 to 4m/s are detected by this method. By measuring the elapsed time of a drop passing through a certain distance the drop speed can be computed. The advantage of this method is that there are no limitations due to repetition frequencies; each drop can be detected below a drop frequency of 2kHz. The drop speed is available in real time. This feature provides the opportunity, to use the measured drop speed for feedback control. As indicated above the drop volume is correlated with the drop speed. If the drop speed is kept constant the drop volume should also stay constant. This way the method provides a tool for drop volume control by measuring the actual drop speed, comparing this value with the set speed and adjusting it - if needed - by changing the voltage for the dispenser. Currently, such a setup in combination with some software algorithms for feed back control are under investigation.

3 Conclusions

For the liquid handling in the pl- and nl-range, a very precise and reliable technique is needed. For this purpose we have different feedback mechanisms for our MicroDrop dispensers under development. It could be shown that with these techniques a drop detection, drop speed measurement and drop volume measurement is possible. The goal is to provide the customer an automated system for calibration and monitoring. The results show that the monitoring of the dispensers is possible and each drop can be detected reliable. It also could be shown that a volume measurement - and therefore a system calibration - is possible. The accuracy are within a few percent; what should be good enough for a large number of applications. Also the investigation of volume control by drop speed measurement is very promising. The investigation will be continued to provide the customer a reliable system for the application of precise amounts of liquids in the pl- and nl-range.

Flow-through microdispenser for liquid handling in a levitated-droplet based analyzing system

Johan Nilsson[1], Thomas Laurell[1], Maria Petersson[2], Sabina Santesson[2], Thomas Johansson[3], Eva Degerman[4] and Staffan Nilsson[2]

[1] Department of Electrical Measurements, Lund Institute of Technology, P.O. Box 118, SE-221 00 Lund, Sweden, E-mail: johan.nilsson@elmat.lth.se
[2] Department of Technical Analytical Chemistry, Center for Chemistry and Chemical Engineering, Lund, Sweden
[3] Department of Atomic Physics, Lund Institute of Technology, Lund, Sweden
[4] Department of Cell and Molecular Biology, Section for Molecular Signaling, Center for Chemistry and Chemical Engineering, Lund, Sweden

Abstract. A flow-through microdispenser actuated by a piezo-electric element is presented. The flow-through design enables droplet ejection from a continuously flowing liquid passing through the device. This also simplifies filling, cleaning and changing liquid in the dispenser since the liquid does not have to pass through the sampling orifice during the process. The total volume of the unit is 400 nL and the droplet volume is normally around 65 pL. The droplets can be ejected with up to 5 kHz frequency. The dispenser is used for sample/reagent handling in a system based on acoustic levitation. The dispenser droplets are supplied to a nodal point of a standing wave in the levitator where a "wall-less" reaction site is created.

Keywords. Flow-through dispenser, acoustic levitator, sample enrichment, evaporation compensation.

1 Introduction

The development of new and miniature tools for liquid/sample handling in analytical chemistry is forced by the demand for faster and less sample/reagent consuming systems, especially found in the area of high-throughput screening (HTS) for drug candidates [1]. For non-contact picoliter sample supply, piezo-electric microdispensing is a method that is more and more frequently used [2]. The commercially available dispensers are, however, all of a single-end type where the sample solution is either supplied from a reservoir or aspirated through the dipenser nozzle. Here, the use of a flow-through dispenser is presented where the sample droplets are ejected from a sampling orifice located between the inlet and the outlet of a flow-through channel. During filling, liquid switching and cleaning none of the liquid is passing through the orifice, which simplifies the use of the device since all these steps may be performed without moving the dispenser away from the target.

The dispenser is used for sample and reagent handling in an analyzing system based on acoustic levitation [3]. The advantage of a levitator is that a reaction site is created containing no walls that may interfere with the on-going reaction or with the selected detection system.

2 The flow-through dispenser

2.1 Design

The principle for the flow through dispenser is shown in figure 1. The dispenser consists of two micromachined silicon dice, which are sandwiched together forming the flow-through channel. The channel dimensions are normally 8 mm x 1 mm x 50 µm resulting in a total volume of 400 nanoliter. With a 50 µm nozzle opening the volume of the droplets ejected is 65 picoliter when water is used. Earlier versions of the dispenser are described in [4, 5]

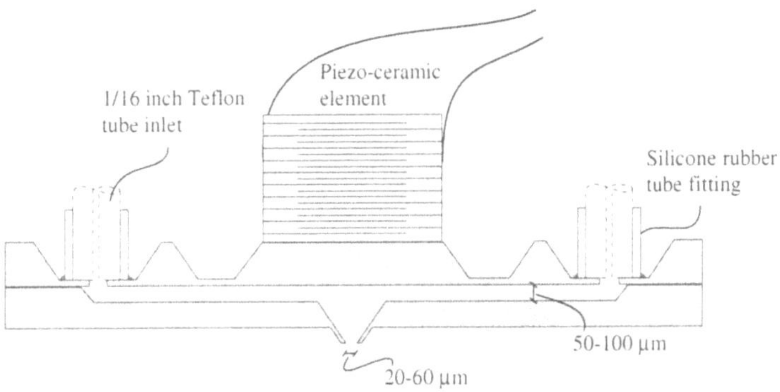

Figure 1. Lengthwise cross-section view of the flow-through dispenser. The total volume of the unit is 400-800 nanoliter depending on the channel height. The nozzle orifice is normally 50 micrometer.

An important design feature is the protruding nozzle with a front surface area of only 8 µm width surrounding the orifice (Figure 2). This minimizes wetting during droplet ejection and reduces potential problems with liquid and particle deposits close to the orifice. As a result, the droplet directivity is stabilized.

Figure 2. SEM picture of the nozzle with minimal front surface area.

2.2 Actuation

The actuation of the dispenser is handled by a multilayer piezo-ceramic element (Ferroperm A/S, Kvistgaard, Denmark) with a Plexiglas™ backing and acting on an integrated push-bar in the channel wall (Figure 3). The piezo-ceramic element elongates in response to a voltage pulse applied across it (typically 10 V amplitude, 100 µs duration). A pressure pulse is thus generated in the flow-through channel and a droplet is ejected (Figure 4). It is important to adjust the shape of the pulse, both amplitude and rise- and falltime, to the liquid used to obtain best possible droplet formation. The pulse amplitude must be reduced when low surface tension liquids are used to avoid formation of unwanted satellite droplets following the main droplet.

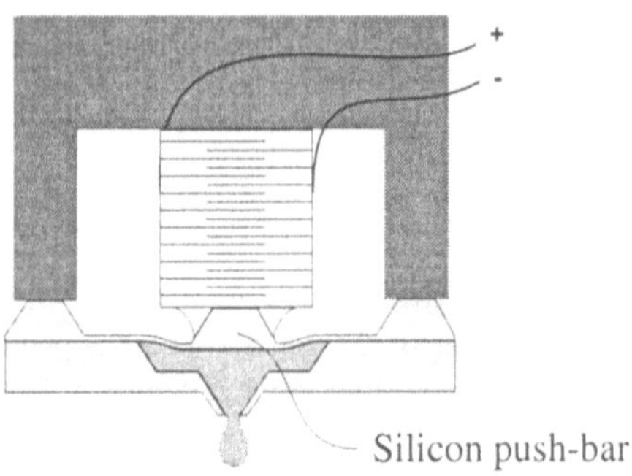

Figure 3. Actuation principle. The piezo-ceramic element elongates when a voltage pulse is applied and generates a pressure pulse that ejects a droplet from the flow-through channel.

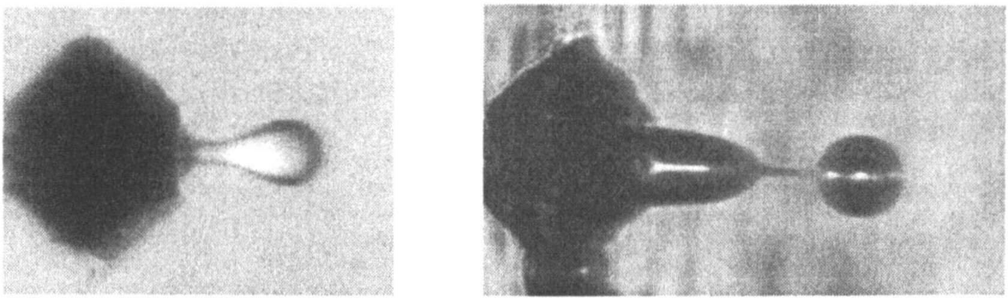

Figure 4. Ejection of water (left) and 1:1 water/glycerol mixture (right).

The dispenser is capable of ejecting both low and high surface tension liquids ranging from ethanol (0.022 N/m) to water (0.073 N/m) and liquids with viscosity up to about 65 mPas. The maximum droplet ejection frequency is around 5 kHz.

3 The acoustic levitator

The acoustic levitator (Martinsson Elektronik AB, Hägersten, Sweden) consists of a 100 kHz ultrasonic transducer and an adjustable reflector. A standing wave pattern is created between the transducer and the reflector and particles or drops may be caught in the nodal points of the standing wave (Figure 5).

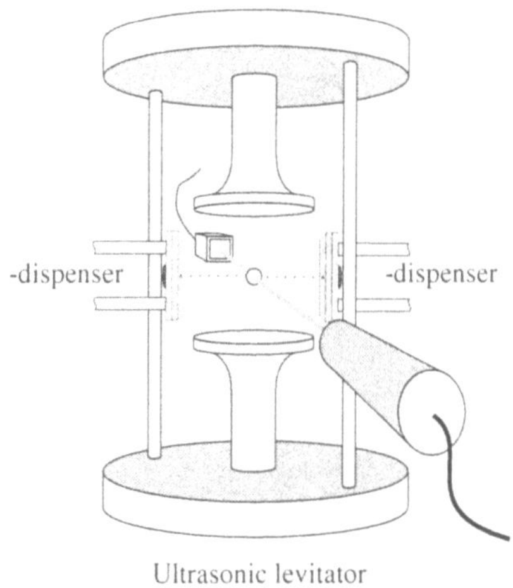

Figure 5. Schematic drawing of a set-up with an acoustic levitator, dispensers and laser-induced fluorescence.

The interesting features of the levitator for use in chemical analysis are that the reaction site contains no interfering walls, which removes the potential problems with analyte adsorption. The reaction site is further readily available for e.g. laser-induced fluorescence detection. Since the dispenser supplies the droplets in a non-contacting fashion it is a perfect candidate for handling the supply of samples and reagents to the levitator.

4 Experiments

When using the acoustic levitator with liquids one must take into account the effect of solvent evaporation. The evaporation may either be an advantage or a problem. In order to analyze low concentration solutions solvent evaporation is beneficial since the sample is enriched.
On the other hand, in experiments involving fluorescence detection it is normally important to keep the reaction volume constant in order not to introduce systematic errors due to a changing fluorophor concentration.

4.1 CE-Sample enrichment

In a sample enrichment experiment the dispenser was used to supply sample droplets containing 0.1 μM Dns-Val and 0.05 μM Dns-Gly in ethanol. The sample droplets were evaporated in the levitator (Figure 6). Up to 36,000 droplets were supplied. The enriched samples were subsequently CE-analyzed and the concentration limit of detection was found to be lowered from 2.5 μM without enrichment to 0.015 μM with enrichment.

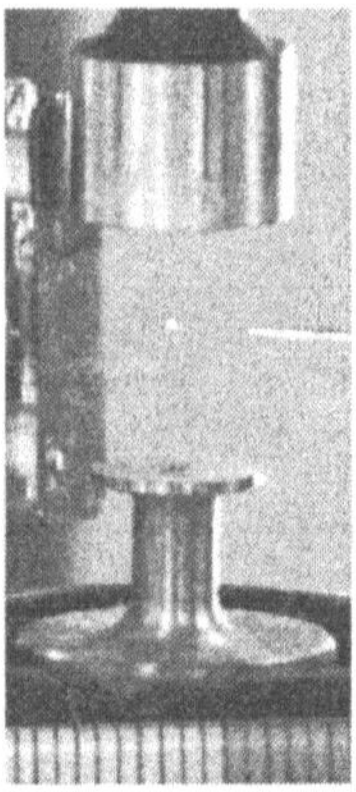

Figure 6. Set-up for sample enrichment. The ultrasonic transducer is seen below and the reflector above. The dispenser to the left supplies sample droplets to the nodal point of the levitator. The enriched sample drop is picked up by the capillary seen to the right and analyzed using CE.

4.2 Evaporation compensation

A 1 µL drop containing a pH-dependent fluorophor (trisodium salt of 8-hydroxypyrene-1,3,6-trisulphonic acid, LabKemi, Lund, Sweden) was placed in the levitator using a 0.8 mm x 40 mm needle. A mercury lamp with interference filters and lenses was used for illumination at the wavelengths 405 nm and 435 nm to allow for ratiometric measurements. The ratio between the emissions for the two different excitation wavelengths was measured during 80 seconds after drop deployment in the levitator with and without evaporation compensation. For compensation the dispenser was supplying water at a constant rate of 10 Hz to the drop in the levitator. Figure 7 shows that it is possible to completely compensate for solvent evaporation maintaining a constant ration level.

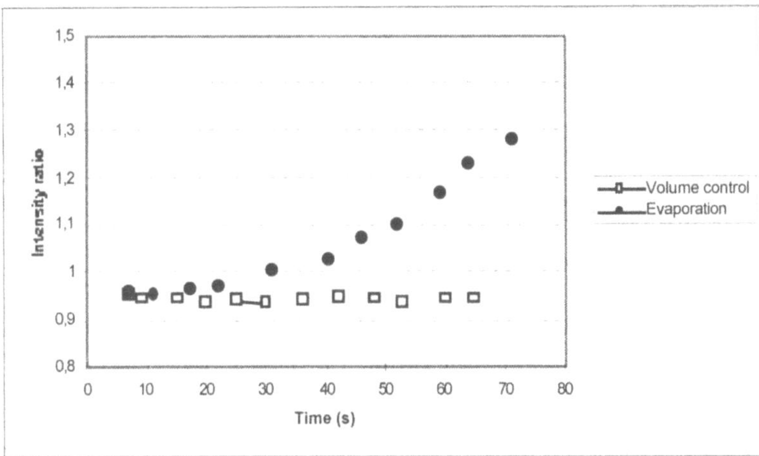

Figure 7. Evaporation compensation by supplying water at a constant rate to the levitator.

5 Conclusions

The combination of one or more flow-through dispensers and an acoustical levitator constitutes a strong set-up for performing analysis involving low volume, low concentration chemistry. The flow-through design simplifies the use of the piezo-electric dispenser and opens up the possibility of connecting the system in an on-line configuration.

6 References

1. Lemmo, A. V., Fisher, J. T., Geyser, H. M. and Rose, D. J., *Anal. Chem.*, 69, 543, 1997.
2. Lemmo, A. V., Rose, D. J., Tisone, T. C., "Inkjet dispensing technology: applications in drug discovery", *Current Opinion in Biotechnology*, 9, 615-617, 1998.
3. Petersson, M., Nilsson, J., Wallman, L., Laurell, T., Johansson, J. and Nilsson, S., J. Chromatogr. B. 714, 39-46, 1998.
4. Nilsson, 1., Wallman, L., Drott, J. and Laurell, T., *Anal. Methods & Instr. Special Issue μTAS'96*. 91, 1996.
5. Nilsson, J., Wallman, L., Önnerfjord, P., Marko-Varga, G. And Laurell, T., Proc. of the μTAS'98 Workshop, Banff, Canada, p. 378, 1998.

Simulation of Microscale Chemical Separation Processes Using the Lattice-Boltzmann Method

David R. Rector[1] and Bruce J. Palmer[1]

[1]Pacific Northwest National Laboratory[(a)]
Box 999
Richland, WA 99352, USA

Introduction

Fluid flow in the micron to hundred micron size range promises to be qualitatively different from fluid flow in large geometries and offers many opportunities for creating and exploiting novel flow configurations that can be used in chemical reactors and separations processes. The most significant difference between flow in conventional macrosized systems and flow at micron length scales is the increase in the relative importance of surface forces, particularly surface tension, as the dimensions of the flow configuration shrink. Surface tension forces between the fluid and the walls and at the interface between immiscible fluids are ordinarily negligable on length scales of a centimeter or so but can dominate the behavior of liquid in a 100 micron channel. The ability to include these effects in simulations of flow in microscale devices is critically important to support the design and testing of micro-chemical components and systems, but conventional fluid dynamics simulations are incapable of modeling the multiphase systems anticipated in many microtechnology applications. A promising new method for simulating fluid flow that can easily be adapted to multiphase flows is the lattice Boltzmann algorithm[1, 2]. One major advantage of the lattice Boltzmann method over other fluid dynamics simulation techniques is the ability to incorporate surface interaction terms directly into the equations of motion[3, 4]. This makes it possible to simulate multiphase-multicomponent systems in a straightforward way, without the introduction of complicated front-tracking techniques.

This paper will present a generalization of the model developed by Swift *et al.*[4] for describing multicomponent-multiphase systems. The model will then be used to simulate two extraction processes in a microscale configuration. The flow geometry in the first process consists of two thin liquid films in direct contact with each other. The stability of the flow is maintained by the relatively high surface tension across the liquid-liquid interface. The second flow geometry is similar to the first, except that a thin plate with pores is used to separate the two liquids.

[(a)]Pacific Northwest National Laboratory is operated for the U.S. Department of Energy by Battelle Memorial Institute under Contract DE-AC06-76RLO 1830.

Multicomponent-Multiphase Model

Lattice Boltzmann simulations represent a discretized version of the Boltzmann equation in which space is divided up into a regular lattice and the velocities are represented by a finite number of vectors instead of a continuous distribution. The original Boltzmann equation modeled the behavior of the one-particle distribution function $f(\mathbf{r}, \mathbf{v}, t)$, which describes the probability of finding a particle at the point $\mathbf{r}$ with velocity $\mathbf{v}$ at time t. Instead of solving for the continuous one-particle distribution function f, the lattice Boltzmann algorithm is based on a discretized distribution $f_i(\mathbf{r})$ where i represents a set of discrete velocities and replaces the variable $\mathbf{v}$ in the continuous distribution. The spatial coordinate $\mathbf{r}$ now refers to a point on a discrete lattice. The discrete velocities are denoted by the displacement vectors towards neighboring lattice sites $\mathbf{e}_i$ where $i = 1, ..., b$ and b is the total number of displacement directions. The magnitude of the displacement vectors is $|\mathbf{e}_i| = c$, where c can be thought of as a mean microscopic particle velocity. A zero displacement vector $\mathbf{e}_0$ is also included in the set to represent particles with zero velocity. For a single component system at constant temperature, only a single distribution f_i is necessary to model the behavior of the system. For a system with M components, $M-1$ additional distributions $g_i^{(k)}$ are needed for each of the extra components labeled by k. The behavior of the final component can always be recovered from knowledge of the total density and the concentrations of the other $M-1$ components.

A total of M sets of distributions, f_i and $g_i^{(k)}$ for $k = 2, ..., M$, are assigned to each site. The distribution f_i represents the transport of the total mass and momentum and satisfies the two relations

$$\rho = \sum_{i=0}^{b} f_i \tag{1}$$

$$\rho \mathbf{u} = \sum_{i=1}^{b} \mathbf{e}_i f_i \tag{2}$$

where ρ is the total mass density and $\mathbf{u}$ is the macroscopic velocity of the fluid. The distributions $g_i^{(k)}$ model the movement of the individual components around the system and satisfy the relations

$$\rho_k = \sum_{i=0}^{b} g_i^{(k)} \tag{3}$$

where ρ_k is the number density of each of the remaining $M-1$ components. The total mass density ρ is related to the number densities ρ_k via

$$\rho = \sum_{k=1}^{M} M_k \rho_k \tag{4}$$

where the M_k are the molecular masses of each of the components. Equation (4) can be used to calculate ρ_1 if ρ and the ρ_k for $k = 2, ..., M$ are known.

The distributions are defined for each of the lattice sites in the system and are updated at each time step by first performing a collision to obtain a new set of distributions. Each of the new distributions f_i and $g_i^{(k)}$ are then displaced along the vector $\mathbf{e}_i$ to get a new set of distributions at each site. The collisions and displacement of the distributions are summarized by the equations of motion

$$f_i(\mathbf{r} + \Delta t\mathbf{e}_i, t + \Delta t) - f_i(\mathbf{r}, t) = -\frac{1}{\tau_\rho}(f_i(\mathbf{r}, t) - f_i^{eq}(\mathbf{r}, t)) \quad (5)$$

$$g_i^{(k)}(\mathbf{r} + \Delta t\mathbf{e}_i, t + \Delta t) - g_i^{(k)}(\mathbf{r}, t) = -\frac{1}{\tau_k}(g_i^{(k)}(\mathbf{r}, t) - g_i^{(k)eq}(\mathbf{r}, t)) \quad (6)$$

where the $\mathbf{r}$ denote lattice sites, t is the discrete time, and Δt is the timestep. The collisions are assumed to take the familiar BGK form[5] and are characterized for the M distributions by the relaxation times τ_ρ and τ_k.

Following Swift *et al.*[4], the equilibrium distributions f_i^{eq} and $g_i^{(k)eq}$ are given by the expressions

$$\begin{aligned} f_i^{eq} &= A + \frac{\rho D}{bc^2} e_{i\alpha} u_\alpha + \frac{\rho D(D+2)}{2bc^4} e_{i\alpha} e_{i\beta} u_\alpha u_\beta - \frac{\rho D}{2bc^2} u_\alpha u_\alpha \\ &+ G_{\alpha\beta} e_{i\alpha} e_{i\beta} \end{aligned} \quad (7)$$

$$f_0^{eq} = A_0 - \frac{\rho}{c^2} u_\alpha u_\alpha \quad (8)$$

$$\begin{aligned} g_i^{(k)eq} &= B^{(k)} + \frac{\rho_k D}{bc^2} e_{i\alpha} u_\alpha + J_\alpha^{(k)} e_{i\alpha} + \frac{\rho_k D(D+2)}{2bc^4} e_{i\alpha} e_{i\beta} u_\alpha u_\beta \\ &- \frac{\rho_k D}{2bc^2} u_\alpha u_\alpha + H_{\alpha\beta}^{(k)} e_{i\alpha} e_{i\beta} \end{aligned} \quad (9)$$

$$g_0^{(k)eq} = B_0^{(k)} - \frac{\rho_k}{c^2} u_\alpha u_\alpha \quad (10)$$

where D is the dimension of the system. The greek indices label spatial components and the Einstein convention of summing over repeated indices is used. The remaining parameters are determined by the requirement that the model generate hydrodynamics that match, as closely as possible, the classical hydrodynamic equations. The parameters $J_\alpha^{(k)}$ and $H_{\alpha\beta}^{(k)}$ are new and are used to eliminate some spurious terms that appear in the continuum hydrodynamic equations generated by the original model of Swift *et al.*[4] Specifically, they remove an extra term proportional to the pressure tensor that appears in the diffusion-convection equation for the individual fluid components. Using the equations of motion and the definition of the equilibrium distributions, the lattice Boltzmann algorithm consists of the following steps:

i) Calculate ρ, ρ_k, and $\mathbf{u}$ at each site.

ii) Evaluate the f_i^{eq} and $g_i^{(k)eq}$ at each site and complete the collision step.

iii) Translate the f_i and $g_i^{(k)}$.

The remaining step is to complete the hydrodynamic analysis and determine the form of the remaining parameters in the equilibrium distributions. This can be accomplished by choosing the f_i^{eq} and $g_i^{(k)eq}$ so that they satisfy the additional moment relations

$$\sum_{i=1}^{b} f_i^{eq} e_{i\alpha} e_{i\beta} = P_{\alpha\beta} + \rho u_\alpha u_\beta \tag{11}$$

$$\sum_{i=1}^{b} g_i^{(k)eq} e_{i\alpha} = \rho_k u_\alpha + \left(\tau_k - \frac{1}{2}\right) P_{\alpha\beta} \partial_\beta \frac{\rho_k}{\rho} \tag{12}$$

$$\sum_{i=1}^{b} g_i^{(k)eq} e_{i\alpha} e_{i\beta} = \mu_k \delta_{\alpha\beta} + \rho_k u_\alpha u_\beta + \frac{\rho_k}{\rho} P_{\alpha\beta} \tag{13}$$

where $P_{\alpha\beta}$ is the pressure tensor and μ_k is the chemical potential of component k. These equations, along with equations (1)-(3) can be used to solve for the unknown parameters appearing in Equations (8)-(10).

If Equations (11)-(13) hold, then the hydrodynamic equations generated by the model have the form

$$\frac{\partial}{\partial t}\rho + \partial_\alpha \rho u_\alpha = 0 \tag{14}$$

$$\begin{aligned} \frac{\partial}{\partial t}\rho u_\alpha + \partial_\beta \rho u_\alpha u_\beta &= -\partial_\alpha p_0 + \partial_\beta \nu \partial_\beta \rho u_\alpha + \partial_\beta \nu \partial_\alpha u_\beta \\ &+ \partial_\alpha \lambda \partial_\beta \rho u_\beta \end{aligned} \tag{15}$$

$$\frac{\partial}{\partial t}\rho_k + \partial_\alpha \rho_k u_\alpha = \partial_\alpha \theta_k \partial_\alpha \mu_k \tag{16}$$

The transport coefficients ν, θ_k, and λ are given by the expressions

$$\nu = \Delta t \left(\tau_\rho - \frac{1}{2}\right) \frac{c^2}{D+2} \tag{17}$$

$$\theta_k = \Delta t \left(\tau_k - \frac{1}{2}\right) \tag{18}$$

$$\lambda = \Delta t \left(\tau_\rho - \frac{1}{2}\right) \left(\frac{c^2}{D+2} - \frac{\partial p_0}{\partial \rho}\right) \tag{19}$$

The coefficient ν is the kinematic shear viscosity, θ_k is the diffusion coefficient for component k, and λ is the bulk kinematic viscosity. The function p_0 is the equilibrium pressure for the system and is part of the pressure tensor $P_{\alpha\beta}$. It will be described in more detail below. Note that because τ_ρ and p_0 are already specified in order to independently choose the kinematic shear viscosity and the equation of state, the bulk kinematic viscosity cannot be independently varied.

Model Equation of State

To test this algorithm, it is necessary to come up with a model equation of state. To simulate a separations process it requires a system with two immiscible liquids along with a soluble third component that is preferentially dissolved in one of the two phases. The model equation of state used for this study is similar to the one used by Swift *et al.* to model two immiscible fluids[4]. Components 1 and 2 are the immiscible components, the remaining components are assumed to be at much lower concentration and are treated as impurities. The free energy functional for this system is chosen to have the Cahn-Hilliard form[6]

$$\Psi = \int \left(\psi(T, \rho, \{\rho_k\}) + \kappa(\nabla\rho_1)^2 + \kappa(\nabla\rho_2)^2\right) d\mathbf{r} \tag{20}$$

where T is the temperature, κ is a parameter that controls the interfacial tension between components 1 and 2, and ψ is the bulk free energy density given by

$$\begin{aligned} \psi(T, \rho, \{\rho_k\}) &= \sum_{k=1}^{M} [kT\rho_k \ln \rho_k - kT\rho_k] \\ &+ \frac{\lambda}{4}(\rho_1 + \rho_2)\left[1 - \left(\frac{\rho_1 - \rho_2}{\rho_1 + \rho_2}\right)\right] \\ &- \sum_{k=3}^{M}\left[\alpha_k\rho_k\left(\frac{\rho_1}{\rho_1 + \rho_2}\right) + \beta_k\rho_k\left(\frac{\rho_2}{\rho_1 + \rho_2}\right)\right] \end{aligned} \tag{21}$$

where k is Boltzmann's constant and λ is a parameter that forces components 1 and 2 to phase-separate. The additional parameters α_k and β_k for each component $k \geq 3$ control the affinity of the impurity for the two phases formed from components 1 and 2. The local chemical potential for each of the components can be calculated from the free energy[7]. For component 2, the chemical potential is

$$\begin{aligned} \mu_2 &= kT \ln \rho_2 + \lambda\frac{\rho_1^2}{(\rho_1 + \rho_2)^2} \\ &- \sum_{k=3}^{M}\left[(\beta_k - \alpha_k)\frac{\rho_k\rho_1}{(\rho_1 + \rho_2)^2}\right] - 2\kappa\nabla^2\rho_2 \end{aligned} \tag{22}$$

Similarly, the chemical potentials for the remaining species $k \geq 3$ are

$$\mu_k = kT \ln \rho_k - \alpha_k \frac{\rho_1}{\rho_1 + \rho_2} - \beta_k \frac{\rho_2}{\rho_1 + \rho_2} \tag{23}$$

The pressure tensor can also be calculated from the free energy functional and has the form

$$P_{\alpha\beta} = p\delta_{\alpha\beta} + 2\kappa\partial_\alpha\rho_1\partial_\beta\rho_1 + 2\kappa\partial_\alpha\rho_2\partial_\beta\rho_2 \tag{24}$$

The scalar function p is given by

$$p = p_0 - 2\kappa(\rho_1\partial_\alpha\partial_\alpha\rho_1 + \rho_2\partial_\alpha\partial_\alpha\rho_2) - \kappa((\partial_\alpha\rho_1)(\partial_\alpha\rho_1) + (\partial_\alpha\rho_2)(\partial_\alpha\rho_2)) \tag{25}$$

and p_0 is the equilibrium equation of state for a uniform system. For this model it is essentially a simple mixture of ideal gases

$$p_0 = \sum_{k=1}^{M} \rho_k kT \tag{26}$$

A partition coefficient K_k for the impurity components can be derived by equating the chemical potentials for the impurities in solutions that are otherwise composed entirely of either component 1 or 2[8]. This gives the relation

$$K_k \equiv \ln(\rho_k^1/\rho_k^2) = \frac{\alpha_k - \beta_k}{kT} \tag{27}$$

where ρ_k^1 is the concentration of component k in the phase composed of component 1 and ρ_k^2 is the concentration of k in the phase composed of component 2. This relation is useful for choosing appropriate values of α_k and β_k.

Results

Two-dimensional simulations of an immiscible fluid pair with a dissolved third component were performed on two micro-flow configurations. The first is a configuration in which two immiscible fluids flow in direct contact with each other along a microchannel. The flow channel configuration is shown in Figure 1. The stability of the flow configuration is maintained by the surface tension between the two fluid phases and the wetting characteristics of the two fluids with the walls. The second configuration, shown in Figure 2, is similar to the first, except that the two fluid phases are separated by a porous contactor plate. The two fluids are only allowed to come into direct contact with each other inside the pores.

The parameters in the equation of state were set at $T = 0.5$, $\lambda = 1.1$, $\alpha_3 = 0.5$, $\beta_3 = -0.5$, and $\kappa = 0.01$. The units of temperature are chosen so that Boltzmann's constant k is equal to 1. The values of α_3 and β_3 favor the partitioning of component 3 in the component 1-rich phase, the partition coefficient for this system is about 7. All components are assumed to have the same molecular mass, equal to 1. The equilibrium compositions of the two phases that form at this temperature are 75% component 1 and 25% component 2 in the component 1-rich phase and 25% component 1 and 75% component 2 in the component 2-rich phase. These two phases are referred to as the 1 phase and the 2 phase in the following discussion. The values of τ_ρ and τ_k were both set equal to 1. The simulations were performed using the orthogonal 2-dimensional lattice.

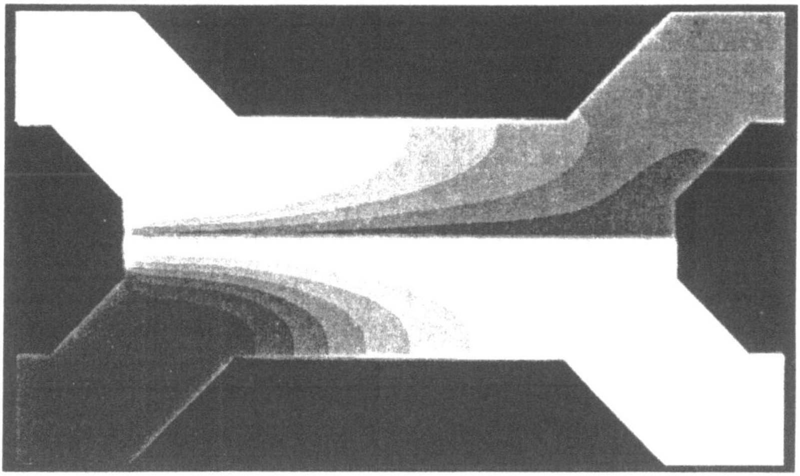

Figure 1. Flow configuration for two immiscible fluids in direct contact with each other. Solid boundaries are shown in black and the concentration of component 3 is shown as grayscale contours.

The simulation shown in Figure 1 has the 1 phase entering the system from the channel in the upper left hand side of the figure and the 2 phase entering from the channel in the lower left hand side of the figure. The pressure is maintained at 0.51 at the boundaries on the left hand side of the figure and at 0.49 on the right hand side. The solid boundaries in the upper half of the flow configuration are set so that they wet the 1 phase, the boundaries in the lower half wet the 2 phase. The concentration of component 3 in the entering 1 phase stream is 0.0001, the concentration in the 2 phase stream is 0.001. The simulation grid is 70×41 lattice units in size. The concentration profile of component 3 is shown as a series of greyscale contours in Figure 1. The sharp horizontal line at the center of the figure marks the boundary between the 1 and 2 phases. The concentration of component 3 gradually increases in the 1 phase as the fluid flows down the channel and decreases in the 2 phase as component 3 moves from the 2 phase to the 1 phase. The third component is always partitioned between the two components at the boundary between the two phases according to the equilibrium partition coefficient, which results in the sharp line in the concentration profile at the center of the figure.

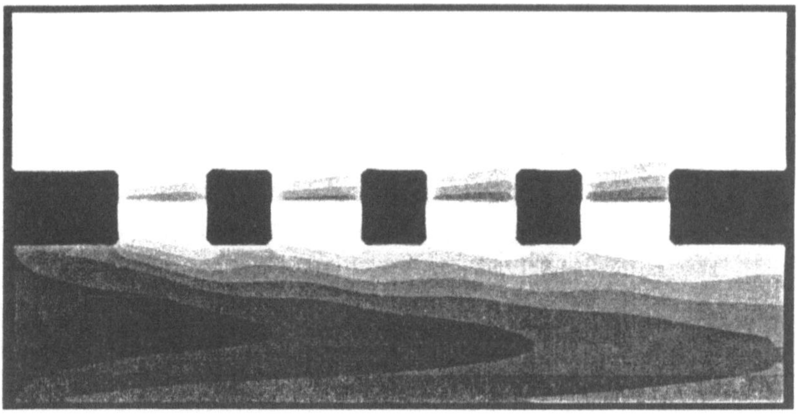

Figure 2. Flow configuration for two immiscible fluids separated by a porous contactor plate. Solid boundaries are shown in black and the concentration of component 3 is shown as grayscale contours. The bottom and top of the figure also represent solid boundaries.

The same boundary conditions described above were used in the simulation shown in Figure 2, except that the simulation size was increased to 100×50 lattice units. The interface between the two phases is pinned in the middle of the pores by the wetting characteristics of the pore boundaries. The concentration of component 3 is shown as a series of grayscale countours. The contours again show the transfer of component 3 from the 2 phase to the 1 phase as the two fluids flow down the channel. The transfer of component 3 appears to be slightly slower in this case due to the reduced surface contact area.

Conclusions

A lattice Boltzmann algorithm for simulating multicomponent-multiphase systems at constant temperature is described and simulations of fluid flow in configurations relevant to microscale separations processes are presented. The algorithm itself improves on an algorithm already in the literature[4] by removing a spurious pressure term in the diffusion equation describing the behavior of minor components. The simulations demonstrate that the algorithm can reproduce the expected transfer of a dissolved minor component from one phase, where it has a relatively low solubility, to another phase for which it has a higher affinity. This is the basic phenomenon that is exploited in many separations processes to extract and isolate particular chemical compounds. The simulations also show that algorithm is capabable of modeling multiphase behavior without specifying the location of the boundary separating the phases in advance. This is a major advantage in simulating less symmetrical systems where the phase boundary may not be linear and its location may not be known *a priori*.

References

[1] H. Chen, S. Chen, and W. H. Matthaeus. *Phys. Rev. A*, **45**:R5339, (1992).

[2] Y. H. Qian, D. d'Humières, and P. Lallemand. *Europhys. Lett.*, **17**:479, (1992).

[3] X. Shan and H. Chen. *Phys. Rev. E*, **47**:1815, (1993).

[4] M. Swift, E. Oflandini, W. Osborn, and J. Yeomans. *Phys. Rev. E*, **54**:5041, (1996).

[5] P. L. Bhatnagar, E. P. Gross, and M. Krook. *Phys. Rev.*, **94**:511, (1954).

[6] J. Cahn and J. Hilliard. *J. Chem. Phys.*, **28**:258, (1958).

[7] L. E. Reichl. *A Modern Course in Statistical Physics*. University of Texas Press, Austin, 1984.

[8] K. Denbigh. *The Principles of Chemical Equilibrium*. Cambridge University Press, Cambridge, 1981.

Modelling of complex reaction-separation processes in a d.c. electric field in microchannels

D. Snita, J. Kosek, H. Sevcikova, J. Lindner, J. Havlica, M. Paces and M. Marek

Department of Chemical Engineering, Center for Nonlinear Dynamics of Chemical and Biological Systems, Prague Institute of Chemical Technology, Technicka 5, 166 28 Prague 6, Czech Republic

Abstract. The paper deals with specific aspects of modelling effects of an electric field applied on microchannels, where also chemical reactions other than dissociation equilibria can take place, and presents three examples of modelling studies that employ software packages developed in our group. We present a set of mass, charge, momentum and heat balance equations that describe reaction-transport systems and form the core of the specialized program code (1D-IONS) that allows to simulate the spatial evolution of species concentrations and of a variety of electric variables, such as charge density, electric field intensity, and conductivity, in spatially 1D systems for the mixtures of up to 999 components. The output of the program is illustrated by solving the isotachophoretic separation. Based on the same set of fundamental equations the mathematical model of the spatially 2D reacting ionic system was developed to simulate the propagation of reaction fronts in the applied electric field. Modification of the 2D model excluding the assumption of electroneutrality was also developed and used to simulate axial dispersion and mixing in the channel with electroosmotic flows.

Keywords. ionic systems, electroseparation, electroosmosis, microreactors

1 Introduction

In liquid - phase microreactors an electric field is often used for dosing and transport of reaction components, migration in electric field affects reaction and electrophoretic separation methods are used to analyse reaction products [1]. Comparison of simulations with experiments on the effects of an electric field on model reactions in 1D and 2D microchannels was discussed at two previous conferences [2, 3]. In the following we shall list an appropriate set of mass, charge, momentum and heat balances used for the description of the system of reacting components in an electric field (a detailed discussion of balance equations was given earlier [4, 5]). Then we shall present three examples of modelling of electrophoretic separations, electroosmosis, and developing patterns of reaction components in an electric field.

2 Balance equations

The *mass balances* of individual components can be written in the form

$$\frac{\partial c_k}{\partial \tau} = -\nabla\cdot(\mathbf{v}c_k) - \nabla\cdot\mathbf{J}_k + \sum_{j=1}^{L} \nu_{k,j} R_j + \sum_{j=L+1}^{M} \nu_{k,j} R_j \ , \quad k = 1, \ldots, N_c \ , \tag{1}$$

where N_c is the number of components, c_k is the molar concentration of the k-th component, τ is time, $\mathbf{v}$ is the local flow velocity, $\mathbf{J}_k$ is the molar flux of the k-th component, $\nu_{k,j}$ is the stoichiometric coefficient of the k-th component in the j-th chemical reaction, $R_j(c_1, \ldots, c_{N_c}, T)$ is the reaction rate of the j-th chemical reaction, T is absolute temperature and M is the number of chemical reactions.

The chemical reactions are divided into two classes: initially far-from-equilibrium reactions ($j = 1, \ldots, L$) and initially near-to-equilibrium reactions ($j = L+1, \ldots, M$). We consider that the initially near-to-equilibrium reactions are mutually independent (The modelling of electrophoretic separations in the case where only dissociation reactions take place is in detail described in [6]).

The Nernst-Planck equation is used for the description of molar fluxes

$$\mathbf{J}_k = -D_k \nabla c_k + z_k D_k \frac{\mathbf{E}F}{RT} \ . \tag{2}$$

Here D_k is the diffusion coefficient of the k-th component, z_k is the charge number of the k-th component, $\mathbf{E}$ is the electric field intensity, F is the Faraday constant and R is the gas constant.

Equations (1) can be transformed into the form

$$\frac{\partial g_i}{\partial \tau} = -\nabla\cdot(\mathbf{v}g_i) - \nabla\cdot\mathbf{F}_i + \sum_{j=1}^{L} \zeta_{i,j} R_j \ , \quad i = 1, \ldots, N_g \le N_c \ , \tag{3}$$

$$\frac{\partial g_i}{\partial \tau} = -\nabla\cdot(\mathbf{v}g_i) - \nabla\cdot\mathbf{F}_i + \sum_{j=1}^{L} \zeta_{i,j} R_j + \sum_{j=L+1}^{M} \zeta_{i,j} R_j \ ,$$

$$i = N_g + 1, \ldots, N_c \ , \tag{4}$$

where

$$g_i = \sum_{k=1}^{N_c} A_{i,k} c_k \ , \quad \mathbf{F}_i = \sum_{k=1}^{N_c} A_{i,k} \mathbf{J}_k \ , \quad \zeta_{i,j} = \sum_{k=1}^{N_c} A_{i,k} \nu_{k,j} \ . \tag{5}$$

The regular transformation matrix $(A_{i,j})_{N_c \times N_c}$ is constructed in such a way that

$$\zeta_{i,j} = 0 \ , \ i = 1, \ldots, N_g \ , \ j = L+1, \ldots, M \ , \ N_c - N_g = M - L = N_e \ . \tag{6}$$

We define the square matrix $(J_{i,n})_{N_c \times N_c}$

$$J_{i,n} = \sum_{k=1}^{N_c} \sum_{j=L+1}^{M} \sum_{m=1}^{N_c} A_{i,k} \nu_{k,j} \frac{\partial R_j}{\partial c_m}(c_1^*, \ldots, c_{N_c}^*) A_{m,n}^{-1} \quad ,$$

$$i = N_g + 1, \ldots, N_c \quad , \quad n = N_g + 1, \ldots, N_c \quad . \tag{7}$$

and consider that the real parts of all eigenvalues of matrix $J_{i,n}$ are negative and have a relatively *high* absolute value for any $c_1^*, \ldots, c_{N_c}^*$ such that

$$\sum_{j=L+1}^{M} \zeta_{i,j} R_j(c_1^*, \ldots, c_{N_c}^*) = 0 \quad , \quad i = N_g + 1, \ldots . N_c \quad . \tag{8}$$

The reaction rates of the near-to-equilibrium reactions can be expressed in the form

$$R_m = \frac{1}{\varepsilon} R_m^*(c_1, \ldots, c_{N_c}) \quad , \quad m = L + 1, \ldots, M \quad , \tag{9}$$

where ε is the *small* parameter. If the first three terms of the right hand side of Eqs. (4) are not very *high*, the pseudo-equilibrium assumption can be used

$$R_m^*(c_1, \ldots, c_{N_c}) = 0 \quad , \quad m = L + 1, \ldots, M \quad . \tag{10}$$

The charge density (q) and the electric current intensity ($\mathbf{i}$) in electrolytes are given by

$$q = F \sum_{k=1}^{N} z_k c_k \quad , \quad \mathbf{i} = F \sum_{k=1}^{N} z_k \mathbf{J}_k \quad . \tag{11}$$

No electric charge is created in the course of chemical reactions.

$$\sum_{k=1}^{M} z_k \nu_{k,j} = 0 \quad , \quad j = 1, \ldots, M \quad . \tag{12}$$

We can set

$$A_{N_g,k} \equiv z_k \quad \Rightarrow \quad g_{N_g} \equiv q \quad , \quad \mathbf{F}_{N_g} \equiv \mathbf{i} \quad . \tag{13}$$

The final set of equations resulting from *mass balances* can be then written in the form

$$\frac{\partial g_i}{\partial \tau} = -\nabla \cdot (\mathbf{v} g_i) - \nabla \cdot \mathbf{F}_i + \sum_{j=1}^{L} \zeta_{i,j} R_j \quad , \quad i = 1, \ldots, N_g - 1 \quad , \tag{14}$$

$$\frac{\partial q}{\partial \tau} = -\nabla \cdot \mathbf{i} \quad , \tag{15}$$

$$\sum_{j=L+1}^{M} \zeta_{i,j} R_j^* = 0 \quad , \quad i = N_g + 1, \ldots, N_c \quad . \tag{16}$$

The laws of electrostatics can be written in the form

$$\nabla \cdot (\epsilon \mathbf{E}) = q \quad , \quad \nabla \times \mathbf{E} = 0 \quad \Leftrightarrow \quad \nabla \cdot (\epsilon \nabla \phi) = -q \quad , \quad \mathbf{E} = -\nabla \phi \quad , \tag{17}$$

where ϵ is the permitivity (dielectric constant) and ϕ is the electrostatic potential.

Eqs. (14 – 17) form (for given field of temperature and velocity) a closed system of equations which can be solved with consistent boundary and initial conditions to obtain fields of concentrations and electrostatic potential.

The *charge balance* equation (15) can be re-written, using Eqs. (2, 11, 17), in the form

$$\frac{\partial q}{\partial \tau} = -\nabla \cdot \mathbf{i} = F \sum_{k=1}^{N} z_k D_k \nabla^2 c_k - \nabla g \cdot \mathbf{E} - \frac{g}{\epsilon} q \quad , \tag{18}$$

where

$$g = \frac{F^2}{RT} \sum_{k=1}^{N_c} z_k^2 D_k c_k \quad , \tag{19}$$

is the electric conductivity. The assumption of a pseudosteady state for (18) results in

$$q = \frac{\epsilon}{g} \left(F \sum_{k=1}^{N} z_k D_k \nabla^2 c_k - \nabla g \cdot \mathbf{E} \right) \quad . \tag{20}$$

As ϵ is small, the assumption of local electroneutrality

$$q = F \sum_{k=1}^{N} z_k c_k = 0 \tag{21}$$

is often used. In principle the assumption (21) is similar to the pseudo-equilibrium assumption (10).

When we apply the electroneutrality condition (21) to Eqs. (17) then, on assuming constant permitivity ϵ, we obtain

$$\nabla \cdot \mathbf{E} = 0 \quad , \quad \nabla^2 \phi = 0 \quad , \tag{22}$$

and the model then consists of Eqs. (14, 16, 21, 22).

However, if we use the electroneutrality assumption in Eq. (15) we obtain

$$\nabla \cdot \mathbf{i} = 0 \quad . \tag{23}$$

We can substitute (2) into the (11) and eliminate $\mathbf{E}$

$$\mathbf{E} = \frac{1}{g}\left(F\sum_{k=1}^{N_c} z_k D_k \nabla c_k + \mathbf{i}\right) \quad . \tag{24}$$

Concentration of a chosen component can be calculated from the concentrations of other components using (21). The model then consists of Eqs. (14, 16, 21, 23).

The actual value of the charge density q is then obtained on applying the divergence operator to Eq. (24) and using Eq. (23)

$$q = \epsilon\nabla\cdot\mathbf{E} = \frac{\epsilon F}{g}\sum_{k=1}^{N_c} z_k D_k \nabla^2 c_k - \frac{\epsilon F}{g^2}\nabla g\cdot\left(\sum_{k=1}^{N_c} z_k D_k \nabla c_k + \mathbf{i}\right) \quad , \tag{25}$$

cf. Eq. (20).

The *momentum balance* (Navier-Stokes equation) and the equation of continuity for non-compressible flow can be written in the form

$$\rho\frac{\partial\mathbf{v}}{\partial\tau} = -\rho\mathbf{v}\cdot\nabla\mathbf{v} - \nabla\cdot\mathbf{T} - \nabla p + \mathbf{E}q \quad , \quad \nabla\cdot\mathbf{v} = 0 \quad , \tag{26}$$

where ρ is the (mass) density, $\mathbf{T} = -\eta(\nabla\mathbf{v} + (\nabla\mathbf{v})^T)$ is the disipative part of the stress tensor where η is the viscosity and p is the pressure.

The *heat balance* can be written in the form

$$\rho c_p\frac{\partial T}{\partial\tau} = -\rho c_p\mathbf{v}\cdot\nabla T - \nabla\cdot\mathbf{q} - \mathbf{T}:\nabla\mathbf{v} + \mathbf{E}\cdot\mathbf{i} + \sum_{j=1}^{M} -\Delta h_j R_j \quad , \tag{27}$$

where c_p is the heat capacity and $\mathbf{q} = -\lambda\nabla T$ is molecular heat flux where λ is the heat conductivity and Δh_j is the reaction heat of the j-th reaction.

3 Spatially one-dimensional problems

Spatially one-dimensional problems of dynamical behaviour of reacting complex ionic systems (described by the above balance equations) can be studied by means of software package "1D-IONS" developed in our research group, installed under UNIX environment (Windows 95 and NT instalations also exist).

Both dimensional and dimensionless form of balance equations with four types of mass flux description (diffusion, diffusion and migration in spatially uniform electric field, controlled spatially constant electric current or potential difference (voltage) between the system boundaries) can be chosen.

For chosen boundary conditions (no flux, known composition, batch or continuous stirred tank reactor at the boundaries) and chosen initial conditions we can follow the evolution of spatial profiles of concentrations $c_k(x,\tau)$,

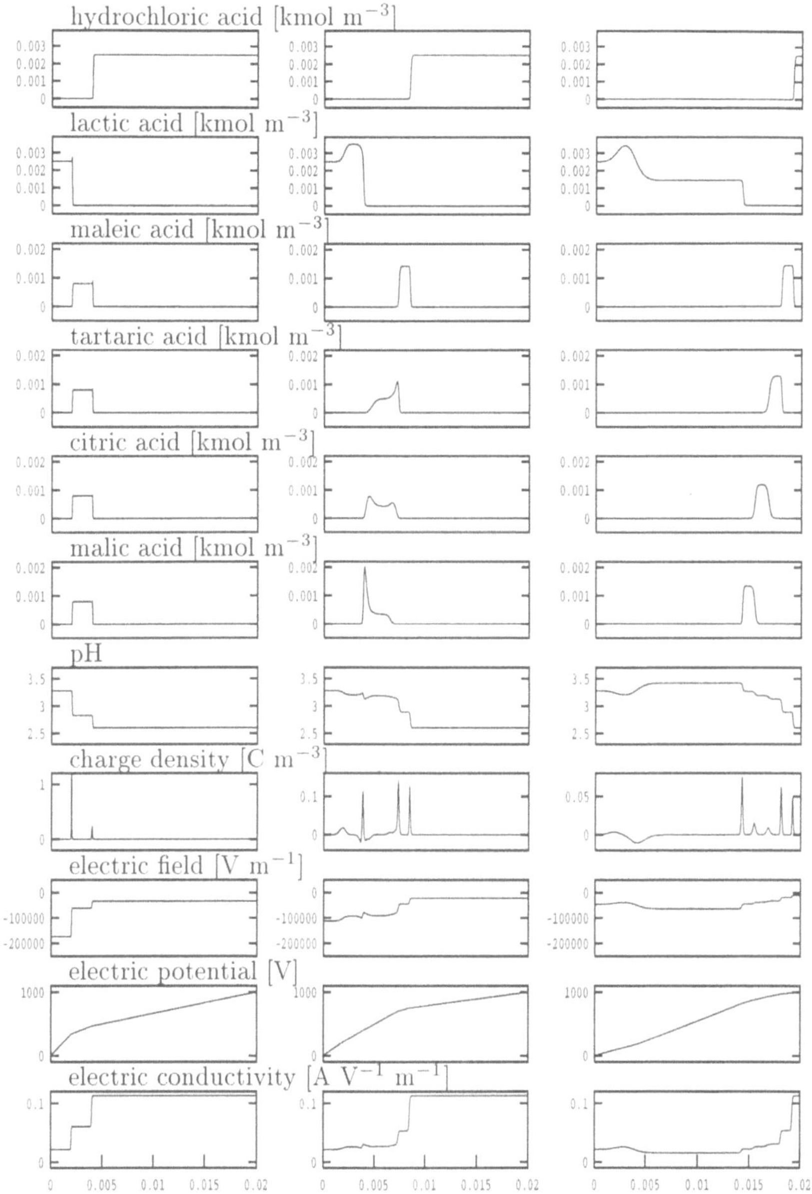

Figure 1: Isotachophoresis: the sample (mixture of four organic acids) is initially placed between the leading (hydrochloric acid) and terminating (lactic acid) electrolytes, the left column - time = 0.01 s, the midle column - time = 1 s, the right column - time = 12 s, capillary of the length 20 mm is exposed to the potential difference 1000 V (the mixture permitivity ϵ is 78.5 $\times$ 8.854 $\times$ 10^{-12} C V^{-1} m^{-1})

and $g_i(x,\tau)$, pH(x,τ), conductivity $g(x,\tau)$, charge density $q(x,\tau)$, electric field intensity $E(x,\tau)$, and electric potential $\phi(x,\tau)$.

Diffusivities (ion mobilities) and equilibrium constants of ionic reactions for common components can be obtained from the connected database. The required set of algebraic equations is generated automatically in a pre-processor stage and an equidistant or adaptive spatial grid is used. An example of the evolution of spatial profiles of concentrations and electric variables in isotachophoresis is shown in Fig. 1. A simple system with small number of components was chosen to enable simple graphical presentation (software package enables to follow far more complex systems containig up to 999 components and the evolution can be also presented via computer animation).

4 Mutual interaction of chemical reaction and an electric field in spatially 2D microreactors

The interaction of chemical reaction and transport can lead to the formation of spatiotemporal patterns, such as fronts and pulse waves. Such patterns can be conveniently used for the comparison of experiments with the results of modelling, cf. [2,3]. Here we study the behaviour of spatiotemporal patterns in the situation where the calculation of the exact 2D distribution of electrostatic potential is important.

A simple autocatalytic reaction

$$A^+ + B^+ \longrightarrow 2\,B^+\ ,\ R_B = -R_A = kc_Ac_B\ ,$$

was chosen as a sufficiently simple yet generic model of propagating reaction fronts. In addition to species A^+, B^+ the reacting electrolyte contains also non-reacting species C^+ and D^-. When the intensity of electric field is low then the reaction and diffusion dominates and kinetic waves propagate in the system. As the intensity of electric field increases the migration of ions becomes more important and strong electric fields lead to complex phenomena [7,8].

Front propagation in our simulation takes place in the rectangular domain in a d.c. electric field. Front propagation is described by equations given in Section 2 supplemented with boundary conditions (fixed concentrations and potential at the left and right boundaries and zero-flux at other boundaries). The evolution of species concentrations, the distribution of electrostatic potential, and the distribution of electric current streamlines were followed during the simulation, cf. Figure 2.

Modelling of ionic reacting systems in 2D is usually performed under the simplifying assumption of either the constant electric field intensity ($\mathbf{E}$ = const) or constant electric current density ($\mathbf{i}$ = const). Our simulations are based on a more general approach to the calculation of distribution of electrostatic potential based on the equation $\nabla \cdot \mathbf{i} = 0$, cf. Section 2.

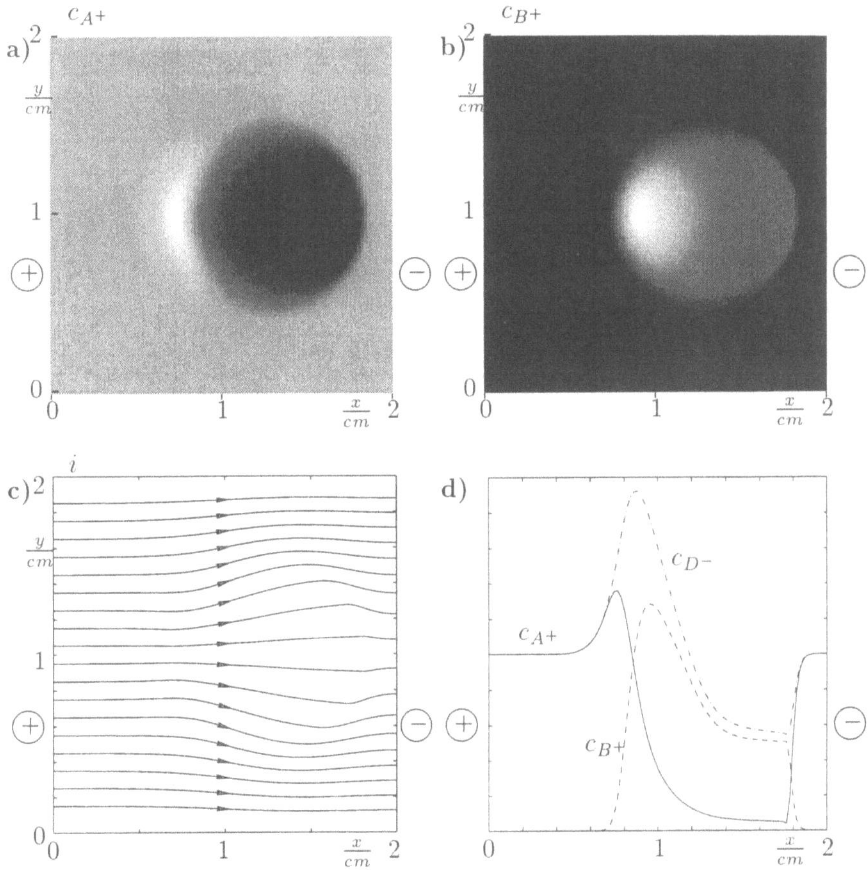

Figure 2: a) Concentration profile of A^+, b) Concentration profile of B^+, (black is low concentration and white is high concentration). c) Electric current streamlines, d) Concentration profiles at $y = 1\,\text{cm}$. Model parameters are $D_A = 2 \times 10^{-5}\,\text{cm}^2\,\text{s}^{-1}$, $D_B = 0.5 \times 10^{-5}\,\text{cm}^2\,\text{s}^{-1}$, $\phi_{right} - \phi_{left} = 5\,\text{V}$, $k = 50\,\text{dm}^3\,\text{mol}^{-1}\,\text{s}^{-1}$.

5 Axial dispersion and mixing in electroosmotic flow

Mathematical modelling of certain reaction-separation processes requires to avoid the assumption of local electroneutrality (21). Examples of such systems are electric-field induced flows along charged surfaces (e.g., electroosmosis), flows around colloid particles, and flows through membranes.

The ζ potential at the capillary wall and the electroosmotic flow in the capillary can be directly controlled either by special chemical treatment of the wall or by the applied radial electric field across the capillary wall [9]. The

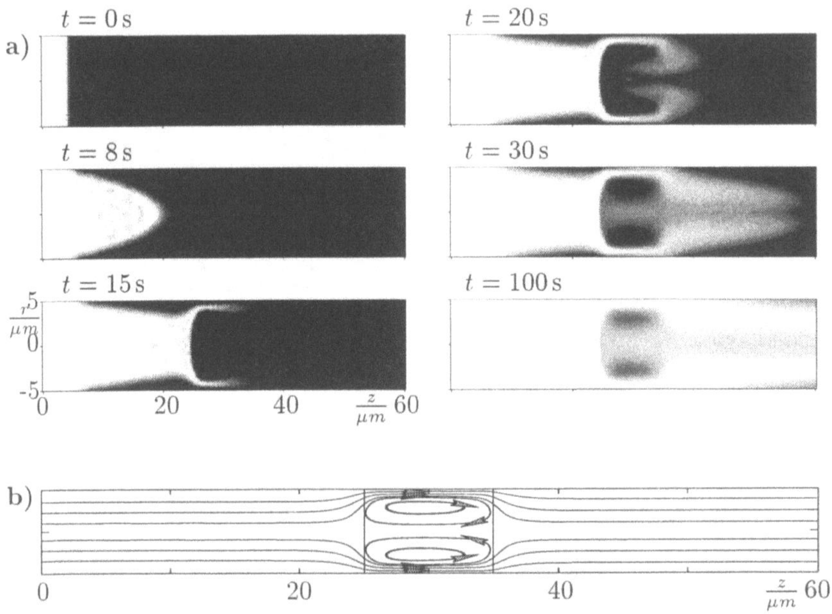

Figure 3: a) Propagation of concentration pulse of uncharged species in the capillary channel (high concentration is white). Concentration pulse is dispersed in the middle region. b) Velocity streamlines in the capillary channel. Electroosmotic flow is suppressed in the left and the right region and maintained in the middle region.

electroosmotic velocity in the controlled region can be decreased or increased and the radial electric field can even reverse the direction of electroosmotic flow. The change in the velocity causes also the dispersion at the boundary between the regions controlled and uncontrolled by the applied radial electric field. This phenomenon can help in the designing of micromixing in a straight capillary channel. Changes in the ζ potential of the order of 50 mV are easily obtained with the external control, e.g., by the radial electric field.

We study electroosmotic flow in the fused silica capillary channel with the surface charge density $\sigma = -8.31 \times 10^{-6}\,\mathrm{C\,m^{-2}}$ corresponding to the complete dissociation of silanol groups. One of the effective mixing configurations in the straight capillary channel is shown in Figure 3. Electroosmotic flow is suppressed at the left and right regions of the shown part of the channel. Regions with the suppressed electroosmotic flows are separated by a zone with the unaffected electroosmotic flow. Since the left and the right regions without electroosmotic flows are relatively wide the parabolic velocity profile develops. Then in the middle region we can observe the development of a vortex and strong mixing.

6 Conclusions

Simulations of transport and reaction in electric field can serve not only for an interpretation of experiments, but also for the design of new types of microreactors and the evaluation of their functions. Similarly to the development of standard CAD software used in the process design, software packages for microreactors and micro-fluidic applications should be further developed and tested. The 1D-IONS is one of available examples, the development of 2D-IONS proceeds in our group just now.

Acknowledgements. Financial support from project VS 96073 (Czech Ministry of Education) and grant 104/99/1408 (Czech Grant Agency) is gratefully acknowledged.

References

[1] Ramsey J.M., Yacobson S.C., Foote R.S., Microfabricated Devices for Performing Chemical and Biochemical Analysis in *Microreaction Technology*, W. Ehrfeld (Ed.), Springer (1998) 204.

[2] Sevcikova H.,Snita D., Marek M.: Reactions in Microreactors in Electric Field, in *Microreaction Technology*, W. Ehrfeld (Ed.), Springer Verlag, Berlin (1998) 47.

[3] Snita D., Lindner J., Sevcikova H., Kosek J., Marek M.: Microreactors for Ionic Reactions in Liquids and Gels in Electric Field, in AICHE Topical Conference Preprints Process Miniatuarization, 2nd Int. Conference on Microreaction Technology (1998) 140.

[4] Snita D., Marek M.: Interaction of Electromagnetic Field with Enzymatic Reaction and pH Effects, in *Horizons of Biochemical Engineering*, University of Tokyo Press (1987) 237.

[5] Snita D., Marek M.: *Physica* **75D** (1994) 521.

[6] Mosher, R. A., Saville, D. A., Thormann, W., *The Dynamics of Electrophoresis*, VCH New York (1992), ISBN: 1-56081-192-7

[7] Snita D., Sevcikova H., Lindner J., Marek M., Merkin J.H., *J. Chem. Soc. Faraday Trans.*, **94**(2) (1998) 213.

[8] Merkin J.H., Sevcikova H., Snita D., Marek M., *IMA Journal of Applied Mathematics* **60** (1998) 1.

[9] Wu C. T., Huang T. L., Lee C.S., *Anal.Chem.* **65** (1993) 568.

Microporous Silica Structures for the Immobilisation of Catalysts and Enhancement of Electroosmotic Flow (EOF) in Micro-Reactors

Natalie G. Wilson and Tom McCreedy

Analytical Science Group, Department of Chemistry, University of Hull, Hull, HU6 7RX, UK

1. Introduction

Since the first work into miniaturisation technology by the micro-electronics industry, many branches of science have continued to follow this trend with the production of micro-sized instruments, one notable example was the manufacture of a miniature GC by Terry *et al* [1] in 1979. However it has only recently become apparent that the name micro-chip does not only apply to those devices found within complex computers and simple wrist watches. The name has also been taken by analytical chemists to describe a miniature appliance capable of performing the functions of a macro-scale system, whilst only a few centimetres in size. This "lab-on-a-chip" concept stems from the idea of the total analysis system (TAS). These complicated instruments which employ the capabilities of an entire lab, i.e. sampling, storage, pre-treatment, analysis, detection and data evaluation initiated the production of the micro-TAS (μ-TAS). These miniature total analysis systems embody the functions of the TAS whilst providing faster sample processing, higher efficiency and lower reagent consumption than conventional TAS. In addition, μ-TAS allow *in-situ* measurement due to their reduced dimensions and portability.

The development and continuing research of these unique miniaturised systems may be contributed to the growing interest in micro-fabrication and the discovery of capillary electrophoresis (CE), and with it electroosmotic flow (EOF) which did not come about until some ten years after the work by Terry *et al* was reported. It is perhaps due to this continued research that the term micro-reactor has been developed. While micro-chip research concerned the separation and detection of compounds on a miniature scale [2], the micro-reactor has surpassed this and has become a platform for the conduction of real chemical reactions. However with this advance comes added problems, including the detection and separation of products, including precipitates, and an overall more complicated set-up. Despite these complications, one of the most interesting areas of micron-scale research remains that of catalytic reactions. In 1998 Senkan [3] developed a micro-synthesis catalytic system to screen catalysts for optimum formulation for any specific reaction. However micro-scale catalytic systems offer advantages other than catalyst screening. Miniature catalytic systems have been shown to give better conversion and selectivity, a fact attributed to the increased surface to volume ratio present. They also offer increased heat transfer, improved process conditions, ease of parallel operation and the ability to be used as test platforms. Additionally, they have the potential to work on the same scale as a macro-reactor with fewer problems. Micro-reactors constructed in banks (scale out compared to scale up) would provide smaller and safer plants. Closer control of thermal reactions could be achieved allowing highly exothermic reactions to be conducted safely. In addition, the ability to switch reactors on and off would provide control of the product output which could be matched to supply and demand and eliminate the need for costly storage and handling.

Despite the rapidly increasing number of applications for these new micro-reactors, the area of heterogeneous catalysis has still not been studied fully. Perhaps this is because the very nature of the reaction causes added complications, namely the immobilisation of a solid catalyst within such small channel dimensions (typically <200 μm). Traditional methods for immobilisation of solid substances have involved using structures called frits [4,5], and have taken the form of metallic or ceramic sinters, thus allowing liquid or gaseous throughput without solid loss. However these traditional methods usually rely on channel dimensions larger than 200 μm in size. A method for the immobilisation of solids within channels of 200 μm and below is required in order to produce a solid catalyst containing micro-reactor.

Heterogeneous catalysis covers a wide area, and as such the choice of reactions is immense. However a growing area at present is the use of solid acid catalysts for industrial reactions. Conventional petrochemical techniques involve the use of strong acids such as HF and H_2SO_4 for reaction of organic substances. However due to their corrosive nature and new laws concerning effluent release, it is becoming increasingly more expensive to use them due to the high costs involved with handling and disposal. A more environmentally friendly approach is to use solid acid catalysts such as sulphated zirconia and silica alumina. These are able to catalyse petrochemical reactions with equal efficiency with the added advantages that a reduced volume is required, reaction conditions are less severe, and disposal and handling is much easier. In addition, solid acid catalysts such as sulphated zirconia have been shown to be applicable to reactions including hydroisomerisation, hydrocracking, alkylation, oligomerisation and dehydration among others.

By combining the areas of micro-reactor and heterogeneous catalytic research, the aim of this project was to produce a solid catalyst containing micro-reactor. Sulphated zirconia was the selected catalyst, and the dehydration of hexanol to hexene (optimum temperature 150°C) was chosen as the test reaction, primarily due to the ease of product collection.

2. Experimental

In 1987 Cortes *et al* [4] reported a method for the formation of porous ceramic material using a mixture of potassium silicate and formamide, which was used for chromatographic bed supports. This led to the study by Christensen *et al* [6] of the formation of these porous ceramic "frit" structures and also optimisation of their production. It was determined that a ratio of 140 μL potassium silicate : 18 μL formamide (or any multiple of this) was optimum, with a mixing time of 30 seconds. Reaction time was optimised to 1h at 100°C. Other significant results included the need for frit priming to remove excess reagents and the formation of a non-porous outer layer when in close contact with the air during the reaction time. A Scanning electron micrograph of the micro-porous frit structure is shown in figure 1. The versatility of this newly found method for producing porous structures meant that solid catalyst immobilisation was not constricted by channel size, in fact frits could be manufactured within channels of 10 mm i.d. and above down to sizes around 50 μm i.d. and less with no adverse effects.

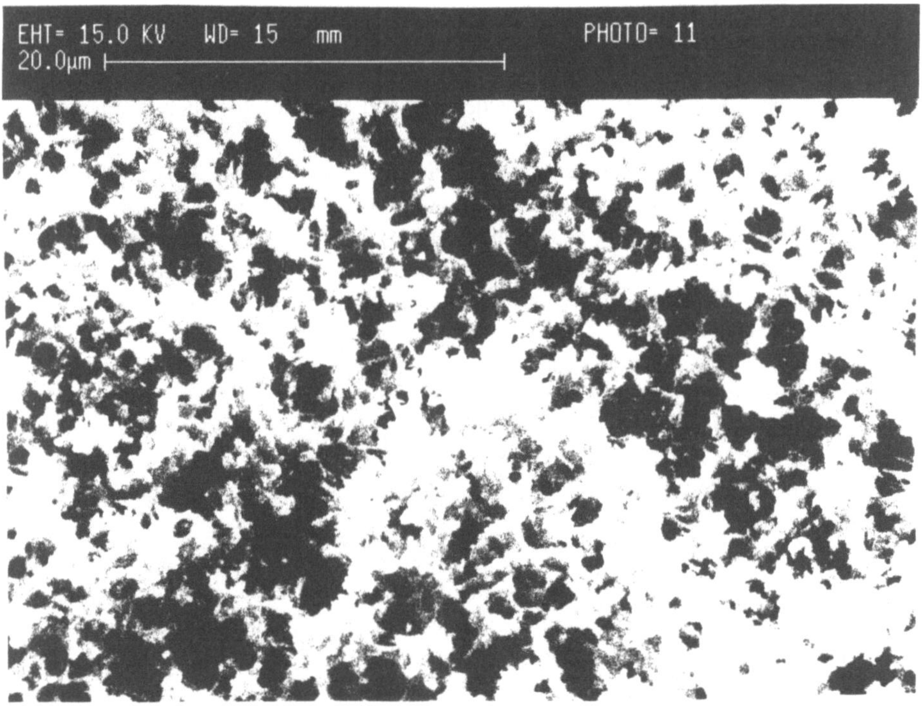

Figure 1: Scanning Electron Micrograph showing the surface of the porous silicate frit.

3. Results

Initial work involved the production of micro-reactors which were not of the conventional chip design. Here a capillary of known i.d. was used to form the channel. Primarily micro-reactors of 1mm i.d. were prepared, but after successful hexanol to hexene conversion within these reactors (approx 80-90% at 0.01 mL min^{-1} flow rate), the channel dimensions were reduced to 200 μm i.d (typical of micro-reactor systems).

Preparation of micro-reactors using 200 μm i.d. fused silica capillaries was achieved by producing a primary frit within the capillary against which a catalyst bed was packed (typically 0.005 g sulphated zirconia). The catalyst was then immobilised fully using a secondary retaining frit. Conversion of hexanol to hexene within these reactor dimensions was achieved with yields within the range 1-5% using a flow rate of 0.01 mL min^{-1}.

For the fabrication of a conventional micro-chip-reator, the immobilisation of the catalyst was more difficult. Two frits were made within the open channel of the micro-reactor bottom plate, this was achieved by blocking the remaining channel area. Frit mixture was added to the open areas and the reaction allowed to go to completion. On removal of the mask (shown in figure 2) the frits were left. Sulphated zirconia catalyst was then packed beween the two frits and the entire frit/catalyst/frit area was made flush to the plate. The bottom channel plate was then fused to a top plate (which contained pre-drilled holes aligned with the ends of the channel) using the programme shown in table 1. An optional heater plate was added and the overall layout is shown in figure 3.

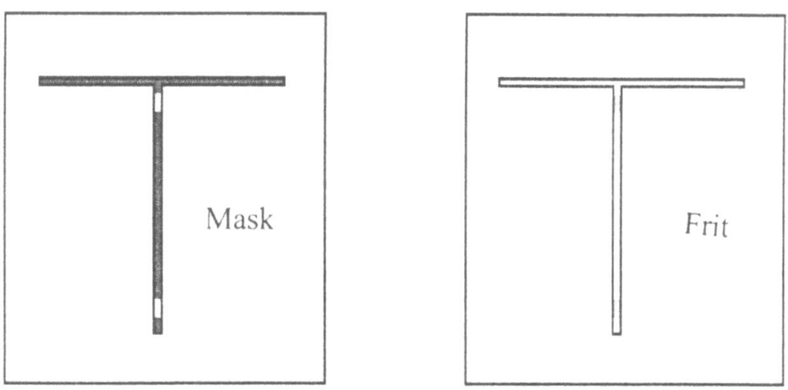

Figure 2: Diagram showing masking of the channel to produce two accurately positioned frits.

Temperature /°C	Time /minutes
75	30
150	30
680	60
650	30
505	30

Table 1: Temperature programme used to fuse the micro-reactor.

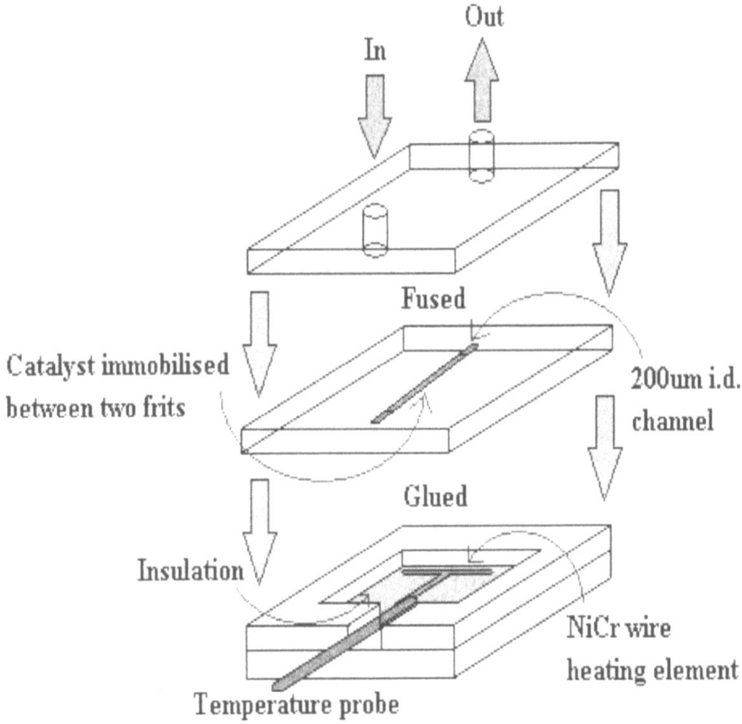

Figure 3: Exploded diagram of the micro-reactor system.

Hence a method for the successful immobilisation of solid phase catalysts within a micro-reactor system was achieved with conversion of hexanol to hexene from the test reaction. From the work by Christensen *et al* [6], it was shown that not only could the frits be used as a means of immobilisation of solids, but when considering a system with reagents moved by EOF, the presence of frits had several distinct effects. EOF can be defined as the movement of a liquid through a capillary or other porous media under the influence of an applied electric field. It arises because the surface of many solids carry a net charge when in contact with aqueous solutions, indeed between pH 4 and 10 surfaces of quartz, ceramics and glass are negatively charged. This negatively charged surface attracts positive ions present within the aqueous buffer solution forming a double layer structure. This diffuse double layer (Guoy layer) is partially mobile and on application of an external electrical field a net migration of the positive ions occurs towards the negative electrode. Due to viscous drag, the whole solution moves in the same direction and this is called EOF.

Using EOF as a fluid propulsion system has advantages, especially for miniaturised systems. The magnitude and direction of the flow can be easily changed by manipulation of the voltage applied and the volumetric flow rate is linearly proportional to the current. However EOF pumping is impractical in situations where analytes are of very high or low pH, (EOF does not occur outside the pH range 4-10), in highly conductive saline solutions or in non-conducting organic medium. In these cases current levels are either excessive or inadequate to support significant EOF. Typically EOF does not occur within channels greater than 300 μm i.d, however due to the nature of the micro-porous frit structures, the presence of frits in larger i.d. channels (>300 μm) allows the movement of liquids by EOF within them. This can be attributed to the physical characteristics of the frit. As they are of silica in origin it is logical to assume that the surface of the frits acts in the same way as any other silica surface, i.e. an electrically induced double layer is formed across the surface which facilitates electroosmotic movement of the solution within. Not only this, but because of the concentrated nature of the Gouy layer within the frit structure i.e. many pores giving many channels within a single area, then EOF rate is increased. Thus the addition of a frit has been found to give increasing flow rate with increasing frit length up to a maximum of 5 cm [6].

This concentrated electrical double layer effect has several distinct advantages, not least the use of the frits as EOF pumps. This means that the addition of a primary frit within a micro-reactor system can facilitate buffer movement through a non-conducting media (e.g. catalyst). Increasing the bed (amount) of non-conducting media was found to decrease the flow rate through the system, however an increase in primary frit size was found to compensate for this. Thus by manipulation of the primary frit size and applied voltage, movement through non-conducting media (eg catalyst) was found to be possible with a controllable flow rate when using frits within a system.

The effect of frits within reduced channel sizes was also studied. It was found that flow rate for capillaries of 1 mm i.d, 0.5 mm i.d. and 0.2 mm i.d was not constant: flow rate was found to increase with increasing i.d. However the unit volume moved over time between the systems followed the same general trend, and was found to increase with increasing voltage up to a maximum (dependent on i.d) where joule heating started to occur. One notable observation however, was the reduction in flow rate between capillaries of 200 μm i.d which were empty and frit filled. There was a pronounced reduction in EOF through the frit filled capillary, with reduction in flow rates from 2.6 to 0.6 μL min^{-1} at 500V. Normally the inclusion of a frit within a system has been found to increase EOF due to increased electrical double layer surface area, however here the opposite was seen. This was attributed to increased friction effects within the system due to the reduced channel dimensions and frit presence. However these effects may still be put to good use. Traditionally within micro-systems homogeneous reactions have been found to occur at the liquid-liquid interface. The addition of frits may serve as a mixing area with which to allow a greater surface area of reaction and thus increase product yields.

The double layer concentration effect of the frits is perhaps most useful when considering the movement of organic substances. Here many are not seen to move readily under normal EOF conditions, or move so slowly that they are not found to be useful. The inclusion of a frit has been found to increase movement of these solvents (propanol, THF etc.).

4. Conclusions

A reliable method for the formation of porous silicate microstructures (frits) has been achieved which has been found to be useful for the immobilisation of solid catalysts as well as for EOF applications. The presence of a frit within micro-reactor systems has been found to increase EOF and also allow movement thorugh non-conducting media. In addition they may also be readily used in order to facilitate mixing within channels.

5. References

1. Terry, S.C., Jarman, J.H., and Angell, J.B, *IEEE Trans Electron. Devices*, 1979, ED-26, 1880.
2. Woolley AT, Mathies RA, Anal. Chem. 67, 1995, 3676
3. Senkan, S.M., *Nature*, July 1998, **394**, 350.
4. Cortes, H.J., Pfeiffer, C.D., Richter, B.E., and Stevens, T.S., *Journal of High Resolution Chromatography and Chromatography Communications*, 1987, **10**, 446.
5. Dean, G.A., and Cardile C.M., Journal of Materials Science Letters, **13**, 872.
6. Christensen, P.D., Johnson, S.W.P., McCreedy, T., Skelton, V., and Wilson, N.G., *Analytical Communications*, 1998, **35**, 341.

A new hydraulic stroke amplifier for microfluidic components

N. Schwesinger, S. Pobering*
TU Ilmenau, *Opal Jena GmbH

A hydraulic stroke amplifier made of silicon is described. The system consists of two membranes made of different materials. Both are connected by a non compressive liquid, which completely fills out a micro cavity. The first membrane act as driving membrane and possesses of a silicon plate and a piezoceramic plate bonded together. Applying a voltage onto the piezoceramic plate a pressure will be generated into all directions of the cavitiy. This pressure forces the output membrane to bend in a direction opposite to the cavity. In contradiction to the first membrane, the output membrane is flabby in bending and very small in cross section. The complete volume shift will be given to the bending of the second membrane due to that and the stiffness of the surrounding walls. Stroke amplifications can be achieved as result. In the experiments we found strokes at the output membrane in a range between 60µm and 120µm depending on the pressure generated at the driving membrane. These values are corresponding with an amplification factor of 20 and more in the idling cycle.
The device is made by silicon technologies and adapted microtechnologies. It can be advantageously used in several applications due to the material properties of the output membrane. New designs of micropumps, microvalves and other devices using this new working principle are described.

1 Introduction

Micropumps and microvalves are favored goals of the investigations of several scientific groups all over the world. Minewhile there exist a huge number of working principles for these devices [1-8]. Although the active parts of the micro components are different one can distinguish between preferred and less important elements. Membrans belong to the group of preferred basic elements. This is caused due to the production techniques which are possible in the micro-area. Especially silicon based microstructures comprises very often thin membranes as key element [9-14]. These membranes consisting of silicon can be produced by dry or wet chemical etching procedures. A direct conversion of electrical energy into mechanical energy is possible using membranes as basic elements.
The flow, the flow rate, the pressure as well as the flow direction of liquids in microstructures can be realized using active controlled membranes. Unfortunatelly these membranes possess in case of silicon of very stiff mechanical properties. Thus the achievable deflection of the membranes respectively the stroke that can be produced is very low. This behavior is independent from the drive that is used to generate a movement of the membrane. Consequently, in all known micropumps using membranes the flow rates are limited. An example of a typical design of a silicon based micropump is shown in fig. 1.

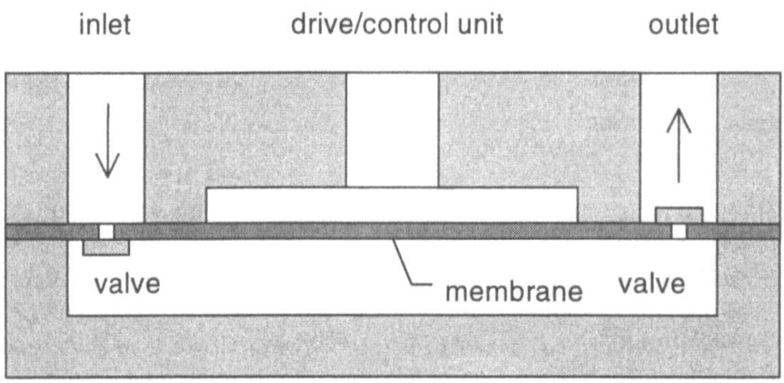

Figure 1: Basic principal of a micropump using a membrane as active driven element

The drive/control unit generate a deflection of the membrane to the bottom. A pressure will be built up that opens the valve at the outlet side, while the valve at the inlet is closed. In the following sequence the membrane will be deflected to the top. A vacuum can be generated that forces the opening of the valve at the inlet and a closing of the outlet valve

Channel diameters must be designed very small in case of microvalves using membranes as active units. Otherwise a defined flow control can not be guaranteed due to the limited deflection of these membranes. Another disadvantage of silicon membranes is based on the high hardness of the material. It is impossible in practice to seal a valve seat perfectly using a silicon membrane. Smallest height tolerances in the surface of seat area can not be balanced by the membrane. A noticeable leakage is the consequence of this behavior.

The goal of our work was focused on the development of a new working principle in microfluidic devices that overcomes the disadvantages of known active microcomponents. We tried to improve the deflection of a membrane applying the advantages of a direct energy conversion in membrane transducers. For that purpose we have designed a new „two-step" drive which contains two types of membranes. Stiff driving membranes should convert the electrical energy into mechanical energy. Membranes made of highly elastic materials should overcome the disadvantages of leakage flow and small deflections. Both membranes should be connected by a hydraulic coupling system. A basic element should be built up using standard silicon microtechnologies and modified technologies.

2 Driving principles

2.1 Basic considerations

The conversion of electrical energy into mechanical energy in technical devices requires a transducer. Such transducers can operate in different ways. Figure 2 shows several typical transducer operating modes.

As shown in figure 2 there exist some transducers which allow the conversion of the

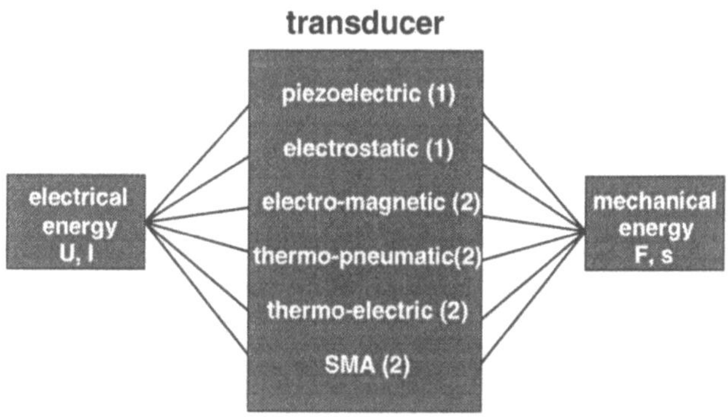

Figure 2: Electromechanical transducer principles

different energy forms in only one step (1). These transducers comprise of advantages in view of the losses which will be generated during the conversion. As more steps required to convert one energy into another as higher the losses in energy. The efficiencies of energy transducers decrease with the number of converting steps. Unfortunately, the application of one of the principles in membrane based micro components lead to not neglectible disadvantages. The results are noticeable leakage rates or low deflections of the membranes. Two step power converters (2) are a compromise to overcome the disadvantages. Unfortunately, the dynamic behavior of these transducers is not comparable to that of the first group. Furthermore, there exist no real standard technologies to produce such transducers in the micro area. Microtechnologies are not well adapted yet to the required design of the transducers of the second group.
Consequently, one has to combine the advantages of these transducer principles (1) with advantages of converting systems which allow higher deflections of membranes and reduced leakage rates. Under these circumstances it is impossible to convert electrical energy into mechanical energy in just one step. A multi-step conversion is required taking into account all disadvantages with energy losses. Membranes must serve as an active mechanical element caused by the requirements of microtechnologies one has to remain on. These elements are relatively easy to handle and to process in comparison to tongs or other mechanical elements.

2.2 Membrane drives

Considering the membranes used in transducers one can distinguish between bending dominated and stretching dominated ones. The difference of the two types is shown in figure 3.

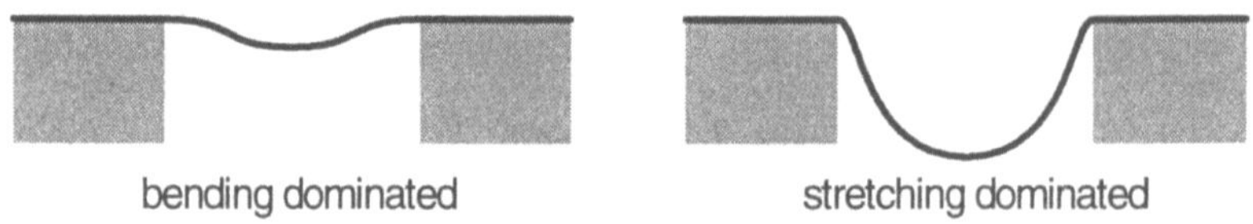

Figure 3: Differences in mechanical properties of membranes

Stretching dominated membranes are week while bending dominated membranes are mainly hard and stiff. Caused by the production technologies membranes in silicon based micro components are mainly bending dominated. Especially piezoelectric drives require relatively stiff membrane materials. Drives are designed often as bimorph structures with all disadvantages described above. Such a drive contains of a piezoceramic plate mounted on a non active plate. Applying a voltage to the piezoceramic in the direction of their polarization one can achieve a stroke ΔV of about

$$\Delta V \approx \frac{1}{3} \cdot A_M \cdot w \qquad (1)$$

- with A_M as the area of the membrane and w as the deflection of the membrane in the center. The deflection w is depending on the thickness of the piezoceramic, the thickness of the non active plate the dimensions of the membranes and mechanical properties of these materials. Typical deflections are in a range of some µm.

Stretching dominated membranes allow much higher deflections but they are not appropriate for piezoelectric drives with high forces. Therfore a combination of both membrane materials seems to be promising for new driving solutions. Such a combination is shown in figure 4.

2.3 Two step membrane drive

The system contains of both types of membranes. They are connected mechanically by means of the non compressive media which is placed inside the arrangement. A displacement or deflection of the stiff membrane can be transferred to the weak one. A stroke ΔV will be produced depending on the size of the membrane although the deflection of the driving membrane is low. The height of the stroke is given in equation 1. The pressure rises up inside the arrangement. In case of stiff housing materials the pressure respectively, the stroke will be transformed to the weak membrane. Depending on their mechanical properties it is possible to achieve a deflection. This deflection depends also on the size of the membrane, the thickness and the height of the stroke. Assuming a similar deflection mechanism of both membranes the simplified behavior of these relations is given in equation 2.

$$a_w = \frac{w_w}{w_d} = \frac{A_d}{A_w} \qquad (2)$$

Where a_w is the amplification in deflection, w_w - the deflection of the weak membrane, w_d - the deflection of the driving membrane, A_d - size of the driving membrane, A_w - size of the weak membrane.

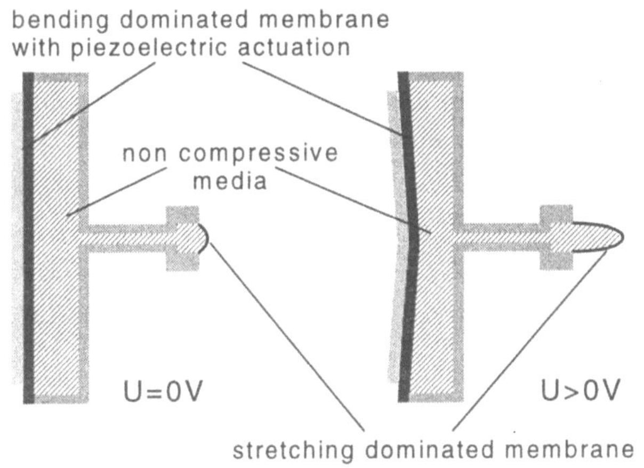

Figure 4: Two step energy converter with two different types of membranes

An amplification in the deflection is possible if the size of the weak membrane much smaller than the size of the driving membrane.

2.4 Operating modes of the two step energy converter

Reactions of the membranes were not taken into account to simulate and understand the behavior of the two step converter. Following simplifications were made:

- The driving membrane is free movable without any counter forces
- The used liquid is ideal non compressive and remains during operation under normal pressure
- Forces at the weak membrane do not influence the forces generated by the driving membrane
- The weak membrane is operating without any losses in force

Dynamic reactions were neglect in the first considerations as well. The setup of the system is shown in figure 5

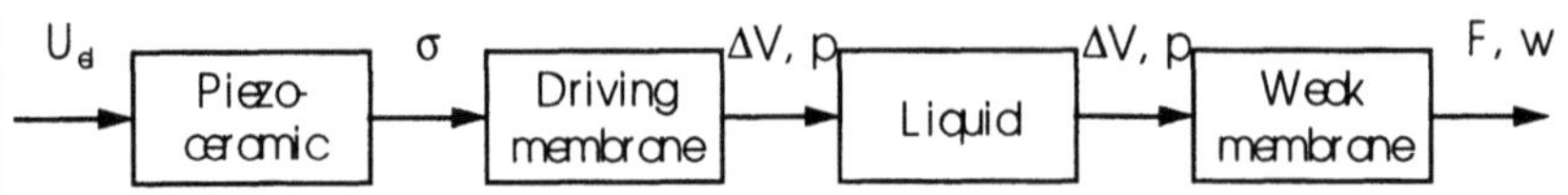

Figure 5: Components of the reaction free two step converter

Due to the inverse transversal piezoelectric effect a contraction of the piezoceramic plate occurs applying a voltage to the electrodes. This contraction is not free caused by the fixing of the piezoceramic with the driving membrane. The whole bimorph system will be bended into the direction of the passive plate. Underneath the driving membrane one can observe a stroke as well as an overpressure. The liquid in the cavity transfers these mechanical quantities to the weak and stretching dominated membrane. Caused by the different deflection effects one can find for the worst case the following relation:

$$a_w = \frac{1}{3} \cdot \frac{A_w}{A_d} \qquad (3)$$

Applying equation (3) one can find for the deflection of the weak membrane in a system free of feedback

$$w_w = \frac{3C_d}{A_d} \cdot U \cdot k_p \cdot a_w \qquad (4)$$

with C_d - mechanical capacity of the driving membrane, k_p - pressure dependent coupling factor of the driving membrane, U - electrical voltage.

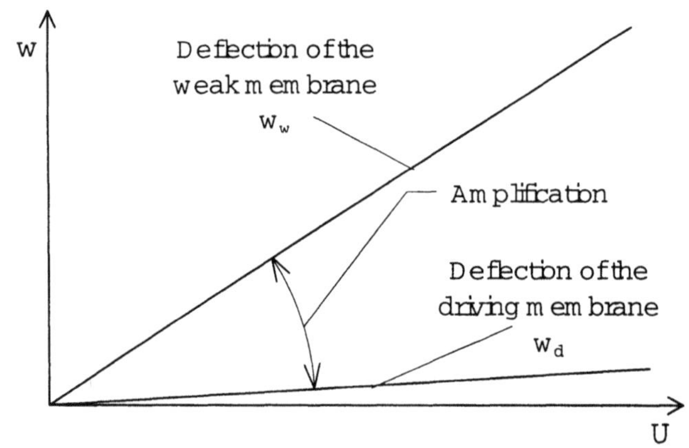

Figure 6: Behavior of a feedback free two step energy converter

The graph of this basic behavior is shown in figure 6.

One has to calculate with a feedback of the driving membrane depending on the weak working membrane under real conditions. Factors which influence the operation of the system are the following:

- External forces at the weak membrane
- Increasing pressure in the liquid
- Forces to deform the weak membrane
- Compressibility of the liquid
- Weak cavity materials

It is possible to take all these factors into consideration applying a complex and quasi static mechanical capacity to the whole system.
The amplification of the deflection is given under these conditions by:

$$w_w = \frac{3C_s}{A_d} \cdot \left(U \cdot k_p - p_0\right) \cdot a_w \qquad (5)$$

With C_S - complex capacity of the system, p_0 - external pressure.
Assuming a linear behavior of the spring represented by the membrane one can find static curves for the two step energy converter depending on the pressure as shown in figure 7.

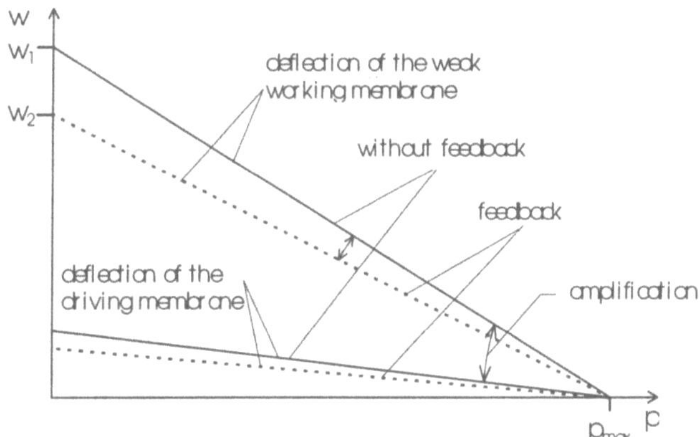

Figure 7: Static feedback reactions in a two step energy converting system

The dynamic behavior of the system is more complicated and will be not discussed in this report.

3 Experimental setup

Experiments were carried out to examine the behavior of a two step energy converting system. Samples were prepared by means of microtechnologies. The structure of a test sample is shown in figure 8.

The first silicon wafer consists partially a cavity that is filled with a liquid. The remaining silicon on the top of this wafer is the driving membrane. It is directly connected with the piezoceramic plate which is mounted by gluing on the first

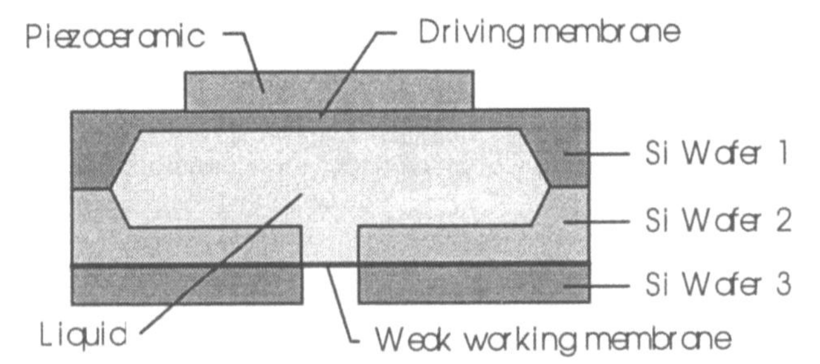

Figure 8: Silicon based test structure of a two step energy converter

wafer. Wafer 2 contains the second part of the cavity as well as a hole etched through the silicon. The hole is made in a dry etching step while the cavity was produced by wet chemical etching procedures. Both silicon wafers are bonded together using a silicon direct bonding technology. Wafer 3 contains just one hole etched through the silicon. A membrane with a high elasticity is arranged between the wafers 2 and 3. A melting- gluing procedure was used as mounting technology for the membrane. Therefore, the material of the membrane was adapted to this process and to the mechanical requirements at this point. We have used LATEX as weak working membrane in our experiments. The dimensions of the driving membrane were fixed at 1cm x 1cm. The size of the holes in the silicon wafers 2 and 3 were 1mm x 1mm and 2mm x 2mm respectively.
Huge problems were found in the filling of the cavity with the liquid. Mainly gas bubbles were generated in this process. Other solutions than water were used. New designs with a system of capillaries were developed. At least we have found a technology which allows a gas free filling of the cavities.

4 Results and discussion

As first we have characterized the driving membrane separately to exclude disturbing influences due to possible external forces.
Deflections up to 6 μm were measured at an operating voltage of 150V at the bimorph structure. The average deformation was calculated as 40nm/V. A volume stroke of about 1,3nl/V was observed without any external pressure. A maximum

pressure of 37.3kPa was achieved at working conditions. A coupling factor k_p=248 Pa/V were found.
The combination of both membranes gives an imagination of the performance of the two step transducers. In Figur 9 is shown the deflection of the weak membrane depending on the pressure produced from the driving membrane. The size of the weak membrane was 2mm x 2mm in square. As to be seen the deflection increases with the pressure nearly linear. The highest deflections were found at about 25 kPa. This value corresponds to a deflection of the driving membrane of about 4µm. Therefore, the amplification of deflection is in a range of about 16. The expected value can be calculated from the relations of the membrane sizes. Theoretical amplifications of 25 should be reached.
The differences can be caused by losses due to the deforming of the second membrane and gas bubbles in the cavity.
Decreasing the size of the second membrane leads to an increase of their deflection. Maximum deflections of about 120 µm were observed while the driving membrane was deflected of about 3 µm. Under these circumstances a stroke amplification factor of 40 was found. The resonance frequency of the whole system was about 21kHz.

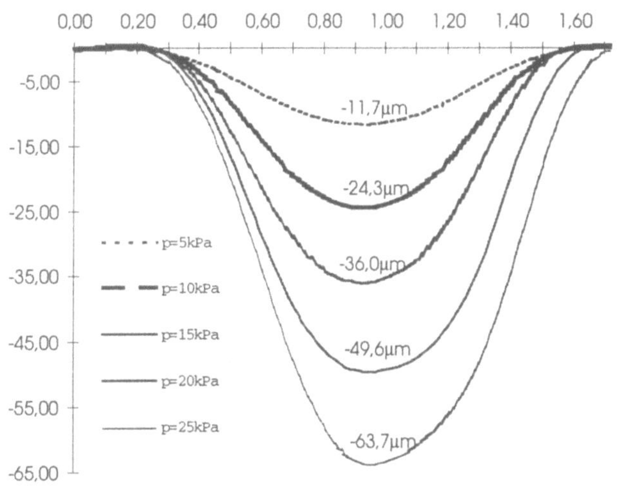

Figure 9: Deflection of the weak stretching dominated membrane depending on the pressure in the cavity

The two step energy converter as shown opens new areas for the application of microcomponents although the amplification factor of the stroke is not as high as possible. Instead of small movements one can operate now with strokes which are acceptable in macro technologies. The handshake" between micro and macro by means of modified microtechnologies should lead to a much higher acceptance of microcomponents in the macroworld. New applications could take profit from this development. Another advantage of the presented solution is the use of a selected membrane material. Due to high elastic properties it is possible to achieve a perfect sealing in fluidic components. New generations of micropumps and microvalves can be designed using this behavior. Other applications could use the linear transducing behavior for way - controlled solutions.

5 References

[1] Bart et al. (1990), Microfabricated electrohydrodynamic pumps, in: Sensors & Actuators A, 21-23(1990), p. 193-197.

[2] Bosch et al. (1993), A silicon microvalve with combined electromagnetic/electrostatic actuation, in: Sensors & Actuators A, 37-38(1993), p. 684-692.

[3] Ahn, Allen (1995), Fluid micropumps based on rotary magnetic actuators, in: MEMS' 95, proceedings , Amsterdam, 408-412.

[4] van de Pol, F.; Breedfeld, P. (1990), Bond-Graph modeling of an electro-thermo-pneumatic micropump, in: MME, Berlin 1990, S. 19-24.

[5] Sato, Shikida (1994), An electrostatically actuated gas valve with an S-shaped film element, in: J. Micromech. Microeng., 4(1994), p. 205-209.

[6] Wagner, B.; Quenzer, H. (1996), Bistable microvalve with pneumatically coupled membranes, in: IEEE [Hrsg.], MEMS' 96, proceedings, San Diego, 384-388.

[7] Schwesinger, N., Planarer Tintenstrahldruckkopf mit piezokeramischem Antrieb, in: Feinwerktechnik Mikrotechnik Meßtechnik F & M, 101. Aufl., 1993, 456-460.

[8] Schwesinger, Bechtel (1998); Micropump for viscous liquids and muds, in: SPIE-proceedings vol. 3515, Santa Clara, 40-45.

[9] Forke (1991), Eine geregelte Membranpumpe als mikrosystemtechnische Lösung, in: Erstes Symp. Mikrosystemtechnik, Regensburg, 1991, S. 131-140

[10] Hirata, S.; Ishii, Y. (1996), An ink-jet using diaphragm microactuator, in: IEEE [Hrsg.], MEMS' 96, proceedings, San Diego, 418-423.

[11] Jerman (1991a), Electrical-Activated, Normally-closed Diaphragma Valves, in: Transducers 1991, Technical Digest p. 1045-1048.

[12] Judy et al. (1991), Surface micromachined micromechanical membrane pump, in: MEMS' 91, proceedings, 182-186.

[13] Richter, Zengerle (1993), Properties and applications of a membrane pump with electrostatic drive, in: "Actuator 93" Bremen 1993, p. 28-33.

[14] Stemme, Stemme (1993); A novel piezoelectric Valve-less Fluid Pump, in: Transducers 1993, digest p. 110-113, Yokohama.

Part 6

Microreactors for Energy Generation and Storage

Microchannel Chemical Reactors for Fuel Processing Applications. II. Compact Fuel Vaporization

Anna Lee Y. Tonkovich,[a] Sean P. Fitzgerald, Jennifer L. Zilka,
Mike J. LaMont, Yong Wang, David P. VanderWiel, Robert S. Wegeng

Pacific Northwest National Laboratory,[b] PO Box 999, Richland, WA 99352

Abstract

A fuel processor is a critical element for the deployment of automotive fuel cell power systems. One component of the fuel processor, the compact gasoline vaporizer, was demonstrated at full-scale using commercial-grade gasoline. The full-scale process volume was less than 0.3 liters; it vaporized nearly 300 mL/min of gasoline, which is sufficient to support a 50-kW$_e$ fuel cell. The reduction in hardware volume was made possible using a microchannel reactor-based design; the compact process hardware is roughly an order of magnitude smaller than conventional technology.

Introduction

A fuel processor produces hydrogen in sufficient quantity and quality for a fuel cell that produces electrical power. The desired fuel for automotive power applications is gasoline, because it reduces costly infrastructure changes to the world's fuel supply system. Furthermore, the constraint of on-board hydrogen generation requires that a process be compact, lightweight, and inexpensive. The microtechnology-based fuel processor meets each constraint, and will enable the commercialization of automotive fuel cell power systems.
The conversion of gasoline to hydrogen in a fuel processor includes a series of process steps, including vaporization, primary conversion of the fuel to synthesis gas, the water gas shift reaction, and carbon monoxide removal. The full-scale gasoline vaporizer, defined at a capacity of 50 kW$_e$, will be discussed in this paper.

Microchemical reactors enable miniaturization of the fuel processor because they minimize heat and mass transfer resistance. Thermal conduction and mass diffusion distances in the described microchannel reactor range between ten and several hundred microns. Conduction and diffusion distances vary from ten to

(a) Author to whom all correspondence should be addressed: aytonkovich@pnl.gov.
(b) PNNL is operated by Battelle for the U.S. Department of Energy.

several hundred millimeters in a conventional reactor. Slow heat and mass transfer dominate the operation of conventional reactor, and thus they will require a larger hardware volume to produce the same quantity and quality of hydrogen for the fuel cell.

Background

Compact fuel cell power systems have potential applications in markets such as residential and commercial power, space exploration and defense. Automotive power systems are one of the most promising applications because of the variety and size of potential markets present in this field. Compared to conventional internal combustion engines, fuel cells are more efficient, can utilize a variety of fuels, result in reduced emissions and can be compact and modular. With the advent of federal and state legislation requiring the implementation of alternative fuels and reduced emissions vehicles into the next century, corporate and government research in this area is rapidly advancing fuel cell power system technologies. Federal initiatives such as the 1992 Energy Policy Act (EPACT) and the 1993 Partnership for a New Generation of Vehicles (PNGV) require the development of efficient and clean technologies that can operate on a wide range of fuels. Adoption of these technologies is required to begin in 2010, with California mandates requiring that 10% of cars sold in the state must be zero-emission vehicles by 2003. Ultimately, the adoption of fuel cell transportation technologies has the potential to decrease the country's dependence on imported crude oil and to decrease nationwide CO_2 emissions, while generating jobs and decreasing consumer fuel costs.

Motivated by this legislation, as well as requirements in foreign countries, the private research sector has responded with a wide range of newly developing technologies. Many automobile manufacturers have invested in electric power systems research, and are preparing to bring fuel cell and hybrid electric vehicles to the market within the next 5 to 10 years. Early fuel cell powered vehicles contained on-board hydrogen tanks, which may not be easily filled by the consumer and which present potential safety concerns. The latest fuel cell vehicles operate on methanol that, while more easily available than hydrogen, does not utilize the existing infrastructure of gasoline distribution and has a much lower volumetric energy content compared to gasoline (or diesel).

The development of compact, efficient fuel processors to generate hydrogen is critical to the deployment of PEM fuel cell power systems. These fuel processors must have the ability to convert conventional fossil fuels, for which a supply and distribution infrastructure already exists. Further, the use of fuel cells in automotive applications requires that the fuel processing system be small, lightweight and modular in nature. Figure 1 is a schematic of a typical fuel processing system for a PEM fuel cell. The first step in this process involves vaporization of a liquid hydrogen fuel such as methanol, ethanol, gasoline, diesel, mixtures of these, or others. Once fully vaporized, the hydrocarbon fuel is converted to syngas, a mixture of carbon monoxide and hydrogen. Options for this primary conversion step include partial oxidation, steam reforming or autothermal reforming. A comparative paper on the attributes of partial oxidation versus steam reforming is reported by Argonne National Laboratory [1]. Since carbon monoxide (CO) poisons the fuel cell, the next step involves secondary conversion of CO to CO_2 in a water-gas shift reactor, which has the added benefit of increasing the hydrogen content of the fuel stream. Additional CO removal to concentrations below 10-ppm is achieved in a third reactor using preferential oxidation or CO absorption. A complete system description of the fuel processor and fuel cell is presented by Delphi [2]. A third fuel processing system configuration is under development at Arthur D. Little [3]. Each of the reported systems is based upon conventional fixed-bed reactor technology, which does not scale linearly with throughput because of inefficient heat and mass transfer. These conventional fuel processors are expected to be at least an order of magnitude larger than the microchannel-based fuel processor. Heat and mass transport limitations, which dramatically increase the size of the conventional technology [1], are mitigated with the emerging technology described in this paper.

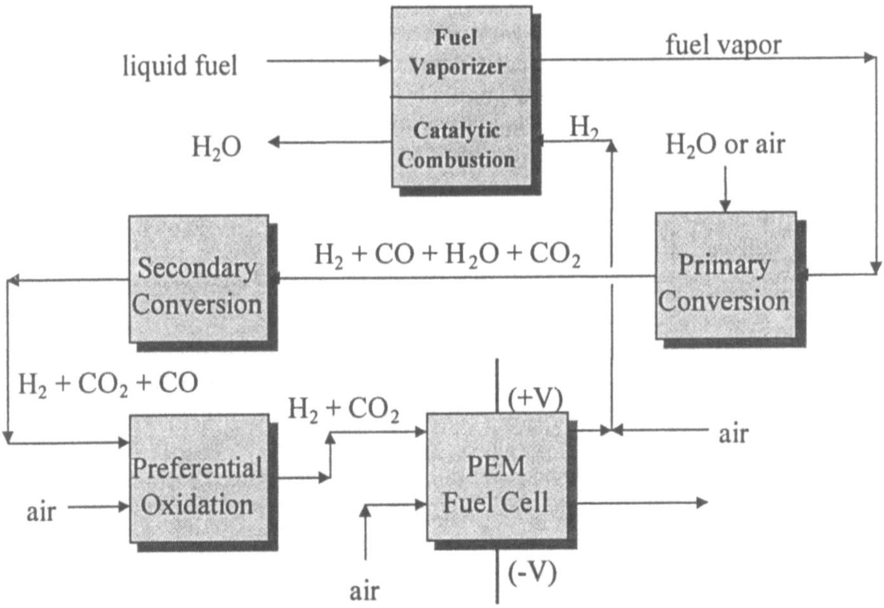

Figure 1. Schematic of Fuel Processor/Fuel Cell Power System

Microchannel chemical reactors [4,5] utilize micron-scale dimensions and integrated engineered catalyst structures to bring reactants, catalysts and heat transfer surfaces into intimate contact. In such systems, the thermal and mass diffusion paths are on the order of a thousand times shorter than those found in traditional packed-bed reactors. The resulting decreased resistances to heat and mass transfer allow the utilization of more highly active catalysts, which take advantage of intrinsically rapid surface reaction kinetics. The application of these principles has been successfully demonstrated for a water-gas shift reactor [6] and microchannel heat exchangers [7]. The characteristics of these systems result in highly efficient and compact reactors. In addition, these systems can be designed without the large pressure drops commonly associated with small or complex flow structures.

A microchannel fuel vaporizer has been developed and demonstrated to supply gasoline vapor at a rate sufficient to feed a 50-kW_e fuel cell power system. Based on early designs [8, 9], this vaporizer utilizes integrated catalytic combustion of a simulated PEM fuel cell anode effluent to supply heat to vaporize liquid gasoline. Weighing only 1.8-kg and displacing about 0.3 L, this compact unit vaporizes gasoline at nearly 300-mL/min. These results show that this component is more than an order of magnitude smaller than a conventional fuel vaporizer [10]. Based upon our results with this and other component investigations, we project a complete system volume of less than 8 L to produce hydrogen at a sufficient rate and quality to produce 50 kW_e from a PEM fuel cell.

Experimental

A fuel processing system requires a vaporizer to prepare liquid fuels for primary conversion to hydrogen and carbon monoxide. The latent heat of vaporization required for a given fuel is provided through combustion of a dilute hydrogen stream from the fuel cell anode effluent.

$$H_2\,(g) + O_2\,(g) \rightarrow H_2O\,(g) \qquad \Delta H_r = -13.4\ \mathrm{kJ/kg}$$

$$\text{gasoline (l, STP)} \rightarrow \text{gasoline (v,215 °C)} \qquad \Delta H_v = 293\ \mathrm{kJ/kg}$$

Catalytic combustion of the dilute hydrogen (6 to 8%) remaining in the anode effluent stream is a facile reaction using palladium. No pre-heating is required because hydrogen combustion initiates at room temperature.

The full-scale vaporizer consists of the body (an integrated reactor and heat exchanger) and one cover plate. The body measures 9.2-cm x 9.9-cm x 2.54-cm (or 3.6" x 3.9" x 1"). The device weighs 1.8-kg. The body is fabricated using a laminated design (patent pending). The reactor section has four parallel engineered-catalysts. The inlet flue-gas stream (anode effluent plus air) enters through the top of the unit and is separated into four parallel catalyst chambers. Each of the four streams of the hot combustion gas is then directed through a slit in the monolith footer to the header of the heat exchange cell below. A cell is defined as an array of parallel microchannels that operates in parallel to another array of parallel microchannels. After passing through the heat exchanger cell, the flue-gas stream exits the top of the vaporizer. Heat conducts through the metal web that separates the two arrays of microchannels. The cooling fluid (evaporating gasoline) enters the unit through the bottom of the vaporizer. The gasoline moves countercurrent to the hot combustion gas stream in the heat exchange cell, before exiting through a port on the bottom side of the unit.

Each catalyst monolith has a cross section (normal to the direction of flow) of 1.61 cm^2 and the heat transfer area between the gasoline vaporization cell and the flue-gas cell is 7.2 cm^2. The channels on the flue-gas side are 254-microns wide and have an aspect ratio of 18:1. The channels on the vaporizing side of the heat exchanger plate are 254-microns wide and have a 10:1 aspect ratio.

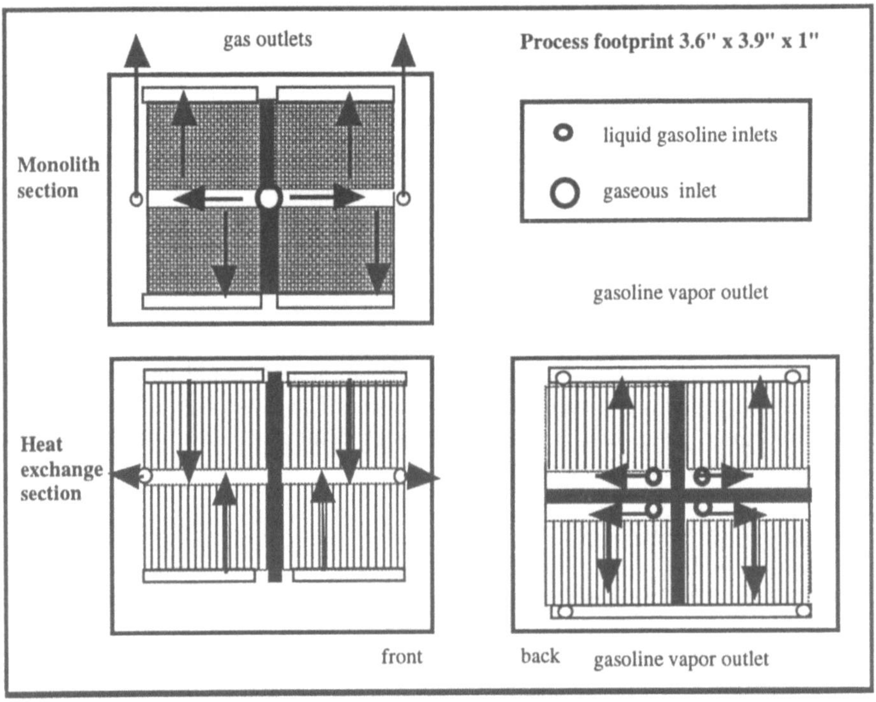

Figure 2. Full-scale gasoline vaporizer (four parallel reactor and heat exchanger cells)

Results and Discussion

The full-scale microchannel vaporizer was evaluated with consumer-grade gasoline. The device occupies a volume less than 0.3-L and is sufficient to vaporize gasoline for a full 50-kW_e fuel processing system. Performance data is given in Figure 3.

The full-scale microchannel vaporizer evaporated nearly 300-mL/min of consumer-grade gasoline through the use of four parallel arrays of microchannels, also referred to as 4 parallel cells. The maximum available flue-gas (anode effluent stream plus air for combustion) for the vaporizer was roughly 1400-

SLPM for a 50-kW$_e$ system. The hydrogen content of this stream was roughly 6.7%. The heat released from the combusted hydrogen was contained within the flue-gas stream, approaching temperatures of 600°C. The heat was then transferred through four parallel arrays (cells) of microchannels to vaporize the gasoline feed stream. Countercurrent flow of flue-gas and gasoline was employed. The pressure drop in the microchannels on the flue-gas side was roughly 2 psi, and well below 1 psi on the gasoline side.

The microchannel gasoline vaporizer was also evaluated under part-load conditions, as shown in Figure 3. The diamond symbol represents the gasoline flow as a percent of the total capacity (~300-mL/min). The device was operated at 25% load, 50% load and 100% load. In all cases, complete vaporization was achieved (as denoted by the exit temperature of the gasoline – square symbols). The highest measured boiling point for the selected commercial-grade gasoline was 209°C. For testing purposes, complete vaporization was taken at 20°C above the highest boiling point, or at a temperature of 229°C.

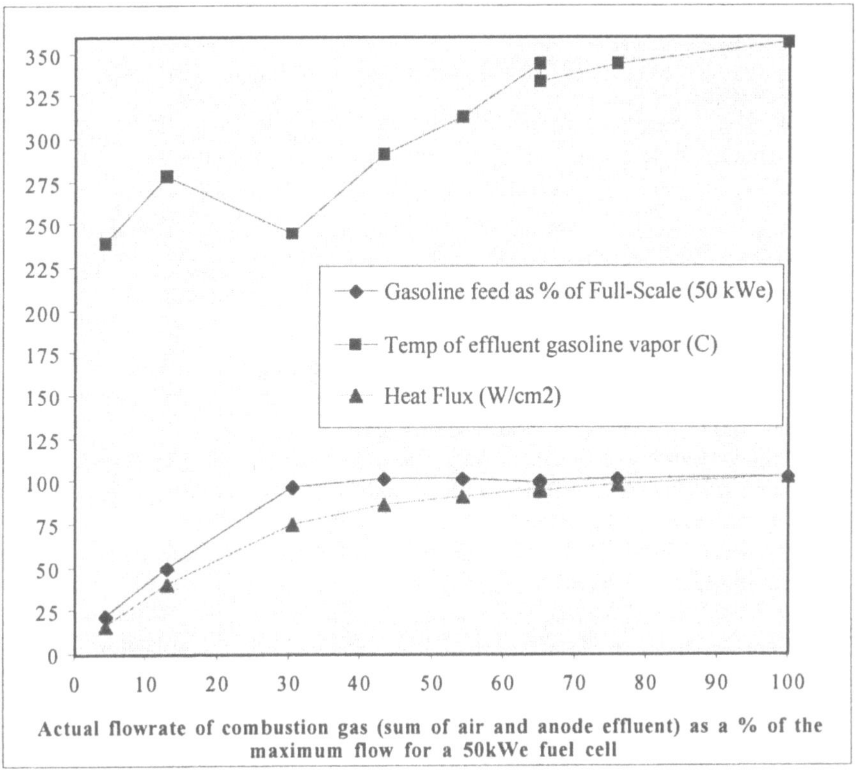

Figure 3. Full-scale gasoline vaporizer performance data

The total amount of flue-gas fed to the vaporizer was also varied between the maximum (~1400-SLPM) and a partial load (5% of maximum). For example, the first data point on Figure 3 reflects a flue-gas feed rate of ~70-SLPM (5% of full load) and a gasoline feed rate of ~75-mL/min (25% of full load). At these conditions, complete vaporization occurred, and a heat flux of roughly 20 W/cm^2 was measured between the arrays of microchannels (shown with the diamond symbols). The addition of extra flue-gas (beyond that needed to vaporize gasoline) superheated the vapor stream. In general, the heat flux during the superheating stage asymptotically approached 105 W/cm^2.

Conclusions

A full-scale gasoline vaporizer was demonstrated for subsequent use in an integrated fuel processing system. The process hardware vaporized sufficient gasoline for a 50-kW_e fuel cell in a volume less than 0.3-L and weighed 1.8-kg. A complete microchannel-based fuel processing system for automotive applications is expected to be less than 8-L.

Acknowledgments

This work was supported by the EE-Office of Transportation Technology in the U.S. Department of Energy. Pacific Northwest National Laboratory is operated by Battelle for the U.S. Department of Energy under contract DE-AC06-76RLO 1830.

References

1. Kumar R and S Ahmed. 1995. "Fuels Processing for Transportation Fuel Cell Systems." First International Symposium on New Materials for Fuel Cell Systems, Montreal. ANL/CMT/CP-84932,.
2. Delphi Automotive Systems Corp 1996. "Research and Development of a Proton-Exchange Membrane (PEM) Fuel Cell System for Transportation Applications." DOE/CH/10435-02.
3. Arthur D Little. 1994. "Multi-Fuel Reformers for Fuel Cells Used in Transportation." DOE/CE/50343-2.
4. Matson DW, PM Martin, AY.Tonkovich, and GL Roberts. September 1998. "Fabrication of a Stainless Steel Microchannel Microcombustor Using a Lamination Process." *Proceedings of the SPIE Micromachined Devices and Components IV Symposium*, Vol. 3514. Santa Clara, California.
5. U. S. Patent pending.
6. Tonkovich AY, JL Zilka, MJ LaMont, Y Wang, and RS Wegeng. 1999. "Microchannel Reactors for Fuel Processing Applications. I. Water Gas Shift Reactor." *Chem. Eng. Sci.*, (in press).

7. Tonkovich AY, CJ Call, DM Jimenez, RS Wegeng, and MK Drost. 1996. "Microchannel Heat Exchangers for Chemical Reactors." *Proceedings of the 1996 National Heat Transfer Conference*, Houston.
8. Wegeng,RS and AY Tonkovich. October 1997. "Microchannel Fuel Vaporizer for Fuel Cells." *Proceedings of the Annual Automotive Technology Development Customers' Coordination Meeting*, Dearborne, Michigan.
9. Tonkovich AY, DM Jimenez, JL Zilka, M LaMont, Y Wang, and RS Wegeng. March 1998. "Microchannel Chemical Reactors for Fuel Processing." *Proceedings of the Second International Conference of Microreaction Technology*, New Orleans.
10. Arthur D Little, private communication, 1997.

Microstructured Catalysts for Methanol-Steam Reforming

P. Pfeifer[1], M. Fichtner[1], K. Schubert[1], M. A. Liauw[2] and G. Emig[2]

[1] Forschungszentrum Karlsruhe GmbH, Herrmann von Helmholtz Platz 1, D-76344 Eggenstein Leopoldshafen, Germany
[2] Lehrstuhl für Technische Chemie I der Universität Erlangen-Nürnberg, Egerlandstr. 3, D-91058 Erlangen, Germany

Abstract. For the application of endothermic methanol-steam reforming e.g. in automobile systems based on hydrogen-driven PEM (proton exchange membrane) fuel cells, active and microstructure-compatible catalyst layers were examined.

As a first approach the catalysts were applied to a structured aluminum foil with microchannels. Nanoparticles were used as the active component and promoter. A porous support based on inert nanoparticles for wet impregnation was also used. Methods of handling and the process of immobilization of the nanoparticles were developed. Additionally the sintering characteristics of the nanoparticles forming the adhesive layer were determined and the morphology of such catalyst layers was examined. Different copper and palladium catalysts were prepared with varying sintering temperatures and mass ratios of the active species, promoter and support. Methanol-steam reforming was carried out in a laboratory setup with a specially designed test reactor for a stack of 1-4 electrically heated foils.

Activity, selectivity as well as deactivation were determined as a function of time and compared for the different catalyst systems. Palladium showed better stability and higher conversion (about 70%) of methanol than copper catalysts at residence times of 0.125-0.25 s. Nevertheless, due to the preparation method the selectivity for methanol-steam reforming was lower for palladium systems.

Keywords. Nanoparticle, Microstructured Foil, Copper, Palladium Catalyst, Methanol-Steam Reforming, Microstructured Reactor, Microchannel Reactor

1. Introduction

Among other alternatives, low-emission passenger transportation in the future is possible by electrically driven vehicles which are equipped with a low-temperature fuel cell (PEM). The fuel cell is fed with hydrogen-rich gas during operation. The hydrogen itself can be provided by on-board conversion of a hydrogen rich liquid. To fulfill this task, methanol-steam reforming to carbon dioxide and hydrogen is discussed as a suitable method [6].

Equation 1 represents the overall formula of the endothermic reforming of methanol with water, whereas for the used catalysts equation 2 is the main secondary reaction [4,13].

$$CH_3OH + H_2O = 3H_2 + CO_2 \qquad (1)$$

$$CO_2 + H_2 \quad = CO + H_2O \qquad (2)$$

First attempts with on-board reforming in prototype cars have already been made by manufacturers of passenger cars. Conventional reformers e.g. tube reactors filled with catalyst pellets have the disadvantage of a slow response to load changes because of the low heat transfer within and between the catalyst particles [2,3,4,5].

An improvement may be achieved by the use of microstructured reactors which offer a high specific surface for heat transfer and an efficient temperature control for the catalyst [10,11,12].

In the first step of such a development, it is necessary to find a microstructure-compatible catalyst with a high activity and stability during operation time. As a first approach catalyst layers based on immobilized and sintered nanoparticles were prepared and studied. In preliminary examinations, sintering of the nanoparticles and the morphology of sintered samples were investigated. In further experiments a process for immobilization of nanoparticle mixtures in microstructures was developed. After immobilization of the particles, a porous and adhesive layer was generated by suitable sintering conditions.

Microstructured foils with catalyst layers prepared with this method were investigated with regard to methanol-steam reforming in a special test apparatus. Conversion, selectivity and the deactivation behavior were examined in particular.

2. Catalyst Preparation

As carrier, microstructured aluminum foils with channel diameters of 100x100 microns separated by fins of 50 microns were used. To provide active layers for the steam reforming of methanol, palladium or copper (reduced or oxidized) were chosen as the active component. Both Pd and Cu are well-known catalysts of the desired reaction [13]. The promoter always was zinc oxide, and in some experiments TiO_2 was used as an inert support. As an alternative preparation method to nanoparticle mixtures, an impregnation of the TiO_2 support with copper and zinc or of the promoter ZnO with palladium was performed.

To handle the nanocrystalline powders and prepare a microstructure-compatible layer of a mixture of oxides, it was necessary to disperse the particles. To obtain stable dispersion which can be used without sedimentation, a polymer such as hydroxyethylcellulose is suitable. In the first step a parent solution of the polymer with a high concentration in deionized water was prepared. After dilution with water the particles were added and the solution was stirred to crush agglomerates. As the following steps and the mass ratios are different for the samples, Table 2.1 shows the chronological procedure for the preparation of the catalysts.

3. Experimental Setup

For experimental catalyst screening a laboratory setup was developed, which allows testing of the active layers at high concentrations of the reactants methanol and water such that an easy upscaling will be possible in the future. The reactant feed system consists of expansion bellows where the liquid is pressurized while being kept separated from gas. Mass flow controllers for the liquids of water and methanol in the range up to

10 g/h control liquid flow to the evaporators. For the vaporization of the reactants a evaporator with precise temperature control was developed so that the concentration of vapor could be held constant within a range of 5-10% of the readings desired. Supplementary mass flow controllers for gases can be used to add other gases such as helium to the feed in the range up to 500 Nml/min.

Table 2.1. Preparation steps of different catalyst samples

Preparation method:	1	2	3	4	5
Sample:	CuO/ZnO	CuO/ZnO	$CuO/ZnO/TiO_2$	Impregnated TiO_2	Pd/ZnO
Step 1	Dispersion of mixture	Dispersion of mixture	Dispersion of mixture	Dispersion of TiO_2	Dispersion of ZnO
Step 2	Drying	Drying	Drying	Drying	Drying
Step 3	Sintering 450°C 5h	Sintering 550°C 5h	Sintering 550°C 5h	Sintering 450°C 5h	Impregnation $Pd(Ac)_2$/toluene
Step 4				Impregnation $Cu/Zn(NO_3)_2$	Drying
Step 5				Calcining 550°C 5h	Calcining 450°C 5h
Step 6					Reduction 500°C 5h 1 vol.% H_2
Mass ratio	1:1 CuO:ZnO	1:1 CuO:ZnO	39:39:22 $CuO:ZnO:TiO_2$	25:2:3 $CuO:ZnO:TiO_2$	1:99 Pd:ZnO
Foil loading	20±2 wt.%	20±2 wt.%	26±2 wt.%	20±2 wt.%	20±2 wt.%

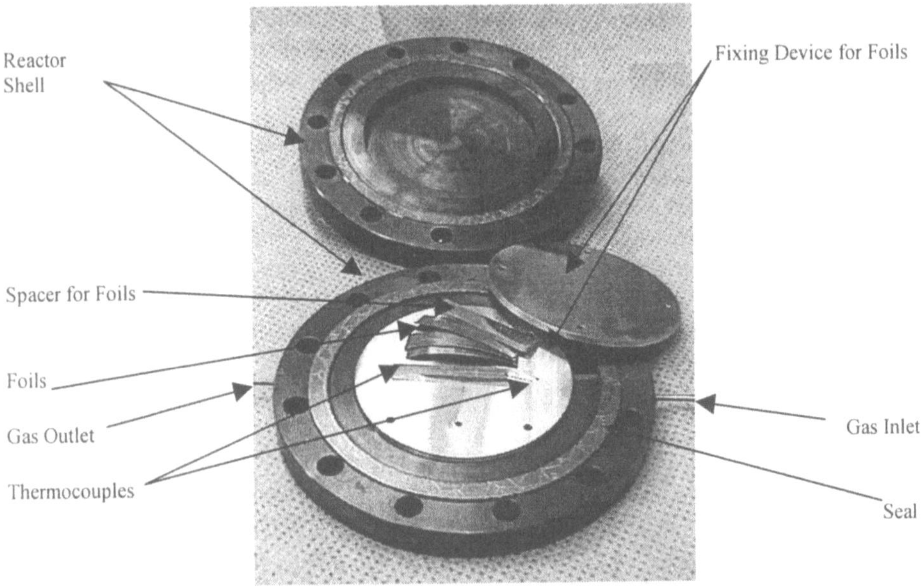

Figure 3.1. Test reactor for 1-4 foils

The test reactor (see Figure 3.1.) was designed for a maximum of 4 stacked foils with 80 channels each and a maximum length of 64 mm. For fast and easy catalyst screening the stack was electrically heated. At the reactor outlet, pressure control up to 2 MPa was possible in the laboratory setup using a pneumatic control valve and a calibrated sensor. For determining the product composition a Micro-GC was used. Vapor calibration of the GC was carried out with a saturator.

The carrier gas for the GC was helium. CO_2, CH_3OH and H_2O as well as other possible components were separated with a Poraplot Q column: The determination of CO and hydrogen was carried out with a molecular sieve column. The reproducibility of the amount of H_2 was found to be low so that the conversion of methanol could only be determined by relating the carbon species detected. Moreover the COcontent was mostly lower than CO_2 by a factor of 10 so that the conversion and CO_2 yield were nearly identical and the CO content was neglected for the determination of the initial molar flow of methanol.

For each catalyst screening experiment 2 foils were prepared simultaneously and tested at a molar ratio of about 1 methanol :2 water with approximately 60 vol.% helium at a residence time of 0.25 s calculated for 250°C and 0.3 MPa.

4. Results

4.1. Properties of the Nanoparticles

The used nanoparticles are commercially available. Their characteristic properties are shown in Table 4.1.

To obtain information about the formation behavior of sintered layers, dilatometer measurements were conducted. Figure 4.1. shows the change of length for rough-pressed samples of 50% theoretical density. For TiO_2 and ZnO there is a sharp change in length at temperatures of 400°C and 450°C, respectively. For copper oxide the length of the sample changes slightly even at low temperatures of 150°C. At a temperature of 550°C the length decreases rapidly with temperature. According to these results, the minimum temperature for sintering of the layers is 450°C.

Table 4.1. Properties of the nanoparticles according to the certificate of analysis

Material	BET Surface [m^2/g]	Average Particle Size [nm]	Density (Gas Pycnom.) [g/cm^3]
CuO	41	23	6.3
ZnO	17	63	5.6
TiO_2	58	26	3.95

High specific surface areas for the catalytic reaction are desirable and so BET measurements with nitrogen of CuO/ZnO mixtures with a mass ratio of 3:5 were carried out. One sample was mixed only, the other samples were pressed before (40% theoretical density) and sintered in the dilatometer at 550°C, 650°C and 750°C for one hour. Table 4.2. shows the density and degree of porosity of these samples together with the specific BET surface.

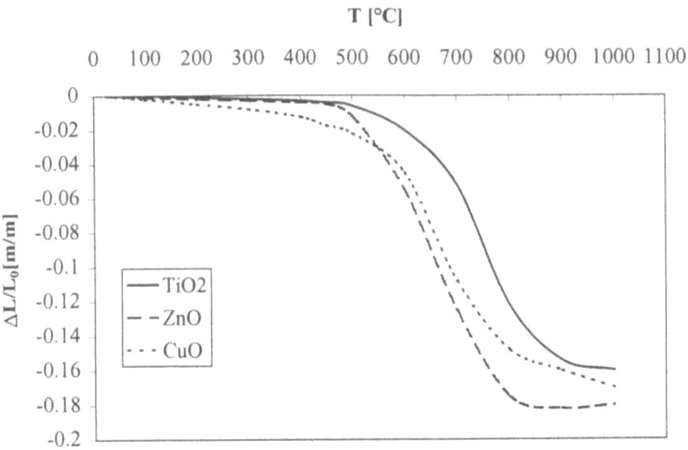

Figure 4.1. Dilatometer measurements for CuO, ZnO and TiO_2 nanoparticles

Table 4.2. BET surface and theoretical density of CuO/ZnO samples sintered in the dilatometer

Temperature for Sintering [°C]	BET Surface [m^2/g]	Density [g/cm^3]	Degree of Porosity [%]
-	19.43	-	-
550	6.36	2.45	58
650	2.59	3.83	35
750	<1	5.05	14

The degree of porosity was calculated on the basis of bulk copper oxide and zinc oxide. XRD measurements of the samples showed good crystallinity of the nanoparticles. Traces of Cu_2O were also detected for the copper particles (see Figure 4.2.)

To investigate the surface area of immobilized and sintered nanoparticles on microstructured aluminum, samples were prepared according to preparation method 1 and BET measurements of these samples were conducted. The BET surface area was 9.3 m^2/g of the coating material so that a surface enhancement factor of about 240 was reached. The morphology of the coating was investigated by Hg porosimetry and physisorption measurements. Only macropores in the range of 100 nm in diameter and no mesopores or micropores were found in the coating.

The regularity of coating was inspected by light-optical microscopy of horizontal cuts. An average thickness of the layer at the bottom of the channels of 20±10 μm was determined.

Figure 4.3. shows a SEM of the distribution of the single particles at the surface.

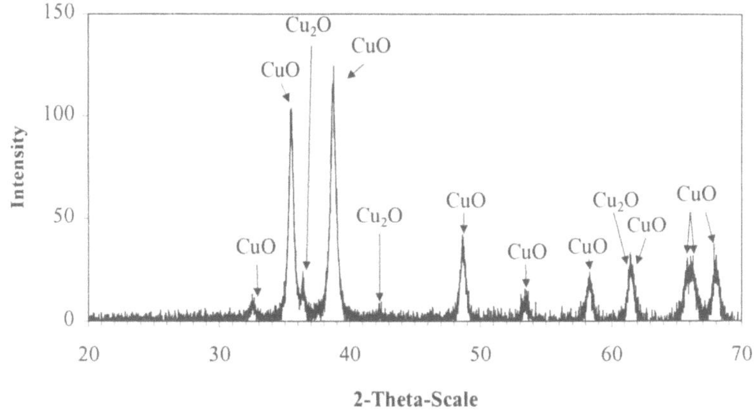

Figure 4.2. XRD of CuO nanoparticles

Figure 4.3. SEM of a CuO/ZnO layer according to preparation method 1

4.2. Copper Catalysts

It is mentioned in literature that pre-reduced copper has a lower activity for methanol-steam reforming than copper reduced in the reaction mixture [7]. The active species in the oxidation state of copper is assumed to be Cu_2O [7].

To investigate this, three catalysts with CuO/ZnO were prepared according preparation method 1. The yield of carbon dioxide was determined for the first catalyst at temperatures between 210°C and 285°C and a heating rate of 0.5°C/min (Figure 4.4). It can be seen that the reaction starts at about 230°C and that there exists a local and reproducible maximum of yield at a temperature of about 235-250°C. The determined CO_2 yield was confirmed by a measurement of hydrogen at 285°C, when the main reaction (1) of methanol and water to CO_2 and H_2 was assumed to take place.

The second catalyst was heated up to the temperature where the local maximum of the yield was reached. Then it was cooled down to a temperature of 210°C. When cooling down the reactor the second catalyst showed neither a local maximum in yield nor the same ignition temperature.

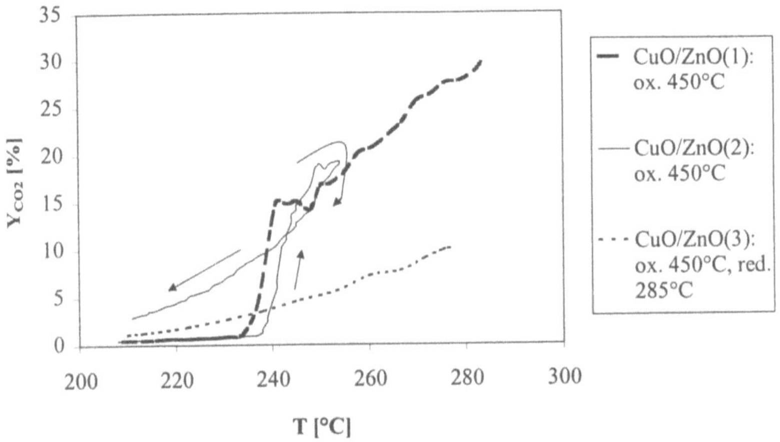

Figure 4.4. Yield of carbon dioxide (Y_{CO2}) for oxidized and reduced copper-zinc catalysts

The third copper catalyst was heated up to 285°C and reduced in situ under hydrogen. After a reduction time of 1 hour it was cooled down to 210°C. With this catalyst the CO_2 yield was found to be much lower than for the pre-oxidized copper catalysts (Figure 4.4.).

According to these results, it can be suggested that the local maximum in the conversion of methanol for the first catalyst is due to the reduction of copper and that there is an active state of copper nanoparticles between fully oxidized Cu(II) and reduced Cu(0), which is in accordance with the observations made in literature for other copper catalysts [7].

In addition to these experiments with different pretreatment procedures, the influence of sintering temperature and catalyst support was investigated (Figure 4.5.). With CuO/ZnO nanoparticles (preparation methods 1 and 2) higher sintering temperatures (450°C => 550°C) led to lower activity, which was probably due to the lower surface area of the layer. The CuO/ZnO/TiO_2 sample had the same copper load and was sintered at 550°C (method 3). The activity at high temperatures was nearly the same as for CuO/ZnO sintered at 550°C (method 2).

The yield of CO_2 for impregnated TiO_2 (method 4) with a smaller copper content was higher than the yield for the other Cu samples. This might be attributed to mixed oxides of copper and promoting [15] zinc. Another explanation could be a higher specific copper surface area which is important for high conversion [15].

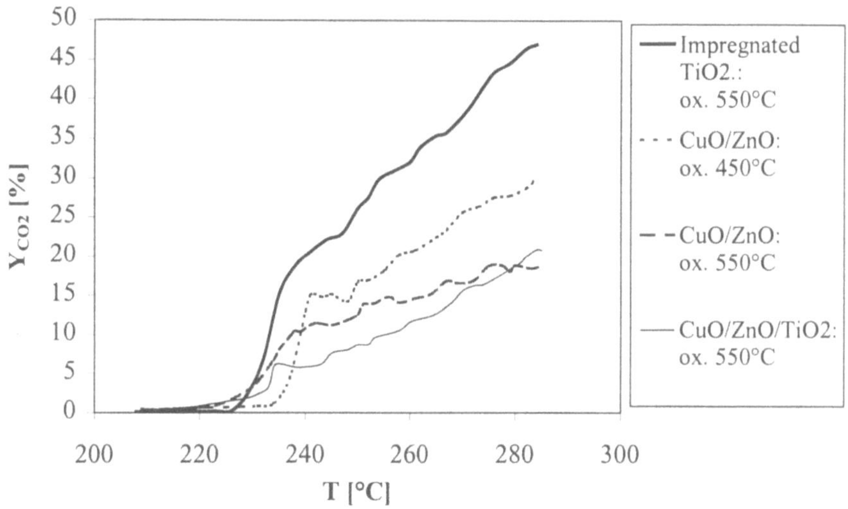

Figure 4.5. Comparison of the CO_2 yields of different copper catalysts

4.3. Palladium Catalysts

The palladium catalysts according to preparation method 5 were found to be more active for the conversion of methanol. Deactivation was less fast than for copper catalysts (see also chapter 4.4.).

In addition to a first optimization of the Pd/ZnO system, methanol-steam reforming was carried out at different vapor velocities and the same residence time in order to examine whether the reaction was limited by diffusion. Moreover possible advantages due to higher velocities in the channels were to be determined.

For these experiments, three catalyst foils of the same kind were prepared. One foil was cut into two pieces and another was cut into 4 pieces to obtain a stack of foils of half and a quarter of the length of one foil, respectively. The foils where then tested under the same conditions of volume flow and reactant concentration. Despite an experimental error for the CO_2 yield of 5%, the results shown in Figure 4.6. indicate a slight tendency to higher yields of CO_2 with increasing vapor velocity. In the case of a regular diffusion limitation, a stronger dependency of the yield would be expected. Hence, the observed weak tendency may also be caused by increased axial diffusion in the case of lower vapor velocities.

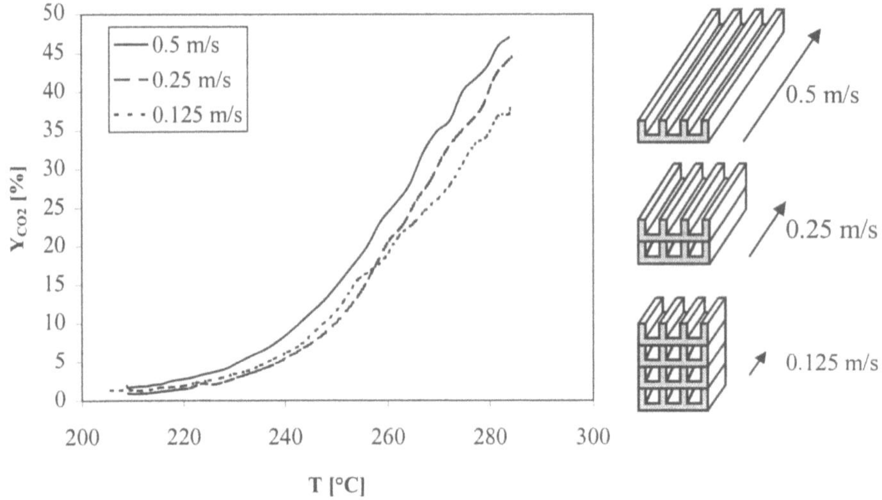

Figure 4.6. Yield of carbon dioxide vs. temperature at a constant residence time of 0.125 s and different channel velocities

4.4. Comparison of Different Catalysts

When examining differences between the catalysts, deactivation is an important factor. Deactivation was carried out at 285°C under the conditions used for catalyst screening (see chapter 3). Figure 4.7. shows the deactivation as a function of time. It can be seen that copper catalysts deactivate rapidly especially in the first few hours. Palladium catalysts show slower deactivation with time.

To reactivate the copper catalysts, reoxidation was carried out, but this did not restore the original activity. The mass of the copper catalysts was determined before and after the reaction in the oxidized state and no mass loss was detected. Hence, decreasing activity cannot be attributed to mass loss. SEMs showed regions where no porosity was left after reaction. For this reason, the strong deactivation at the beginning could be due to an initial reduction of copper and copper catalysts could be subject to sintering during reaction. However, this conjecture has not yet been fully proven (see also [1]).

The palladium catalysts were stable when a small amount of oxygen was added to the reactants. A possible explanation for this behavior could be the removal of adsorbed carbon species from the surface by oxidation.

This interpretation is supported by the results shown in Figure 4.8. where the actual CO concentrations at the reactor outlet are compared to the equilibrium for the water gas shift reaction (data see [14]). Due to the preparation technique, palladium also crystallizes on the aluminum carrier where no contact to zinc oxide exists. Consequently, no alloy can be formed between Pd and Zn when reducing the sample in flowing hydrogen. Having this amount of elementary palladium, the reaction then proceeds with a decomposition to CO [8].

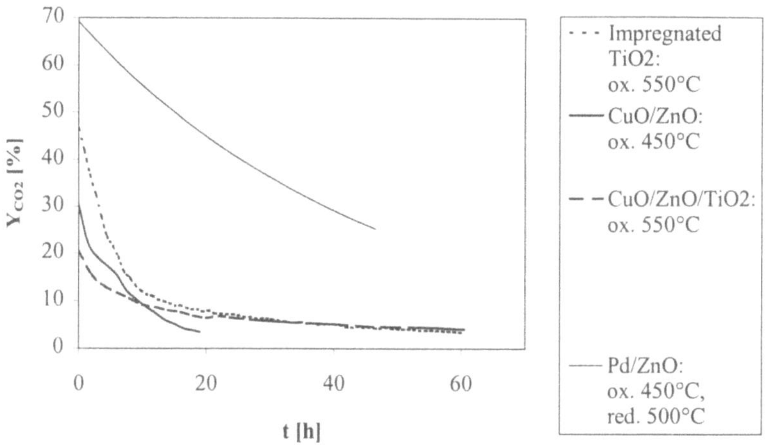

Figure 4.7. Deactivation of Pd and copper catalysts as a function of time, residence time 0.25s

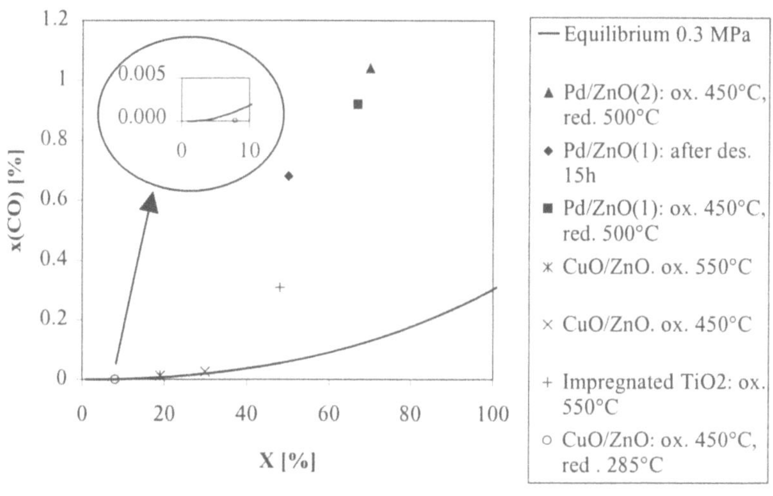

Figure 4.8. CO content in the reactor outlet as a function of the conversion of methanol to CO_2 and H_2 and comparison with equilibrium compositions for the water gas shift reaction, residence time 0.25 s, temperature 285°C

Copper catalysts exhibited a lower selectivity for CO. The selectivity depended on the oxidation state and on whether or not TiO_2 was used as support. For reduced copper the CO content was lower than the equilibrium concentration, which is in agreement with the statement that the reaction does not proceed with the formation of carbon monoxide.

Because of the reduction of CuO to Cu(0) during the reaction (which was confirmed by XRD measurements), small amounts of methanol can be partially oxidized (CO formation) in the case of pre-oxidized copper catalysts.

Pure TiO_2 layers were investigated and no conversion to CO or CO_2 was found.

5. Conclusion

Microstructure-compatible catalytic coatings based on immobilized and sintered nanoparticles can be applied to convert a mixture of methanol and water vapor into hydrogen and carbon dioxide. The active component can be provided either by adding catalytically active nanoparticles to the particle mixture or by wet impregnation of support particles or the sintered layer. Conversions of up to 70% were obtained at residence times of only 0.25 s. Due to the strong deactivation and the high CO contents, the main work for the future will focus on the improvement of selectivity and stability of the catalysts.

Acknowledgement
The authors thank the Institute for Materials Research III (Karlsruhe Research Center) for the dilatometer and XRD measurements. They also express thanks to HVT-HZ (Karlsruhe Research Center) for the SEMs. A Liebig grant (FCI/BMBF) to one of the authors (M.L.) is gratefully acknowledged.

References

[1] Autenrieth-R, Strobel-B; DE patent 19727589A, 1999
[2] Colsman-G; Diss. KFA Jülich, Report Jül-3127, 1995
[3] Düsterwald-HG; Diss. KFA Jülich, Report Jül-3452, 1997
[4] Düsterwald-HG, Höhlein-B, Kraut-H, Meusinger-J, Peters-R, Stimming-U, Chem. Eng. Technol., Vol.20 (1997), p.617-623
[5] Höhlein-B, Boe-M, Bogild-Hansen-J, Bröckerhoff-P, Colsman-G, Emonts-B, Menzer-R, Riedel-E; J. Power Sources, Vol.61 (1996), p.143-147
[6] Höhlein-B, Stimming-U; Chemie Ingenieur Technik, Vol.67 Iss.10 (1995), p.1306-1309
[7] Idem-RO, Bakhshi-NN; Ind. Eng. Chem. Res., Vol.33 Iss.9 (1994), p.2056-2065
[8] Iwasa-N, Masuda-S, Ogawa-N, Takezawa-N; Appl. Catal. A-General, Vol. 125 Iss.1 (1995), p.145-157
[9] Iwasa-N, Masuda-S, Takezawa-N; Reaction kinetics and catalysis letters, Vol.55 Iss.2 (1995), p.349-353
[10] Peters-R, Düsterwald-HG, Höhlein-B, Meusinger-J, Stimming-U; Proc. 1st Inter. Conf. Microreaction Technol. 1997, p.27-34
[11] Schubert-K, Bier-W, Brandner-J, Fichtner-M, Franz-C, Linder-G; Proc. 2nd Inter. Conf. Microreaction Technol. 1998, p.88-95
[12] Schubert-K, Bier-W, Keller-W, Linder-G, Seidel-D; Symp. Vol., DSC-Vol.19 (1990), ASME, p.189-197
[13] Takezawa-N, Iwasa-N; Catal. Today, Vol.36 Iss1 (1997), p.45-56
[14] Twigg-MV; Catalyst Handbook, Sec. Ed. 1996
[15] Wang-D, Ma-L, Jiang-CJ, Trimm-DL, Wainwright-MS, Kim-DH; Stud. Surface Sci. Catal., Vol.101 (1996), p.1379-1387

Fuel Cells for Low Power Applications

C. Hebling, A. Heinzel, D. Golombowski, T. Meyer, M. Müller, M. Zedda

Fraunhofer Institute for Solar Energy Systems, Oltmannsstr. 5, 79100 Freiburg, Germany

1. Introduction

Fuel cells are under discussion for a large variety of applications as efficient and clean power generators today. In general, fuel cells are easily scalable due to their modular design and furthermore, the storage capacity and the output power can be combined independently. This enables the integration into portable devices or even in micro systems, which is favourable in those cases when batteries do not meet the respective demands. In particular, fuel cells offer advantages like variable geometry, high storage capacity, no self discharge and the potential for low production costs which make them interesting candidates for an environmentally friendly energy source [1].
Two main options for energy conversion, both of which are based on the membrane technology, are available for fuel cell systems: hydrogen as an energy supply from hydride cartridges or methanol from an appropriate methanol container feeding the direct methanol fuel cell (DMFC). If a simple design with air-breathing electrodes and a hydride cartridge is chosen as hydrogen supply, nearly no auxiliary devices are necessary. Therefore, the system efficiency does not depend on the power and is even high at partial load of small cells in the Watt-range. Both concepts will be described in this paper.
To obtain higher output voltages compared to a single cell, two approaches for a fuel cell system were realized at Fraunhofer ISE: First, a banded structure with 25 internally series-connected fuel cells delivering up to 20 V was constructed. This design is an interesting alternative to the conventional stack design, since a certain number of cells can be interconnected in one plane. In principle, it is possible to combine the banded structure with the stack principle resulting in a banded structure stack, thus allowing a very compact geometry. The second approach is a single large fuel cell combined with a special DC/DC-converter in order to transform for example 0.7 V input voltage to 12 V output voltage. The efficiency of such a 10 W DC/DC-converter is between 80 and 90%, depending on the input power. With this design, various controlling schemes such as charge controlling, motor controlling or fixed output voltage supply may be implemented additionally.

2. Hydrogen-based fuel cells

2.1. Design of the fuel cell

The operating principle as well as the main components of a hydrogen/air membrane fuel cell are schematically depicted in figure 1. Basically, one of the most important components is the ion conducting membrane, which is coated with the electrocatalysts for electrochemical hydrogen oxidation and oxygen reduction. Such membrane/electrode assemblies (MEAs) have to achieve a very good electrical conductivity even at room temperature which can be realized by humidified gases. Therefore, the water management is one important subject, which is solved preferably by including the water produced by the electrochemical reaction.

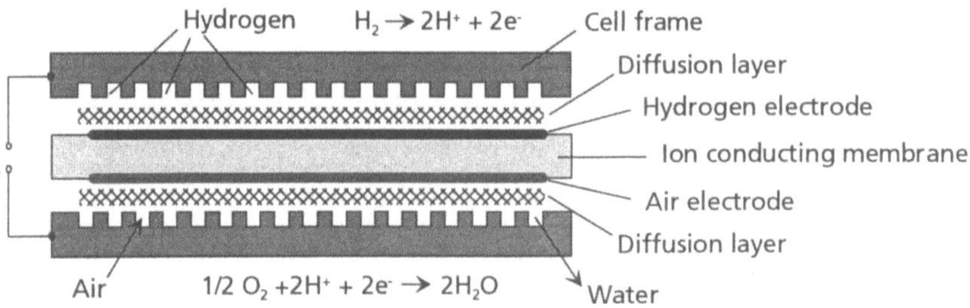

Fig. 1: Operating principle of a membrane fuel cell

The membrane thickness ranges from 30 - 100 μm, the electrodes (catalyst layer, diffusion layer and porous current collector) add another 150 μm on each side. However, the dimensions of the cell are dominated by the cell frame, as the gas supply of the active area of the cell and the removal of the product water require gas channels with dimensions which are usually in the mm range in order not to show a significant pressure drop for the gas flow.

2.2. Computational Fluid Dynamics

For low power applications and smaller active areas, all dimensions should be reduced to a minimum since low weight, flat design and a small volume are the most important requirements. In order to evaluate the crucial parameters with respect to flow and transport phenomena for such a miniaturized cell, three-dimensional flow-simulations were carried out with computational fluid dynamics (CFD).

Hereby, the flow area is divided into small control volumes – in this case about 200.000. The Navier-Stokes-equations, that describe the conservation of mass, momentum and energy, are integrated over each control volume. Subsequently, the set of difference-equations have to be solved iteratively for given boundary conditions, either for stationary flows with constant boundary conditions or transient flows with varying boundary conditions.
Additional transport equations e.g. for the vapour-pressure of the water or the oxygen component of air can be set up and solved simultaneously, which results in time- and space-resolved information of these parameters. Models, e.g. for the formation of water on the electrodes and its re-diffusion into the gas flow can be implemented. Furthermore, the diffusion layers can be modelled as porous media. The geometry of the four simulated meander-like flow channels as well as the results of the first simulations are shown in fig. 2; the isolated four cross-sections in the foreground are shown in fig. 3.
First calculations concentrated on the transport of the oxygen component of air from the flow channel through the diffusion layer to the electrode. Using a constant oxygen supply, different diffusion-

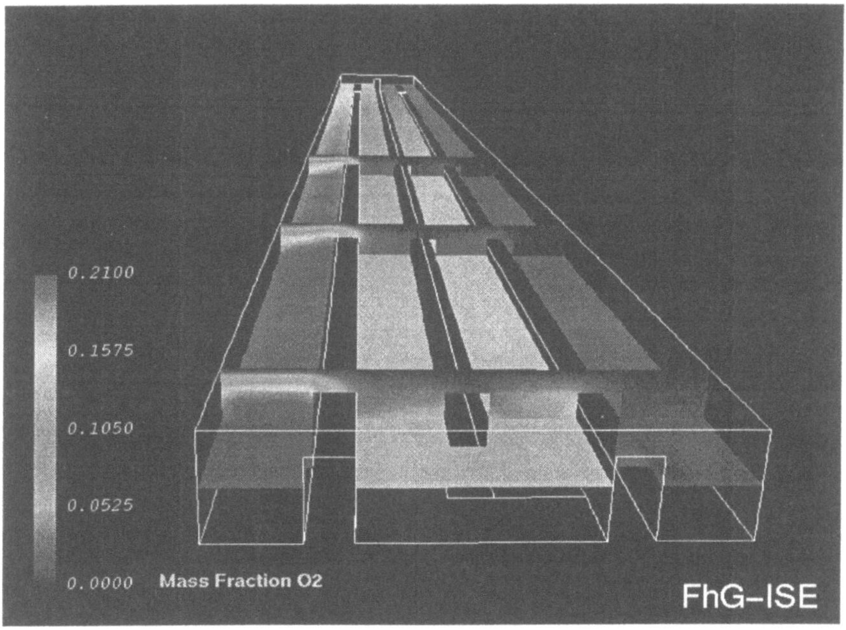

Fig. 2: Three-dimensional simulations of a flow-field including a diffusion-layer. The oxygen-inlet is located at the beginning (bottom) of the left flow channel, the outlet at the end of the right flow-channel. Three planes perpendicular to the gas flow are showing the lateral oxygen distribution due to it's transfer through the diffusion-layer.

layer thickness and flow channel cross sections were simulated. The boundary conditions were set as follows:
Mass flow: 3.37E-6 kg/s, Fluid: air with an initial oxygen mass fraction of 21 %; Electrode kinetics: infinite i.e. oxygen mass fraction on electrode surface is zero; Diff.-layer: carbon paper (Toray TGP-H-120, 150 µm thickness); Flow channel: 1x1 mm^2 (0.5 x 1 mm^2); Total length 240 mm (4 x 60 mm^2).
Various cross sections featuring the results of simulations on the influence of the diffusion layer thickness and the flow channel cross-section on the lateral oxygen transition are shown in fig.3. The depletion of the oxygen is reduced when a larger cross-section is chosen. This, of course, also depends on additional boundary conditions like e.g. gas pressure and current density, which corresponds to the oxygen consumption. The lateral oxygen transition via the diffusion layer is reduced with a thinner diffusion layer and a larger cross-section of the flow channel.

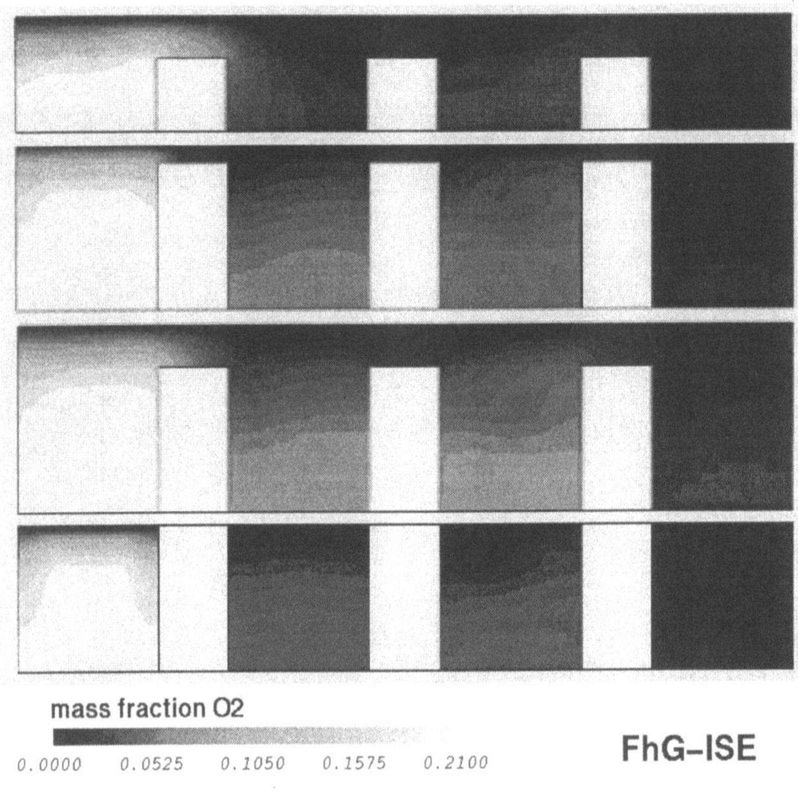

Fig. 3: Flow-simulations of the 4 gas channels with cross sections of 0.5 x 1 mm^2 and 1x1 mm^2, diffusion layers of 300 µm, 150 µm thickness and no diffusion layer, respectively.

With this powerful simulation tool, the optimal design of flow fields for various geometries depending on application will be calculated with respect to the crucial parameters like the gas composition, pressure, temperature, gas kinetics or the smoothness of the surfaces.

2.3. Hydrogen storage

For hydrogen supply in the small power range, low temperature hydride materials are the best choice. An adsorption isotherm of a commercial available hydride material is shown in fig. 3. A flat pressure plateau is suitable for selecting a simple pressure regulation device. The level of this plateau can be tailored by modifying the AB_5-alloy with A=Lanthanum, B=Ni.

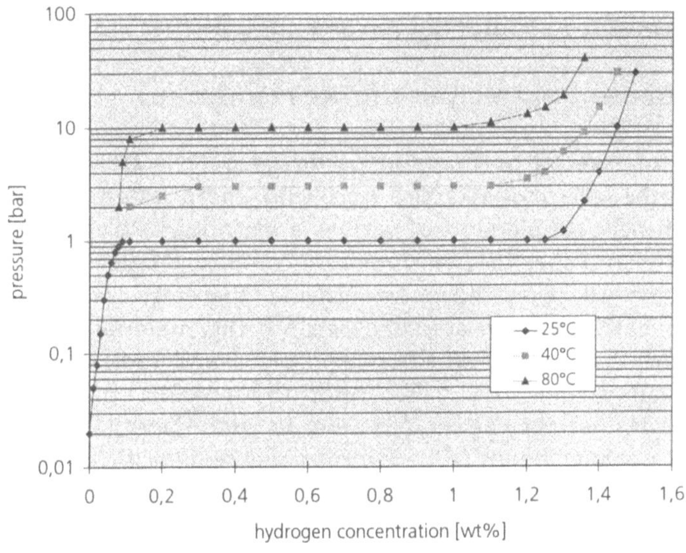

Fig. 4: Concentration/pressure isotherms for Hydralloy E60/0 (GfE comp., Nürnberg)

3. Direct methanol fuel cell

The direct methanol fuel cell offers the possibility of a simple storage system: liquid fuel and air are directly fed into the cell, and power is generated even at room temperature. This type of fuel cell can be realized today as well, but the power density is still quite low. The main reasons for the significant difference in performance compared to hydrogen as a fuel are:

- the low catalytic activity of all known electrocatalysts with respect to methanol oxidation at low temperature
- the methanol crossover from the fuel side to the air side due to the permeability of the membrane materials. This methanol loss also influences the fuel conversion efficiency
- the polarisation loss of the air electrode which is even worsened by the methanol, leading to a mixed potential

In the future development the goals are to increase the efficiency and the power density. In addition to the cell design, the main subjects for R&D activities are related to the methanol crossover through the membrane. Membranes for high operating temperatures are one approach for larger systems, offering the advantage of better methanol oxidation kinetics, lower methanol transfer and thus higher cathode potentials. Diffusion barriers (metallic layers) are one of many approaches pursued in order to overcome the crossover problem.

3.1. Development of a polyelectrolyte membrane

New types of layered polymers in particular polyelectrolytes with a higher degree of crosslinks are receiving more attention as an important functional element of various applications. Polyelectrolytes have to withstand specific environmental conditions especially in fuel cells. First of all, they have to provide thermal, chemical and electrochemical long term stability. Second, the material has to be inert towards oxidation, reduction, hydrolysis and radical reactions which prevail within fuel cells. Third, the polymer backbone of polyelectrolytes needs to be rigid in order to endure mechanical stress. Fourth, the polymer needs to be hydrophobic so that in combination with ionic side-groups, the polyelectrolytes attain only a small degree of swelling and stay insoluble.
As temperatures above the boiling-temperature of water in the DMFC were approached, polyelectrolytes showed worsening stability and decreasing ionic conductivity, and the existing polymers did not meet the required properties. At present, the commercially available polymer Polybenzimidazole (PBI) is a promising candidate for improvements on a physical and chemical basis. This is due to it's stabilizing effects of aromaticity and rigidity. Although PBI was known for it's good material characteristics, the overall performance in fuel cells still has to be improved [2-4]. In addition, proton conductivity has to be incorporated into the polymer.
In order to improve the various properties of PBI with respect to fuel cell application, a new technique has been examined at Fraunhofer ISE. Recently, layer-by-layer absorbed polyelectrolyte films are gaining interest in many areas [5-9]. Based on electrostatic attraction between

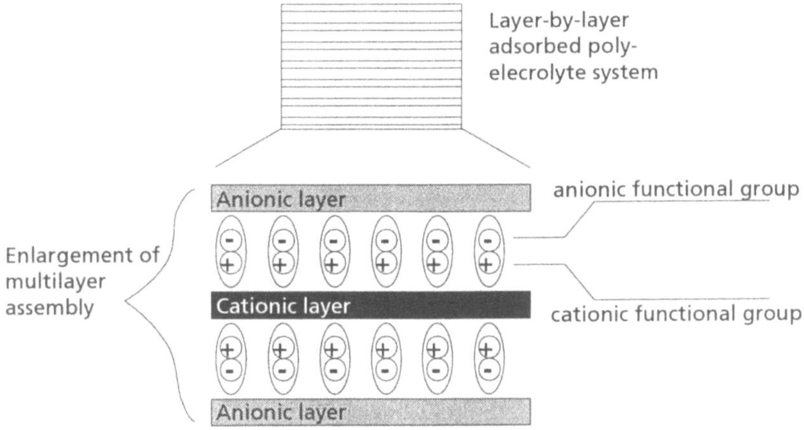

Fig. 5: Schematic drawing of multilayer system

opposite charges, polyanions and polycations can form multilayered films by alternating deposition. Figure 5 depicts the principle of film-forming between anionic and cationic layers of polyelectrolytes. It is expected, that multilayer assemblies of bipolar polyelectrolytes may offer promising effects on methanol crossover of DMFC-PEMs [5]. Furthermore, the general concept of multilayer formation may improve the physical endurance of fuel cell membranes with respect to stress and temperature endurance.

In order to establish a multilayer system we modified PBI to form a cationic and anionic functionalised polymer. In general, PBI is a high performance polymer which is known for its resistance to oxidation, reduction, hydrolysis and temperature. The introduction of functional groups must not alter those properties toward impairing qualities. Thus, the introduction of stable cationic and anionic side groups to the polymer backbone is mandatory.

The idea is to make a multilayer system by a specific dipping process [7]. A layer-by-layer deposition may result in the electrostatic adsorbation of polyelectrolytes to form alternating layers of opposite charge which is shown in fig. 5 and 6. Fig. 6 describes the structure of a multilayered membrane in detail. A three dimensional view upon the cross-section of three layers of anionic and cationic polyelectrolytes depict the structure within the membrane alongside each layer. Ionic crosslinking may culminate in a very strong interaction between the layer-structure. In addition, such strong interactions may support better film-forming characteristics.

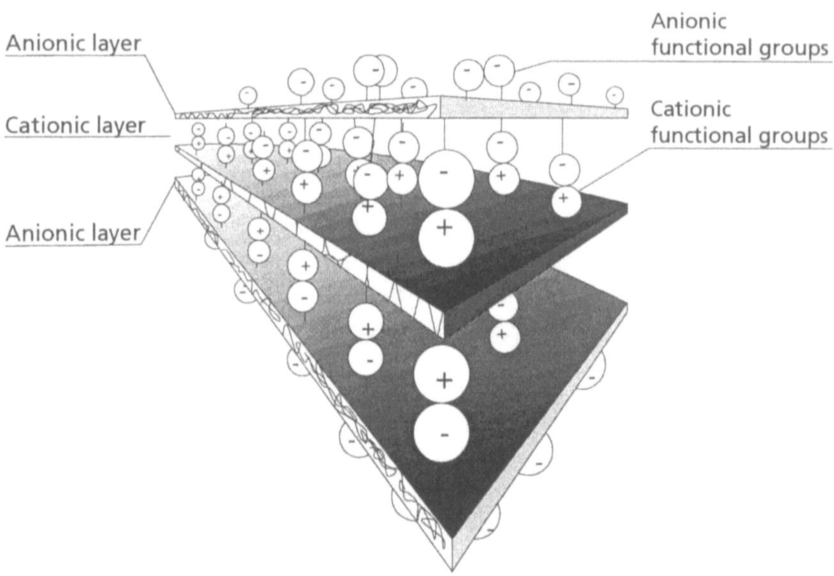

Fig. 6: Three-dimensional view depicting a multilayer assembly

3.2 Energy density of fuel cell - systems compared to Li-Ion battery

As fuel cells are a competitor to batteries, advantages and disadvantages have to be taken into consideration for this new development:

The best system in terms of rechargeable batteries on the market is the Li-Ion battery. The fuel cell system is advantageous if a high energy to power ratio is required. The properties of the system are dominated by the hydrogen storage system, hence, an improvement in storage capacity of the hydride material leads to an extension of the possible application field. Today, remote sensors and signals, control systems for the transport of valuable goods and portable computers and cameras are interesting products.

The storage capacity of methanol is attractive - though a stochiometric mixture of water and methanol (1:1) has to be used for the acid system of membrane fuel cells even in the ideal case. In reality, even values of 10:1 are used. The energy density is given in Table 1 in comparison to the rechargeable Li-Ion battery and the primary Li/$SOCl_2$-System.

Tab. 1: Energy densities of fuel cell - systems and Li-Ion- systems

	Wh/l, chemical energy	Wh/l, electrical energy	Wh/kg, chemical energy	Wh/kg, electrical energy
Methanol	4 300		5 300	
Methanol/water 1:1	2 150		2 650	
Methanol/water tank, + fuel cell	~2 100	800*	~2 300	900*
Li-Ion battery		210		100
primary Li.battery		1280		650

* 40% conversion efficiency is assumed for the direct methanol fuel cell

The values for the energy density show that competition with primary batteries is possible only in cases where other advantages like flexible choice of energy and power, environmental advantages compared to batteries, easy recycling of the plastic components of DMFC-systems after removal of the traces of noble metal catalyst, or the expected very low fabrication cost in mass production count more than energy density.
Future developments will show if the mentioned advantages will bring the DMFC into market for the low power and low cost segment of energy supply. An interesting field of application emerges for micro-energy systems since more and more decentralized intelligence is required in future products, ranging from telecommunication to smart cards.

4. Fuel cell systems for high output voltages

4.1 The banded structure fuel cell

A fuel cell system with a so-called banded structure fuel cell was realized as one approach for a power supply for an electronic device with a typical power consumption of 10 to 20 W [10]. The main advantages of this design are the compactness compared to a fuel cell stack and the high output voltage compared to a single cell. The construction principle is depicted in fig. 7 [11]. It could be shown that the banded structure concept is able to reduce the necessary stack volume significantly. Furthermore, the series resistance of the total system is on the same order of magnitude compared to a normal stack.

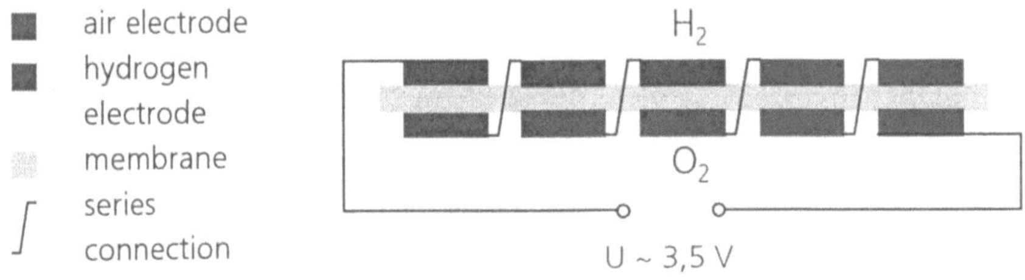

Fig. 7: Series connection of several cells in a plane

This approach enables an easy H_2-supply for e.g. 5 cells in 1 plane since the cross flow through open air channels can be maintained by low power fans (nearly air-breathing). The optimized cell geometry is described in [12]. Any combinations whatever of banded structure and stacked fuel cells are conceivable - the system design only depends on the available space and the necessary output voltage.

4.2 The combination of a single fuel cell with a DC/DC - converter

Another possibility of gaining the appropriate voltages is the combination of a single or asymmetrical twin-cell with a highly efficient DC-converter [13]. This simplifies the system design and increases the reliability in the long term. Additionally, the problem of the water management in a complex combination of fuel cells is avoided. The thickness of such a fuel cell is in the several mm-range and the area is defined by the needed final output power.

The DC/DC-converter shown in Fig.8 which was designed for high power density transforms 0.7 V (at 15 A) to 12 V output voltage with an efficiency of over 90 %. The overall size still can be scaled down and fitted to an existing empty space in an electronic device. Furthermore, additional electronic functions such as charge control, motor control of a small fan or a fixed output voltage can be implemented. This approach is an interesting alternative to the banded structure when only extensive and flat cavities are available.

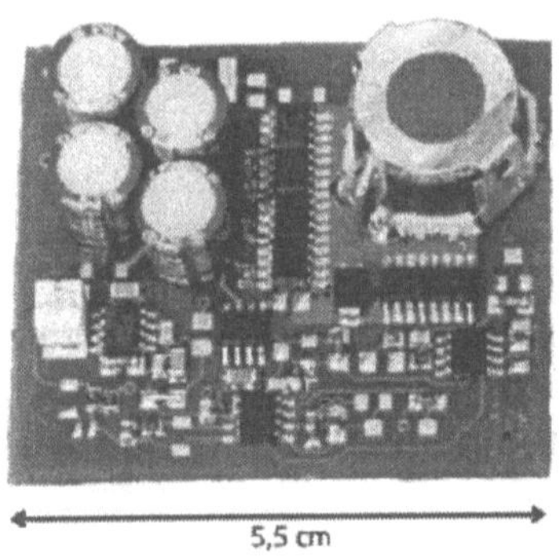

Fig. 8: 10 W DC/DC converter which was designed for high power density.

5. Conclusions and Outlook

Two fuel cell systems using the membrane technology were described. One is based on hydrogen and the other on methanol for the energy supply. Both technologies can be scaled down into the sub-watt range and have their respective advantages in certain applications.
The activities at Fraunhofer ISE range from material development to system design. In the field of membrane development, for example, a method to modify polybenzimidazole (PBI) is under investigation. This method aims to increase resistance to oxidation, to reduction, to thermal degradation and to hydrolysis and it aims to improve ductility and film-forming properties. In this context a multilayer assembly of polyelectroyte with PBI is of major interest.
The total system is defined mainly by geometry, power density, storage media and output voltage. Therefore we developed two systems with high output voltage (banded structure vs. single cell with DC/DC converter), each of these having their respective advantages.
The challenge for further development deals with the robustness for all required operating conditions. This is connected mainly with membrane properties, as the conductivity of the state-of-the-art membranes strongly depends on the water content. There are two possibilities: either to design a cell housing and system with intelligent water management, or to look for new membrane materials with a sufficient proton conductivity even under "dry" conditions. Future efforts will focus on the miniaturization of all components, since portable electronic applications in the low power range are judged to be a rapidly expanding market for small and adaptable energy supply systems.

---

1. K. Kordesch, G. Simader, *Fuel Cells and Their Appl.*, VCH Weinheim 1996.
2. J.-T. Wang, R.F. Savinell, J. Wainright, M. Litt, H. Yu *Electrochimica Acta,* 41 1996, 193.
3. J.-T. Wang, S. Wasmus, R.F. Savinell J. Electrochem. Soc., 143 1996, 1233.
4. W.-F. Lin, J.-T. Wang, R.F. Savinell, J. Electrochem. Soc., 144 1997, 1917.
5. L. Krasemann, B. Tieke, *J. Membr. Sci.,* 1998 (in press).
6. G. Decher *Science,* 277 1997, 1232.
7. J. Schmitt, T. Grünewald, G. Decher, P.S. Pershan, K. Kjaer, M. Lösche, *Macromolecules,* 26 1993, 7058.
8. Y. Lvov, F. Essler, G. Decher, *J. Phys. Chem.,* 97 1993, 13773.
9. Y. Lvov, H. Haas, G. Decher, H. Möhwald, *J. Phys. Chem.,* 97 1993, 12835.
10. K. Ledjeff, R. Nolte, DE 4329 819, EP 0711461, DE 4443945, DE 19502 391.
11. R. Nolte, A. Kolbe, K. Ledjeff, *Proc. 37th Power Sources Conf.,* 1996, 77.
12. A. Heinzel, R. Nolte, K. Ledjeff, M. Zedda, *El. chim. Acta,* 43 1998, 3817.
13. A. Heinzel, M. Zedda, T. Meier, H. Schmidt, *Fuel Cell Seminar* 1998, 686.

Recent Developments in Microtechnology-Based Chemical Heat Pumps

M. Kevin Drost, Michele Friedrich, Christine Martin, Jerry Martin, Rick Cameron

Pacific Northwest National Laboratory,[a] P.O. Box 999, Richland, WA 99352 USA

Abstract

Pacific Northwest National Laboratory (PNNL) has an ongoing program focused on the development of compact chemical heat pumps based on the absorption cycle. Design and laboratory data suggest that, by taking advantage of the high rates of heat and mass transfer available in microstructures, we are able to radically reduce the size of a heat-actuated heat pump based on the absorption cycle. Current estimates suggest a size reduction of a factor of 60 compared with a conventional heat pump. High performance is achieved by using microchannel combustor technology, microchannel heat exchangers and ultra-thin-film micromachined contactors. Currently all of the components of the system have been demonstrated and PNNL is assembling a complete bench-scale version of the device. Successful development of this technology will enable a new class of compact, heat-actuated space conditioning systems that can be used for portable or distributed heating and cooling.

Introduction

In a number of microclimate control applications individuals must wear protective clothing, which significantly reduces heat transfer from the body. Examples include workers exposed to hazardous materials, police wearing body armor, and military personnel exposed to nuclear, biological, or chemical (NBC) warfare agents. The threat from NBC warfare agents on the modern battlefield imposes a substantial burden on U.S. military forces. Present and future individual protective equipment for military personnel includes a protective suit. While such uniforms provide protection against hazards, they significantly decrease an individual's military effectiveness. Personnel performing labor intensive tasks in a hot environment are susceptible to heat stress, especially when wearing protective clothing. The time that can be spent performing essential tasks before succumbing to heat injury is limited under these conditions.

---

[a] Pacific Northwest National Laboratory is operated by Battelle for the U.S. Department of Energy under Contract DE-AC06-76RLO 1830.

Supplemental cooling will permit tasks to be performed under hazardous conditions in hot climates with enhanced efficiency and reduced heat stress. Although cooling systems can be integrated with protective suits, presently available systems are too heavy to carry for extended periods. Typically, a complete conventional system sized for eight-hour operation with a cooling capacity of 350 W_t weighs more than 10 kg.

Based on research at PNNL, a compact heat pump using a chemical compressor can provide 350 W_t of cooling and weighs approximately 1 kg. When fuel, batteries, and an air-cooled heat exchanger are included, the complete system is projected to weigh between 4 and 5 kg, less than half the weight of a conventional microclimate control system. The system is shown in Figure 1.

The device will be primarily driven by thermal energy produced by the combustion of liquid hydrocarbon fuels. The energy storage density of liquid fuels exceeds the energy storage density attainable in conventional batteries by a factor of 130 (13,000 W_t/kg for liquid hydrocarbon fuel compared with 100 W_e/kg for batteries). A cooling system based on a chemical compressor still requires electric power to operate a liquid pump and fan, but its overall electric power requirements are approximately an order of magnitude lower than those of a conventional vapor compression system. The combination of reduced demand for electric power and using liquid fuels significantly reduces the weight of the primary energy source compared with cooling schemes that require significant shaft work, such as vapor compression cycles.

In addition to manportable cooling, the miniature heat-actuated chemical heat pump (Figure 1) can meet a number of important microclimate control and space conditioning requirements, including 1) vehicle space conditioning; 2) distributed

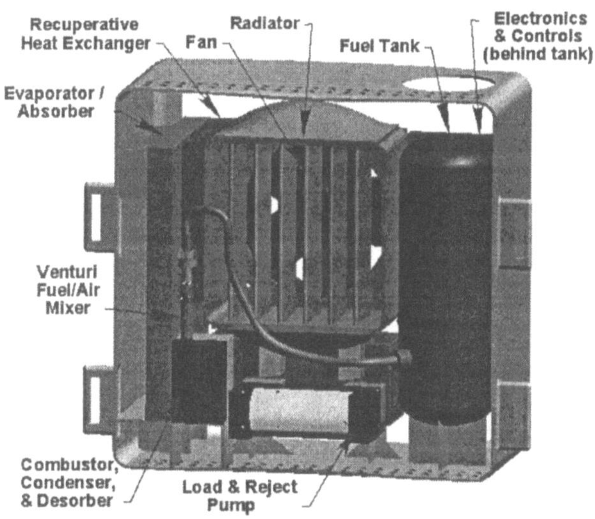

Figure 1. Manportable Cooler

cooling of buildings where the use of multiple small heat pumps eliminates the need for ducting and cycling of oversized HVAC equipment (which typically waste 50% of the cooling produced by a central cooling system), 3) lightweight air-transportable space conditioning, and 4) autonomous cooling for shipping containers.

Concept Description

Although the chemical compression and vapor compression cycles differ in the way compression is provided, both systems take the same approach to heat absorption and rejection. In both cycles, superheated refrigerant enters the condensing heat exchanger, where it undergoes constant-pressure heat rejection. The resulting condensate or mixture of condensate and vapor is then adiabatically expanded through either a throttling valve or a capillary. The mixture is then routed to an evaporating heat exchanger for constant-pressure heat absorption.

Compression is accomplished in the chemical heat pump system with a single-effect thermochemical compressor consisting of an absorber, a solution pump, a regenerative heat exchanger, and a desorber (gas generator), as illustrated in Figure 2. The cycle presented in Figure 2 is also described as an absorption heat pump.

A conventional absorption heat pump relies on gravity to form falling films, which provide liquid-to-gas contact in the absorber and desorber. This approach has two decisive disadvantages for manportable microclimate control applications. First, the heat pump must be oriented so that the solution will fall over heat exchanger tubes and form a thin film. Deviations from the proper orientation will prevent the heat pump from working. Second, falling films have a film thickness on the order of 1 mm. This becomes a significant barrier to mass diffusion and results in a physically large absorber and desorber. The miniature absorption heat pump avoids both disadvantages by relying on the high rates of heat and mass transfer available in microstructures.

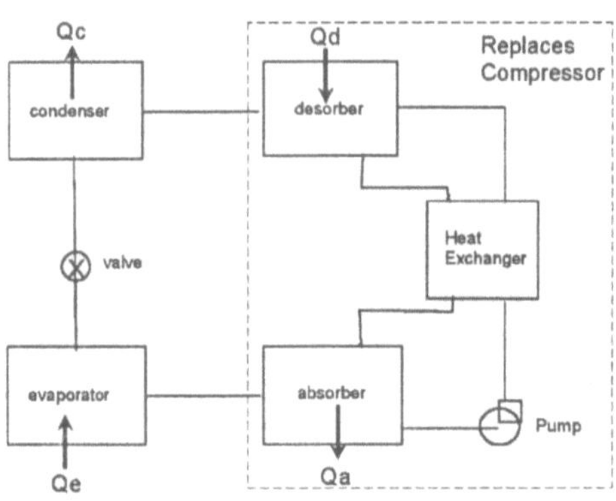

Figure 2. Single-Effect Absorption Chiller

The miniature chemical heat pump depends on the extraordinarily high heat and mass transfer rates available in microstructures to radically reduce its size while maintaining cooling capacity and efficiency. Its performance ultimately depends on a microstructure with individual features as small as 5 microns. The heat pump is a miniature device, but it is large enough to use a small but conventional solution pump.

A number of chemical heat pump cycles have been evaluated. Cycles can be classified based on the fluid combination and on cycle arrangements. The most widely used fluid combinations are lithium bromide (LiBr) and water, where water is the refrigerant; and water and ammonia (NH_3), where ammonia is the refrigerant. Cycle arrangements range from the single-effect cycle described above to progressively more efficient but complicated multiple effects. Previous research at PNNL simulated three cycles, 1) single-effect $LiBr/H_2O$, double-effect $LiBr/H_2O$, and single-effect H_2O/NH_3.

The single-effect $LiBr/H_2O$ cycle requires a low pressure solution pump with a 41 kPa (6 psi) pressure rise, but the cycle is inefficient compared to the double-effect cycle. While more efficient, the double-effect $LiBr/H_2O$ cycle requires a higher pressure pump [410 kPa (60 psi) pressure rise] and is more complicated than the single-effect cycle. The pressure rise required for a H_2O/NH_3 solution pump (2400 kPa, 350 psi) is too high for currently available small pumps and results in a heavy and inefficient system. Based on this screening, both the single-effect and the double-effect $LiBr/H_2O$ cycles are candidates for cooling applications where weight and size are key issues.

Heat Pump Components

An ongoing research project at PNNL is focused on developing a miniature $LiBr/H_2O$ heat pump for manportable cooling applications. This research for developing a manportable cooler is funded by the Defense Advanced Research Projects Agency (DARPA). Each component of the manportable heat pump is described; Figure 3 shows typical examples.

- Absorber – LiBr and water solution enters the absorber and forms an ultra-thin absorbing film. The ultra-thin film is maintained by a proprietary micromachined structure. Refrigerant vapor passes through the micromachined structure and is

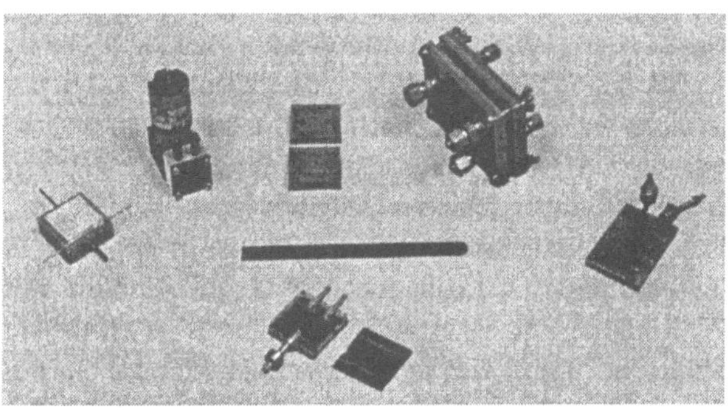

Figure 3. Absorption Heat Pump Components

absorbed into the weak solution. The absorption process results in a high rate of heat generation, which is subsequently removed from the solution by a microchannel heat exchanger. Figure 4 presents an overview of the absorption process. The water vapor pressure, at the interface where the vapor and solution are in contact, is the driving force for the absorption process and must equal the partial pressure of water in the solution at the interface. As the solution passes through the absorber, water is diffused through the thin film. It is important to keep the pressure drop across the micromachined contactor low so that the maximum vapor pressure is available to drive the absorption process.

Absorber performance is strongly dependent on the thickness of the mechanically constrained, ultra-thin film. As Figure 5 shows, the reduction in the thickness of the thin film from 200 microns to 50 microns would reduce the length of the absorber by a factor of 16 while keeping the absorption rate constant. In addition, mechanically constrained absorption means that the absorber performance is independent of its physical orientation.

PNNL has developed and demonstrated the absorber, and preliminary results show that ammonia could be absorbed in water at a rate that generates between 10 and 30 W/cm^2. This is an extraordinary absorption rate, which exceeds the performance of conventional absorbers by more than a factor of 10. We are currently developing LiBr and water absorbers.

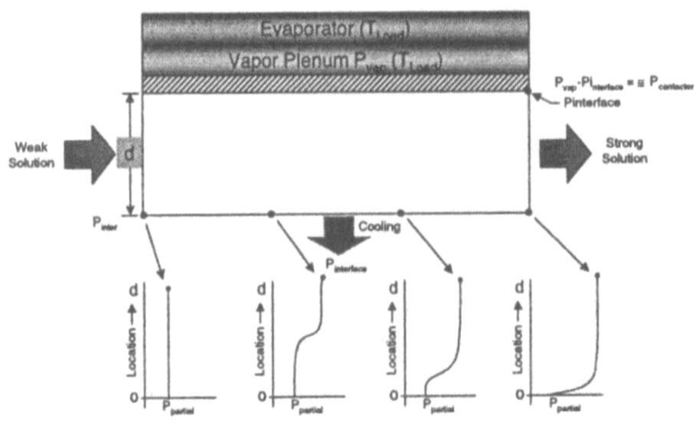

Figure 4. Ultra-Thin Film Absorption Process

- Evaporator - The evaporator also consists of an array of microchannels with channel widths between 100 and 300 microns and depths up to 1 mm. Extensive experimental investigations of microchannel evaporation have been published by PNNL [1]. Results show that convective heat transfer coefficients of 1.0 to 2.0 W/cm^2-K are readily attainable, and heat transfer rates up to 100 W/cm^2 can be obtained with a small temperature difference. These heat transfer coefficients and rates exceed those of conventional evaporators by a factor of 4.
- Desorber – Water and LiBr solution enters the desorber and forms an ultra-thin desorbing film. The ultra-thin film (with a film thickness of approximately 100 microns) is maintained by a micromachined contactor, which allows refrigerant vapor to evolve from the thin film. Thermal energy from a combustor can be used to heat the strong solution in the ultra-thin film, evaporating the refrigerant vapor

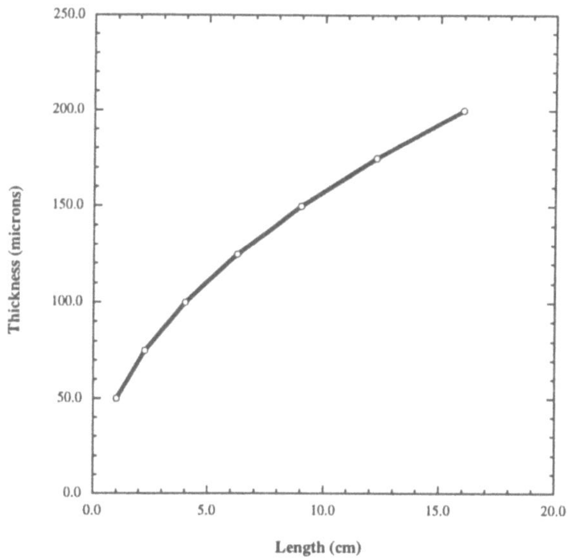

Figure 5. Impact of Film Thickness on Absorber Length

from the solution at a faster rate. The refrigerant vapor passes through the micromachined structure and is transferred to the condenser. PNNL has developed and demonstrated the desorber. Results show that water is desorbed from the strong solution at a rate of 0.2 g/cm^2/min. This exceeds the performance of conventional desorbers by a factor of 10.

- Condenser - The condenser consists of an array of microchannels with channel widths of 100 to 300 microns and depths up to 1 mm. The microchannel condenser has been demonstrated by PNNL. Heat transfer rates in excess of 30 W/cm^2 were attained with a small temperature difference and low pressure drop.
- Regenerative Heat Exchanger - The regenerative heat exchanger consists of arrays of microchannels with channel widths of 100 to 300 microns and depths up to 1 mm. PNNL has published extensive experimental investigations of single-phase microchannel heat transfer [2]. Convective heat transfer coefficients of 1.0 to 1.2 W/cm^2-K are attainable. These heat transfer coefficients exceed conventional regenerator performance by a factor of between 3 and 6.
- Combustor - PNNL has demonstrated a microchannel combustor for the DARPA. The microchannel combustor can produce thermal energy at a rate of at least 30 W_t/cm^2 with a thermal efficiency between 82 and 85% [3]. We plan to use the microscale combustor as the heat source for the miniature absorption heat pump.

Contactor Development

The desorber and absorber depend on the availability of micromachined contactors that will prevent liquids from passing though the contactor while having a minimum impact on the diffusion of water vapor. PNNL has evaluated a wide range of micromachined and other types of contactors. Data have been collected on breakthrough pressure (the pressure at which solution will pass through the contactor) and permeability (the mass transfer through the contactor as a function of pressure difference across the contactor). Typical data are shown in Figure 6 (numbers refer to different contactor designs). Contactors in the upper right hand corner of Figure 6 are desirable contactors with high breakthough pressure and high permeability.

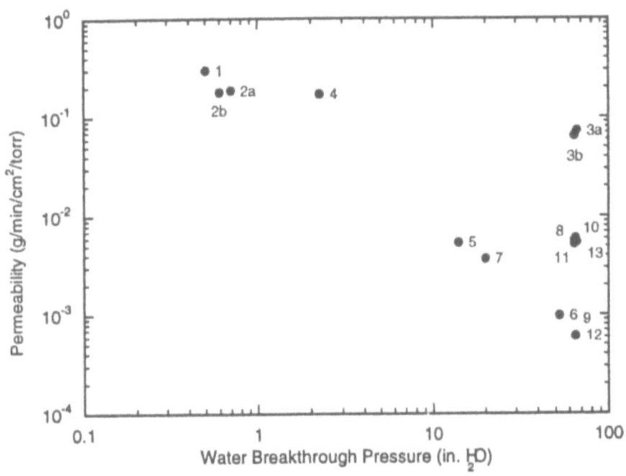

Figure 6. Permeability versus Supported Liquid Pressure

Heat Pump Performance

Based on experimental data we have collected to date, a prototype manportable cooler has been designed. This section presents the predicted performance of a single-effect, LiBr/H_2O heat pump sized to provide 350 W of cooling. The system includes a water-cooled condenser and absorber, a water heat source for the evaporator, and a microchannel combustor with exhaust gas at 250°C as the desorber heat source. In many applications, water cooling will be available (from the vehicle radiator for vehicle cooling and from a cooling tower for space conditioning). Other applications, such as manportable cooling, will require an air-cooled heat exchanger for ultimate heat rejection from the system.

Table 1 summarizes the weight and performance characteristics of two manportable coolers. The first system uses a battery to provide electric power for pumps and the fan, while the second uses a thermoelectric generator (TEG) installed between the combustor and the desorber to provide electric power. As Table 1 shows, the heat pump will weigh approximately 1000 g and have a coefficient-of-performance (COP) of 0.68. COP is the ratio of the amount of cooling provided divided by the thermal energy supplied to the desorber. For the manportable design, 1 W of heat provided by combusting fuel would provide 0.68 W of cooling. An electrically driven heat pump would have a COP of 2.8 if we assume that fuel must be consumed in a 25% efficient power generation system to provide compressor power. As shown in the table, a thermoelectric generator significantly reduces the weight of the system by eliminating the need for batteries.

Table 1. Miniature Chemical Heat Pump Characteristics

	NiCd Battery	PbTe TEG
Heat Pump	0.92	1.18
Radiator	0.85	0.83
Fan	0.65	0.65
Battery	1.07	0.03
TEG	0.00	0.18
Structure	0.47	0.47
Pumps/Fluids	0.43	0.43
Fuel/Tank	0.54	0.68
Electronics	0.22	0.22
Total, kg	**5.144**	**4.675**

These device characteristics are compelling examples of the advantage of devices in the miniature size range. The specific cooling (cooling per unit volume) of the miniature heat pump is higher than that of a conventional macroscopic absorption heat pump by a factor of 60 (1.25 W/cm^3 compared with 0.02 W/cm^3) for a macroscale device. The chemical heat pump also requires considerably less electric power than a vapor compression heat pump (20 W compared with 120 W). This will reduce the need for manportable power generation or batteries for powering microclimate control systems.

Conclusions

By taking advantage of the high rates of heat and mass transfer attainable in microstructures, PNNL is developing a miniature chemical heat pump with a cooling capacity of 350 W that weighs only 1 kg. This device achieves a reduction in volume of a factor of 60 over a macroscale absorption heat pump. A complete manportable cooling system, including the heat pump, an air-cooled heat exchanger, batteries, and fuel, is estimated to weigh between 4 and 5 kg, compared to the 10-kg weight of alternative systems.

References

1. Cuta JM, CE McDonald, and A Shekarriz. 1996. "Forced Convection Heat Transfer in Parallel Channel Array Microchannel Heat Exchangers." *Advances in Energy Efficiency, Heat/Mass Transfer Enhancement.* PID-Vol.2 HTD-Vol.338, American Society of Mechanical Engineers, New York.
2. Ravigururajan TS, J Cuta, C McDonald, and MK Drost. 1996. "Single Phase Flow Thermal Performance of a Parallel Micro-Channel Heat Exchanger." Presented at American Society of Mechanical Engineers 1996 National Heat Transfer Conference.
3. Drost MK, CJ Call, JM Cuta, and RS Wegeng. 1996. "Microchannel Integrated Evaporator/Combustor Thermal Processes." Presented and 2nd U.S Japan Seminar in Molecular and Microscale Transport Phenomena, August 8-10, Santa Barbara, California.

Miniaturised direct methanol fuel cell with a plasma polymerised electrolyte membrane

Laurent Mex[1], Jörg Müller

Technical University Hamburg-Harburg, Department of Semiconductor Technology, D-21073 Hamburg, Germany

Introduction. Polymer electrolyte fuel cells are well established as power sources for e.g. vehicle applications or plants. For the area of applications in the low power regime (commonly battery powered devices) we have evolved a new concept for a direct methanol fuel cell (DMFC) based on the utilisation of thin-film technologies, e.g. plasma enhanced chemical vapour deposition, sputtering etc.

In principle a polymer electrolyte fuel cell consists of two electrodes separated by an ion conductive membrane, the so called electrode membrane assembly. One electrode is fed with oxygen or air, and the other with a methanol/water mixture in the case of a direct methanol fuel cell. The methanol and water react in the presence of a platinum/ruthenium catalyst to electrons, hydrogen ions and carbon dioxide. The hydrogen ions penetrate through the membrane and the electrodes move along an outer circuit to the opposite electrode, where together with oxygen and a platinum catalyst the reaction to water takes place.

The membrane electrode assembly in our miniaturised DMFC consists of an electrolyte and porous graphite electrodes which are deposited with plasma enhanced chemical vapour deposition techniques.

A silicon-wafer is used as a substrate for the membrane electrode assembly consisting of porous silicon areas for the gas feed of the lower electrode. The encapsulation is realised by pyrex-wafers on top and bottom side of the silicon substrate using anodic bonding.

In a first step towards the realisation of this miniaturised fuel cell a deposition process of the plasma polymerised electrolyte was developed. The resulting membranes exhibit an ion conductivity of $\sigma = 15 mS/cm$ at $T = 81°C$.

[1] e-mail: laurent@tu-harburg.de

1 Structure of the miniaturised DMFC

1.1 Substrate and gas feed system

The silicon substrate shown in the center of the DMFC scheme in figure 1.1 requires permeable and conductive areas for the gas feed to the lower graphite electrode. These regions can be realised by the etching of a porous silicon layer and a KOH etching process for the removal of the substrate material.

Several fuel cells can be structured on one substrate connected in series for a desired voltage. In this case an additional p^+-silicon region between neighbouring fuel cell compartments prevents an electrical short circuit along the silicon substrate.

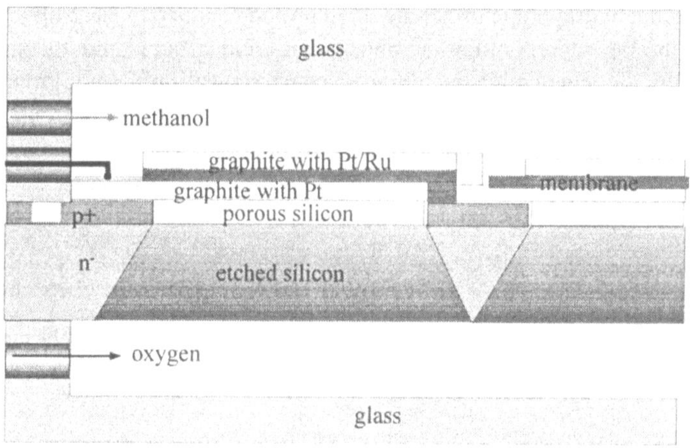

Fig. 1.1 Scheme of a miniaturised fuel cell

The central gas in- and outlet for both electrodes is made of capillaries inserted at the edges of pyrex coverages anodically bonded to top and bottom of the silicon substrate. The pyrex contains etched grooves or caverns for a homogeneous gas distribution across the substrate which may be covered with a teflon-like polymer to prevent adhesion of water.

1.2 Electrodes

Following the configuration of common polymer electrolyte fuel cells porous graphite is used as the electrode material. For the reaction of methanol resulting in the formation of hydrogen ions and electrons a combination of platinum and ruthenium is needed as a catalyst, and on the other electrode it is platinum which maintain the reaction of oxygen, hydrogen ions and electrons to form water. These catalysts have to be incorporated in the upper area of the electrodes adjacent to the ion conductive membrane between them.

1.3 Plasma polymerised ion conductive membrane

In contrast to conventionally polymerised membranes plasma polymerised films exhibit a high degree of crosslinkage and are pinhole free even for thin films of a few hundred nanometer thickness. From these benefits a drastic reduction of the methanol cross-over in direct methanol fuel cells is expected.

In the PECVD-process for the ion conductive membrane the monomers Tetrafluorethylene (TFE) and Vinylphosphonic acid (VPA) are chosen. TFE composes the hydrophobic backbone of an proton conductive electrolyte and VPA is used for the incorporation of phosphonic acid groups into the membrane. Commonly ion conductive membranes comprise sulfonic acid groups as ion conductors. However, due to their dibasic nature phosphonic acid groups will lead to a higher conductivity compared to sulfonic acid groups.

(a) (b)

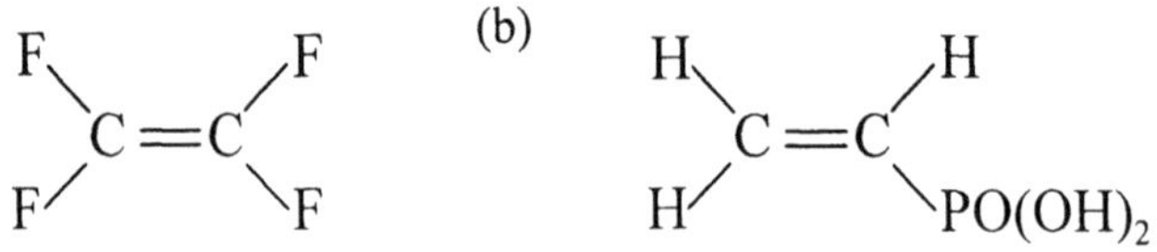

Fig. 1.2 Tetrafluorethylen (a) and Vinylphosphonic acid (b)

Looking at the structures of TFE and VPA (fig. 1.2) the polymerisation should occur via the π-bindings to prevent a distortion of the phosphonic acid group and ensure a chemical binding of these groups into the polymeric backbone of the membrane. Hence, low power densities are required for the plasma polymerisation. Furthermore VPA is a liquid and has to be volatilised. For this the VPA is stored in a container consisting of a oil-bath heater and a gas in- and outlet for a carrier gas, e.g. Argon or TFE.

2. Results

Since the choice of the substrate material and the encapsulation of the membrane electrode assembly can be varied, e.g. to polymers or ceramics, we started the development of the miniaturised DMFC by evaluating a suitable process for the polymer membrane.

2.1 Characterisation of TFE/VPA plasma polymers

The composition of the TFE/VPA plasma polymers is analysed by infrared spectroscopy in attenuated total reflection measuring technique (Perkin Elmer 1800 FTIR; 45° KRS-5 crystal). Figure 2.1 shows a typical TFE/VPA and for comparison a TFE spectrum as well as their difference. The phosphonic acid groups exhibit three characteristic absorbance bands corresponding to vibrational modes of the P = O binding (1220cm^{-1}), the P – O binding in P-OH (1032cm^{-1}) and the O – H bindings (1700cm^{-1} – 1500cm^{-1}).

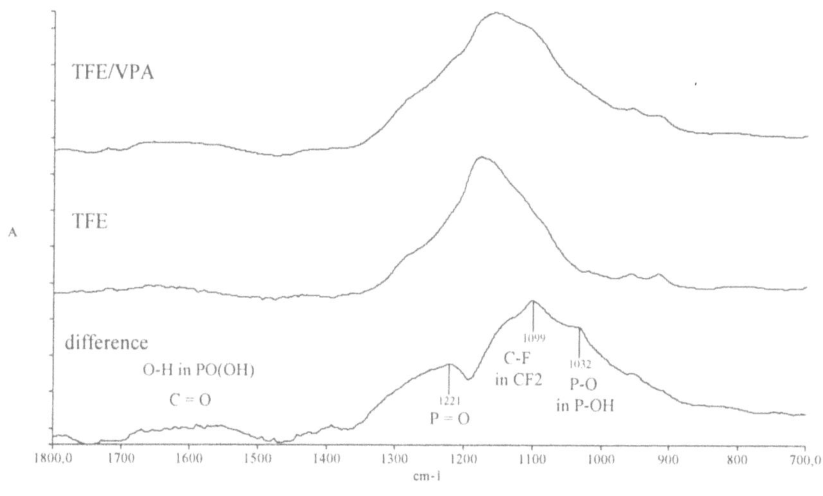

Fig. 2.1 Attenuated total reflection infrared measurements of TFE/VPA (4W, 100μbar, 5sccm TFE carrier gas, 120°C) and TFE plasma polymers (4W, 100μbar, 5sccm TFE)

To evaluate the corresponding rf-power in the PECVD-process, for which the phosphonic acid groups remain unfragmented in the plasma, several polymers have been deposited at identical process parameters except for a variation of the rf-power from 6W to 50W. Looking at the characteristic absorbance bands of the phosphonic acid groups in the IR-spectra of these films (fig. 2.2) it is obvious that the fragmentation of the phosphonic acid group increases with increasing power and the IR spectra at high powers are identical to TFE polymer spectra. Only at a rf-power of less than 10W an incorporation of complete groups is apparent.

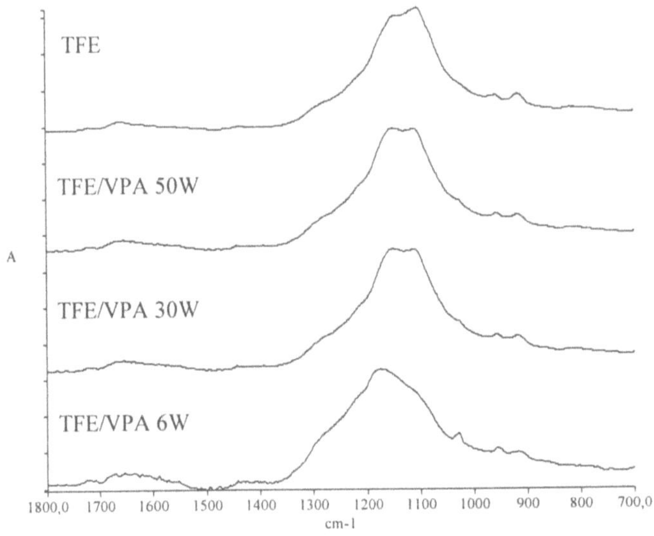

Fig. 2.2 IR-spectra of TFE/VPA polymers with different applied rf-power (10sccm TFE carrier gas, p = 100μbar, T = 70°C) and TFE polymer spectrum (10sccm TFE, p = 100μbar, P = 50W)

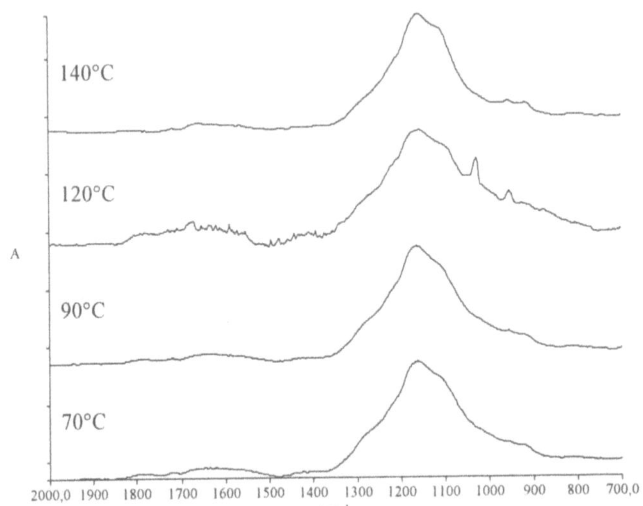

Fig. 2.3 IR-spectra of TFE/VPA polymers deposited at different VPA reservoir temperatures (10 sccm TFE carrier gas, p = 100 μbar, P = 6 W)

To increase the amount of phosphonic acid groups in the TFE/VPA polymers the temperature of the VPA reservoir was increased up to 140°C. The IR-spectra obtained from TFE/VPA polymers deposited at 70°C, 90°C, 120°C and 140°C show an increasing amount of VPA up to 120°C (fig.2.3). At 140°C the VPA content decreases again, due to a spontaneous polymerisation of the VPA in the heated gas inlet (≈ 150°C).

2.2 ac-impedance measurements

For the determination of the ion conductivity of the plasma polymerised TFE/VPA films ac-impedance measurements were carried out with a HP 4192A LF impedance analyser in the frequency range between 150Hz and 13MHz.

The test assembly consists of two TFE/VPA polymer films with 300nm in thickness embedding two parallel gold tracks (2.5mm x 17mm x 90nm) separated by 3.6mm on a glass substrate. These tracks were contacted in four point measuring technique and immersed in deionised H_2O. The temperatures was varied between 23°C and 90°C.

Figure 2.4 shows a typical impedance plot at T = 81°C. The slope at high frequencies ($Z' \rightarrow 0\Omega$) is caused by a parasitic impedance from the leads. From this measurement an ion conductivity of σ = 15mS/cm was estimated.

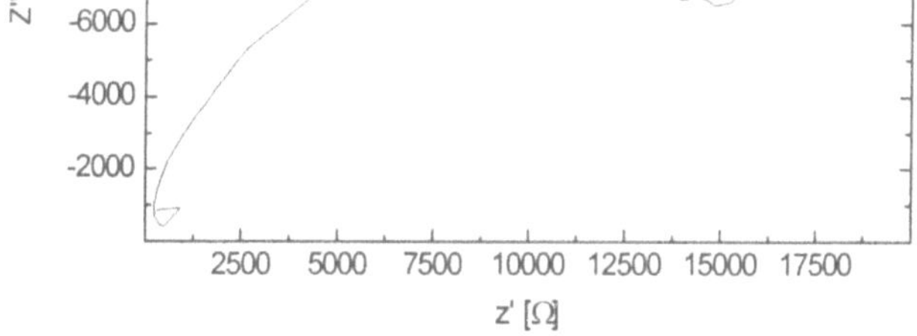

Fig. 2.4 Impedance plot from the ion conductivity measurement of a TFE/VPA polymer

The dependence of the ion conductivity on the temperature is shown in figure 2.5. The dashed line was measured after the sample was stored for 3h in water and the solid line after a storage of 24h, resulting in an increase of its ion conductivity by a factor of 2.5 at 80°C.

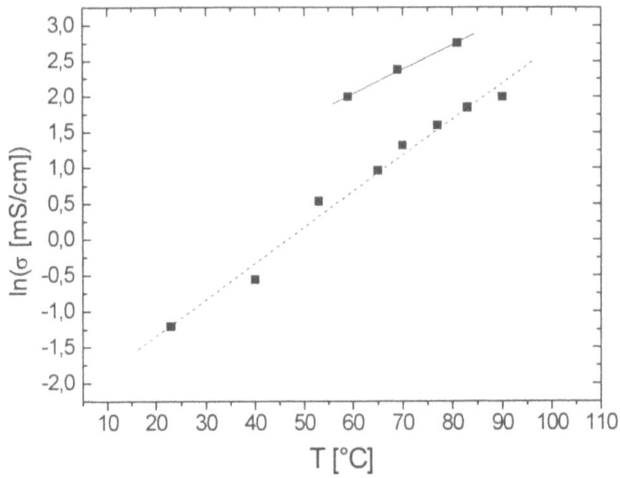

Fig. 2.5 Temperature vs. ion conductivity from a TFE/VPA polymer

3 Conclusion

A concept for a miniaturised direct methanol fuel cell using thin-film technologies and a plasma polymerisation process for an ion conductive membrane were presented. An ion conductivity of this membrane of $\sigma(81°C) = 15mS/cm$ was determined by ac-impedance measurements. Thus an integration of a plasma polymerised membrane into a miniaturised DMFC is very promising since the necessary membrane thickness of 1μm leads to a lower resitance compared to membranes used in commercial DMFC (e.g. Nafion $\sigma = 75mS/cm$, 175μm thick).

Part 7

Application of Microreactors in Drug Development and Biotechnology

Rapid PCR in flow-through Si chip thermocyclers

S. Poser*, R. Ehricht*, T. Schulz, S. Uebel, U. Dillner, J.M. Köhler
Institute for Physical High Technology Jena, Helmholtzweg 4, D-07743 Jena, Germany
*present address: Clondiag Chip Technologies GmbH, Löbstedter Str. 105, D-07743 Jena, Germany

Abstract. Rapid micro flow PCR can be performed in chip devices made by silicon and glass and including heat transfer barriers for thermal decoupling of the three temperature zones. We tested the system in the PCR of 106 base pairs as template. The number of temperature cycles is given by the construction of chip or experimental arrangement, resp. Continous flow experiments were done using different chips and setups with 16 up to 32 temperature cycles within 16, resp. 32 minutes. The PCR product was characterized by gel electrophoresis and was found to be identically with PCR products of the same template amplified slowlier in a conventional thermocycler. Gel electrophoresis showed a certain cross talk in products in case of a serial flow process caused by template dispersion, which can be minimized by the segmentation of aqueous samples by oil.

Keywords. DNA-analysis, polymerase chain reaction, PCR, thermocycling, continous flow, serial flow, micro fluid reactor, segmented flow

1. Introduction

The amplification of nucleic acids outsite living cells became one of the key methods of molecular biology during the last decade. The most favorited method for DNA amplification is the polymerase chain reaction (PCR), which allows to amplify DNA specifically and with high precision [1]. The PCR needs the periodical application of three temperatures to the reaction mixture. The process needs between about 2 and 4 hours in conventional table devices. Several types of chip thermocyclers were developed for miniaturized rapid polymerase chain reaction (PCR) during the last few years [2-8]. They use the quick heat transfer between liquid and walls caused by the low heat diffusion distances. In addition, rapid PCR becomes possible at low heating and cooling power due to the reduction of parasitic thermal masses. But, PCR in micro wells suffers from wetting and evaporation problems during thermocycling. Flow-through thermocycling is favorited for miniaturized PCR in chip devices in order to avoid such problems [9], whereby chip cyclers made from polymer material are tested to be suited for quick PCR [10].

Silicon chips should be particularly suited for flow-through thermocyclers due to the high heat conductivity of Si and the technological compatibility with thin film processes, which allows the simple integration of thin film heaters and sensors in the device. Here, the preparation of new types of flow-through Si chip thermocyclers and PCR results are reported.

2. Chip thermocyler design

The principle design of flow thermocyclers has to be orientated on the functional demands of PCR. Therefore three temperature zones are necessary, which has to be passed by the process liquids several times. A temperature of about 47 - 55 °C is necessary for the selective hybridization of the short starter molecules (primers) on the recognition ranges of the template. At temperatures of about 70 - 75 °C, the polymerase is working, which elongates the chain of starter molecules in the order of the complementary nucleotides of the template. The third temperature of about 95°C is necessary for the so-called melting, the separation of the both single strands of the double strand templates at the beginning of the process and after finishing each elongation step. These three processes - annealing, extension and denaturation - have to be performed between about 15 and 30 times for getting a sufficient enhancement of the concentration of the template. The principle construction of a Si-flow-through thermocycler was presented in [9].

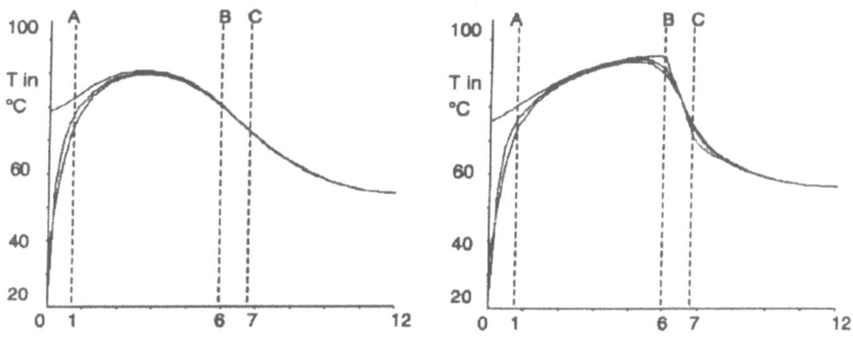

Fig. 1 Comparison of temperature distribution over two temperature zones for 2 different material combinations in μ-fluid thermocyclers for 3 different flow rates, left: Si only, right: Si and glass, optimized (Result of FEM simulations)

The combination of two wall materials with different heat conductivities is advantageous in miniaturized PCR devices (Fig. 1). This way low heat transfer rates can be realized between the different temperature zones and a high degree of thermal equalization in each zone. A chip device with vertical heat flow control was realized previously by isolating three silicon chips - one for each temperature zone- by a glass chip [9]. For the new miniaturized flow-through thermocycler chips, we used the principle of horizontal heat flow control, which was found to be suited for stationary chip thermocyclers, too [5,6]. The three temperature zones were thermally separated by holes etched into the Si substrate in order to prevent a fast heat transport by the highly conductive Si between the zones. The fluid channel is formed by a second chip of glass, in which the heat flow rate is low due to the low heat conductivity.

FEM simulation showed the effect of material choice in case of the use of glass or silicon alone or in combination. It is found, that the temperature history of a sample flowing through the device is marked by only very soft transitions between annealing and extension and between extension and melting in the case of a homogenous material. The profiles are much sharper, if glass and Si are combined, so that Si dominates in each temperature zone, while glass is dominant between the temperature zones.

3. Device constructions and preparation

Two types of flow-through thermocyclers were designed, prepared and tested. Both designs are based on the horizontal heat flow control principle. The first type with 16 integrated channel loops ("FTC16") uses an arrangement, in which the annealing zone is placed in the center and the primer extension zone and the denaturation zone at the edges of the chip. The extension and the denaturation zone are heated by integrated resistors, the annealing zone is cooled by a cooling block.

Fig. 2
Flow-through Si chip thermocycler FTC 16. (Front side with flow channel etched in Si for 16 temperature cycles)

This arrangement possesses the advantage of comparatively steep temperature transitions between the three zones and a good thermal control behaviour. Disadvantageous is the necessity of passing the annealing zone between extension and denaturation, which contradicts the usual PCR protocol. Therefore, the residence time of liquid was minimized in this phase by the reduction of the cross section of the channel (Fig. 3). Chip devices for 4, 8 and 16 cycles were designed and prepared.

Fig. 3
FTC 16: Si chip with inlet port and thermal gaps etched in Si

The second design with 25 channel loops ("FTC25") was stronger orientated on the temperature order of PCR protocol: The extension zone was arranged in the center, annealing and denaturation zones at the edges of the chip. The disadvantage of this cycler consists in the fact that the liquid has to pass the extension zone after the denaturation. That means that there is no sharp temperature transition between denaturation and annealing.

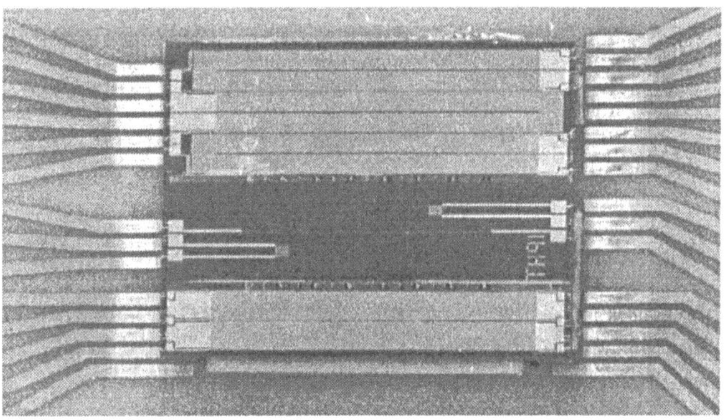

Fig. 4
FTC 16: back side with thin film heaters, resistive sensors and wiring

In order to get optimal conditions of liquid transport in heterogenous systems, sharp edges and rough surfaces are avoided in this design. Therefore, the flow channel is not etched into the Si, but into the glass chip. A rounded shape of channel loops should support not only a continous flow PCR but a serial flow processing of different templates, too.

Silicon wafers were micropatterned using a $Si_3N_4/SiO_2/Si_3N_4$ sandwich system [9]. Thermal gaps were etched anisotropically using an anisotropic NaOH etchant. The glass chips were etched isotropically using a HF etchant and micropatterned chromium film as etch mask. Thin platinum films (0.2 μm thick) were deposited by sputtering and microlithographically etched for the preparation of thermal transducers acting as heaters and resistive sensors. Contact pads were made by sputtered thin aluminium films. The prepared chips were connected by anodic bonding. Fluid connections were prefabricated mechanically and attached by glue bonding.

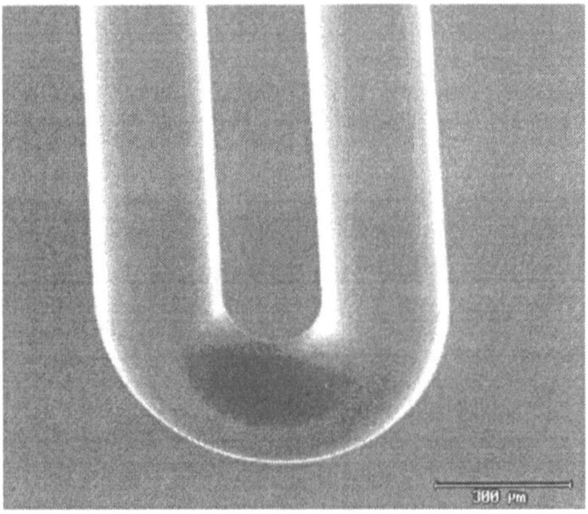

Fig. 5
Flow channel of FTC25, etched in glass,
channel width: 0.25 mm;
channel depth: 0.1 mm

4. Flow experiments

Both μ-PCR devices were tested in micro flow systems. Syringe pumps were used for the fluid transport. Single flow experiments were performed at first. They show that flow rates between about 0.5 μl/min up to 100 μl/min can be realized in both type of cyclers.

In a second experiment series, the dispersion behaviour of dyes was studied in the flow-through cyclers. The observed dispersion is important and leads to the conclusion that a serial PCR amplification needs sufficient liquid volumes, e.g. template-free buffer, between the single injected template solution volumes or that the PCR has to performed in separated process volumes, which can be realized by oil segmentation like shown previously [8].

Then, both types of flow cyclers were tested then for the flow-through behaviour of water segemented by small droplets of oil. It was found that the flow behaviour of segmented liquid columns in the FTC 25 was much better than their flow behaviour in the FTC16. Problems during the transport of smaller water in oil segements were obviously caused by sharp edges in the fluid channels and the changes in cross section. A very stable liquid transport with small segment volumes was realized in FTC25 (Fig. 6).

Fig. 6 Segmented flow in the FTC 25 device

5. PCR experiments

5.1. PCR setup

PCR experiments were performed in continous flow by use of the FTC16 device. Two different arrangements were used: a single chip and a double chip variant. The template was injected in a stream of buffer solution after a pretreatment of the cycler by flowing buffer solution containing the primers, the NTPs, and additions.

The temperature control was realized by a feed back circuit, which was controlled by a programmable power source. The temperature protocols were stored in a PC by using an A/D interface. Products were measured on a Li-Core sequencing device by addition of IRO-41 labelled oligonucleotide primers [11,12].

5.2. Continous flow PCR process and products

Three different templates were tested in the PCR experiments. We chose DNA segments of a length of 61 base pairs (bp), 106 bp, and 160 bp.

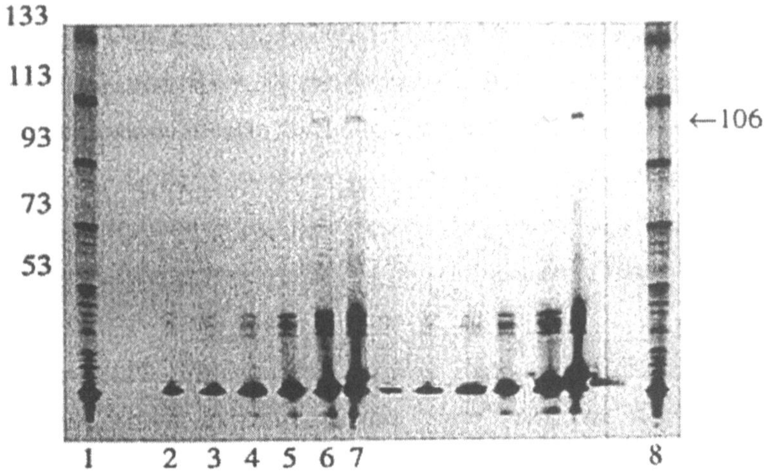

Fig. 7
Gel electrophoresis of μ-flow-PCR products after 16 cycles; Track 1 and 8: length standards; tracks 2-7: products from μ-flow-PCR in 6 different concentrations.

The flow cyclers arrangements were tested at a flow rate of 62 nl/s corresponding to a residence time of 15 s in each melting zone, of 15 s in the annealing zones and of 30 s in the primer extension zones. The temperatures were adapted to 94, 43 and 72 °C. The time need for each cycle was about one minute, that means the overall process took 16 minutes in the single chip, and 32 minutes in the double chip arrangement.

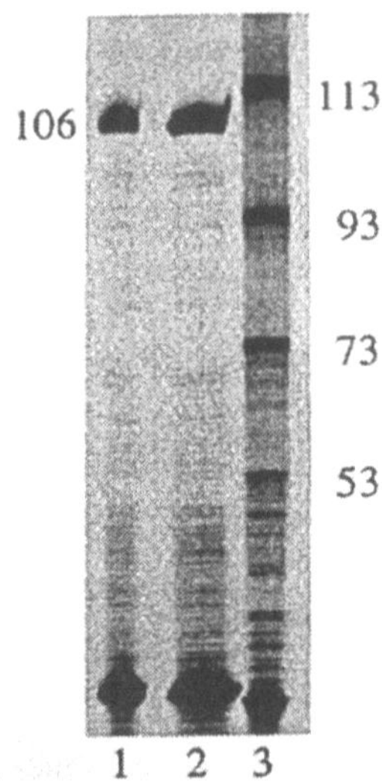

Fig. 8
Gel electrophoresis of μ-flow-PCR products after 32 cycles; Track 3: standard; tracks 1 and 2: products from μ-flow-PCR labbeled with 2 different primers

The PCR product was characterized by denaturing gel electrophoresis. The comparison with markers and with the product obtained in a conventional thermocycler prooved the selectivity of the rapid PCR process. The concentration of the PCR product was comparatively low in case of a 16 cycle process (Fig. 7) . But the product could be identified unambiguously. The expected product with a length of 106 bps was clearly found at the three higher concentrations in the gel electrophoresis of a concentration series of 6 different values.

High concentrations of the right product were obtained, if the reaction mixtures passed not only one but two cycler chips, that means 32 temperature cycles (Fig. 8). It is important to remark that in the case of the 106 bp template during 32 cycles, too, no side products were found. This shows a high specifity of the amplification procedure.

5.3. Serial procedures

Flow-through thermocyclers are particularly interesting for carrying out serial amplifications for DNA analytics. Therefore, the cross-talk between serially applied templates was studied.

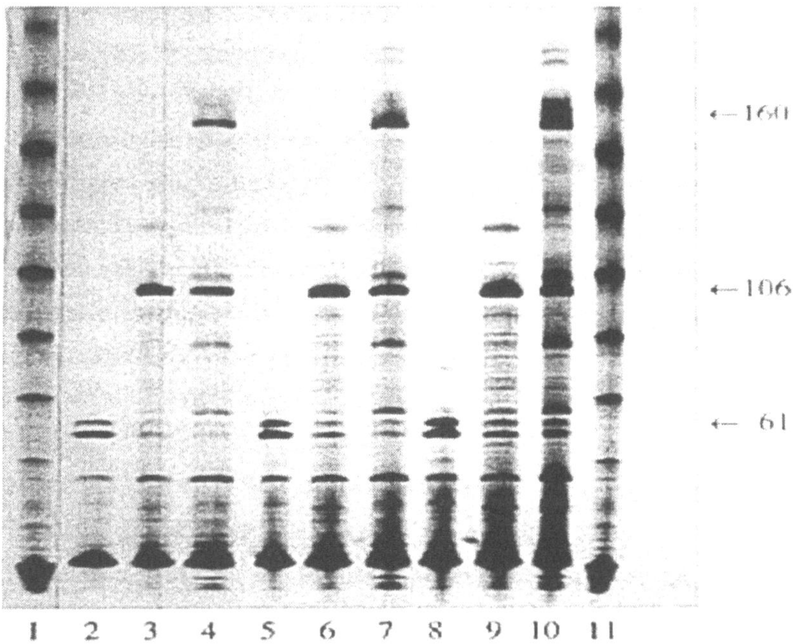

Fig. 9
Gel electrophoresis of μ-flow-PCR products after an unsegmented serial flow experiment with three different templates and three different concentrations for each template: tracks 1 and 11: standard; tracks 2,5,8: PCR product from μ-flow-PCR after injection of first template (61 bps) tracks 3,6,9: PCR product from μ-flow-PCR after injection of the second template (106 bps); tracks 4,7,10: PCR product from μ-flow-PCR after injection of the third template.

The test started with the amplification of the shortest sequence fragment of the DNA (61 bp), followed by the mediate DNA segment (106 bp) and finished with the amplification of the longest DNA segment (160 bp). Between the first and the second and between the second and the third template the thermocyclers were cleaned only by air. Flow rates of 3 μl/min for the both first amplifications and of 2μl/min for the third were chosen. In all cases the same sets of primers were chosen.

The PCR succeeded for all runs. The expected PCR products were obtained in all tests independently of the length of templates. The result is shown by the electrophoresis gel (Fig. 9). But, the electrophoresis shows also clearly that a small, but significant amount of the first template (61 bp) is amplified in the second and the third run, too. Much more important is the cross-talk from second to third run. The 106 bp DNA segment from second PCR run was found in a large extend after the third PCR run with the amplification of the 160 bp segment of the template. This observation underlines the need of a safe separation of PCR jobs in a μ-fluid process on chip. Either, there is necessary a decontamination step after each run or there has to be applied a safe segmentation of fluid column elements [13].

6. Conclusions

Two new designs of chip flow-through thermocyclers were presented, in which the temperature management was supported by the combination of a silicon and glass as a good and a poor heat conductive material in order to realize thermally equalized zones, which are separated by thermal gaps. The amplification of three different artificial sequence segments of a DNA templates by microflow PCR in a silicon chip flow-through thermocycler was demonstrated.

The selectivity of priming was proved by denaturing gel electrophoresis on a sequencing device. It was possible to get the right amplified product after passing a chip with 16 temperature cycles within 16 minutes. High yields of products were obtained after 32 cycles.

Experiments with serial injection of templates showed that amplificates of previous runs lead to a contamination of following runs. This contamination problem remains, if the flow channel of the chip device is blown out and dried by air between the runs. The best solution to overcome the cross-talk problem seems to be to process the templates in a segmented liquid column.

The realization of a stable flow of aqueous solution segments in an immiscible liquid succeeded in a micro flow thermocycler design with flow channels etched in the glass half sphere. The adaption of PCR protocols for μ-flow procedures in Si chip devices for serial amplification in micro segmented fluids is under progress.

Acknowledgements

Fruitful discussions and support in PCR-tests by S. Wölfl and H.-P. Saluz is gratefully acknowledged. We like to thank M. Sossna, H. Porwol, F. Jahn and H. Fischer for support in the microtechnical preparations. The cooperation and financial support of the Biometra GmbH is gratefully acknowledged.

References

[1] Mullis, K.; Faloona, F.; Scharf, S.; Saiki, R.; Horn, G.; Erlich, H.: Cold Spring Harb. Symp. Quant. Biol., 51, Pt. 1 (1986), 263-273

[2] Northrup, M.A., Gonzales, C., Hadley, D., Hill, R.F., Landre, P., Lehew, S., Saiki, R., Sninsky, J.J., Watson, R., Watson, R.jr., 8th Internat. Conf. On Solid State Sensors and Actuators (Yokohama 1993), 924-926

[3] Wilding, P., Mann, A., Shoffner, L., Kricka, L., Oak Ridge Conf. 1994

[4] M.A. Northrup, C.Gonzales, D.Hadley, R.F.Hills, P.Landre, S.Lehew, R.Saiki, J.J.Sninsky, R.Watson, R. Watson jr.: 8th Internat. Conf. on Solid-State Sensors and Actuators (Stockholm 1995), 764-767

[5] Baier, V., Dillner, U., Poser, S., Schulz, T., Köhler, J.M., Bodner, U., Schimkat, D., Adelhelm, K., Jäger, R., Proc. Conf. On Micro System Technology (Chemnitz 1995), 184-189

[6] Poser, S., Schulz, T., Dillner, U., Baier, V., Köhler, J.M., Mayer, G.; Siebert, A.: Eurosensors X (Leuven 1996), 1197-1199; Sensors and Actuators A 62 (1997), 672-675

[7] Poser, S.; Schulz, T.; Dillner, U.; Baier, V.;Köhler, J.M.; Mayer, G.; Siebert, A. Schimkat, D.: (Frankfurt/M. 23.-25.2.1997), O.25 Proc. Ed. W. Ehrfeld (Springer 1997), 294

[8] Daniel, J.H.; Iqbal, S.; Millington, R.B.; Moore, D.F.; Lowe, C.R., Leslie, D.L.; Lee, M.A.; Pearce, M.J.: Sensors and Actuators A 71 (1998), 81-88

[9] Köhler, J.M.; Dillner, U.; Mokansky, A.; Poser, S.; Schulz. T.: 2nd Internat. Conf. on Mircoreaction Technology (New Orleans, März 1998), 241-247

[10] Koop, M.U.; DeMello, A.J.; Manz, A.: Science 280 (1998), 1946

[11] Ehricht, R.; Kirner, T.; Ellinger, T.; Foerster, P.; McCaskill, J.S.: Nucl. Acids Res. 25,22 (1997), 4697-4699

[12] Ellinger, T.; Ehricht, R.; McCaskill, J.S.: Chem. Biol. 5,12 (1998), 729-741

[13] Ehricht, R.; Ellinger, T.; Poser, S.; Schulz, T.; Uebel, S.; Mokansky, A.; Koehler, J.M. : BIOTechniques, submitted

Disposable electrophoresis chip for high throughput analysis of biomolecules

R. Konrad[1], W. Ehrfeld[1], H.-J. Hartmann[1], P. Jacob[1], M. Neumann[2], R. Pommersheim[1], I. Sommer[1], J. Wolfrum[2]

[1] Institut für Mikrotechnik Mainz GmbH, Carl-Zeiss-Str. 18-20, D-55129 Mainz, Germany
Phone: +49-6131-990-432, E-mail: konrad@imm-mainz.de
[2] Physikalisch-Chemisches Institut, Ruprecht-Karls-Universität, Im Neuenheimer Feld 253, D-69120 Heidelberg, Germany

Abstract.

An electrophoresis chip for high-throughput applications within the life sciences will be presented. The chip has been developed as a disposable and enables the parallel application and rapid separation of 96 samples on an area similar to a credit card. A transfer system in the well-to-well dimensions of a 384-well titer plate allows the direct transfer of 96 samples from the reaction well into the separation channels of the chip. This reduces analysis time and sample consumption drastically. Analyses which require several hours and several microliters of substance in conventional systems can be reduced to minutes and pico- or nanoliter consumption with the new system. Preliminary tests concerning sample transfer and separation efficiency in a one-channel chip verify the good functionality of the system. Simply by adsorption at the pin picoliter amounts of sample can be transferred, which are sufficient for subsequent fluorescence-lifetime analysis. All eleven fragments of the double-stranded DNA Φ κ 174 Hae III could be separated within four minutes with baseline resolution.

Keywords: Capillary electrophoresis chip, sample transfer system, highly parallel, time-resolved fluorescence detection

1 Introduction

Since the introduction of micromachined capillary electrophoresis systems in 1992 [1] different microchip geometries have been applied to DNA separations [2]. Microfabricated glass chips were designed to separate DNA restriction fragments [3-5], PCR products [3], short oligonucleotides [6] and DNA sequencing fragments [7-8]. Further improvements have been made to run multiple samples on capillary array electrophoresis microplates [5,9].

Commonly those chips have been fabricated as reusable, expensive, and handcrafted microstructures made of silicon or glass [2]. Recently, polymer microchips have been used to separate DNA restriction fragments [10,11] and to detect single DNA molecules [12]. The application of inexpensive mass production for microstructured components and systems for analytical purposes together with the extension of individual channels to a high density array format will help to solve the demands of high-throughput fields like the Human Genome Project.

The project described here is part of a joint project funded by the Federal Ministry for Education, Science, Research and Technology (BMBF) [13] for the development of a new approach for the high throughput sequencing of DNA. IMM's role within this project was the development of a microstructured electrophoresis chip for the rapid analysis of the huge sample amounts as they arise from the total analysis of the human genome or of other organisms. The chip was designed as a disposable.

As given in figure 1 [14] an electrophoresis chip commonly consists of four reservoirs for sample and buffer, four electrodes, which are plugged into these reservoirs from the outside, an injection channel, which crosses a linear or meandric separation channel, and a detection area near the anode. This arrangement requires much space so that parallelization, as demanded for a high throughput, is only possible on a huge area. Pipetting the sample into the reservoir requires volumes in the microliter range. Richard Mathies and coworkers [8] were the first to present a new design at the μ-TAS ´98 in October 1998 in Banff, Canada (fig. 2). Here, only 7 electrodes (6 cathodes and 1 anode) and 127 reservoirs (5/4 N + 1) are required for 48 separation channels. The microstructure is placed on a 4" boro-

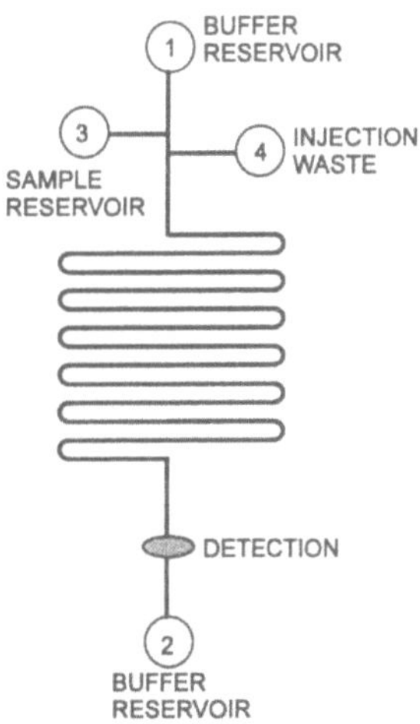

Fig. 1: Electrophoresis chip with injection channel (source: Soper et al. [14])

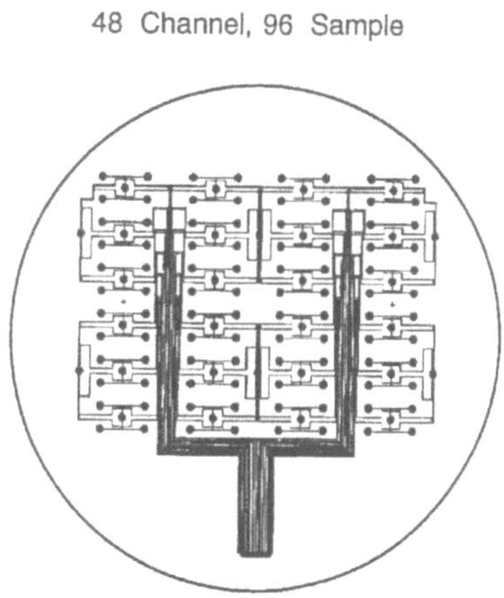

Fig. 2: 48-channel chip on a 4" borosilicate wafer (source: Mathies et al. [8])

silicate wafer and still uses crossed injection channels. A further reduction of the required space is impossible with this kind of injection and so the high throughput given by closely packed titer plates (384 wells and more) can not be achieved in an appropriate time scale.

2 New Design

The chip presented here comprises a bundle of new features. The sample injection has been changed from crossed channels to direct transfer, which results in the need for only two electrodes in total, and it is fabricated as a disposable from inexpensive plastic material. 96 samples can be applied in parallel and can be separated in 96 parallel separation channels (fig. 3). Analysis is achieved by a new detection method developed at the University of Heidelberg [15,16]. After excitation of the dye-labeled biomolecule with a pulsed laser diode at 640 nm the emitted fluorescence light was collected by an avalanche photodiode. Data acquisition was performed by a time-correlated single-photon counting PC interface card.

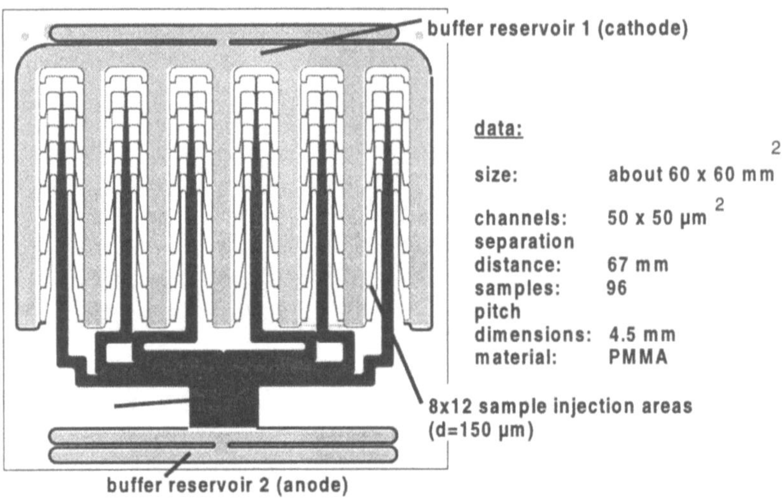

Fig. 3: Design of the 96-channel electrophoresis chip

With outer dimensions of 60 x 60 mm^2, the chip is slightly smaller than a credit card comprising 96 separation channels of 50 µm depth and 50 µm width and a total length of 67 mm. The sample application areas having a diameter of 150 µm are microstructured as circular widenings of the separation channels in the pitch dimensions of a 384-well microtiter plate. They are connected to both the cathode and the separation channel. The cathode´s comb-like shape ensures an uniform electrical field for each separation channel. Cathode and anode have been integrated into the chip by vapor deposition of a 100 nm gold layer through a mask. For faster detection the separation channels are bundled above the anode.

3 Fabrication

Commonly used replication processes for polymers are hot embossing and injection molding. Whereas hot embossing is a method appropriate for prototyping and small series, injection molding is a very fast process for mass production. For the production of microstructured disposable chips a metallic mold insert, preferably of stainless steel, nickel and nickel alloys, is required. Fabrication of a nickel mold insert (fig. 4) was achieved by advanced silicon etching (ASE) of a 5"-silicon wafer in the positive tone followed by the transformation into the complementary structure in nickel by electroplating. We chose the hot embossing method for the manufacture of a small series of the 96-channel electrophoresis chip in Polymethyl methacrylate (PMMA). Replication of the microstructures into the polymer material was achieved with high precision.

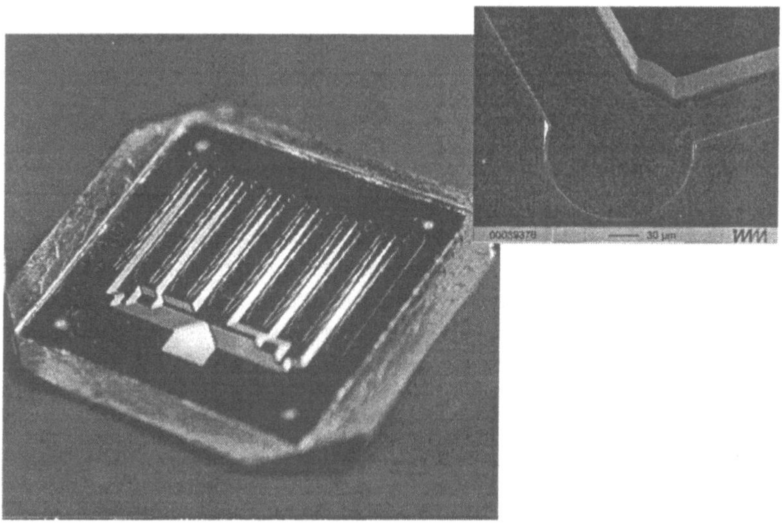

Fig. 4: Nickel mold insert with SEM-image of design details

Integration of the electrodes has been achieved by vapor deposition of a 100 nm gold layer through a mask. In figure 5 the chip is shown with SEM-images of prominent details.

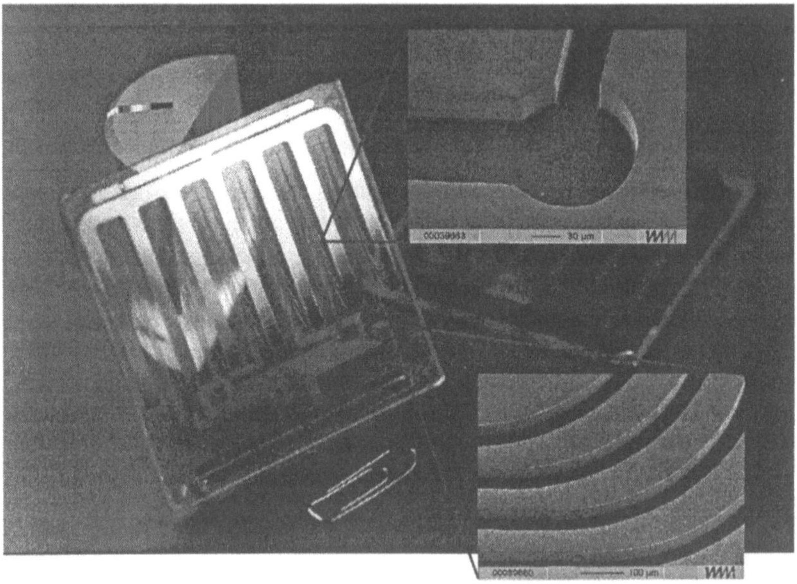

Fig. 5: PMMA-chip with integrated electrodes and SEM-images of sample application area and separation channels

Application of a structured 125 μm thick PMMA foil on top of the electrophoresis chip not only enables the fast and easy filling of the channels with buffer and gel, but also prevents cross talking and evaporation. From a choice of methods only lamination at room temperature supported by organic solvents was applicable. The geometries and dimensions of the channels remained unchanged. After filling with an aqueous dye solution no cross contamination is observed as shown in figure 6.

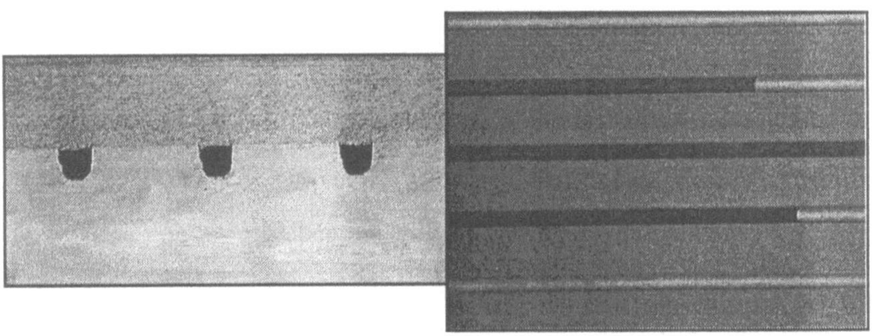

Fig. 6: Bonding of the channels with a 125 μm top layer. Microscopic images of the cut channels (left) and after filling with an aqueous dye solution (right). Channel width 50 μm, distance between channels 150 μm.

4 Sample Transfer

To obtain the parallel transfer of 96 samples from a 384-well titer plate to the separation channels of the chip in small volumes, a sample transfer system consisting of a plate with 96 pins has been fabricated. The pins having a base of 2 mm and tops of 80 μm^2 (fig. 8) are arranged in the pitch dimensions of the plate as well as of the sample application zones of the chip. Thus, according to fig. 7 the sample transfer is easily achieved by dipping the pins into the wells, resulting in adsorption of the samples at the pin surface, and subsequent deloading of the pins by injection into the slightly widened separation channels of the chip. Application of a low voltage at the metallic pins during the loading process leads to higher sample concentrations.

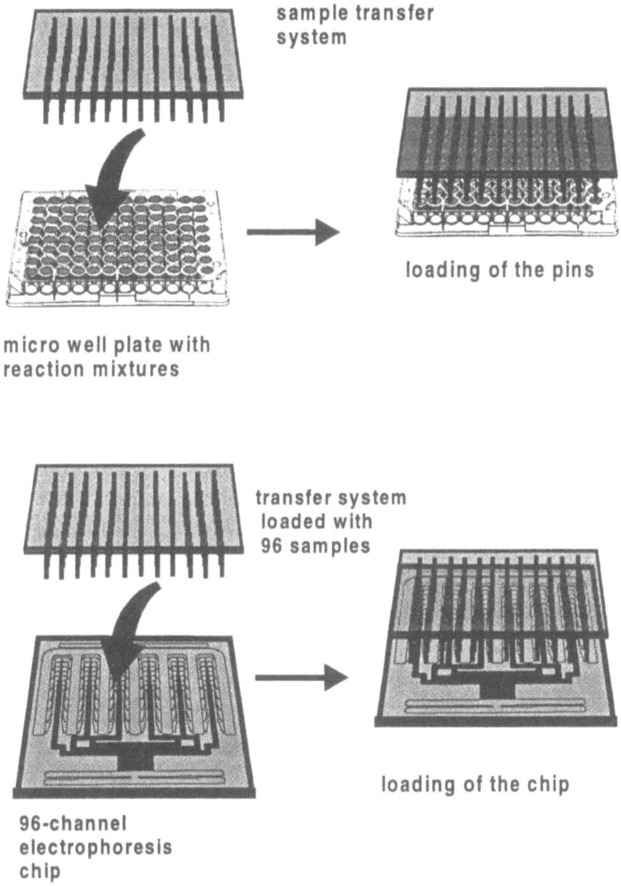

Fig. 7: Scheme of sample transfer from titer plate to chip

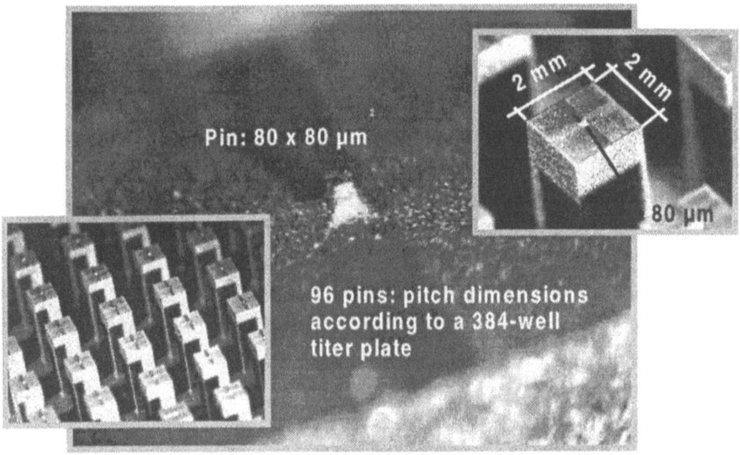

Fig. 8: Microscopic image of the pin array

5 Experimental

The appropriate polymer material for an electrophoresis chip has a low autofluorescence at the emission wavelength of the fluorescent marker molecule and an electroosmotic flow comparable to coated quartz capillaries. From all the different materials we tested, we chose PMMA. In addition to the properties required above, it is a good insulator with a reasonable low price and good fabrication properties.

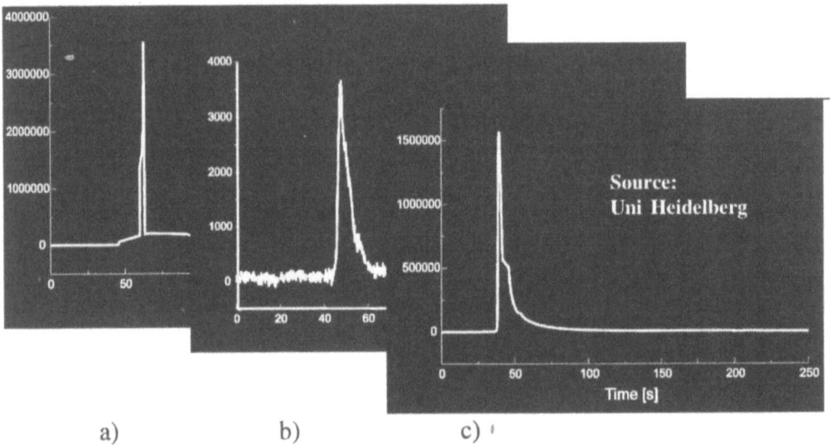

Fig. 9: Comparison of the different injection methods: a) cross channel, b) pin adsorption (5 sec.), c) pin adsorption with low voltage (5 sec. 3.8 V)

The different methods for sample injection had to be tested and evaluated. For these tests two different chips comprising only one channel have been structured. The channel was straight with 50 μm each for depth and width, a total length of 60 mm and an effective length of 50 mm. A tightly bonded 125 μm thick PMMA foil prevented the channel from leaking. The difference between the two kinds of chips was the sample injection. The first kind of chip had the "classical" crossed channel design according to fig. 1, while the second one has the 150 μm application hole. Conditions were the same: 0.75 % HEC (Hydroxy ethyl cellulose) + 0.5 % PVP (Polyvinyl pyrrolidone) in 0.89 M TBE (Tris-borate-EDTA) buffer at pH 8,4 as medium and 400 V/cm as separation voltage. Detection of the fluorescence lifetime has been achieved with a confocal microscope after excitation at 640 nm by a laser diode and single photon counting by an avalanche diode according to [15]. As test substance we chose a single stranded DNA-primer M13 mp21, 5′-end labeled with Cy5 in concentrations from 25 fmol/μl to 2.5 pmol/μl. Sample transfer has been achieved a) by cross channel, b) by pin adsorption (5 seconds) and c) by application of 3.8 V during pin adsorption (5 seconds). Fig. 9 a), b) and c) reflect the different transfer methods. The amounts of transferred samples have been estimated by comparison of the peak intensities.

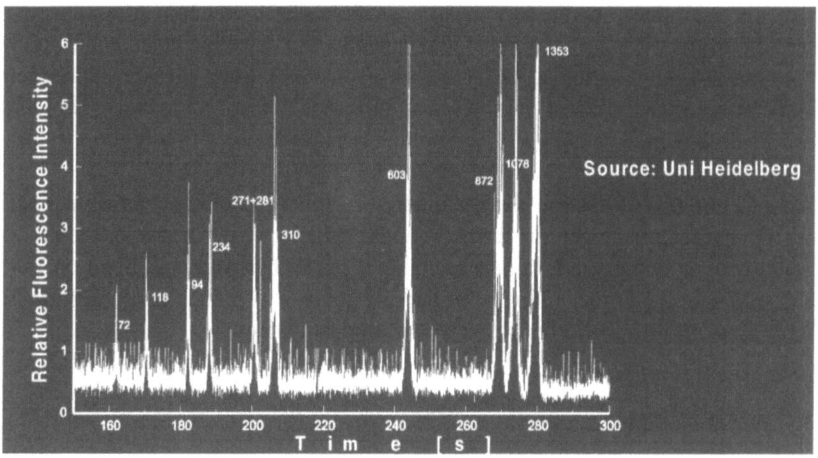

Fig. 10: Fragment separation of Hae III Φκ 174 DNA in a PMMA chip

Compared to the well-defined sample volume of 125 pl given by the cross section of injection channel and separation channel in version a) consisting of 30 fmol DNA, methods b) and c) result in the transfer of 3 and 120 fmol respectively. It could be shown that simple adsorption of aqueous biomolecule samples leads to a transferred amount which is sufficient for a sensitive detectioı method and that this amount could be enhanced by a factor of 40 by applying a low voltage at the pins during loading. The sample consumption could be reduced drastically, because we did not have to fill a reservoir as in a) with 20 µl of sample. Only picoliter amounts have to be transferred.

For the tests of the separation efficiency of polymeric electrophoresis chips we chose the well known fragment separation of the double-stranded Φ κ 174 DNA after restriction with Hae III. After labeling with the intercalating agent To-Pro-3-iodid (Molecular Probes) the mixture was injected for 50 seconds with a voltage of 400 V/cm and a concentration of 10 ng/µl using 5 % Genescan in 2 mM TTE (Tris-TAPS-EDTA buffer, pH 8.4) as sieving agent. As shown in fig. 10 the separation was achieved within four minutes with baseline resolution of all eleven fragments. The concentration of the single peaks is estimated to be 750 zmol.

6 Conclusions and Outlook

Design, fabrication and preliminary tests of an disposable electrophoresis chip with 96 separation channels and of a pin array for the direct transfer of samples from a 384-well plate onto the chip has been described. The parallelization of separation channels was achieved by integration of the injection area in the channels instead of using space-filling crossed channel systems. Preliminary tests confirmed the usefulness of the new design. It allows the reduction of sample consumption from microliters to picoliters, enables the baseline resolved fragment separation of double-stranded DNA and saves analysis time. Further experiments will be done with different samples and different detection methods, including labelfree detection. The chip will be applicable in all areas with a high throughput of samples.

As a follow-up of this one-dimensional electrophoresis chip a two-dimensional chip for application in the proteome analysis will be constructed. The chip will allow the medical diagnosis of auto immune diseases by pattern recognition. As is the case for the genomics chip the proteomics chip will reduce analysis time and sample consumption. A fully automated version is planned.

7 Acknowledgement

This work has been supported by the Bundesministerium für Bildung, Wissenschaft, Forschung und Technologie (BMBF) under reference 0311088.

8 References

[1] Manz A., Harrison, D., Verpoorte E., Fettinger J., Paulus A., Ludi H., Widmer H. Planar chips technology for miniaturization and integration of separation tecniques into monitoring systems. Capillary electrophoresis on a chip. J. Chromatogr. 593 (1992) 253

[2] Effenhauser C. Integrated chip-based microcolumn separation systems. Topics in Current Chemistry, Vol. 194. Springer Verlag Berlin Heidelberg 1998, 52-82

[3] Wooley, A., Mathies, R. Ultra-high-speed DNA fragment separations using microfabricated capillary array electrophoresis chips. Proc. Natl. Acad. Sci. USA, 91 (1994) 11348-11352

[4] Jacobson S., Ramsey, J. Integrated microdevice for DNA restriction fragment analysis. Anal. Chem. 68 (1996) 720-723

[5] Wooley, A., Sensabaugh, G., Mathies, R. High-speed DNA genotyping using microfabricated capillary array electrophoresis chips. Anal. Chem. 69 (1997) 2181-2186

[6] Effenhauser, C., Paulus A., Manz A., Widmer H. High-speed separation of antisense oligonucleotides on a micromachined capillary electrophoresis device. Anal. Chem. 66 (1994) 2949-2953

[7] Wooley, A., Mathies, R. Ultra-high-speed DNA sequencing using capillary electrophoresis chips. Anal. Chem. 67 (1995) 3676-3680

[8] Mathies R., Simpson P., Woolley A. DNA analysis with capillary array electrophoresis microplates. Kluwer Acad. Publishers, Dordrecht/Boston/London. Proc. of the μ-TAS 98 Workshop, Banff, Canada 13-16 Oct. 1998, 1-6

[9] Simpson P., Roach D., Woolley A., Thorsen T., Johnston R., Sensabaugh G., Mathies R. High-throughput genetic analysis using microfabricated 96-sample capillary array electrophoresis microplates. Proc. Natl. Acad. Sci. USA 1998, 95, 2256

[10] Roberts, M., Rossier, J., Bercier, P., Girault, H. UV laser machined polymer substrates for the development of microdiagnostic systems. Anal. Chem. 69 (1997) 2035-2042

[11] McCormick, R., Nelson, R., Goretty Alonso-Amigo, M., Benvegnu, D., Hooper, H. Microchannel electrophoretic separations of DNA in injection-molded plastic substrates. Anal. Chem. 69 (1997) 2626-2630

[12] Effenhauser C., Bruin G., Paulus A., Ehrat M. Integrated capillary electrophoresis on flexible silicone microdevices: Analysis of DNA restriction fragments and detection of single DNA molecules on microchips. Anal. Chem. 69 (1997) 3451

[13] Joint project "DNA-sequencing with multiplex dyes and capillary electrophoresis – part 3: Micro electrophoresis chip". Funded by the Bundesministerium für Bildung, Wissenschaft, Forschung und Technologie (BMBF) as 0311088.

[14] Soper S., Ford S., Davies J., Williams D., Cheng B., Klopf M., Calderon G., Saile V. Micro-electrophoresis devices with integrated fluorescence detectors and reactors for high-throughput DNA sequencing ultrasensitive biochemical diagnostics II, SPIE Vol. 2985 (1997) 36-46

[15] Lieberwirth U., Arden-Jacob J., Drexhage K., Herten D., Müller R., Neumann M., Schulz A., Siebert S., Sagner G., Klingel S., Sauer M., Wolfrum J. Multiplex Dye DNA Sequencing in Capillary Gel Electrophoresis by Diode Laser-Based Time-Resolved Fluorescence Detection. Anal. Chem. 1999, 70, 22, 4771-4779.

[16] Neumann, M; Ehrfeld, W.; Han, K; Jacob, P., Konrad, R.; Siebert, S.; Wolfrum, J.; Sauer, M. 1999. Biomed. Chromatogr. 13, 109-110. Electrophoretic sizing of DNA fragments and detection of single mononucleotide molecules in microstructures.

A Micromachined Analysis System for Rapid Protein Identification.

[1]Simon Ekström, [1]Patrik Önnerfjord, [2]Martin Bengtsson, [2]Johan Nilsson, [2]Thomas Laurell and [3]György Marko-Varga

[1] Dept. Analytical Chemistry, Lund University,
P.O. Box 124, 221 00 Lund, Sweden
[2] Dept. Electrical Measurements, University of Lund,
P.O. Box 124, 221 00 Lund, Sweden
[3] Astra Draco AB, Bioanalytical Chemistry,
Preclinical Research and Development,
P.O. Box 34, 221 00 Lund

Abstract

A microfabricated flow-through analysis system for rapid protein identification is presented. The system facilitates rapid enzymatic digestion of proteins, on a porous silicon immobilized protease microreactor and on-line picoliter sample preparation, by micro-dispensing of the sample into microchip vials for matrix-assisted laser desorption/ionization time of flight mass spectrometry (MALDI-TOF MS) analysis. The generated peptide map is used for database identification of the parent protein. This system enables automated analysis of 100 protein samples to be performed in 5 hours. Kinetic studies on the micro-reactors are reported and application of this system for identifying model proteins, such as lysozyme, with high sequence cover is shown.

Keywords. Micromachining, microdispensing, mirovials, silicon, enzyme immobilization, trypsin, MALDI-TOF MS, peptide mapping, protein identification, biotechnology

1 Introduction

The explosive progress in the fields of protein science and bioinformatics during the latest decade has led to an increasing interest in the development of analytical technology that allows proteins to be subjected to mass screening, as identification of proteins is central in all proteome projects. Miniaturization is generally considered to be one of the best ways to increase process speed and keep the cost of analysis down. By utilizing the versatility of silicon micromachining to fabricate efficient minute volume microstructures it is possible to make analysis systems that are extremely small [1-3]. The advantage of miniaturization arises from the fact that reaction kinetics as well as sample handling procedures can be performed at a high speed in micro-nanoliter systems. Also, the sample amounts in life science research fields often are very small and precious. It is therefore desirable that every analysis consumes as little sample as possible while providing the desired information. In this perspective mass spectrometry has evolved as a very important and widely used analytical method in the recent years [4, 5]. Ideally the small sample consumption of MALDI-TOF MS [6] also allows other complementary analyses to be performed on the remaining sample, extracting the maximum amount of information that is obtainable from each sample.

2 Materials and methods

2.1 Chemicals

Chymotrypsin, bovine pancreas (Lot: 85060820-64) from Boehringer Mannheim (GmbH, Mannheim, Germany). Glutaraldehyde, grade II, 25% (aq), (Lot: 56H0107); 3-Aminopropyltriethtoxysilane (ATPES), (Lot 127H1073); Guanidine HCl (Lot: 12H5725); Dithiothreitol (DTT), (Lot 45H007215); Iodoacetamide (Lot: 25H8452); α-N-Benzoyl-L-Arginine Ethyl Ester (BAEE) (Lot: 17H0861); Trizma® Base (Lot:12H56001); Trizma® HCl (Lot: 35H5705); Trypsin type IX, porcine pancreas, (Lot 25H0359); myoglobin, horse heart, (Lot: 76H7210); and α-cyano-hydroxycinnamic acid, (CHCA) (Lot: 51229) were all purchased from Sigma Chem. Co. (St. Louis, MO, USA). Lysozyme (Lot: 56518) was purchased from ICN Chemicals (Costa Mesa, CA, USA). HCl, acetone, 2-propanol, NH_4HCO_3, KOH, HF, $CaCl_2$, KH_2PO_4, Na_2HPO_4, NaN_3, NaCl, acetonitrile (ACN) from Merck (Darmstadt, Germany). Urea (Lot: 51459); ACTH (Lot: 91007); trifluoro acetic acid (TFA) and sinapinic acid were obtained from Fluka Chemicals (Buchswald, Switzerland), and used without any further purification. Angiotensin I, substance P, ACTH clip 1-17, ACTH clip 18-39, ACTH clip 7-3 were all part of mass standard kit no: 2-3143-00 (Lot: 7022801), Perseptive Biosystems, Framingham, MA, USA). Nitro Cellulose Trans-Blot® transfer medium, 0,45 micron from Bio-Rad laboratories (Richmond, CA 94804, USA). The water was purified using a Millipore apparatus (Bedford, MA, USA). All chemicals were used without further purification and all buffers used were freshly prepared, pH checked and if used in a critical application autoclaved.

2.2 Micromachining

The micromachined parts of the analysis system described in this paper includes a μ-chip IMER, wherein enzymatic protein digestion takes place, the microdispenser, which was used to deposit samples and the μ-vial target plate.

2.2.1 The μ-IMER

The μ-chip IMER was fabricated in <110> -oriented silicon, which allows vertical anisotropic etching, to generate a surface enlarging high aspect ratio parallel channel structure[7]. To further enhance the surface area in the μ-chip IMER it was treated by anodization in an HF-solution yielding a thin porous surface layer on the channel walls. The optimal porosity and manufacture procedure for this layer have been determined for a glucose monitoring systems in an earlier work by Drott et al. [8, 9]. Porous silicon was in that application shown to give a 170-fold increase in enzyme activity compared to non-porous reactors. The internal-volume of the μ-chip IMER used in this work was 6 μl.

2.2.2 The micro-dispenser

The microdispenser was fabricated by anisotropic etching of monocrystalline silicon as described elsewere [10]. In this work the in-house developed piezoactuated flow through microdispenser was used to deposit sample volumes ranging from 50 pL to over 100 nanoliter. The dispensing principle is based on a piezo-ceramic element, which elongates when a voltage pulse is applied, creating a pressure pulse that results in droplet ejection. The unique dispenser construction features flow-through capability, and a pyramid-shaped nozzle giving stable droplet directivity. The internal volume of the microdispenser used was 3 μl. The use of this microdispenser for MALDI-TOF MS sample preparation has previously been described in work by Önnerfjord et al [11]. Sample deposition on high density MALDI-targets was accomplished by adjustment of the target plate using an in-house built computer controlled x-y stage.

2.2.3 μ-vials

Collecting the sample in micro-vials is another approach that can be used in order to increase the sensitivity. This has been reported by Jespersen et al. [12], who presented picoliter vials etched in silicon wafers. The vials were inverse pyramids of 100x100 μm with a volume of 250 pL and low attomole levels of peptides and proteins were detected with MALDI-TOF MS. The increased sensitivity found with these vials can probably be attributed to the fact that the vial geometry allows a larger fraction of the whole sample to be analyzed. In this work similar vials ranging in size from 100-400 μm were used. The use of vials has the advantage over planar surfaces that a greater amount of sample can be collected on a smaller surface area, due to surface tension effects at the vial edges, and the exact positioning of the samples in the vials makes automation of the MALDI analysis simpler.

2.3 Enzyme immobilization

The immobilization of proteolytic enzymes onto the μ-chip IMER was made according to standard procedures for enzyme coupling to silica matrices [13]. This was done in three steps, first by silanization in 10% v/v 3-aminopropyltriethoxysilane, pH 3,5 at 75°C for 3 hours, followed by extensive washing. Glutaraldehyde activation was subsequently performed in 0,1 M PBS, pH 7,0 containing 2,5% glutaraldehyde for 4 hours at room temperature. Finally the enzyme was coupled, 2-4 mg/ml enzyme in PBS for 24 hours. Enzymes that were immobilized in this work, were trypsin and chymotrypsin.

2.4 System set-up

The integrated total analysis system is depicted in Figure 1, with photo inserts of micromachined parts. The internal volume of the total analysis system is 13 μl.

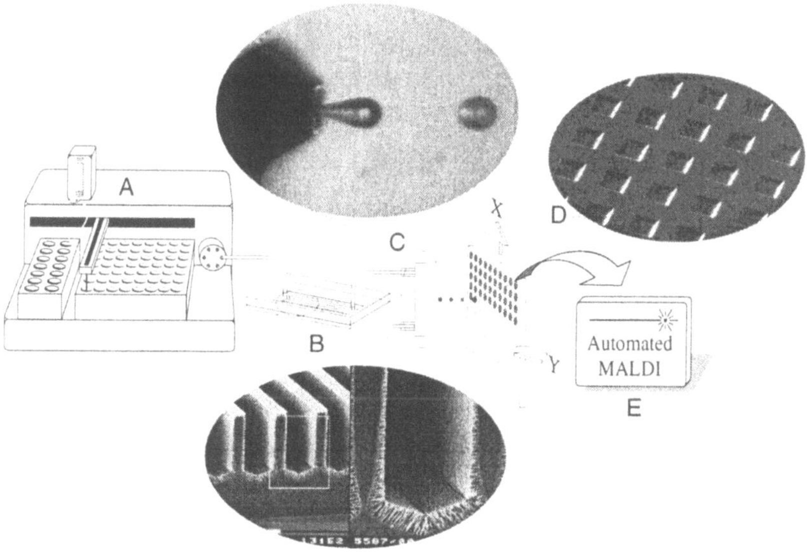

Fig. 1. The total analysis system and its different parts; A = automated sample handling and injection, B = μ-chip IMER, photo insert shows a SEM picture of the lamella structure and porous layer, C = microdispenser, the magnified photo shows an ejected droplet in the air approaching the MALDI target plate, D =ground vials (300x300x20 μm) on the MALDI-target plate and E = automated MALDI-TOF MS analysis.

The protein samples are first subjected to pre-treatment in an autosampler (A in Figure 1) where it is made amenable for digestion and then injected into a flow. This flow carries the sample to a μ-chip IMER (B in Figure 1) made of porous silicon wherein the enzymatic digestion takes place, and then further to a piezo-electric flow-through microdispenser (C in Figure 1). The microdispenser ejects an adjustable fraction of the sample onto an MALDI-target plate (D in Figure 1). Finally the deposited sample is analyzed by MALDI-TOF MS (E in Figure 1) and the resulting peptide map is used for database search.

2.4.1 Pre-treatment

The pre-treatment robot, shown as part A in Figure 1, was a Gilson (Villiers-le-Bel, France) ASPEC XL autosampler equipped with a Rheodyne (Berkeley, CA, USA) four-port injection valve with an internal loop of 0,5-5 μl volume. The sample rack was adapted for this application, giving a maximum capacity of 576 samples per run. This pre-treatment system was controlled with Gilson 719 software from a Pentium PC. The pre-treatment was performed by adding a denaturating agent, urea (1-8 M), or guanidine HCl (0,5-6 M) and a reducing agent, dithiothreitol (4-8 mM) to the protein sample. All reagents and samples were dissolved in digestion buffer (25 mM NH_4HCO_3, pH 7,8 or 25 mM Tris, pH 8,2). The sample was then heated to 75°C for 20 minutes. In cases where refolding of the protein was to be avoided an alkylating agent, iodoacetamide (10-30 mM) was added after the denaturation/reduction step. The carboxyamidomethylation reaction took 30 minutes at room temperature.

2.4.2 Automated digestion and sample preparation

Normally a digestion time of 1 minute was sufficient to achieve the desired degree of digestion for the studied proteins. To aid the digestion the μ-chip IMER was heated to a temperature of 40-60° C. At a flow rate of 6μl/min, the digestion buffer carries a sample through the system and completes the process of digestion and sample preparation in 2 minutes.
The time axis below (Figure 2) describes the analysis of 100 samples. The time between two consecutive injections of sample into the μ-chip IMER is currently around 2 minutes. This was determined by the need to avoid carry-over effects.

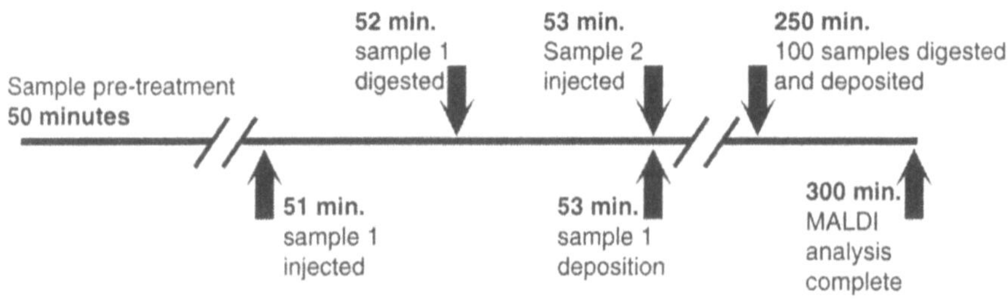

Fig. 2. Time axis describing the analysis of 100 samples with the total analysis system.

For peptide analysis the sample was deposited directly onto a pre-deposited thin-layer matrix surface [15, 16] using the microdispenser. Alternatively the matrix was deposited after the sample. To increase the speed of on-target pre-concentration the MALDI plate was heated to a temperature of 50°C. Sample clean up can be performed in many ways, the most commonly used method is washing the prepared sample spot. Washing of the sample spot surface was done by depositing cooled acidic water (0.1 % TFA) over the entire target plate, and removing it by an air stream.

2.4.3 MALDI-TOF MS analysis

The MALDI-TOF instrument used was a Voyager DE-RP/PRO (Perseptive Biosystems Inc., Framingham, MA, USA) mass spectrometer. The instrument was equipped with a delayed extraction ion source, and used a nitrogen laser at 337 nm, which could be operated in both linear (L) and reflector (R) mode. For high accuracy peptide mapping reflector mode was used. The high degree of mass accuracy, 5-100 ppm obtained with instruments using the delayed extraction principle [17-20] and reflectron detectors makes sequencing and identification amenable [21]. Identifications done in this work were performed with the aid of internet accessible programs, MS-Fit, ProFound and PeptideSearch [22].

2.5 Enzyme activity determination

The catalytic efficiency of the trypsin μ-chip IMER was measured using a spectrophotometric assay [14]. This was done to obtain absolute values of the reactor performance for theoretical calculations and comparison with future μ-chip IMER designs. As the substrate was injected into the μ-chip IMER the enzymatic reaction yielded a colored compound as product. A solution containing α-N-benzoyl arginine ethyl ester (BAEE) was continuously injected directly into the absorbance detector, Waters 486 Tunable Absorbance Detector (Millipore Corp., Milford, MA,USA) to obtain the baseline level. Subsequently a valve was switched and the reagent was passed through the μ-chip IMER before entering the absorbance detector. The observed shift in absorbance reading was the result of enzymatic cleavage of the substrate (BAEE).

3 Results

3.1 Enzyme activity determination

By applying the Lambert-Beer Law to the observed absorbance readings , the actual BAEE turn-over rate, V_e can be calculated [7]. The obtained turn-over rates was plotted against flow for BAEE concentrations of 0,075-0,5 mM, Figure 3 shows the plot for the BAEE concentration of 0,5 mM. To ensure that first order enzyme kinetics prevailed, i.e. that the turn-over rate was independent of the flow, a value for the maximum enzyme activity at each BAEE concentration was estimated at a flow rate where the turn-over had leveled. By plotting these values against the substrate concentration, estimation of apparent V_{max} and K_m were obtained by fitting the theoretical equation of a first order enzyme kinetic system [23] to a plot of the data. Additional estimation of apparent V_{max} and K_m were obtained by making an Eadie-Hofstee plot and a Lineweaver-Burk plot . All these methods treat errors in the data points differently i.e. in a Lineweaver-Burk plot an error at low concentration have greater impact on the curve fit than at high concentration, and the opposite is true for an Eadie-Hofstee plot.

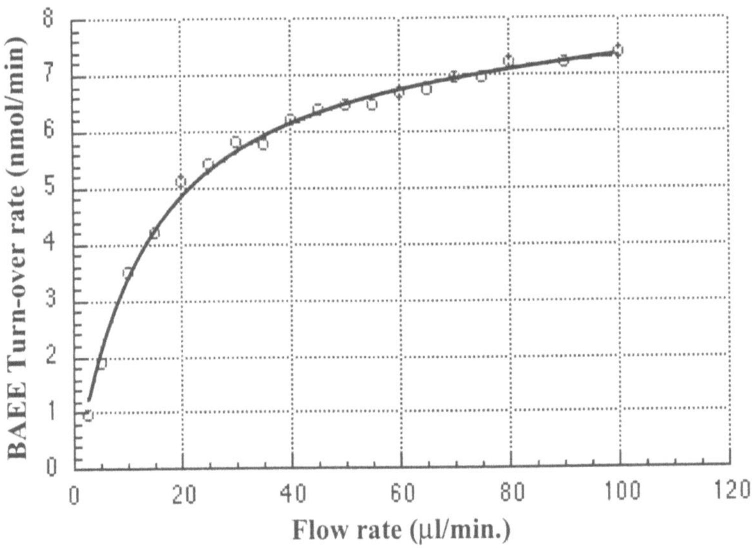

Fig. 3. BAEE turn-over rate versus flow rate plot for the BAEE concentration of 0,5 mM. The maximum enzyme activity was estimated at the highest deployed flow rate.

The estimated apparent V_{max} and K_m values are shown in Table1, as can be seen there is very little variation between the different estimates, which indicates that there is a good estimation. All measurements were made at 25°C. Although, the turn-over rate is also affected by temperature and to get an estimation of the gains that may be obtained by increasing the temperature, measurements were also performed at 50° C, giving an increase in turn-over-rate of 25%.

Table 2. Estimated values of apparent V_{max} and K_m for a trypsin m-chip IMER obtained from the different plot methods.

	V_e versus S_0	Lineweaver-Burke	Eadie-Hofstee
Apparent V_{max} (nmol/min)	8,9	8,8	8,9
Apparent K_m (mM)	0,11	0,10	0,11

3.2 Protein digestion

Model protein samples (1 pmol/μl), cytochrome c, myoglobin, lysozyme, ribonuclease A, and albumin, digested on a trypsin μ-chip IMER was identified by database search. In all cases 8-20 peptides could be assigned to the parent protein, resulting in a sequence coverage of 50-100%, Figure 5 shows spectra of lysozyme obtained by μ-chip IMER digestion

Comparison of traditional in-solution digestion and μ-chip IMER digestion revealed that under equal conditions the μ-chip IMER gave better or equal identification results to that of in-solution digestion. But μ-chip IMER digestion was completed in 1-3 min. This should be compared to the in-solution digestion times of 8-24 hours to get comparable results. Also, the mass spectra resulting from μ-chip IMER digestion of proteins were devoid of trypsin autolysis peaks.

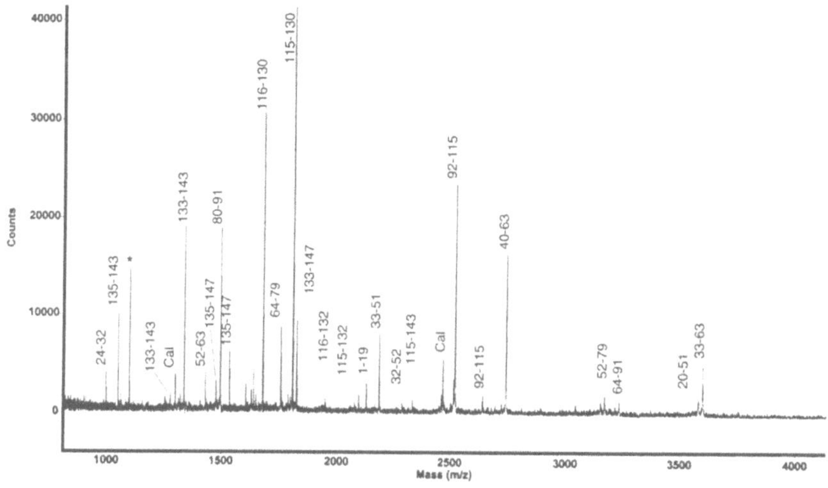

Fig. 5. The mass spectrum shows the resulting digest of lysozyme on a trypsin immobilized μ-reactor. Peaks are marked according to which amino acid residues the peptides correspond to in the parent protein.

4. Discussion

The system presented here is designed to be used as a first step in protein identification processes, were it is desirable to quickly screen, large amounts of purified protein samples. The goal is to avoid time-consuming manual steps and to make analysis as cost-effective as possible. The use of miniaturized micromachined structures in this application has many advantages, such as the ability to handle small amounts of

3.2 Protein digestion

Model protein samples (1 pmol/μl), cytochrome c, myoglobin, lysozyme, ribonuclease A, and albumin, digested on a trypsin μ-chip IMER was identified by database search. In all cases 8-20 peptides could be assigned to the parent protein, resulting in a sequence coverage of 50-100%, Figure 5 shows spectra of lysozyme obtained by μ-chip IMER digestion

Comparison of traditional in-solution digestion and μ-chip IMER digestion revealed that under equal conditions the μ-chip IMER gave better or equal identification results to that of in-solution digestion. But μ-chip IMER digestion was completed in 1-3 min. This should be compared to the in-solution digestion times of 8-24 hours to get comparable results. Also, the mass spectra resulting from μ-chip IMER digestion of proteins were devoid of trypsin autolysis peaks.

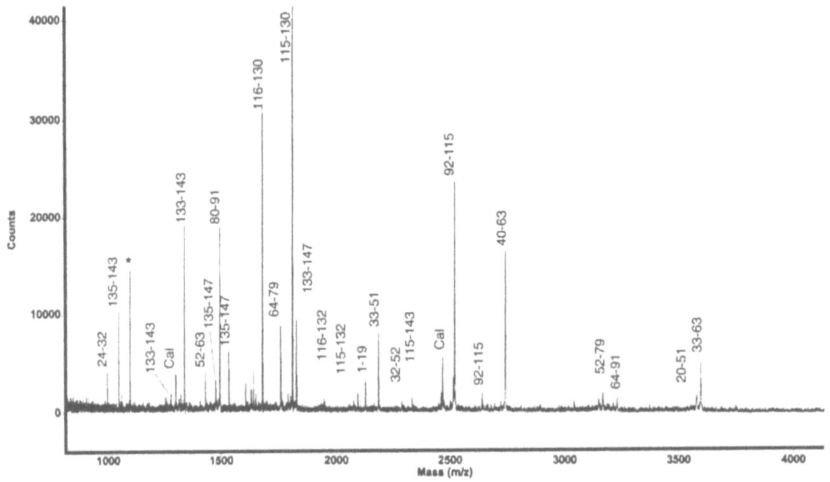

Fig. 5. The mass spectrum shows the resulting digest of lysozyme on a trypsin immobilized μ-reactor. Peaks are marked according to which amino acid residues the peptides correspond to in the parent protein.

4. Discussion

The system presented here is designed to be used as a first step in protein identification processes, were it is desirable to quickly screen, large amounts of purified protein samples. The goal is to avoid time-consuming manual steps and to make analysis as cost-effective as possible. The use of miniaturized micromachined structures in this application has many advantages, such as the ability to handle small amounts of sample and achieving rapid digestion of proteins. The reduced and automated sample handling also leads to minimized sample loss. The immobilization of the protease also has the advantage of avoiding autolytic interferences from the proteolytic enzyme in the mass spectra. By using porous silicon as a carrier for the immobilized enzyme in the μ-chip IMERs a large surface area and thus a high catalytic turn-over is obtained. The micromachined reactor does not quite offer the same surface area as a micro-column with packed material, but it has better flow characteristics, lower hydrodynamic pressure drop, is less sensitive to carry-over effects and much simpler to manufacture in the desired small sizes. The long-term operational stability of the μ-chip IMER was very high. This was proven by using one of the reactors for performing more than 400 digestions of samples containing high concentrations of pre-treatment reagents conditions, during a time period spanning over two months, with no observed loss in catalytic efficiency. Another big advantage of the system set-up is the sample deposition step. Microdispensing of the samples easily allows several 1000 samples to be deposited on each MALDI-plate. The use of the flow-through microdispenser allows on plate sample enrichment at the same time as not all of the sample is deposited on the target. This allows collection of the remaining sample at the flow outlet. Unidentified samples or proteins of special interest can then be further analyzed by other methods, such as immunoassays, or in the case of key regulating proteins MS/MS sequencing by use of nanospray technology.

5. Conclusions

This report presents a new step in the development of tomorrows totally integrated and miniaturized analysis systems. The next generation of this microsystem, which already is under construction in our laboratory, will be even smaller with higher degree of integration. Smaller μ-chip IMERs in combination with even smaller micro-dispensers will be used to increase sensitivity and process speed. Making the system even smaller is important, because this would allow small sample volumes that contains low concentrations of protein to be injected continuously and not as a plug in a flow. Thus avoiding dispersion effects and gaining valuable sensitivity, especially together with sample enrichment from micro dispensing.

6. Acknowledgements

The authors would like to express thier gratitude to Carl Tryggers Stiftelse, Crafordska Stiftelsen, NUTEK (KOFUMA), Swedish Reasearch Council for Engineering Sciences, Kungliga Fysiografiska Sällskapet i Lund, Svenska Läkare-sällskapet, ELFA Forskningsstiftelse for financial support. Martin Hjelmtvedt at Pretech Instruments is acknowledged for assistance with the Gilson instrumentation.

7. References

1. B. He, N. Tait, and F. Regnier, Anal. Chem. 70:3790-3797 (1998).
2. J. D. Harrison and A. van den Berg, μ-TAS´98 , Kluwer Academic Publishers, Banff, Canada , (1998).
3. Y. Fintschenko and A. v. d. Berg, J. Chromatogr. A 819:3-12 (1998).
4. G. Siuzdak, Proc. Natl. Acad. Sci 91:11290-11297 (1994).
5. J. R. Yates, III, J. Mass Spectrom. 33:1-19 (1998).
6. G. Allmaier, Rapid Commun. Mass Spectrom. 11:1567-1569 (1997).
7. T. Laurell, L. Rosengren, and J. Drott, Biosensors & Bioelectronics 10:289-299 (1995).
8. J. Drott, L. Rosengren, K. Lindström, and T. Laurell, Thin Solid Films 330:161-166 (1998).
9. J. Drott, L. Rosengren, K. Lindström, and T. Laurell, Microchimica Acta in press (1999).
10. J. Nilsson, L. Wallman, P. Önnerfjord, G. Marko-Varga, and T. Laurell, μ-TAS´98 Kluwer Academic Publishers, Banff, 1998, p. 387-390.
11. P. Önnerfjord, J. Nilsson, L. Wallman, T. Laurell, and G. Marko-Varga, Anal. Chem. 70:4755-4760. (1998).
12. S. Jespersen, W. M. A. Niessen, U. R. Tjaden, J. v. d. Greef, E. Litborn, U. Lindberg, and J. Roeraade, Rapid Commun. Mass Spectrom. 8:581-584 (1994).
13. H. H. Weetall, Methods in Enzymology 44:134-148 (1976).
14. G. W. Shewert and Y. Takenaka, Biochim. Biophys. Acta:570 (1955).
15. A. Shevchenko, M. Wilm, O. Vorm, and M. Mann, Anal. Chem. 68:850-8 (1996).
16. M. Kussmann, E. Nordhoff, H. Rahbek-Nielsen, S. Haebel, M. Rossel-Larsen, L. Jakobsen, J. Gobom, E. Mirgorodskaya, A. Kroll-Kristensen, L. Palm, and P. Roepstorff, J. Mass Spectrom. 32:593-601 (1997).
17. M. L. Vestal, P. Juhasz, and S. A. Martin, Rapid Commun. Mass Spectrom. 9:1044-1050 (1995).
18. R. M. Whittal and L. Li, Anal. Chem. 67:1950-1954 (1995).
19. T. B. King, S. M. Colby, and J. P. Reilly, Int. J. Mass Spectrom. Ion Processes 145:L1-L7 (1995).
20. R. S. Brown and J. J. Lennon, Anal. Chem. 67:1998-2003 (1995).
21. O. N. Jensen, A. Podtelejnikov, and M. Mann, Rapid Commun. Mass Spectrom. 10:1371-1378 (1996).
22. Ms-Fit@http://prospector.ucsf.edu/cgi-bin/msfit.exe
    ProFound@http://prowl.rockefeller.edu/cgi-bin/ProFound
    PeptideSearch@http://193.175.249.95:80/CGI
23. C. R. Cantor and P. R. Schimmel, W.H. Freeman & Co., San Francisco, 896-8 (1980).

Simulation of Biochemical Reaction Kinetics in Microfluidic Systems

Vinod B. Makhijani[1], J. Raghavan, Agata Przekwas and Andrzej Przekwas,

CFD Research Corporation, Huntsville, Alabama 35805, USA

Abstract. Bench-top microfluidic bioassay systems often use biosensors to detect analytes. The efficacy of the bioassay systems is influenced by the mass transport of analytes as well as the biochemical reactions that occur within the system. This paper describes computational design capabilities that are being developed at CFD Research Corporation for bio-microfluidic assay devices which simulate convective-diffusive mass transport as well as biochemical reaction kinetics in a coupled manner. Simulation results from two sample test cases, one involving enzyme-catalyzed reactions and the other involving cooperative ligand-binding kinetics for an allosteric protein (hemoglobin), are presented.

Keywords. Computational fluid dynamics, bioassay systems, enzyme-catalyzed reactions, oxygen-hemoglobin binding

1 Introduction

Several microfluidic devices are being developed worldwide to conduct bench-top biological and chemical assays directly in a miniaturized automated manner. The bio-sensors typically used in such devices convert observed chemical changes into measurable signals through interaction of a biological element, usually an enzyme, with a electrochemical or optical transducer (Kovacs, 1998). The enzyme acts as a biological catalyst for certain reactions and can bind itself to specific substrates to form a product. In bio-sensors used for blood-gas assays, the reactions involve proteins such as hemoglobin which display cooperativity with complex ligand-binding kinetics. In order to develop computational design capabilities for bioassay devices, one needs to be able to simulate the associated biochemical reactions.

This paper describes numerical simulation capabilities for enzyme-catalyzed reactions and for oxygen-hemoglobin reaction kinetics in micro-flow systems being developed at CFD Research Corporation (CFDRC). CFDRC's commercial multi-physics simulation code, CFD-ACE+MEMS, was used for modeling the underlying convection-diffusion-reaction processes. Two sample studies are presented here. The first study involves modeling the conversion of the zymogen, factor X (FX) to the serine protease, factor Xa (FXa), in the presence of the tissue

[1]Manager/Biomedical Technology, 215 Wynn Dr., Huntsville, AL, 35805; vbm@cfdrc.com

factor-factor VII catalytic complex (TF-FVIIa) in a parallel-plate flow channel (100 micron height). The second study analyzes the mixing of co-flowing streams of oxyhemoglobin (HbO_2) and deoxyhemoglobin (Hb) in a T-shaped microchannel (500 microns width). The simulations accounts for convection and diffusion of Hb, HbO_2 and oxygen, as well as the oxygen-hemoglobin reaction kinetics

2 Simulation of Enzyme-Catalyzed Reactions

2.1 Problem Description

In this study, the conversion of the zymogen, factor X (FX) to the serine protease, factor Xa (FXa), in the presence of the tissue factor-factor VII catalytic complex (TF-FVIIa) in a microfluidic channel (100 micron height) was simulated. The catalyst was assumed to be present in a layer of rat vascular smooth muscle cells along the lower surface of the flow channel. This process was modeled as a convection-diffusion-reaction process. Figure 2.1 shows the schematic of the flow domain, which is essentially flow between parallel plates, where the lower plate represented the reacting surface. The enzymatic conversion of FX to FXa was modeled by simulating a biochemical surface reaction along the lower wall.

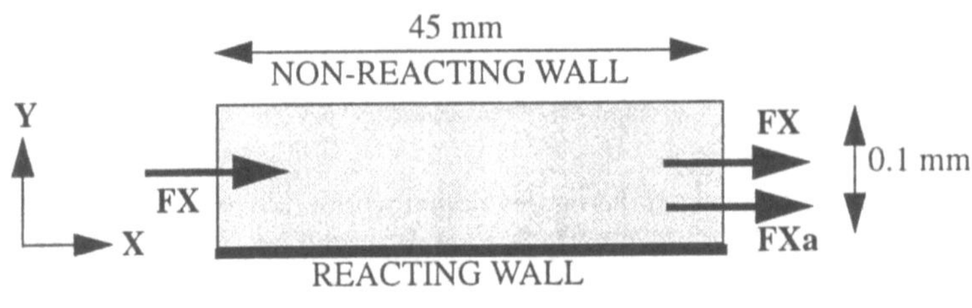

Fig. 2.1. Schematic of flow domain for enzyme-catalyzed reaction problem

2.2 Solution Procedure

A two-dimensional geometric model was created for the flow domain shown in Figure 2.1 using CFDRC's geometric modeling/mesh generation software, CFD-GEOM. A body-fitted structured grid was used to discretize the flow domain into approximately 4,000 computational cell volumes. Because mass transfer occurs by both convective and diffusive transport, the flow velocities needed to describe convective transport of the species were obtained by solving the Navier-Stokes equation shown in vector form in Equation 2.1. The SIMPLEC algorithm (Van Doormal and Raithby 1984) in CFD-ACE+MEMS along with the second order upwind-differencing scheme for spatial discretization was used to obtain the flow solutions. Flow was modeled as steady laminar incompressible, Newtonian flow.

The mass fractions of the two species, FX and FXa, were computed by solving the conservation of mass equation (Equation 2.2). The conversion ratio of FX to FXa was taken as 1:1. Since there is a reaction along the lower plate where there is conversion of FX to FXa, the effect of this reaction on bulk flow mass transport was incorporated through appropriate source and sink terms for FXa and FX, respectively, at the boundary cells along the wall. The reaction was modeled based on the Michaelis-Menten kinetics (Marshall, 1978).

$$\rho\vec{U}\cdot\nabla\vec{U}+\nabla P-\mu\nabla^2\vec{U}=0 \qquad (2.1)$$

$$\vec{U}\nabla C-D\nabla^2 C=0 \qquad (2.2)$$

where ρ is the fluid density (taken as 1 g/cm^3), $\vec{U}$ is the velocity vector, C represents the species concentration and D is the species diffusivity (5 x 10^{-7} cm^2/s).

2.3. Boundary Conditions

The flow boundary conditions for the problem include uniform inlet velocity, fixed exit pressure (taken as zero) and zero-slip conditions along the bounding walls. The mass transport boundary conditions include uniform inlet FX concentration ($C=Co$ = 100 nM)), zero flux condition at the upper wall (Equation 2.3) and the reaction at the lower plate (Equation 2.4).

$$D\frac{dC}{dy}=0 \qquad \text{at } y=y_{max} \qquad (2.3)$$

$$-D\frac{dC}{dy}=V_{max}\left[\frac{C_w}{K_m+C_w}\right] \qquad \text{at } y=0 \qquad (2.4)$$

where K_m is the Michaelis constant (= 36 nM), V_{max} is the maximum rate of FXa formation (in fmole/min/cm^2), C_w is the near wall concentration of FX and y_{max} is the chamber height. All the model parameters used here were based on a previous study by Hall *et al.* (1998).

2.4 Model Validation and Parametric Studies

Simulations were carried out for various inlet flow conditions corresponding to physiologic wall shear rates (10 - 1300 s^{-1}) encountered in the human vasculature. To study the influence of reaction kinetics on the biochemical process occurring in the flow system, different values of the reaction rate constant V_{max} (124 - 4000 fmole/min/cm^2) were used in the study. A grid sensitivity study was also done to ensure that the numerical solutions were not influenced by the grid density.

Sample Simulation Result. Fully developed flow solutions corresponding to the different wall shear rates were obtained prior to solving for mass transport. Figure 2.2 shows the velocity vectors for a wall shear rate of $100s^{-1}$. Figure 2.3 shows the FXa concentration contours in the flow domain. (The figure has been scaled 150 times in the y direction).

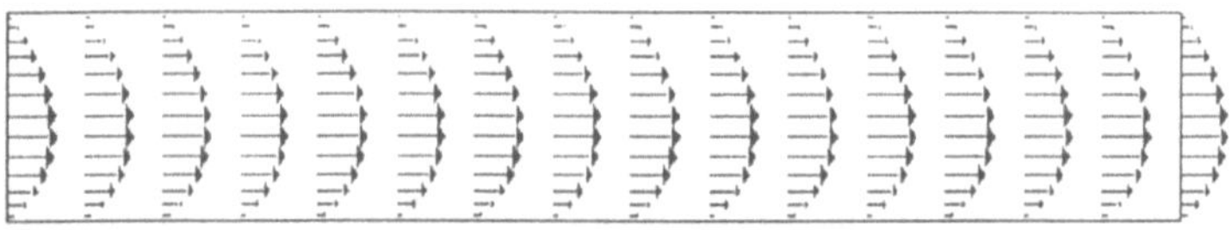

Fig. 2.2. Velocity Vectors in the Flow Domain

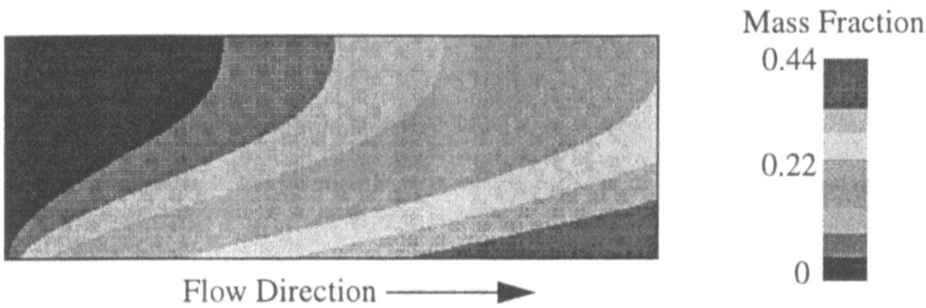

Fig. 2.3. Mass fraction contours of FXa in the flow domain (figure scale = 150y)

Model Validation. Table 2.1 shows a comparison of numerical predictions on the mean value of FXa outflux for different values of V_{max} and wall shear rates with corresponding experimental data from Hall *et al* (1998). The numerical fluxes predicted by CFD-ACE+MEMS are in close agreement with the experimental data,

Effect of Wall Shear Rate on FX Conversion. V_{max} was set constant at 4000 fmole/min/cm^2 and the wall shear rates were varied from 10 to 1000 s^{-1}. Figure 2.4a shows the normalized mass fraction profiles of FX at the chamber midpoint (22.5 cm). A very high V_{max} value implies larger reaction rates which causes increased generation of FXa from FX. At high shear rates, as we move away from the wall, convection forces become more dominant and the concentration of FX becomes more or less constant.

Effect of Reaction Rate on FX Conversion. Figure 2.4b shows the effect of surface reaction rate on the FX concentration at the chamber mid point at a fixed value of wall shear rate (20 s^{-1}). As expected, the plots show a higher consumption, and hence lower concentration of FX near the reacting wall at higher reaction rates.

Table 2.1. Comparison of numerical and experimental data

V_{max} (fmole/min/cm^2)	Wall shear rate (s^{-1})	Numerical flux (fmole/min/cm^2)	Experimental mean flux (fmole/min/cm^2)
792	1280	532.5	593 ±32
200	80	138.9	138 ±33
124	20	85.5	95 ±20

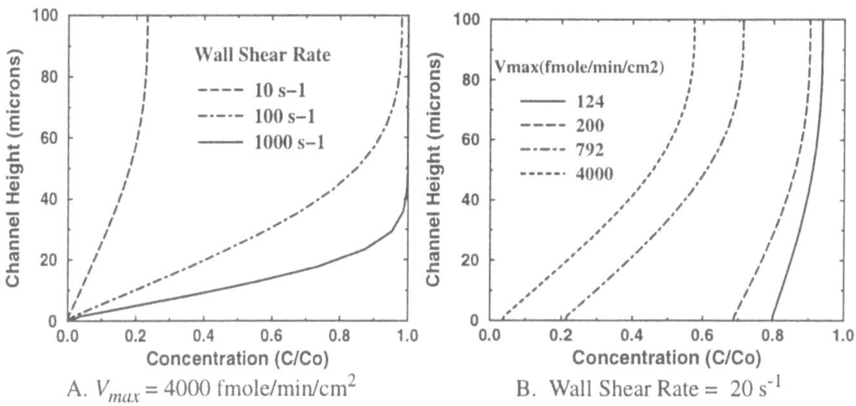

Fig. 2.4. Normalized FX concentration versus channel height

3. Simulation of Oxygen-Hemoglobin Reaction Kinetics

3.1. Problem Description

The second study involved simulating the kinetics of oxygen binding to hemoglobin, an allosteric protein in which binding of the ligand (oxygen) is cooperative. Cooperativity refers to the facilitation of binding of subsequent ligand molecules after the first ligand has bound to the oligomer. A hemoglobin molecule consists of four peptide chains, each with an incorporated heme group to which an oxygen molecule binds. Binding of an oxygen molecule results in a conformational change in the hemoglobin molecule and affects its ability to further bind oxygen molecules. Oxygen-hemoglobin kinetics have been defined by either the complex four-step Adair reaction scheme or by the simpler one-step approximation in which each hemoglobin molecule is replaced by four independent heme groups (Popel, 1989) . The latter approach was used in the present study.

The sample problem chosen to demonstrate this simulation capability was the mixing of co-flowing streams of oxyhemoglobin (HbO_2) and deoxyhemoglobin (Hb) in a T-shaped microchannel. The simulation accounted for convection and diffusion of Hb, HbO_2 and oxygen in addition to the oxygen-hemoglobin reaction kinetics, which reduce to the Hill equilibrium model when the rate term is set to zero (Makhijani *et al.*, 1990). The model described a facilitated diffusion system with association-disassociation reactions involving the three species.

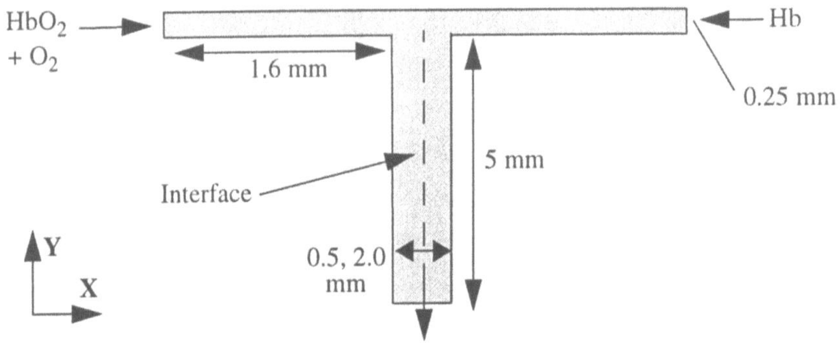

Fig. 3.1. Schematic of the T-channel flow domain

3.2. Solution Procedure

A two-dimensional geometric model was generated for the flow domain shown in Figure 3.1 using CFD-GEOM and discretized into approximately 2100 cells using a body-fitted structured grid. Steady-state flow solutions were obtained using CFD-ACE+MEMS using the same procedure described in Section 2.2. Time-varying mass fractions of the three species (HbO_2, Hb and O_2) at different locations in the flow domain were obtained by solving the conservation of mass equations given in Equations 3.1 - 3.3. N_i, and F_i, are the concentrations, and fluxes (convective + diffusive), of species "i", respectively, and "t" refers to time. Γ is the reaction term which is described by Equation 3.4.

$$\frac{\partial N_O}{\partial t} = -\nabla F_O - \Gamma \qquad 3.1$$

$$\frac{\partial N_{HO}}{\partial t} = -\nabla F_{HO} + \Gamma \qquad 3.2$$

$$\frac{\partial N_H}{\partial t} = -\nabla F_H - \Gamma \qquad 3.3$$

$$\Gamma = K\left[N_H\left(\frac{N_O}{N_{50}}\right)^n - N_{HO}\right] \qquad 3.4$$

N_{50} is the concentration of oxygen in equilibrium with hemoglobin at 50% fractional oxygen saturation, which can be expressed as $Bu * P_{50}$ where Bu is the Bunsen solubility coefficient of oxygen and P_{50} is the oxygen partial pressure at 50% fractional oxygen saturation. K is the dissociation rate constant for oxyhemoglobin and n is the dimensionless Hill coefficient.

A zero-flux mass boundary condition and no-slip velocity condition was applied along all walls. A fixed exit pressure condition was used for the flow equations.

3.3. Model Validation and Parametric Study

Grid sensitivity studies were carried out to ensure that the model predictions were insensitive to grid density. Simulations were then done for the case in which the common branch of the T-channel had a width of 2 mm. Model validation was done through comparison of the results with published experimental data from Vay (1986). Finally, the model was used to analyze the effects of flow rate and reaction rate on the diffusional mixing of HbO_2 and Hb.

Model Validation. This simulation was done under no flow conditions. The following model parameters were chosen to enable comparison of simulation results with published experimental data (Vay, 1986): $P_{50} = 7.0$ Torr, $Bu = 1.45\text{x}10^{-3}$ mM/Torr, $n = 2.5$, $K = 39.49\ \text{s}^{-1}$, hemoglobin diffusivity = oxyhemoglobin diffusivity = $9.375\text{x}10^{-7}\ \text{cm}^2/\text{s}$, oxygen diffusivity = $1.51\text{x}10^{-5}\ \text{cm}^2/\text{s}$, total heme concentration = 7.21 mM. The initial condition chosen was 100% fractional oxygen saturation in the left-half of the flow domain ($N_H = 0$) and 0% on the right side ($N_{HO} = 0$). The corresponding initial equilibrium concentrations of O_2 in the domain were calculated from the Hill equation by setting $\Gamma = 0$ in Equation 3.4. The simulation was carried out for 420 seconds with a time step of 0.03 sec.

The transient variation predicted in the oxygen saturation (oxyhemoglobin/total hemoglobin concentration) 100 microns to the left of the channel midpoint shows close agreement with the experimental data (see Figure 3.2). The saturation at this location decreases with time as oxyhemoglobin diffuses to the right side of the channel and dissociates to form unbound O2, which also diffuses to the right

Effect of O2-Hb Reactions on the Diffusive Mixing Process. The remaining simulations were done for a common channel width of 500 microns. The total duration of the transient simulations was 25 seconds with a time step of 0.005 sec. The effect of the oxygen-hemoglobin reaction terms on the diffusive mixing of oxygen in the T-channel was studied by performing simulations with and without the reaction terms. The convective terms were turned on in these simulations. The density and kinematic viscosity of the hemoglobin solution was taken as 1170 kg/m^3 and $2.99 \times 10^{-6}\ \text{m}^2/\text{s}$, respectively.

Figure 3.3 shows oxygen concentration profiles (non-dimensionalized with respect to the total heme concentration) at end of 20 seconds in the domain. (The common channel has been scaled in the X-direction for clarity). In absence of an association reaction with hemoglobin (which essentially acts as a sink term), we see higher concentrations of unbound oxygen transported across the interface.

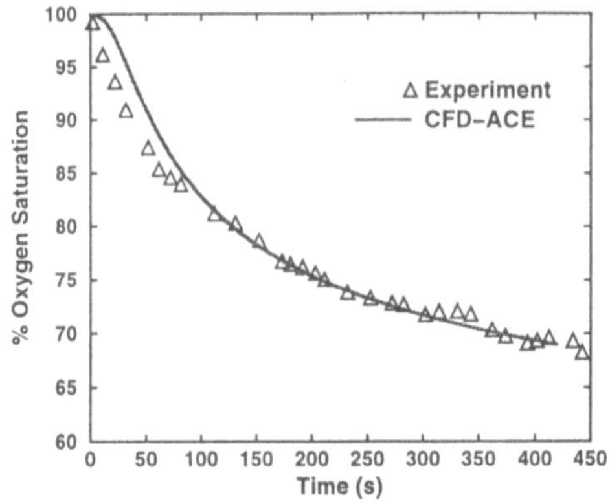

Fig. 3.2. Transient variation in oxygen saturation at a fixed location in the common channel

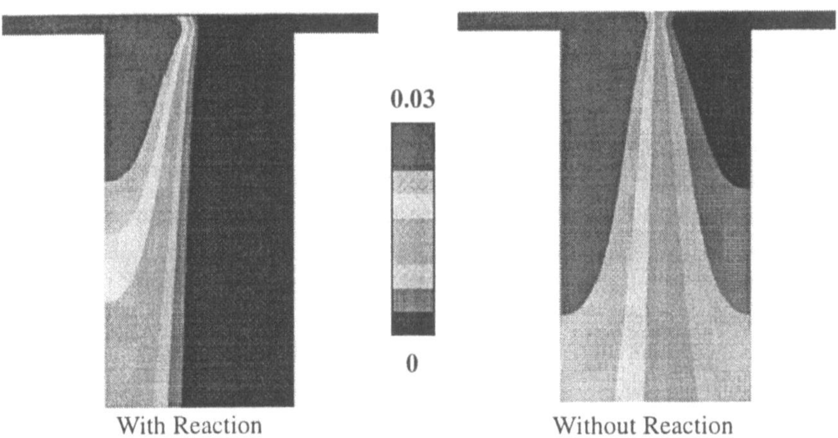

Fig. 3.3. Non-dimensionalized concentration contours of unbound oxygen in the T-channel

Effect of Convection on the Diffusive Mixing Process. The effect of the convective terms on the diffusive mixing process was studied by performing simulations with and without the convective terms. Here, the initial conditions specified were 95% oxygen saturation on the left side and 5% saturation on the right side. The corresponding equilibrium values of HbO2 and Hb were also used as the initial conditions.

Non-dimensionalized concentration contours of free oxygen in the channel at t=20 seconds are shown in Figure 3.4. There is clearly more diffusive mixing of free oxygen on the left and right sides of the channel in the absence of convection.

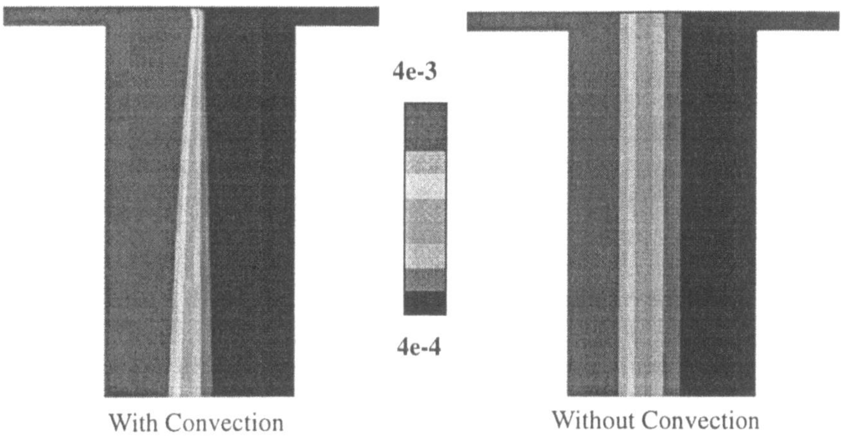

Fig. 3.4. Non-dimensionalized concentration contours of free oxygen in the T-channel

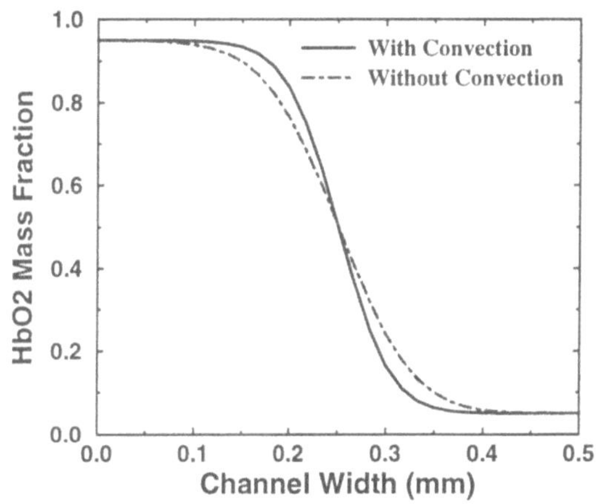

Fig. 3.5. Spatial variation in HbO_2 concentration along channel width at the exit

Figure 3.5 shows the effect of convection on the spatial profile of HbO_2 along the channel width at the exit of the T-junction at t=20 seconds. HbO_2 concentrations are higher in the oxygenated (left) side of the T-junction and lower in the deoxygenated side. This indicates that the convective washout of HbO_2 from the flow channel decreases the residence time of this species in the channel which results in a smaller amount of HbO_2 dissociation, and Hb and O_2 association, on the left, and right sides, respectively.

4. Conclusion

The two studies presented here demonstrate the applicability of CFD-ACE+MEMS for simulating microfluidic problems involving biochemical reaction kinetics. Two different types of biochemical reaction systems were simulated. The first involved enzyme-mediated surface reactions based upon the Michaelis-Menten kinetics. The second involved more complex non-linear reaction kinetics in the bulk flow.

Efforts are currently underway to incorporate simulation capabilities in CFD-ACE+MEMS for modeling the reaction kinetics associated with competitive multi-protein adsorption (reversible and irreversible) on biomaterials surfaces in microflow systems. This will further enable the application of CFD-ACE+MEMS for design and optimization studies on microfluidic bioassay/biosensor devices.

Acknowledgements

This work was funded under DARPA/ETO Composite CAD Program. The authors acknowledge the technical assistance on the passive scalar transport model provided by Drs. S. Krishmoorthy and S. Li of CFDRC during the course of this study.

Citations

Hall, C. L., Slack, S. M., and Turrito, V. T. (1998) A computational analysis of FXa generation by TF:FVIIa on the surface of rat vascular smooth muscle cells. *Ann. Biomed. Eng.* **70**, 129-134.

Kovacs, G. T. A. (1998) *Micromachined Transducers Sourcebook.* McGraw-Hill, New York.

Makhijani, V. B, Cokelet, G. R., and Clark, A. Jr. (1990) Dynamics of oxygen unloading from sickle erythrocytes. *Biophys. J.* **58**, 1025-1052.

Marshall, A. G. (1978) *Biophysical Chemistry - Principles, Techniques and Applications.* John Wiley & Sons, New York.

Popel, A. S. (1989) Theory of oxygen transport to tissue. *Crit. Rev Biomed. Eng.* **17(3)**, 257-321.

VanDoormal, J. P. and Raithby, G. D. (1984) Enhancements of the SIMPLE method for predicting incompressible fluid flows. *Numer. Heat Transf.* **7**, 147-163.

Vay, J. R. (1986) *Measurement of Oxygen Saturation and Hemoglobin Diffusivity in Hemoglobin Solutions with a Microscope Spectrophotometer.* MS Thesis, Univ. of Rochester, Rochester, New York.

Microfluidic Devices on Polymer Substrates for Bioanalytical Applications

Yuehe Lin*, Dean W. Matson, Dean E. Kurath, Jenny Wen, Fan Xiang, Wendy D. Bennett, Peter M. Martin, Richard D. Smith

Pacific Northwest National Laboratory, PO Box 999, Richland, WA 99352 USA

1. INTRODUCTION

Development of capabilities to miniaturize analytical devices and components offers a number of potential benefits. Among these are the ability to reduce sample sizes, development of low cost, single-use disposable devices, and improved device portability. Extensive work has been done on producing such microanalytical systems on silicon or glass substrates using processes commonly employed in electronic chip manufacturing [1-2]. For many analytical applications, however, common polymeric materials provide acceptable substrates from which to produce components or complete analytical systems [3-4]. The low flow rate characteristic of these microfluidic devices is compatible with the electrospray ionization/mass spectrometry (ESI-MS) [5-7]. In this paper, development of microfluidic analytical devices fabricated on polymer substrates using excimer laser micromachining technology will be described. These include a microfluidic motherboard, dual-stage microdialysis chip, and a micro capillary isoelectric focusing device. The applications of these microdevices for cleanup, fractionation, and separation of biological samples are discussed.

EXPERIMENTAL

2.1. Fabrication Methods

UV excimer laser direct-write micromachining systems (Potomac LMT-4000 and LMT-5000) were used to produce microchannels of various geometries in polymeric materials. The systems contained Potomac TGX-1000 KrF (248 nm) excimer lasers configured for focused (direct write) operation at the machining surface. The minimum spot size of an unapertured laser beam at the work surface was approximately 10 microns. The maximum pulse repetition rate was 2 KHz with a maximum pulse energy of 45 μJ. Machining was performed on set of 4-inch (LMT-4000) or 12-inch (LMT-5000) computer-controlled X-Y stages.

* Corresponding author. Email: yuehe.lin@pnl.gov

Microchannels in polymer devices were produced using two distinct methods. One approach was to use the focused excimer laser beam to directly mill channels into bulk polycarbonate or polyimide substrates. Because the unapertured laser beam width at the substrate was approximately 10 microns, multiple parallel passes were required to produce typical channels having widths of 50 or 100 microns. The depths of channels produced by this method, commonly 50 to 150 microns, were controlled by adjusting the feed rate of the part being machined or by changing the repetition rate of the laser. Microchannels produced by this method were sealed by applying a thin polyethylene terephthalate (PET) sheet directly to the face of the part with a heat activated adhesive, or by using a gasket compressed between the machined face and a second piece of flat polymer. Fluid connections to the microchannel were commonly made by mechanically machining the female part of a small chromatographic fitting (e.g., Upchurch Scientific) through the part from the back, and connecting to the microchannel with a small drilled hole.

A second approach to the formation of microchannels involved use of a polymer (e.g., polyimide) gasket with the channel machined through the thickness of the material. This method was used to produce microchannels up to 500 microns wide and 225 microns deep. For this approach, only the perimeter of the channel needed to be machined, with the remaining center portion of the channel being stripped out and discarded. When sandwiched between two solid pieces of polymer, the gasket formed a well-defined microchannel whose depth was determined by the thickness of the gasket material. These gaskets could be produced from sheet material having adhesive on one or both sides to improve sealing capabilities.

When feasible, devices were designed such that they could be assembled in a stacked geometry. In doing so, fluid flows were directed through a series of layers in the device, each layer having a distinct functionality. Minimum dead volume in the device was achieved by directing the fluid flow from one layer to the next using small diameter drilled vias.

2.2. Mass Spectrometric Analysis

A Finnigan LCQ (San Jose, CA) quadrupole ion trap mass spectrometer with a modified micro-ESI source was used for all experiments. The heated inlet capillary was maintained at 160^{0}C. For on-line microdialysis/ESI-MS experiment, the high voltage for ESI was applied to the metal syringe needle used for the buffer delivery and was carried to the sample side through the electrical conductivity of the buffer solution. Details of the membranes and flow rates used in specific experiments are discussed in the following section. A typical electrospray voltage was 2.2 kV. The maximum sample injection time was 40 msec, and 3 microscans were summed for each scan with a m/z range of 150 to 2,000.

3. RESULTS AND DISCUSSION

3.1. System Integration Platform - Microfluidic Motherboard

Dramatic gains in detector system performance can be expected through the seamless integration of miniature and microfluidics sample handling components with microscale chemical and biosensors through a novel motherboard technology: the microfluidics system integration motherboard. The objective of this work was to develop a suitcase-sized portable standard platform which integrates microfluidics sample handling components and standard connectors to interface the other miniturized biosensors or pathogen detectors based on "plug-and-play" modules. All components including micropumps, microvalves, microfluid channels and electronics were tightly integrated in the same microfluidic motherboard.

The microfluidic motherboard borrowed the "functionality on a chip" concept from the electronics industry and was intended to be the heart of a complex microfluidic device. The motherboard platform was designed to be tightly integrated and solid-state (i.e., liquid flows were all confined within machined microchannels), reducing the need for tubing for fluid distribution and connectivity. This concept greatly facilitated system integration and miniaturization. The first-generation motherboard consisted of three fluid reservoirs connected to three embedded piezoelectric micropumps (IMM) by microchannels fabricated from adhesive-backed polyimide [8]. The fluids could either be pumped independently or mixed in microchannels before being directed to exterior analytical components via outlet ports. Acrylic was used for all polymer parts of the fluidic motherboard except the laser-machined polyimide gaskets used to form the microchannels.

A second-generation microfluidic motherboard was designed and fabricated that incorporated valves in addition to the pumps (Figures 1-2). Initial tests indicate that the second-generation microfluidic motherboard is much more convenient to use compared with the previous version. Because all components in motherboard are based on "plug-and play" design, it is much easier for part replacement and repair.

It has been demonstrated that the microfluidic motherboard can deliver sample solutions and reagents to other devices (e.g., chemical and biosensors) at flow rates of between 10 and 300 μl/min. Integration of microfluidic motherboard with other micro biosensors and detectors is currently being explored in our lab.

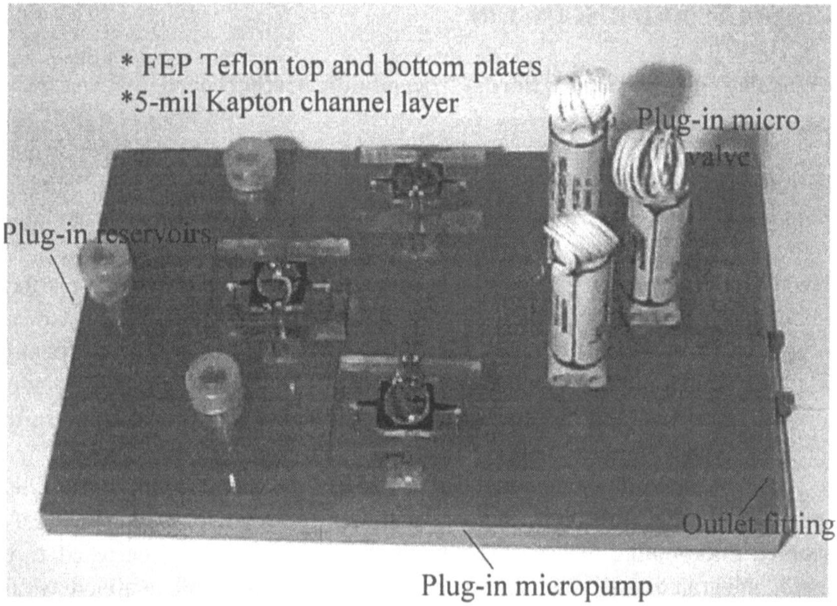

Fig. 1. Microfluidic Motherboard

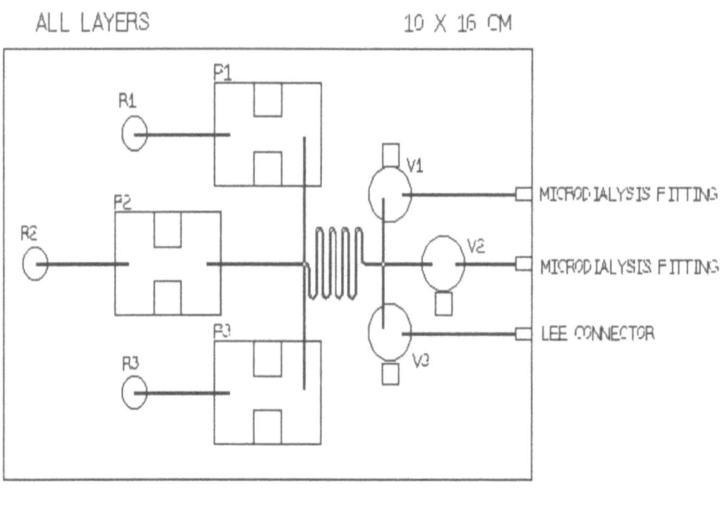

Fig.2. Microchannels in Microfliuidic Motherboard

3.2 Microfabricated Dual-Stage Microdialysis Devices for On-Line Fractionation and Cleanup in ESI-MS Analysis of Proteins

A microfabricated dual-stage microdialysis device was developed for fast and efficient fractionation and cleanup for ESI-MS analysis of complex biological samples. Figure 3 shows the structure and flow diagram of the dual-microdialysis device. The sample channel (60 μm deep, 150 μm wide) was machined directly into the polymer chip (30x30x6 mm) by using multiple parallel laser passes. A 500 μm wide buffer channel was machined from 225 μm thick polyimide with a silicone adhesive on one side. The adhesive was used to attach the buffer channel to a flat polymer chip. Two microdialysis membranes are sandwiched between three polymer chips with microfabricated serpentine channels. This novel design makes the dual-microdialysis device highly compact and rugged, eliminating the tubing for connection of two stages of microdialysis units. The fluid channels were designed such that the raw sample flows in a microchannel along one side of the first stage membrane so that a portion of sample flows through the membrane and enters the second stage. This partial flow in first stage provides a mechanism for sweeping out cellular residues in the sample channel and results in a reduced possibility of membrane clotting. Alternatively, the sample channel outlet (b) can be completely sealed if the life-time of the membrane is not of concern. The sample flow rate entering the second stage of the device can be adjusted by controlling the flow rate at outlet (b) by throttling the sample flow through fused silica tubing of various length and inner diameters. Sample entering the second stage of the device is stripped of the small molecules and salts by a dialysis buffer flowing in a microchannel on the opposite side of the second stage membrane.

The motivation for use of a microfabricated dual-microdialysis approach to MS sample clean-up was an attempt to improve on the shortcoming of similar larger-scale devices described previously [9]. A protein mixture consisting of 30 μM BSA, 8 μM cytochrome *c* and 2.4 μM ubiquitin in 10 mM PBS was used to evaluate the performance of the dual microdialysis device. A 50 kDa MWCO membrane and a 8 kDa MWCO membrane were used in the first and the second stage microdialysis, respectively. The sample was injected at 0.5 μL/min, and the clean-up buffer at 10 μL/min. As shown in Figure 4A, direct infusion of the protein mixture produced a no useful ESI-MS spectrum. ESI-MS performance is well known to be poor for such solutions, presumably due to the excess BSA and NaCl. After the on-line dual dialysis in the microfabricated device, the spectrum in Figure 4B clearly showed peaks from cytochrome *c*, ubiquitin, and removal of the unresolved envelope of peaks at higher m/z. The improvement in spectral quality using the dual-dialysis device enabled effective peak assignments and accurate molecular weight determinations.

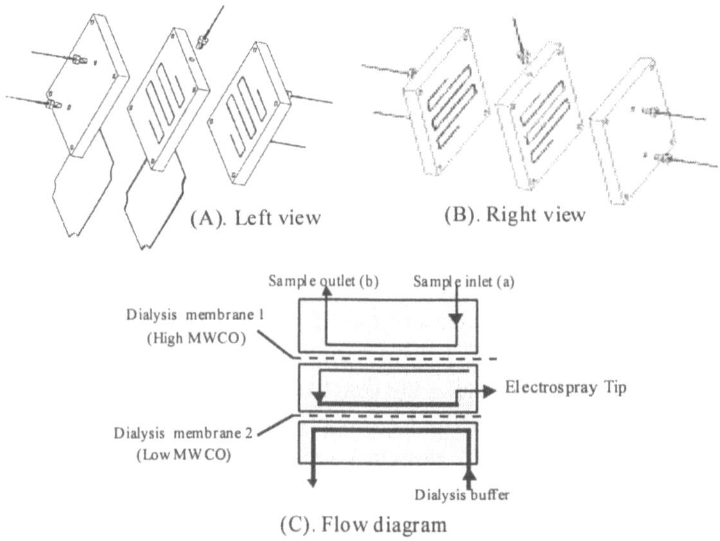

Fig. 3. Dual-Stage Microdialysis Device

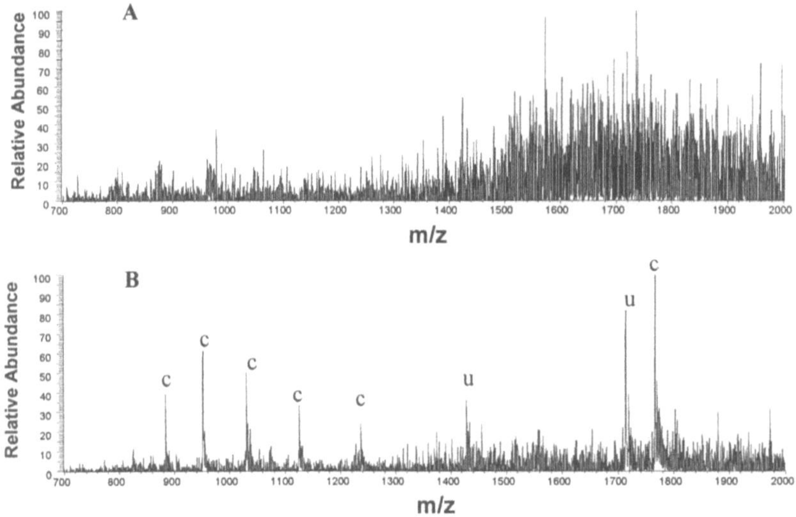

Fig. 4. Dual-Dialysis On-line ESI-MS Spectra of a Protein Mixture Consisting of 30μM Bovine Serum Albumin (BSA), 8 μM Cytochrome *c* and 2.4 μM Ubiquitin in 0.01 M PBS.
(A) 100 kDa (MWCO) membrane used on the first stage of the dual dialysis device;
(B) 50 kDa (MWCO) membrane used on the first stage of the dual dialysis device.

For analyzing complex biological samples, initial sample fractionation and clenup can simplify analytical procedure. On line dual-dialysis using the microfabricated device is a fast and efficient means to achieve do this. The molecular weight range selected for analysis by dual-microdialysis can be controlled by selection of appropriate MWCO membranes at both stages. Application of dual-microdialysis/ESI/MS for the identification of "biomarks" of crude *E.coli* whole cell lysate has been successfully demonstrated in our lab [10].

3.3. Microfabricated Capillary Isoelectric Focusing (CIEF) Device for Protein Separations

CIEF is a high-resolution technique for the separation of protein mixtures based on the difference in their isoelectric points. A major drawback of conventional CIEF technique is poor reproducibility because of inconsistent sample handling and instability of the coating required on the silica capillary. Successful on-line interfacing of the microfabricated isoelectric focusing device with ESI-MS may provide a rapid and automated tool to concentrate, separate, and analyze complex protein mixtures. In this work, microfabricated CIEF chips on polymer substrates (polycarbonate, polyimide) were used for the separation of proteins. Electroosmotic flows in microchannels directly laser-machined on these polymers are typically lower than those in fused silica capillary. This suggests that isoelectric focusing experiments may be performed directly in these microfabricated CIEF chips without the need for coating of the microchannels, allowing for improved reproducibility in this microchip CIEF mode.

To reduce the dead volume in the interface between microchip and the mass spectrometer and improve both speed and sensitivity, an electrospray emitter was directly fabricated into the end of the device (Figure 5). Extensive experience with ESI at PNNL has demonstrated that sharper emitter "tips" give higher electric fields at the tip and result in better electrospray performance. The laser-machined microchannel (50 μm wide, 60 μm deep, and 16 cm long) in the CIEF chip (11 x 3 x 0.5 cm) was fabricated directly on the polycarbonate substrate, similar to the sample channel on the microdialysis device described above. A dialysis membrane or ion exchange membrane was integrated on-chip at both the cathode and anode reservoirs, which allowed electric conduction in the microchannel but prevented the fluid from backflowing through the membranes.

CIEF separation with mass spectrometric detection was conducted in two steps as shown in Figure 6. In the first step, sample solution containing proteins and 1 5 pharmalyte (PI 3-10) was injected and filled the entire channel, across which electric field was applied. Isoelectric focusing was performed at a constant potential(-8 kV) for approximately 5-10 min. Proteins were focused on different zones based on difference in their isoelectric points (PIs). In the second step, separated proteins were moved into the MS detector by applying pressure from a syringe pump at a flow rate of 0.05 μL/min. During operation, the electrospray emitter tip was positioned in front of the sampling orifice of the mass spectrometer. The applied high voltage (2 kV) and nebulizer gas (nitrogen) flowing around the electrospray emitter tip create a fine spray of charged particles for the MS analysis.

Preliminary experiments have been conducted for the separation of a protein mixture with a concentration of 0.05 mg/mL for each protein. The separation profile is shown in Figure 7. Peak 1 and 2 in Figure 7 were from carbonic anhydrase and peak 3 from myoglobin. Figure 8 shown the positive ESI mass spectrum of peak 2 and 3. The results indicate that electrospray directly from the sharp-edge microchip is feasible for separation and detection of proteins with microfabricated CIEF/ESI-MS.

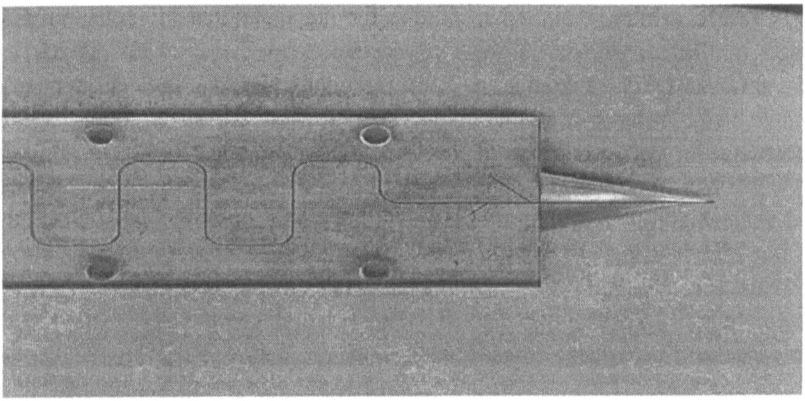

Fig. 5 CIEF microchip with a sharp electrospray emitter tip

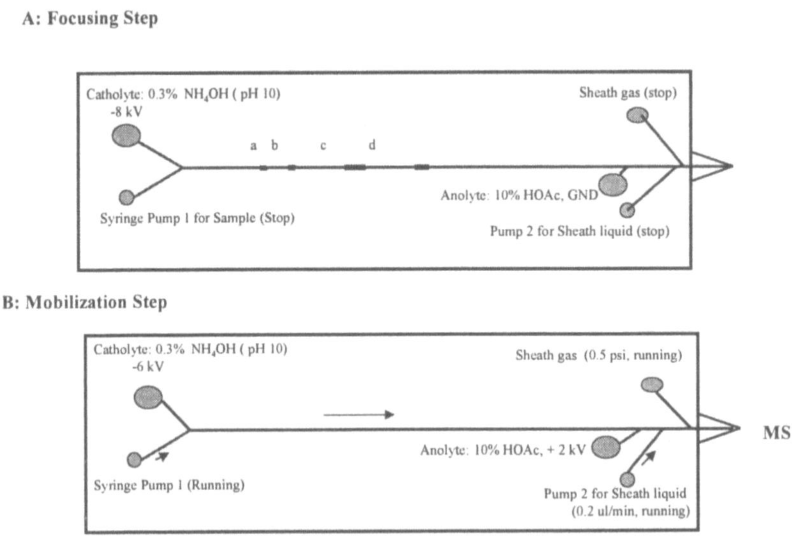

Fig. 6. Two step in CIEF separation

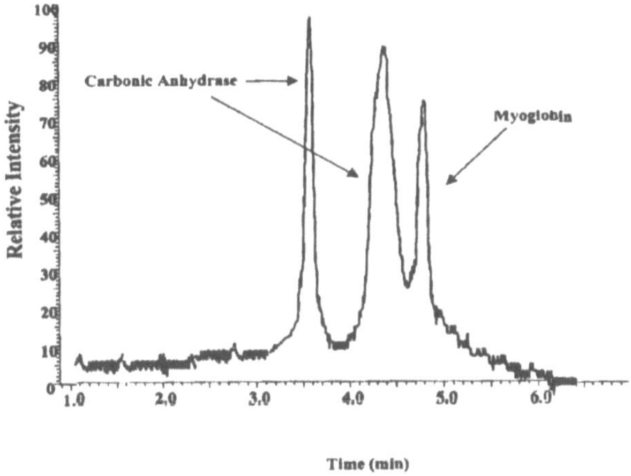

Fig. 7. CIEF/ESI-MS Separation Profile on CIEF Microchip for Protein Mixture with a Concentration of 0.05 mg/ml for Each Protein. Peak 1 &2, carbonic anhydrase (pI 5.9 & 6.8); peak 3, myoglobin (pI 7.2).

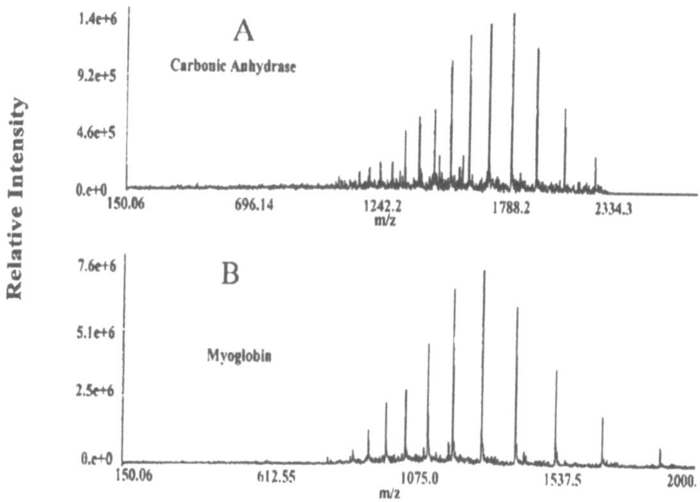

Fig.8. Positive ESI Mass Spectrum of Carbonic Anhydrase (A) and Myoglobin (B)

4. CONCLUSIONS

There are many advantages in the use of polymeric materials for producing microanalytical components and devices. We have demonstrated the fabrication of such devices using laser machining and lamination processes amenable to low cost, large-scale production. The designs used for these devices offer potential for additional size reduction with optimization of flow paths and support materials. Efficient sample cleanup, fractionation, and separation of protein samples were demonstrated using ESI with an ion trap mass spectrometer. Integration of these microfabricated devices with other nanoscale manipulations is expected to lead to more powerful, sensitive, less expensive, and rugged instrumentation for biomolecule analysis.

ACKNOWLEDGMENT

This project was supported by a LDRD program at Pacific Northwest National Laboratory (PNNL). PNNL is operated by Battelle Memorial Institute for the U.S. DOE under Contract DE-AC06-76RLO 1830. Helpful discussions with Drs. C. J. Call and D. J. Hammerstrom are gratefully acknowledged.

REFERENCES:

1. Harrison, D.J.; Fluri, K.; Seiler, K.; Fan, Z.; Effenhauser, C.S.; Manz, A. *Science*, 1993, 261, 895
2. Kop, M.U.; deMello, A.J.; Manz, A., *Science*, 1998, 280, 1046-1048
3. McCormick, R.M.; Nelson, R.J.; Alonso-Amigo, M.G.; Benvegnu, D.J.; and Hooper, H.H., *Anal. Chem.*, 1997, 69, 2626-2630.
4. Martin, P.M.; Matson, D.W.; Bennett, W.D.; Lin, Y., Hammerstrom, D.J., *J. of Vacuum Science and Technology*; In press
5. Xue, Q.F.; Foret, F.;. Dunayevskiy, Y. M.; Zavracky, P. M.; McGruer, N. E. and Karger, B. L., *Anal. Chem.*, 1997, 69, 426-430
6. Xu, N.; Lin, Y.; Hofstadler, S.A.; Matson, D.W.; Call, C.J. and Smith, R.D. Anal. Chem., 1998, 70, 3553-3556
7. Lin, Y.; Xu, N.; Matson, D.W.; Smith, R.D., in the *Micro Total Analysis Systems*; Ed.: Harrison, D.J. and den Berg, A.V., pp343-346, Kluwer Academic Publishers, Boston, 1998
8. Matson, D.W.; Martin, P.M.; Kurath, D.E.; Lin, Y.; Hammerstrom, D.J., in the *Micro Total Analysis Systems*, Ed.: Harrison, D.J. and den Berg, A.V., pp371-374, Kluwer Academic Publishers, 1998
9. Liu, C. L.; Hofstadler, S. A.; Bresson, J. A.; Udseth, H. R.; Tsukuda, T.; Smith, R. D., *Anal. Chem.* 1998, 70, 1797-1801.
10. Xiang, F.; Lin, Y.; Wen, J.; Matson, D.W.; Smith, R.D, *Anal. Chem.*, In press

Transfer of plasmid DNA into cells with microelectroporation arrays on a chip

Robert Poppe

Johannes Gutenberg-Universität Mainz, Institut für Genetik, Becherweg 32, 55099 Mainz, Germany. *E-mail: robert.poppe@mainz.netsurf.de*

The possibility to transfer pure DNA into bacterial cells forms the basis for the genetic engineering of the cell. Electroporation is a powerful and easy technique to introduce plasmid DNA into cells. Its drawback for use with high-throughput approaches is that with standard electroporation chambers the reactions have to carried out one after the other and the electroporation cuvettes are expensive. To obtain the possibility of high-throughput electroporation reactions *Escherichia coli* cells were electroinjected in parallel with different plasmids in reactions as small as 100 nl on a microstructured array of electrodes, forming hundred separate electroporation units on a chip of a square inch. The electroporation reactions reached about 1000 transformants per reaction. In addition microstructured electrode arrays in standard 96-, 384- and 1536-well microtiter plate format, which are compatible with robot arraying technology, were fabricated by means of photolithographically patterning and etching of photoresist coated copper films deposited on epoxy resin plates. Microelectroporation reactions were investigated with bacteria, and yeast.

Keywords: Microelectroporation arrays, DNA transfection, electropermeabilization, high-throughput electroporation, yeast two-hybrid assay, drug screening

1 Introduction

The cytoplasm of cells is surrounded by a semipermeable membrane that prevents escape of cellular constituents to the environment. Transport systems in the cell membrane facilitate the selective uptake and release of specific molecules. As a result the cell membrane represents a formidable barrier for most molecules. The metabolic and genetic engineering of the cell is a major task in biotechnology. Therefore the possibility to increase the membrane permeability at will, to introduce selected molecules into the cell, is of significant interest.

Electroporation - the use of high-voltage electric shocks to temporarily render the membrane permeable- has gained widespread use to introduce water-soluble molecules into cells because of its ease of use, reproducibility, and high efficiency to transfer DNA into a diverse variety of cell types [1-3]. The increasing demand for high-throughput operation in the field of molecular biology requires parallel processing, automation and hence miniaturization of standard laboratory techniques. Standard electroporation chambers are expensive and incompatible with automation and parallel sample processing. To address the need for high-throughput bacterial transformation reactions, an electroporation chip has been developed on which the samples can be spotted on a microstructured array of electrodes, forming up to hundred separate electroporation units on a chip of a square inch.

To facilitate automation of sample handling microelectroporation units have been arrayed in standard microplate formats which can be loaded with multi-channel pipettes or robot pipetting machines. This electroporation arrays allow the parallel transformation of 96, 384 or 1536 separate samples. The samples can be recovered from the chip surface by pipettes or can be stamped directly with the electroporation chip on an appropriate selection media surface. Cleaning of the electroporation chips is performed easily and allows their reuse.

2 Methods

2.1 Microelectroporation chip fabrication

Arrays of microelectrodes were fabricated by standard circuit board processing techniques using photoresist coated epoxy resin plates with a 35 μm copperoverlay (Bungard, Germany). The positive layout of the electrode array was printed on transparency film and used as a mask for the photolithographically patterning of the board. The board was exposed with about 1.5 mJ/cm^2 with a UV-lamp, subsequently the photoresist was developed for 90 sec in 200 mM NaOH followed by a brief rinse in water. Etching at 45 °C in 25% sodiumpersulphate was performed to develop the photolithographically defined electrodearray followed by thorough rinsing in water. The copper electrodes were galvanically covered with silver and subsequently with a gold overlay.

The epoxy resin carrier is naturally hydrophilic, causing aqueous solutions to spread on the surface. Therefore the surface of the chip was modified by soaking in dilute (1% in isopropanol) dichlordimethylsilane for 10 min. at room temperature. After this treatment the epoxy resin surface is hydrophobic.

Several electroporation chip designs were produced. The smallest chip consists of 100 separate electroporation units on a square inch. Each electroporation unit is composed of two opposing electrodes in 700 μm distance. The electrodes are 800 μm in diameter and 35 μm thick. Center to center distance between neighboring electrodes is 1.5 mm. Microelectroporation reactions on this chip were carried out with volumes between 100 and 500 nl. Electroporation chips in 96-well microplate format have opposing electrodes in 1 mm distance The electrodes are 3 mm in diameter and 35 μm thick. Center to center distance between neighboring electrodes is 9 mm Microelectroporation reactions on this chip were carried out with volumes between 1 and 5 μl . Chips in 384-well microplate format have opposing electrodes in 1 mm distance The electrodes are 2 mm in diameter and 35 μm thick. Center to center distance between neighboring electrodes is 4.5 mm. Microelectroporation reactions on this chip were carried out with volumes between 0.5 and 2 μl. Chips in 1536-well microplate format show opposing electrodes in 500 μm distance The electrodes are 1 mm in diameter and 35 μm thick. Center to center distance between neighboring electrodes is 2.25 mm. Microelectroporation reactions on this chip were carried out with volumes between 100 and 500 nl.

2.2 Cleaning protocol

Cleaning of electroporation chips was performed by rinsing in water followed by washing for 5 min. in 100 mM sodiumphosphate, 1 mM Na-EDTA, 0.1% SDS, pH 7.0 and thorough rinsing in deionized water. The chips were sterilized by incubation with 70 % ethanol for 10 min and air dried. To prevent cross contamination with DNA the chips were exposed in a UV-crosslinker(Stratagene) for 5 min.

2.3 Transformation of *Escherichia coli*

Electrocompetent cells were prepared as described by Dower et al. with modifications [4]. In brief, 1000 ml SOB-broth (2% Bacto-Tryptone, 0.5 % yeast extract, 10 mM NaCl, 2.5 mM KCl) were inoculated with 10 ml of a fresh overnight culture of *E. coli* DH10B. Cells were grown at 37 °C with vigorous shaking to an OD_{600} of 0.8, subsequently chilled in an ice-water bath and centrifuged in a cold rotor at 5000 x g_{max} for 10 minutes. The cell pellet was resuspended in a total of 500 ml of ice-cold sterile 10% glycerol and recentrifuged. The cell washing step was repeated twice. Finally the cell pellet was resuspended to a final volume of 3 ml in ice-cold sterile 10% glycerol and stored at -70 °C in aliquots.

The electroporation chips and the socking chamber were prechilled on ice. In a microcentrifuge tube on ice cells were mixed with plasmid DNA to a final concentration of 5 ng/μl and transferred in the appropriate volume on the electroporation chip. The socking chamber was connected with a square wave pulse generator type EPI 2500 (Intergen, Heidelberg, Germany) and electroporation was performed with the settings 15 kV/cm, 100 μF, 5 msec.

To determine the transformation efficiency the electroporation reactions were incubated for 1h at 37 °C in 100 μl SOB-medium supplemented with 10 mM $MgCl_2$, 25 mM glucose and titered on ampicillin agar plates.

2.4 Transformation of *Saccharomyces cerevisiae*

Yeast cells were prepared by inoculation of 1000 ml YP-medium (1% Bacto-peptone, 1 % yeast extract, 1%) with 10 ml of a fresh overnight culture of *S. cerevisiae* Y190. Cells were grown at 30 °C with vigorous shaking to an OD_{600} of 1.3, subsequently chilled in an ice-water bath and centrifuged in a cold rotor at 5000 x g_{max} for 5 minutes. The cell pellet was resuspended in a total of 500 ml of ice-cold sterile water and recentrifuged. The cell washing step was repeated twice. Finally the cell pellet was resuspended with one volume sterile 2M sorbitol.

In a microcentrifuge tube on ice the cells were mixed with plasmid DNA to a final concentration of 10 ng/μl and transferred in the appropriate volume on the pre-chilled electroporation chip. Electroporation was performed with the settings 7.5 kV/cm, 100 μF, 5 msec. The transformation efficiency was determined by spreading the electroporation reactions onto selective plates containing 1M sorbitol and lacking the aminoacid leucine to select for pACT2 transformants or lacking the aminoacid tryptophan to select for pAS2-1 transformants. Double transformants were selected on plates lacking leucine and tryptophan [5].

3 Results

3.1 Fabrication of microelectroporation arrays and handling

Arrays of microelectrodes were easily fabricated by standard circuit board processing techniques (Fig. 1). An advantage of this low-end technique is the high flexibility because computer generated layouts, printed on transparency film can be used directly for the photolithographically patterning of a coppersurface. Initial experiments with copper-electrodes revealed corrosion of the electrodes because of electrochemical oxidation. To prevent the release of toxic copperions into the electroporation reaction the electrodearrays were galvanically refined with a silver- and a final goldsurface. Gold-plated microelectroporation arrays have been cleaned and reused several times without any visible damage. To recover small reaction volumes and to prevent cross-contamination between neighboring microelectroporation units the surface of the epoxy resin plate which carries the microelectroporation arrays was silanized with dimethydichlorsilane. After this treatment aqueous solutions can be placed without spreading as small droplets between the electrodes (Fig 1. b). Evaporation of small sample volumes for the time between sample application, electroporation and sample recovery was prevented by chilling the electroporation chip and the socking chamber on ice.

3.2 Transformation of *E. coli*

3.2.1 Uniformity of parallel microelectroporation reactions

E. coli transformation tests were conducted by using electrocompetent cells from the strain DH10B. To investigate the uniformity of parallel microelectroporation reactions *E. coli* cells were mixed with plasmid DNA pBluescript-KS (pKS) to a final concentration of 2 ng/µl and placed as 10 separate electroporation reactions with a volume of 100 nl on a chip with a density of 100 electroporation units on a square inch. After electroporation (1200 V, 250 µF, 5 msec) cells were resuspended in SOB-medium and titered onto ampicillin agar plates. Per reaction between 1440 and 750 transformants with a mean of 1100 were obtained. The experiment was repeated once with the same electroporation chip. No correlation between the position of the electroporation reaction and the transformation efficiency was observed.

3.2.2 Investigation of possible cross-contamination

A sensitive test was performed to measure cross-contamination between neighboring electroporation reactions and between subsequent electroporation reactions after cleaning the chip.

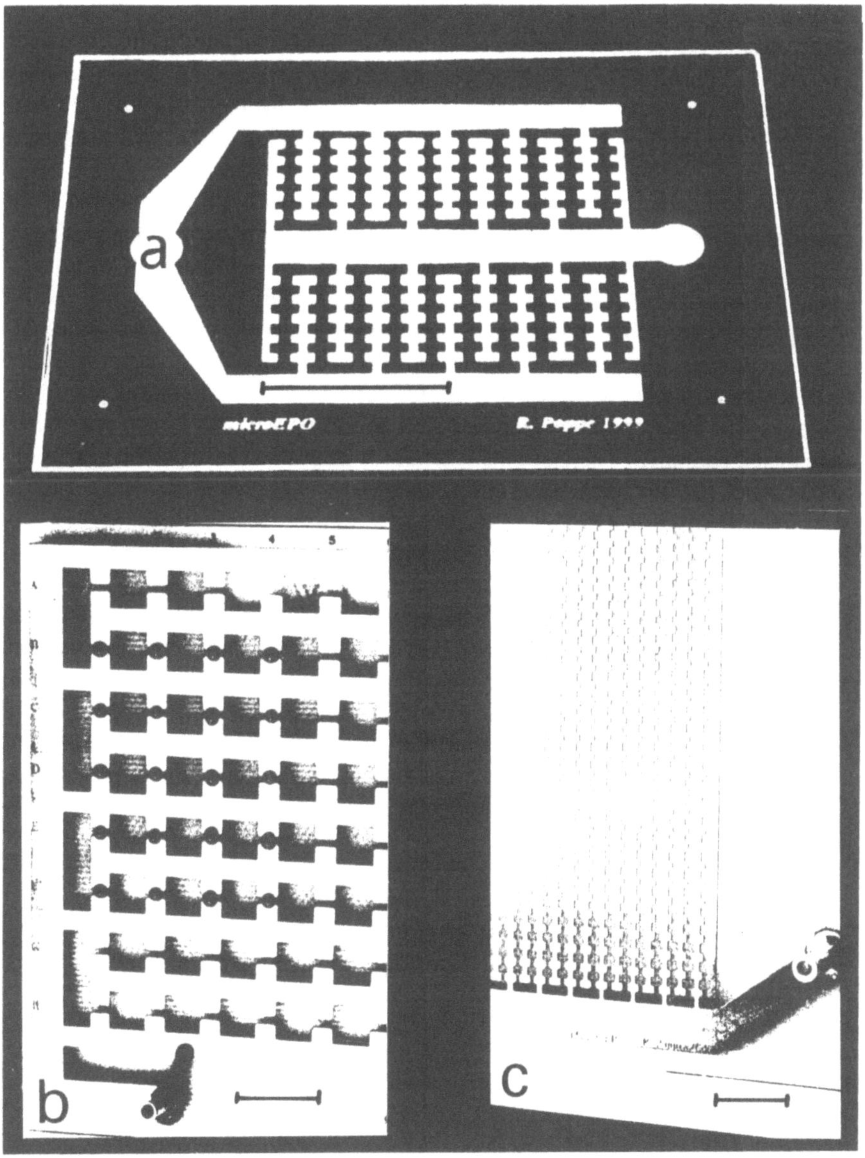

Fig. 1. **Microelectroporation arrays build in circuit board etching technique**
The layouts of the electrode arrays were printed on transparency film photolithographically patterning of a coppersurface. The copperstructure was galvanically refined with a goldsurface. A microelectroporation array with 100 electroporation units on a square inch (a) an array in standard 96 well format (b) and in 1536 well format (c) are shown. The hydrophobic modification of the epoxy resin plate was visualized with red ink droplets (5 μl) in b. Bars: 1cm.

E. coli cells were mixed with pKS or with pKS-A containing a 1.5 kB insert to a final concentration of 2 ng/μl and placed as 10 separate 500 nl electroporation reactions neighboring each other directly on a chip with a density of 100 electroporation units on a square inch. After electroporation (1200 V, 100 μF, 5 msec) cells were resuspended in SOB-medium and plated on ampicillin agar plates containing 5-bromo-4-chloro-3-indolyl β-D-galactoside (X-gal) to assay for β-galactosidase expression. Cross-contamination should be indicated by the occurrence of blue colonies in the pKS-A transformation and white colonies in the pKS containing reaction.

No cross-contamination between neighboring colonies could be observed if the reactions are placed carefully in position. However pipetting with air bubbles could cause significant cross-contamination because of aerosol production. In principal arcing of reactions because of high ion concentration could be an additional risk for cross-contamination, but arcing was never observed under the conditions indicated.

To investigate the cross-contamination between subsequent electroporation reactions *E. coli* cells were mixed with pKS-A and electroporated as previously indicated. The electroporation chip was cleaned and reused with *E. coli* cells which were mixed with pKS. These second reactions were plated on ampicillin agar plates containing X-gal. No contamination was observed.

3.2.3 Parallel transformation of *E. coli* with plasmids of different size

E. coli transformation tests were conducted with different plasmids in parallel reactions. Separate aliquots of cells were mixed with plasmid-DNA in the size range of 3.0 - 13 kB to a final concentration of 2 ng/μl. As a control an aliquot of cells without DNA was prepared. Electroporation reactions were performed in parallel with 500 nl of each reaction on a chip with a density of 100 electroporation units on a square inch (1200 V, 250 μF, 5 msec). Cells were titered on ampicillin agar plates. Between 500 and 1800 transformants were obtained with the DNA-containing reactions (Tab.1.).

Tab. 1. Efficiency of microelectroporation of *E. coli* DH10B with different plasmids

Plasmid	Size (kB)	transformants/reaction
pAS2-1	8.4	900
pLAM5′-1	9.0	500
pACT2	8.1	600
pKS-Insert	8.0	900
pKS	3.0	1200
pVA3-1	9.4	500
pTD-1-1	10.0	700
pGAD10-Insert	13.1	700
pGAD10	6.6	1300
no DNA	/	0

3.2.4 Direct plasmid transfer from yeast to *E. coli* cells

To perform a plasmid rescue fresh colonies of yeast cells which were transformed with pAS2-1 were picked from agar plates and suspended into 30 µl water. 5 µl *E. coli* cells and 1µl yeast cells were placed between the electrodes of the 96X electroporation chip. A first pulse (1000 V, 100 µF, 5 msec) was performed to release plasmid DNA from yeast cells. After 1 min a second pulse (1500 V, 100 µF, 5 msec) was performed and the reaction mixture was plated on ampicillin agar plates. Between 10 and 220 bacterial clones were obtained.

3.3 Transformation of yeast cells

Electrocompetent yeast cells were mixed on ice with the yeast two-hybrid plasmids pAS2-1, pACT2 or both plasmids in mixture to a final concentration of 50 ng/µl. Reactions with a volume of 1 µl were placed on the electroporation and were pulsed (600 V, 100 µF, 5 msec). Spreading on agar plates selected for pAS2-1 or pACT2 about 2300 transformants per reaction. About 50 double transformants were obtained with the plasmid mixture.

4 Conclusions and outlook

These results show that transfer of plasmid DNA into cells with electroporation can be performed effectively in parallel with microelectroporation arrays. The reaction volume in standard electroporation cuvettes is in the size of 50 µl [6]. Because of the high efficiency of electroporation this volume is uneconomic for most standard applications. With the electroporation chips the plasmid transfer reaction have been scaled down 100 fold, still yielding a sufficient amount of transformants. A small sample volume allows economical use of DNA and cells. An additional benefit of miniaturization is that with a smaller sample cross-section, it provides the higher resistance for high voltage operation. Small devices allow easy remove of heat from the system despite the thermal conductivity of water [7]. This may be the reason why arcing never occurred in the experiments with microelectroporation chips.

The microelectroporation chips have been successfully applied with *E. coli* and yeast transformation techniques and enable parallel processing and automation of this standard laboratory techniques. This may be hintful for large scale functional proteomic investigations with the yeast two-hybrid technique [8] and for new transfected-cell assay systems for high-throughput screening to identify drug leads [9]. One can envision future microelectroporation devices for drug screening which integrate the electroporation arrays into the bottomplate of microtiterplates and hence enable the controlled transfer of DNA and drugs into cell cultures.

In conclusion the new developed microelectroporation chips have several advantages over conventional electroporations chambers, including design flexibility, parallel sample processing, size, cost and automation of sample handling.

5 References

1. Chang DC, Chassy BM, Saunders JA, Sowers AE, (eds) (1992) Guide to electroporation and electrofusion, Academic Press, New York
2. Zimmermann U, Urvonitz HB (1987) Principles of electrofusion and electropermeabilization, Methods in Enzymology, Vol. 151, 195-221
3. Miller JF (1994) Bacterial transformation by electroporation, Methods in Enzymology, Vol. 235, 375-385
4. Dower WJ, Miller JF, Ragsdale CW (1988) High efficiency transformation of *E. coli* by high voltage electroporation, Nucleic Acids Res. 16:6127-6145
5. Clontech yeast protocols handbook (1996) Clontech Laboratories, Inc., Palo Alto, USA
6. Methods in electroporation (1993) Bio-Rad Laboratories, Hercules, USA
7. Fuhr G, Shirley SG (1998) Biological applications of microstructures. In: Manz A, Becker H (eds) Microsystem technology in chemistry and life sciences, Springer Verlag, Berlin, Heidelberg
8. Wanker E. (1998) An automated yeast two-hybrid system for analysis of protein-protein interaction networks. In:, International symposium on genomics and proteomics, Abstracts Vol. 6, Deutsches Krebsforschungszentrum, Heidelberg
9. Okun I, Veerapandian P (1997) New methods to mimic nature in high-throughput screening, Nature Biotechnology 15:287-288

On-line Monitoring of Enzyme Activities using Microreactor Heat Power Meters

Johannes Lerchner[1], Antje Wolf[1], Annekatrin Weber[1], Regina Hüttl[1], Gert Wolf[1], J. Michael Köhler[2], Martin Zieren[2]

[1] TU Bergakademie Freiberg, Inst. Phys. Chem., Leipziger Straße 29, D-09596 Freiberg,
[2] Institute of Physical High Technology, Microsystems Dept., Winzerlaer Str. 10, D-07745 Jena

Abstract: To study the influence of the design parameters of miniaturized flow-through calorimeters on their signal generation behavior four different versions of flow-through calorimeters have been prepared and tested. For chemical reactions with low rate constants such as enzyme catalyzed reactions the calorimeter volume needs to be larger than a threshold volume of about 20 µl provided sufficient mixing. The modular arrangement of a highly efficient micromixer with a low-volume calorimeter of 2 µl appears to accomplish a considerable signal increase compared to using an integrated mixer of poor efficiency.

Keywords: enzyme, catalysis, calorimeter, flow injection analysis FIA, on-line monitoring

1 Introduction

The heat power measurement is a simple and direct way for the monitoring of primary biochemical processes. Because of the universal nature of the heat power production of chemical reactions no sophisticated cascades of reaction steps are necessary to generate electrical signals which are related to the rate of the primary reactions under investigation. Further, a single heat power meter can be applied to different reactions without modifications. Therefore, analytical arrangements for sequential flow-injection techniques should be possible on the basis of heat power measurement.
With the design of calorimeters in the scale of microreactors the essential conditions for their practical applications are given. Recently different constructions and applications of integrated circuit (IC) calorimeters have been described in the literature [1]. In general, the main part of an IC-calorimeter consists of a silicon chip with integrated temperature transducer, heater and heat sink. As an important advantage calorimeters with very low reaction volumes (only a few microliters) and time constants result in turn with the miniaturization.
To study the influence of the design parameters of miniaturized flow-through calorimeters on their signal generation behavior four different versions of flow-through calorimeters have been prepared and tested. They differ in the reactor volume, the

layout and the size of the inlet and outlet channels and the type of the used temperature transducers. In all cases the basic parts of the calorimeters are silicon chips with integrated thermopiles.

2 Design of miniaturized Flow-Through Calorimeters

2.1 Calorimeters based on p-type-Si/Al-Thermocouples

The basic parts of the first two calorimeters (calorimeter 1 and 2) are monolithic integrated silicon thermopile chips manufactured by Xensor Integration (Delft, Netherlands) [2]. The sensitive area for the heat detection is located in the center of a free-standing silicon membrane of up to 30 μm thickness and contains a heater resistance suitable for calibration purposes. Between the perimeter of the sensitive middle section (4x4 mm) and the silicon frame a radially arranged thermopile is integrated into the membrane to measure the temperature difference given as consequence of the heat flow from the membrane center through the membrane with the thermocouples towards the Si-frame. The thermopile consists of an array of 160 p-type silicon – aluminum thermocouples connected in series and integrated into an n-type epilayer grown on the silicon wafer (Fig.1). The silicon chips are assembled in a standard ceramic chip carrier (PGA).

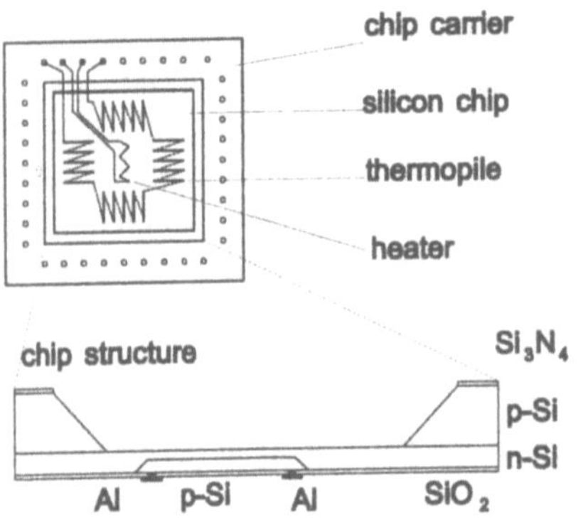

Fig.1: Silicon chip with integrated Si/Al-Thermopile

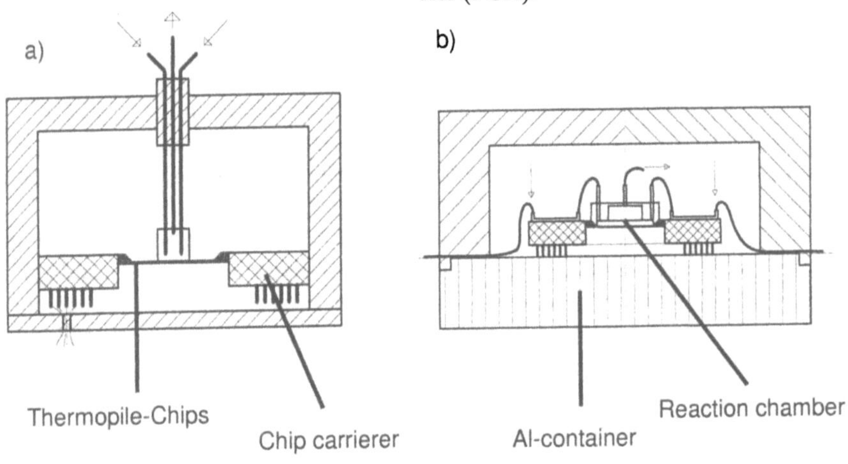

Fig.2: Schemes of calorimeter 1 (a) and calorimeter 2 (b). Both calorimeters are based on the thermopile chip LCM 2524 . Reaction chamber and input channels of calorimeter 2 are milled into a plastic disc.

Figures 2 depicts the schemes of both calorimeters. The reaction chamber of calorimeter 1 is made of a stainless steel capsule of 40 µl which is glue-bonded onto the sensitive area of the silicon membrane. Capillaries of the same materials serve as inlet and outlet for the liquid flow. Due to the poor heat conductivity of the capillaries in comparison with that of the membrane the decrease of the sensitivity is minor. In the case of the more compact calorimeter 2 a plastic disc with a thickness of 2 mm is attached to the surface of the silicon chip. The disc contains a cylindrical reaction chamber with a volume of 18 µl and two inlet channels. To improve the thermostatting of the input liquid flows the connecting tubes are contacted to the chip carrier. To establish defined temperature conditions both calorimeter arrangement are mounted in a temperature controlled aluminium jacket.

2.2 Calorimeters based on BiSb/Sb-thermocouples

The third calorimeter design combines a free-standing membrane within a silicon frame and a glass chip with wet-etched reaction channel, connected by glue-bonding.
The stress-compensated multilayered SiO_xN_y-membrane of about 1 µm thickness is prepared by anisotropic wet etching, whereupon 144 BiSb/Sb-thermocouples in series are deposited by electron beam evaporation and photolithographically patterned with active junctions on the membrane area and reference junctions in close contact with the silicon frame. The 0.1 mm deep glass channel is 2 mm in width and features a Y-shaped static mixer structure (Fig. 3) [3]. The mixer structure allows only diffusive mixing with a maximum diffusion path of half the channel width (i.e. 1 mm) and a diffusion area of channel height times length (i.e. 0.1 mm x 10 mm). Mounting and electrical connection is made to a printed circuit board (PCB).

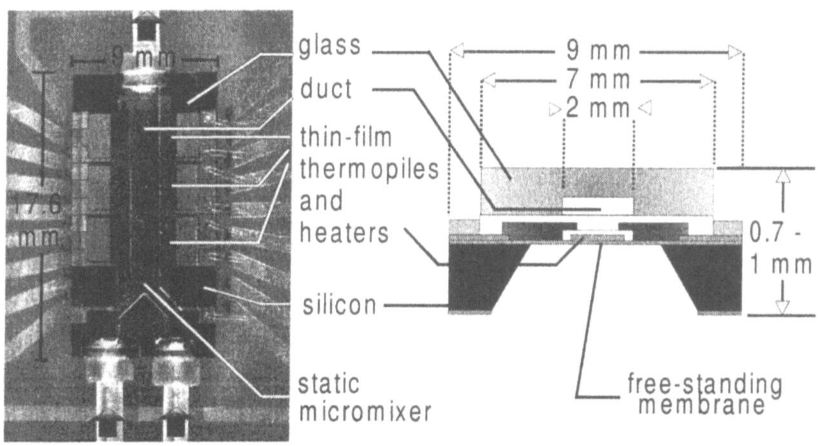

Fig. 3: Flow-through calorimeter made of one deep-etched Si-chip with thin-film arrangement consisting of three spatially separated thermopiles (thermocouples in series) plus attributed heaters deposited onto a free-standing membrane glue bonded to one glass chip with etched reactor channel of some 2 µl featuring a Y-shaped static micromixer

The fourth type of calorimeter is prepared of a silicon chip featuring a deposited thermopile and a reaction chamber made from plastic. Here, the thermopile consists of up to 130 BiSb/Sb-thermocouples in series (Fig. 4), arranged radially between a circular sensitive area with 3.6 mm radius and the Si-frame with 12 mm x 12 mm outer dimensions [4]. The metallic layers are deposited onto a 8 mm x 8 mm free-standing membrane either deep-etched into silicon with a thickness of a few μm or prepared of the stress-compensated SiO_xN_y-multilayer already mentioned. The cylindrical reaction chamber of 6.5 mm in diameter is milled 1 mm deep into a 1.2 mm thick PMA-disk measuring 7.5 mm x 7.5 mm, creating a volume of some 33 μl. To accomplish sufficient mixing of two educts within the chamber, a matrix of 38 holes (0.2 mm in diameter, 0.2 mm apart) is drilled into the 0.2 mm thick PMA-membrane (Fig. 5) in order to yield micro-plumes similar to those already described [5]. The chamber is equipped with one additional inlet and one outlet fluid port for the second educt and the product, respectively. Finally the chamber is glue-bonded into the cavity of the Si-chip directly onto the membrane. Calorimeter 4 is glue-bonded and electrically connected to a standard ceramic chip carrier (PGA) of through-hole type (Fig. 6).

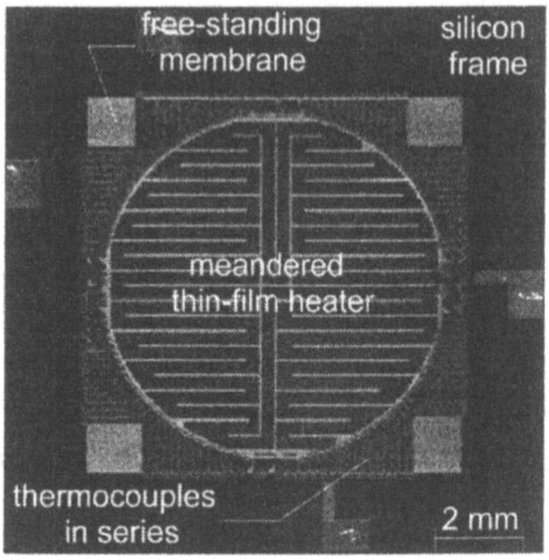

⇑ **Fig. 4**: Thermoelectric chip for caloric measurements: BiSb/Sb-thermopile in circular arrangement, meandered thin-film heater for calibration

⇓ **Fig. 6**: Assembled flow-through calorimeter with 33 μl volume

Fig. 5: Scheme of PMA fluid chamber with integrated hole matrix mixer structure

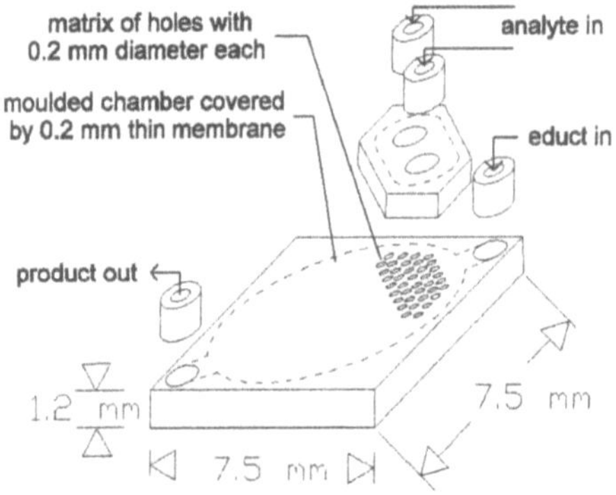

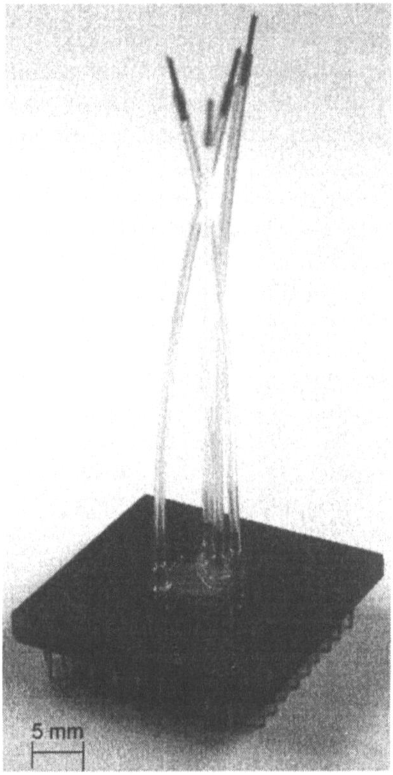

3 Characterization of the Calorimeters

3.1 Analysis of Heat Transfer

The sensitivity of the discussed flow-through calorimeters depends on the temperature coefficient of the thermopiles, the heat conduction across membrane, tubes and chamber material and it is influenced by mass flow through the calorimeter. To understand the flow rate dependency of the sensitivity and to apply optimal working conditions the relation between heat transfer by conduction and by forced convection should be known.

The sensitivities can be calculated from the linear slope of the steady-state signal vs. heat power curves. To generate heat power at various flow rates the integrated electrical heaters were used. Assuming concentrated thermal resistances R_{th} the heat power balance equation

$$\dot{q}_{el} = \frac{\Delta T}{R_{th}} + 2 v \rho c_p \Delta T \tag{1}$$

yields the expression

$$S_{el} = \frac{dU}{d\dot{q}_{el}} = \frac{\varepsilon R_{th}}{1 + 2 v \rho c_p R_{th}} \tag{2}$$

for the electrical sensitivity S_{el} with temperature difference ΔT, thermoelectric output voltage $U = \varepsilon \Delta T$, fluid density and specific heat capacity ρ and c_p, respectively, thermopower ε, and volume flow rate v.

According to the linearized form of Eq. (2)

$$\frac{1}{S_{el}} = \frac{2 \rho c_p}{\varepsilon} v + \frac{1}{\varepsilon R_{th}} \quad , \tag{3}$$

R_{th} and c_p of the fluid should be available from the linear relationship between experimentally determined reciprocal sensitivities and flow rates. Figure 7 depicts this relationship for calorimeter 1. Slope and intercept yield R_{th}=14.3 K W^{-1} and c_p = 5.6 J K^{-1} g^{-1}, respectively. The calculated R_{th} agrees with the value determined at zero flow (14.1 K W^{-1}), i. e. the heat conduction is not influenced by the fluid flow. This is in contrast to the behaviour of calorimeter 3. As shown in Fig. 8 the above-mentioned assumptions are not realistic at small flow rates. The sensitivities are calculated from data of the first thermopile. Further, the calculated R_{th} (150 K W^{-1}) is markedly higher than that at zero flow (116 K W^{-1}). Apparently, there is a heat flow through the glass plate in an opposite direction to the input liquid flow which is more and more eliminated by forced convection.

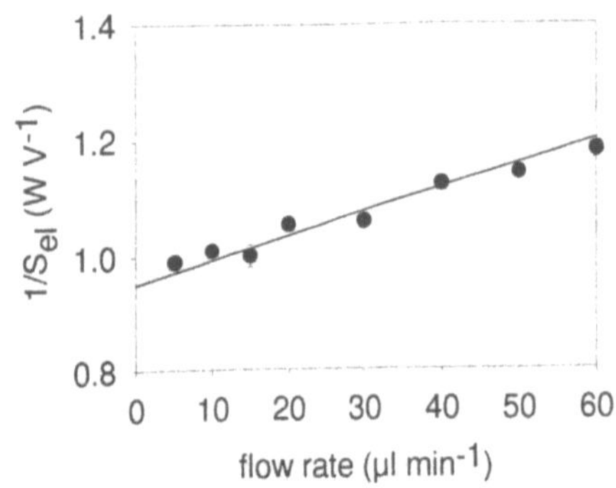

Fig. 7: Flow rate dependency of the sensitivity of calorimeter 1

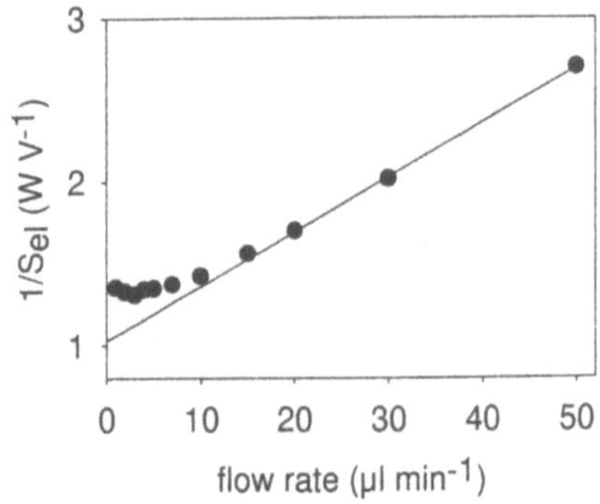

Fig. 8: Flow rate dependency of the sensitivity of calorimeter 3

3.2 Chemical Reaction Sensitivities

One of the most important problems of the application of chemical reactions in miniaturized flow-through calorimeters evolves from the reduced mixing efficiency of the input liquid flows. The relation of the virtual sensitivity determined using chemical reactions and the joule heating based sensitivity is suitable to quantify the degree of mixing at any given flow rate.

Assuming a 'fast' reaction with immediate conversion, the degree of mixing can be defined as

$$D_{mix} = \frac{\left(\frac{dn}{dt}\right)_r}{\left(\frac{dn}{dt}\right)_{in}} = \frac{\dot{q}_r}{\dot{q}_{\max}} \tag{4}$$

with the rate of reaction $(dn/dt)_r$ corresponding to the generated heat power q_r and the input rate of reactant $(dn/dt)_{in}$ corresponding to the maximum heat power

$$\dot{q}_{\max} = \nu c_0 \Delta H_r \tag{5}$$

assuming total mixing and total conversion. Here, c_0 denotes initial concentration, $\Delta_r H$ denotes molar reaction enthalpy. By defining a virtual chemical sensitivity

$$S_{chem} = \frac{dU}{d\dot{q}_{\max}} = \left(\frac{\dot{q}_r}{\dot{q}_{\max}}\right) \frac{\varepsilon R_{th}}{1 + 2\nu \rho c_p R_{th}} \quad , \tag{6}$$

the degree of mixing can be determined experimentally by measuring S_{el} and S_{chem}. :

$$D_{mix} = \frac{S_{chem}}{S_{el}} \tag{7}$$

The precondition is, that in joule heating and chemical reaction experiments the thermal resistances are matching as can be seen by dividing Eq. (2) and Eq. (6).
As a fast test reaction the protonization of tris-(hydroxymethyl)-aminomethane was used. Figures 9 and 10 show, that the virtual chemical sensitivity is always lower than the joule heating based sensitivity and that the difference increases with increasing flow rates, i.e. the degree of mixing is a function of the flow rate. As one can see from Figure 11 there are big differences in the degree of mixing between the low-volume calorimeter 3 and both calorimeters 1 and 2 with the larger chamber volumes. This is due to the fact that the rate of mixing is controlled by diffusion only and therefore the volume dependent retention time has a considerable influence.

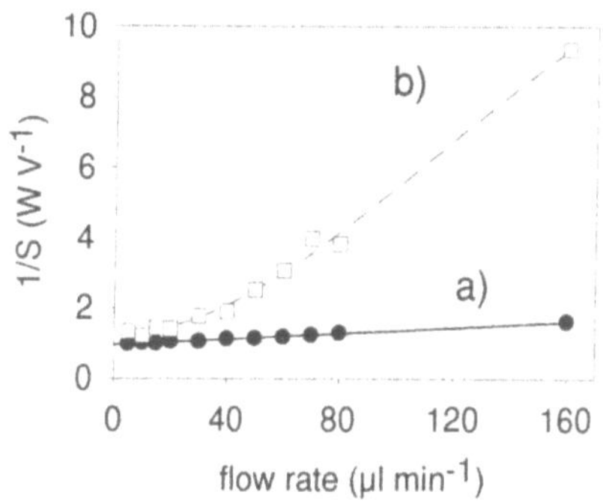

Fig. 9: Comparison of joule heating (a) and virtual chemical (b) sensitivities of calorimeter 1

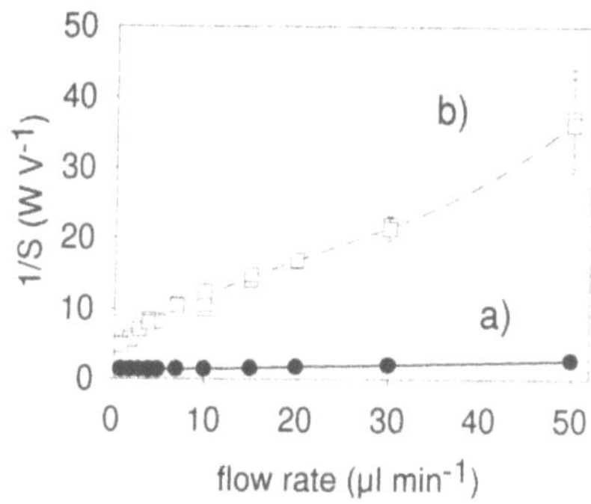

Fig. 10: Comparison of joule heating (a) and virtual chemical (b) sensitivities of calorimeter 3

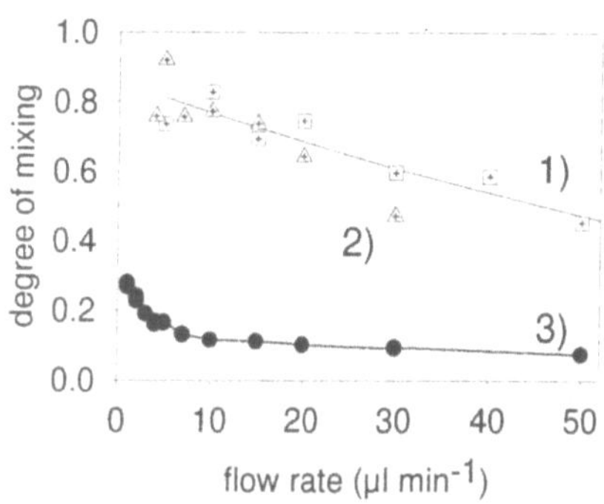

Fig. 11: Degree of mixing of the calorimeters 1 (squares), 2 (triangles) and 3 (circles)

4. Thermochemical Detectors for Flow Injection Analysis

Due to the universal nature of the heat power production of chemical reactions a single heat power meter can be applied as detector to different sensor reactions without modifications. Therefore, analytical arrangements for sequential flow-injection techniques should be possible on the basis of heat power measurements. Because of their high selectivity enzyme catalyzed reactions possess great potentials for applications as sensor reactions in sequential flow injection analysis [6].

To apply enzyme catalyzed reactions as sensor reaction the substrate solution which is under investigation has to be combined in the flow-through calorimeter with enzyme solutions of a defined activity. Then the detected heat power is a measure of the concentration of the substrate.

Figure 12 shows results of applications of the saccharose inversion reaction to the miniaturized flow-through calorimeter described above. The compared signals are generated by enzyme solution pulses injected into saccharose solution flows of constant concentrations. They correspond to equal concentrations of saccharose (100 mmol l^{-1}) and enzyme activities (0.4 mg ml^{-1} enzyme with an activity of 220 µmol min^{-1} mg^{-1}). Obviously, the signal strongly depends on the volume of the reaction chamber. Because the reaction is kinetically controlled the reduction of the volume and hence the decrease of the retention time not only reduces the mixing efficiency but also the degree of conversion. Therefore, it is very important to optimize the volume of the reaction chamber.

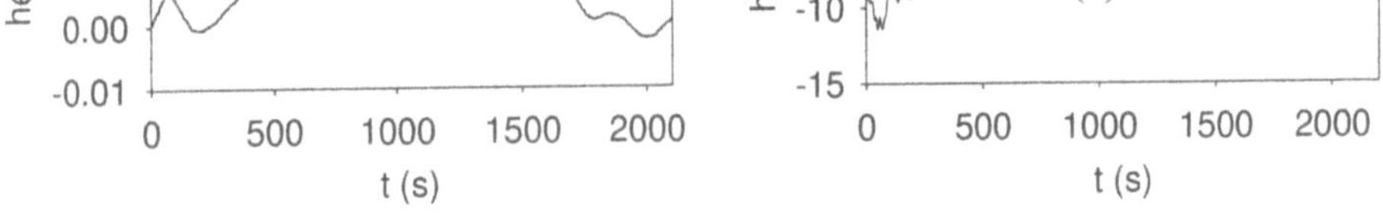

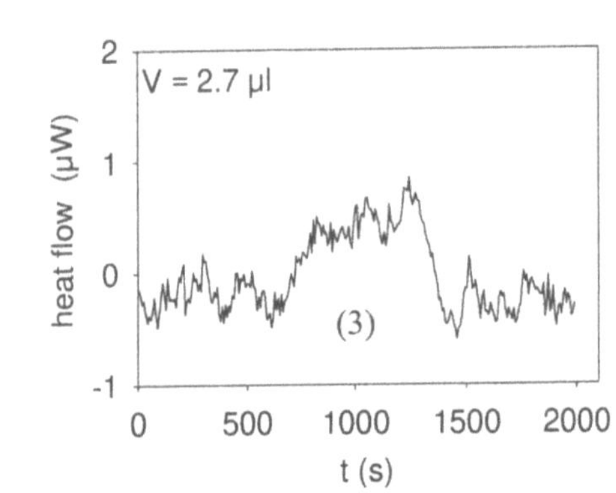

Fig. 12: Heat power signals generated by enzyme solution pulse injected into a continuos flow of saccharose solution. 1: calorimeter 1; 2: calorimeter 2; 3: calorimeter 3

5 Kinetic-adapted Miniaturized Flow Injection Analysis (μFIA)

In order to overcome the strong signal dependency on the retention time in calorimeter 3 (see 3.2) a modular set-up of highly efficient static micromixer (Fig. 13) [7] and calorimeter 3 has been tested [8]. For 'fast' chemical reactions with immediate conversion of reactants, i.e. reactions controlled by the degree of mixing such as protonization, it is expected that a considerable part of the reaction will have evolved before the sample volume enters the calorimeter associated with a non-desirable heat loss. Hence, the set-up is considered non-feasible. For kinetically controlled, i.e. rather 'slow' reactions the signal yield should be comparably higher than using the mixer integrated into the calorimeter 3. For testing this hypothesis, the cerium catalyzed redoxreaction of malonic acid and sodium bromate was used as an examplary 'slow' reaction.

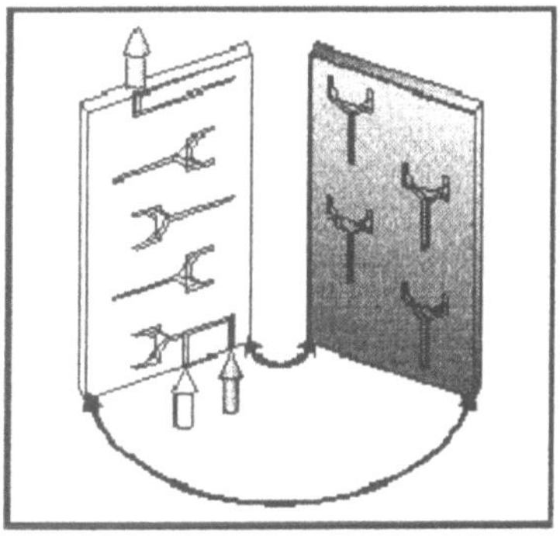

Fig. 13: Working principle of the static micromixer (patented by and courtesy of TU Ilmenau, Germany): chemically deep-etched silicon chips anodically bonded to form closed ducts, causing two injected fluids to be alternatingly stacked in thin layers [7]

In plotting both signal peak height and integral signal (i.e. detected chemical heat) vs. flow rate, the typical graph known from traditional FIA is yielded. An increase of concentrations of educts leads to a rise of signal heights and integral signal. Moreover, it leads to a shift of the maximum towards higher flow rates, indicating a 'faster' reaction. The comparison of signal heights and detected heat with data obtained using the integrated mixer indicates, that an increase by a factor of ten is accomplished (Fig. 14). Further investigations especially on enzymatically catalyzed reactions are currently conducted.

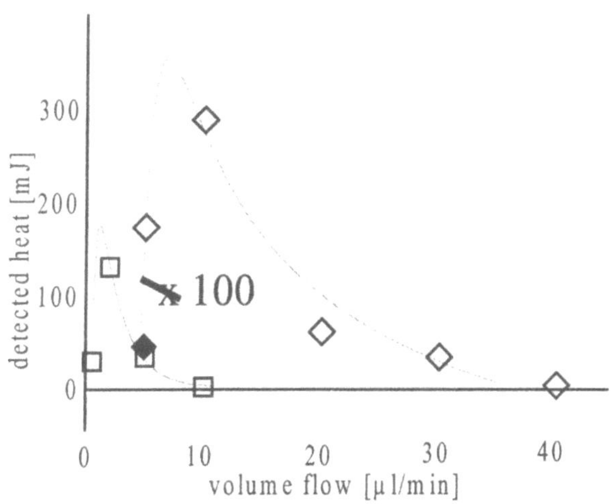

Fig. 14: Detected heat vs. volume flow for two different educt concentrations: malonic acid and sodium bromate both 0.3 mol/l (hollow squares, sample volume 0.25 μl, enhanced 100-fold) and 1 mol/l (hollow diamonds, sample volume 2 μl). Data using the integrated mixer added (filled diamond, concentrations like those for hollow diamonds) [buffer: 2 mol/l sulfuric acid, 50 mmol/l cerium]

6 Summarizing Conclusions

Three miniaturized flow-through calorimeters different in reactor volume, reactor design, chosen materials, and degree of mixing have been prepared and characterized both by Joule heating and chemical reactions. A fourth type has been introduced and is yet to be evaluated. Retention time and degree of mixing evolved as the crucial parameters in signal formation with significantly better performances of calorimeters with volumes of a few 10 µl compared to one with only 2.7 µl and poor mixing efficiency. It appears, that for chemical reactions with low rate constants the modular arrangement of a highly efficient micromixer with the latter low-volume calorimeter 3 accomplishes a considerable signal increase.
Integration of a highly efficient mixer structure into a calorimeter with kinetic-adapted reactor volume for sufficiently increased retention time will be the next step towards a totally integrated µFIA arrangement. Reactor volumes of more than 10 µl in combination with small heat conduction paths seem to be achievable only in glue-bonded hybrid devices or in stacked-wafer devices made of material with low heat capacitance.
For a more versatile modular set-up of mixer and calorimeter dead-volume reduction seems to be necessary to minimize heat losses and retention time between mixer and calorimeter, urging the adaptation and standardization of fluid connectors.

Acknowledgement

This work was partly funded by the german research council (Deutsche Forschungsgemeinschaft) under grants Wo 576/5-1 and Ko1403/3-1, 3-3, 10-1. We thank H. Fischer, F. Jahn, I. Menzel, K.-H. Wagner, and M. Wiedemann (IPHT Jena) for technical support, N. Schwesinger (Technical University of Ilmenau, Germany) for providing the micromixer.

References

[1] Recent developments in integrated circuit calorimetry; J. Lerchner, A. Wolf, G. Wolf; J. Thermal Anal. 55 (1999) 212

[2] Thermal sensors based on the Seebeck effect; A.W. Herwaarden, P. Sarro; Sensors and Actuators 10 (1986) 321

[3] Chip reactor for microfluid calorimetry; J.M. Köhler, M. Zieren; Thermochimica Acta, 310 (1998) 25-35

[4] First measuring results concerning a new thin-film absolute radiometer; E. Kessler, J. Müller, F. Völklein, M. Pawlak, K. Möstl, F. Brandt; SPIE, 1712 '14th Symp. on Photonic Measurements' (1992)

[5] Towards integrated microliquid handling systems; M. Elwenspoek, T.S. Lammerink, R. Miyake, J.H.J. Fluitman; J. Micromech. Microeng., 4 (1994) 227-245

[6] Sequential flow injection analysis based on calorimetric detection; J. Lerchner, A. Wolf, A. Weber, R. Hüttl, G. Wolf; Thermochim. Acta, 318 (1999), in press

[7] A modular microfluid system with integrated micromixer; N. Schwesinger, T. Frank, H. Wurmus; J. Micromech. Microeng., 6 (1996) 99-102

[8] Detection of catalytic reactions in a miniaturized FIA set-up using a static micromixer and a flow-through chip calorimeter; M. Zieren, R. Willnauer, J.M. Köhler, N. Schwesinger; Abstract collection '13. Ulm-Freiberger Kalorimetrietage', 17.-19.03.1999 (Freiberg, Germany) 110; to be published

Isothermal Biochemical Amplification in Miniaturized Reactors with Integrated Microvalves

R. Bräutigam[1], D. Steen, R. Ehricht, J.S. McCaskill[2]

Institute of Molecular Biotechnology Jena, Department of Molecular Information Processing, Beutenbergstr. 11, D-07745 Jena, Germany

1 Abstract

A microreactor of 64 nl volume with integrated micro membrane valves of 1.7 mm^2 active area was designed for examination of biochemical reaction diffusion systems. The goal was to create a reactor with integrated flow control to improve the system response by reducing delay times from minutes to seconds compared to the use of micro syringe pumps for fluid control. To demonstrate the functionality of this microsystem a cooperatively coupled amplification system (CATCH) [1] was run in the reactor. The course of the reaction was observed by CCD imaging using an intercalating dye which monitors the concentration of the amplified double stranded DNA molecules.
The microvalves allow actively switchable mixing of the reactants at the entrance to the reactor and the regulation of the flow rates. The reactor also demonstrates switchable inoculation of the amplification mixes with DNA template and a spatially resolved course of the reaction by fluorescence imaging.

2 Introduction

Recently, there has been much activity and a rapid development in the field of microreactors and microsystems for their application in biotechnology. But one of the most important prerequisites for a serious and automatic handling of the biochemical solutions on the nanolitre scale is the availability of microvalves for a controlled supply of these solutions which should be integrated on a chip. Different proposals were made for such microvalves in microfluidic systems. For the investigation of DNA amplification as a reaction-diffusion system in microreactors we decided to use the principle of pneumatic actuated membrane valves presented by Vieider [2] in 1995.
We chose this particular concept since the required methods to prepare this type of valve are easily available, the valves are compatible with the biochemicals which we use and it is possible to integrate these valves into our existing reactor designs.

---

[1] corresp. author, current address: IPHT Jena, Winzerlaer Str. 10, D-07745 Jena
[2] presenting author, current address: GMD, Schloß Birlinghoven, D-53754 Sankt Augustin

3 Materials and Methods

3.1 Design and preparation of the microstructured reactor with integrated valves

The microvalve controlled reactor consists of a glass wafer where liquid supply channels and the reactor are etched with a depth of 16 μm and a silicon wafer with membranes 10 μm thick [3] which serve as mobile parts for the valves (fig. 1). The valve is compatible with our established reactor design where two layers are sufficient to create a closed system. To form a microvalve, a small barrier is left in each channel separating it into two parts. A silicon wafer is anodically bonded to the glass wafer to cover the etched structures. The silicon wafer is prepared in such a way that only a thin silicon membrane covers the channels at the position of the barriers. When the membrane is lifted using pneumatic actuation the liquid flows over the barrier. This means that the membrane displacement determines the fluidic cross section of the channel (fig. 2).

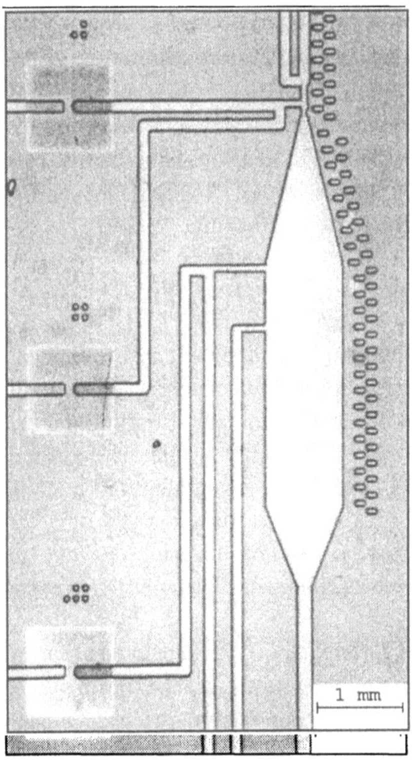

Fig. 1: Microscopic image of the microstructured reactor controlled by a total of 8 microvalves for separate supply of different solutions used in isothermal amplification of DNA

To connect the microreactor channels to the solution vessels outside, the glass wafer is ultrasonically drilled at the end of the channels. PTFE microtubes are attached to the holes using an UV curable adhesive.

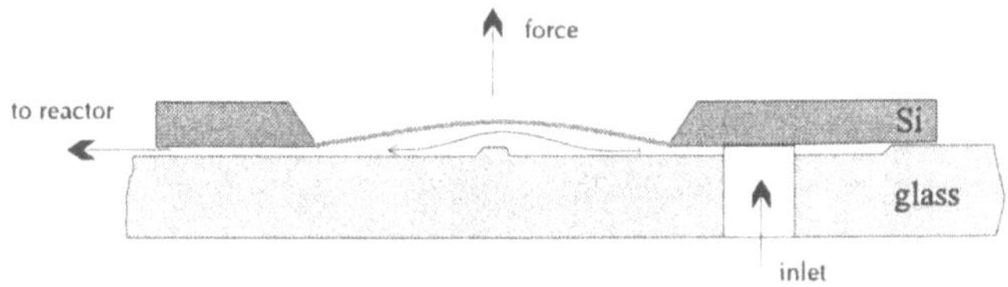

Fig. 2: Schematic side view of the microvalve composed by two channel pieces (etched in the pyrex glass) separated by a barrier and a Si membrane 10 μm thick which covers the barrier from the opposite side

3.2 Preparation of biochemical solutions and biochemical system

The details of designing the isothermal amplification system CATCH and for preparing the necessary solutions are described in [1, 4, 5].
Here, we list only all the components with their final concentrations in the separately added solutions:
Enzymes
4 U/μl T7 RNA polymerase (prepared as described in [5]),
0.16 U/μl RNAseH (MBI Fermentas),
5 U/μl HIV-1 reverse transcriptase (prepared according to [6] with p6HRT-PROT)
Nucleic acids
0.8 nM 106 bp dsDNA [1],
4 μM primer P1, 4 μM primer P2, 0.08 μM P1*-IRD-41 (MWG Biotech),
2 mM NTP, 1 mM dNTP (Pharmacia)
Other solutions
30 mM $MgCl_2$, 10 mM KCl, 40 mM Tris/HCl (pH 8.1), 5 mM Dithiothreitol, 2 mM Spermidin, 0.6 μM TO-PRO-1 (Molecular Probes), 0.4 mg/μl BSA, 1x Solution Q (Qiagen)

To prevent the reaction from starting in the supply channels, the biochemicals are divided into three different solutions: The solution which is refered to as enzyme solution contains all of the enzymes in their listed concentrations with all components mentioned in 'other solutions' except the dye TO-PRO-1. The second introduced solution is called resources and consists of all of the components from 'other solutions' and all of the listed nucleic acids except the dsDNA. Only the third solution, the so-called template solution, contains all of the components above but with the half of the given concentrations except for the dsDNA and 'other solutions'. Enzymes and resources are mixed directly at the entrance to the reactor whereas the template solution comes only into contact to this mixture to initiate the amplification at the end of the triangular shaped part of the reactor.
After this initiation, the isothermal amplification of the dsDNA takes place according to the schematic drawing in figure 3.

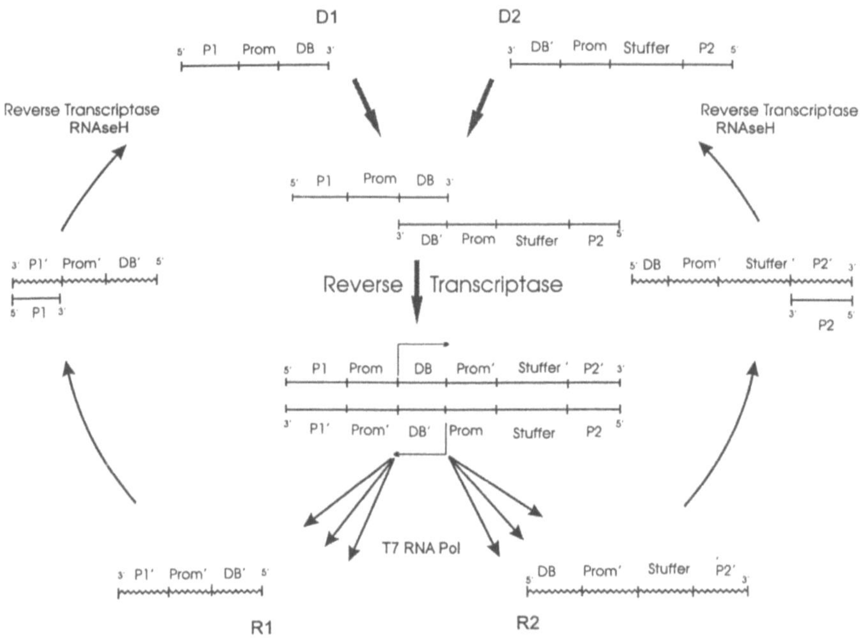

Fig. 3: The 3SR [7] amplification cycles of the single stranded RNAs R1 and R2 are mutually cooperatively coupled by the joint dsDNA intermediate. The hybridization of the primers P1 and P2 to each complementary region of the RNAs R1 and R2 starts the cycle. Reverse transcription and digestion by RNAseH produce the ssDNA intermediates D1 and D2. Their 3' ends hybridize via the complementary region DB/DB' and are elongated to the joint DNA double strand by the reverse transcriptase. Both RNAs R1 and R2 are recreated after reverse transcription by T7 RNA polymerase starting from the promotor (Prom) regions.

3.3 Experimental setup and handling

A fluorescence based setup (fig. 4) using the DNA intercalating dye TO-PRO-1 (Molecular Probes, absorption 514 nm, emission 531 nm) is applied to detect the concentration changes of the amplification system in the reactor. The 514 nm line of an argon ion laser (Spectra Physics) in cw mode (1W) is directed to the silicon/glass wafer with the microvalve reactor by optical fibers to excite the dye. To prevent photodegradation of the reaction solution a shutter is placed in the excitation beam. The fluorescence images are collected by a cooled CCD camera CH250 (Photometrics) with the Kodak KAF 1400 chip and equipped with two objectives (Nikon 105 and 50 mm) in tandem arrangement and a custom-designed interference filter (AHF). The spatial resolution of the images of 1317x1035 pixels realized in the experiments with the microvalve reactors was 3 μm per

pixel. The images were taken every minute with an exposure time of 0.3 seconds during a period of at least one hour by a custom extended version of the PC program PMIS (Photometrics).
Continuously monitored pressure is applied to the solutions to pump them through the channels into the reactor while microvalves control the fluidic cross sections of each channel. The reaction chamber is temperature controlled from the rear side of the reactor and kept at a constant temperature of 42 °C.

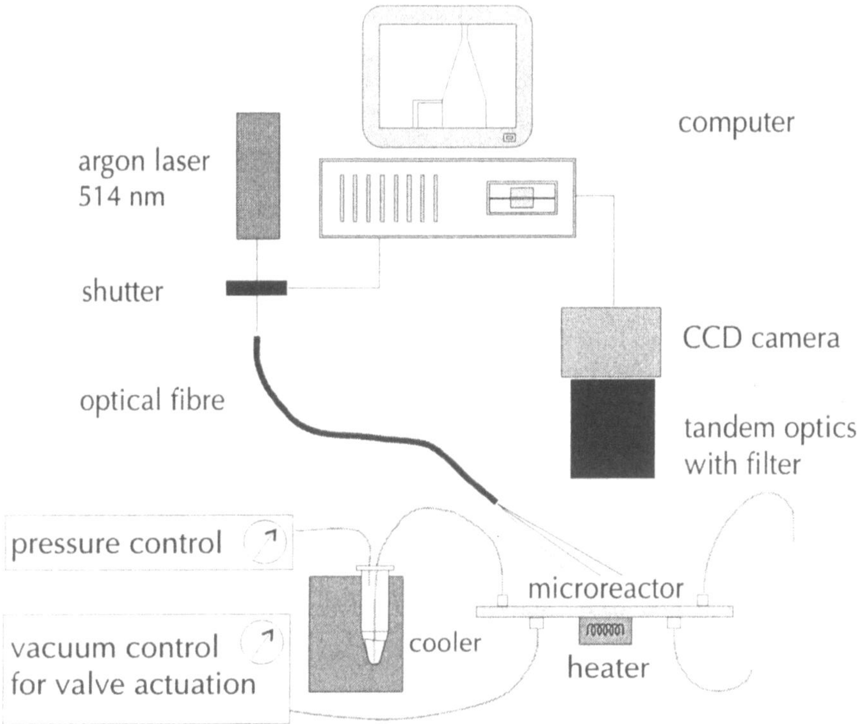

Fig. 4: Schematic experimental setup of the CCD fluorescence imaging for observation of the DNA amplification in the microvalve controlled reactors

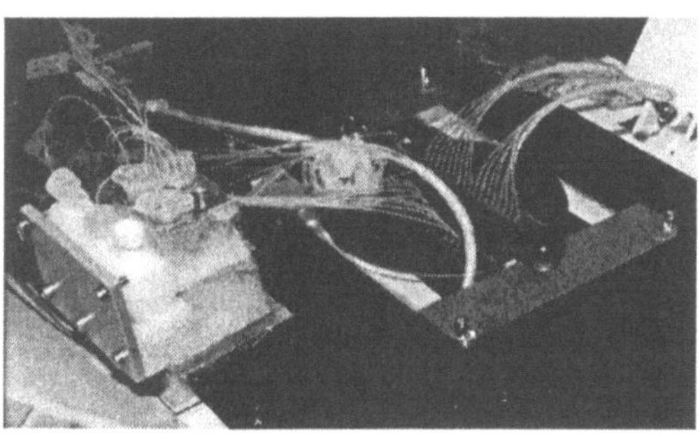

Fig.5: The image shows the wafer connected via PTFE tubing to the supply vessels which are stored in a cooler at 1°C to prevent the enzymes from losing activity. The pressure system is mounted on the back side of the cooler. The pressure is also fed into the vessels through PTFE capillaries.

4 Results

4.1 Membrane displacement as a function of the applied vacuum

The variation in the fluidic resistance of the valve is due to membrane lift. The membrane displacement can be measured by using monochromatic light with a microscope and counting the number of Newtonian rings. Examples of these measurements are shown in the image sequence of fig. 6. The graph in fig. 7 shows the determined pressure dependence of the displacement for two membranes of different sizes, 1300 μm (used here) and 1000 μm.

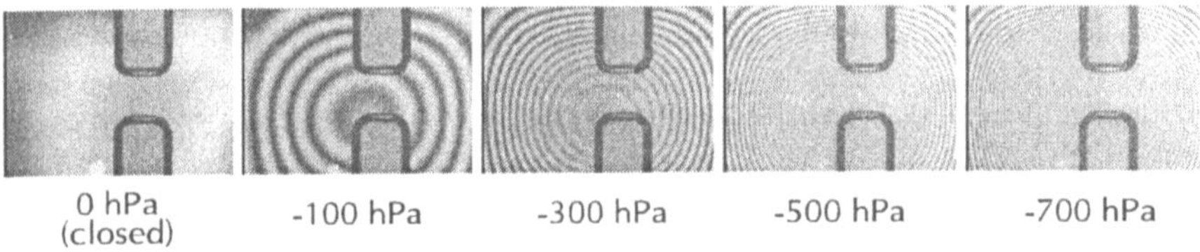

Fig. 6: Examples of determination of the membrane displacement as a function of the applied pressure by counting the number of Newtonian rings from microscopic images with monochromatic light (514 nm)

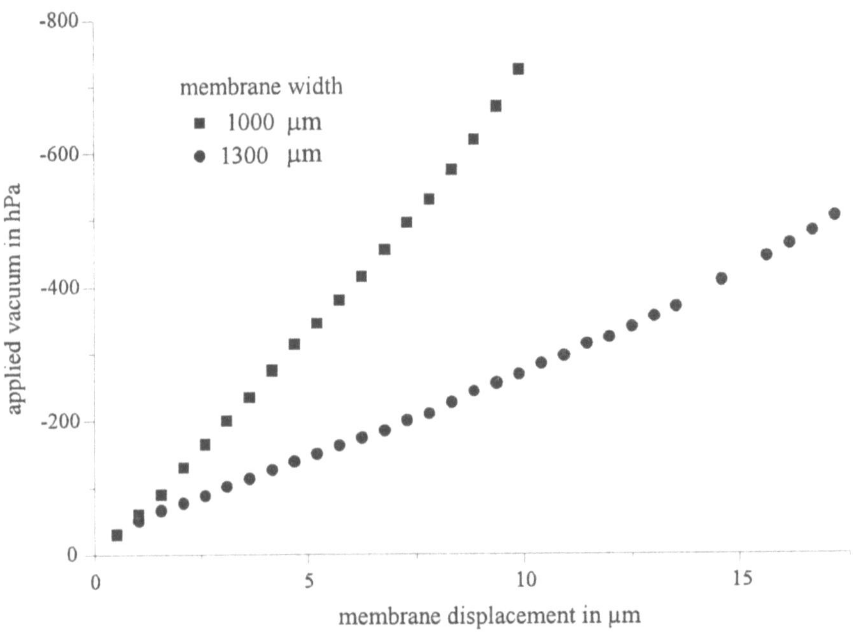

Fig. 7: Pressure dependence of the determined membrane displacement for two membranes of different sizes (1000 and 1300 μm)

4.2 Amplification in the microvalve controlled microreactor

The coupled amplification reaction ran successfully in the microchamber as shown in fig. 8. The flow control system employed made a precise liquid management on the subnanolitre scale possible, thus the reaction could be started from a predetermined spot in the reactor. After this initiation, amplification of the dsDNA concentration was observed by fluorescence intensity increase of the intercalated dye (fig. 8, 9 and 10). This was proved by gel electrophoresis where the amplification of the template molecule could be confirmed using the dye labeled primer P1*-IRD-41 (not shown).

We also started with an inhomogeneous distribution of the resources (at the beginning enzymes and NTPs were kept separately) so that the solutions had to mix by diffusion before the amplification could take place. Further experiments were carried out applying an amplification system that uses the RNA-Z mechanism [8].

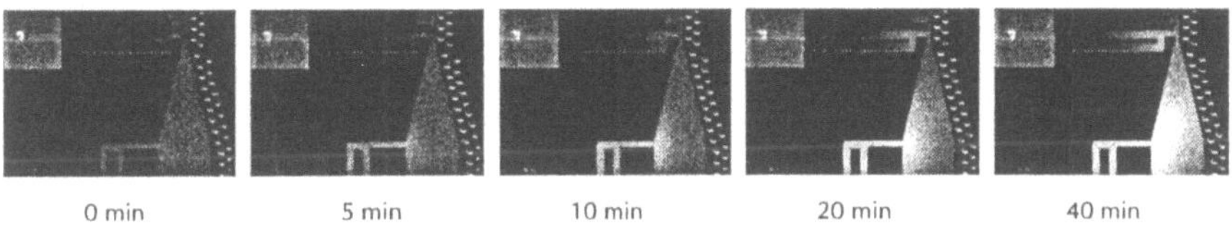

Fig. 8: CCD camera images of the DNA amplification in the microreactor at different times (0, 5, 10, 20, 40 min) detected using the fluorescence intercalator TO-PRO-1 (excitation 514 nm, exposure time 0.3 s). Each single image corresponds to a real region of 3.95 x 3.1 mm.

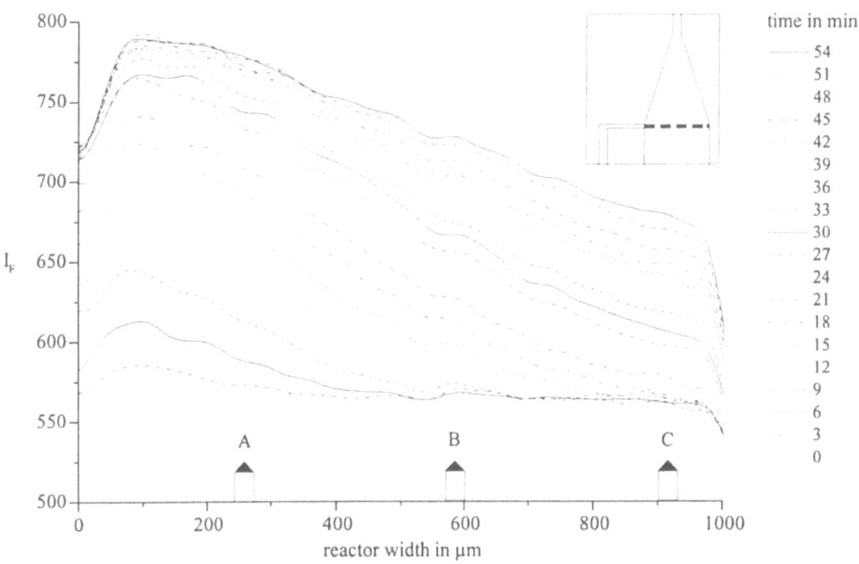

Fig. 9: Fluorescence intensities measured in the microreactor at the entrance of the template channel (dashed line of the insertion) at time intervals of 3 minutes

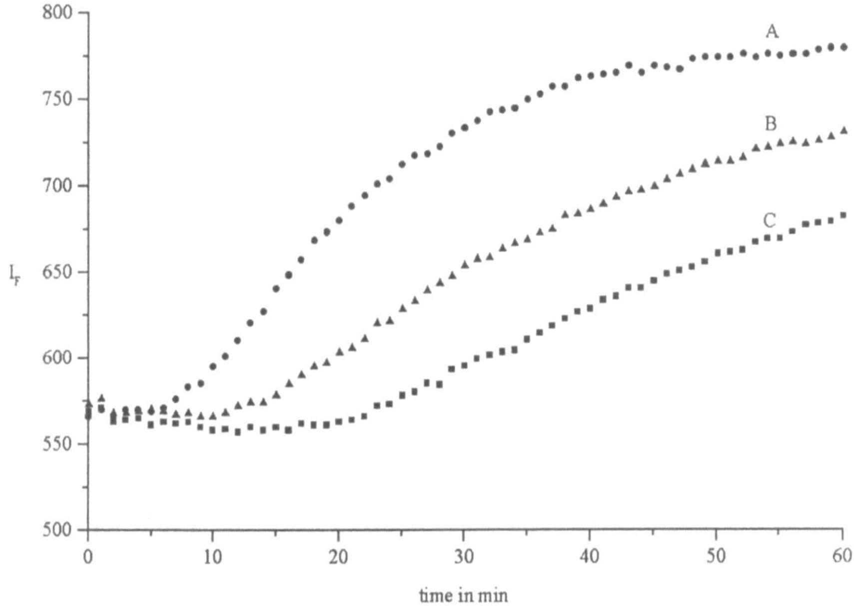

Fig. 10: Time course of the amplification in the microvalve reactor at three chosen positions A, B, and C shown by the corresponding marks in fig. 9

5 Conclusions

To our knowledge, we have shown the first successful biochemical reaction in a microreactor whose supply channels are controlled by microvalves on the same wafer. Our results demonstrate that the microvalve controlled reactor which we created can improve the investigations of biochemical reaction-diffusion systems which are important in evolutionary studies. It seems likely that further biochemical rections and amplifications will be performed using integrated microvalve reactors.

The next steps for routine applications and a more automatic and precise handling would be the computer control of the pneumatic system, its miniaturization and extension to more valves. It would be also desirable to integrate the pressure control in the microsystem itself.

6 Acknowledgments

We would like to thank T. Kirner for valuable discussions, A. Wagenhaus for biochemical assistance and M. Gohlke for assistance in microstructuring. This work was supported by grants from the German Ministery of Education and Technology (BMBF 03 10799).

7 References

[1] R. Ehricht, T. Ellinger, J.S. McCaskill, Eur. J. of Biochem. 243 (1997), 358
[2] C. Vieider, O.Ohman, H. Elderstig, Digest of Technical Papers, Vol. 2, 8th Int. Conf. on Solid State Sensors and Actuators (1995), 284
[3] B. Kloeck, S.D. Collins, R.L. Smith, N.F. de Rooij, Reproducible Micromachining by Electrochemical Etch-Stop, University of Neuchâtel, Institute of Microtechnology, Rue A.-L. Breguet 2, CH-2000 Neuchâtel
[4] R. Ehricht, Dissertation, Friedrich-Schiller-Universität Jena (1998)
[5] T. Ellinger, R. Ehricht, BioTechniques 24 (1998), 718
[6] S.F. Le Grice, G.E. Cameron, S.J. Benkovic, Methods Enzymol. 262 (1995), 130
[7] J.C. Guatelli, K.M. Whitefield, J.K. Barringer, D.Y. Kwoh, D.D. Richman, T.R. Gingeras, Proc. Natl. Acad. Sci. USA 87 (1990), 1874
[8] R.R. Breaker, G.F. Joyce, Proc. Natl. Acad. Sci. USA 91 (1994), 6093

Microfluidic Filtration Chip for DNA Extraction and Concentration

Agata Przekwas, DeMing Wang, Vinod B. Makhijani,
and Andrzej J. Przekwas

CFD Research Corp., 215 Wynn Dr. Huntsville AL 35805, USA
{ ph: 256-726-4815 , e-mail: ajp@cfdrc.com}

Abstract. In the last few years there has been tremendous interest in developing a complete biomedical/biochemical intelligent microsystem for extraction, concentration, amplification, analysis and processing of DNA. This paper presents multi-disciplinary computational tool, CFD-ACE+MEMS, for simulation and design of microfluidic bioMEMS. It is demonstrated on 3D high-fidelity simulations in microfluidic channels and on a complex DNA filtration chip.

Keywords. Microfluidics, DNA, separation, computational fluid dynamics, MEMS

1 Introduction

In the last few years significant advances have been made in the development of multifunction bioMEMS chips. An integrated microchip for Capillary Electrophoresis (CE) and a postcolumn reactor on a glass substrate were presented by Jacobsen [1] in 1994. Shortly after Wooley and Mathies [2] showed ultra high speed DNA sequencing using CE chips with four color fluorescent photodetection. In 1996, the UC Berkeley and Lawrence Livermore Laboratory team [3] showed the first fully integrated Polymerase Chain Reactor (PCR) and CE chip for complete DNA analysis. The first silicon based commercial (Cepheid) version of an integrated actively controlled PCR and real time optical fluorescence monitoring of product DNA was demonstrated by Northrup [4] in 1998. Michael Ramsey's team [5] at Oakridge presented a multiple sample PCR and Gel Electrophoresis (GE) single monolithic chip. A Cepheid-Stanford University team [6] has recently explored DNA-sold wall binding methods to extract and concentrate pure DNA samples from "real life" samples in Si micromachined microstructure arrays. In 1999, the Oakridge team presented a biochip with an array of DNA microprobes and 220 on-chip phototransistor detectors [7].

These bioMEMS devices are being developed in a very conventional experimental trial and error manner with little computational design support. There are no multidisciplinary CAD tools with coupled microfluidic, biochemical and electrodynamic modeling capabilities. This paper presents a new software tool, CFD-ACE+MEMS, which is specifically developed for multidisciplinary, multiscale design of bioMEMS. It describes the computational physics involved in modeling bio-microfluidic devices, and demonstrates it on a biochip for extraction and concentration of DNA from fluidic samples.

2 Virtual BioChip Concept for DNA Processing

2.1 DNA in External Fields and Forces

Long DNA strands in solutions adopt random orientations due to thermal agitation. Their internal configuration is determined by internal forces/fields (hydrogen bonding, electrostatics, bending, torsion tension) and by external fields such as flow shear, temperature gradients, electric fields, etc. DNA dynamics can be computationally investigated at several levels ranging from molecular dynamics, Monte Carlo, finite element, integral potential methods, etc [8,9]. Molecular dynamics is limited to very short times and small number of base pairs (bp) and does not account for hydrodynamic interactions. The most suitable model for unsteady multiscale DNA-environment fields interaction is a finite element model (FEM) using flexible beams to model a spiral staircase whose supports are made of flexible springs. Chemical, electrostatic and mechanical loads can be accounted for in such models.

In many applications (e.g., CE) DNA structures are exposed to electric fields. In an electric field (E), negatively charged DNA moves in the opposite direction of its counterions. The motion of counterions induces a hydrodynamic flow which in turn exerts shear and drag forces on DNA. Most of the friction between the DNA molecule and the solvent occurs in the double layer around the DNA backbone. A "macroscopic" electrohydrodynamic mathematical model of DNA dynamics in external fields has been recently presented by Long et al., [10]. Figure 1 schematically illustrates DNA structures in the presence of external fields.

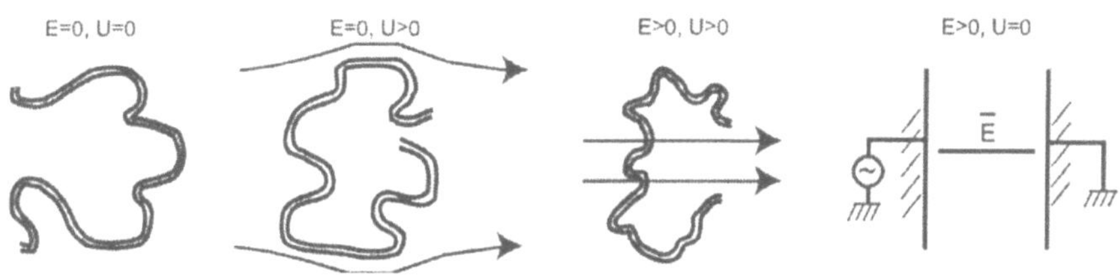

Fig. 1. Schematics of DNA behavior in external hydrodynamic and electric fields

CFDRC is exploring utilization of an Eulerian Lagrangian formulation to study coupled DNA dynamics and hydro-thermo-electric fields in microfluidic channels.

2.2 DNA Amplification in PCR

Computational simulation of DNA dynamics in varying thermal environments associated with the PCR process is very difficult. Only few studies have been reported presenting the thermochemistry of PCR. The most complete modeling approach including hybridization, thermodynamics and second order kinetics has been discussed by Wetmur [11]. Most studies are limited to simple heat transfer simulations of temperature uniformity and ramping rates between the denaturation process (shown in Figure 2), annealing and the elongation steps.

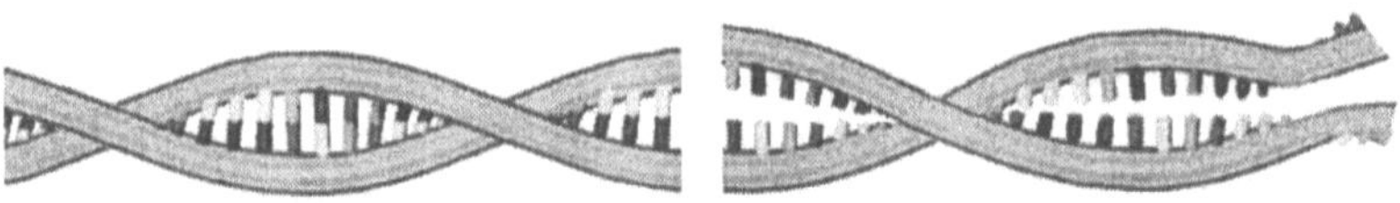

Fig. 2 Denaturation of a double strand DNA in the first step of PCR process

2.3 DNA Separation and Detection in Capillary Electrophoresis (CE)

CE is performed in narrow capillaries (~25 um) filled with aqueous buffer and exposed to an axial electric field (~100-500V/m). Separation of different sizes and charges in a mixture is obtained by differential migration of DNA and solutes in an electrical field. If the sample is placed near the anode and the electric field is applied, the ions in the capillary will start to migrate to electrodes with different speed depending on their charge, size, and shape. The migration velocity, v, is proportional to the electric field: $v=\mu E$ (μ is the electrophoretic mobility). Figure 3 schematically illustrates the CE separation concept.

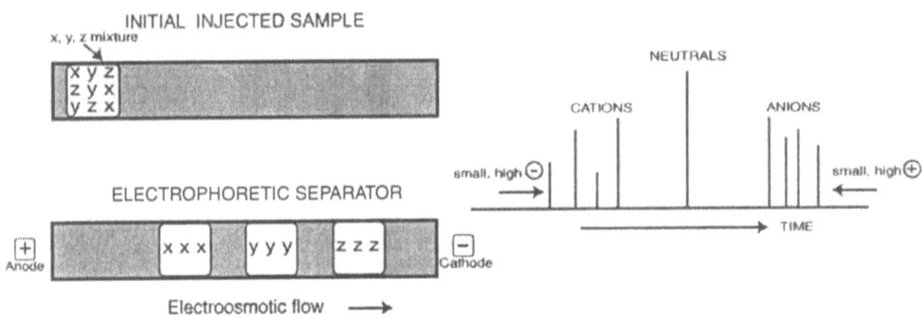

Fig. 3. Schematics of capillary zone electrophoresis and a typical electropherogram

Computational modeling of CE in bioMEMS has been recently demonstrated by Ermakov [12], and by Giridharan [13] using CFD-ACE+MEMS.

2.4 DNA-Wall Interaction

DNA can be both covalently and noncovalently linked to solid walls of glass, metals, silicon, and organic polymers The attachment chemistries have been developed for several materials (proteins, DNA). The DNA molecule preserves its native double helix when adsorbed onto a negatively charged surface, and undergoes unfolding when the surface is charged positively. This wall affinity property can be used to extract, separate, and concentrate DNA [6]. A DNA solution is passed through a binding matrix (e.g., silica powder) in which DNA binds to the material surface. Undesired pollutants (salts, PCR inhibitors) are washed with an ethanol-based solution. Finally an elution reagent is passed through the system releasing DNA into the fluid solution. Computational simulations of the DNA-wall binding are presented in the next chapters of this paper.

3. Description of Computational Models

Computational models are presented in the form of governing equations for the various phenomena that influence DNA dynamics on a biochip. These models have been uniquely incorporated for coupled simulations for the first time ever in the software package CFD-ACE+MEMS.

3.1 Fluidics

The fluid flow model in CFD-ACE+MEMS solves the continuity and the pressure-based Navier-Stokes equations assuming that the fluid is a continuum.

$$\frac{\partial \rho}{\partial t} + \nabla \cdot (\rho \vec{u}) = \dot{m} \tag{1}$$

$$\frac{\partial}{\partial t}(\rho \vec{u}_i) + \nabla \cdot (\rho \vec{u}\vec{u}) = \nabla p + \nabla \cdot \underline{\underline{\tau}} + \rho \vec{f} \tag{2}$$

where ρ is the fluid density, $\vec{u}$ is the instantaneous velocity, $\dot{m}$ is the mass generation rate, p is the static pressure, $\underline{\underline{\tau}}$ is the viscous stress tensor, and $\vec{f}$ is the body force.

3.2 Heat Transport and Thermodynamics

The thermal model solves the total enthalpy form of the energy equation

$$\frac{\partial}{\partial t}(\rho H) + \nabla \cdot (\rho \vec{u} H) = \nabla(k \nabla T) + \frac{\partial p}{\partial t} + \nabla \cdot (\tau - \vec{u}) - \nabla \cdot (J_i h_i) + S_a + \rho \vec{f} \cdot \vec{u} \tag{3}$$

where J_i is the total (concentration-driven+temperature-driven) diffusive mass flux for species i, h_i represents the enthalpy, S_a represents additional sources due to surface reaction, radiation, joule heating, and liquid spray, and q_j is the j-component of the heat flux. Fourier's Law is employed to model the heat flux: $q = -k\nabla T$, where k is the thermal conductivity and T is the temperature

3.3 Mixing and Chemistry

The species transport equation may be written as

$$\frac{\partial \rho Y_i}{\partial t} + \nabla \cdot (\rho \vec{u} Y_i) = \nabla \cdot (D_i \nabla Y_i) + \dot{w}_i \qquad i = 1,\ldots, N_{sp} \tag{4}$$

where N_{sp} is the number of species involved. Y_i is the species mass fraction D_i is the speues diffusivity and w_i represents the reaction term.

3.4 Electrostatics

The electrostatic model solves the Poisson's equation for the electric potential.

$$\nabla \cdot \varepsilon_r \nabla \phi = -\frac{\rho}{\varepsilon_0} \tag{5}$$

where ϕ is the electric potential, ε_r is the electric relative permittivity, ε_o is the permittivity of free space, ρ is the space charge. From the electric potential, the electric field $\mathbf{E}(\mathbf{E} = -\nabla\phi)$, the pressure force, and the capacitance of the geometry are calculated. Two different techniques are available for solving Poisson's equation: finite volume method and boundary element method.

4. Demonstration Simulations

4.1 Surface Tension Driven Liquid Filling in Microchannels

Liquid filling occurs when the DNA solution is being introduced through a microfluidic circuit into the separation chamber. Two forces are available to induce the filling: surface tension and an externally applied pressure or vacuum. In the present demonstration case, the externally applied force does not exist, and transport is due to surface tension. This transfer is possible only when the contact angle is less than 90°. On the other hand, with a contact angle greater than 90°, the fluid must be pressurized to induce transport. The pressure force has to be higher than the pressure jump across the interface. Detailed analysis of surface tension driven flows has been presented by Yang and Przekwas [14].

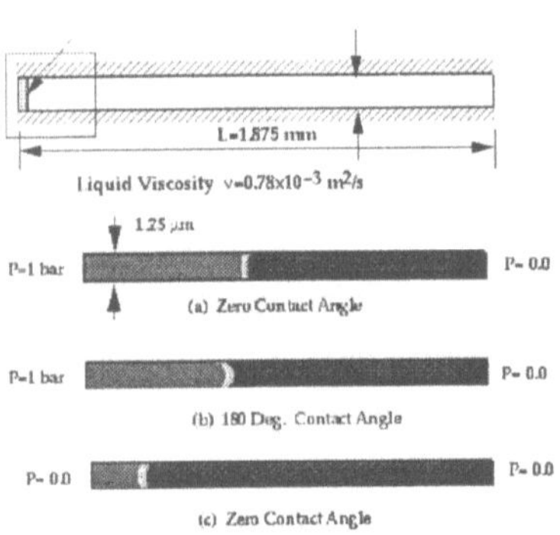

Fig. 4 Surface tension effect on liquid flow in a microchannel

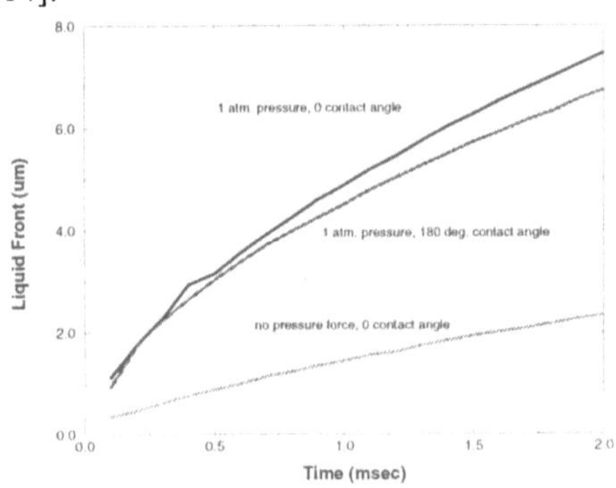

Fig. 5. Front propagation due to pressure and surface tension forces

The numerical model of liquid transport in a microchannel is shown in Figure 4. The boundary conditions and physical properties are also shown in this Figure.

Cases (a) and (b) correspond to vacuum transport. Two extreme contact angles are presented: (a) $\alpha = 0°$ and (b) $\alpha = 180°$. The front shapes are very different for these two cases. At $\alpha = 0°$, the surface tension assists, whereas at $\alpha = 180°$, the surface tension force resists the motion. As a result, one sees slower transport of the front. For comparison purposes, Figure 4(c) shows the front driven by surface tension alone. The liquid front positions with time are shown in Figure 5. Due to the fact that the driving force is constant, the front position is proportional to the square root of time, as explained in reference [14].

4.2 Simulation of DNA Extraction

Problem Description

In this study, the flow and surface reaction processes governing the extraction of DNA from a sample flowing through a microfluidic DNA filtration chip were analyzed using CFD-ACE+MEMS. The filtration chip consisted of inlet and outlet ports and an interior reaction/extraction chamber with multiple pillars. The top and bottom walls of the chamber as well as the pillars were made of a DNA binding material. The geometry of the device was similar to that of the Cepheid chip [6]. The pillar diameters, however, were larger, and the number of pillars were also much lower than in the Cepheid chip. The resulting DNA binding surface area was 0.16 cm^2 as compared to 0.36 mm^2 in the Cepheid chip.

Flow was modeled as steady, laminar, incompressible and Newtonian. Equations 1, 2, and 4 were solved on an unstructured mesh of 27,440 cells (see Fig. 6). The mass fractions of the DNA in different regions of the flow domain were computed by solving the conservation of mass equation for a passive scalar. The DNA diffusivity, D, was taken approximately equal to 10^{-7} cm^2/s, based upon molecular weight considerations for a DNA mixture with an average strand length of 500 base pairs, and the corresponding diffusivity for other macromolecules (e.g. hemoglobin) in water. Since the binding reaction results in DNA removal from the flow sample, the effect of this reaction on the bulk flow DNA transport was modeled by adding appropriate sink terms to the transport equation at the boundary cells along the reacting walls.

Boundary Conditions

The flow boundary conditions for the problem include uniform inlet velocity, fixed exit pressure (taken as zero) and zero-slip conditions along the bounding walls. The mass transport boundary conditions include uniform inlet DNA concentration ($C=Co$ = 1000 ng/ml)) and zero flux condition at the non-reacting side walls. As a first approximation, the surface binding reaction was modeled as a first-order reaction which is proportional to the near wall concentration of DNA.

$$-D\frac{dC}{dn} = K_d C_W \qquad \text{at binding surface} \tag{7}$$

where K_d is the binding rate constant, n describes the normal direction to the binding surfaces and C_w is the near wall DNA concentration.

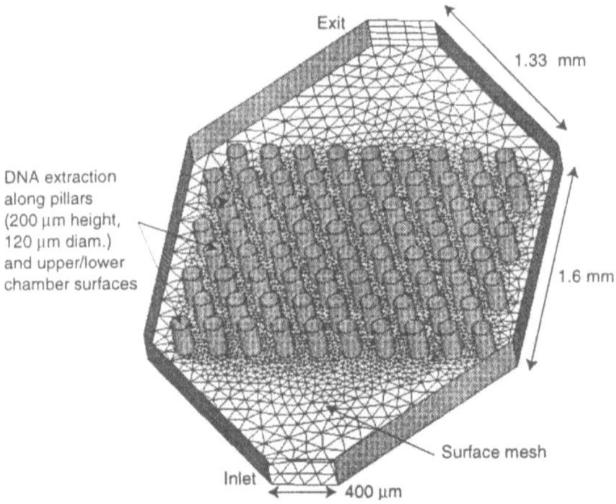

Fig. 6. Geometric model of DNA extraction chip with unstructured surface mesh

Simulation Results

Simulations were initially carried out to determine K_d through comparison of simulation results with DNA deposition data measured by Christel et al. [6]. The inlet flow rate and DNA concentration used in these baseline simulations were 0.5 μl/s and 1000 ng/ml, respectively. Once K_d was obtained, parametric studies were carried out to study the effect of DNA flow rate on the chip capture efficiency.

Flow Patterns in the Extraction Chamber. Figure 7a shows the flow velocity contours along the mid-plane of the extraction chamber. The velocity vectors are shown in the vicinity of the pillars in Figure 7b. Results show a symmetric distribution of fluid velocities (and pressures) due to the system layout. Because the Reynolds number is very low, there is no recirculation around the pillars.

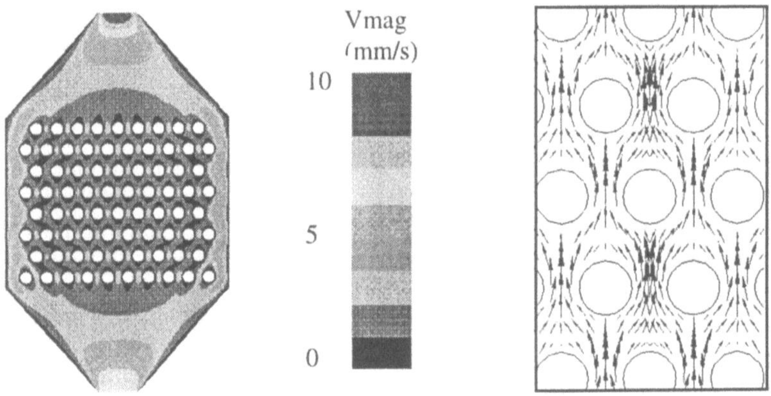

(a) velocity contours in chamber (b) velocity vectors around pillars

Fig. 7. Flow patterns along the central plane in the DNA extraction chip

Determination of DNA Binding Rate Constant. For the chosen baseline conditions, the DNA capture efficiency of the Cepheid chip was 2.8 [6]. However, their capture surface area was approximately 2.25 times larger. Assuming an approximately linear relationship between surface area and capture efficiency, the corresponding capture efficiency in the present study should be around 1.25%.

The binding rate constant was varied in the simulations until the computed DNA capture efficiency matched the expected value. The resulting K_d was 7 x 10^{-5} cm/s. The corresponding concentration contours for the solution DNA, normalized with respect to inlet concentration, along the lower surface of the chamber are shown in Figure 8a. The plot shows two streams of DNA solution bypassing the pillars along the side walls in which the DNA concentration is relatively higher than in the center of the chamber. In the solution that goes through the pillars, the DNA concentration reduces progressively as the solution moves towards the exit.

Effect of Chamber Geometry on Capture Efficiency. To demonstrate the enhancement of capture surface area due to the pillars, a simulation was carried out under the baseline flow conditions for the "flat chip" configuration in which the pillars were absent. The normalized DNA concentration contours along the lower surface of the chamber are shown in Figure 8b. The model predicts a lower capture efficiency of 0.84% for the flat chip configuration, which is consistent with its reduced capture surface area (0.12 cm^2 as compared to 0.16 cm^2). Similar results have been reported by Reference 6.

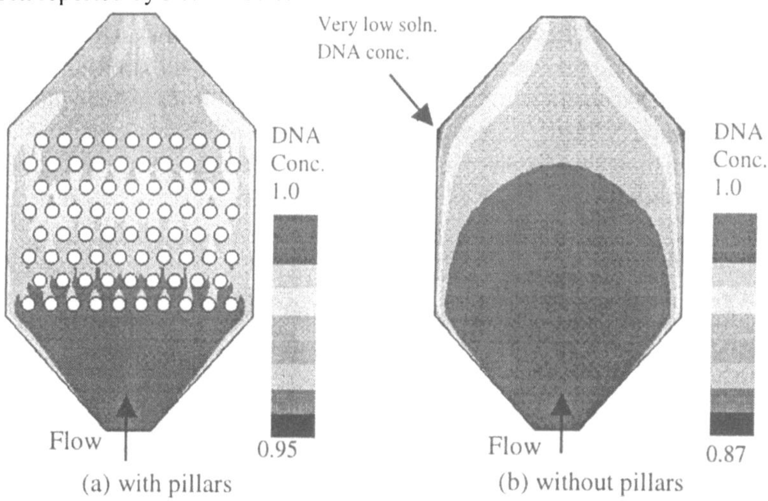

Fig. 8. Normalized DNA concentration contours in the solution flowing through the chip

What is interesting to note, however, is that there is a greater amount of localized DNA capture in the flat chip (lowest DNA concentration is 87% of inlet value) as compared to the chip with pillars (lowest DNA concentration is 95% of inlet value) as shown in Figure 8 legends. This can be attributed to the significantly different flow field in both chips as well as the locations of the reaction surfaces. In the flat chip, there are larger areas of low flow velocity in the converging region of the extraction chamber. This results is a large amount of localized DNA deposition in these regions. The overall capture efficiency, however, is still lower than that for the chip with the pillars because the solution DNA in regions away from the lower and upper surfaces exits the chamber virtually "untouched".

Effect of DNA Flow Rate on Capture Efficiency. Parametric studies were carried out to determine the effect of sample flow velocity on the DNA capture efficiency. Two different flow rates were used: 5 ml/s and 0.05 ml/s. The DNA concentration in the sample was maintained at 1000 ng/ml.

The model predicted a decrease in DNA capture efficiency from 1.25% to 0.128% when the flow rate was increased. Christel et al. [6] have similarly reported a decrease in capture efficiency from 10 to 5% when they increase their flow rate from 0.5 μl/s to 5 μl/s for a DNA concentration of 100 ng/ml. This can be attributed to the decreased residence time of the DNA strands in the extraction chamber due to increased convective transport, which reduces the potential for completion of the surface binding reaction.

For the case with the decreased flow rate, as expected, the model predicted an increased capture efficiency of 11.2%. In the measurements reported by Christel et al. [6], a decrease in DNA flow rate from 0.5 μl/s to 0.1 μl/s resulted in the same capture efficiency. The discrepancy between their experimental data and our simulation results could be attributed to factors such as differences in the chip geometry and the assumption of first-order binding kinetics in our model. The model is currently being refined to account for second-order surface kinetics in which the binding process also depends on the "capture sites" available on the surface in addition to the near wall DNA concentration.

5 Conclusions

The paper presents a novel concept of virtual prototyping of a microfluidic DNA processing chip. A multi-disciplinary simulation environment and computational techniques have been presented for several biofluidic functions of an integrated bioMEMS chip. The paper presents simulation results, obtained with CFD-ACE+MEM for grid filling with surface tension within microchannels and DNA filtration and concentration in a micromachined MEMS structure. Detailed 3D flow and DNA transport and wall binding have been investigated for a DNA extraction chip with and without pillar arrays. Simulation results were compared to experimental observations at Cepheid. The model correctly predicted the influence of flow rate, surface area and device geometry on the DNA capture efficiency. Future model extensions are in progress and will incorporate Lagrangian modeling of DNA transport, electrodynamics, wall binding, DNA amplification in PCR devices and separation in electrophoetic devices.

6 Acknowledgments

This study was funded by a DARPA/ETO contract (F30602-98-2-0152). The authors acknowledge the contributions of Mr. J. Raghavan, Drs. H.Q. Yang and S. Mazumder, and Mr. Milind Talpallikar of CFDRC.

7 References

1. Jacobsen C.J., et al., "Microchip Capillary Electrophoresis with an Integrated Postcolumn Separation," Anal. Chem. 66, p. 3472, 1994.
2. Wooley A.T. and Mathies R.A., "Ultra-High-Speed DNA Sequencing Using Capillary Electrophoresis Chips," Anal. Chem. 67, p. 3676, 1995.
3. Wooley A.T., et al, "Functional Integration of PCR Amplification and Capillary Electrophoresis in a Microfabricated DNA Analysis Device," Anal. Chem. 68, p. 4081, 1996.
4. Northrup M.A., "A Miniature Analytical Instrument for Nucleic Acids Based on Micromachined Silicon Reaction Chambers," Anal. Chem. 70, p. 918, 1998.
5. Waters L.C. et al. "Multiple Sample PCR Amplification and Electrophoretic Analysis on a Microchip," Anal. Chem. 70, p. 5172, 1998.
6. Christel L.A. et al "High Aspect Ratio Silicon Microstructures for Nucleic Acid Extraction," Solid State Sensors and Actuators Workshop, Hilton Head, SC, June 7-11, 1998
7. Vo-Dinh et al. "DNA Biochip Using a Phototransistor Integrated Circuit," Anal.Chem. 71, p. 358, 1999.
8. Olson W. K. "Computational Approaches to Nucleic Acid Structure," In Computational Biomolecular Structures, ed. By Soumpasis D.M. and Jovin T.M. Springer-Verlag, 1993.
9. Robinson B.H. and Drobny G.P. "Site Specific Dynamics in DNA: Theory," Ann Rev. Biophys. Biomol. Struct. V24, p. 523, 1995.
10. Long D. Viovy J.L., and Ajdari A., "Simultaneous action of Electric Fields and Nonelectric Forces on a Polyelectrolyte," Phys. Rev. Lett. 76, p. 3858, 1996
11. Wetmur J.G., "DNA Probes: Applications of the Principles of Nucleic Acid Hybridization," Crit. Rev. in Biochem. And Mol. Biol. 26(3/4), p. 227, 1991.
12. Ermakov S.V., et al., "Computer Simulation of Electrokinetic Transport in Microfabricated Channel Structures," Anal. Chem. 70, p. 4494, 1998.
13. Giridharan M.G., et al., "Computational Simulation of Microfluidics, Electrokinetics, and Particle Transport in Bilogical MEMS Devices," Symp. On Design, Test and Microfabrication of MEMS/MOEMS, Paris, March 30, 1999.
14. Yang, H.Q. and Przekwas, A.J., "Computational Modeling of Microfluid Devices with Free Surface Handling," MSM-98, Santa Clara, CA, 1998.

Part 8

Commercialization of Microreaction Technology

The commercial use of multilayer microcapsules for biotechnological applications

R. Pommersheim; A. Noack; S. Scholz

CAVIS GmbH, Kupferbergterrasse 17-19, D-55116 Mainz, Germany,
e-mail: pommersheim@cavis-mainz.de

1. Introduction

Cost optimization has become a major demand in every field of economic activity, so biotechnology, a new and very fast growing sector makes no exception. Therefore new technologies like the immobilization of substances and living cells have been developed and constantly improved. These techniques can provide an almost complete recovery of expensive additives like enzymes and living cells from the production process increasing its effectiveness and reducing substantially the production costs.

New developments make now immobilization techniques available for the food industry too, commonly used only in other biotechnological areas: The encapsulation of living cells (e.g. yeast, bacteria) or enzymes in 0.5-2 mm microspheres, covered with a stable, semipermeable membrane made of natural polymers. Because of the filtering property of this capsule membrane, additional filtering steps are no longer required. The encapsulated cells or enzymes are not hindered in their production capacity by the presence of this multilayer shell. Sensitive processes can be conducted easier due the possibility of a very precise timing of the removal of the encapsulated additives. These microcapsules are made entirely of food approved substances and can by easily disposed as fodder.

2. Structure of the membrane capsule

In extension of classical immobilization methods based on alginate, here gelspheres acting as a matrix for the immobilization of substances or living cells are covered in consecutive steps with several polymer layers forming a membrane of about 0.8 μm thickness (fig. 1).

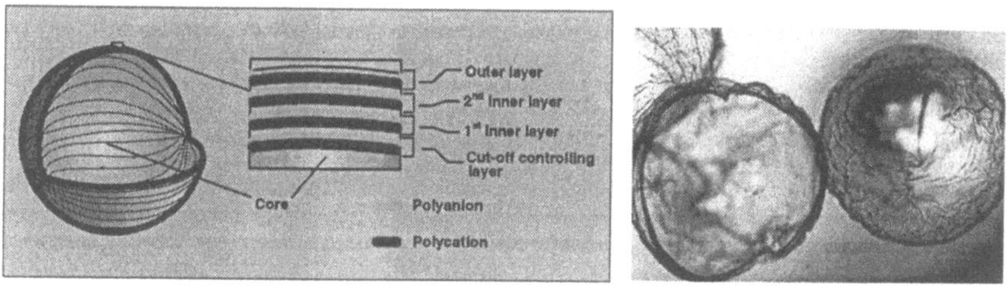

Fig. 1: Structure of the membrane capsule.
Left: Schematic representation,
Right: Membrane of 800 µm capsules after destruction by osmotic shock

Each of these layers is optimized with respect to a different property of the capsule's shell. So, one layer regulates the permeability, one the mechanical stability and an outer layer its storage properties. The materials used are all food approved natural polymers. The tuning of the membrane for a specific application is made by the choice of the polymers. The core contains the immobilized material either as an aqueous solution (e.g. enzymes) or as a suspension (e.g. living cells). If requested, this core can be gelled or liquefied.

3. Properties of the MLC

The capsules are available in a size range from 0.6 to 2 mm in correspondence to the users requirements. They can be stored either in a dried form or in liquids. The membrane resists to an inner pressure of up to 3.8 atmospheres. They should be used in liquid media, in a fluidized bed. MLC can be integrated in existing processes without any modification of installations.

4. Fields of application

Considering the large possibilities in adapting the capsules membranes to the practical demands, the simplicity and the low costs of the production process these microcapsules have large applications in various fields. Thanks to the very gentle encapsulating process substances like sensitive enzymes can be immobilized without significant loss of activity, as well as living cells or microorganisms.

MLC are mainly designed for applications where a clear liquid has to be transformed during the production process into a clear product or where additives such as enzymes or microorganisms have to be recovered or removed from the product. In the last few years in Mainz/Germany many experiments with these capsules gave very good results demonstrating the feasibility of industrial processes done with these microspheres.

Encapsulation of Enzymes

Enzymes are in general very expensive substances, therefore a recovery of them in industrial processes is frequently desired. To lower the costs of the recuperation steps these substances must be immobilized. Using classical methods for this purpose such as bonding to surfaces or adsorption on porous materials up to 70 % of the enzymes initial activity may get lost in a denaturation process of the molecule.

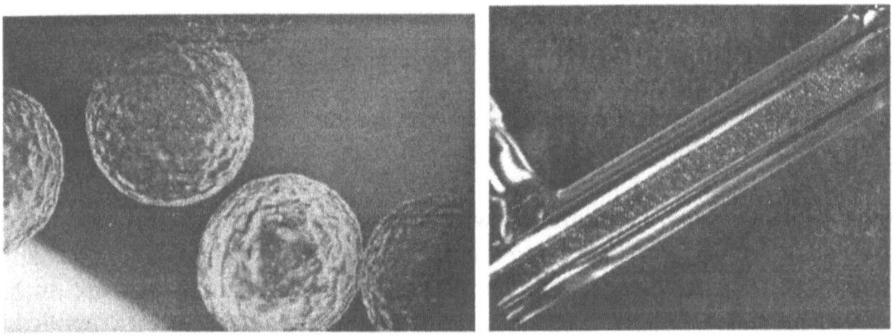

Fig. 2: Encapsulated enzymes (capsule diameter 0.8 mm)
Left: Microscopic image of enzyme-containing MLC;
Right: MLC-encapsulated enzymes in a tube-reactor

By the immobilization of the enzymes in multilayer microcapsules (Fig. 2) the loss of activity is generally below 5 %. The capsules having a liquid core, the activity of the enzymes in this alginate solution is nearly the same as in any other liquid media because they are free floating and capable to adopt their native conformation.

Encapsulation of microorganisms:

Encapsulating microorganisms like fungi or bacteria, this technique can be used to lower costs in fermentation or other production processes because no special filtering systems or reactors are needed.

To demonstrate the potential of these multilayer microcapsules at the Institut für Mikrotechnik Mainz/Germany a new continuous process for alcohol production in a miniplant scale has been developed (Fig. 3).

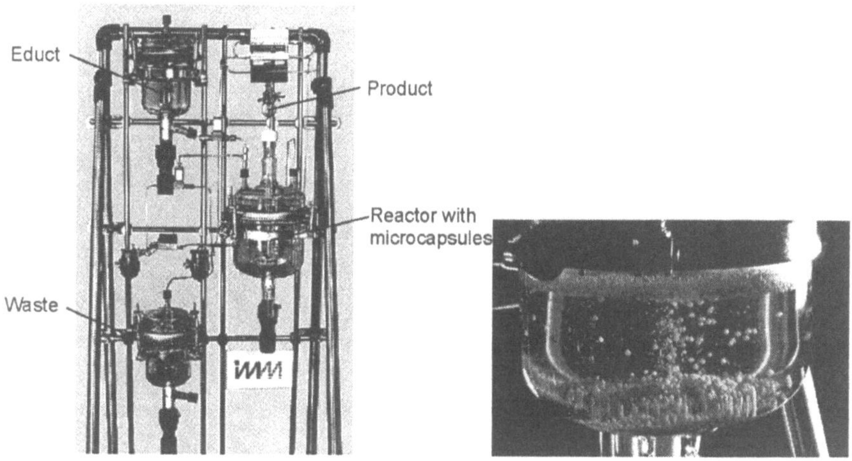

Fig. 3: Continuous alcohol production process with microcapsules (Source: IMM)
Left: Test-installation in a miniplant scale
Right: Fermentation process with encapsulated yeast

Enzymes, yeast and bacteria used in this process are recovered completely without any filtering elements. By the continuous extraction of the resulting alcohol the productivity is much higher than in classical fermentation.

Using only natural polymers for the preparation of the capsules they can be very easily disposed without producing any environmental damage.

Encapsulation of living cells

For medical use, there is a great need of encapsulated living cells: If animal (e.g. porcine) cells are protected from the patients immune system by the capsules membrane, they can be xenotransplantated into humans. So, for example they could be used in diabetes therapy (fig. 4).

A similar approach can by done in liver therapy by using encapsulated liver cells. First experiments performed at the clinics of the University Mainz with MLC-encapsulated porcine cells have given very promising results.

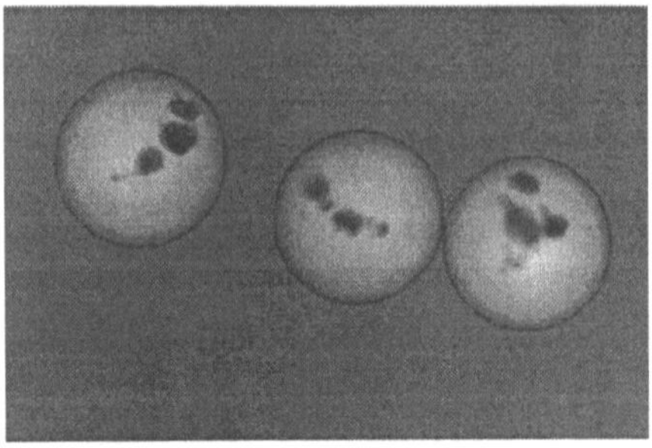

Fig. 4: Encapsulated living cells (capsule diameter 0.4 mm)
Left: Immobilized Langerhan's Islets for diabetes therapy (source: University Mainz)
Right: MLC-encapsulated living cells in the well of titer-plate

5. Market potential

In order to illustrate the economic potential which lies in this immobilization technique, the world market for enzymes can be a good example. This market is estimated to be in the range of 2 billions US $/ year, with an annual growth of approximately 10 %. Not considering the 50 % of enzymes used for the fabrication of paper, leather, textiles or cleaners, which can not be recovered from these products, the other 50 % of the enzymes can be recycled. These are enzymes mainly utilized in the food-, pharma-, or chemical industry. Assuming that only 2 % of these enzymes will by immobilized in MLC' s, the potential for these capsules can be estimated above 40 Mio. US $ / p.a.

Considering that the immobilization of enzymes is only one of various other applications of the MLC's, one can estimate the great market potential of these microcapsules.

6. Conclusions

The multilayer microcapsules presented here can be used for the immobilization of enzymes and living cells. The membrane is permeable for substances of low molecular weight. Applications of the capsules in organic and biotechnological syntheses are possible. Using the multilayer technique, capsules coated with membranes of defined permeability, mechanical stability and storage properties can be produced. First industrial processes with these capsules have been developed and are now tested with very good results.

Part 9

Microreaction Devices

A Capillary Force Filled Auto-mixing Device

R. U. Seidel*, D. Y. Sim⁺, W. Menz*, M. Esashi⁺

* Institut für Mikrosystemtechnik, Albert-Ludwigs-Universität Freiburg, Germany
⁺ Tohoku University, Sendai, Japan

Abstract

A new mixing device for chemical analysis was developed and fabricated using silicon bulk micromachining. All liquids are supplied by capillary forces and mixing is done by diffusion. Hence, particularly for medical and environmental analysis, easy systems without the need of damageable micropumps can be realised. The chips have two channels of 200 µm and 300 µm depth, respectively, and of 10 mm length. They are separated by a 2 µm thick one-sided clamped plate made of SiO_2. The mixing volumes of both channels were 0.22 µl and 0.86 µl, respectively. The separate filling by capillary forces and the mixing by diffusion was demonstrated successfully.

Keywords: micro, mixer, mixing, capillary force, diffusion

Introduction

In most fluidic analysis devices mixing plays a major role. Effective mixing is crucial for the reproducibility of every analysis, because generally the result depends strongly on the concentration of the reagent in the sample. In addition, economical usage of expensive reagents is requested by industry and researchers. Thus the minimum possible amount of reactants should be employed [1]. Furthermore, in medical analysis, the most valuable part of an analysis is the sample drawn from the patient, because it is usually only available in small amounts. So, small mixing volumes are a major advantage of micro mixers.

Apart from the mixing volume, all modern approaches to mixing consider mixing speed as a second important parameter [2]. Especially the chemical industry is interested in fast mixing of micro-volumes e.g. in analytical or micro reactor applications. However, mixing of liquids in micro-dimensions is a tricky task, because turbulent flow can hardly ever be achieved. The existence of laminar flow means, that except for diffusion, no mixing can occur between two fluid layers [3]. Unlike with all our experiences of the macro-world, no vortices will speed up mixing in this case. One approach to reduce mixing times is to split the common flow into a number of small interleaved channels of some micrometers in diameter. Hence, diffusion distances are reduced drastically and mixing times of a few seconds down to less than one second become feasible.

To achieve both, small mixing volumes and fast mixing times, conventional analysis systems have been reduced in size by replacing macro with micro components [5]. However, pumps, valves and pipes contribute to the system's internal volume, increasing the overall liquid consumption and decreasing the reliability of the analysis. Furthermore, pumps and valves can be clogged by particles unless all liquids are filtered intensely. In turn, filters increase the pressure drop in the system, necessitating stronger pumps and system components. Therefore, pumps and valves should preferably not handle those fluids but only clean, sterile liquids, limiting the scope of applications.

Motivation for a new approach to mixing

For every-day use, an analysis system should be able to deal with 'dirty' liquids containing particles and bacteria. Therefore, not the maximum mixing speed was decisive for the development of the new device, but ruggedness, prevention of clogging and contamination, as well as simplicity of the analysis systems. We propose self-filling capillaries to circumvent the problems mentioned above.

In the following, we present a novel approach to mixing based on the use of capillary forces for pumping. Unfortunately, *one capillary* can only be filled with *one liquid* by capillary forces. Therefore, the problem definition was how to use capillary forces to fill *one capillary* with *two liquids* and at the same time to keep the diffusion distances small.

The mixing chip with no need of external forces and its use

To solve the problem conventionally, two long channels are placed parallel, next to each other, either separated by a membrane or a thin plate with small holes. We propose two capillaries separated by a one-sided clamped plate (in the following abbreviated to plate) leaving a small gap for diffusion. The gap is closed by capillary forces of the first reactant being filled into one of the capillaries and re-opened by the second reactant.

Figure 1 shows schematically the principle structure of such a capillary force mixer. Two capillaries anisotropically etched into a silicon wafer are separated by a thin plate of silicon dioxide. This structure is closed on both sides by anodically bonded glass wafers. The two reactants are supplied to the capillaries by holes on either side. Mixing occurs by diffusion from one capillary to the other through the small gap left between the plate and the top glass wafer. In *figure 2 a*, a cross section of the virgin structure can be seen before filling. In *figure 2 b*, fluid is drawn into channel 1 by capillary forces from the top supply hole. Surface tension

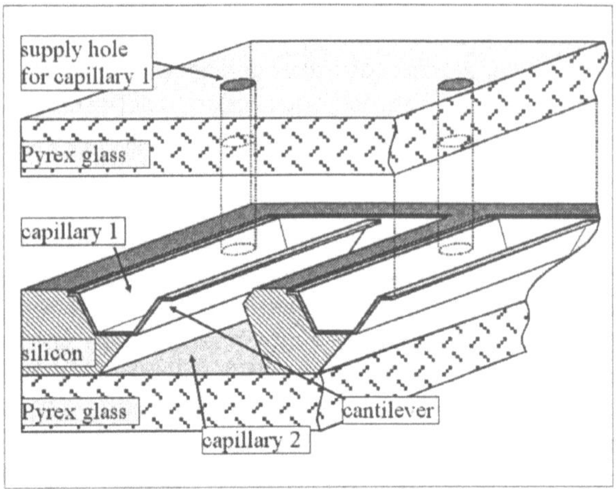

Figure 1: Exploded view of the capillary force mixer's Pyrex-Si-Pyrex - structure

Figure 1: Exploded view of the capillary force mixer's Pyrex-Si-Pyrex - structure

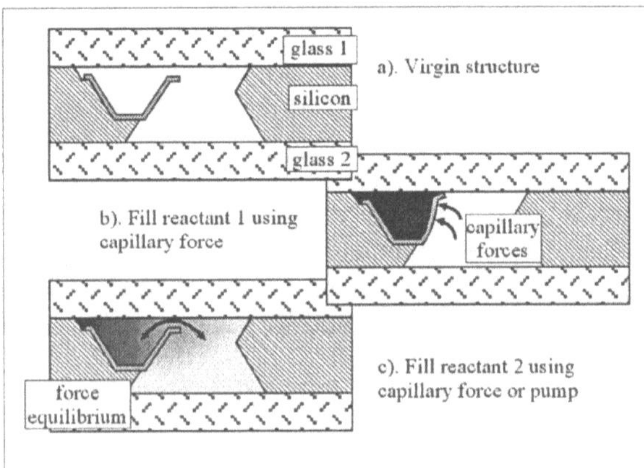

Figure 2: Filling and mixing principle of the capillary force mixer

attracts the separating plate to the cover wafer and thus closes the gap to channel 2. Then, the latter is filled from the bottom, also with the help of capillary forces (*figure 2 c*). Due to equilibrium, the separating plate is released and the gap re-opens allowing mixing to take place. The mixed reactants can be analysed in several ways, e.g. optically or electrically using electrodes integrated into the covering glass substrates.

By changing the channel geometries, the mixing ratio of both reactants can be precisely pre-determined in a wide range. Hence, user interference is excluded. Exact dosage and very small mixing volumes are guaranteed within the tolerances of silicon micromachining. Even environmental, non-filtered liquids can be used, because the capillaries are not blocked easily.

As far as filling is concerned, both reactants can either be added manually and the chip can then be inserted into a spectrometer. Because the fluid inlets are placed at opposite sides, the danger of confusing the two reactants is minimised. Alternatively, a modular analysis system could handle the reactants. In this case, the 'dirty' liquid is added manually by placing one drop onto the supply hole, while the clean reactant is pumped from a reservoir. Such a system using the capillary force mixer could consist of a reagent container, a pump, the replaceable mixing chip and an optical analysis device. In such a modular system, neither pumps nor any other parts can be clogged or contaminated and only the mixing chip has to be replaced after use.

One drawback of the capillary force mixer is, that mixing is a little slow compared to other micro mixers due to the small diffusion area. But this is not necessarily a drawback, as especially environmental and medical analyses always need some further preparation, e.g. labelling of the samples.

Fabrication of the capillary force mixer

The mixer's process chart is shown in *figure 3*. First, a silicon wafer is thermally oxidised on both sides (200 nm). Next, openings for the capillaries are defined by standard photolithography. After removing the oxide, the gap-defining groove is

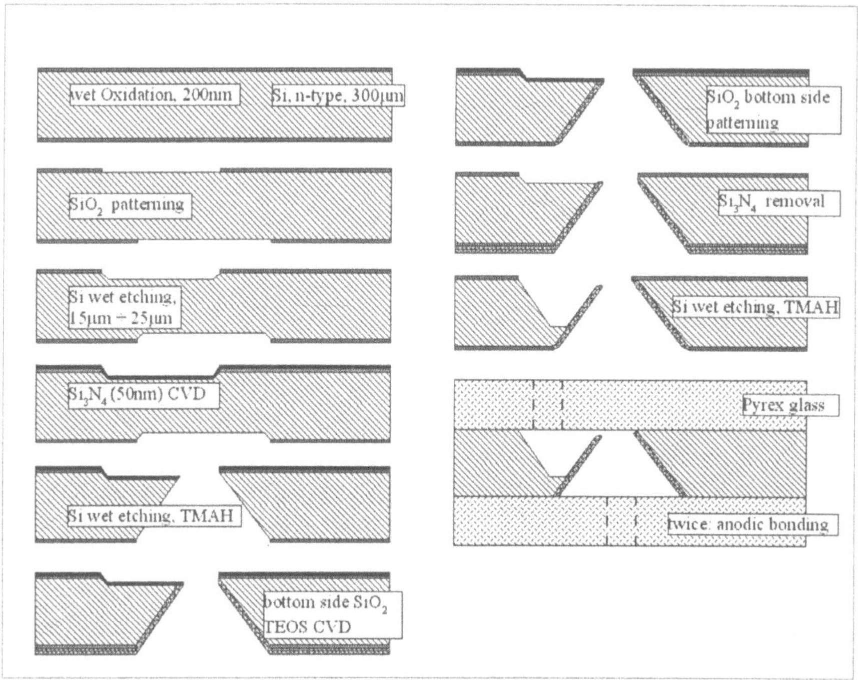

Figure 3: Simplified process chart of the capillary force mixing chip

anisotropically etched using TMAH [6] to a depth of approx. 15 - 25 µm. Then a thin CVD silicon nitride layer (50 nm) is deposited on the top of the wafer, whereupon the deep grooves are anisotropically etched in TMAH (300 µm) from the bottom side.

A thick TEOS-CVD silicon dioxide layer (e.g. 2 µm) is deposited on the bottom and patterned in such a way, that only the side walls of the etched grooves remain coated with SiO_2. After removing the top nitride, the second channel is again etched anisotropically from the top side using TMAH, leaving the thin silicon dioxide plate separating both channels.

After stripping the thermal oxide, Pyrex glass substrates containing the liquid supply holes are bonded anodically to the silicon chip's top and bottom sides.

Results

The mixing chips have a total size of 20 mm x 20 mm with an array of 7 pairs of mixing channels of 10 mm length each *(figure 4)*. Channel 1 measures 300 µm width and 200 µm depth, while channel 2 measures 500 µm width and 300 µm depth. The total volumes of channel 1 and channel 2 are calculated to be 0.22 µl and 0.86 µl, respectively. To render the mixing chip transparent (needed for non-reflecting optical analysis), the depth of channel 2 is determined by the wafer thickness.

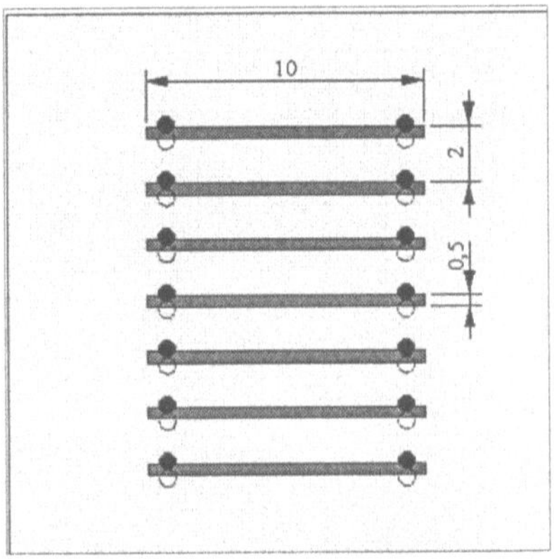

Figure 4: Layout of one mixer chip containing 7 pairs of channels; dimensions in millimetres

The plate consists of a 2 µm thick deposited silicon dioxide layer with a length and height of 10 mm and 245 µm, respectively. It is transparent for optical analysis. The gap for diffusion is 20 µm.

For preliminary testing, all but one of the array's liquid supply holes were covered with adhesive tape to prevent accidental filling of the other channels (*figure 5*). In a future study, the mixers will be separated by dicing or sawing.

Red and blue comestible colours were chosen as *liquid 1* and *liquid 2*, respectively. After filling of channel 1 (*figure 6*), the chip was turned round for filling of channel 2. This does not need to be done particularly carefully because the liquid is kept save in channel 1 owing to the closed gap. Then, channel 2 was filled (*figure 8*). Unfortunately, the microscope's light was not strong enough to resolve the process of diffusion. Only a colour change of the initial blue of *liquid 2* could be seen immediately after filling channel 2.

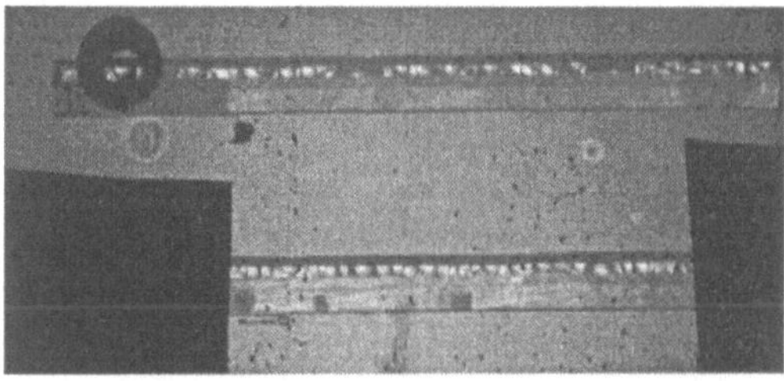

Figure 5: Empty channels, holes covered with adhesive tape; the pair of channels at the bottom will be filled. The ground of channel 1 can be recognised by hillocks stemming from old TMAH solution

Figure 6: Channel 1 filled with blue colour (change of colour especially of the plate)

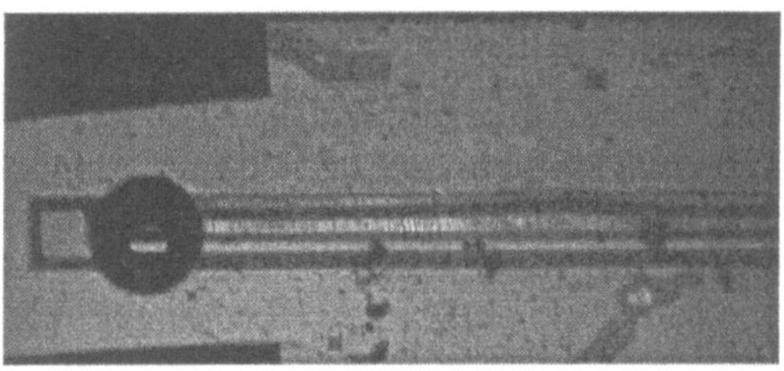

Figure 7: Channel 1 filled with blue colour, shown from the bottom side

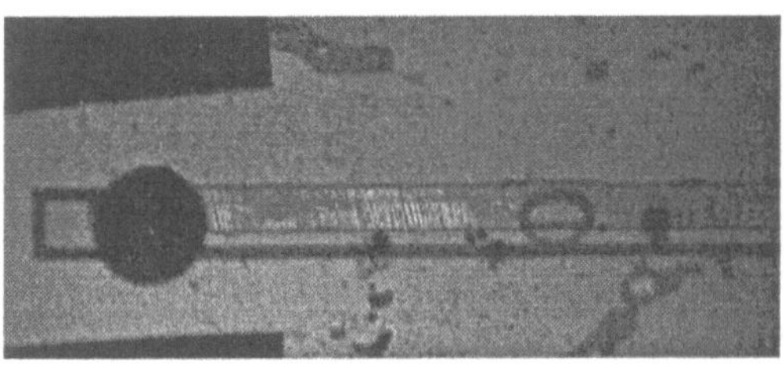

Figure 8: Channel 2 filled, mixing already finished; an air bubble was caught in the middle of the mixer

Conclusion and outlook

A new capillary force mixing chip has been produced and tested. Results presented show that pumping and mixing of two liquids can be achieved by capillary force and diffusion only. One major advantage of this device is its robustness to particles, another the simplicity of use making way to simple modular systems: no pumps or valves are needed.

However, mixing might be much slower than in highly sophisticated micro mixers. In aqueous fluids at room temperature, the diffusion speed is some 5 μm per second [7]. Therefore, the mixing time of the capillary force mixer is supposed to be in the range of seconds to minutes. High-speed photographic work is currently under way to evaluate the mixing properties of the capillary force micro mixer.

The mixing chips can be designed as disposable parts or cleaned using alcohol or hot vapour cleaning for multiple usage. In a further development of the capillary force mixer, sensors can be integrated into the design for electrochemical analysis. Electrodes on either bottom or top side could also detect successful diffusion.

In case of medical analysis, a needle could be integrated into the capillary force mixer. Hence, a very small amount of blood could be taken directly from the patient into channel 1 and economically mixed by subsequent filling of channel 2. Sucking blood like a mosquito, such a system could get along with the minimum amount of sampled blood. Even multiple blood tests could be performed simultaneously by a capillary array. Capillary forces combined with blood pressure will be the driving forces to fill the channel.

Acknowledgements

This publication would not have been possible without the extensive help of Dr. Michael A. E. Wand.

The authors would also like to thank Dr. Yoichi Haga for friendly advice about medical analysis.

This work was supported by the Japanese Ministry of Education, Science and Culture under grant-in-aid 100 44 112.

References:

[1] J. Ruzicka, E.H. Hansen, 'Flow injection analysis', Chemical Analysis Vol 62, John Wiley & Sons, New York, 1987

[2] A.v.d.Berg, 'Technologies and Microstructures for chemical microsystems', 1st Intenational Conference on Microreaction Technology, 23. - 25. Feb. 1997

[3] J. Zierep, 'Grundzüge der Strömungslehre', G. Braun, Karlsruhe, 1987
[4] P. Gravesen, J.Branebjerk, J.P.Krog, C.R.Nielsen, 'Fast mixing by lamination', Proceedings of IEEE MEMS'96 1996, San Diego, USA
[5] M.Richter, A.Prak, J.Naundorf, M.Eberl, H.Leeuwis, P.Woias, A.Steckenborn, 'A chemical microanalysis system as a microfluid system demonstrator', Proc. Transducers '97
[6] O. Tabata, R. Asahi, H. Funabahsi, K. Shimaoka, S. Sugiyama, 'Anisotropic etching of silicon in TMAH solutions', Sensors and Actuators A, 34 (1992) 51-57
[7] J.T.Davies, E.K.Rideal, 'Interfacial Phenomena', Academic Press, New York 1961

Contacting the author:

Tel: +49-761-203-8289, Fax: 8282
E-mail: rseidel@informatik.uni-freiburg.de

Fast Heating and Cooling for High Temperature Chemical Microreactors

Ch. Alépée[1], R. Maurer[2], L. Paratte[1], L. Vulpescu[1], Ph. Renaud[1] and A. Renken[2]

[1] Institute of Microsystems, Swiss Federal Institute of Technology, 1015 Lausanne, Switzerland

[2] Laboratory of Chemical Reaction Engineering, Swiss Federal Institute of Technology, 1015 Lausanne, Switzerland

Abstract. Many chemical reactions, like the dehydrogenation of gaseous methanol to formaldehyde, require fast heating and cooling of the gases in order to control precisely the residence time at the reaction temperature (on the order of hundreds of ms) to minimise consecutive reactions. For these applications, microsystems are of great interest thanks to their characteristic high surface-to-volume ratio and low thermal inertia [1-3]. Heater and cooler units were developed with compatible micromachining technologies in view of a future integration in a complete microreactor system.

The heater is composed of a silicon duct, capped with a silicon nitride membrane heated by a platinum filament [4]. Within the mean residence time of approximately 4 ms nitrogen was heated to a maximum temperature of 600°C. Numerical simulations are in agreement with the performed temperature measurements. However, the large temperature gradient of this prototype due to single-sided heating is not well suited for chemical reactions. A solution with two heated membranes face-to-face is being investigated.

The cooler is a co- or counter-current flow heat exchanger machined into silicon by deep plasma etching. The measured overall heat transfer coefficients are lower than expected. This can be explained by the longitudinal heat conduction in the fins, which is neglected in the basic theory. New exchangers with thinner fins are therefore being developed but the cooling needed for our particular application (from 700°C to 200°C in a few ms) is already possible with the current structures.

Keywords. Microreactor, micro-heater, platinum thermoresistors, micro heat exchanger, longitudinal conduction

1 Introduction

Currently, the main industrial production path to formaldehyde is the oxidative dehydrogenation of methanol [5]. During this reaction water is inevitably produced in stoichiometric amounts, which is a major drawback for some applications like the production of POM (Polyoxymethylen). For these cases, the direct dehydrogenation of methanol to formaldehyde (Eq. 1) becomes an interesting alternative. This reaction takes place at temperatures above 600°C [6]. Under those conditions, the formed formaldehyde is thermodynamically unstable and decomposes in consecutive reactions to hydrogen and carbon monoxide (Eq. 2).

$$CH_3OH \rightarrow CH_2O + H_2 \quad (1)$$

$$CH_2O \rightarrow CO + H_2 \quad (2)$$

In order to minimise these consecutive reactions it is necessary to quench the reaction products by cooling rapidly and to control precisely the residence time at the reaction temperature (on the order of hundreds of ms). Therefore the heating and cooling times must be reduced to some ms. Microsystems, which are characterised by high surface to volume ratios and low thermal inertia, are therefore well-suited for this application. Short paths from one section of the reactor system (heater, reactor, cooler) to the other can be obtained by integrating these functions on a single chip (Fig. 1.1).

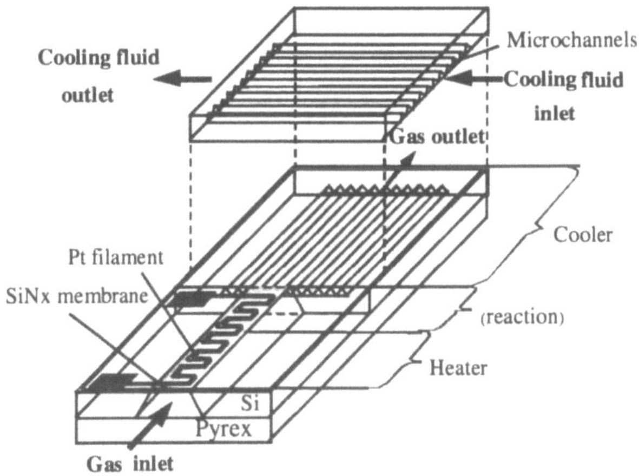

Fig. 1.1. Schematic view of a high temperature microreactor.

A heating and a cooling unit were fabricated with compatible silicon technologies in view of future integration. In the first part of this paper the heater is described while the second part deals with the cooling unit.

2 The heating unit

2.1 Description of the heating unit

Similar to [4, 7], the heater consists of a micro-structured silicon duct capped with a silicon nitride membrane. The other side is sealed with a Pyrex plate attached with standard epoxy (Fig. 2.1, Fig. 2.2). Pt-Ta filaments to heat the membrane and thermoresistors to sense the temperature of membrane and substrate are defined by a lift-off technique. The fluidic channels are formed by KOH-etching on the back side of the silicon. The membrane above the access hole (Fig. 2.2.b) is pierced and a steel tube is glued on the front side (Fig. 2.2 a).

Preliminary calculations showed that heating a nitrogen flow of 30 ml/min (STP) to a mean temperature of 650°C needs an electrical power of 1.7 W with an assumed substrate temperature of 130°C. The flow is laminar in the 1.3 x 0.38 mm^2 trapezoidal cross-section and 7.4 mm long heating duct (Re=64), leading to an average residence time of approximately 4 ms.

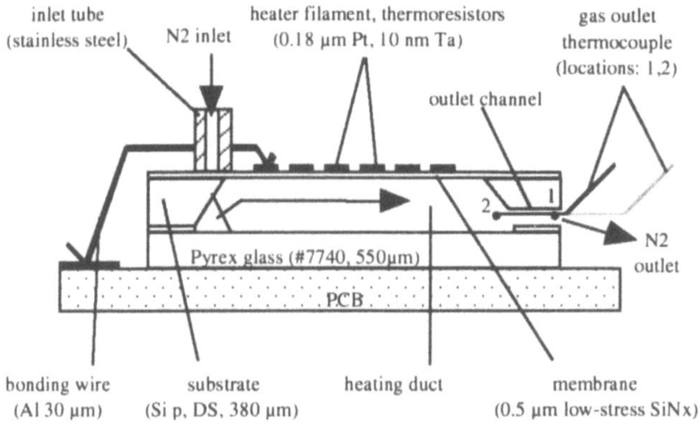

Fig. 2.1. Heater's longitudinal cross-section.

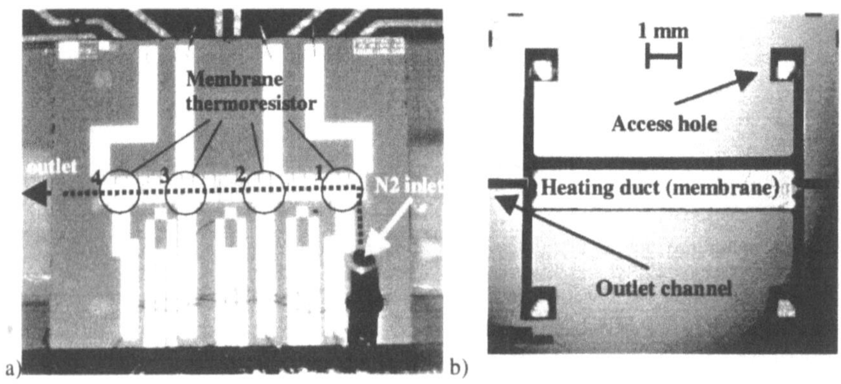

Fig. 2.2. a) Top view of a heater chip (10 x 10 mm^2); the membrane thermoresistors are labelled 1-4. b) Back side, showing the fluidic paths and the heating membrane (still covered with 10 µm of Si).

2.2 Description of the test set-up

The test set-up is similar to the one of the heat exchanger (Fig. 3.3). It allows controlling of the nitrogen flow rate and the electrical supply to the filaments. The pressure drop across the microstructure as well as the temperature with thermocouples and on-chip thermoresistors (±7%) can be measured. To avoid microstructure fouling, 0.2µm filters are placed upstream.

A thermocouple was used to measure the temperature of the gas. To limit the disturbance caused by the insertion inside the triangular outlet channel (Fig. 2.1, width 300 μm, height 214 μm) the smallest available thermocouple (type K, Thermocoax AG, CH) with a head of 80μm in diameter has been selected. The heat loss of the gas to the thermocouple and the increase of fluidic resistance are estimated at 10% and 15% respectively.

2.3 Results

The thermocouple was placed either 120 μm inside the outlet channel or at the end of the heating duct (locations 1 and 2 in Fig. 2.1). The substrate temperature (TS) and the membrane temperatures at four equidistant locations (Fig. 2.2. a, T1-T4) were recorded using calibrated thermoresistors. It was found that the substrate temperature was uniform within 2°C over the whole chip.

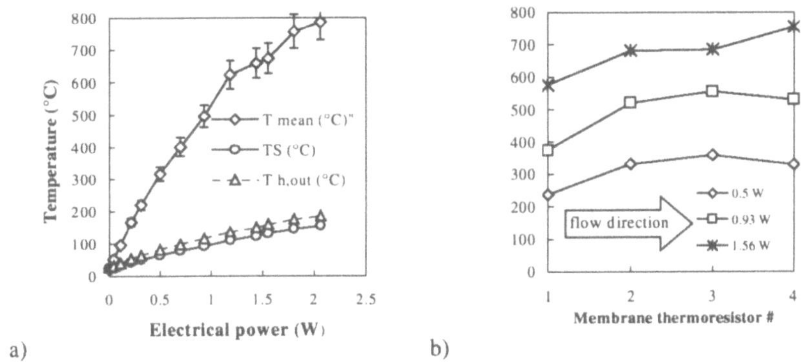

Fig. 2.3. a) Temperatures of the membrane (Tmean), the substrate (TS) and the gas at channel outlet (Th,out, location 1, Fig. 2.1) as function of the applied electrical power, at 30 ml/min (STP) of nitrogen. b) Membrane temperature profile (T1-T4).

The temperature dependence of the membrane and the substrate for the applied power at a flow rate of 30 ml/min (STP) and a pressure drop of 15 mbar are given in Fig. 2.3. Around 600°C (mean membrane temperature) the filament lights red and its resistance changes abruptly as shown in Fig. 2.3 a) indicating a membrane degradation, which is consistent with the observations of [4,7]. Below 600°C, the temperature profile along the membrane length (Fig. 2.3 b) behaves as expected; T1 and T4 are lower than the central T2 and T3, because they are "surrounded" by three silicon walls and the maximum is shifted downstream. Again, this curve becomes irregular above temperatures of 600°C.

With the thermocouple placed at location 1 in Fig. 2.1, the measured gas temperature is very close to the substrate temperature (Fig. 2.3 a) due to rapid cooling at the channel walls. Hence, a measurement was performed with the thermocouple fully inserted (location 2 in Fig. 2.1) giving a higher temperature of 200°C with a mean membrane temperature at approximately 600°C.

2.4 Finite element modelling

In order to determine the temperature profile in the heating duct finite element modelling (FEM) was done using the ANSYS - FLOTRAN 5.4 package. The flow was considered to be laminar at 30ml/min (STP). The temperatures of the membrane and substrate (lateral surfaces) were held constant and the thermal losses through the Pyrex were neglected.

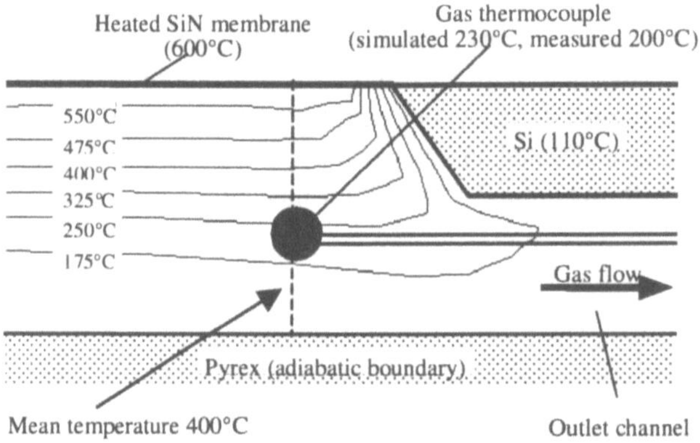

Fig. 2.4. Simulation results of the temperature distribution (longitudinal cross section).

A simulation with a mean membrane temperature of 495°C and a substrate temperature of 92°C resulted in a gas temperature of 150°C at the channel outlet location. The measured value was only 115°C, which is due to the heat losses through the Pyrex plate and the thermocouple wires.

Another simulation was performed with a mean membrane temperature of 600°C and a substrate temperature of 110°C. The results are depicted in Fig. 2.4 showing a large temperature gradient in the heating duct. The calculated gas temperature of approximately 230°C at the duct outlet corresponds well to the measured value of 200°C, taking in account the relatively large temperature gradient in combination with the incertitude on the exact location of the thermocouple head.

2.5 Discussion

With this preliminary heater design it was possible to heat a 30 ml/min (STP) nitrogen flow up to an approximate mean temperature of 400°C at the end of the duct. The FEM simulations were in agreement with the performed temperature measurements. Simulations show the flow patterns within the heater and the temperature distribution, which makes FEM very useful as a design tool. The single-sided heating displays a large axial temperature gradient. Positioning two heating membranes face-to face will reduce this. Based on the actual concept a double-sided heater is currently being developed.

3 The cooling unit

3.1 Description of the heat exchanger

The cooler is a gas/gas or gas/liquid heat exchanger, which can be operated either in co- or counter-current mode. It consists of three layers: a silicon chip, in which a row of channels is machined by deep plasma etching (Fig. 3.1), and two Pyrex covers, which are anodically bonded on both sides of the silicon chip (Fig. 3.1.b). The stainless steel and glass tubes used as fluid connectors are glued into the inlets and outlets, which were drilled into the covers with diamond grinding pins.

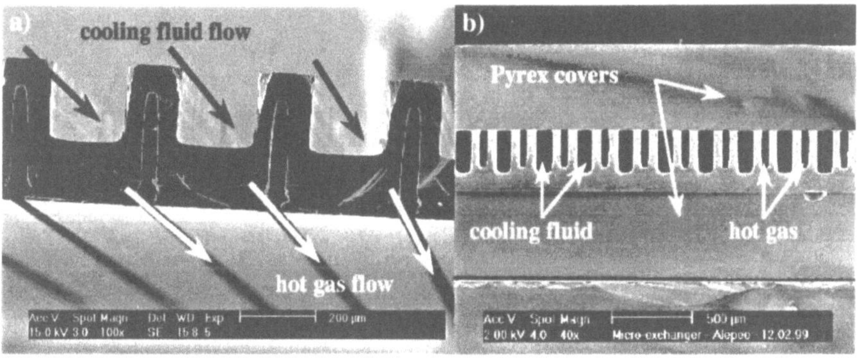

Fig. 3.1. Cross sections of the heat-exchangers: a) "double-sided" structure, silicon chip alone; b) "one-sided" structure, silicon chip with the two Pyrex covers

In the following, the fluid to be cooled and the cooling fluid will be called the hot and cold gas respectively. The channels for the two fluids are intercalated and are about 300µm deep and 2cm long. They are separated by 50µm thick walls. The cold gas channels are 100 or 200µm wide. The hot gas channels (50 or 100µm wide) are either machined on the same or on the opposite side of the chip (Fig. 3.1.b and Fig. 3.1.a respectively). The "one-sided" structure is easier to fabricate, but there is risk of leakage between the channels of the two different fluids, whereas with the "double-sided" structure, the two fluids are perfectly isolated.

The fluid distribution channels for the cold fluid are etched by HF into one of the Pyrex covers, while those for the hot gas are machined by deep plasma etching of the silicon layer. The design of the fluid distribution (Fig. 3.2) guarantees that the resistances of all the paths are equal and the flows in all the channels (for one fluid) are identical.

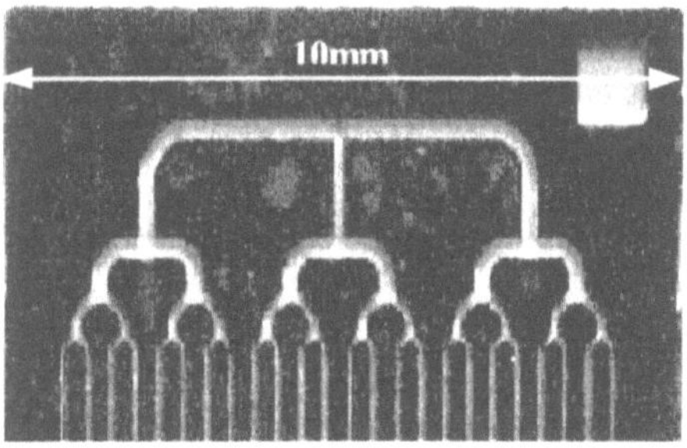

Fig. 3.2. Fluid distribution channels in the Si chip

3.2 Description of the test set-up

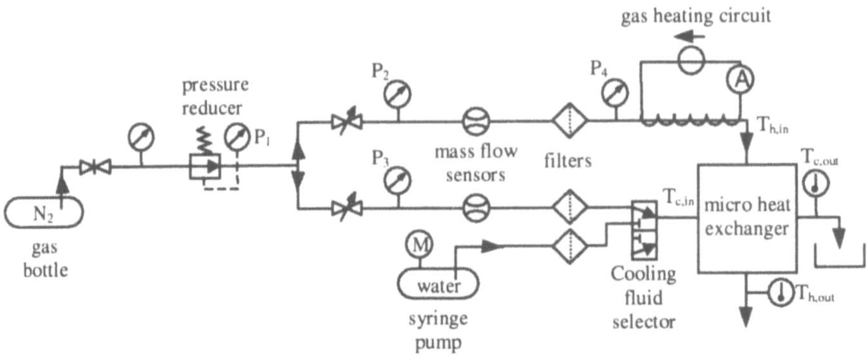

Fig. 3.3. Schematic view of the test set-up for the micro heat exchangers

A feed gas set-up (see Fig. 3.3) allows to control the pressure and flow rate of the gas. Water as coolant is supplied with a syringe pump and a fluidic switch allows to choose between liquid and gas as the cooling fluid. In order to avoid fouling of the microchannels, 0.2μm filters were inserted in each feed line.

The hot gas is brought to the desired inlet temperature (up to 500°C) by a glass tube heated with a copper filament coiled around it. The outlet temperatures of both fluids are measured by miniature thermocouples (see section 2.2).

Before the heat exchangers were connected, the pressure drop in the gas heating circuit was measured (P_4) for various $T_{h,in}$ and gas flows in order to calculate the pressure drop of the exchanger alone during the test.

3.3 Results

The tested exchangers are one-sided types with 32 50μm wide hot gas channels and 33 100μm wide cold gas channels. The tests were carried out in the gas to gas counter-flow configuration with nitrogen. This was chosen as in the final application nitrogen will be used as carrier gas.

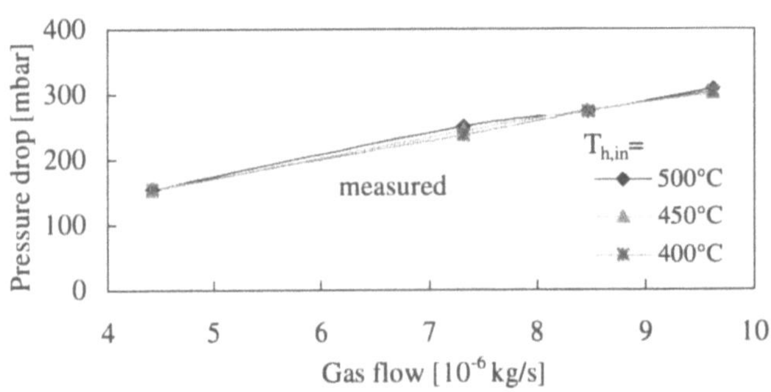

Fig. 3.4. Hot gas pressure drop in the micro-heat exchanger

For the final application aimed at, the hot gas pressure drop in the heat exchanger has to be as low as possible. Indeed, it will determine the pressure in the reactor and thus may influence the chemical reaction. The measured pressure drops are plotted in Fig 3.4 for several flows and inlet temperatures. They are perfectly acceptable for the aimed application. The linear dependence of the pressure drop on the flow rate implies a well-developed laminar velocity profile. In fact, all the measured Reynolds numbers are lower than 70.

Heat transfer

The exchangers were tested with gas flows up to 500ml/min (STP) ($9.6 \cdot 10^{-6}$kg/s) and inlet temperatures of the hot gas varied between 400°C and 500°C. The inlet temperature of the cold gas was held constant at 20°C.

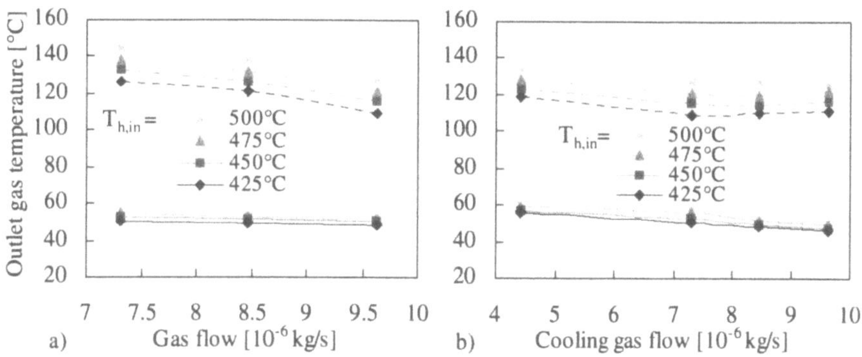

Fig. 3.5. Outlet temperatures of the gas to be cooled (hot gas, plain lines) and the cooling gas (cool gas, broken lines); a) equal gas flows; b) hot gas flow fixed at 380ml/min (STP) ($7.6 \cdot 10^{-6}$ kg/s)

The measured outlet temperatures of both fluids as a function of the gas flow and the inlet temperature are plotted in Fig. 3.5. As expected for the counter-flow mode, low outlet temperatures for the hot gas were achieved. Surprisingly, the hot gas outlet temperatures tended to decrease with increasing gas flows. This is due to the effect of the longitudinal conduction (see section 3.4), which becomes more important at lower flow rates.

The heat exchanger's overall heat transfer coefficients are calculated from the measured inlet and outlet temperatures according to [8]:

$$U = q / (A \, \Delta T_m) \qquad (3)$$

where $q = (q_h - q_c) / 2$ $\qquad \Delta T_m = (\Delta T_{in} - \Delta T_{out}) / \ln(\Delta T_{in} / \Delta T_{out})$

$q_h = \dot{m}_h \, c_{p,h} (T_{h,in} - T_{h,out})$ $\qquad \Delta T_{in} = T_{h,in} - T_{c,out}$

$q_c = \dot{m}_c \, c_{p,c} (T_{c,out} - T_{c,in})$ $\qquad \Delta T_{out} = T_{h,out} - T_{c,in}$

The measured coefficients are plotted in Fig. 3.6 where it appears that they are much smaller than expected with the basic theory.

3.4 Heat transfer theory

The overall heat transfer coefficient of a counter-flow heat exchanger can be calculated with [8]:

$$1/U = 1/\alpha_h + 1/\alpha_w + 1/\alpha_c \qquad (4)$$

where $\alpha_h = Nu_h\,\lambda_h/d_{h,h}$, $\alpha_w = \lambda_w/d_w$ and $\alpha_c = Nu_c\,\lambda_c/d_{h,c}$.

The Nusselt number depends on the boundary conditions of the flow and on the channel's cross-section shape. For a counter-current heat exchanger, the assumption of a constant axial wall heat flux and constant peripheral wall temperature is adequate [9]. For this boundary condition and rectangular sectional channels Shah et al. [9] give the Nusselt number as a function of the channel's aspect ratio (p. 206). In our case we found, $Nu_h = 6{,}4$ and $Nu_c = 5{,}2$.

The overall heat transfer coefficients calculated with this basic theory are plotted in Fig. 3.6.a) where the inadequacy of this simplified theory clearly appears. The basic theory is based on the assumption that the longitudinal heat conduction is negligible. This is not valid any more in the studied case because of the relative thickness of the fins and their high thermal conductivity.

The influence of longitudinal heat conduction

When the gas velocities in a heat exchanger become low, the heat conduction in the flow direction becomes important. In [10], Bier et al. proposed several methods for calculating a correction factor for U in order to take into account the longitudinal conduction. They showed that the longitudinal conduction in the gases is negligible but the longitudinal conduction in the fins can be important.

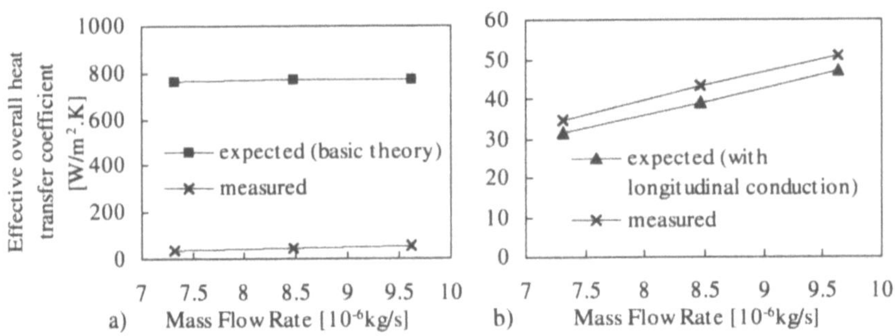

Fig. 3.6. Comparison of the measured and expected overall heat transfer coefficients (equal hot and cold gas flows)

It was shown that a two-phase homogeneous model is sufficient to estimate the effective overall heat transfer coefficient [10]. In this model the boundaries of the two gas phases are extended to half of the wall's thickness on each side of the channel. By using the modified heat conductivity λ_{2Ph} instead of λ_f the model takes into account the thermal conductivity of the walls and their thickness compared to the width of the channels. The two characteristic numbers of longitudinal conduction N and Λ are defined as follows:

$N_i = U A / (\dot{m}_i \, c_{p,i})$ (number of transfer units) $i = h, c$

$\Lambda_i = (\lambda_{2Ph,i} / L) \, S_i / (\dot{m}_i \, c_{p,i})$ $i = h, c$

where $\lambda_{2Ph,i} = \lambda_i + \lambda_w \, (1 - \Psi) / \Psi$ and $\Psi = S / S_{tot}$.

In [10] Bier et al. proposed an approximation of the correction factor which is valid for $\Lambda << 1$ and also gave the exact solution for the correction factor for the case where $N_h = N_c$ and $\Lambda_h = \Lambda_c$. Unfortunately, neither can be used in our case. Therefore the solution for the general case has to be found. With the help of [8] and [11] (chaps. 10 and 14) the following expressions are obtained:

$$U_{eff} = \frac{C_h}{A} \frac{1}{C_r - 1} \ln \frac{\varepsilon - 1}{\varepsilon \cdot C_r - 1} \quad \text{with} \quad C_r = C_h / C_c \quad \text{and} \quad C_i = \dot{m}_i \, c_{p,i} \quad (i = h, c).$$

$$\varepsilon = \frac{\sum_{i=1}^{3} \left(g_{i+1} \, g_{i+2} \, w_i + g_i \, w_{i+1} \, w_{i+2} \right) b_i \left(b_{i+1} - b_{i+2} \right)}{\sum_{i=1}^{3} \left(g_{i+1} \, g_{i+2} \, w_i + g_i \, w_{i+1} \, w_{i+2} \frac{N_h}{N_c} \right) b_i \left(b_{i+1} - b_{i+2} \right)}$$

$$g_i = \frac{1 - \Lambda_h b_i}{1 + \Lambda_c b_i} \qquad b_{3+i} = b_i \qquad w_i = e^{b_i}$$

where $b_{1,2,3}$ are the three solutions of the following cubic equation :

$$b^3 + b^2 \cdot \left(\frac{1}{\Lambda_c} - \frac{1}{\Lambda_h} \right) - b \cdot \left(\frac{1}{\Lambda_h \Lambda_c} + \frac{N_h}{\Lambda_h} + \frac{N_c}{\Lambda_c} \right) + \frac{N_c - N_h}{\Lambda_h \Lambda_c} = 0$$

All these equations were solved analytically in MathCad and the resulting overall heat transfer coefficients are plotted in Fig. 3.6.b). As can be seen the experimental results are in very good agreement with the theory, the maximum error being 5%.

3.5 Discussion

Cooling of a 500ml/min nitrogen flow from 500°C to 54°C in less than 1 ms was accomplished, which shows that the cooling needed for our final application (from 700°C to 200°C in a few ms) is achievable with the tested heat exchangers. Based on the highly accurate model used above, new geometry and operation conditions can be defined to increase the effective overall heat transfer.

The other exchangers that were fabricated have respectively 100µm and 200µm wide hot and cold gas channels. According to the more accurate model the influence of longitudinal conduction decreases, thanks to the broader channels, so that the effective overall heat transfer of these exchangers will be nearly twice as high as for the tested structures.

Both higher cooling gas flows as well as thinner fins will increase the heat transfer further. Therefore, new heat exchangers with thinner fins are currently being developed. These exchangers combined with a doubled cooling gas flow are expected to show an increase of the overall heat transfer by a factor 3. Increasing the hot gas flow would also enhance the heat transfer, but then the pressure drop in the hot gas would rise as well and would become unacceptable for the final application.

With water as a cooling fluid the heat transfer is expected to increase considerably too. Indeed the higher heat conductivity and heat capacity of water decrease drastically the relative influence of heat conduction along the walls.

4 Conclusions

According to the results of these first tests, both the heating and the cooling unit are currently being improved and the next step will be their integration into a complete microreactor system. It is also important to note that the project aims at an industrial application where the microsystems will have to be "scaled-out"[12], which will be easily feasible because the standard silicon micromachining technologies used, allow for batch processing.

5 Acknowledgements

The authors would like to thank Harald van Lintel, Peter Cousseau and Olivier Wolfrath for their precious help and advice on the theoretical, numerical and experimental part of this project and during the elaboration of this paper. The financial support from EPFL is gratefully acknowledged.

6 Nomenclature

A	[m^2]	total heat exchange area
c_p	[J/kgK]	specific heat at constant P
d_h	[m]	hydraulic channel diam.
d_w	[m]	wall thickness
L	[m]	channel length
$\dot{m}$	[kg/s]	mass flow rate
Nu		Nusselt number
P	[bar]	pressure
q	[W/m^2]	heat flux
Re		Reynolds number
S	[m^2]	Channel cross-sect surface
S_{tot}	[m^2]	S + half wall cross-section on each channel side
STP		@ Std T,P (0°C, 1 atm)
T	[°C]	temperature
TS	[°C]	substrate temperature
U	[W/m^{2}K]	overall heat transfer coeff.
u	[m/s]	fluid velocity

Greek letters:

α	[W/m^2.K]	heat transfer coefficient
ΔT_m	[°C]	Log. mean T difference
ε		heat exchanger efficiency
λ	[W/m.K]	thermal conductivity
μ	[Pa.s]	dynamic viscosity

Indices:

c	cold fluid = cooling fluid
eff	effective
f	fluid
h	hot fluid = fluid to be cooled (except for d_h)
in, out	heat exchanger inlet, outlet
m	mean
th	theoretical
w	wall

7 References

[1] W. Ehrfeld, V. Hessel, H. Möbius, T. Richter, and K. Russow, "Potentials and Realization of Microreactors", *Workshop on Microsystem Technology for Chemical and Biological Microreactors*, 1996, VCH Verlagsgesellschaft: Frankfurt am Main, p. 1-28.

[2] K.P. Jäckel, "Microtechnology: Application Opportunities in the Chemical Industry", *Workshop on Microsystem Technology for Chemical and Biological Microreactors*, 1996, VCH Verlagsgesellschaft: Frankfurt am Main, p. 29-50.

[3] T. Schaller, W. Bier, G. Linder, and K. Schubert, "Mechanische Microtechnik für Abformwerkzeuge und Kleinserien", 1995, Forschungszentrum Karlsruhe.

[4] R. Srinivasan, S.L. Firebaugh, I.-M. Hsing, J. Ryley, M.P. Harold, K.F. Jensen, and M.A. Schmidt, "Chemical Performance and High Temperature Characterization of Micromachined Chemical Reactors", *Transducers '97*, 1997.

[5] G. Reuss, W. Diesteldorf, and O. Grundler, "Formaldehyde", *Encyclopedia of industrial Chemistry*, 1988, Ullmann: Berlin, p. 619-651.

[6] S. Su, P. Zaza, and A. Renken, "Catalytic dehydrogenation of methanol to water-free formaldehyde", Chemical Engineering and Technology, 1994, **17**: p. 34-40.

[7] M. Weber, P. Lerch, and P. Renaud, "Improved design for fast modulating IR sources", J. Micromech. Microeng., 1997, **7**: p. 210-213.

[8] F.P. Incropera and D.P.D. Witt, "Fundamentals of Heat and Mass Transfer", 4th ed., 1996, New York: John Wiley & Sons.

[9] R.K. Shah and A.L. London, "Laminar Flow: Forced Convection in Ducts", Advances in Heat Transfer, ed. J. Thomas F. Irvine and J.P. Hartnett, Vol. Supplement 1, 1978, New York: Academic Press.

[10] W. Bier, W. Keller, G. Linder, D. Seidel, K. Schubert, and H. Martin, "Gas to gas heat transfer in micro heat exchangers", Chemical Engineering and Processing, 1993, **32**: p. 33-43.

[11] J.C. Mecklenburgh and S. Hartland, "The Theory of Backmixing", 1975, London: John Wiley & Sons.

[12] J.J. Lerou, M.P. Harold, J. Ryley, J. Ashmead, T.C. O'Brien, M. Johnson, J. Perrotto, C.T. Blaisdell, T.A. Rensi, and J. Nyquist, "Microfabricated Minichemical Systems: Technical Feasibility", *Workshop on Microsystem Technology for Chemical and Biological Microreactors*, 1996, VCH Verlagsgesellschaft: Frankfurt am Main, p. 51-69.

Gas / Liquid Microreactors for Direct Fluorination of Aromatic Compounds Using Elemental Fluorine

V.Hessel, W. Ehrfeld, K. Golbig, V. Haverkamp, H. Löwe, M. Storz, Ch. Wille
Institut für Mikrotechnik Mainz GmbH, Carl-Zeiss-Str. 18-20, D-55129 Mainz, Germany
Tel.: +49-6131-990-450, Fax: +49-6131-990-475, e-mail: hessel@imm-mainz.de

A. E. Guber
Institut für Mikrostrukturtechnik, Forschungszentrum Karlsruhe GmbH, D-76021 Karlsruhe, Germany

K. Jähnisch, M. Baerns
Institut für Angewandte Chemie Berlin-Adlershof e.V., Rudower Chaussee 5, D-12489 Berlin, Germany

Abstract

Two different concepts of contacting gas and liquid media were realized in miniaturized reaction devices. The first one, a micro bubble column, utilizes dispersed systems consisting of small gas bubbles in a liquid phase. A large specific interfacial area of about 9,000 m^2/m^3 was determined by applying the sulfite oxidation method. Flow equipartition of a high number of parallel operating channels was achieved by using a pressure barrier. First visual observations of the liquid and gaseous flows indicate that this principle leads to a uniform flow equipartition. The second concept was based on the generation of thin stable films in a falling film microreactor. Very narrow residence time distributions of the flows in the channels were achieved also by using a pressure barrier.

The direct fluorination of toluene was feasible in both types of microreactors as proven by the synthesis of quantitative amounts of monofluorinated products with yields up to 20%. The yields, and in particular the conversion, of the microreactors were much larger compared to a laboratory bubble column which was operated at similar process conditions, e.g. equal ratios of fluorine to toluene, temperature, toluene and fluorine concentration. The falling film microreactor showed a very high conversion up to 50%, whereas the bubble column gave high selectivities reaching 49%. An analysis of the substitution pattern obtained in both microreactors is consistent with an electrophilic substitution mechanism.

1. Fluorination of aromatic compounds

Fluorinated aromatic compounds are of high commercial interest, e.g. as pharmaceuticals, reactive dyestuff, isotope labels and liquid crystals [1]. For their synthesis two routes are favoured, the Schiemann reaction [1,2], replacing an amino by a fluoro group via a diazotization reaction, and the Halex process [1,3], substituting chlorinated precursor molecules using metal fluorides.

Although both processes are widely used in industry, it is aimed since more than sixty years to perform the direct fluorination process using elemental fluorine [4]. To prefer the direct route, a number of reasons can be listed, concerning process efficiency in terms of yield, process variability for different reactants, energy consumption, and waste production. A major disadvantage of the presently used processes is their low overall yield of about 25 to 30% [4,5]. In case of the Schiemann process this is a consequence of the multi-step synthesis route applied, while the selectivity of the Halex process is decreased by side-reactions due to the long reaction times needed. Moreover, the Halex process is strongly limited regarding the choice of reactants, since due to its nucleophilic character only deactivated aromatic compounds can be used [4]. Furthermore, multi-step syntheses as the Schiemann process produce considerable waste, in this case including toxic solutions of nitric acid, sodium nitrite and sodium tetrafluoroborate [1,5].

The first twenty to thirty years of investigations dealing with the direct fluorination of aromatic compounds only seemed to show that this reaction is not feasible [6], in contrast to all other types of electrophilic halogenations [7]. Mainly undefined addition as well as precipitated polymeric products, i.e. tar, were found and, in some cases, even explosions made further research in this field unattractive [8,9]. However, since the late sixties it was shown by laboratory experiments that the direct fluorination can be performed with reasonable yields in case of an effective heat management [9-11]. This was either achieved by operation at extremely low temperatures [9], processing of very low concentrations or, most often, a combination of both [10].

Microreactors using thin fluid layers have proven for a number of liquid as well as gas phase applications that significant process improvements can be achieved due to the highly efficient heat transfer [12,13], even for highly concentrated solutions. In addition, microreactors strongly increase mass transport by providing thin fluid layers and, hence, high specific interfacial areas [14,15]. This is particularly important, since fluorine is insoluble in nearly all solvents commonly used.

In this context, the use of microreactors for the highly exothermic fluorination of toluene is described in this publication. For the generation of thin fluid layers two concepts have been followed. A disperse system consisting of small gas bubbles separated by thin liquid layers is generated in a micro bubble column. In a falling film microreactor a thin liquid film is contacted with a gas phase.

2. Disperse Contacting of Gas and Liquid in a Micro Bubble Column

The central component of the micro bubble column is a static micromixer constructed analogous to a design previously published [16]. However, two distinct differences from the original design have to be mentioned. First, the width of the gas feed channels is much smaller compared to the liquid feed channels in order to guarantee flow equipartition by applying a large pressure barrier. Second, instead of multilaminating gas and liquid streams in one common discharge slit, each gas/liquid bilayer stream is fed into a separated reaction channel (see figure 1).

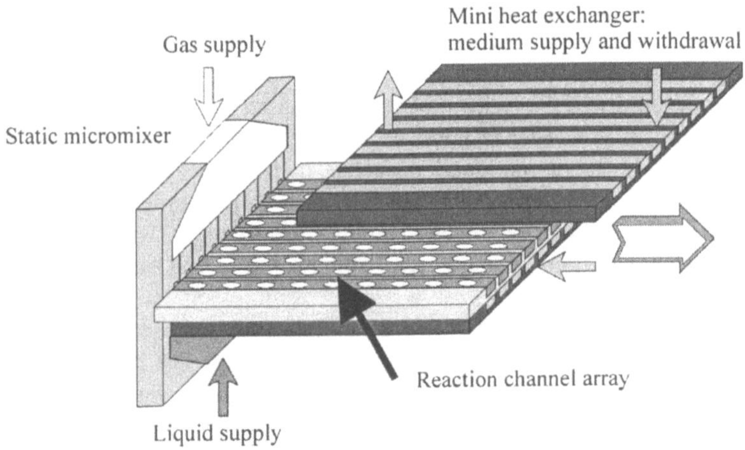

Fig. 1 Schematic of a static micromixer with an interdigitated channel configuration with gas and liquid feed channels of different size.

These parallel oriented reaction channels form a channel array on a platelet. Highly efficient thermal control is provided by heat transport to a heat transfer medium flowing in countercurrent mode in minichannel arrays adjacent to the reaction channel array on both sides. The reactor components of the micro bubble column are shown in figure 2.

Fig. 2 Components of the micro bubble column.

The design criteria for the microbubble column were based on the demands already defined (see section 1). The flow pattern, which is expected to establish in the microchannels, is characterized by gas bubbles dispersed in a liquid, the so-called segmented or bubble-train flow [17,18]. It was further assumed that the diameter of the gas bubbles is in the range of the channel width. Estimations of the bubble diameter and length as well as the film thickness between bubble and wall were made on the basis of measurements done in square or circular capillaries [18]. From these data a large specific interfacial area of 14,000 m^2/m^3 was calculated. To meet the criteria for conversion, flow rate, and hot-spot, rectangular reaction microchannels were designed with a width, depth and length of 300 μm, 150 μm and 50 mm, respectively. If the microchannel cross-section is further decreased to 50 x 50 μm^2, the specific interfacial area can be even higher (40,000 m^2/m^3).

Flow equipartition of gas and liquid streams was achieved by using a pressure barrier. Pressure loss calculations show that the width of the gas mixing channels has to be as small as 5 μm while the width of the liquid mixing channels should not exceed 20 μm (figure 3).

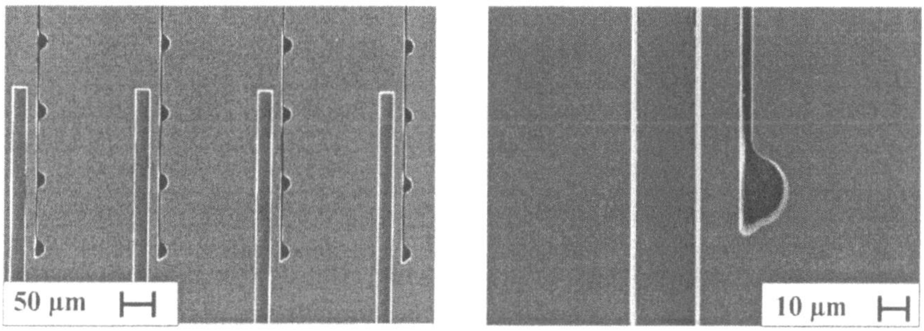

Fig. 3 Scanning electron micrographs of the interdigitated gas and liquid feed channels of the micromixer.

The micromixer platelets were realized in nickel on copper by using a UV-LIGA process. The reaction platelets with the larger cross-section of 300 x 150 μm^2 were fabricated by means of wet chemical etching in nickel and a high-alloy stainless steel. Smaller microchannels of 50 x 50 μm^2 size were achieved by applying a UV-LIGA process. The housing was made of high-alloy stainless steel by precision engineering and micro electro discharge machining using a micro die sinking process.

First measurements using the reaction platelets of large cross-section, based on the widely used sulfite oxidation method [19], indicate that a high specific interfacial area is generated by this type of contacting. Values of up to 9,000 m^2/m^3 were measured which is in the same order of magnitude as the calculated value of 14,000 m^2/m^3. Thus, it can be concluded that bubbles with a diameter and length in the order of 300 μm and 1200 μm, respectively, were generated. A visual observation of the gaseous and liquid flows of each channel at the reactor outlet indicates that flow equipartition was achieved for the majority of the channels.

3. Continuous Gas / Liquid Contacting in Open Parallel Channels

Another way of achieving high specific surface area for enhancement of mass and heat transfer is the generation of thin films. The principle of the formation of these thin films by contacting of liquid and gaseous reactants in a falling film microreactor is shown in figure 4.

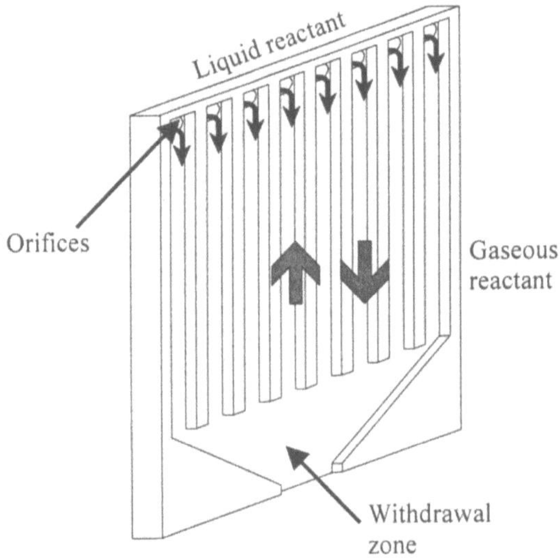

Fig. 4: Principle of contacting liquid and gaseous reactants in a falling film microreactor.

The liquid reactant is distributed onto a reaction plate, which is vertically oriented to the horizontal plane. Each substream enters a reaction channel through a single orifice at the top of the plate and flows downward as a fluid film to the withdrawal zone at the bottom.

The film thickness can be calculated by equation (1) [20]. Whereas the

$$\delta = \sqrt[3]{\frac{3 \cdot \dot{V} \cdot \eta}{\rho \cdot b \cdot n \cdot g}} \qquad (1)$$

gravity constant, g, and physical parameters of the liquid phase e.g. viscosity, η, and density, ρ, are fixed parameters for a given reaction system, the film thickness is determined for a given volume flow, $\dot{V}$, by width, b, and number, n, of the microchannels. Thus, the film thickness can be easily adapted to special demands of mass controlled reactions in solution by variation of geometrical parameters of the microdevices.

While the flow direction of the liquid reactant corresponds to the direction of gravity, the gaseous phase can be guided co- or countercurrently relative to the liquid phase. Due to the strong capillary forces acting in microchannels it is expected that high countercurrent gas loads are allowed, before flow reversal of the liquid occurs.

4. Measuring of Thin Films Using a Microfocus Measurement System

To verify the film thicknesses calculated by equation (1), non-contact surface measurements were performed using Microfocus-UBM® [21]. These experiments were carried out with a dead gas load.

In Figure 5 five graphical representations of profile measurements of the streaming fluid front are shown. This collection includes a profile across the

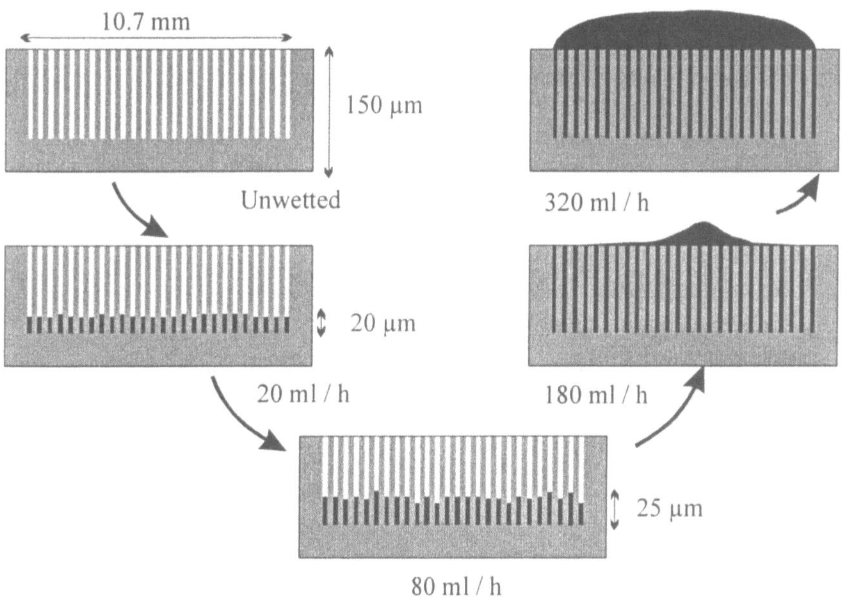

Fig 5: Graphical representation of profile measurements of the reaction plate at different operation modes.

unwetted reaction plate and two profiles of only partially filled microstructures at different stationary volume flows. Additionally, two examples for an operation mode with partially and totally flooded reaction plate are given. Comparing the first four plots, it is evident that film thickness increases with volume flow. The measured film thicknesses are in the range of the predicted values derived from equation (1). However, the data set is to small to allow a judgement if the measurements are in accordance to equation (1).

5. Falling Film Microreactor for Highly Exothermic and Diffusion Controlled Reactions

The complete falling film microreactor consists of four components, shown in Fig. 6, namely the bottom plate with an integrated heat exchanger, the reaction plate, the contact zone mask and the top plate with the gas chamber. The central

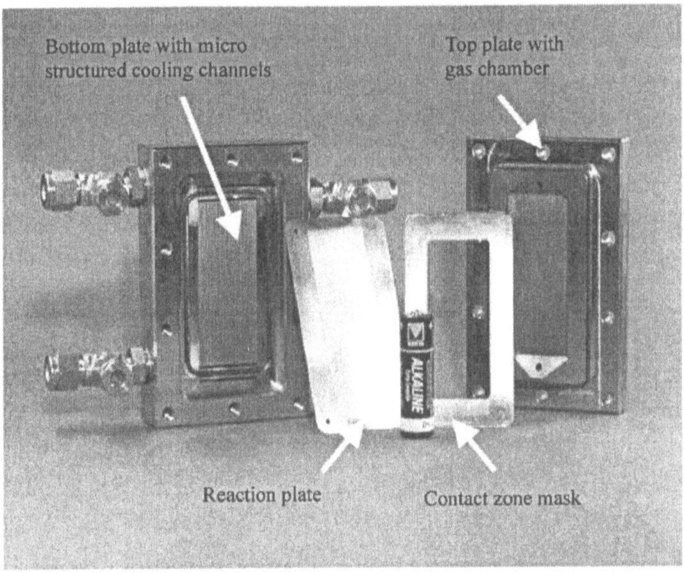

Fig. 6: Components of falling film microreactor.

element of the microreactor is the reaction plate comprising the microchannels with the liquid reactant. Supply and withdrawal of the liquid reactant are each performed in boreholes which are connected via one large slit to numerous small orifices, see figure 7.

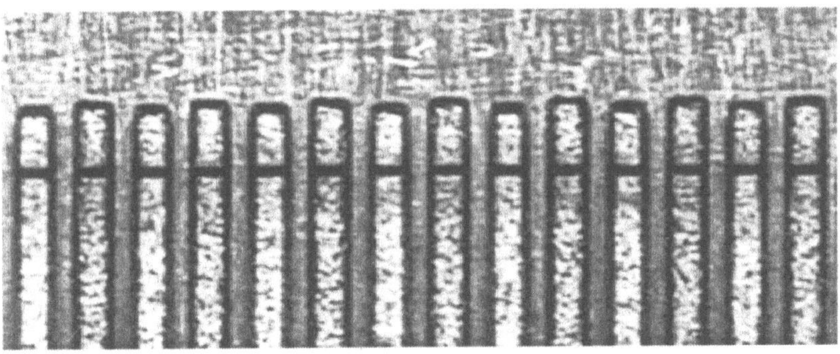

Fig. 7: Supply orifices in the reaction plate connected via one large slit.

The same manifolding principle is used for distributing the heat transfer medium among the cooling channels in the bottom plate. The reaction plate is inserted on top of the cooling channels in such a manner that each reaction channel covers a single cooling channel. The cooling channels have the same width and depth as the reaction channels, but are shorter due to the chosen positions of the feeding and withdrawal boreholes. In order to restrict the reaction to the cooled zone, a contact zone mask has to be added. The reaction channels are covered by a rectangular-shaped gas chamber in the top plate whereby the inlet and outlet of this gas chamber act as diffusors.

The reaction and cooling channels have a width of 300 μm and a depth of 100 μm, the length amounts to several ten millimeters. Microfabrication of the reaction plate was performed by means of micro electro discharge machining (μ-EDM) and wet etching of thin metall foils. In case of μ-EDM, the channels were realized by a die sinking process using electrodes with inverse shapes of the microstructures, fabricated by thin-wire microerosion. The orificies and the slit have a height of 80 μm and 1 mm, respectively. They were generated by die sinking as well. The contact zone mask was realized using laser machining, the fabrication of the gas chamber was performed by die sinking.

6. Fluid Equipartition for Uniform Residence Times

The quality of equipartition was measured by imaging the starting wetting behaviour of the reaction plate with a dead gas load, i.e. the plate was not covered. It was found that the starting behaviour is representative for stationary flow conditions as well. Figure 8 shows the time behavior of the flowing fluid front for acetonitrile with a volume flow of 20 ml/h at five time intervals, each separated by 4 s.

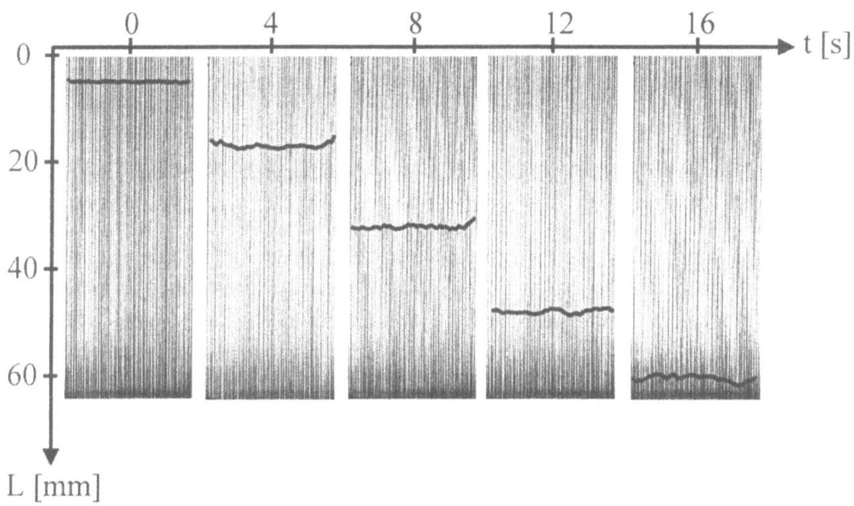

Fig. 8: Streaming fluid front (acetonitrile) in open channel system at different times.

It is obvious that very uniform residence time distributions were achieved. A cumulative plot shows that 90 % of all substreams, before leaving the reaction zone, have residence times in the range of 17 to 18 s. The mean residence time was measured to be 17.5 s (see figure 9).

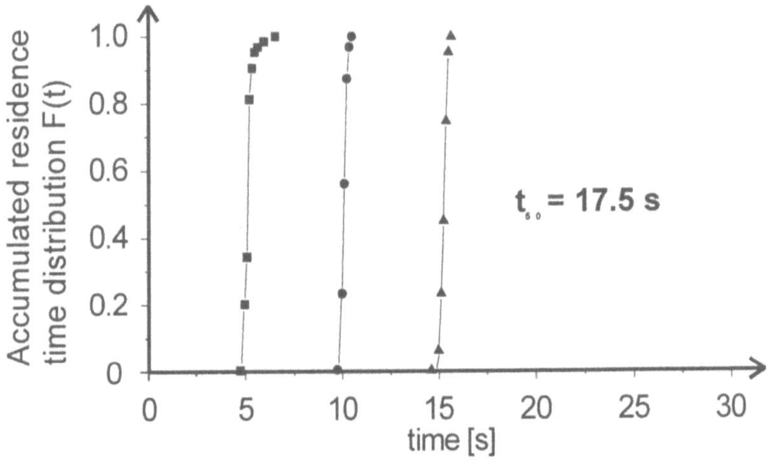

Fig. 9: Cumulative frequency distribution at different times.

During the experiments it became obvious that microfabrication parameters have an effect on the quality of equipartition. The distribution of the residence times is stochastical, thus, obviously revealing small, statistic deviations of the channel geometries from ideal. When comparing reaction plates manufactured by different fabrication techniques, the precision of the microstructuring of the channels has, as expected, a prominent influence on the efficiency of the manifolding.

Moreover, it turned out that the quality of the sealing in the entrance region is another important factor for flow equipartition. In particular, a strong effect related to the bending of the reaction plate was observed. Based on this experience, microfabrication and sealing techniques could be significantly improved.

7. High Yield and Selectivity for Direct Fluorination of Toluene

The direct fluorination of toluene with elemental fluorine (10 % F_2 in N_2) was performed using the falling film microreactor and the micro bubble column. The results were compared to fluorination results of a standard macroscopic contacting device, a laboratory bubble column. This column consists of an U-tube with a fine porous frit filled with a reaction volume of 10 ml. Similar devices were used in a number of experimental investigations published in [9-11].

Toluene was dissolved in the solvent acetonitrile with a molar concentration of 10 %. The temperature was kept at –17°C and molar concentration ratios of fluorine to toluene ranging from 40 % to 100 % were used. The gas volume flow of the laboratory bubble column was 20 ml/min. The liquid volume flow was set to 13 ml/h for the micro bubble column and to 11.6 ml/h for the falling film microreactor. The gaseous reactant was guided cocurrently to the falling film.

Since in the experiments only fluorine-to-toluene ratios below or equal to one were used, conversion, yield and selectivity usually are referred to the under-stoichiometric component, i.e. fluorine. However, using the present experimental set-up the consumption of fluorine in the gaseous phase could not be measured. Instead, the consumption of toluene and the generation of reaction products, i.e. monofluorinated toluenes, could be quantified. Therefore, conversion, yield and selectivity were defined according to the following equations:

$$X = \frac{n_{\text{Toluene},0} - n_{\text{Toluene},R}}{n_{\text{Toluene},0}} \cdot 100 \quad [\%] \tag{2}$$

$$Y = \frac{n_{\text{mono-F-toluene}}}{n_{\text{Toluene},0}} \cdot 100 \quad [\%] \tag{3}$$

$$S = \frac{n_{\text{mono-F-toluene}}}{n_{\text{Toluene},0} - n_{\text{Toluene},R}} \cdot 100 \quad [\%] \tag{4}$$

X denotes the conversion of toluene, Y the yield of monofluorinated toluenes and S the selectivity of monofluorinated toluenes referred to the amount of converted toluene; $n_{\text{Toluene},0}$ denotes the amount of toluene in the educt mixture, $n_{\text{Toluene},R}$ the amount of toluene in the reaction product and $n_{\text{mono-F-toluene}}$ the amount of ortho-, meta-, and para-monofluorinated toluene.

In tables 1a to 1c quantitative High Pressure Liquid Chromatography (HPLC)-data regarding conversion, yield and selectivity of the fluorination of toluene in dependence on the molar fluorine ratio are given for the three types of reactors.

Molar ratio mol F_2 / mol C_7H_8 [%]	Gaseous flow (10% F_2 in N_2) [ml/min]	X [%] *	Y [%] **	S [%] ***
40	20	5	3	45
54	20	13	4	27
100	20	34	8	22

Table 1a: Quantitative HPLC data of fluorination experiments performed in the laboratory bubble column.

Molar ratio mol F_2 / mol C_7H_8 [%]	Liquid flow (1.1 mmol C_7H_8 in 1 ml CH_3CN) [ml/h]	X [%] *	Y [%] **	S [%] ***
40	13	12	6	49
54	13	26	11	46
83	13	28	6	24

Table 1b: Quantitative HPLC data of fluorination experiments performed in the micro bubble column.

Molar ratio mol F_2 / mol C_7H_8 [%]	Liquid flow (1.1 mmol C_7H_8 in 1 ml CH_3CN) [ml/h]	X [%] *	Y [%] **	S [%] ***
40	11.6	30	11	36
54	11.6	38	14	36
92.5	11.6	50	20	41

Table 1c: Quantitative HPLC data of fluorination experiments performed in the falling film microreactor.

* see equation (2), ** see equation (3), *** see equation (4)

Nearly all experiments performed in both types of microreactors gave higher conversions and yields of the monofluorinated products, i.e. ortho-, meta- and para-fluorotoluene. Yields of up to 20% for the falling film microreactor and 11% for the micro bubble column were found, whereas the best yield of the laboratory bubble column did not exceed 8%. It has to be taken into account that in nearly all experiments the yield could not reach 100% due to the excess of toluene relative to fluorine. E.g., at a molar ratio of 40% a yield based on toluene of 11% was measured for the falling film microreactor. Actually, 27% of the fluorine molecules were converted to the desired product.

The conversions reached in both microreactors, in particular in the falling film reactor, were much higher compared to the laboratory reactor. Conversions of up to 50% were achieved in the falling film microreactor and, additionally, high selectivities were found. Only for the molar ratio of 40%, corresponding to a low conversion, the laboratory bubble column gives comparable results. Thus, direct fluorination using this reactor only led to either high conversion or high selectivity, while the microreactors combine high conversion at reasonable selectivities.

Another, often used graphical representation to characterize the performance of reactors is a selectivity-conversion diagram (see figure 10). Although the data set available is small, the selectivity-conversion diagram clearly indicates the better performance of the microreactors, in particular of the falling film reactor. The high conversions achieved in the microreactors demonstrate that mass transport is extremely facilitated by the generation of thin liquid films. The high selectivities measured show that a similar influence of the heat transfer was found. The low selectivity of the micro bubble column at high conversion is not understood at present. This may be caused by a change of the flow regime. Further investigations are necessary to clarify this result.

In addition to the analysis of reactor performance, experiments were performed to reveal the reaction mechanism. Gas chromatographic (GC) analysis allowed to efficiently separate the fluorinated products. A typical GC-plot is shown in figure 11.

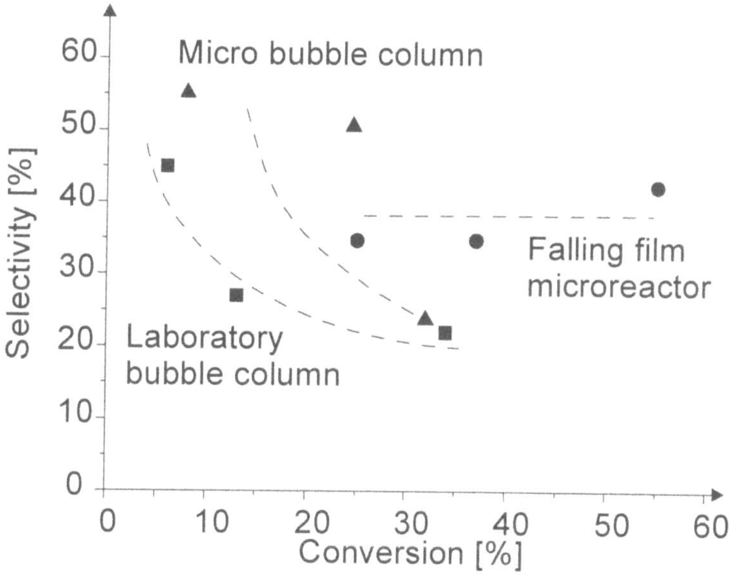

Fig. 10: Selectivity-conversion diagram for the three reactors.

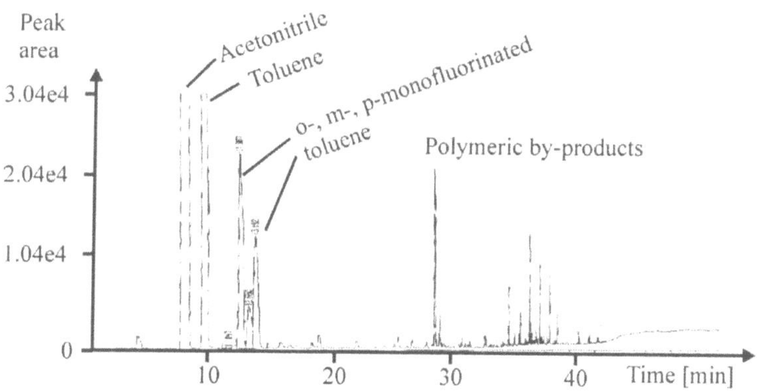

Fig. 11: Product spectrum of direct fluorination using acetonitrile.

A substitution pattern of 5 : 1 : 3 regarding ortho, meta and para isomers, respectively, was determined by integration of the peak areas. This is consistent with an electrophilic substitution mechanism which has been proposed for this reaction by mechanistic studies using highly diluted solutions [10]. Furthermore, the GC-measurements, in accordance to the HPLC results of tables 1a-c, indicate a significant generation of monofluorinated products relative to the undesired by-products.

In addition to GC and HPLC analysis, reactants and products can be analysed and quantified using IR-spectroscopic methods. A variant of this technique, attenuated total reflection (ATR), is particularly suitable to analyze samples of small volume. First IR-spectra corresponding to fluorination experiments with low conversion are shown in figure 12.

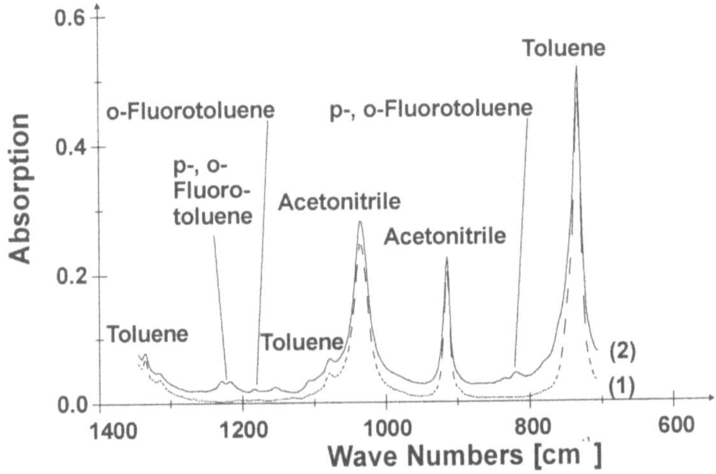

Fig. 11: Product spectrum of direct fluorination using acetonitrile.

The spectra of the reactants (1) and the products (2) can be distinguished easily, although the total amount of reaction products is low. The IR-absorption bands of toluene were found at 1339.6, 1081.9 and 737.0 cm^{-1}, for acetonitrile at 1038.7 cm^{-1}, for o-fluorotoluene at 1232.4 and 837.9 cm^{-1} and for p-fluorotoluene at 1220.8 and 822.8 cm^{-1}.

At present, fluorination experiments using high fluorine contents, up to 100% fluorine, are performed at the FZK. The expected high conversions may allow an even more accurate quantitative analysis by the IR-ATR-technique compared to the measurement shown in figure 11.

8. Conclusions

The results presented here clearly demonstrate that significant process improvements can be achieved for the mass transport limited, highly exothermic fluorination reaction by enhancing mass and heat transfer. Further measurements are needed to yield a comprehensive analysis of the performance of both reactors. However, the results indicate a huge potential for further optimization of reaction parameters which should lead to a faster reaction process and, hence, higher throughputs. Moreover, numbering-up by operating a higher number of channels as well as the combined action of several reaction plates with multichannel arrays can be easily achieved. In this context, the application of microreactors to the production of fine chemicals seems to be feasible in principle.

Moreover, the overall yields of the microreactors are close to the yields of industrially applied processes for the synthesis of monofluorinated products, e.g. the Balz-Schiemann or Halex route. These routes are usually fixed to special substitution patterns, depending on the substituents on the aromatic ring. The direct fluorination may enable the synthesis of fluorinated aromatic compounds with substitution patterns which can not be realized using the established process pathways.

Based on the knowledge gained about the hydrodynamics and reaction performance, the application of these microreactors may be extended easily to other hazardous gas/liquid reactions with similar process features as the fluorination. Moreover, an adaptation of the reactor design should allow to use them as analytical measuring tools for process development. In particular, the field of catalysis in multiphase reactions, either homogeneous or heterogeneous, seems to be a promising field for application of these versatile applicable falling film micoreactors and micro bubble columns.

9. Acknowledgments

We would like to thank Acordis, BASF, DuPont de Nemours, Aventis and Merck for financial and technical support of this project. This work was supported by the DECHEMA and the Bundesministerium für Bildung, Wissenschaft, Forschung und Technologie (Contract No.: 03 D 0067).

10. References

[1] G. Siegmund, W. Schwertfeger, A. Feiring, B. Smart, F. Behr, H. Vogel, B. McKusick, *Fluorine compounds, organic* in Ullmann's Encyclopedia of Industrial Chemistry, 5th edition, volume A11, pp. 349 - 383

[2] G. Schiemann, B. Cornils, *Chemie und Technologie cyclischer Fluorverbindungen*, Ferdinand Enke Verlag, Stuttgart, pp. 188 –193 (1969)

[3] L. Dolby-Glover, Chem. & Ind. (London) **1986**, p. 518 (1986)

[4] S.T. Purrington, B.S. Kagen, Chem. Rev. **86**, pp. 997 - 1018 (1986)

[5] Organikum, 16th edition, chapter P.8, pp. 530 - 563, VEB Verlag, Berlin (1986)

[6] R. D. Chambers: *Organofluorine Chemistry;* Springer Verlag, Berlin, p. 3 (1997)

[7] J. March, Advanced Organic Chemistry, 4th edition, Wiley-Interscience, New York (1992)

[8] N. Fukuhara, L.A. Bigelow, J. Am. Chem. Soc. **60**, pp. 427 - 429 (1938)

[9] V. Grakauskas, J. Org. Chem. **35** (3), pp. 723 - 728 (1970)

[10] F. Cacace, P. Giacomello, A.P. Wolf, J. Am. Chem. Soc. **102** (10), pp. 3511 - 3515 (1980)

[11] L. Conte, G.P. Gambaretto, M. Napoli, C. Fraccaro, E. Legnaro, J. Fluorine Chem., **70**, pp. 175 – 179 (1995)

[12] O. Wörz, K.P. Jäckel, Th. Richter, A. Wolf, in: Proc. of 2nd International Conference on Microreaction Technology, W. Ehrfeld, I.H. Rinard, R.S. Wegeng, eds., Topical Conference Preprints, March 9th - 12th 1998, New Orleans, U.S.A., pp. 88 - 95 (1998)

[13] K. Schubert, W. Bier, J. Brandner, M. Fichtner, C. Franz, G. Linder, in: Proc. of 2nd International Conference on Microreaction Technology, W. Ehrfeld, I.H. Rinard, R.S. Wegeng, eds., Topical Conference Preprints, March 9th - 12th 1998, New Orleans, U.S.A., pp. 88 - 95 (1998)
[14] V. Hessel, W. Ehrfeld, K. Golbig, V. Haverkamp, H. Löwe, Th. Richter, in: Proc. of 2nd International Conference on Microreaction Technology, W. Ehrfeld, I.H. Rinard, R.S. Wegeng, eds., Topical Conference Preprints, March 9th - 12th 1998, New Orleans, U.S.A., pp. 259 - 266 (1998)
[15] I. Robins, J. Shaw, B. Miller, C. Turner, M. Harper, in: Microreaction Technology, Proc. of the 1st Int. Conf. on Microreaction Technology, ed.: W. Ehrfeld, Springer-Verlag, Berlin, pp. 35 - 46 (1997)
[16] W. Ehrfeld, K. Golbig, V. Hessel, H. Löwe, Th. Richter, Ind. Chem. Res., Accepted for publication (1999)
[17] V. Hatziantoniou, B. Andersson, Ind. Eng. Chem. Fundam. **23**, pp. 82 – 88 (1984)
[18] T.C. Thulasidas, M.A. Abraham, R.L. Cerro, Chem. Eng. Science **50** (2), pp. 183 – 199 (1995)
[19] W.-D. Deckwer, *Reaktionstechnik in Blasensäulen;* Otto Salle Verlag, Frankfurt /M, Berlin, München (1984)
[20] R. H. Perry, D.W. Green: *Perry's Chemical Engineers' Handbook*; McGraw-Hill, New York, p. 6-43 (1997)
[21] UBM Microfocus System by UBM Messtechnik GmbH, Ettlingen, Germany

Solvent Extraction and Gas Absorption Using Microchannel Contactors

W.E. TeGrotenhuis, R.J. Cameron, V.V. Viswanathan, R.S. Wegeng

Pacific Northwest National Laboratory, PO Box 999, Richland, WA 99352 USA

Abstract

Microchannel devices offer many technical advantages over conventional technologies for chemical separations. The overall objective for microchemical separations is to achieve extremely high throughput per unit hardware volume by engineering devices with high contact area to volume ratios, micro-thin films, and uniform flow distribution. Pacific Northwest National Laboratory's proprietary sheet architecture is being used to construct microchannel devices consisting of channels that are 50 to 500 μm deep and separated by porous contactor plates. Rapid mass transfer occurs between immiscible fluids, both gases and liquids, as they flow through the channels. By immobilizing the interface, both co-current and counter-current flow are achieved. The resulting improvements in heat and mass transfer rates through the devices result in dramatic improvements in productivity and efficiency.

Devices have been fabricated and tested for both solvent extraction and gas absorption. Data are presented for cyclohexanol extraction from water, carbon dioxide absorption into diethanolamine, and carbon monoxide and carbon dioxide absorption into copper ammonia formate solutions. Mass transfer models based on first principles have been developed for co-current and counter-current flow for solvent extraction and gas absorption. Such predictive models are useful for evaluating the relative importance of resistances to mass transfer, predicting full-scale performance, and comparing microchannel separators with conventional technologies. Experimental data are compared with theoretical prediction.

Applications for microchannel separations technology include radiochemical processing; cleanup of fuel-cell feed gases; in situ resource utilization for space exploration; recovery and sequestration of carbon dioxide for carbon management; and other separations of interest to the chemical, environmental, and pharmaceutical industries.

1 Introduction

Concepts are being developed to provide capabilities for engineering and fabricating devices containing channels and wall-thicknesses that are 100 μm deep or less with features as small as 1 to 10 μm. These capabilities allow engineering of chemical processes in devices that are orders of magnitude smaller than conventional chemical process equipment. The advantage arises from highly uniform flow distribution at micron length-scales, giving extremely rapid heat and mass transport rates that result in very short residence times and high throughput per unit hardware volume [1-3]. Progress in developing microtechnology for all traditional chemical processing unit operations, including heat exchangers [4], reactors [5] separators [6,7], and actuators [8,9], will facilitate miniaturization of entire chemical processes. Although traditional economies of scale are not realized with microtechnology, the potential exists for economies of mass production [10,11] analogous to the dramatic reductions in manufacturing costs experienced in the microelectronics industry.

Microchannel contactors are being developed that allow for rapid contacting of two immiscible fluids to facilitate chemical separations. Figure 1 illustrates the concept for solvent extraction in a simple two-channel geometry. The solvent and raffinate streams flow through separate channels separated by a micro-porous

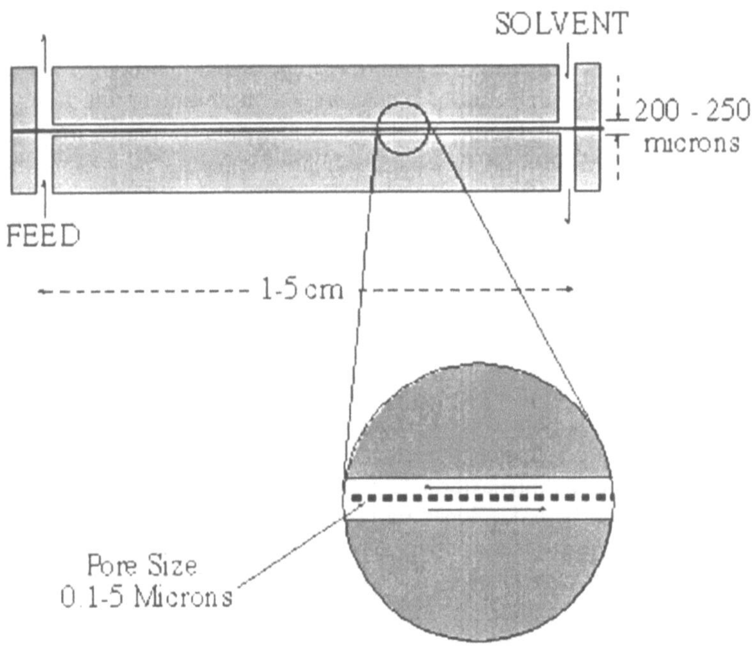

Figure 1. Representation of microchannel architecture for chemical separations

contactor plate, which allows for either co-current or counter-current operation. The channel heights are on the order of 100 μm, the micro-porous plate is 10 to 50 μm thick, and the channel length is typically 1 to 20 cm. The concept is scaled-up to multi-channel devices with high overall volume efficiency using Battelle's proprietary sheet architecture [12,13].

Although the contactor plate facilitates counter-current operation, it also adds mass transfer resistance, which can be minimized by reducing thickness, increasing porosity, or reducing tortuosity. Tortuosity is eliminated by microfabricating cylindrical or conical holes in thin substrate materials. Interfacial forces immobilize the fluid-fluid interface within the pores of the contactor plate, and the maximum pressure difference that the contactor plate can support without breakthrough is a function of pore size, contact angle, and interfacial tension. Pressure differentials ranging from several inches to tens of inches of water column are typical. Breakthrough pressure constrains the operating range of the microchannel contactors.

The most promising near-term applications for microchannel based separations are for cases where size and/or weight are extremely significant. NASA's In Situ Resource Utilization (ISRU) program is planning near-term missions to Mars that will include chemical processes for converting the carbon dioxide and possibly water in the Martian environment to propellants, oxygen, and other useful chemicals [14]. The launch mass of future missions will be dramatically reduced by not having to bring all of the resources necessary for the visit and the return flight from earth. Obviously, extraterrestrial chemical plants will need to be compact, lightweight, efficient and able to operate reliably for prolonged periods in reduced gravity environments. A second example is the development of fuel-cell powered vehicles by the Office of Transportation Technology of the U.S. Department of Energy. One approach for supplying hydrogen to the fuel cell is to reform gasoline or methanol on board. Two important challenges for microchemical separations are scrubbing of sulfur and carbon monoxide, which are catalyst poisons. A final example where size is important due to the impact on costs is in the treatment of high-level radioactive wastes, such as the waste stored in tanks at the Hanford site in Eastern Washington State. Reducing the size of the equipment required to process the waste has enormous cost savings potential due to infrastructure and other scaled costs.

2 Theoretical

Three parameters are useful for characterizing the performance of microchannel contactors for a given application. Mass transfer performance is characterized by the inverse of the mass transfer Peclet number defined as

$$\frac{1}{Pe_m} = \frac{w\,L\,D_s}{Q\,h} \tag{1}$$

that is used for both the solvent and raffinate channels. Here, w is channel width, L is channel length, D_s is the solute diffusion coefficient in the solvent or raffinate phase, Q is flowrate of the solvent or raffinate phase, and h is channel height. The inverse of the Peclet number is used because a larger value corresponds to greater mass transfer.

The second performance parameter is residence time, which is a metric for the size of the device;

$$T_R = \frac{w\,h\,L}{Q} \tag{2}$$

The final performance parameter reflects the constraint on the maximum pressure drop across the contactor plate, referred to as the breakthrough pressure. For counter-current flow, the minimum breakthrough pressure required is equal to the sum of the pressure drops along the lengths of the raffinate and solvent channels. Using the pressure drop equation for 2-dimensional Pouiseille flow in a closed channel, the pressure drop scales by the relationship

$$\Delta P \propto \frac{L\,Q}{w\,h^3} \tag{3}$$

These three scaling parameters are used to size and scale the microcontactor architecture for a given application, as described below.

Performance of microcontactor devices can also be predicted more precisely using numerical solutions of the transport equations with the appropriate geometry and boundary conditions [7]. These equations are solved as an initial value problem for co-current flow using the Crank-Nicholson [15] implicit method of finite differences, and as a boundary value problem for counter-current flow. Solutions are used for comparison to experimental results, predicting full-scale performance, and comparing microchannel separators to conventional technologies.

The design objectives for applying the microchannel technology to solvent extraction are to achieve the desired separation requiring a given number of theoretical stages and minimize equipment size while not exceeding the breakthrough pressure constraint for the contactor plate. To allow size comparison between microchannel hardware and conventional technologies, a size metric defined as the residence time divided by the theoretical number of stages is adopted that is analogous to the height equivalent of a theoretical plate (HETP) [16].

A design procedure is followed where first an operating point is selected at a given channel height, channel length, and flow rate that will accomplish the required separation while meeting the ΔP constraint. From this operating point, the flow rate (Q) and the inverse of the channel length (1/L) are varied proportionally while keeping the channel height constant to maintain a constant pressure drop. A flow rate and channel height combination is selected that will accomplish the required separation. The second step of the design procedure is to vary the parameters such that the pressure drop and Peclet number are kept constant. This is achieved by decreasing the flow rate (Q), the square root of the channel length ($L^{1/2}$) and channel height (h) proportionally in order to further miniaturize final device. The mass transfer resistance of the contactor plate will become more significant as the other dimensions are reduced, causing a decrease in the number of theoretical number of stages, which can be corrected by returning to step one of the procedure.

The procedure has been applied to the extraction of acetone from water using 1,1,2-trichloroethane. Theoretical predictions result in a design point for 8 theoretical stages of separation that corresponds to a channel height of 50 μm, a channel length of 4.5 cm, and a flow rate of 0.07 ml/min per cm of channel width. This design point gives a 2.4 second residence time per theoretical stage for the raffinate stream, and an HETP of only 0.6 cm.

Table 1 compares the predicted performance of microchannel devices to conventional solvent extraction technologies for acetone extraction into 1,1,2-TCA. The flow capacity decrease is more than offset by the reduction in the HETP, resulting in a potential order of magnitude reduction in hardware volume when compared to high-efficiency structured packing. As much as two orders of magnitude size reduction is possible over conventional sieve tray columns.

3 Experimental Data

Both solvent and gas absorption experiments have been performed using microchannel contactor test devices. Experimental data have been collect for the extraction of cyclohexanol from water using cyclohexane as the solvent, and for extracting acetone from water using 1,1,2-trichloroethane. Gas absorption experiments experiments have included the removal of carbon dioxide from a mixture of CO_2 and N_2 using diethanolamine (DEA) solutions as the absorbent, and for scrubbing CO from a H_2 rich stream using copper ammonia formate solutions.

Table 1. Comparison of microchannel contactor performance to conventional technologies

	Capacity [$cm^3/cm^2/s$]	HETP [cm]	T_{res}/Stage [sec]
Sieve Trays	1	100-250	100-250
Structured Packings	0.8-5	20-180	10-80
Microchannels	0.2	0.25-5	1-20

Data are presented in Figure 2 for the extraction of acetone from water using 1,1,2-TCA. The same device and contactor plate are used in both co-current and counter-current flow configuration. As the flow rates are decreased, equilibrium effluent concentrations are approached in co-current operation, while counter-current flow is able to achieve greater than one theoretical stage. The data are also compared to predicted performance in Figure 2. In both flow configurations, actual performance does not meet expected performance, possibly due to non-

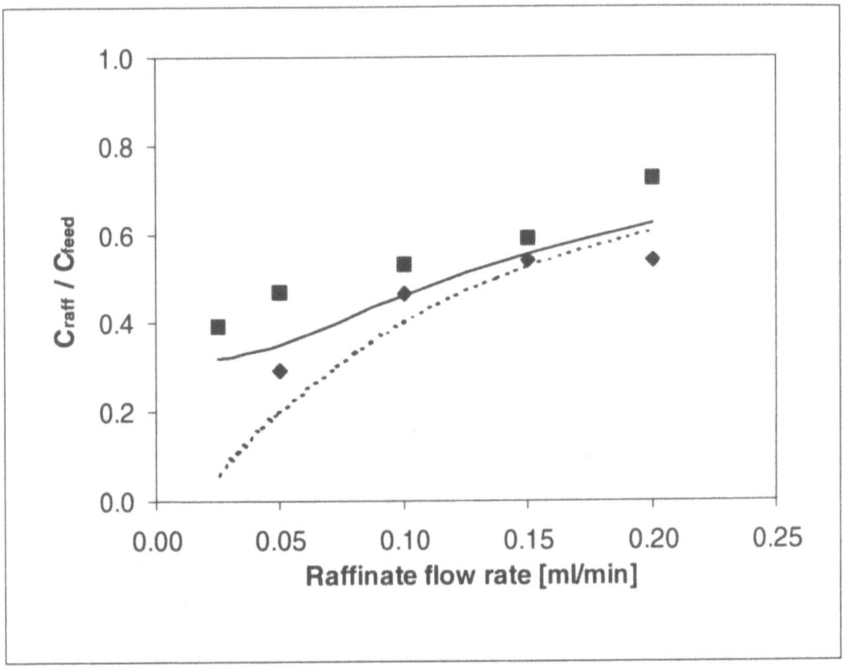

Figure 2. Raffinate to feed concentration ratio as a function of flow rate in the extraction of acetone from a 1 g/L aqueous solution using 1,1,2-trichloroethane as the solvent; equal flow rates; 240 μm deep, 8 cm long flow channels; and a 180 μm thick contactor plate in co-current flow (■) and counter-current flow (◆). Finite difference solutions of the transport equations are shown for co-current (—) and counter-current (- - -) flow.

idealities, such as poor flow distribution, or errors in physical parameters, such as diffusivities, channel dimensions, or in the equilibrium distribution coefficient.

Figure 3 illustrates the performance of microchannel devices in gas absorption. Over 90% of the carbon dioxide is removed in less than 10 seconds from a stream containing 25% CO_2. Increasing the concentration of absorbent molecule, in this case DEA, is shown to have diminishing returns. Similar results have been obtained using copper ammonia formate solutions to scrub CO from a H_2 rich stream, where CO is reduced from 1% to less than 10 ppm in 1.3 seconds gas residence time from a stream also containing 25% CO_2 and 4% H_2. Over 99% of the CO_2 was also removed with only 10% loss of H_2 hydrogen.

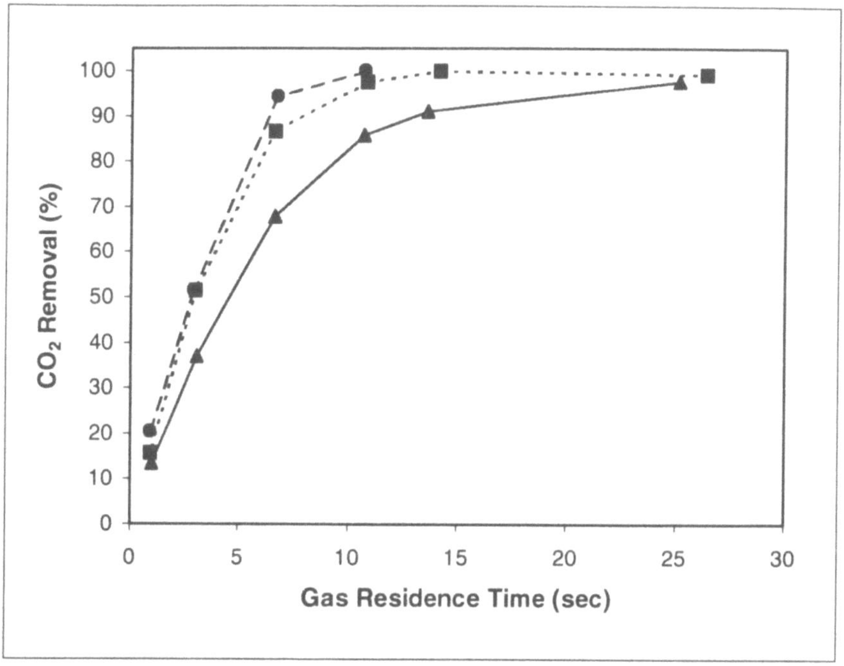

Figure 3. Absorption effectiveness as a function of gas residence for removal of CO_2 from a 25% CO_2 and 75% N_2 gas stream into a 10% diethanolamine solution (▲), 20% diethanolamine solution (●), and 40% diethanolamine solution (■) at a liquid residence time of 26 seconds.

4 Summary

Miniaturizing chemical separations equipment by an order of magnitude or more is likely through the development of microchannel contactor technology. Progress is being made to reduce the residence time per theoretical stage for solvent extraction and other applications to 1 second, with values for the height equivalent of a theoretical plate (HETP) approaching a fraction of a centimeter. Data have been presented illustrating performance of microchannel test devices for solvent extraction and gas absorption. The most likely applications for near-term market penetration are cases where reduced size and weight can have an enormous impact.

5 References

1. Wegeng RS, CJ Call, and MK Drost. 1996. "Chemical System Miniaturization." PNNL-SA-27317, Pacific Northwest National Laboratory, Richland, Washington.
2. Bowers M and I Mudawar. 1993. "High Flux Boiling in Low Flow Rate, Low Pressure Drop Mini-Channel and Micro-Channel Heat Sinks." *Int. J. Heat Mass Trans. 3*, 321–332.
3. Kleiner M, K Stefan, and K Haberger. September 1995. "High Performance Forced Air Cooling Scheme Employing Microchannel Heat Exchangers." *MicroSim 95*, Southampton.
4. Cuta J, W Bennett, and C McDonald. 1995. "Fabrication and Testing of Microchannel Heat Exchangers." *Proceedings of the SPIE Conference on Micromachining and Microfabrication Process Technology.*
5. Ehrfield W, V Hessel, H Mobius, T Richter, and K Russow. February 1995. "Potentials and Realization of Microreactors." Presented at the DECHEMA Workshop on Microsystem Technology for Chemical and Biological Microreactors.
6. Becker E. 1986. "Development of the Separation Nozzle Process for Enrichment of Uranium." *Ger. Chem. Eng. 9*, 204–208.
7. TeGrotenhuis WE, RJ Cameron, MG Butcher, PM Martin, and RS Wegeng. 1999. "Microchannel Devices for the Efficient Contacting of Liquids in Solvent Extraction." Accepted for publication in *Sep. Sci. and Tech.*
8. Zbedlick M, R Anderson, J Jankowski, B Kline-Schoder, L Christel, R Miles, and W Weber. 1994. "Thermopneumatically Actuated Microvalves and Integrated Electro-Fluidic Circuits." Presented at the Solid-State Sensor and Actuator Workshop.
9. Barth P, C Beatty, L Field, J Baker, and G Gordon. 1994. "A Robust Normally Closed Silicon Microvalve." Presented at the Solid-State Sensor and Actuator Workshop.
10. Friedrich C and B Kikkeri. 1995. "Rapid Fabrication of Molds by Mechanical Micromilling: Process Development." *Proceedings of the SPIE Conference on Micromachining and Microfabrication Process Technology.*
11. Arnold J, U Dasbach, W Ehrfeld, K Hesch, and H Lowe. May 1994. "Combination of Excimer Laser Micromachining and Replication Processes Suited for Large-Scale Production (Laser-LIGA)." Presented at the EMRS Spring Meeting, Strasbourg, France.
12. Wegeng RS, MK Drost, and CE McDonald. March 1997. "Microcomponent Sheet Architecture." U.S. Patent No. 5,611,214.
13. Wegeng RS, MK Drost, CJ Call, JG Birmingham, CE McDonald, DE Kurath, and M Friedrich. September 1998. "Microcomponent Chemical Process Sheet Architecture." U.S. Patent No. 5,811,062.
14. Frankie BM and R Zubrin. 1999. "Chemical Engineering in Extraterrestrial Environments." *Chem. Eng. Prog.*, *95*(2):45-54.
15. Crank J. 1975. *The Mathematics of Diffusion*, Second Edition. Oxford University Press, New York.

16. Humphrey JL and GE Keller II. 1997. *Separation Process Technology.* McGraw-Hill, New York.
17. Lo TC, MHI Baird, and C Hanson, eds. 1983. *Handbook of Solvent Extraction.* Krieger Publishing Company, Florida.

Acknowledgment

Pacific Northwest National Laboratory is operated by Battelle for the U.S. Department of Energy under Contract DE-AC06-76RLO 1830.

Polymer membranes for product enrichment in microreaction technology

B. Schiewe, A. Vuin, N. Günther*,
K. Gebauer*, Th. Richter*, G. Wegner

MPI für Polymerforschung, Ackermannweg 10, 55128 Mainz, Germany
*Institut für Mikrotechnik Mainz GmbH, Carl-Zeiss-Str. 18-20,
55129 Mainz, Germany

Abstract

For product separation of the off-gas components of an ethylene oxide microreactor new concepts need to be developed. Therefore, instead of using the technical process of low temperature distillation, the performance of gas separation through a dense polymer membrane was investigated. In order to characterise several membrane materials, a special set-up for low pressure permeation was realised utilising micro-structured membrane supports. With this device, different polymer membranes were investigated with respect to the permeability of ethylene oxide as a function of partial pressure. The work reported here focused on the screening of different membrane materials. Simultaneously, the development of a prototype membrane module to be used with the microreactor has been initiated.

Keywords

ethylene oxide synthesis, membrane separation

1. Introduction

Microreaction technology offers many advantages for a large number of applications in chemical engineering, pharmacy, medicine and biotechnology.

Much effort is made in order to develop microreactors for point-of-use production of explosive or toxic chemicals on-site and on-demand. Since a huge variety of microfabrication techniques is available today, the production of suitable microreactors has become possible /1,2/. As a consequence, improvement in microreaction technology requires separation techniques capable of fast and effective separation steps that can be miniaturised. The use of thin polymer membranes for gas separation is currently a fast growing field enabling fast product separation at low costs.

Progress has been made in the preparation of ultra-thin polymeric membranes and the characterisation of these membranes for separation of gases and liquids. It has been shown that polymer layers with a thickness of only 50-100 nm can be used very effectively in separation processes and are stable enough to overcome strong pressure changes /3/.

This project deals with the development of miniaturised separation modules for on-line separation and enrichment of gaseous products from the ethylene oxide microreactor. The focus of this study is on the formation and processing of gas separation membranes and their physico-chemical characterisation.

2. Aim and motivation

The ethylene oxide microreactor produces small amounts of a gas mixture comprising ethylene, ethylene oxide, carbon dioxide, oxygen and nitrogen. Therefore the aim of product separation is enrichment and separation of the product ethylene oxide as well as the isolation of the reactant ethylene for re-use in the reactor.

The large scale technical process, currently employed in industry, is low temperature distillation after scrubbing of ethylene oxide in a wash process with water /4/. For gas separation of the microreactor off-gas this technology is not feasible. An alternative is the gas separation employing dense polymer membranes enabling the continuos withdrawal of a small product stream. The principle is shown schematically in figure 1.

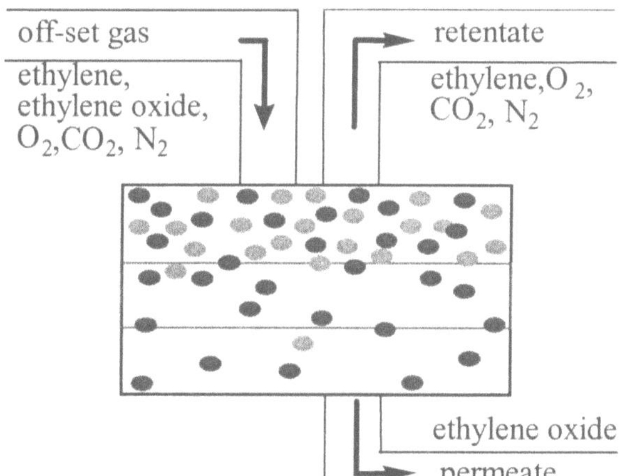

Figure 1: Scheme of the principle of gas separation through a dense polymer membrane.

3. Results and Discussion

3.1. Set-up of the permeation device

Since the partial pressures of ethylene and ethylene oxide in the product gas mixture are very low, usually in the range of 100 to 200 mbar at a total pressure of 5 bar, a specific permeation device was developed in order to perform measurements below 1 bar. The construction of this device is based on the design of a standard permeation device. The schematically view of the set-up developed is shown in figure 2.

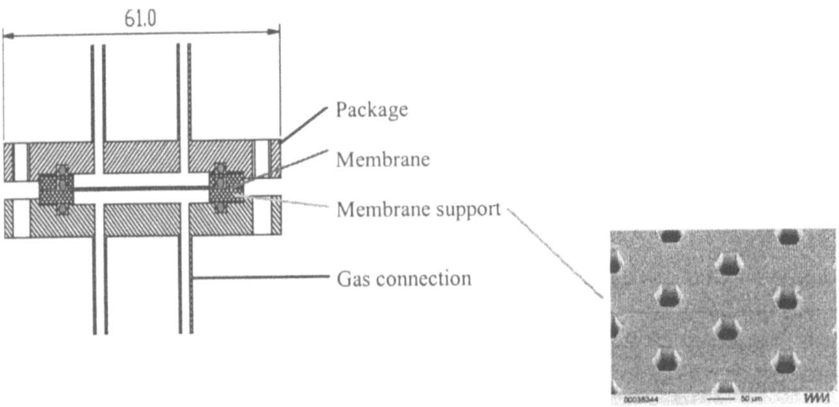

Figure 2: Schematic of view of the permeation device developed.

In order to realise short permeation times, the use of ultra-thin polymer membranes is necessary. Therefore, a membrane support with a correspondingly small thickness, which can be realised by micro-structuring techniques, has to be employed. The material chosen for this support was a Nickel structure assembled in a stainless steel housing, which is inert to the very reactive ethylene oxide. Calculations of the support dimensions based on the use of 200 nm thin polymer membranes resulted in a pore size of 60 µm and a interpore distance of 80 µm. A regularly structured Nickel support with these dimensions was realised at the Institut für Mikrotechnik Mainz using a UV-LIGA process (LIGA is the German acronym for lithography, electroforming and moulding).

All measurements were performed at 35°C and in a pressure range 0 to 20 mbar of the permeate. Therefore, specifically calibrated pressure sensors were used. In order to analyse binary gas mixtures, a differential refractometer was built up to measure the gas density of the permeate /5/. This method is based on optical spectrometry requiring only small gas amounts which is advantageous when measuring highly reactive and toxic substances, like ethylene oxide.

3.2. Membrane screening

Permeability data for ethylene and ethylene oxide through dense polymer membranes are not known from the literature. The gases to be separated are very similar in their molecular size and structure, thus, a high selectivity is necessary to achieve efficient separation. Therefore, polymer membranes which are used in other difficult separation processes, such as the separation of C2 and C3 olefins from paraffines, were used as starting materials /6/.

Another limitation to be considered is the reaction of ethylene oxide with the polymer membranes. In order to avoid this, various substituted polyimides as well as polyethylene were tested but showed significant adsorption of ethylene oxide at the polymer, regardless of their chemical structure. Even at low pressures, considerable swelling of the membrane material was observed (see figures 3 and 4).

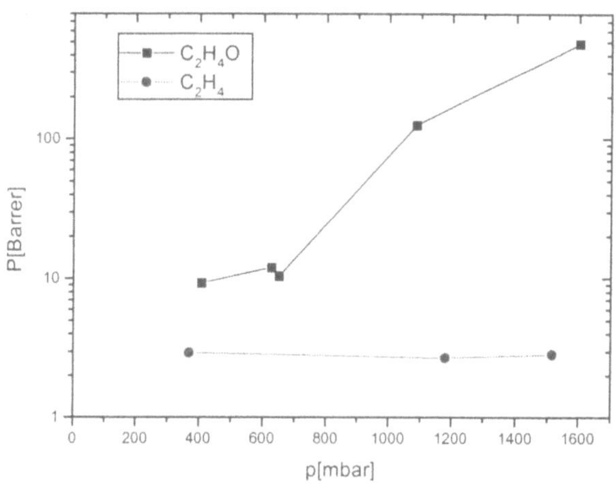

Figure 3: Permeability plot as a function of pressure for ethylene and ethylene oxide.

Figure 3 shows an almost linear behaviour of the permeability of ethylene, whereas the curve for ethylene oxide increases in a stepwise function of the pressure, indicating polymer swelling.

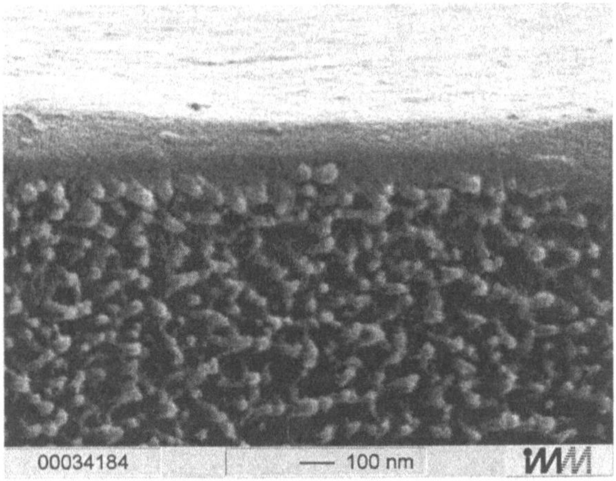

Figure 4: SEM picture of a swollen polyimide membrane (6FDA-mPD) exposed to ethylene oxide for 4 min at 1 bar.

The SEM picture of a dense polyimide membrane (figure 4) shows considerable swelling after exposure to ethylene oxide for only 4 minutes at 1 bar leading to microporosity.

These results lead to the conclusion that membranes to be used for the separation of ethylene oxide have to be crosslinked by means of physical or chemical methods. The investigation of crosslinked polyimide membranes showed that swelling decreases with increasing degree of crosslinking. On the other hand, increasing the degree of crosslinking also leads to a decrease in selectivity with respect to the separation of ethylene from ethylene oxide /7/. Nevertheless, at a degree of 20% crosslinking an ideal selectivity of 3 could be achieved. However, this is to low for a complete separation.

4. Conclusions

We report the development of a specific permeation device in order to perform gas permeation measurements below 1 bar. Therefore, a membrane support with regular microstructures was realised using a UV-LIGA process. With this equipment, various substituted polyimides as well as polyethylene were examined in terms of their selectivity for the separation of ethylene and ethylene oxide. Regardless of their chemical structure all membranes showed significant adsorption of ethylene oxide at the polymer, leading to considerable swelling. The permeability function of ethylene shows linear behaviour, whereas the curve for ethylene oxide increases in a stepwise function of the pressure.

Therefore, crosslinked polymer membranes should be used for ethylene oxide enrichment. The investigation of crosslinked polyimide membranes showed that swelling decreases with increasing degree of crosslinking, whereas the selectivity decreases.

5. References

/1/ W. Ehrfeld, K. Golbig, V. Hessel, R. Konrad, H. Löwe, Th. Richter, In: Microreaction Technology, Proceedings of the First International Conference on Microreaction Technology, Ed.: W. Ehrfeld, Springer-Verlag, Berlin (1997), p. 72 - 90.

/2/ W. Ehrfeld, V. Hessel, V. Haverkamp, Microreactors, Ullmann`s Encyclopedia of industrial chemistry, sixth ed., 1999

/3/ M. Seufert, C. Fakirov, G. Wegner; Adv. Mater., 7, 52-55, 1995.

/4/ W. Gerhartz, Ullmann`s Encyclopedia of industrial chemistry, vol. A10, 117-135, 1991

/5/ A. Becker, W. Köhler, B. Müller, Ber. Bunsenges. Phys.Chem. 99(4), 600-608, 1995

/6/ C. Staudt-Bickel, Dechema-Jahrestagungen `98, Kurzfassungen, Band 2, S. 255

/7/ B. Schiewe, C. Staudt-Bickel, G. Wegner, Gas Sep. Purif., in progress

Compact Heat Exchangers

L. A. Luo, U. D'Ortona, D. Tondeur

Laboratoire des Sciences du Génie Chimique, CNRS-ENSIC
1 rue Grandville, BP 451, F-54000 Nancy, France

Abstract. Three cross flow compact heat exchangers manufactured using classical workshop techniques are presented. Their heat-transfer coefficient are measured and compared to various correlations. Numerical simulations are performed and show that the introduction of mixing significantly improves exchanger performances. A new compact heat exchanger that practically implement this idea of mixing is presented.

Keywords. Compact Heat Exchangers, Microdevices, Heat Transfer, Mixing, Chaotic Advection

1 Introduction: From Macro to Micro

Micro devices raise a number of new questions in relation with mass and energy transfer phenomena [1]: Does the Navier-Stokes equation still hold for very small channels ? Does viscosity need a correction for micro channels ? For gas flow, how to introduce a Knudsen number correction ? In the case of heat exchangers: Do classical heat transfer correlations still hold or do we need new correlations ? The channel size being small, the flow is laminar. Does this limit heat exchange enhancement ? Can we reduce the size "indefinitively" or is there a lower limitation ? From which size down do specific effects appear due to small sizes ?

The present work addresses some of these questions, but also that of ease of manufacture. For these purpose, we have manufactured several "compact" heat exchangers, compact meaning here millimeter channel size. These exchangers are interesting for several reasons: they are much easier to fabricate than micro devices; they allow rather complex flow channel organization, as we shall illustrate; moreover, they exhibit behaviors usually expected only with smaller sizes. For example, Adams and Coll. [2] have studied circular copper micro-channels of diameter 0.76 to 1.09mm, Reynolds numbers between 3000 and 20000, and Prandtl numbers between 4.2 to 6.4. Their results show that for a channel of 0.76mm, the Nusselt number has values significantly higher than the ones predicted by the following classical correlations [3]:

$$Nu = 1.86\,(Re\,Pr)^{1/3}\Big(\frac{D_h}{L}\Big)^{1/3}\Big(\frac{\mu}{\mu_p}\Big)^{0.14} \qquad Re < 2200 \qquad (1)$$

$$Nu = 1.116\,(Re^{2/3} - 125)\Big[1 + \Big(\frac{D_h}{L}\Big)^{2/3}\Big]\Big(\frac{\mu}{\mu_p}\Big)^{0.14} \qquad 2200 < Re < 10^4 \qquad (2)$$

$$Nu = 0.023\,Re^{0.8}Pr^{0.4} \qquad Re > 10^4 \qquad (3)$$

For the channel of 1.09mm, the difference is much less important.

Zhang and Coll. [4] have observed no differences with classical correlations in the case of channel size larger than 0.8mm, but for smaller diameters, they suggest a new correlation

$$Nu = 0.021\,(Re\,Pr)^{0.7} \quad Re > 600 \tag{4}$$

Peng and Coll. [5, 6, 7] have used water and methanol to study rectangular channels of hydraulic diameter D_h between 0.133 and 0.367mm. By changing channel height H and wall thickness between two channels W_c they show that transfer is improved for a large D_h/W_c ratio and a small H/W ratio (W is the channel width). They propose the following correlation

$$Nu = 0.1165\left(\frac{D_h}{W_c}\right)^{0.81}\left(\frac{H}{W}\right)^{-0.79} Re^{0.62} Pr^{1/3} \quad \text{laminar} \tag{5}$$

$$Nu = 0.072\left(\frac{D_h}{W_c}\right)^{1.15}\left[1 - 2.421(Z-0.5)^2\right] Re^{0.8} Pr^{1/3} \quad \text{turbulent} \tag{6}$$

with Z=min$(H,W)/\max(H,W)$. They observed that the heat transfer is optimal for Z=0.5 and a large ratio D_h/W_c.

These results show that the increase of performance found in micro heat exchangers can already be observed for millimeter size channel.

To be complete, we have to cite the work of Bier and Keller [8] who have studied a 1cm^3 cross-flow heat exchanger made with 4000 channels of $100\mu m \times 78\mu m$ rectangular section. They compared the global transfer coefficient H and find a good agreement with classical correlations (eq. 2-3 in the laminar case, but an improvement for larger flow rates. They think this might be related to local turbulence in the hot fluid.

2 An experimental study

Two compact cross flow heat exchangers have been fabricated using classical mechanical means. The first is made of 33 brass plates where 16 channels 2mm wide and 1mm deep were grooved (see fig. 1a), the solid thickness between two channels being 0.5mm. The plates were assembled with a small amount of tin and welded using an oven at 350°C. Two plates of 5mm thickness were added at the top and the bottom to fix the liquid collectors. Thus the external size is 50×50×60 mm, for a compactness of 550 m^2/m^3. The second exchanger is made of 23 plates of graphitized carbon with 20 channels 1.5mm wide and 1mm deep. The external size is the same as the brass one, the compactness is 535 m^2/m^3. To give a typical example, for a cold water flowrate Q_c=404.3g/s with t_{ci}=20.8°C and hot water flowrate Q_h=465g/s with t_{hi}=83.7°C, the temperatures variation is Δt_c=8.0°C for the cold liquid and Δt_h=-7.9°C for the hot one. The power is caculated using

$$P = \frac{P_c + P_h}{2} = \frac{Q_c C_{pc}(T_{co} - T_{ci}) + Q_h C_{ph}(T_{hi} - T_{ho})}{2} \tag{7}$$

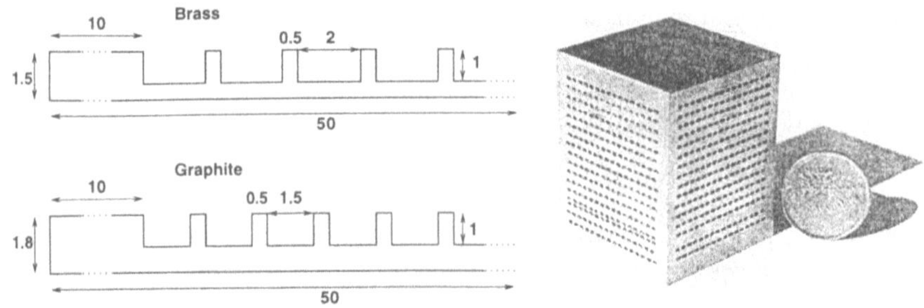

Figure 1: The compact heat exchangers: a) geometry of one plate of brass and one of graphite, b) the copper compact heat exchanger and a 5 francs French coin

and we find in this case P=14.2 kW.

To compare compact heat exchangers, the global transfer coefficient H is used. It is obtain using DTLM (logarithmic mean temperature difference) method:

$$H = \frac{P}{S\,F\,\Delta T_{lm}} \qquad \text{with} \qquad \Delta T_{lm} = \frac{(T_{hi} - T_{co}) - (T_{ho} - T_{ci})}{\ln\left(\frac{T_{hi}-T_{co}}{T_{ho}-T_{ci}}\right)} \tag{8}$$

For cross flow exchangers with the conditions studied, the correction factor F is very close to one

Figures 2 show the global transfer coefficient H for the brass (fig. 2a) and graphite (fig. 2b) heat exchangers versus Reynolds number for the cold fluid. The series of dots are experimental values obtained for 2 different Reynolds

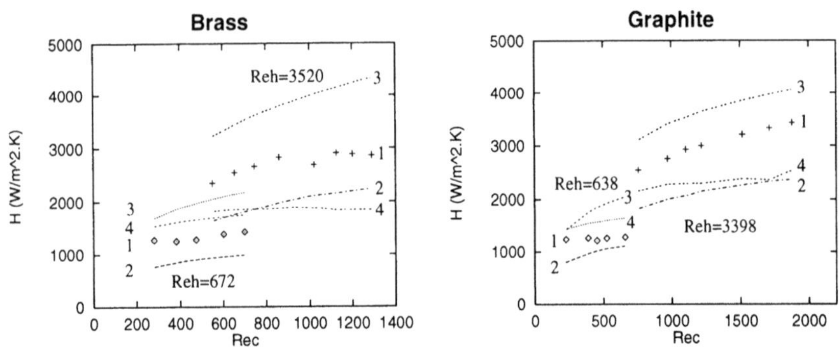

Figure 2: Global transfer coefficient H versus Reynolds number on the cold fluid; labels 1 to 4 correspond respectively to experimental measurements, Zhang and Coll. correlation, Peng and Peterson correlation, and Beek and Muttzal classical correlation

number of the hot fluid. These values are compared to classical correlation [3] (eq. 2-3), and the correlations of Zhang and Coll. [4] (eq. 4) and Peng and Peterson [5] (eq. 5-6) obtained for micro heat exchangers.

The results presented in these graphs show some discrepancy with correlations presented in the literature. The correlation of Zhang should be the more adequate since it was established for channels close to one millimeter size. Nevertheless, the predictions are 30% too low. In the case of Peng and Peterson, the correlation values are clearly higher than the experimental ones. This was expected since these correlations are for hydraulic diameter less than 0.3mm. For classical correlations, we must distinguish two regimes: For $Re < 800$, the experimental values are 20% lower and $Re > 800$ gives 30% higher performances.

These results show that even for millimeter size channels, improvement can be observed compared to classical heat transfer correlations.

3 A numerical study

In this part, we present the results of a direct simulation of the compact heat exchangers. The aim is to have a better understanding of basic phenomena, and determine where the limitations to heat exchange are. By direct simulation we mean here, 1) solving Navier-Stokes equation to obtain the stationary velocity profile in one channel, 2) solving the advection-diffusion equation of heat in the liquid and 3) the diffusion equation of heat in the solid. To present the model and the hypotheses we use, we shall first study the case of a single channel.

3.1 Single Channel Case

In our heat exchangers, the liquid flows through rectangular channels. In the simulation, we suppose there is no inlet effect, i.e. infinite channel length. This is not unreasonable since the length over thickness ratio is about fifty. With this hypothesis, we no longer have a 3D flow, but simply a flow in the Z direction of a X-Y plane. Thus, temperature is the only 3D variable to be stored in the computer memory. The physical properties of the liquid do not change with temperature. This may seem oversimplified, but we shall see that the simulation brings enough information to show how to improve heat exchange. Table 3.1 gives the characteristics of the liquid and solids (at 20°C) we have used in the simulations:

Table 3.1 Physical properties of the materials [9, 10]

	specific mass ρ (Kg m^{-3})	specific heat c_p (J Kg^{-1}K^{-1})	thermal conductivity λ (W m^{-1}K^{-1})	viscosity μ (W s m^{-2})
Water	998.2	4181	0.598	1.005
Copper	8960	385	401	-
Brass	8530	377	121	-
Zinc	7140	388	116	-
Bakelite	1300	1464	1.4	-

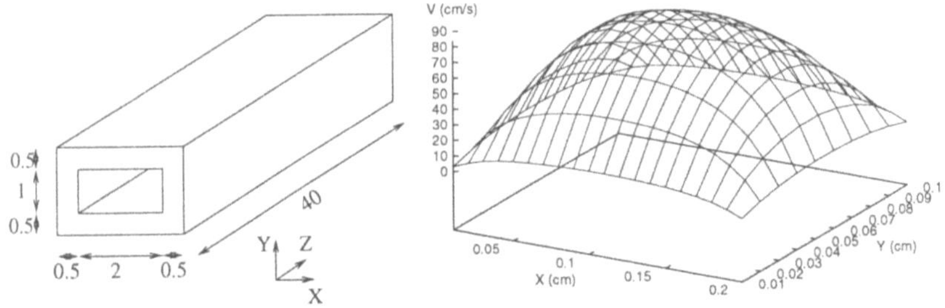

Figure 3: Single channel, a: geometry of the channel (in mm), b: flow profile inside the channel

To obtain the cross section flow profile (see figure 3b), we have used a rather unusual numerical technique: Lattice Boltzmann Automaton. The reason is that LBA allows two-phase-flow [11] and very complex geometries [12] pertinent for future studies. But identical results would be expected with other techniques or commercial software. Once the flow profile is obtained, the thermal advection-diffusion equation

$$\frac{\partial T}{\partial t} + \vec{v} \cdot \nabla T = D_l \Delta T \tag{9}$$

is solved within the fluid. For the solid, only the diffusive part is used

$$\frac{\partial T}{\partial t} = D_s \Delta T \tag{10}$$

The numerical technique adopted is finite differences (see for example [13]). We have tested upwind and central scheme for the advective part. Both give the same results, thus we have kept the upwind scheme which allows larger time steps. The spatial discretisation is done with a rectangular lattice. In the

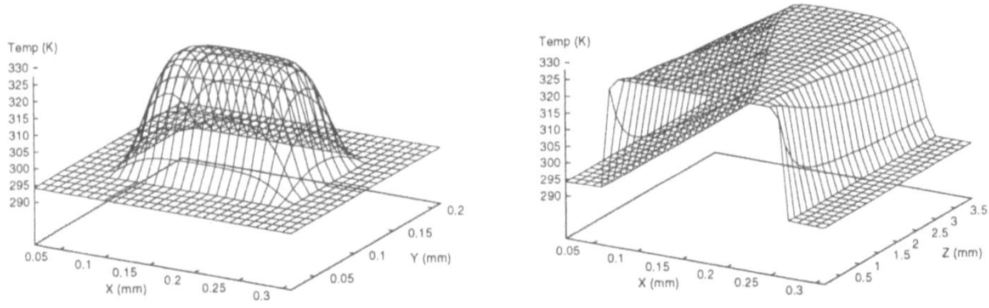

Figure 4: Temperature profile in liquid and solid, a: at the end of the channel (X-Y plane), b: in a longitudinal median (X-Z) plane

case of a single channel, the lattice is $30 \times 20 \times 100$. One node has been added in each direction to apply boundary conditions.

Figure 4 shows the stationary temperature profile for a flowrate Q_c=0.33 g/s, an inlet temperature T_i=289.6K and a fixed solid temperature T_s=327.7K. This flowrate in one channel corresponds to a flow of Q_c=85.44g/s in the heat exchanger studied previously. In this simulation, the liquid is water and the solid is copper. The outlet mean temperature is T_o=303.9K. The ΔT=14.3K is about twice the one obtained experimentally on a real heat exchanger. This comes from the fact that in this simulation, heat is transferred directly from the wall to the liquid, while in the exchanger, it is transferred from one liquid to the other, thus meets twice a fluid-side resistance. In both figures, the bottom flat parts show the temperature within the solid. As we can see, the thermal conductivity of copper is so high that there is nearly no temperature gradient in the solid. Figure 4a shows the temperature profile at the last node of the lattice in the Z direction i.e. where the liquid leaves the channel. At the center of the channel, the temperature is maximum, as expected, and only 0.3K lower than the inlet temperature. This is easy to understand because of the conductive transfer mechanism in a poorly conductive fluid and in laminar flow. This can also be seen in Figure 4b where the temperature in a plane along the channel only decreases near the wall.

These two figures point out the bottleneck of heat exchange: the conductivity of the liquid (if the flow is laminar which is usually the case in compact and micro heat exchangers). To overcome this limitation, we need some mixing mechanism or chaotic advection. In the simulations, we have done this by adding a fictitious perfect mixing of infinitesimal thickness. After crossing such a "mixing" plane, the X-Y temperature profile is averaged. Figure 5 shows the

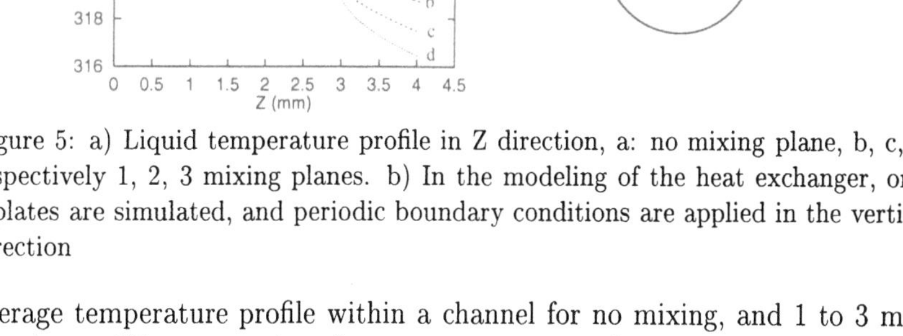

Figure 5: a) Liquid temperature profile in Z direction, a: no mixing plane, b, c, d: respectively 1, 2, 3 mixing planes. b) In the modeling of the heat exchanger, only 2 plates are simulated, and periodic boundary conditions are applied in the vertical direction

average temperature profile within a channel for no mixing, and 1 to 3 mixing planes. We can see that the temperature strongly decreases after the inlet as well as after a mixing plane, in fact, wherever the transverse temperature gradient is large (see fig. 4).

In the present case, increasing the channel length by a factor of two gives the same temperature decrease as 3 mixing planes.

3.2 Complete Heat Exchanger Case

To simulate a complete heat exchanger, we have modeled only two plates, a cold and a hot one, and applied periodic boundary conditions (5b). A plate is composed of 16 channels (see part 3.1), and the boundary conditions of each channel are adjusted to realize the contact between channels. Reproducing the experimental conditions found in [14], the simulation gives temperature differences of ΔT_c=7.88°C and ΔT_h=-3.01°C for a power of P=2.82kW while experimentaly, a power of P=3.33kW was observed. These values are extremely satisfactory if we remember the simplifying hypothesis we have adopted in the simulations.

If mixing planes are introduced in each channel, we obtain a power of P_{mix}=4.16kW, thus an improvement of heat transfer of about 50% . This confirms that breaking the temperature profile in the liquid is a very simple and efficient way to improve heat exchangers performances.

To check the influence of exchanger material, we have done simulations with copper, brass, zinc and a very poor heat conductor: bakelite. For the 3 metals, the power obtained is the same (within 0.3%) while for bakelite, P=1.75kW. This shows that a conductivity ratio of 4 between brass and copper (see table 3.1) has no influence since the conductivities of metals is large compare to water. But in the case of bakelite, the conductivity are comparable, and power is significantly reduced. The brass and graphite exchangers studied experimentally have shown very fast dynamic behavior, such that it is difficult to quantify with our instruments. A simulation was performed for a step change in the liquid inlet temperature. First, the system evolves until it reaches a stationary state. The physical conditions are: flow rate Q_h=Q_c=231g/s, inlet temperature T_{ci}=289.8K and T_{hi}=327.9K. The flow rates are chosen equal to avoid the interference of several relaxation times. After 3 seconds (in physical time), inlet temperatures are switched to T_{ci}=269.8K and T_{hi}=347.9K. Figure 6a shows the time evolution of both outlet temperatures T_o. There is a small delay between the inlet temperature change and the outlet one (zone a-b). This simply corresponds to the residence time of the liquid ($t \sim 0.4s$) . Then there is a strong temperature relaxation (zone b-c) where outlet temperatures almost reach their final values. After this, there is a second relaxation period (zone d) which can be seen in figure 6b where temperature minus final temperature is plotted on a semi-logarithmic scale. We have no clear explanation for this second relaxation time. Nevertheless, the temperature variation involved in this second relaxation is so small that it will be difficult to observe experimentally. If we consider that at point c the system has reached its new stationary state, the relaxation phenomenon has taken around 0.1s.

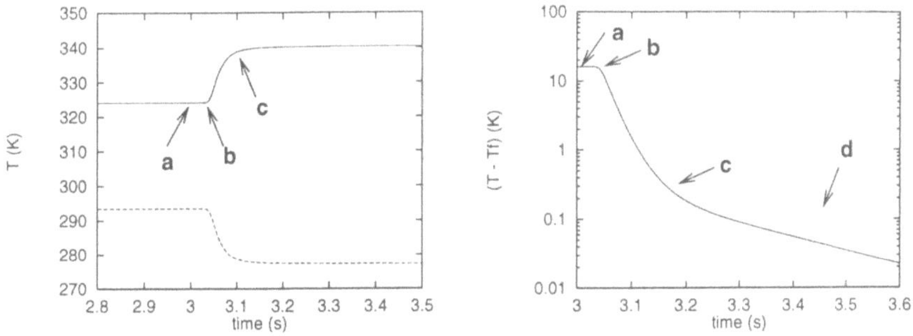

Figure 6: Dynamic of heat exchanger: a) outlet temperature versus time, b) difference between hot outlet temperature and its final value $T_{h\,o} - T_{h\,f}$ versus time

4 A new compact heat exchanger

In order to implement practically the idea of cross mixing, a new cubic heat exchanger has been designed (fig. 7a). It is composed of two networks, hot and cold, of 3mm diameter channels grooved in a $4\sqrt{2} \times 4\sqrt{2} \times 5$ cm duralumin block. In each network, the channels are organized in such a way that they divide and remix the liquid, insuring 8 mixing points through the trajectory of the liquid. Of course, these mixing points can not be as efficient as the mixing plane in the simulations.

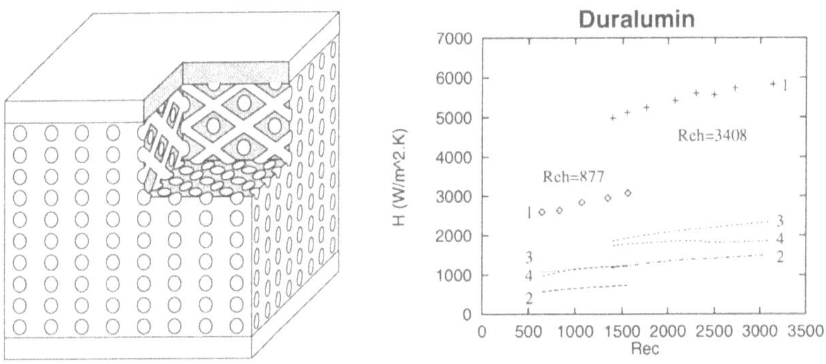

Figure 7: A new cross-flow compact heat exchanger: a) geometry of the system, b) global transfer coefficient H; label 1 to 4 correspond respectively to experimental measurements, Zhang and Coll. correlation, Peng and Peterson correlation, and Beek and Muttzal classical correlation

Figure 7 shows the global transfer coefficient versus Reynolds number on the cold fluid, obtained for 2 different Reynolds number in the hot fluid. The experimental values are compared with classical correlations, and Zhang and Coll. and Peng and Peterson correlations, both obtained for micro-exchangers. This new exchanger presents higher performances compared to these correlations, and compared to the two previous exchangers (see fig. 2).

5 Conclusions

Two cross flow compact heat exchangers made of grooved plates were studied. They present overall performances one decade larger than usual industrial exchangers, but one decade lower than micro-exchangers. Compared to classical commercial heat exchangers, they achieved high compactness, low thermal inertia and close temperature pinch. Numerical simulations have shown that limitation to heat transfer comes from the poor conductivity of the fluid. A way of solving this problem is to introduce mixing or chaotic advection. A new compact heat exchanger applying this idea of mixing was designed, and its performances are very promising. This mixing enhancement could be applied to any system where the flow is mainly laminar, from macro to micro-heat-exchangers. Being fabricated by classical mechanical means, these exchangers present a low cost of production, and very interesting features. There may be some practical and economic "niches" for these "intermediate scale" devices.

Symbols

Roman letters

c_p	specific heat
D	thermal diffusivity
D_h	hydraulic diameter
H	channel height
H	global transfer coefficient
L	channel length
Nu	Nusselt number
Re	Reynolds number
P	power
Pe	Peclet number
Pr	Prandtl number
Q	massic flow rate
t	time
T	temperature
W	channel width
W_c	wall thickness

Greek letters

λ	conductivity
μ	viscosity

index letters

c	cold
f	final
h	hot
i	inlet
l	liquid
o	outlet
s	solid

References

[1] C.-M. Ho, Y.-C. Tai, Micro-electro-mechanical-systems (MEMS) and Fluid Flows, Annu. Rev. Fluid Mech., **30** (1998) 579

[2] T.M. Adams and Coll., An experimental Investigation on Single-Phase Forced Convection in Microchannels, AIChe Symp. Ser., **314** (1997) 87

[3] Beek and Muttzal, Transport Phenomena, Ed. Wiley (1980) Chap. 3

[4] N. Zhang and M.D. Xin, Liquid flow and convective heat transfer in microtube, Proc. China National Heat Transfer, Beijing, China (1991) 11189

[5] Peng and Peterson, The effect of thermofluid and geometrical parameters on convection of liquids through rectangular microchannels, Int. J. of Heat and Mass Transfer **38** (1995) 755

[6] Peng and Wang, Experimental investigation of heat transfer in flat plates with rectangular microchannels, Int. J. of Heat and Mass Transfer, **38** (1995) 127

[7] Wang and Peng, Experimental investigation on liquid forced-convection heat transfer through microchannels, Int. J. of Heat and Mass Transfer, **37** (1994) 73

[8] Bier and Keller, Manufacturing and testing of compact micro heat-exchanger with high volumetric heat transfer coefficients, Microstructure, Sensor and Actuators, DSC-Vol 19 (1990) 189

[9] D.R. Lide, Handbook of Chemistry and Physics, 74th edition, 1993-1994

[10] R.H. Perry and D.W. Green, Perry's Chemical Engineers Handbook, 6th edition, 1984

[11] U. D'Ortona, D. Salin, M. Cieplak, R. Rybka and J. R. Banavar, Two-Color Nonlinear Boltzmann Cellular Automata: Surface Tension and Wetting, Phys. Rev. E, **51** (1995) 3718

[12] S. Chen, K. Diemer, G. Doolen, K. Eggert, C. Fu, S. Gutman and B. J. Travis, Lattice Gas Automata For Flow Through Porous Media, Physica D **47** (1991) 72

[13] B.A. Finlayson, Numerical Methods for Problems with Moving Fronts, Seattle, Rovenna Park, 1992

[14] L.A. Luo and D. Tondeur, Performance d'un échangeur thermique compact à courant croisés, Acte Société Française de Thermique, Elsevier editeur, 1997

Optimizing the Geometry of a Catalytic Enzyme Microreactor in Porous Silicon

Martin Bengtsson, Johan Drott and Thomas Laurell
Department of Electrical Measurements,
Lund Institute of Technology, Lund University Sweden
P.O. Box 118, S-221 00 Sweden, email: thomas.laurell@elmat.lth.se

Abstract

The optimization of the catalytic performance of porous silicon enzyme microreactors with respect to the reactor geometry is reported. The microreactors were fabricated in <110> p-type silicon with anisotropic KOH-etching. To increase the effective area the silicon was made porous by anodisation in a HF-ethanol solution. The performance of the different reactors was evaluated by coupling glucose oxidase onto the porous matrix and measuring the catalytic turn-over rate.

Keywords

porous silicon; biosensor; microreactor; immobilized enzyme; glucose

1 Introduction

Micromachining is today used to develop miniaturized chemical analysis systems where decreased analyte consumption, analysis time and price are the driving factors, [1,2]. In this development enzyme microreactors have proven to be stable components for selective detection or conversion of biochemicals [3]. The use of porous silicon (PS) as an enzyme coupling matrix in microreactors has been shown to provide a tremendous increase in the effective area thus increasing the catalytic turn-over rate with up to a factor of 350 [4].

Previous research has shown that the most favourable porous depth for microreactors is about 5 - 10 μm [5]. However due to the electric field distribution in the bulk silicon, the pores becomes deepest at the bottom of the channels and gets more shallow closer to the top of the channelwalls, fig 1. This might decrease the catalytic performance since only a small part of the channel actually has an optimized porous depth. When anodizing long enough to achieve a deep porous layer at the top, the porous depth at the base instead becomes so large that the bulk silicon supporting the channelwalls all becomes porous and a very fragile structure is obtained, fig 2.

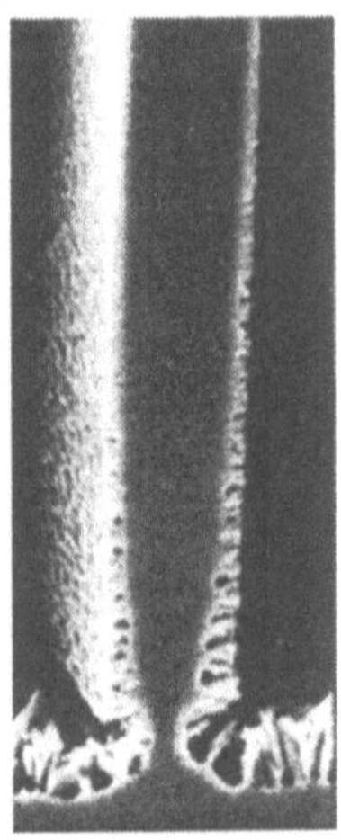

Fig 1. Cross section of porous channel wall

In this paper, reactors with different spacing between the channels and with different wall thickness are compared in order to learn the criterias for obtaining more homogeneous PS-layers at the channel wall. The porosity is dependent of the hole migration in

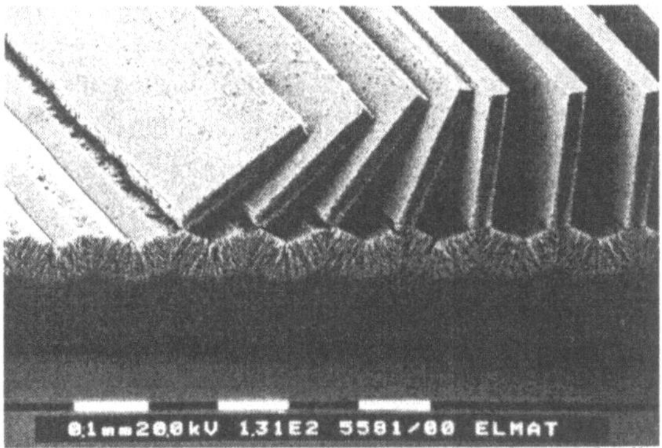

Fig 2. A too deep porous layer make the reactor walls very fragile.

the silicon and thus the electric field. Increased wall thickness between the channels changes the electric field, allowing more charge carriers to migrate to the top of the channelwalls.

In order to decrease analyte consumption and dispersion, the dead volume of the reactors must be minimized. This calls for fewer and more narrow channels in the reactor, but smaller cross-section flow area has its drawbacks in increasing the hydrodynamic pressure drop. With narrow channels there is also a risk that particles in the analyte get stuck and block a part of, or the whole channel. With fewer open channels the performance of the reactor totally changes.

This paper compares different designs of microreactors with different channel width, channel length and dead volume, all with the same internal area, were fabricated and tested, in order to optimize catalytic turnover and minimize dispersion.

2 Materials and methods

2.1 Anisotropic etching

The microreactors were fabricated from <110> p-type silicon, 20 - 70 Ωcm. The reactor structures were patterned with standard lithography methods and the SiO_2 was etched with buffered HF (28ml 40% HF, 170ml H_2O, 113g FNH_4 etched anisotropicly in 30% KOH at 80 °C for 100 min, with a resulting channel depth of 225μm, fig 3.

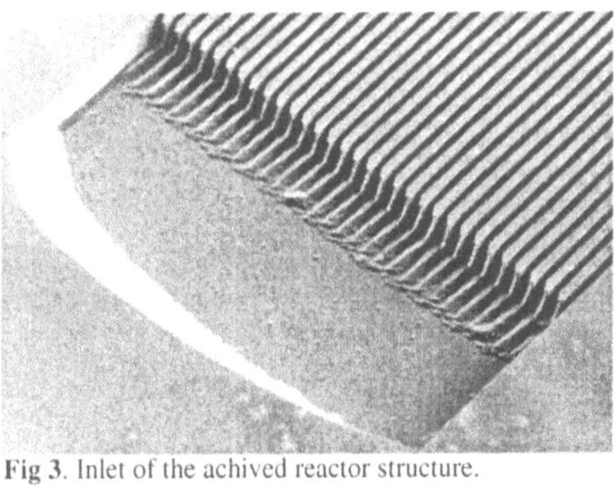
Fig 3. Inlet of the achived reactor structure.

2.2 Porous silicon fabrication

To obtain a porous layer on the channel walls, the silicon was anodized in an ethanol (96%) - HF (48%) solution mixed (1:1), during illumination, fig 4. The current density was held constant at 50mA/cm^2 for 5 min. To ensure sufficient contact with the electrolyte, the wafer backside was previously doped with boron. After anodization the reactors were rinsed in H_2O to clean the pores from HF.

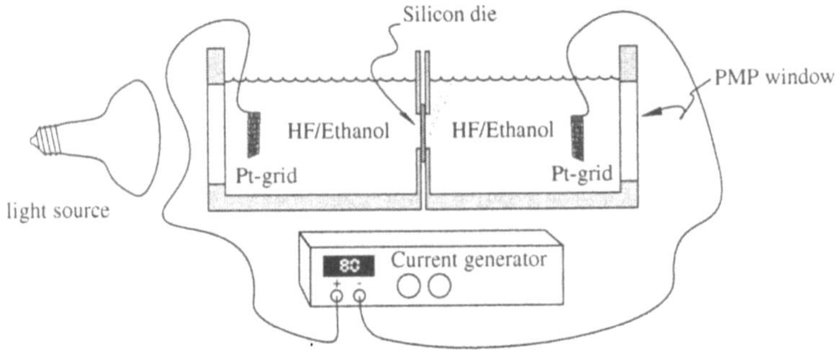

Fig 4 The setup for the anodization process

2.3 Enzyme coupling

Glucose oxidase was immobilized onto the silicon using a standard silane based coupling method [6]. The silicon was first immersed in 75°C 10% 3-aminopropyltrietoxysilane for 2 h, and after that in 20°C 2.5% glutaraldehyde for 2 h. In the last step glucose oxidase was coupled onto the matrix in room temperature for 12 h. Afterwards the reactors were stored in phosphate buffer solution, PBS (pH 7), at 8°C.

2.4 Enzyme activity monitoring set-up

The catalytic performances of the different reactors were tested using a colourometric reagent, Trinder reagent. The Trinder reagent yields a colour shift proportional to the H_2O_2 concentration change due to the enzyme catalyzed glucose oxidation. The colourometric shift was monitored by a flow through spectrometer as depicted in fig 5. and the corresponding enzyme activity was calculated from the recorded absorbance shift.

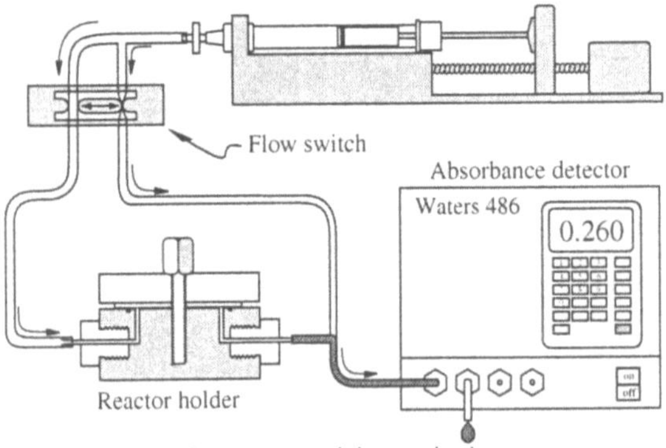

Fig 5. The set-up for enzyme activity monitoring

2.5 Wall/channel width variation

For the wall thickness study, reactors were fabricated with channelwidth/wall thickness proportions varying from 75μm/25μm to 25μm/75μm. To measure the porous depth on the walls, the different reactors were cut in half and photographed in a SEM. The average porous depth was registered at the top of the channels, 1/3 down, 2/3 down, at the base and in the bottom of the channels.

2.6 Width/length variation of the reactor

For the reactor geometry study, reactors were made with between 18 channels (0.96μl dead volume) and 45 channels (1.6μl), all with an internal area of 56.4 mm^2, prior to the PS-anodisation process. All these reactors had a channel width of 20μm and a wall thickness of 50 μm.

3 Results & Discussion

3.1 Wall/channel width variation

The measured average porous depth is shown in table 1.

Wall Thickness/ μm	Channel Width/ μm	Porous Depth/ μm				
		Top	1/3	2/3	Base	Bottom
25	75	2.0	2.3	2.5	5.5	22.5
25	25	2.5	2.4	2.7	5.5	9.5
50	50	3.0	3.4	4.9	8.9	14.3
50	25	3.9	4.1	4.6	6.5	11.1
75	25	6.3	4.7	6.3	9.7	14.2

Table 1. The resulting porous depths

The porous depth in the channel walls seems to be dependent only of the wall thickness, walls with the same thickness has the same porous gradient. Comparing the porous profiles of the different reactors, fig 6., it is obvious that the profile becomes more homogeneous the thicker the walls get.
The bulk silicon of the thinner wall has almost been cut off at the base and in the case of the 75μm wide channel (b), the channel has become significantly deeper due to the large loss of silicon at the bottom. The measured porous depth at the base is low, caused by the high current density in the corner wall/bottom, which has undermined the wall and made it thinner. The thickest wall has a porous depth that only decreases by 35% over the wall height and actually lies within the desired range for optimum performance, 5-10μm.
The actual influence of the porous profile with respect to the catalytic performance of the reactor is difficult to say without further studies. It is easy to see that the porous layer becomes more homogeneous with thicker walls (d), but for these the dead volume of the inlet/outlet basins may lead to increased dispersion in the system.

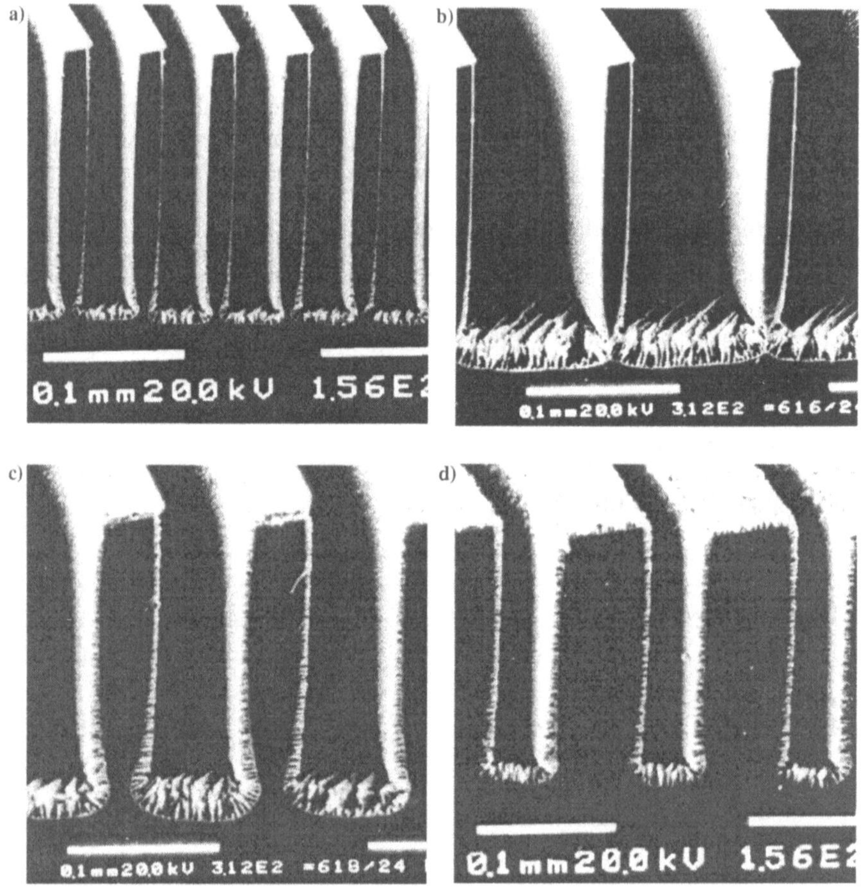

Fig 6. SEM of cross section from reactors with wall thickness / channel width (a) 25μm/25μm, (b) 25μm/75μm, (c) 50μm/50μm, (d) 75μm/25μm

3.2 Width/length variation of the reactor

The turnover dependency of glucose concentration for the different reactors defined in table 2, is shown in fig 7.

From these data the apparent maximum turn-over rate ($V_{max\ app}$) and the apparent Michaelis constant ($K_{m\ app}$) was extracted by fitting the theoretical equation of a first order enzyme kinetic system, to the measured data, table 2.

No. of channels	Reactor length / mm	Reactor width / mm	Reactor volume / μL	Vmax / nmol/min	Km / mM
18	8.2	1.2	0.95	17.5	5.7
30	5.5	2.0	1.2	8.4	7.5
45	4.2	3.1	1.6	3.3	11.6

Table 2. Maximum turnover and Michaelis constant of the different reactors

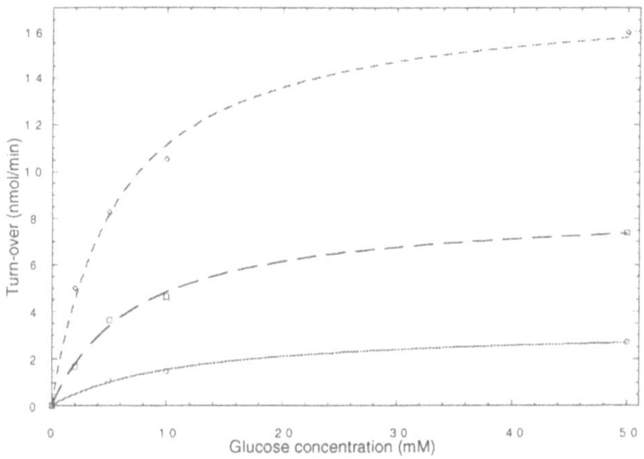

Fig 7 The turn-over rate for different glucose concentrations. Reactor with 18 (upper), 30 (middle) and 45 (lower) channels.

Fig 7. clearly shows the influence of the reactor geometry, wide and short compared to narrow and long, having the same channel density and depth. The narrow reactor with only 18 channels displays a more than 5 times increase in $V_{max\ app}$ compared to the wide reactor with 45 channels, both having the same internal surface area. This increase is observed as the cross-section surface area is reduced by 60%, indicating that a majority of the flow passes through the centre channels and the flow in the side channels presumably is very low. The proposed non-uniform flow distribution in the wider reactors will be verified by imaging the channel array as fluorescent or luminescent samples are injected into the reactor.

4 Conclusions

Enzyme microrectors with different geometries has been fabricated and studied.
It was found that channels with partition walls of 75μm thickness are sufficient to overcome minority carrier depletion at the higher channel wall regions while performing the porous silicon etching. Thereby, the desired porous depth could be obtained over the full wall height.
A further improvement, ~5 times, of the reactor performance was obtained by reducing the number of channels and increasing their length. The porous non-uniform flow distribution across the larger channel arrays not only lowers the reactor efficiency but also an increased fluid dispersion could be expected for these components. Further studies will now focus on the flow profile and the dispersion of the developed microreactors.

5 Acknowledgements

This work was most kindly supported by The Swedish research Council for Engineering Sciences, Carl Tryggers Stiftelse and Crafoord Stiftelsen.

6 References

[1] Fintchenko Y, van den Berg A, Silicon microtechnology and microstructures in separation science, J of Chromatography A, 819 (1998), 3-12

[2] van den Berg and Harrison J (eds.), Proceedings of the µTAS'98 Workshop, Banff, Canada, Oct. 13-16, 1998.

[3] Drott J, Lindström K , Rosengren L., Laurell T, Porous silicon as the carrier matrix in microstructured enzyme reactors yielding high enzyme activities, J of Micromech. Microeng. 7 (1997), 14-23

[4] Drott J, Rosengren L, Lindström K, Laurell T, Pore morphology influence on catalytic turn-over for enzyme activated porous silicon matrices. Thin Solid Films, 330 (1998), 161-166

[5] Drott J, Rosengren L, Lindström K, Laurell T, Porous silicon carrier matrices in micro enzyme reactors - influence of matrix depth, Microchimica Acta, (1999), In press

[6] Weetall H, Covalent Coupling Methods for Inorganic Support Materials, Methods in enzymology, 44, Mosbach K (ed.), Academic Press N.Y. 134-148

High aspect ratio silicon micromachined heat exchanger

Jonas Bengtsson, Lars Wallman and Thomas Laurell

Department of Electrical Measurements
Lund Institute of Technology
P.O. Box 118, S-221 00 Lund
Sweden
email:thomas.laurell@elmat.lth.se

Abstract

The fabrication process for a silicon micro machined heat exchanger is descibed. A thin walled, 7 µm, channel structure defined by vertical channels, 360 µm deep and 220 µm wide, anisotropically etched in <110>-silicon followed by a pn-etch stop process provided a heat exchanger module in the shape of a "square wave" corrugated membrane. The double sided processing provided vertical channels interdigitated from one side to the other giving a counter flow component with rear side fluid access from the front side. The design allows the stacking of multiple elements to obtain a unit with the desired heat sink capacity.

Keywords: micro heat exchanger, PN-etch stop, anisotropic, micro machined, silicon

1. Introduction

Microscaled heat exchangers require thin partition walls with a large contact area and a high thermal conductivity to offer high heat transfer capabilities. Earlier work on micro heat exchangers describe parallel channel, sub 100 micron, structures in metal foils (e.g. stainless steel, copper, aluminum and titanium) either fabricated by diamond sawing [1-3] or LIGA-EDM [4]. A key parameter in designing an efficient heat exchanger is the thickness of the partition wall. Also the surface roughness on the channel wall is an important parameter for an optimal heat transfer. It would thus be desirable to obtain channel structures that offer a finer surface roughness and thinner partition walls, issues earlier addressed in [5]

This paper presents the design and the fabrication of a silicon micromachined counter-flow heat exchanger, realizing high aspect ratio micro channels having a partition wall thickness of 7 µm. The partition walls are fabricated by means of anisotropic etching and PN-etch stop. The heat exchanger is constructed as a modular sandwich device enabling the stacking of an arbitrary number of elements.

2. Device fabrication

The heat exchanger was fabricated in 3", p-type, <110>-oriented silicon using anisotropic KOH-etching to realise vertical flow channel arrays. The obtained structure was subsequently phosphorus doped, generating an n-type layer in the channels. To achieve the desired interdigitized counter flow channel design the wafer was PN-etched from the rear side allowing the removal of bulk material down to the n-doped layer. The partition walls were thus defined by the phosphor dopant depth, leaving a 7 μm thick corrugated membrane structure Fig. 1.

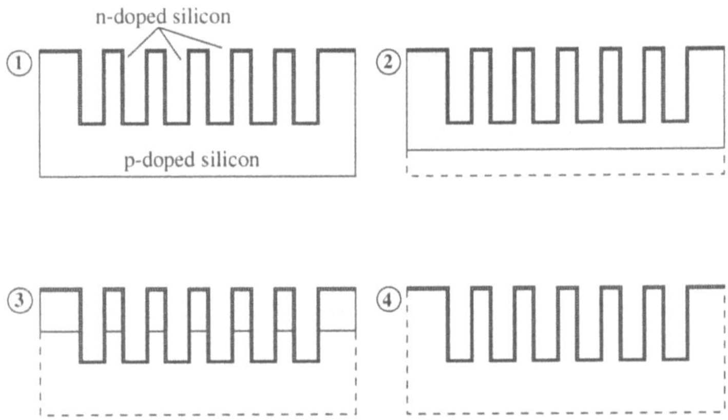

Fig. 1. The proccess steps for generating the micro heat exchanger. 1) The anisotropically etched channels are n-doped. 2) KOH etching of the bulk silicon. 3) The etching is stopped at the doped regions due to the PN-etch stop. 4) The etching stops at the n-layer leaving the desired thin corrugated membrane structure.

2.1 PN-etch stop process

The silicon wafer potential is controlled via a four terminal measurement set-up comprising two dry, o-ring protected, electrode contacts on the n-doped side. One electrode passes the current and the other measures the n-region potential. The n-layer potential was set to +0.6 Volt versus a platinum reference electrode in the KOH solution. A second platinum grid acted as the counter electrode during the etchprocess, Fig. 2. The anodic potential applied to the n-layer protects this region from the KOH. Due to the reversed poled PN-junction the p-doped bulk silicon is on its open circuit potential and thus etches normally.

To separate the cold side from the warm side a thin standing wall was defined as depicted in Fig 4. To avoid a cavity formation in the separating wall from the wafer rear side the wall thickness was set to be only 50 μm. When two n-doped regions are close together in a PN-etch configuration, the undoped region inbetween will not etch if the distance is typically below 100 μm. Thus, when performing the rear side PN-etching the etch will not form a cavity in the separating wall in spite of the fact that this bulk silicon is not n-doped as seen in Fig 5.

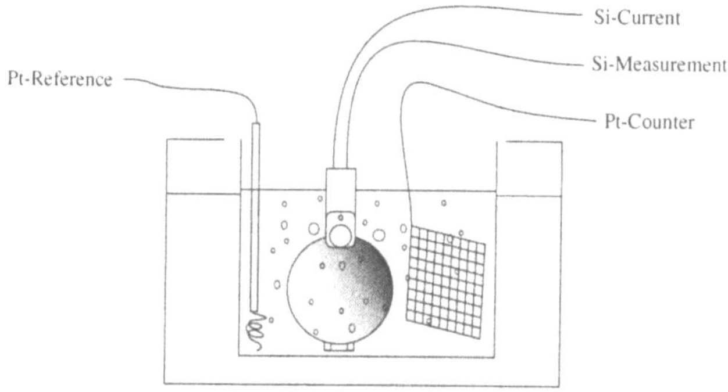

Fig 2. PN-etch stop equipment comprising a temperature controlled ultrasonic stirred etch bath, with the Teflon wafer holder and the Pt counter and reference electrodes.

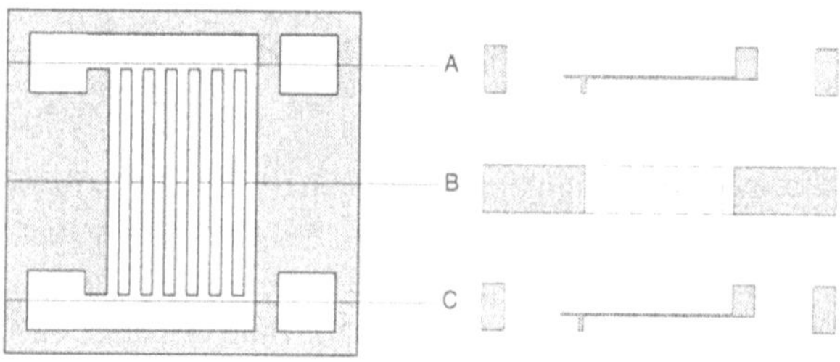

Fig 3. Schematic layout of the heat exchanger. A) Flow inlet of the top side and oulet of the bottom side. B) The folded heat exchanger structure. C) Flow outlet of the top side and flow inlet of the bottom side.

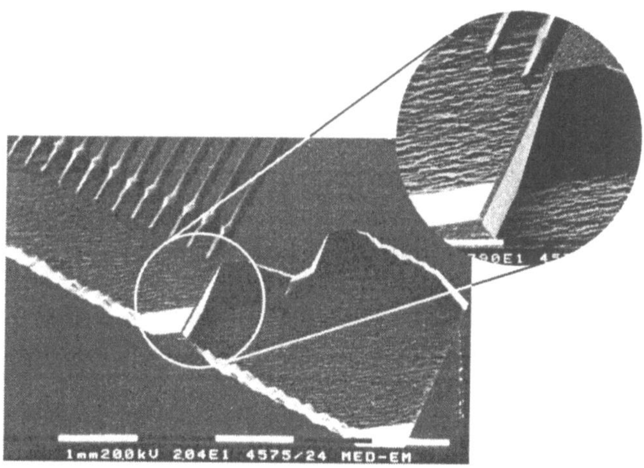

Fig 4. The thin wall separates the cold and warm inlet/outlet flows and thus allows a modular sandwich stacking of multiple heat exchanger elements.

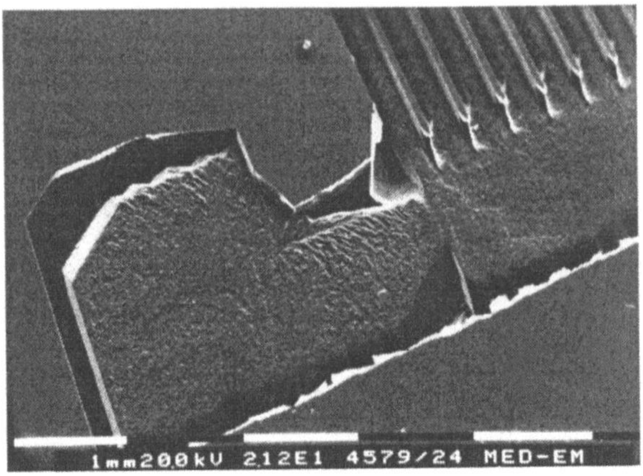

Fig 5. The rear side of Fig 4, here shows the intact bulk silicon opposing the separating wall after the PN-etch. No cavity is formed in spite of the undoped bulk silicon i the separating wall.

The finalised heat exchanger component is seen in Fig. 6. The corrugated membrane wall achived by the PN-etch stop is a very thin but yet strong membrane of 7 µm, which yields a exellent heat transfer capacity. The high aspect ratio channels further improve the lateral heat transfer by increasing the fluid contact area to the partition wall. The heat conductivity coefficient of silicon is comparable to aluminum, table.1

Material	**Heat conductivity coeff. W/mK**
Copper	390
Chromium	43
Silicon	170
Aluminum	220
Glass	0.9
Silver	420
Steel	75
Zink	110

Table 1. Heat conductivity coefficients for some materials.

The rectangular silicon studs seen on top of the corrugated membrane, Fig 6., serve as a sealing rim to the bottom of the above module when multiple elements are being stacked to a complete heat exchanging unit.

Conclusions

The described microfabricated heat exchanger holds promise of a high heat transfer thanks to the high aspect ratio channel structure and the extremely thin partition walls. Furthermore, the anisotropically etched channel walls ensure an almost perfectly planar surface which improves hydrodynamic conditions and eliminates stagnant boundary layers with high heat resistance. Further work will focus on the thermal properties of the device and investigate improved geometries.

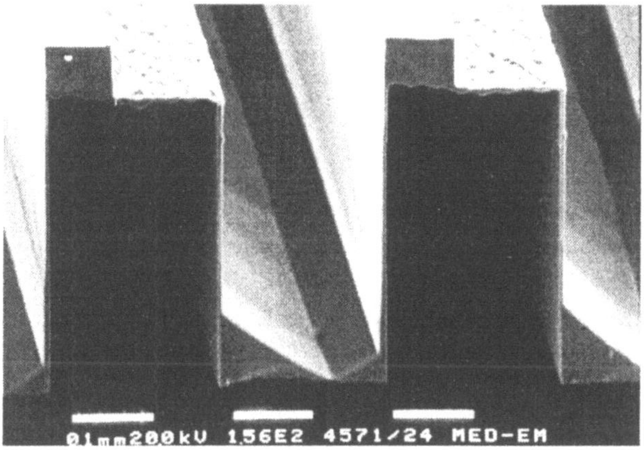

Fig 6. A cross-section of the heat exchanger which shows the counter flow channels, separated by a 7μm thick wall.

Acknowledgements

This work was funded by the Crafoord Foundation, The Carl Trygger Foundation and Kungliga Fysiografiska Sällskapet i Lund

References

[1] Hönicke D., Weissmeier G., Heterogeneously Catalyzed Reactions in a Microreactor, in DECHEMA Monographs Vol. 132-VCH Verlagsgesellschaft 1996, Ed. Ehrfeld W., 93-107

[2] Bier W., Keller W., Linder G., Seidel D., Schubert K. and Martin H., Gas to gas heat-transfer in micro heat-exchangers. Chemical Engineering and processing, 32,1, 1993, 33-43

[3] Friedrich C. D. and Kang S. D., Micro-heat exchangers fabricated by diamond machining, J. Am. Soc. Precision Engineering, 16, 1, 1994, 56-59

[4] Ehrlfeld W., Gärtner Cl. Golbig K., Hessel V., Konrad R., Löwe H., Richter Th. and Schultz Ch., Fabrication of components and systems for chemical and biological microreactors, in DECHEMA Monographs Vol. 132-VCH Verlagsgesellschaft 1996, Ed. Ehrfeld W., 73-90

[5] Bengtsson J., Development of a fabrication process for silicon micromachined heat exchangers, Master of Science Thesis, Department of Electrical Measurements, Lund Institute of Technology, Lund Sweden, No. 3, 1996

Comparison of Two Microvalve Designs Fabricated in Mild Steel

Brian K. Paul[1] and Tyson Terhaar [1]

[1] Department of Industrial and Manufacturing Engineering, Oregon State University, Corvallis, Oregon 97331-2407, USA

Abstract. Researchers at Oregon State University (OSU) are currently interested in creating a microcompressor for various miniature energy and chemical systems (MECS) applications. One approach to microcompression involves the use of miniature check valves. Requirements for a check valve design include one-way flow and major dimensions less than 3 mm in size. Two microvalves were created using a microlamination process. A prerequisite to the creation of a microflapper valve was a capability for selective bonding so as not to bond the flapper mechanism to the valve orifice. A microprojection welding process was developed to meet this requirement. A microfloat valve was also fabricated consisting of five laminates. A prerequisite for creating the microfloat valve was the capability for preassembly. A capacitive dissociation process was developed to separate the small disk from within the valve body. Pressure drop tests were performed across each valve as a function of mass flow rate. Results indicate that the microfloat valve performed best with a theoretical orifice size of 0.629 mm for a 1.5 mm valve opening and an average diodicity ratio of 11.2. A variety of reasons are presented for improper sealing within the microflapper valves including the formation of an ejecta ridge around the valve seat during the laser forming process.

Keywords: ***Micromanufacturing, microlamination, microvalves, microprojection welding, microassembly***

1. Introduction

Over the past three decades, sheet lamination techniques have been used to fabricate a variety of complex geometric structures which were difficult to manufacture with more conventional techniques. Sheet lamination techniques involve the forming, alignment, and bonding of thin laminates or sheets. Each laminate is machined with a two-dimensional pattern so that when the laminates are bonded together, they form a functional solid geometry. Laminates may be either metal or glass and can be machined, photoetched, or laser cut. The laminate stack may be diffusion bonded, fusion bonded, or even adhesive bonded. Sheet lamination techniques were the forerunners to the automated rapid prototyping technologies used today.

Sheet lamination technology was first introduced as platelet technology in the 1960's. Initially it was used to reduce the costs associated with embedding cooling channels into the sidewalls of rocket engine thrust chambers.[1] More recently sheet lamination technology has been used to fabricate fluidic control systems,[2] plasma arc electrodes with integrated cooling channels for mixed waste vitrification,[1] and corrugated feed horn arrays for spaceborne wave sensing systems.[3,4] Platelet technology is still used in the aerospace industry, for example, to fabricate the engine injector assembly within the engine thrust chamber of the space shuttle orbital maneuvering system.[5]

More recently, these sheet lamination techniques have begun to be applied to the fabrication of microfluidic devices. Goldberg[6] fabricated and tested microchannel, forced air heat sinks constructed of copper. Goldberg employed a microlamination approach involving shearing of copper shims and heli-arc welding. Microchannel dimensions varied from 125 to 625 micrometers. Krause et al.[7] fabricated a water-cooled, copper microchannel heat exchanger by registering five pre-formed copper laminates and diffusion bonding. Microchannel dimensions varied from 60 to 100 micrometers. Kleiner et al.[8] assembled both copper and aluminum air-cooled, microchannel heat sinks with mechanical fasteners. Microchannel dimensions were on the order of 500 micrometers.

Most of the microlamination work to date has involved the fabrication of microchannel devices such as heat exchangers. However, researchers at Oregon State University are currently involved in microlaminating one-way microvalves for use in microcompression applications. This paper will discuss the microlamination of two microvalves; a microflapper valve and a microfloat valve. The microflapper valve works the same way as an ordinary flapper valve. A flapper mechanism is used to seal a small orifice. When the pressure is reversed, the flapper pivots allowing fluid passage. A prerequisite to the microlamination of the microflapper valve was a capability for selective bonding so as not to bond the flapper mechanism to the valve orifice. A microprojection welding process was developed to meet this requirement. Mild steel was chosen because of its excellent welding behavior as well as its resistance to fatigue. Projections as small as 125 microns in height were used to fabricate the microflapper valve.

The microfloat valve consists of a small disk that floats up and down within a valve cylinder. When the fluid flows through in the forward direction, the disk moves against a ring orifice and does not impede flow. When the flow is reversed, the disk moves to the other end of the cylinder covering a small orifice and thereby impeding flow. A prerequisite for creating the microfloat valve was the capability for preassembly. A capacitive dissociation process was developed to separate the small disk from within the valve body. A comparison of the valves based on theoretical orifice size and diodicity is presented.

2. Device Designs

Currently, there are many microvalves that have been created. These valves range in size from as little as 60 microns, up to several inches.[9,10,11,12,13,14] However, most of the microvalves created have been fabricated in silicon. One goal of this effort was to fabricate the microvalves out of mild steel so to be compatible with the remainder of the microcompressor.

2.1. Microflapper Valve

The microflapper valve design consists of two 1" x 1" x 0.010" laminates that were bonded together. Figure 1 shows the two laminates used to fabricate the valve. Laminate A contains the valve seat. The orifice opening on this laminate is 1.5 mm in diameter. Laminate B contains the flapper mechanism. As shown in Figure 2, this laminate consists of a 3 mm hole with a 2.2 mm disk on the interior held in place with a hinge that is approximately 0.3 mm across. The flapper mechanism is surrounded by a projection that is 5 mm in diameter with a thickness of 0.127 mm, and an average height of 0.152 mm. The projection is used in the microprojection welding

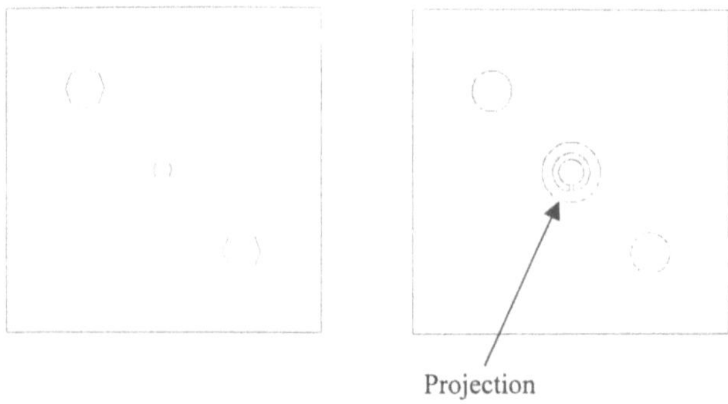

Figure 1. The microflapper valve design.

process and will be described more fully in the device fabrication section. The holes in the upper left and lower right of each of the laminates are used for alignment.

2.2. Microfloat Valve

The microfloat valve was designed to have a free floating disk that would seal against the valve seat when pressure was applied in one direction, and float up to a ring nozzle that would permit fluid flow when pressure was applied in the reverse direction. The microfloat valve design consisted of five 1" x 1" x 0.010" laminates that were bonded together. Figure 3 shows the five laminates used to fabricate the valve. Laminate A is the valve seat, which has an orifice of 1.5 mm in diameter. Laminate C contains the float mechanism. Figure 4 show that the float disk is held in place via the use of three 0.1 mm wide fixture bridges. The purpose of the bridges were to hold the valve disk in place during bonding. After bonding, the bridges were removed with a capacitive dissociation process to permit the disk to be free-floating. The capacitor discharge procedure will be discussed in further detail in the device fabrication section. The three tabs on the edge of the disk were designed to center the valve disk in the channel created by laminates B, C, and D. Laminate E is a ring nozzle that has a ring-shape orifice consisting of a 2.2 mm

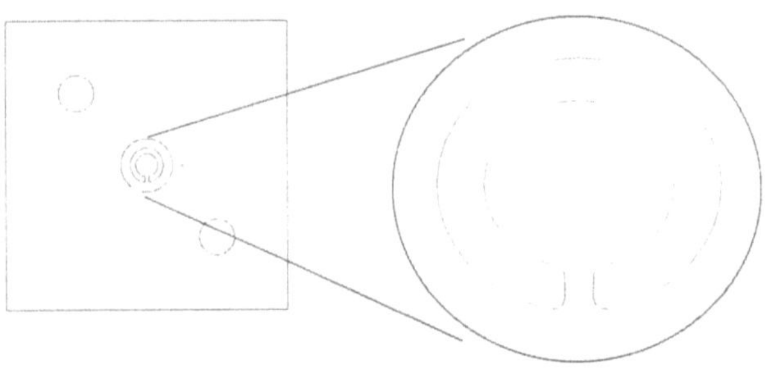

Figure 2. Exploded view of flapper mechanism.

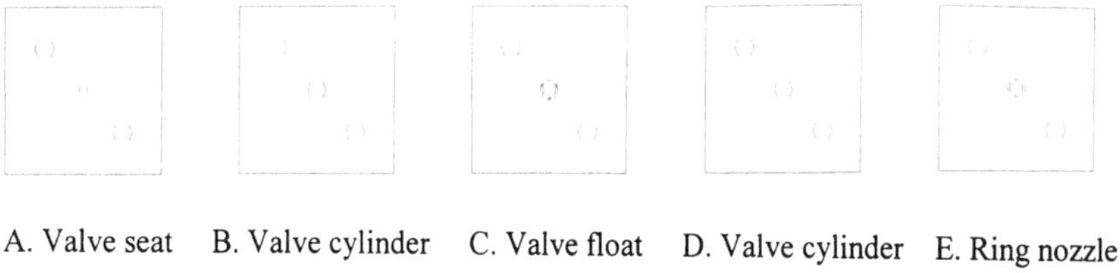

Figure 3. The microfloat valve design.

diameter disk inside a 3 mm diameter hole, and is held in place by two bridges that are 0.3 mm across (see Figure 5). Laminates B and D contain 3 mm diameter holes which together form a valve cylinder within which the float disk moves. Laminates A, B, D, and E all were all given 5 mm diameter projection rings with a thickness of 0.127 mm, and an average height of 0.152 mm. (Projection rings not shown to reduce drawing clutter.) Again, the holes in the upper left and lower right of each of the laminates are used for alignment.

One consideration given to the valve design was the orifice openings of the valve seat and the ring nozzle. These orifices were designed to be the same cross sectional area to minimize flow restrictions through the valve. Another consideration in the design was length of the float disk tabs. Since laminate registration can be problematic in microlamination, a tolerance was built into the length of the fixture bridges. Since the alignment precision was unknown, an allowance of 0.0025" was designed into the fixture bridges in order to allow for misalignment of the laminates.

3. Device Fabrication

3.1. Microflapper Valve

The microflapper valve was fabricated in two steps: laminate formation and laminate bonding. Laminate formation included formation of projections for microprojection welding and formation of through-hole and through-slot geometry. Laminate bonding was accomplished by microprojection welding.

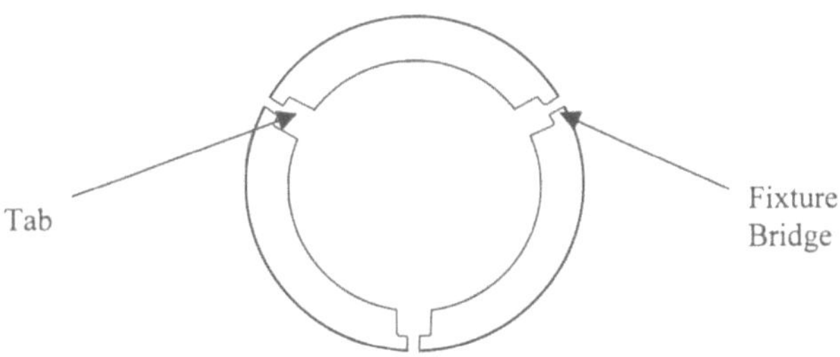

Figure 4. Exploded view of valve float mechanism.

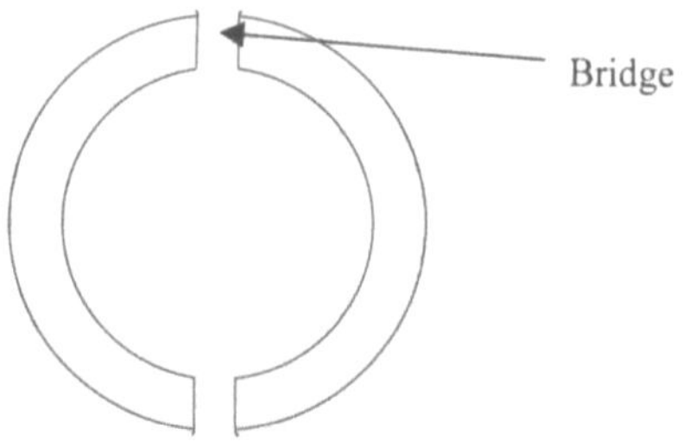

Figure 5. Exploded view of ring nozzle.

3.1.1. Laminate Formation

Acid etching was used to create the weld projections and laser micromachining was used to form the necessary two-dimensional geometry. Laminates were pre-cut out of 0.010" AISI-1010 mild steel shim stock and cleaned with Alconox and water, and the residues were removed by washing with acetone (Ashland M368), methyl alcohol (Fisher 946-4), and de-ionized water. The laminates were blown dry with compressed air (Kensington Dust Blaster), and placed in a clean dish. The laminates were then spin coated (1000 rpm for 10 seconds) with 9 drops of System 812 Photoresist and placed in a soft bake oven (Blue M, model OV-8A) at 100°C for 10 minutes to cure the positive photo resist. The laminates were then spin coated and baked on the opposite side using the same process. Photoresist exposure was performed using an ultra-violet exposure machine (Kulicke and Soffa, model 521) that had been pre-loaded with the projection mask. The contact-aligned mask was visually centered on the laminate, and the laminate was exposed for 20 seconds. The laminates were then placed in a 2.5:1 solution of Micropost M319 developer and water. The developing solution was agitated for 3 minutes, and the laminates were removed and washed with de-ionized water. Laminate etching was performed in an etching bath containing a 1:1:2.5 solution of 39% HCL solution (Mallinckrodt 2612), 95% Nitric acid solution (Fisher Scientific A200-500), and de-ionized water, heated to 65°C. The laminates were etched in the solution for 2 minutes 20 seconds. The laminates were removed and washed with acetone, methanol, and de-ionized water. The average projection height was 0.152 mm, with an average projection thickness of 0.127 mm.

Laser micromachining was performed at ElectroScientific Industries, Inc., in Portland, Oregon. The machine used was an ESI-4420, which employed a Q-switched Nd:YAG laser. The settings of the machine were experimentally determined for steels. The laser frequency used was 1000 kHz, using the third harmonic of the Nd:YAG laser (532 nm wavelength). The laser bite size was 2 μm, with an average power of 2.735 watts. The center of the projection was obtained by indexing the controller from edge to edge of the projection, and then dividing by two. This was done both horizontally and vertically in order to find the exact center of the projection. The lens of the machine was then focused, and the geometries and aligning holes were cut by the laser. The resulting laminate was trimmed to 1" X 1". As shown in Figure 6, inspection of the laminates revealed a 21-micron high ridge that formed on both sides of the laser path. This non-uniform ridge is caused by the deposition of molten material ejected during laser ablation.

Figure 6. Ejecta ridge formed as vaporized material is redeposited after laser scan.

3.1.2. Microprojection Welding

The projection welding was performed via modification of a Miller MPS-10 spot welder. The welder was modified by removing the welding tongs, and replacing them with four #40 welding leads. The leads were then fastened to the welding platens, which in turn were fastened to an arbor press. Aligning pins in the welding platens were used to align laminates to one another.

The first step in the process was to insert the laminates onto the aligning pins. The current and weld times were then set to 8,500 amps and 1.5 seconds respectively. The platens were brought together via the arbor press, and approximately 388 psi of pressure was applied. The welder was then activated. Laminates were welded to the stack, one at a time, using the previous procedure. Due to wear, the platens were re-machined every 20 welds to maintain flatness.

3.2. Microfloat Valve

The microfloat valve was fabricated in three steps: laminate formation, laminate bonding, and component dissociation. The first two steps involved the same procedures suggested above. Component dissociation was accomplished via capacitor discharge.

3.2.1. Component Dissociation

The microfloat valve required a component dissociation procedure. A large capacitor bank was used to provide a high current for the vaporization of the steel fixture bridges. After bonding, the laminated device was placed into a fixture. A small electric probe was manufactured out of a 10 gauge copper wire and inserted through the valve seat nozzle, onto the float disk. The other end of the circuit was connected to the outside of the device. A 0.07 Farad capacitor bank was charged to 11 volts. The capacitor bank was discharged across the fixture bridges which vaporized the fixture bridges. As a result, the float disk was dissociated from the rest of the valve, and was free to float.

3.3. Device Testing Procedure

Each of the valves were characterized by the volumetric flow rate as a function of pressure drop. The valves were tested using: 1) a nitrogen gas cylinder and pressure gauges; 2) a set of three rotometers; 3) a low pressure transducer (0 to 10 Torr) and associated equipment; 4) a high pressure transducer (0 to 760 Torr) and associated equipment; 5) a manometer (for calibration); 6) two stop valves; and 7) a fixture. The setup is diagramed in Figure 7. All connections were tested for hermetic seal under water or with a stream of flowing detergent and water. A manometer was used to calibrate the low and high pressure transducers. The nitrogen tank was turned on to a pressure of 15 psi. Volumetric flow rate was manipulated and pressure drop across the valve was measured for every 10 scalar increase in rotometer flow. The procedure was then repeated for reverse flow for each valve.

4. Results and discussion

Four valves were fabricated using the procedures explained above. Micrographs of the first microflapper valve and first microfloat valve are shown in Figure 8. In particular the micrograph in Figure 8b shows warpage in laminate 3. It was determined that this was caused by volumetric expansion during the microprojection welding process.[15]

In addition, all valves were tested for weld hermetic seal using the previously mentioned test procedure. All the valves were found to be airtight. Device performance was determined using the flow rate vs. pressure drop results. Specifically, performance was determined by two measures: 1) diodicity; and 2) theoretical orifice size.

4.1. Diodicity

Diodicity is defined as the ratio of flow (intended in one direction) to leakage (undesired in the opposite direction) at a given pressure. The resulting diodicity ratios for valves manufactured are shown in Table 1. As shown, the diodicity ratio for the first flapper valve was barely better than unity. This suggests that almost as much fluid passes through the valve in the open direction than in the closed direction. This suggests very poor performance. It was expected that this poor performance was due to the ejecta ridges formed during the laser micromachining procedure. (See Figure 6.)

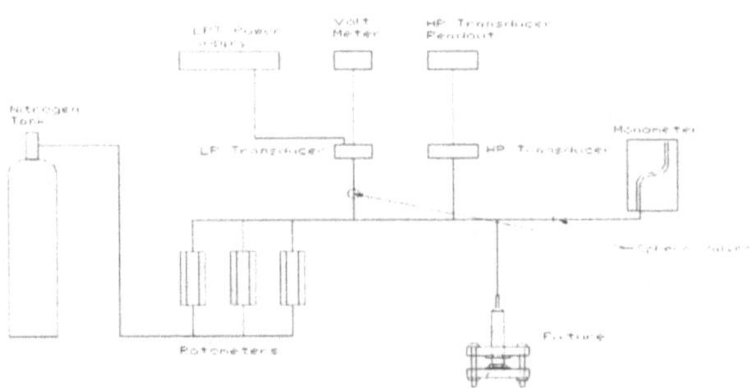

Figure 7. Valve test loop.

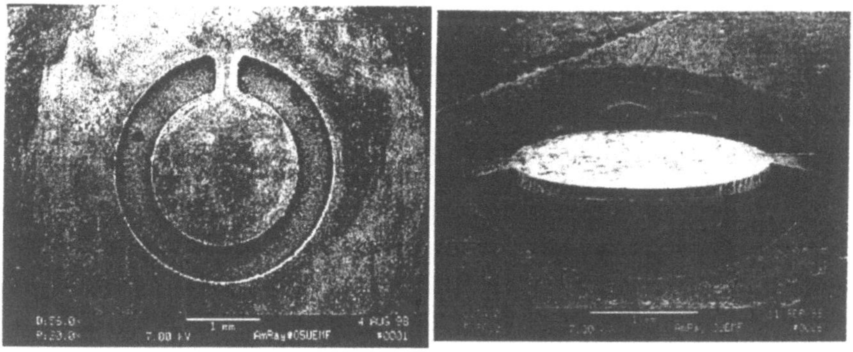

a) b)

Figure 8. a) Top view of microflapper valve; b) isometric view of microfloat valve showing warpage due to volumetric expansion caused by microprojection welding. Note that the ring nozzle has been bent to show the warpage in Laminate C.

To improve the performance of the flapper valve, a polyimide resist (DuPont Pyralin™) was applied to subsequent microflapper valves prior to microprojection welding to permit better sealing. The polyimide was applied to two different surfaces on the laminates: the valve seat and the back of the flapper (side toward the valve seat). Application of the polyimide was accomplished with the aid of a 3 ml dropper. The polyimide was placed on the surface, and planarized with the dropper. The laminate was then placed in the oven for 10 minutes at 100 °C.

Table 1 suggests that application of the polyimide to the back of the flapper did indeed improve the sealing of the microflapper valve. Application of the polyimide to the valve seat degraded performance. It is expected that this was due to uneven application of the polyimide resist to the valve seat laminate. Of more importance, is the finding that the microfloat valve significantly outperformed the microflapper valves. It is expected that this is due to the greater degree of freedom allowed by the float mechanism compared with the flapper mechanism. As a result, the float mechanism is able to reorient to compensate for asparities along the valve seat.

4.2. Theoretical Orifice Size

Theoretical orifice size is an important measure for valve performance. The larger the theoretical orifice size, the more efficient the design of the valve. Since the largest feature on all of the valves is the projection, the actual size of each valve is 5 mm. Table 2 shows the

Table 1. Diodicity values for various valves.

Diodicity		
	Average	Maximum
Flapper Valve (Polyimide on the Back of Valve)	4.08	6.32
Flapper Valve (Polyimide on the Valve Seat)	1.22	1.78
Flapper Valve (No Polyimide)	1.71	2.90
Float Valve	11.19	17.10

Table 2. Valve Theoretical Orifice Size

Theoretical Orifice Size	
	Average Diameter (mm)
lapper Valve (Polyimide on the Back of Valve)	0.611
lapper Valve (Polyimide on the Valve Seat)	0.587
lapper Valve (No Polyimide)	0.557
loat Valve	0.629

theoretical orifice size for the four valves. The best performing valve based upon the theoretical orifice size is the microfloat valve. It has the largest theoretical orifice size at 0.629 mm.

5. Conclusions

Two unique valves were created using two novel processes. The results indicate that the best performing valve was the microfloat valve, with a theoretical orifice size of 0.629 mm and an average diodicity ratio of 11.19. Reasons for this include the ability for floats to reorient for sealing against surface asparities along the valve seat. Several reasons exist for improper sealing of the valve seats. The formation of an uneven ejecta ridge from the laser ablation process may not allow the flap or float to be coplanar to the valve seat. Furthermore, a polyimide seal applied to the flapper valve was not made flat.

Both valve designs have their advantages. While the microfloat valve seals better and allows for larger mass flow rates than the microflapper valve, it does require an additional process step. Further investigation into the elimination of surface asparities and improving sealing methods would be beneficial.

6. References

1. Young, M. "Platelet-Cooled Plasma Arc Torch", DOE Contract No. DE-AR21-93MC30361, Technology Development Data Sheet, (Morgantown, WV: Federal Energy Technology Center, 1997), pp5-6.
2. Esposito, A. Fluid Power with Applications, (Englewood Cliffs, NJ: Prentice-Hall, 1988), pp380-381.
3. Haas, R.W., D. Brest, H. Mueggenburg, L. Lang, and D. Heimlich, "Fabrication and performance of MMW and SMMW platelet horn arrays," Intl. J. Infrared and Millimeter Waves, (v14 , n11, 1993), pp2289-93.
4. Haas, R.W., "Further development of MMW and SMMW platelet feed horn arrays," Astron. Soc. Pac. Conf. Ser., (v75, 1995), pp99-105.
5. Kennedy Space Center, National Space Transportation System (NSTS) Shuttle Reference Manual, (NASA, 1988).
6. Goldberg, N., "Narrow channel forced air heat sink," IEEE Transactions on Components, Hybrids, and Manufacturing Technology, (v7, n1, 1984), pp154-9.
7. Krause, V., H.-G. Treusch, P. Loosen, T. Kimpel, J. Biesenbach, A. Kosters, F. Robert, H. Oestreicher, M. Marchiano, B. DeOdorico. "Microchannel Coolers for High Power

Laser Diodes in Copper Technology." SPIE Proceedings, (v2148, 1995), pp351-358.
8. Kleiner, M. B., S. A. Kuhn, K. Haberger, "High Performance Forced Air Cooling Scheme Employing Microchannel Heat Exchangers," IEEE Transactions on Components, Packaging, and Manufacturing-Part A, (v18, n4, 1995), pp795-804.
9. T. Ohnstein, T. Fukiura, J. Ridley, and U. Bonne. "Micromachined Silicon Microvalve." Proceedings of IEEE Micro Electro Mechanical Systems. An Investigation of Micro Structures, Sensors, Actuators, Machines and Robots, 1990, pp. 95-98.
10. A. Meckes, J. Behrens, W. Benecke, and K. Strasse. "Electromagnetically Driven Microvalve Fabricated in Silicon." *Proceedings of IEEE International Conference on Solid-State Sensors and Actuators*, June 16-19, 1997, pp. 821-824.
11. M. Stehr, H. Gruhler, H. Straatmann, S. Messner, H. Sandmaire, and R. Zengerle. "Self Priming VAMP." *Proceedings of IEEE International Conference on Solid-State Sensors and Actuators*, June 16-19, 1997, pp. 351-352.
12. H. Jermand, and M. Dunbar. *Sensors*. Vol. 11, No. 9, September, 1994, pp. 26-36.
13. E. Sevick, and D. Williams. *Macromolecules*. Vol. 27, No. 19, August, 1994, pp. 5285-5290.
14. I. Stoev, M. Syrzycki, M. Parameswaran, G. Chapman, and A. Rawicz. *Canadian Journal of Physics*. Vol. 70, No. 10, August 11, 1992, pp. 881-885.
15. Paul, B.K. and T. Terhaar. "Comparison of Microlamination Methods for Fabricating Two Microvalve Designs," to be published in *Proceedings of 1999 Industrial Engineering Research Conference*, May, 1999.

Integrated Microfluidics/Electrochemical Sensor System for Field-Monitoring of Toxic Metals

Yuehe Lin*, Dean W Matson, Wendy D Bennett, Karla D Thrall, Charles Timchalk

Pacific Northwest National Laboratory, PO Box 999, Richland, WA 99352 USA

1. Introduction

Because of the toxicity of environmental contaminants, a reliable analytical system is desired for monitoring them in complex environmental matrices. For metal species, such assays are usually carried out in central laboratories using ICP-AES or ICP-MS. However, in view of the huge labor and analytical costs or time delays associated with centralized laboratory analyses there are immediate needs for developing portable analytical systems for on-site monitoring. Development of a portable system for analysis of environmental contaminants will greatly improve the quality and efficiency for on-site characterization of chemical contamination and will allow real-time monitoring of toxic metal concentrations in the tissues of waste-site clean-up workers.

Microfabrication technology has been utilized to integrate whole laboratory systems onto microchips. These systems have been termed microscale total analytical systems (μ-TAS) [1], and their development has become an active research area. Numerous μ-TAS systems have been developed and reported in the recent literature, and have shown very promising capabilities [1-5]. The advantages of the μ-TAS approach include the ability to analyze extremely small volume samples, increased speed of analysis, reduction in reagent consumption, and consequent reduction in waste disposal. One potential application of μ-TAS is their integration into micro-chemical or biological systems for real-time/on-site monitoring.

Most fabrication processes for microfluidic chemical and biological systems involve photolithography and the associated wet etching processes to form the microchannels, valves, and interconnects. Historically, silicon and glasses have been the most widely used materials [1-2]. Most devices constructed from these materials are planar, which makes it difficult to incorporate a number of functions into the microfluidic component and to efficiently direct fluid flow to the required site. This paper describes a significantly different design and fabrication approach, the use of functional

* To whom correspondence should be addressed. Email: yuehe.lin@pnl.gov

laminates arranged in a stacked architecture. Each laminate layer performs a specific function, such as fluid reservoirs, mixing chamber, or a reaction chamber. Microsensors, micropumps, and valves can easily be integrated into such a laminated structure.

Analyte detection remains an important issue for microchip devices. Most analyses have been conducted using large detectors, such as laser-induced fluorescence (LIF) [6] and mass spectrometry [7]. For applications where appropriate, electrochemical detection is sensitive, compact and easily integrated into a small size [8]. Furthermore, electrochemical detectors can be microfabricated at low cost relative to the above detectors. In this paper, development of miniaturized analytical systems based on a microfluidics/electrochemical detection scheme is described. The design, fabrication process, and testing of the microfluidic devices are discussed.

2. Fabrication Process

An UV excimer direct-write laser micromachining system (Potomac LMT-4000) was used to produce microchannels of various geometries in polymeric materials. The system contained a Potomac TGX-1000 KrF (248 nm) excimer laser configured for focused (direct write) operation at the machining surface. The minimum spot size of an unapertured laser beam at the work surface was approximately 10 microns. The maximum pulse repetition rate was 2 KHz with a maximum pulse energy of 45 μJ. Machining was performed on set of 4-inch computer-controlled X-Y stages with the channel geometry defined by inputting a CAD file.

Microchannels for the microfluidic devices described here were produced by machining the channel into a thin polymer (e.g., polyimide) gasket. For this approach, the laser cut was made through the gasket material. Only the perimeter of the channel needed to be machined, with the remaining center portion of the channel being stripped out and discarded. When sandwiched between two solid pieces of polymer, the gasket formed a well-defined microchannel whose depth was determined by the thickness of the gasket material. This method was used to produce microchannels up to 500 microns wide and 225 microns deep. These micromachined gaskets were produced from sheet material having adhesive on both sides to improve sealing capabilities. Microfluidic devices were designed such that they can be assembled in a stacked geometry. In doing so, fluid flows were directed through a series of layers in the device, each layer having a distinct functionality. Minimum dead volume in the device was achieved by directing the fluid flow from one layer to the next using small diameter drilled vias. Where required for electrical conductivity, appropriate patterns were generated using standard photolithographic processes and conductive metal or insulator layers were applied by magnetron sputtering.

3. Descriptions and Evaluation of the Integrated Microfluidics / Electrochemical Sensor Systems

Two integrated systems were fabricated based on the different designs and fabrication methods for the electrochemical sensors. In one system, the electrochemical sensors were fabricated using standard photolithographic processes and sputtered leads were applied to a glass plate using magnetron sputtering. In the second system, electrodes were embedded in Teflon or PEEK polymer blocks and form micro flow cells based on a thin-layer or wall-jet design.

3.1 Microfluidic device with electrochemical sensor fabricated using standard photolithographic process

The fabrication scheme for the first device was based on a mutilayer lamination method utilizing stacking modules, each having a distinct functionality, to achieve a compact microanalytical device (Figure 1). With the exception of the detector module, all components were fabricated from FEP Teflon or polyimide. The detector module was produced on BK7 glass, with sputtered electrodes and insulator layers. All microchannels including the mixing/reaction channel and detector channel, were produced by sandwiching laser-machined adhesive-backed polyimide gaskets between layers of the device. Individual components, such as the microfabricated piezoelectrically actuated pumps (modified from IMM pump) and the sample and reagent reservoirs were designed as plug-in parts, allowing for rapid change and repair if necessary. The base dimensions of the fabricated part were 5.0 x 8.0 cm as shown in Figure 1A. The overall height of the device, including pumps, but without reservoirs, was approximately 2.5 cm.

The sensor consisted of four functional components comprised of five laminations shown in Figure 1B. Components contained within the laminated sensor device included: 1). Sample and reagent reservoirs; 2) Micropumps; 3). Mixing and reaction channel; 4). Electrochemical sensors module.

Two fluid reservoirs are shown in Figure 1 (One held sample and the other, an electrolyte solution or reagent). Fluid was pumped from the reservoirs by two integrated plug-in micro piezoelectric pumps. Typical pumping flow rates were 10–300 μl/min. The two fluids were then pumped into a serpentine mixing/reaction channel, which facilitated initial mixing of the fluids and allowed time for the reactions to occur between sample and reagent. The mixing/reaction channel was laser machined from 125-μm polyimide with adhesive on both faces and the channel was sealed between adjacent solid polymer sheets using the gasket process described above. The channel width was 200 μm. The sample was next pumped through a 200-μm-wide detector channel where it flowed over iridium, gold, and silver electrodes for the voltammetric measurements. The detector base was polished 0.16-cm-thick BK7 glass

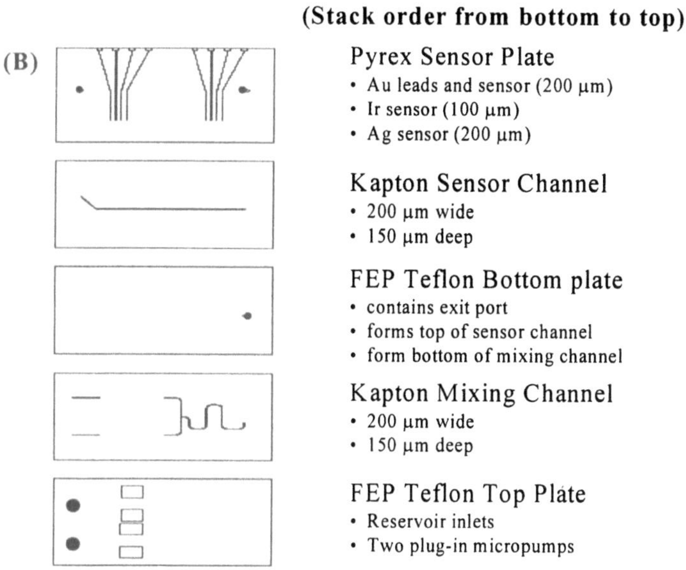

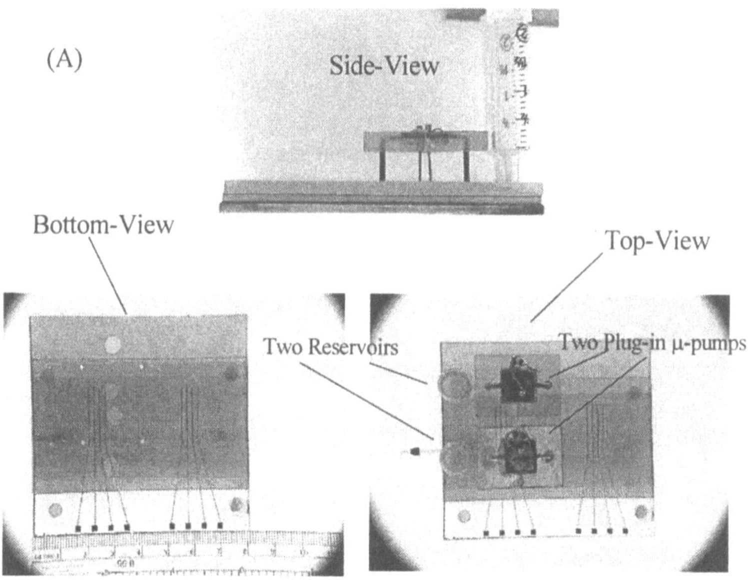

Figure 1. Photos of Integrated Microfluidics/Electrochemical Sensor Array (A) and Its Stacking Scheme (B)

(A)

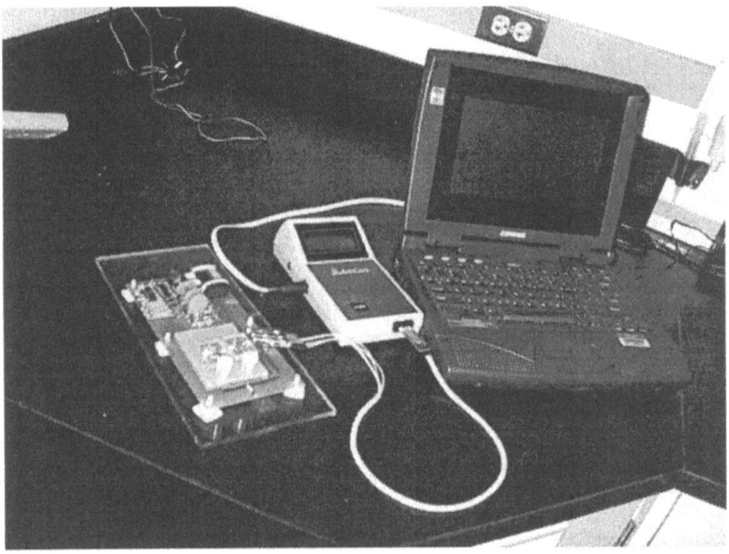

(B)

Figure 2. (A) Photo of the Prototype of Microfluidic Device with Electrochemical Sensors and Micropump Controller
(B) Photo of the Prototype of Portable Analyzer

All metallization was deposited by magnetron sputtering. Gold connecting leads and sensor bases were first deposited on the substrate and formed using the lift off technique. A miniaturized power supply was integrated on an electronic board which is located on the bottom of the microfluidic device. This design made the entire system highly compact. The entire analytical system was control using a laptop computer as shown in Figure 2.

The microfabricated electrochemical cell consisted of an Ir working electrode, a Pt auxiliary electrode, and an Ag electrode. In order to make the device fully functional, the sputtered Ag electrode had to be converted into an Ag/AgCl reference electrode. Conversion of thin-film Ag into Ag/AgCl reference electrode presented a challenging technical issue. Complete oxidation of Ag would damage the Ag electrode. Optimally, the Ag was oxidized under a low constant current (1 μA) in 0.1 M KCl solution for 5 min. Evaluation of the micro electrochemical system was conducted by studying the cyclic voltammetric property of 0.5 mM $K_3Fe(CN)_6$ solution. In order to form the Ag/AgCl reference electrode, 0.3 M KCl was added to the solution. For comparison purpose, cyclic voltammogram of 0.5 M $K_3Fe(CN)_6$ solution was also measured with an Ir microelectrode (10 μm) in a conventional electrochemical cell (5 mL) with a 3 M Ag/AgCl as a reference electrode. As shown in Figure 3, a well-defined sigmoidal response (Figure 3) was obtained for both the microfabricated Ir electrode and the Ir microelectrode in a conventional electrochemical cell. These results indicate that a diffusion controlled process and a steady-state current were achieved in the microfabricated electrochemical cell. Since lower KCl concentration was used in the microfabricated electrochemical system, a negative potential shift was observed.

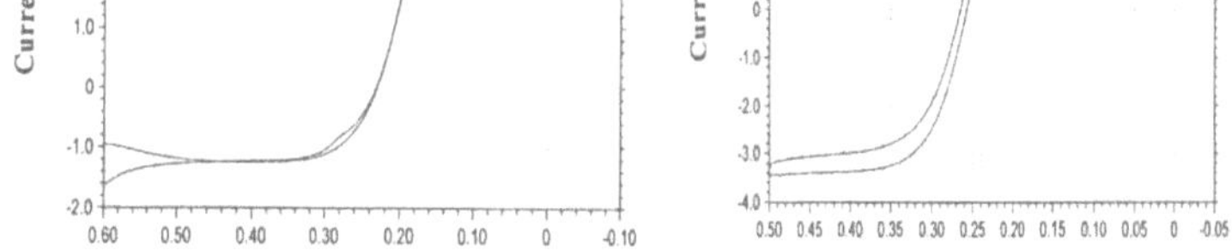

Figure 3. Cyclic votammogram of 0.5 mM ferricyanide in 0.4 M KNO_3 with a microfabricated Ir electrode (A) and Ir microelectrode in a conventional cell (B). Scan rate: 10 mV/s.

3.2. Microfluidic device with electrochemical sensor fabricated based on thin-layer or wall-jet design

In the second scheme, all individual modules were fabricated on the surface of a standard microfluidic platform based on a plug-and-play design (Figure 4C). In this design, the microelectrochemical flow cell was incorporated as a plug-in module rather than being laminated on the bottom layer of the microfluidic device. The microelectrochemical flow cell which integrates three electrodes, was based either on a thin-layer (Figure 4A) or wall-jet (Figure 4B) design and was fabricated on polymer substrates. The microelectrochemical flow cell could then be plugged directly into the microfluidic platform. Although less compact that the multilayer device described above, this fabrication scheme allows for more analytical flexibility. Specifically, different electrochemical cells having different analyte sensing capabilities can be quickly replaced and plugged into the microfluidic platform. Microelectrochemical cells with different working electrodes, such as glassy carbon, micro Ir, Pt, and Au, have been fabricated for different applications.

The integrated microfluidic device with the on-surface microelectrochemical cell was used to detect lead in water using Square Wave Anodic Stripping Voltammetry (SWASV). For lead detection, a constant potential of –0.80 V is applied on a glass carbon mercury film electrode. When sample solution flowed through the electrode surface, Pb^{2+} was reduced and accumulated. After accumulation for 200 s, the potential of the mercury film electrode was scanned from –0.80 V to 0 V. Pb was measured with square wave voltammetry. After measurement, the Pb was stripped from the electrode surface and the sensor could be reused. Figure 5 shows the SWASV votammograms obtained for 1 ppb and 10 ppb Pb^{2+} in aqueous solutions using this device. These results indicate that 1 ppb Pb^{2+} in a water sample could be detected in a 200s preconcentration time period. The micro analytical system has also been used for the analysis of Pb^{2+} in water samples collected from rivers, analytical results were comparable with those obtained using ICP/MS.

4. Conclusion

The successful integration of microfluidics/electrochemical sensor arrays in compact devices has been demonstrated. Individual modules, such as sample and buffer reservoirs, microfabricated piezoelectrically actuated pumps, and a microelectrochemical cell were integrated onto portable platforms. This allowed rapid change-out and repair of individual components by incorporating "plug and play" concepts now standard in PC's. Preliminary fluidic and electrochemical experimental results from the integrated devices show a feasibility of using the portable analyzer developed in this work for detection of trace-level toxic metals. This type of portable analyzer can provides a low-cost and rapid method for on-site and *in situ* determination of toxic metals in groundwater. Demonstration of the portable analyzer for *in-situ*/on-site characterization of chemical

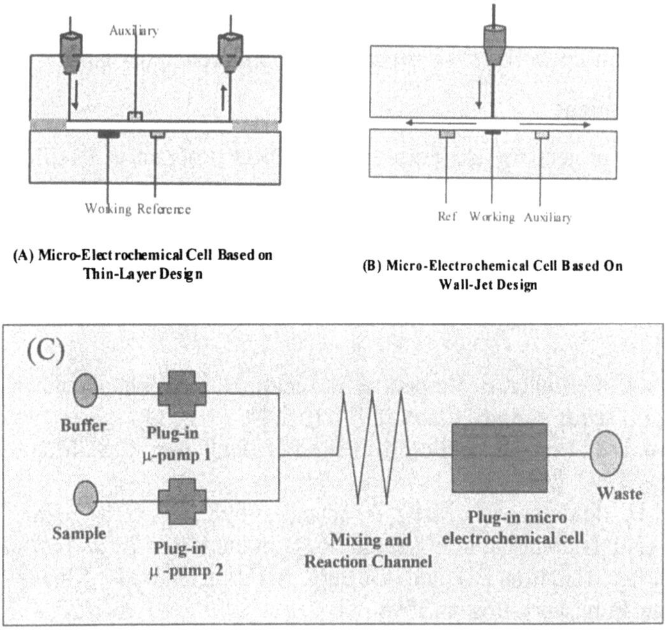

Figure 4. Integrated Microfuidics Device with Microelectrochemical Cell Based on Thin-Layer or Wall-Jet Design

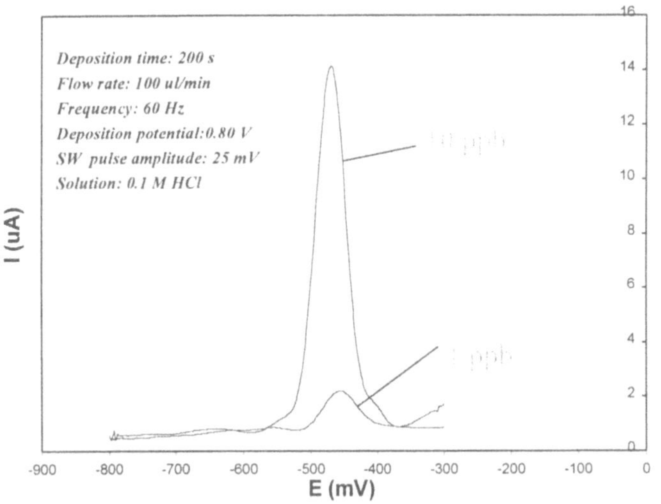

Figure 5. The SWASV for Pb in 0. 1M HCl solution

contamination and real-time monitoring of toxic metal concentrations in tissues of waste-site clean-up workers is currently being explored in our lab.

Acknowledgement

This project was supported by a LDRD program at Pacific Northwest National Laboratory (PNNL). PNNL is operated by Battelle Memorial Institute for the U.S. DOE under Contract DE-AC06-76RLO 1830.

References:

1. Manz, J. C. Fettinger, E. Verpoorte, H. Lundi, H. M. Widmer, and D. J. Harrison, *Trends in Anal. Chem.*, 1991,10, 144.
2. Harrison, D.J.; Fluri, K.; Seiler, K.; Fan, Z.; Effenhauser, C.S.; Manz, A. *Science,* 1993, 261, 895
3. Kop, M.U., deMello, A.J., Manz, A. *Science*, 1998, 280, 1046-1048
4. Lin, Y.; Xu, N.; Matson, D.W.; Smith, R.D., in the *Micro Total Analysis Systems*; Ed.: Harrison, D.J. and den Berg, A.V.; pp3343-346; Kluwer Academic Publishers, Boston, 1998
5. Ramsey, R.S. and Ramsey, J.M., *Anal. Chem.*, 1997, 69, 1174-1178
6. Woolley, A.T., and Mathies, R.A. *Proc. Natl. Acad. Sci. USA*. 1994, 91, 11348-11352
7. Xue, Q.F.; Foret, F.; Dunayevskiy, Y.M.; Zavracky, P.M.; McGruer, N.E. and Karger, B.L., *Anal. Chem.*, 1997, 69, 426-430
8. Kounaves, S.P.; Deng, W.; Kovacs, G. T. A., *Anal. Chem.*, 1994, 66, 418-423.

Thermoelectrical Measurement System for Chemical Instrumentation

(Topic: 3. Microreaction devices, Sensors)
(Poster Presentation)

Stefan Beißner[1], Thomas Elbel[1], J. Michael Köhler[2], Martin Zieren[2]

1 Fachhochschule Hannover, Forschungsschwerpunkt 'Angewandte Mikrosysteme (AMIS)', Ricklinger Stadtweg 120, 30459 Hannover, Tel.: 0511/9296-381, Fax: 0511/9296-99-381

2 Institut für Physikalische Hochtechnologie IPHT Jena, Winzerlaer Straße 10, 07743 Jena, Tel.: 03641/657715, Fax: 03641/657700

Keywords: Microcalorimetry, microfluidics, microcontroller, reaction enthalpy

1 Introduction

We report on a measurement system for determination of the concentration of liquid chemicals and the quantity of substance using the thermoelectric measurement principle.
The application of the thermoelectric measurement principle with wire thermocouples for contact temperature measurements and with thin film radiation thermopiles for non-contact temperature measurements is well known [1, 2].
The measurement principle of the described system is comparable to the detection of radiation heat with thin-film thermopiles. In comparison our system uses similar thin-film thermopiles but detects induced chemical reaction heat. Applications for this measurement system are medical and environmental analysis.
The system measures the heat of exothermic reactions in the liquid phase. For this purpose we use a chip calorimeter integrated in a sequential flow-injection analysis arrangement. Two chemicals are provided via inlets, mix by diffusion only and react in the sensor chip [3]. The reaction heat causes a temperature difference between the active and the passive junctions of three thin-film thermopiles. By integrating the voltage signals of the sensor over time the released heat energy is calculated, which is proportional to the quantity of substance. By evaluating the voltage signals of the three thermopiles separately, conclusions about the reaction kinetics can be drawn [4].
The measurement system consists of analog amplifiers for the signals from the thermopiles, a microcontroller for signal conditioning and for the control of two serial interfaces, microcontroller driven current sources for electrical sensor calibration via integrated thin-film heaters, a display and a keyboard for operating the device. A software package for downloading, visualisation and evaluation of the measured data on a personal computer has been developed.

In each measurement cycle the sensitivity of the thermopiles together with the amplifiers, which are arranged in series, is determined. Therefore a defined power in the μW-range is transferred to one of the integrated heaters using a high precision current source, which is controlled by the microcontroller. The syringe pumps used for providing the chemicals are adjusted and controlled by the system and can easily be substituted by micropumps. The voltage signals of the thermopiles are recorded for further calculations and an optional download. Using the sensitivity previously determined, the heat of reaction is calculated and the quantity of substance is determined based on the molar reaction enthalpy.
The system can be used for automated investigations on reaction kinetics as well.

2 The sensorchip

A sensorchip described in [3, 4, 5] has been used for the measurement system (Fig.1).

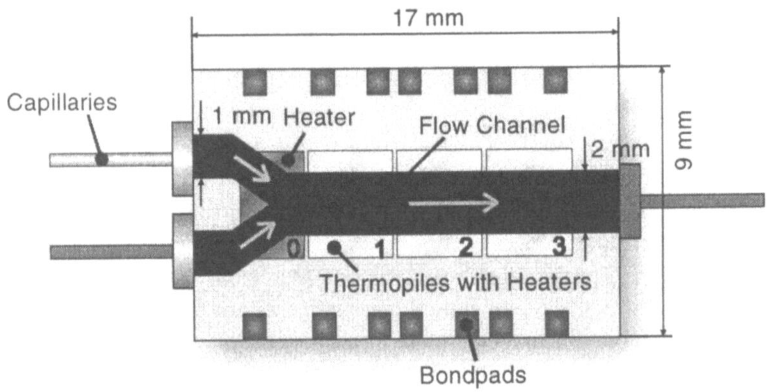

Figure 1: Outline of the sensor chip.

The chip is built by means of micromachining. It provides a Y-shaped flow channel with a volume of approximately 2 μl, three thin-film thermopiles which serve as transducers and four thin-film heaters for calibration purposes.
In former investigations free-standing thermoelectric sensors were used as transducers [7, 8]. The quantity of substance of glucose has been determined with this measuring device.
The advantages of the sensorchip used in the presented system are the insensitivity against environmental influences like air humidity and better potential for process control.

3 Chemical Fundamentals

The measurement system is used for analytical investigation of exothermic chemical reactions in the liquid phase. An application for the system might be the determination of the concentration of glucose in blood. Measuring is done according to the principle of flow injection analysis [6] (Fig.2).

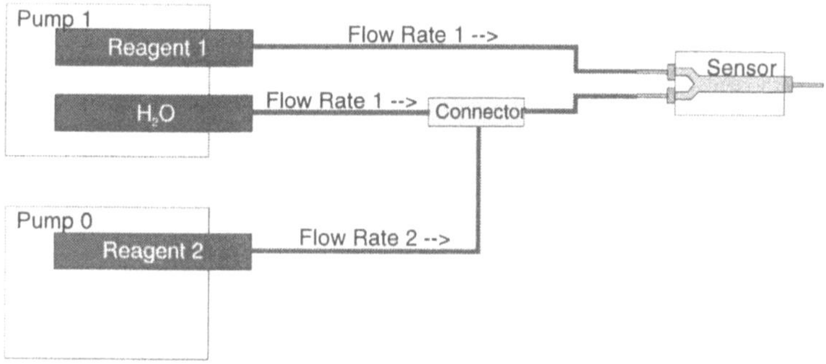

Figure 2: Measurements according to the principle of flow injection analysis

A constant flow of liquid is established by pump 1. One reagent and a carrier stream of distilled water are merged in the Y-shaped flow channel of the sensor chip.
Driven by pump 0 a defined volume of the second reagent is injected into the water stream.
The two reagents merge and react in the flow channel.
During the reaction a reaction enthalpy ΔH_R [9]

$$\Delta H_R = \sum \Delta H_{B,products} - \sum \Delta H_{B,reagents}, \qquad (1)$$

which equals the difference between the enthalpy of formation ΔH_B of the reaction products and the reagents, is emitted.
The change of enthalpy is

$$dH = dU + pdV + Vdp, \qquad (2)$$

where U is the internal energy, p is the pressure and V is the volume. Therefore the change in the internal energy of the liquids and of parts of the sensorchip between the moments t_1 and t_2 in time can be written as:

$$U_{12} = \int_{t1}^{t2} dU = \Delta H_{R,12} - \int_{t1}^{t2} pdV - \int_{t1}^{t2} Vdp. \qquad (3)$$

The reaction enthalpy subtracted by the energy which is used to increase the volume and the pressure of the fluids causes a change in the internal energy.
Considering the thermal material parameters of the sensor chip, the fluids and the air, temperature gradients on the chip, which induce a voltage signal on the thermoelectric sensors, can be calculated numerically[4].
Using the sensitivity of the amplified thermoelectric sensor

$$S = \frac{U_{el}}{P}, \qquad (4)$$

where U_{el} is the voltage and P is the power, the reaction enthalpy can be calculated from the voltage signal. A factor K has to be taken into account for the efficiency of mixing and degree of completeness of the chemical reaction considering chemical kinetics and residence time.
The reaction enthalpy is

$$\Delta H_{R,12} = \frac{K}{S} \cdot \int_{t1}^{t2} u_{el}(t)\,dt. \qquad (5)$$

The quantity of substance n which has been converted in the region of the sensor can be calculated using the molar reaction enthalpy:

$$n = \frac{\Delta H_{R,12}}{\Delta h_R^{mol}}. \qquad (6)$$

4 Hardware Architecture

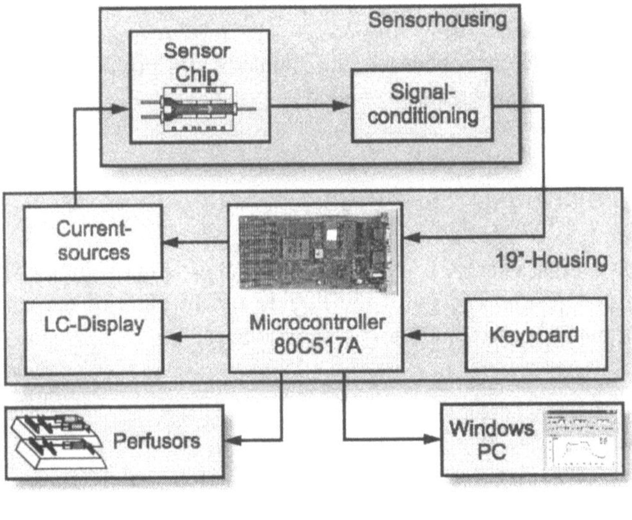

Figure 3: Structure of the measurement system.

All components of the measurement system are controlled by a siemens 80C517A microcontroller.

Two syringe pumps are providing the reagents according to the principle of flow injection analysis. These pumps can be configured and controlled via a RS – 232 interface of the microcontroller.

Current sources (Fig. 4) with an accuracy in the μA – range are provided for calibration purposes. They are used to transfer a power in the μW – range (e.g. 70 μW) to the heaters of the sensor chip [10].

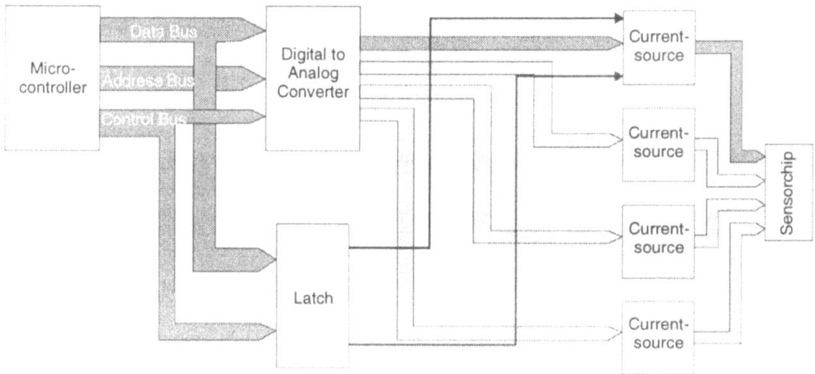

Figure 4: Structure of the current sources.

To achieve galvanic insulation between the current sources and the microcontroller, optocouplers are installed in the data-, address- and the control bus. Furthermore the current sources are powered by their own dedicated potential transformer.

A digital to analog converter with a resolution of 8 bits is used to convert a data byte sent by the microcontroller into an analog voltage. This analog voltage regulates one of the voltage controlled current sources. The output current of the current source powers the correspondent heater of the sensor chip. Analog switches are provided to ensure that no disturbance from the current sources can be coupled into the thermopiles during the measurement.

The analog voltage signals of the thermopiles are conditioned by three two-stage operational amplifier circuits to match the input voltage range of the analog to digital converter (Fig. 5).

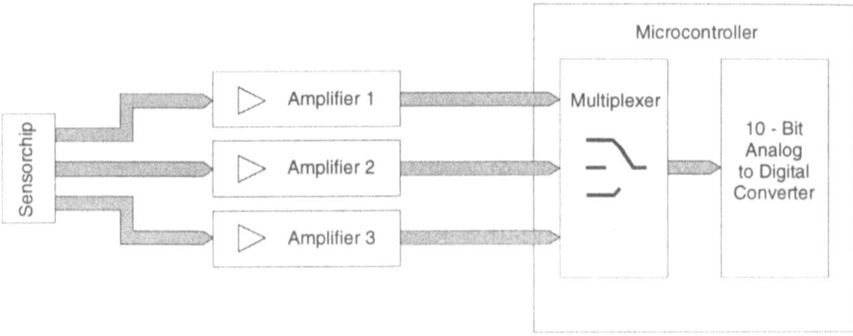

Figure 5: Signal conditioning concept.

The circuit diagram of the two-stage amplifiers is shown in figure 6.

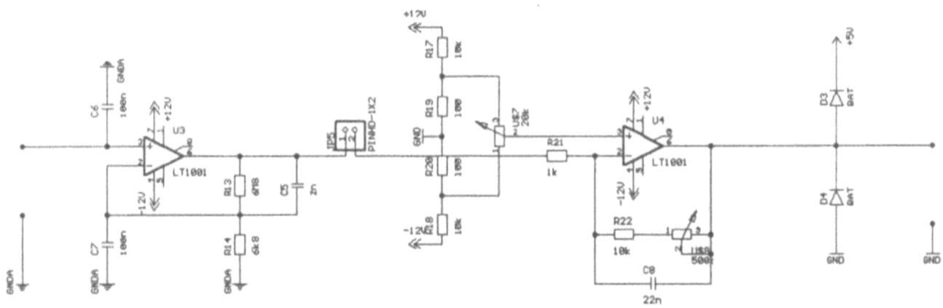

Figure 6: Circuit diagram of the two stage amplifier.

A noninverting amplifier has been used for the first stage because of its high input resistance. Care has been taken to minimise current flow on ground tracks of the first amplifier stage which could tamper with the sensor signal.
An inverting amplifier with an adjustable gain serves as the second amplifier stage. The noninverting input of the second stage operational amplifier is put on an adjustable voltage for offset compensation.
The LT1001 operational amplifier has been chosen for its low noise, low offset and high precision. Two schottky diodes prevent overloading of the analog to digital converter.
We are using the 10 – bit analog to digital converter which is implemented in the 80C517A microcontroller. The amplified signals are converted and they are stored in a random access memory (RAM) by the microcontroller.
The stored data can be downloaded to a personal computer (PC) via a RS – 232 interface of the microcontroller. A comfortable windows program is provided for taking the data up on the PC, visualising and evaluating them [11].
A liquid crystal display and a twelve key keyboard serve as the user interface of the measurement system.

5 Software

The software for the microcontroller of the measurement system has been written in the high level language 'C'. Therefore maintenance and servicing of the software is easy and the system can easily be adapted to different tasks.
The software performs an initialisation for the current sources, the serial ports of the microcontroller, the analog to digital converter and the liquid crystal display of the system.
Parameters of the measurement, like flow rates or the volume of the chemical compound which shall be injected, are read from the keyboard. The output offset of the amplifiers is adjusted.
The sensitivity of the thermopiles together with the amplifiers is determined by the software (Fig. 7). Therefore the output voltage of an amplifier is measured while the heaters are turned off. Then a defined power P of 70 μW is transferred to the heater which is spatially attributed to the thermopile under investigation. After a delay time of 10 seconds the voltage is measured again. To minimise the influence of noise each voltage is taken five times and the mean value is calculated by

$$U_{low} = \frac{1}{5} \cdot \sum_{i=0}^{4} U_{low,i} \quad , \quad U_{high} = \frac{1}{5} \cdot \sum_{i=0}^{4} U_{high,i} .$$

The sampling frequency here is 2 Hz. The sensitivity of the thermopile and the amplifier in series is calculated to

$$S = \frac{U_{high} - U_{low}}{P} .$$

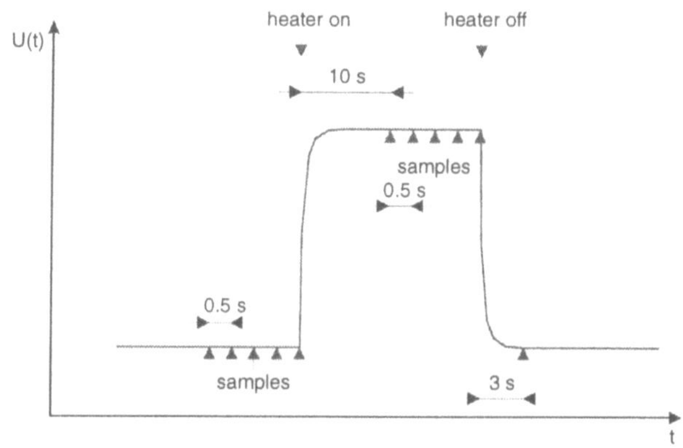

Figure 7: Measurement principle for determining the sensitivity of thermopile and amplifier in series.

To minimise disturbance by heated fluids these measurements are taken against the flow direction in the channel. This means the sensitivity of thermopile 3 is determined first, then measurements on thermopiles 2 and 1 follow.
The amplified voltage signal of the thermoelectrical sensors are gathered over several minutes, and then they are stored in the systems random access memory. Using the sensitivity determined before, the detected enthalpy can be calculated by the microcontroller.
A download of the measured data to a personal computer over a serial port is supported. A comfortable windows™-program for data visualisation is provided (Fig. 8).
The program for data visualisation reads the data from the serial port of a personal computer. The data is displayed both as table and as diagram. A comfortable function for integration is provided. This function takes the offset and the drift of the thermopile – amplifier arrangement into account.

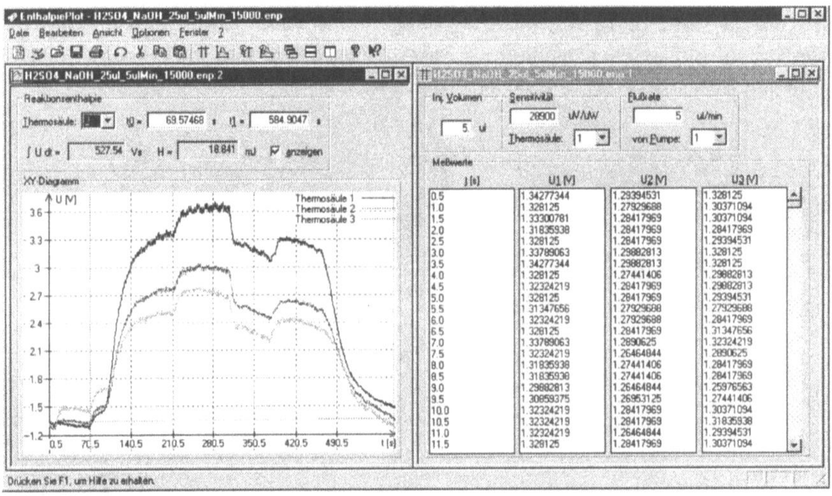

Figure 8: Typical view on the personal computer program for data visualisation. The diagram view on the left shows the integration limits and the integration result for one of the three thermopiles. The right side shows the table view of the data.

6 Results

A measurement system for automated investigation of chemical reactions has been developed. A new generation of thermoelectric sensorchips for investigations in a flow injection arrangement has been used. Electronics have been developed which use this chip's potentials for calibration and measurement.
A personal computer program for the evaluation of the measured data has been developed.

The suitability of the thermoelectric measurement principle for chemical measurements has been proven via the neutralisation reaction between sodium hydroxide and sulphuric acid.
Figure 9 shows an example of a neutralisation reaction measured with this system.

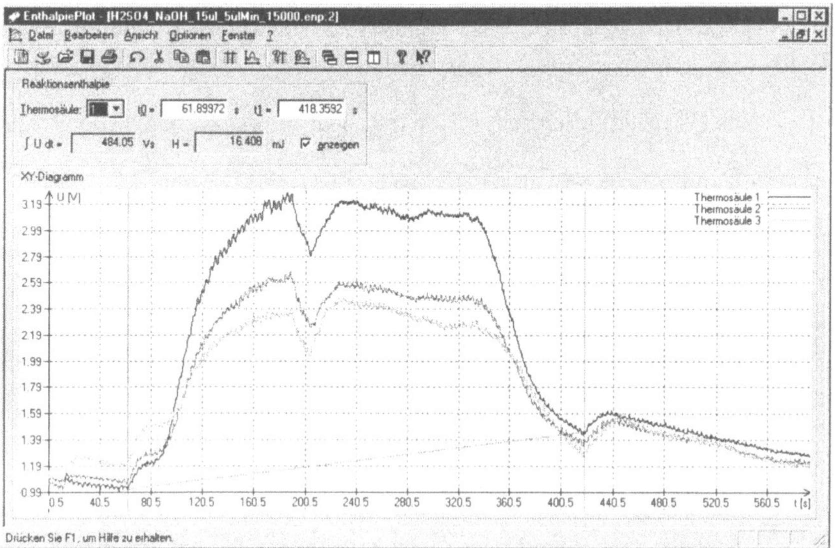

Figure 9: Neutralisation of 15 μl 0.1 molar sodium hydroxide with 0.5 molar sulphuric acid. Flow rate on both pumps: 5 μl/min.

The molecular formula of the neutralisation reaction shown in Figure 9 is:

$$H_2SO_4 + 2\,NaOH \rightarrow Na_2SO_4 + 2\,H_2O + 148.5\frac{kJ}{mol_{NaOH}}.$$

The quantity of sodium hydroxide converted in this reaction is calculated to:

$$n_{NaOH} = c \cdot V = 1.5\,\mu mol.$$

Therefore the reaction enthalpy of this reaction is

$$\Delta H_R = \Delta h_R^{mol} \cdot n_{NaOH} = 223\,mJ.$$

The enthalpy which is detected by the sensors is calculated via their sensitivity. Using the integration limits shown in Figure 9 the following values are calculated:

$$\Delta H_{R,1} = 16.4\,mJ \quad , \quad \Delta H_{R,2} = 12.1\,mJ \quad , \quad \Delta H_{R,3} = 10.5\,mJ,$$

where 1, 2 and 3 indicate the first, second and third thermopile. About 17 percent of the reaction enthalpy is detected by the sensors. Thus we calibrate our system by determining the factor K in equation (5).
In further investigations the system can be used to draw conclusions about reaction kinetics.

This work has been supported by the Ministry of Science and Culture of Lower Saxony.

1 Elbel T (1996) Mikrosensorik, Wiesbaden: Vieweg-Verlag, ISBN 3-528-03377-0
2 Elbel T (1991) Miniaturized Thermoelectric Radiation Sensors, Sensors and Materials 3, No.2, P. 97 - 109
3 Köhler JM, Zieren M (1998) Chip reactor for microfluid calorimetry, Thermochimica Acta 310:25-35
4 Zieren M, Dillner U, Köhler JM (1998) Time-Resolved Calorimetry in a New Type of Micro Fluid Reactor Using Spatially Seperated Thin-Film Thermopiles and FIA-Technique, IMRET II: 2nd Int. Conf. on Microreaction Technology, New Orleans
5 Köhler JM, Zieren M (1997) Micro flow calorimeter for thermoeletrical detection of heat of reaction in small volumes, Fresenius J Anal Chem 358:683-686
6 Steinhage G (1996) Dünnschichtthermosäulen für die chemische und biochemische Analytik, Ph.D. Thesis, University of Münster, Germany
7 Lodziak D, Meinken A (1996) Entwicklung eines Meßsystems für einen thermoelektrischen Biosensor, Diplomarbeit, Fachhochschule Hannover, Germany
8 Schimkus T, Küchel H (1997) Mikrocontrollergesteuertes Meßsystem für einen thermoelektrischen Biosensor, Diplomarbeit, Fachhochschule Hannover, Germany
9 Schröter, Lautenschläger, Bibrack (1991) Taschenbuch der Chemie, Fachbuchverlag Leipzip ISBN 3-343-00760-9
10 Arkenberg L (1998) Mikrocontrollergesteuertes Kalibriersystem für thermoelektrische Chemosensoren, Diplomarbeit, Fachhochschule Hannover, Germany
11 Pralle K (1999) Meßwertvisualisierung eines mikrocontrollergesteuerten Meßsystems für thermoelektrische Mikrokalorimeter mit Visual C++ 6.0 unter Windows NT, Diplomarbeit, Fachhochschule Hannover, Germany

Electrically Heated Microstructure Heat Exchangers and Reactors

J. Brandner, M. Fichtner and K. Schubert
Forschungszentrum Karlsruhe,
Hauptabteilung Versuchstechnik, Postfach 3640, D-76021 Karlsruhe

Abstract

Keywords: micro heat exchangers, electrical heaters, heating fluids, evaporation of liquids, heat transfer, temperature control

In this paper, the designs and properties of electrically heated microstructure heat exchangers shall be presented. As material, stainless steel is used. The maximum electrical power achievable with those laboratory prototype devices is 600 W and 1800 W. The power to be transferred is adjusted by either a voltage controller or a conventional programmable controller. The devices may be applied for the temperature control of fluids up to several hundred degrees Kelvin as well as for the evaporation of liquids or endothermic chemical reactions. Experiments were performed with water and gases as test fluids. Results of heating rates, response behavior etc. will be given. Furthermore, the results of first evaporation tests will be presented.

1 Introduction

Controllable heating of fluids in process technology is normally done by either electrical heaters or heat exchangers driven by heating fluids. Conventional electrical heating elements can be rather voluminous. Moreover, they may have a high temperature difference between the fluid and the surface of the heater, which may cause problems with fluids that are sensitive to overheating.

Heat exchangers driven by heating fluids have an upper temperature limit depending on the fluid used.

Because of the limited heat transfer capabilities of conventional electrical heaters such as heating plates, the heating rates are lower than those realizable with microstructure devices [1]. Hence, the combination of microchannel structures with electrical resistance heaters should lead to compact heater devices. They should be easily controllable by a power control.

This paper will give an overview of the properties of electrically driven microstructure heat exchangers like heating rates, temperature range and the possibility of evaporation of liquids.

2 Design of the Laboratory Prototypes

At the Central Experimental Department of the Karlsruhe Research Center first prototypes of electrically driven microstructure heat exchangers and reactors have been developed and tested.

The prototypes have the same core structure. Between two metal plates used as top and bottom of the heat exchanger block, spacers to insert the heating elements and microstructured metal foil packages are arranged in alternating order.

The microstructures have been made of mechanically microstructured stainless steel foils with thicknesses of 200 μm [2,3]. The spacers were made of the same material as the microstructure foils.

All parts were connected by diffusion bonding, electron beam welding or other conventional welding technologies.

As heating elements, small electrical resistors have been used. Their advantages are mainly their fast response time and good controllability of the heating power by, e. g., a voltage controller.

The heating elements are small compared to normal heating resistors used, for example, in heating plates. The heaters used were either resistors built by thick film technology on ceramic plate substrates, or simple cartridge resistor heaters. Heating power varied between 60 W up to several hundred watts per heater. The surface power density was between 18 W/cm^2 and 165 W/cm^2 depending on the heater.

2.1 Electrical Microstructure Heat Exchanger Type 1

By diffusion bonding of mechanically structured stainless steel foils with non-structured stainless steel foils, microstructure foil-doublets were made. Each of these doublets contained 80 microchannels. The microchannels were 150 μm wide, 150 μm deep and had a length of 22 mm. The hydraulic diameter was 133 μm. Eleven doublets were welded together with ten spacers and two plates as bottom and top of the micro heat exchanger. This was also done by diffusion bonding.

After completing the heat exchanger block with fluid adapters, ten ceramic heaters were set into the spacers, and the electrical circuit was built. The ceramic heaters had a maximum electrical power of 60 W each, so a maximum overall power of 600 W was reached.

Because the electrical connection to the ceramic heaters was done by soldering, the maximum temperature of the heat exchanger is limited to the melting point of the soft solder.

Figure 2.1 shows a schematic drawing of this type of electrically driven microstructure heat exchanger.

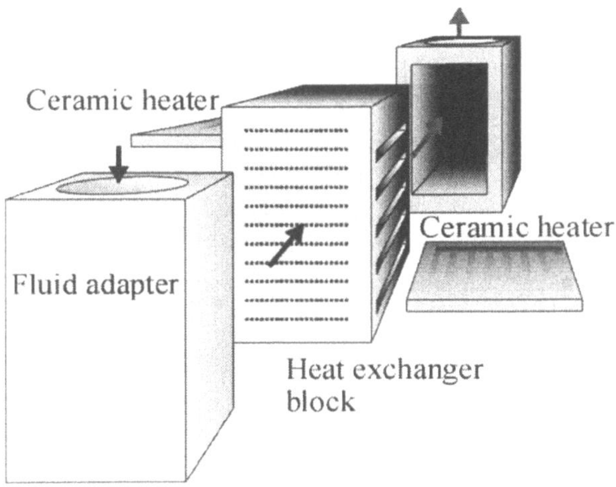

Fig. 2.1: Schematic drawing of an electrically driven microstructure heat exchanger type 1. Eleven microstructured stainless steel foil doublets and ten stainless steel spacers for the heating resistors were arranged in alternating order. Two stainless steel plates were used as top and bottom of the device body. The resistors are fitted in, and the heater core is adapted for fluids.

2.2 Electrical Microstructure Heat Exchanger Type 2

For this laboratory prototype cartridge resistor heaters were chosen as heating elements.

Between two stainless steel plates as top and bottom of the microstructure heat exchanger, ten microstructured stainless steel foils and two spacer blocks were arranged in alternating order and then welded by diffusion bonding. The microstructured foils contained 153 microchannels each. The channels were 200 μm wide, 100 μm deep and had a length of 27 mm. The hydraulic diameter was 133 μm.

The heater cartridges were inserted into holes of the spacer blocks. Then the micro heat exchanger was adapted to stainless steel tubes with standard fittings, which served as fluid connectors.

Depending on the number and the power of the heater cartridges, the overall power of a type 2 device is in the range of several 100 W up to several kW. With the present heat exchanger device having 6 cartridges, a maximum electrical power of 1800 W can be reached.

In Figure 2.2, a schematic drawing of the micro heat exchanger type 2 is shown.

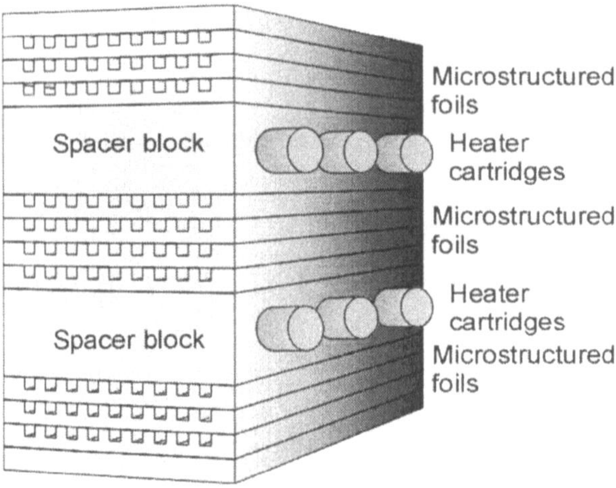

Fig. 2.2: Schematic drawing of a type 2 device.

Figure 2.3 shows a photo of the heat exchanger before it is adapted to the standard fittings.

Fig. 2.3: Photo of a type 2 device before it is welded to the fluid adapters and without heater cartridges inserted.

To obtain an idea as to which heat transfer coefficients α could be achievable, the well-known equation (1) was used for calculation [4]:

$$(1) \qquad Nu = \frac{\alpha \cdot d_h}{\lambda} = \left\{ 3.66^3 + 0.7^3 + \left[1.615 \cdot \sqrt[3]{Re \cdot Pr \cdot d_h / l} - 0.7 \right]^3 \right\}^{1/3}$$

In this equation, d_h is the hydraulic diameter of a microchannel, l is the length of the channel, and λ is the heat conductivity of the fluid.

First the Reynolds number Re was calculated from the geometric data of the microstructures, the mass flow and temperatures of the fluid at the inlet and the outlet of the device. The Prandtl number was taken from [4,5]. Then the Nusselt number Nu was calculated using (1) and the equation was rearranged to calculate the heat transfer coefficient α. For a water mass flow range of 6 kg/h to 60 kg/h, the heat transfer coefficient α of a type 2 device was calculated. The heat transfer coefficient varies between 17200 W/m^{2}K and 17600 W/m^{2}K. Conventional electrical heaters have heat transfer coefficient values of, e. g., 2440 W/m^{2}K, as mentioned in [6].

3 Experimental Results

The test program for both types of laboratory prototype devices included measuring of the maximum throughput, system responses to changes of the electrical power, system responses to changes of the throughput and measuring the heating rate. With type 2 devices, also evaporation experiments were made. Furthermore, experiments with gases were performed with a type 2 device.

3.1 Microstructure Heat Exchanger Type 1

Figure 3.1 gives an example of the test results achieved with a type 1 device. The data shown in Figure 3.1 have been measured at a constant water throughput of 9 kg/h and fast change of the electrical voltage between zero volt and 200 volt. The voltage was switched to zero volt again after about 45 seconds. Figure 3.1 shows the system response in this experiment.

Figure 3.1 reveals a time delay of approximately 3 seconds between the application of the electrical voltage and the system response. This is because the relatively large mass of the micro heat exchanger is acting like a heat sink for heating up or heat source for cooling down the device. It takes roughly 22 seconds to reach a value of 90% of the end temperature of 45°C and the same time to cool the device down to 10% above the inlet temperature after the voltage is switched off.

3.2 Microstructure Heat Exchanger Type 2

For the experiments made with a type 2 device some results shall be shown by way of example. In Figure 3.2, the temperatures of the water at the inlet and the outlet as well as the housing temperature of a heat exchanger device type 2 are shown as

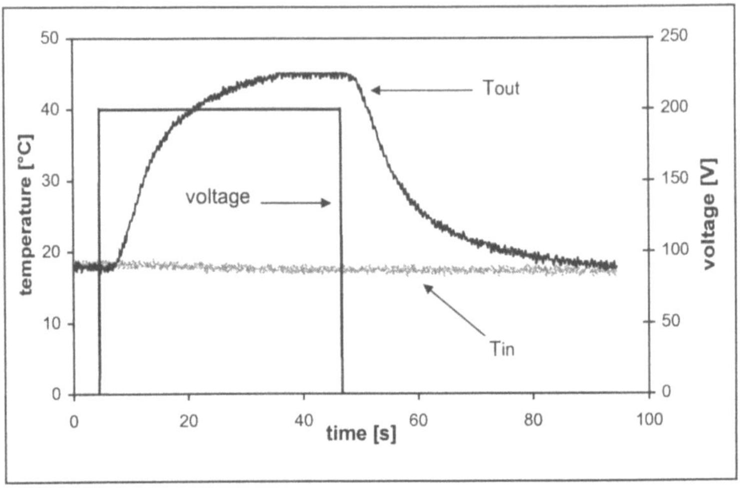

Fig. 3.1: System response of a type 1 device to a fast change of the electrical voltage between 0 V and 200 V. The water throughput was 9 kg/h.

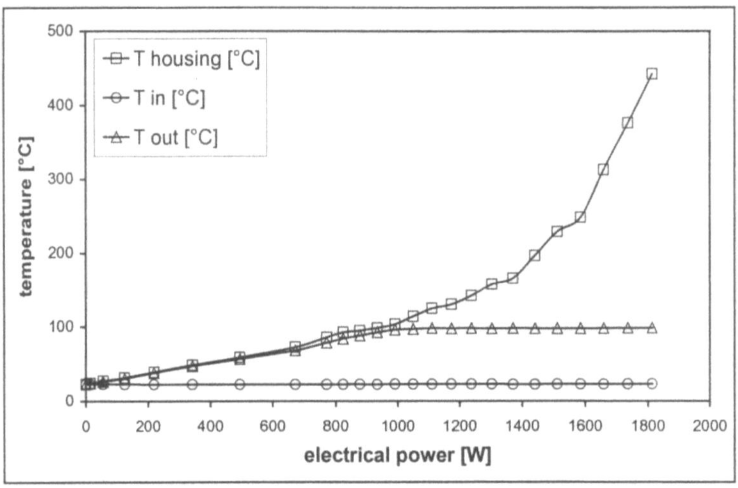

Fig. 3.2: Temperatures of the housing, at the inlet and at the outlet of a type 2 device. The electrical power was slowly increased to reach the evaporation point precisely. .Evaporation took place under atmospheric pressure. The outlet reached a maximum temperature of 100°C. The water throughput was 9 kg/h.

a function of the electrical power. The water throughput was held constant at 9 kg/h.

When the power was increased, the outlet temperature was close to the housing temperature until evaporation started. Evaporation temperature was reached with an electrical power of about 1000 W. A further increase of the power to 1800 W led to a high housing temperature of 450°C at the maximum, while the outlet temperature remained constant at the evaporation temperature of water. The constant outlet temperature of 100°C indicates that the evaporation of the water throughput has not yet been completed.

In Figure 3.3, the electrical power, the thermal power transferred and the ratio between thermal power and electrical power are shown as a function of the applied voltage. Thermal power was calculated from the constant mass throughput of 9 kg/h, the difference between inlet and outlet temperature of the water throughput and the specific heat capacity of water. Figure 3.3 shows the results up to the evaporation point of water. The results indicate that the micro heat exchanger is very efficient with a power ratio of thermal power / electrical power of up to 0.93 when liquid water is heated.

In Figure 3.4, the water temperatures at the inlet and the outlet and the transferred power of a type 2 device are shown as a function of the water throughput. The applied electrical power was held constant at 1100 W, the throughput was varied between 12 kg/h and 60 kg/h. At a constant inlet temperature, the outlet temperature is decreasing with increasing throughput. Furthermore, the transferred power is slightly reduced.

To calculate the heating rate $\Delta T/\Delta t$ [K/s], equation (2) was used:

$$\frac{\Delta T}{\Delta t} = \frac{T_{out} - T_{in}}{\Delta t} \quad (2)$$

T_{out} is the fluid temperature at the outlet, T_{in} the fluid temperature at the inlet of the device. Δt is the residence time of the fluid.

In the experiment shown in Figure 3.4, a residence time of 247 ms was achieved at the lowest water throughput of 12 kg/h. A temperature difference of 75 K was measured at this point, which corresponds to a heating rate value of 304 K/s. A water throughput of 60 kg/h led to a residence time of 50 ms and a temperature difference of 15 K. For this, a heating rate of 300 K/s can be calculated. The electrical power applied in this experiment was kept at 1100 W.

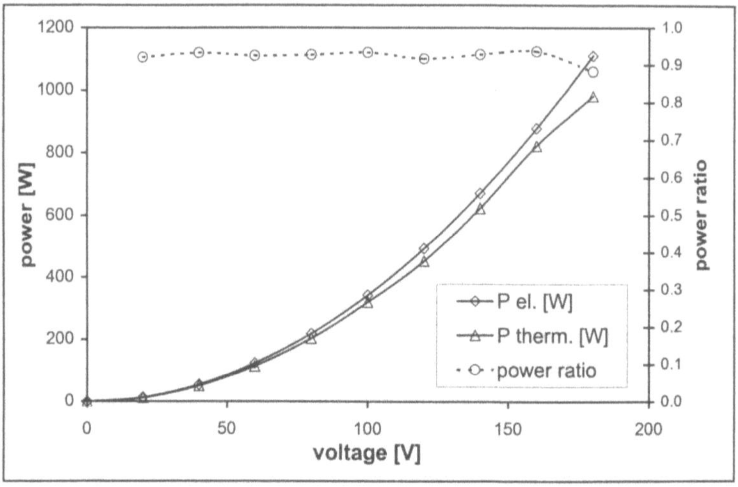

Fig. 3.3: Applied electrical and transferred thermal power as a function of voltage. This figure also shows the power ratio between applied and transferred power. As the maximum, a power ratio of 0.93 was reached at a water throughput of 9 kg/h.

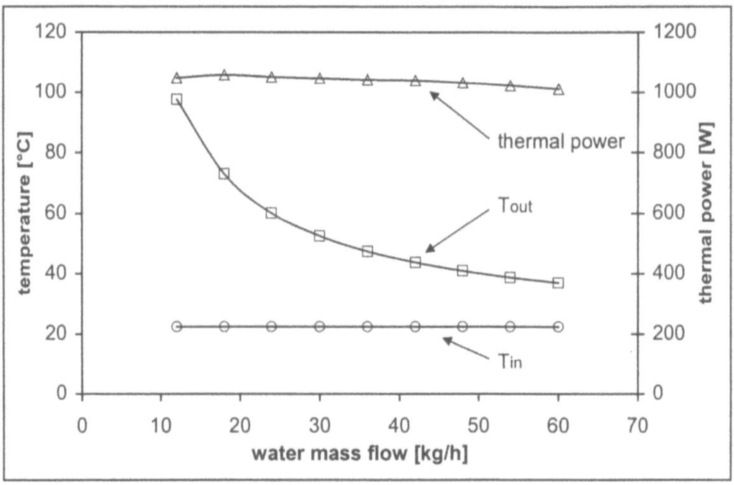

Fig. 3.4: Temperatures at the inlet and the outlet and the transferred power of a type 2 device as a function of the water throughput. The electrical power was held constant at 1100 W.

Besides testing the devices with water, tests to heat up gases and gas mixtures have been made.

Within the framework of the BMBF project MIKREAK [7] an electrically driven microstructure heat exchanger type 2 is used for preheating an inflammable CH_4/O_2 gas mixture safely up to 450°C before it enters the reaction chamber. The use of non-microstructure electrical heaters might cause problems because of the explosion risk of the gas mixture. A liquid-driven microstructure heat exchanger cannot be used because the highest temperature that can be reached with conventional heat transfer liquids is about 400°C.

In this application, the temperature is controlled by a simple program controller, the electrical power can be adjusted by a controllable voltage transformer.

With a flow of 360 l/h of the gas mixture at a pressure of 1 bar, a gas temperature of 25°C at the inlet of the device and an electrical power of 76 W, an outlet gas temperature of 450°C could be reached. Taking into account the expanded volume of the heated gas, the mean residence time Δt_m of the gas was 5.9 ms. The mean heating rate achieved was 72000 K/s.

As a maximum temperature, 600°C was reached at a flow of 1500 l/h N_2 as test fluid and a pressure of 1 bar. The electrical power was maintained at 650 W. Due to the expanded volume of the heated gas, the mean residence time in this experiment was 1.4 ms. The temperature of the gas at the inlet was 25°C. A heating rate of 410700 K/s was reached.

4 Summary

Prototypes of electrically heated microstructure heat exchangers have been developed, which can be used to gain fast and easy control of the process temperature of fluids. As material, stainless steel is used. The maximum electrical power achievable with those laboratory prototype devices is 600 W and 1800 W. The power to be transferred is adjusted by either a voltage controller or a conventional programmable controller. The devices may be applied for the temperature control of fluids up to several hundred degrees Kelvin as well as for endothermic chemical reactions.

Water throughputs of up to 60 kg/h and gas throughputs of up to 1500 l/h have been applied.

The internal structure of the prototype devices is determined by an alternating order of heating elements and microchannel structures. The differences between the device types are the arrangement and type of the electrical resistor heaters used.

Under certain operating conditions, the electrically driven heat exchanger devices can be used for evaporation of liquids. The precise controlling of the evaporation temperature with a high-resolution voltage controller is possible. This is a prerequisite for a good access to vapors of various fluids.

In the experiments with water as test fluid, a heating rate along the microchannels of 300 K/s was reached. Experiments with N_2 have shown that fluid temperatures of up to 600°C can be reached. In these tests, a heating rate of 410700 K/s at the maximum was achieved.

One of the devices is currently used as a preheater for an inflammable mixture of CH_4 and O_2. The mixture can be heated up safely to a target temperature of 450°C.

Acknowledgement

The authors thank Mr. H. Schlenker and Mr. S. Zimmermann for their precious contribution to the construction of the experimental setup.

5 References

[1] K. Schubert, W. Bier, J. Brandner, M. Fichtner, C. Franz, G. Linder: Realization and Testing of Microstructure Reactors, Micro Heat Exchangers and Micromixers for Industrial Applications in Chemical Engineering; Proceedings of the 2nd International Conference on Microreaction Technology, March 9-12, 1998, New Orleans, Louisiana, USA; pp.88-95

[2] K. Schubert, W. Bier, G. Linder, D. Seidel: Herstellung und Test von kompakten Mikrowärmeübertragern; Chem. Ing. Tech. 61, No. 2, P. 172 (1989)

[3] Th. Schaller, W. Bier, G. Linder, K. Schubert: Mechanische Mikrostrukturierung metallischer Oberflächen; F&M 102 1994/5-6

[4] VDI-Wärmeatlas; 7., erweiterte Auflage 1994; VDI-Verlag

[5] K. Stephan: Wärmeübergang beim Kondensieren und beim Sieden; Springer Verlag, 1988

[6] The wte / cps – system: Information to electrically driven heater systems for fluids HTT Hoch – Temperatur – Technik GmbH, P.O. Box 1922, D – 32009 Herford, Germany

[7] J. Mayer, M. Fichtner, K. Schubert, D. Wolf: A microstructured reactor for the catalytic partial oxidation of methane to syngas; 3rd International Conference on Microreaction Technology, April 18-21, 1999, Frankfurt, Germany

Analytical Module for In-line IR Spectroscopy of Chemical Reactions in Microchannels

A. E. Guber and W. Bacher

Forschungszentrum Karlsruhe GmbH
Institut für Mikrostrukturtechnik
Postfach 3640, D-76021 Karlsruhe, Germany
e-mail: andreas.guber@imt.fzk.de

Abstract

In-line spectroscopy in microchannels can be accomplished using a specially developed analytical module. By means of the micro electrical discharge machining technique (μEDM technique), the respective channel forms are first cut into metal substrates and then covered by transparent disks on both sides. IR spectroscopic irradiation of the partially transparent microchannels takes place in an IR microscope and yields any number of data along and across a reaction channel depending on the focus of the measurement IR beam.

Keywords. transparent microchannels, μEDM technique, IR microscope, In-line IR spectroscopy

1. Introduction

Microreaction systems are mainly qualified by practical tests. For this purpose, the entering and leaving substance flows are balanced precisely at the inlet and outlet of the system with particular regard to the reaction products formed and the possibly existing educts. However, it may be advantageous for an exact design of the microreactor to carry out in advance an in-line analysis of the chemical reaction taking place in the individual microchannels. By special channel forms, partially transparent microchannels are obtained, which may be subjected to e.g. IR irradiation for in-line measurement inside these channels [1]. Fabrication of different channel forms takes place by means of the micro electrical discharge machining technique (μEDM technique) based on cutting or sinking microerosion. The IR microscopic measurement method has been further developed and is now well suited for concentration measurements of substance flows at nearly any point of the microchannels.

2. Fabrication of Microchannels by Means of the Micro Electrical Discharge Machining Technique (μEDM Technique)

The μEDM technique is characterized by the material being removed in a contact-free manner by a number of successive spark discharges. The mechanical, physical and chemical properties of the materials are not changed. Hence, hardened or mechanically hardly machinable materials can be machined relatively easily. Complex microstructures, for instance those having difficult profiles, can be achieved with high geometric accuracy, good reproducibility and an absence of burrs. Machining may be accomplished by means of microcutting erosion with minimum cutting wire diameters of 30 μm [2] or by microsinking erosion with miniaturized electrodes [3].

Nearly any geometrical forms of microchannels can be produced by the microcutting EDM technique in thin metal platelets made of stainless steel. Fig. 1 shows a channel system with five educt channels that run into a mixing zone. All channels are cut into a 2 mm thin stainless steel platelet as profiles using the microcutting EDM technique. The five educt channels have a width of only 200 μm each, whereas the mixing zone and the downstream reaction channel are 1 mm wide.

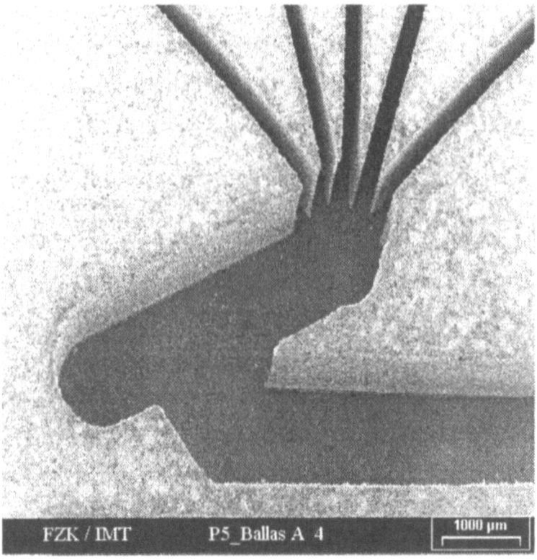

Fig. 1: SEM of a microchannel structure with five educt channels and a downstream mixing zone which runs into the actual reaction channel.

Stepped microchannels can be cut into stainless steel platelets by means of micro-sinking erosion. The tool electrode used for this purpose is made of copper and provided with the inverse structures of channel shapes desired (cf. Fig. 2). The walls between the five educt channels are about 300 μm thick, wall thickness of the reaction channel is about 900 μm. The SEM in Fig. 3 presents a meandering microchannel structure with 3 stepped educt channels that are about 400 μm wide and 400 μm deep each.

Fig. 2: Microstructured tool electrode made of copper for sinking erosion (left) and structured stainless steel platelet with cut-in microchannel structures (right).

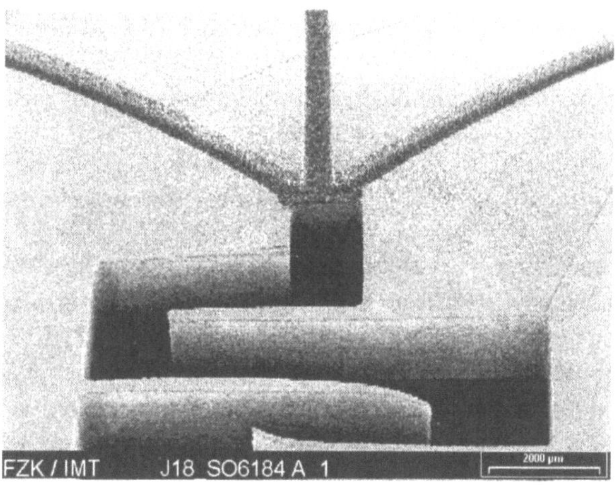

Fig.3: SEM of a stepped microchannel structure with three educt channels and a downstream meandering reaction channel.

They run into a meandering reaction channel, into which 200 μm wide metal ridges protrude. Additional microchannels may be cut into the rear of the stainless steel platelet by positioned sinking microerosion (cf. Fig. 4). The lateral walls and the base of the channels produced by sinking EDM technique have relatively rough surfaces.

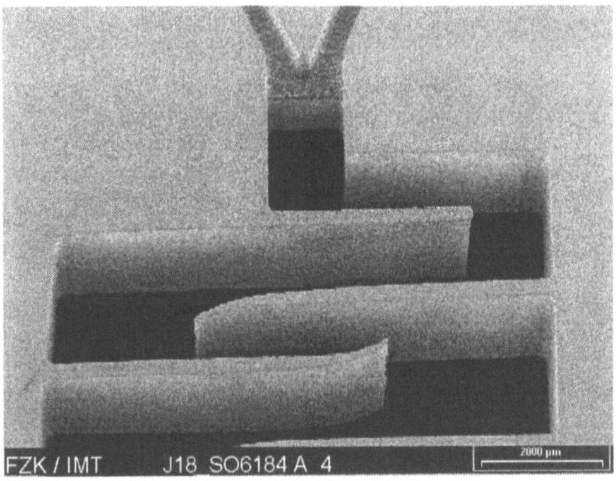

Fig. 4: SEM of a stepped mircochannel structure with two educt channels cut into a stainless steel platelet by positioned sinking microerosion.

3. Analytical Module for the Investigation of Gas Reactions

Fig. 5 (top) shows the modular analytical system with all components required for assembly. Prior to the assembly of the analytical module, the microstructured metal platelets with the cut-in microchannels have to be sputtered on both sides with a very thin teflon layer and then covered by two IR-transparent disks. The AgCl windows used for this purpose are fixed in teflon holding units specially manufactured for this purpose. Vacuum tightness is ensured by Viton O-ring seals. The assembled sandwich of the AgCl windows and the microstructured metal platelet is fixed between two CF flanges (cf. Fig. 5; bottom). Here, a Y-mixer with two educt channels and a reaction channel is shown. Both supply lines as well as the discharge line for the educts or the products are integrated in the lower CF flange. The entire analytical module can be integrated very easily and quickly into the existing test setups.

The analytical module with the partially transparent microchannels can be fixed on the object carrier of an IR microscope for in-line measurements at all points of the transparent sections of the microchannels (cf. Fig. 6). By coupling the IR microscope with an FTIR spectrometer (type IFS 88; BRUKER company), a high-performance microanalytical system for spectroscopic measurements in microchannels is obtained.

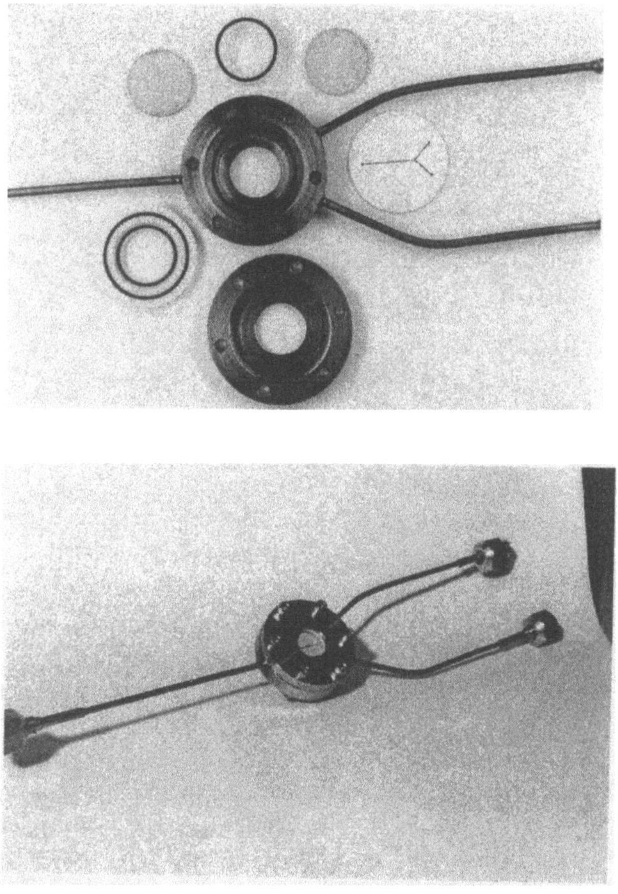

Fig. 5: Modular analytical system in the disassembled (top) and completely assembled state (bottom).

Routine IR measurements are performed in the spectral range between 4500 and 600 cm^{-1} with a spectroscopic resolution of 4 cm^{-1}. The measurement spot of the IR beam can be adjusted variably between 10 and 230 μm using the microscope optics. Micrometer-precise movement of the object carrier of the IR microscope allows to approach nearly any number of measuring points in the microchannels. Preliminary tests have demonstrated that the measurements are to be performed in the transmission mode, as reflection measurements were found to be associated with a significant signal loss.

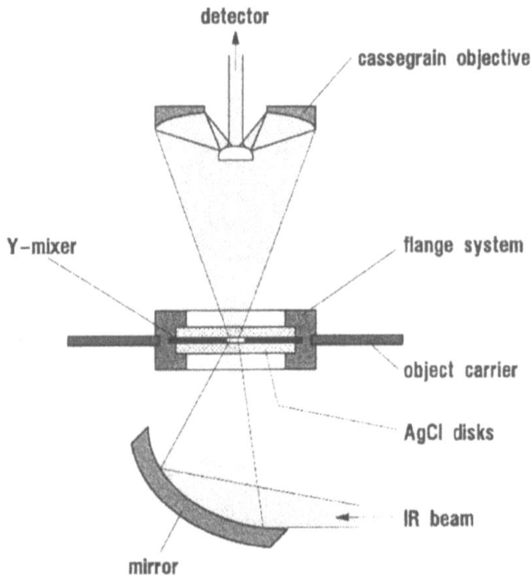

Fig. 6: Schematic representation of in-line measurements in microchannels using IR microscopy.

The measurement method presented here allows to determine the concentrations of all IR-active components in the substance flows at practically any point of a transparent microchannel. The above analytical modules have an optical length of 2 mm and, hence, are very well suited to study gas/gas reactions in microchannels [1].

The model chemical exchange reaction between the two halogen-containing gases $SiCl_4$ and F_2 was studied again [1]. The educt $SiCl_4$, the intermediate reaction products of the type $SiCl_xF_y$ ($SiFCl_3$, SiF_2Cl_2 and SiF_3Cl) as well as the product SiF_4 are IR-active, whereas the F_2 supplied and the Cl_2 formed are not IR-active. The three-dimensional representation in Fig. 7 shows the development of the SiF_4 concentration along a 1 mm wide and 5 mm long linear microchannel. The inlet of the initial substances of $SiCl_4$ and F_2 takes place via a total of five educt channels (2 educt channels are supplied with $SiCl_4$ gas and 3 educt channels with F_2 gas). $SiCl_4$ and F_2 are supplied stoichiometrically: The $SiCl_4$ and F_2 flow rates amount to 2 st.cm^3/60 s.

All partial pressure data were determined by IR microscopy on the basis of a total of 260 measurement points (10 measurement points across and 26 measurement points along the microchannel). The diameter of the IR measurement spot was about 33 μm. Fig. 8 shows the decreasing $SiCl_4$ signal in a linear reaction channel.

It is demonstrated by the measurements that in case of a pure gas/gas reaction a complete mixing of all educt flows in the reaction channel is achieved after a few micrometers already. The gas exchange reaction between $SiCl_4$ and F_2 is characterized by rather rapid SiF_4 formation. The reaction yield can be controlled through the educt flows supplied and the residence times in the microchannel.

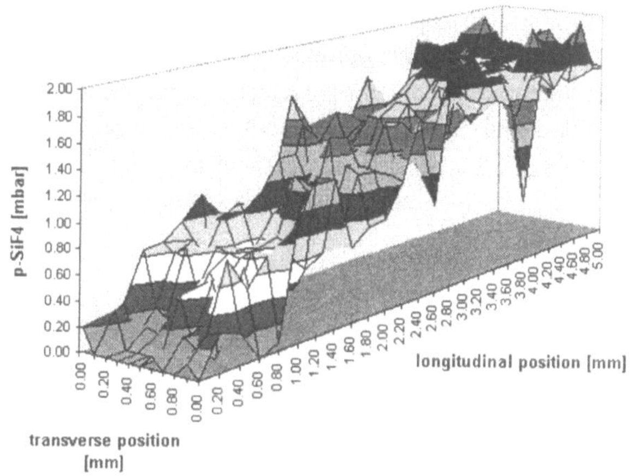

Fig. 7: 3D representation of the SiF_4 concentration and partial pressure development inside a linear microchannel.

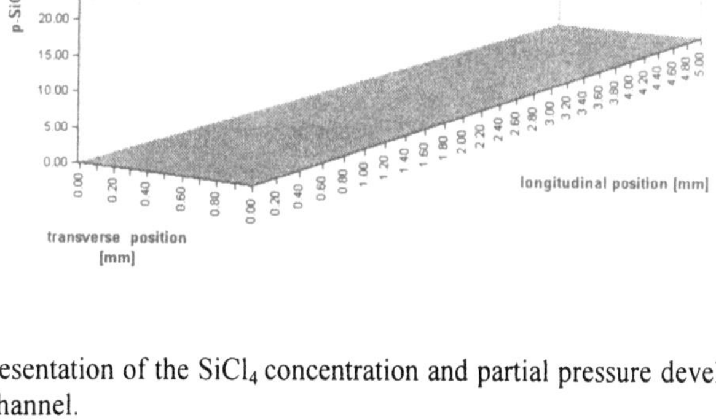

Fig. 8: 3D representation of the $SiCl_4$ concentration and partial pressure development inside a linear microchannel.

4. Summary and Outlook

The micro electrical discharge machining technique allows to cut profiles into thin metal platelets. By coverage on both sides, partially transparent microreaction systems are obtained. They can be irradiated by a specially developed in-line measurement method based on IR microscopy and FTIR spectroscopy. Thus, a detailed study of a pure gas/gas reaction in a microchannel system is possible. The measurement data obtained can also be used for the verification of simulation calculations and, hence, for an optimization of the channel geometries of microreactors. At the moment, this in-line spectroscopic measurement technique is adapted to gas/liquid and liquid/liquid reactions for application in the field of biotechnology.

5. Acknowledgments

We would like to thank Prof. Dr. V. Saile for discussions and the constant support of our work. Our special thanks go to Mr. P. Abaffy for the performance of numerous IR measurements and their evaluation. Furthermore we would like to thank Mr. A. Mayer for the rapid fabrication of various microchannel structures using the μEDM technique.

6. References

[1] A. E. Guber, W. Bier, K. Schubert: IR Spectroscopic Studies of a Chemical Reaction in Various Micromixer Designs, in Proc. of 2nd International Conference on Microreaction Technology, March 9 – 12, 1998, New Orleans, U.S.A., pp. 284 – 289 (1998)

[2] A. E. Guber, N. Giordano, M. Loser, P. Wieneke: Mikroinstrumente aus Nickel-Titan, F&M Feinwerktechnik, Mikrotechnik, Mikroelektronik **104 (4)**, pp. 247 – 251 (1997)

[3] W. Ehrfeld, C. Gärtner, K. Golbig, V. Hessel, R. Konrad, H. Löwe, Th. Richter, Ch. Schulz: Fabrication of Components and Systems for Chemical and Biological Microreactors, in Proc. of 1nd International Conference on Microreaction Technology, February 23 – 25, 1997, Frankfurt, Germany, pp. 72 – 90 (1997)

A Microstructure Reactor for Gas Purification

R. Wunsch, M. Fichtner and K. Schubert

Forschungszentrum Karlsruhe, 76021 Karlsruhe, Germany

Abstract. For certain laboratory applications, continuous flows of hydrocarbon-free gases are required. Microstructured reactors are suited for this task, because their small operating volume combined with the high mass transfer rates in the microchannels may result in high space-time yields.

In the present paper, first results are reported regarding the oxidation of small amounts of methane in air by a heterogeneously catalyzed reaction in a microstructure reactor which was especially developed for that purpose. In this reactor the oxidation takes place at the walls of the microchannels. A large active surface area for the catalytic reaction was provided by the anodic oxidation of the aluminum microstructures. An impregnation with diluted platinum nitrate solutions, followed by the reduction of the metal precursors, served to generate catalytically active sites on the aluminum oxide substrate.

Testing of the catalyst performance focussed on the oxidation of methane to evaluate the efficiency of the catalytic microstructure reactor as well as to determine the impact of the operating conditions on the catalytic activity.

The experimental results obtained so far show that the microstructure reactor may be a suitable and compact device for hydrocarbon removal from gas flows.

Keywords. Methane Conversion, Gas Purification, Platinum Catalyst, Microstructure Reactor, Microchannel Reactor

1 Introduction

One aspect in the development of laboratory equipment with ever increasing precision and sensitivity is the demand for highly pure gases. To achieve a high reproducibility of the measured results it is mandatory to supply highly purified gases of constant quality. Purified gases are available with less than 2 ppm impurity levels from gas cylinders. However, existing safety regulations require the use of complicated and expensive gas supply systems. Furthermore the operational costs are high due to the price of these highly purified gases.

Therefore it is desirable to provide highly purified gases without the storage requirements of cylinders through the use of gas purifiers.

Hydrocarbon-free air is one of the most often required gases in analytics and emission monitoring. It is used for example as carrier gas or oxygen source in flame ionization detectors (FID's). A major characteristic of this purified air is a typical hydrocarbon and carbon monoxide content of only a few ppm.

One possibility to remove hydrocarbons from air is their catalytic combustion. The products of this heterogeneous catalysis should only be carbon dioxide and water.

According to the literature, precious metal catalysts are mostly used in the conversion of hydrocarbons to the desired products at temperatures ranging from 400 to 550 °C [1-4]. At present, the full conversion of all hydrocarbons in the air requires quite large reactors even at low volume flows. The reactor size could be reduced through the use of microstructures which should possess much larger space time yields than conventional fixed-bed reactors. The short diffusion distances and large heat transfer coefficients in the microchannels affect the mass and heat transfer characteristics of these reactors positively [5,6]. Furthermore, miniaturization shortens start-up times and the overall power consumption of these devices .

In the following contribution the development of such a microstructure reactor for the removal of hydrocarbons from air is described. One of the main prerequisites for this purpose is the formation of catalytically active sites at the surface of the microchannels [7]. Therefore the anodic oxidation of the aluminum support was chosen to create high surface area in microchannels [8]. By means of anodic oxidation, the surface was converted into thin porous alumina shell containing regular mesopores [9,10]. The catalytically active component was immobilized by a wet impregnation technique. The microstructured and catalytically activated foils were piled up and accommodated in a housing.

In order to test the activity of the microstructured reactors for gas purification we chose methane as the most unreactive hydrocarbon in an oxygen-rich atmosphere to simulate a real feedstream.

In the following paper the conditons and the experimental results of operation of the laboratory device are described.

2 Experimental

2.1 Microchannel reactor configuration

The layout of the microstructure reactor includes microstructured foils with the catalytical coating, electrical heat cartridges and the housing with gas inlet and outlet.

For this study, aluminum was used as base material for the microstructured foils. After preparation of the catalyst, the microstructured and catalytically activated foils were stacked and clamped in the reactor housing. A stack with 820 reaction channels with lengths of 220 mm each was formed. Typically the microchannels have hydraulic diameters of about 120 µm. The total geometrical surface of all microstructured foils is 0.19 m^2. The microstructured reactor was heated with electrical cartridges.

2.2 Preparation of the catalytic coating

For preparation of the catalytic active surface the microstructured foils were degreased in acetone, stained in an aqueous solution of NaOH/NaF (1.0/1.0 wt.%) for 10 min and anodically oxidized in an aqueous solution of 1.5 wt.% oxalic acid. The anodic oxidation was carried out under conditions of constant voltage (50 V) and temperature (12 °C) for 240 minutes. The anodized samples were rinsed with deionized water and dried at 500 °C for 5 h.

The noble metal catalysts were prepared by wetness impregnation. Platinum was incorporated by adsorption from a solution using platinum nitrate and water as the solvent. To prepare different metal loadings, solutions of appropriate concentration between 0.001 up to 0.2 mol/l $Pt(NO_3)_2$ were used. To obtain an optimum adsorption, impregnation was carried out at 50 mbar and 40 °C for 1 h.

The activation of the noble metal precursors was done by drying at 100 °C for 1 hour. Finally the temperature was increased at a rate of 1.7 °C/min up to 500 °C in a H_2/Ar stream and held constant at 500 °C for 5 additional hours.

2.3 Determination of textural coating parameters

The specific surface area of the catalyst systems was determined by the BET method from the nitrogen adsorption isotherm at 77 K. Mean pore size distribution was calculated by applying the BJH method. In addition, the metal dispersion and the particle size were determined by hydrogen chemisorption at 40 °C. Field emission scanning electron microscopy was applied to estimate textural parameters of the pore system. The platinum distribution in the alumina layer was measured by X-ray fluorescence (WDX). The total amount of platinum in the layer was determined by ICP-emission spectrometry.

2.4 Test equipment

The experiments have been performed in an apparatus which consists of a gas mixing section, a reactor section and an analysis section. In the gas mixing section a model gas mixture consisting of 20.5 vol.% O_2 and 79.5 vol.% N_2 was prepared and 70 to 1000 ppm CH_4 were added employing thermal flow meters for each gas component. Upstream of the mixing section, water vapor from 0 to 20 vol.% H_2O can be mixed to the model gas. The total pressure was varied in the range between 110 kPa and 510 kPa. Thus, the residence time varied from 200 to 800 ms. The methane turnovers were measured in the temperature range between 100 and 520 °C at a heating rate of 1.5 °C/min. It was impossible to perform activity measurements above 520 °C due to the use of aluminum as carrier material.

At the entrance and exit of the microstructured reactor, on-line analysis was performed. A flame ionization detector was applied to monitor CH_4. CO and CO_2 were detected under dry conditions with dedicated nondispersive IR detectors.

The CH_4 conversion was calculated based on the following equations:

$$x(CH_4) = \frac{c(CH_4)_i - c(CH_4)_f}{c(CH_4)_i}$$

where c_i and c_f describe the initial (at the inlet to the reactor) and final (at the exit of the reactor) concentrations.

3 Results and discussion

3.1 Catalyst characterization

The alumina coatings prepared by anodic oxidation were regular and homogeneous. Figure 1 shows a field emission scanning electron microscope view on the top of the surface of an anodized foil.

By using the conditions of anodic oxidation presented in chapter 2.2, the thickness of the alumina layer is 20 µm as measured by SEM.

Pore size distribution was calculated from nitrogen desorption isotherms. It was found that the average pore diameter of the samples is 41 nm. As can be seen in Figure 2 the diameters of the mesopores are in a narrow range between 30 and 80 nm with a maximum at about 39 nm. Furthermore, the pore structure of impregnated alumina was almost the same as that of not impregnated alumina.

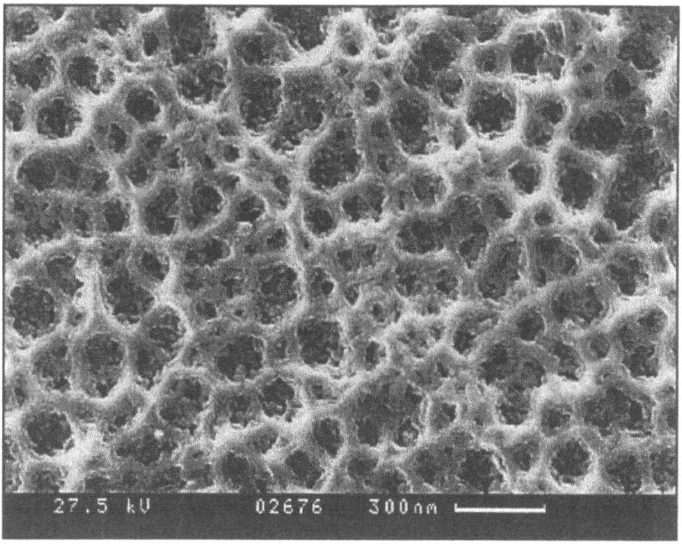

Fig. 3.1: FEM image of the top of the surface of an anodized aluminum foil. Conditions of anodic oxidation: oxalic acid (1.5 wt.%); U = 50 V; t = 4 h; T = 12 °C; electrode distance = 8 cm.

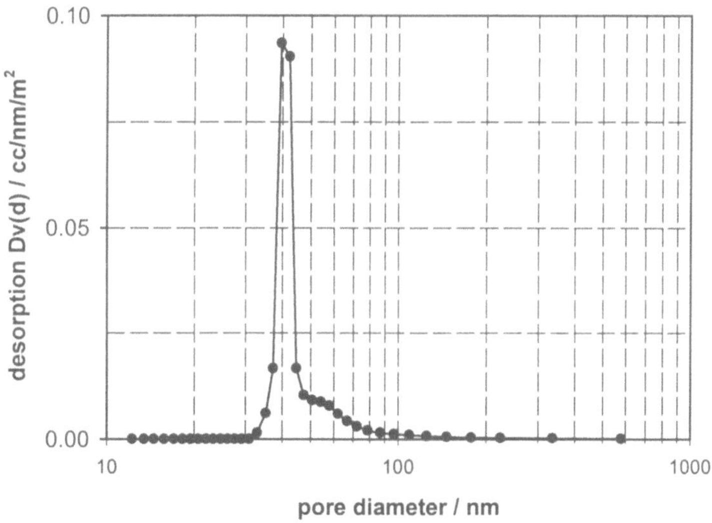

Fig. 3.2: Pore size distribution by applying the BJH method of a surface of an anodized alumina layer coating. Conditions of anodic oxidation: oxalic acid (1.5 wt.%); U = 50 V; t = 4 h; T = 12 °C; electrode distance = 8 cm.

Due to the formation of mesopores in the oxide layer and the chemical treatment of the microstructured foils, the surface area increased. The specific surface area of the non-oxidized microstructured foils with a surface of $2.9 \cdot 10^{-3}$ m^2/g was increased up to 7.0 m^2/g. For practical reasons the real surface area after anodization was related to the geometrical surface area of the microchannels, i.e. a magnification factor of the geometrical surface of 2400 m^2/m^2 was reached.

As well, the magnification factor of the impregnated alumina was determined. It was observed with a platinum metal loading in the range between 0.02 and 1.6 wt.% that the magnification factor is almost the same.

The results of the hydrogen chemisorption showed a small platinum particle size. For example, in the catalytic system with 0.02 wt.% Pt, the average crystallite size was found to be 2.6 nm. The metal dispersion which is defined as the fraction of metal atoms found on the surface of active metal particles was 43%.

Figure 3 shows a WDX linescan (aluminum, oxygen and platinum) for one of the catalysts in which a small particle size distribution was observed. Similar micrographs were obtained for other catalysts, with metal loadings between 0.02 and 1.6 wt.% .

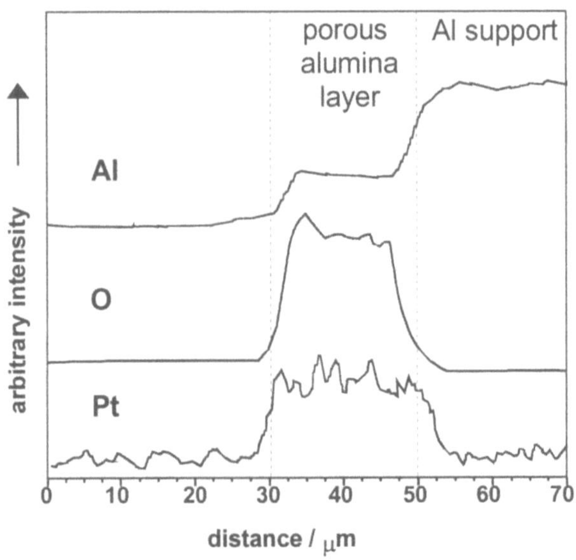

Fig. 3.3: WDX linescan of the cross section of microstructured and coated aluminum foils (20 μm porous alumina layer, 0.4 wt.% Pt)

From the intensity distributions in Figure 3 it can be concluded that the platinum was homogeneously distributed in the oxide layer. Obviously, the wetness impregnation described in chapter 2.2 is a suitable method for a uniform deposition.

3.2 Studies of catalytic activity

Figure 4 shows activity data for methane oxidation as a function of the inlet temperature and the system pressure as parameter for the catalytic microstructure reactor with 0.4 wt.% Pt in the catalytic coating.

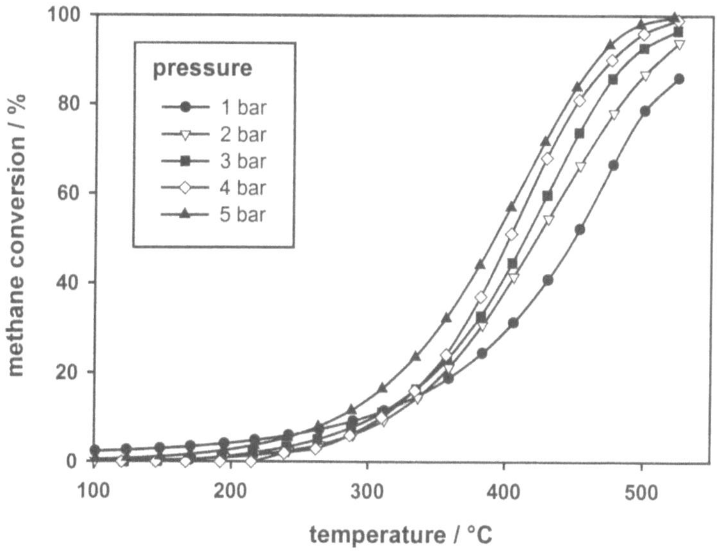

Fig. 3.4: Activity data of methane oxidation in a catalytic microstructure reactor with 0.4 wt.% Pt; operation conditions: 120 ppm methane; 20.5 vol.% O_2 in N_2; flow rate = 1000 sccm/min

Typical sigmoidal curves were observed. The reaction started at about 250 °C; first the conversion increased slowly then drastically as the temperature increased. As well, the methane conversion increased as the system pressure increased. The complete conversion (>99.8%) was reached by a system pressure of 5 bar. Up to full conversion, only CO_2 and H_2O were observed as products and CO was not formed in the investigated temperature range.

The effect of different methane concentrations on the catalytic activity was also investigated. In Fig. 5 the results for 3 different catalyst temperatures are plotted.

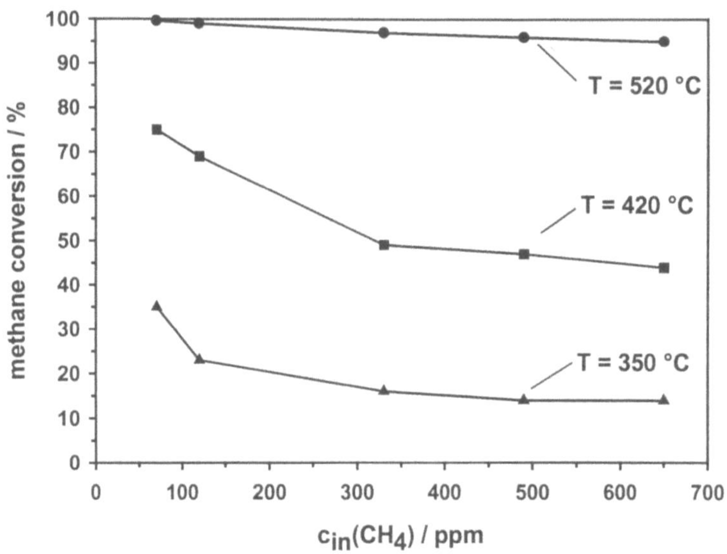

Fig. 3.5: The dependence of the catalytic activity on the methane concentration in a catalytic microstructure reactor with 0.4 wt.% Pt; operation conditions: 20.5 vol.% O_2 in N_2; pressure = 4 bar; flow rate = 1000 sccm/min

Figure 5 shows that the methane conversion decreased with rising methane concentration. The reduction at a catalyst temperature of 520 °C was found to be only a few percent, while a stronger decline was observed at 420 and 350 °C.

For practical application of a microreactor as a catalytic scrubber, it is important to know the effect of water vapor on the activity. Therefore, the effect of water vapor on the methane conversion was investigated at three reaction temperatures (Fig. 6). At a reaction temperature of 520 °C, the conversion is reduced by only 9% as the water vapor content increases from 0 to 20%. The observed deactivation in the range up to 20 vol.% water vapor was reversible.

For gas purification applications in the region from 0 to 5%, the conversion drops from 99.8 to 98.2%. Again, at catalyst temperatures lower than 520 °C, a much stronger decline of the methane conversion as a function of the gas composition was observed.

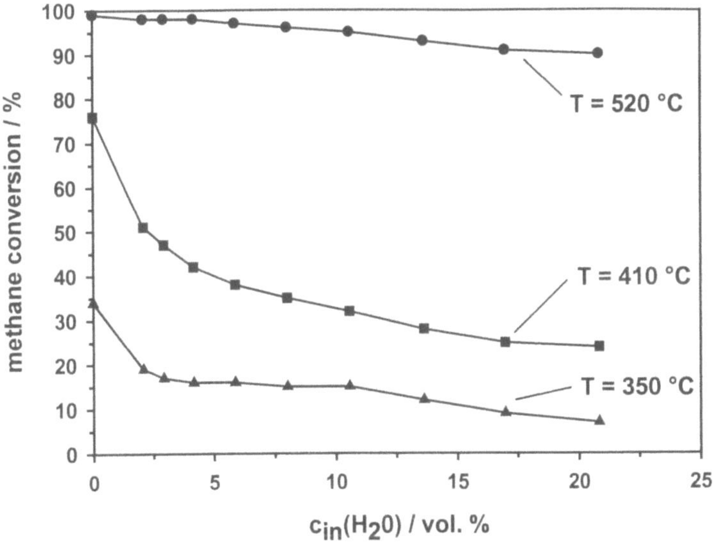

Fig. 3.6: Effect of water vapor on the methane conversion; catalytic microstructure reactor with 0.4 wt.% Pt; operation conditions: 70 ppm methane; 20.5 vol.% O_2 in N_2; pressure = 4 bar; flow rate = 1000 sccm/min

Finally, the effect of precious metal loading on the methane conversion was investigated. Figure 7 shows the methane conversion at 520 °C as a function of metal loading.

The conversion increases from 4 to 92 % when increasing the loading from 0.02 to 0.4 wt.% Pt. Above 0.4 wt.% metal loading, the conversion begins to level out with the metal loading. However, for high methane conversion (>98%) at 2 bar the metal loading has to be higher than 1.4 wt.% Pt.

In order to study the durability of the manufactured catalysts, they were subjected to an ageing process consisting of consecutive measurements by heating the microreactor from 100 °C to 550 °C with a temperature rate of 1.5 °C/min under steady feed stream for 100 days. The catalytic behavior of the aged microstructured reactor with a 0.3 wt.% Pt catalyst was compared with that of the fresh catalysts.

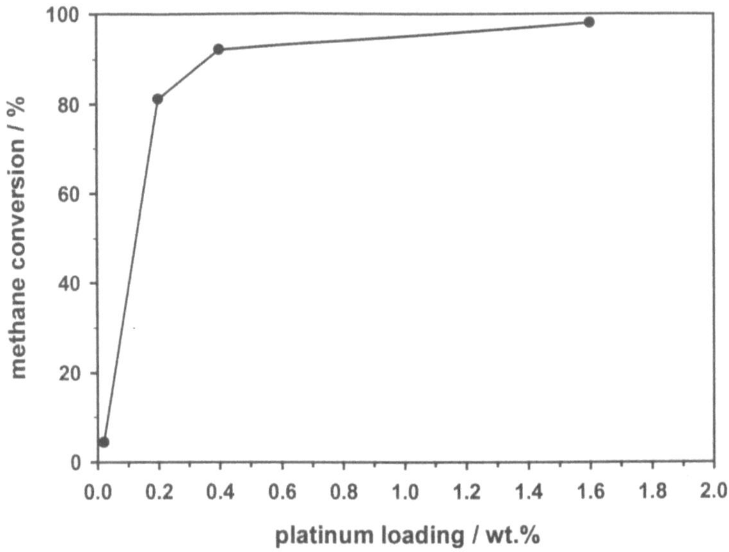

Fig. 3.7: Effect of metal loading on the methane conversion; operation conditions: 120 ppm CH_4; 20.5 vol.% O_2 in N_2; pressure = 2 bar; flow rate = 1000 sccm/min; temperature = 520 °C

4 Conclusion

For the combustion of small amounts of hydrocarbons in air, a microchannel reactor device equipped with catalytically active microstructured aluminum foils was developed. The catalytic coating was manufactured by anodic oxidation of the aluminum foils, followed by an optimized wet impregnation procedure with diluted platinum nitrate. To attain complete turnover of methane, the microreactor should be operated at temperatures higher than 500 °C and a system pressure of more than 2 bar. Furthermore noble metal loading of more than 1.4 wt.% should be envisaged. In such a system complete conversion to carbon dioxide and water can be achieved.

This catalytic microstructured reactor will serve as a basis for the development of a small compact processing unit for the removal of hydrocarbons in an air stream. Future work will focus on the integration into analytical devices.

5 References

[1] R. Anderson, K. Stein, j. Freenan, and L. Hofer, Ind. Eng. Chem., 53,(1961) 809

[2] P. Norton, R. Tapping and J. Goodale, J. Vac. Sci. Technol., 14 (1977) 446

[3] C. Cullis, D. Keene and D. Trimm, J. Catal., 19 (1970) 378

[4] K.R. Thampi, J. Kiwi and M. Gratzel, Catal. Lett., 4 (1990) 49

[5] K. Schubert, W. Bier, J. Brander, M. Fichtner, C. Franz and G. Linder, Proc. 2nd Inter. Conf. Microreaction Technol., (1998)

[6] R. S. Wegeng, C. J. Call and M. K. Drost, Proc. Spring National Meeting of AICHE, (1996)

[7] M. Fichtner, W. Benzinger, R. Wunsch, K. Schubert, K. Haas-Santo, this issue

[8] G. Wießmeier, Doctoral Thesis, University of Karlsruhe, (1996)

[9] G. Wießmeier, K. Schubert and D. Hönicke, Proc. 1st Inter. Conf. Microreaction Technol., (1997)

[10] F. Keller, M. S. Hunter and D. L. Robinson, J Electrochem. Soc., 100 (1953) 411

A Flexible Multi-Component Microreaction System for Liquid Phase Reactions

Th. Richter, W. Ehrfeld, V. Hessel, H. Löwe, M. Storz, A. Wolf

Institut für Mikrotechnik Mainz GmbH, Carl-Zeiss-Str. 18-20, D-55129 Mainz, Germany, e-mail: richter@imm-mainz.de

Abstract

First microreaction systems, introduced only a few years ago, mainly were tailored to very specific needs or process requirements. Meanwhile, due to the increasing interest in microreaction systems and their applications, there is a rising demand regarding flexible microreactors which are easy to adjust to various reaction conditions. Enhanced variability regarding process parameters like temperature and residence time as well as simple changes in reaction sequences are important features in order to explore the specific qualities of microreactors and their potential for process optimization.

Here, a microreaction system, available in small lot production and suited especially for liquid multiphase reactions, is introduced. With respect to heat management and variability of residence times this microreactor enables novel reaction conditions not applicable in macroscopic devices. At the same time its specific order of process steps can be changed easily. All components are constructed in a highly alloyed stainless steel suitable even for extremely corrosive reactants and solvents respectively.

1 Introduction

The development of a huge variety of microstructuring techniques has widened significantly the choice of materials available for microreaction technology in recent years [1]. Especially the use of microstructures resistant to high pressures and temperatures as well as corrosive reactants notably sped up the development of microreaction technology. Materials used include metals, special alloys as well as ceramic materials. Meanwhile, microstructured components are available for various unit operations [2].

Today, first microreactor components as well as integrated systems are being used in industrial laboratories as well as for production purposes. The main prospects of microreaction technology cannot only be found in the mere reduction of the system size but in availability of completely new reaction regimes not accessible when using conventional equipment. More recently, these advantages, which lead to improved yields and selectivities, were utilized for industrial production processes as well. For example, miniaturized components within a production line of a metallo-organic compound at Merck KGaA led to a 20 % improvement to a yield of 90 % [3]. At Aventis Research & Technologies the amount of side reactions within a polymerization process was minimized significantly by assuring a defined mixing process of the reactants using micromixers [4].

The recent advances in microreaction technology lead to a high demand with respect to flexible microreaction systems for process development and exploration of new reaction regimes. Effective heat management within microreactors often turns out to be the major issue for reactor design as well as for process optimization and development. There are some basic characteristics of microreaction systems leading to favorable and improved reaction conditions as briefly discussed in the following.

Large surface to volume ratios

Miniaturization with respect to the third dimension provides a significant enlargement in surface to volume ratio as compared to large scale processes. Thus, even within small reaction volumes, huge transfer areas become accessible for improved exchange of heat or matter leading to efficient heat management and mixing processes in microreaction systems. In case of multiphase reactions, large surface to volume ratios enable process optimization in case of intense transport of matter e.g. between immiscible phases [5, 6, 7].

Short reaction pathways

Down-scaling along the flow axis enables large gradients in temperature, pressure, reactant concentration or density of the reactants [2]. Thus, integration of several unit operations within small volumes provides rapid changes in reaction conditions, e.g. in temperature [8]. This way, even highly unstable intermediates may be used for subsequent reaction sequences before their deactivation or decomposition [9]. Moreover, small dimensions of fluid channels in microreaction systems often lead to laminar flow behavior and, therefore, defined residence times. Residence time distributions tend to be well defined assuring unchanged reaction conditions regardless of the actual flow path.

Defined process characteristics

The above mentioned effects of miniaturization lead to defined process characteristics in microreactors. This in particular holds for isothermal processing even of highly exothermic processes enabling a save performance of potentially hazardous reactions [10]. Meanwhile, a variety of applications in reaction engineering, both on lab and production scale, depend on the potential benefits of microreactors such as significantly enhanced yields and selectivities [3, 4]. Additionally, there is experimental proof with respect to the advantages of isothermal processing [7, 10] and the use of explosive reaction regimes in microreactors as well [11, 12]. These should lead to widespread applications of microreactors in chemical processing in the near future.

2 System description

The integrated microreaction system discussed here can be flexibly adapted with respect to varying demands of liquid / liquid multiphase reactions. It has been designed for fast exothermal reactions, where benefit can be expected from isothermal processing. It is especially suited for fast liquid multiphase reactions which show fast side reactions as well. The reaction process is expected to mainly take place at the interfacial area but not to be extremely sensible to mass transport through the interfacial area.

The construction material, a highly alloyed stainless steel (1.4539; X1NiCrMoCuN 25 20 5) is suited even for highly corrosive reactants. The systems plate-like architecture enables easy exchange or addition of system components and allows various combinations of the components with respect to each other. Therefore, reaction sequences as well as residence times can be varied easily. The main components comprise channel structures either microstructured by a combination of micro electro discharge machining techniques (μ-EDM) or realized by precision engineering like milling or grinding.

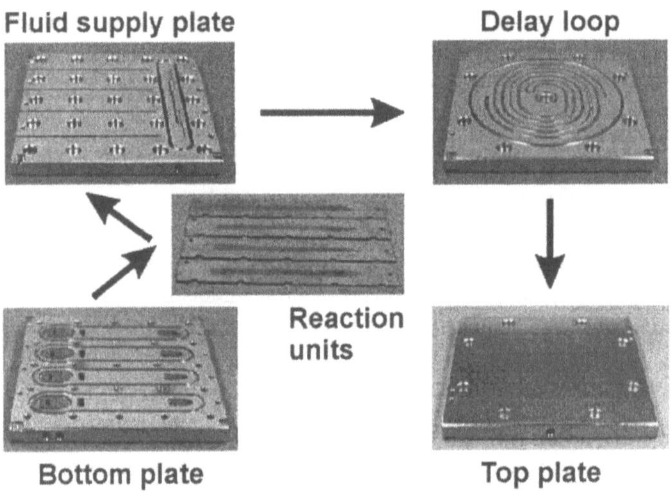

Fig. 1: Overview on the standard system components. Flexible arrangement of the reactor components is assured by variable stacking.

2.1 Reaction unit

The system's main component, a reaction unit of 1.2 mm thickness and featuring a combined micro-contactor / heat exchanger component, is crucial for assuring an efficient heat transfer within the reaction zone. A similar system already proved to assure isothermal reaction conditions even for highly exothermic reactions [7].

The 32 reaction channels, showing a length of 61.4 mm and an overall volume of 0.12 ml, are evenly distributed over four independent reaction units, each comprising 8 reaction channels. Thus, the system's volume flow can be adjusted to lower values at constant reaction conditions by replacement of any reaction unit by an unstructured platelet, using only 8, 16 or 24 reaction channels.

The reactants enter the reaction zone from opposite sides at the beginning of each reaction unit. The fluid phase showing the higher density is supposed to enter each reaction unit from the upper side, whereas the less dense fluid is introduced from the lower side. Both fluids are guided to the contactor area through narrow but deep channels (width 70 μm, depth 450 μm). Within the contact zone, they enter a microchannel of unchanged width but doubled depth (900 μm) located at the upper side of the reaction unit. Intense contact between the reactants is assured by gravity, forcing the more dense liquid to move to the bottom of the channel. Therefore, a difference in density of the reacting liquids as well as high flow velocities are recommended in order to assure large contact areas between both fluid phases.

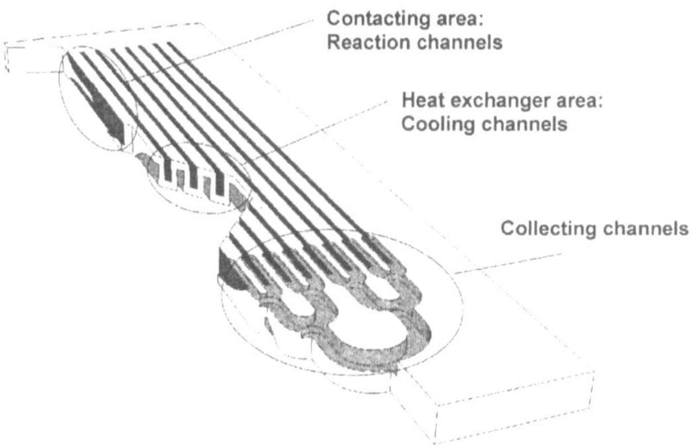

Fig. 2: Schematic view of the reaction unit showing reaction channels (upper side) and cooling channels (lower side).

Microchannels for the cooling liquid (width 70 μm, depth 900 μm) are introduced from the lower side of each reaction unit. These are located in between the reaction channels. Thus, both types of microchannels are separated only by a thin layer of metal (approx. 100 μm) leading to an efficient heat transfer between reaction and cooling channels.

The system is equipped with eight thermocouples for thermal control of the chemical process at the very beginning as well as at the end of each individual reaction zone. Therefore, differences in reaction performance of each of the 32 reaction channels can be determined by measuring possible differences in reaction temperature of the individual channels.

2.2 System specifications and possible combinations of components

Figure 3 shows a set of possible combinations with respect to the sequence of the microreactor components. Delay loop plates are available for variation of residence times of the reaction process ranging from minutes to hours depending on the actual flow rate. Residence times can be changed by adjusting the volume flow or inserting delay loops as well. Based on an overall volume flow of 750 ml / h, residence times between 1 s and 10 s can be achieved for aqueous solutions. For a minimum flow of 1 ml / h, residence times vary between 14 min and 2 h respectively. The data are given for systems with and without one delay loop, respectively.

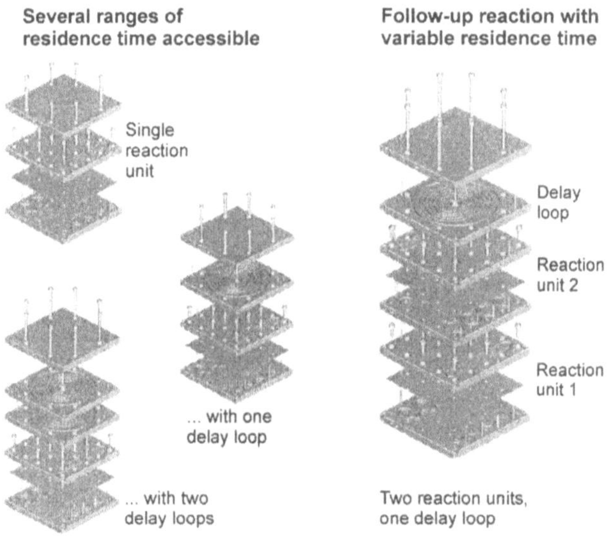

Fig. 3: Possible combinations of the system components along the reaction pathway for easy variation of reaction sequences and residence times.

Moreover, combination of two reaction units allows the performance of a consecutive reaction sequence, for instance immediate processing of unstable reaction products within a second reaction step as indicated in figure 3. Different temperature regimes can be maintained by inserting insulation layers (not shown in figures 2 and 3) between almost any system layer. This thermal de-coupling of single reactor components is suitable in case of any deviation in thermal conditions within the microreactor, e.g. different temperature levels in case of consecutive reactions.

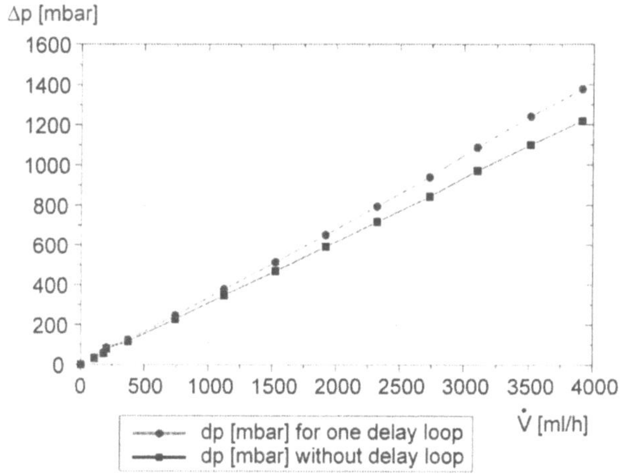

Fig. 4: Pressure drop as a function of volume flow with and without delay loop.

As expected, the system's volume as well as the pressure drop depend on the actual number of delay loops. System volumes are in the milliliter range (< 1 ml using one reaction unit and no delay loop, < 3 ml using one and < 5 ml using two delay loops). Figure 4 indicates the system's pressure drop at ambient conditions using water. Assuming a limit in pressure drop of 1 bar, volume flows of about 3.2 l / h are feasible without delay loop (2.8 l / h using one delay loop).

2.3 Fabrication technique

In order to assure repeatable and less time consuming production of the microstructured components, a new microstructuring process was developed and adapted to the specific needs of small lot production. The process previously used depends on a μ-EDM grinding technique employing rotating disc electrodes [13]. This relatively slow process requires a very accurate pre-shaping of the electrodes as well as several consecutive process steps which lead to a certain loss in precision. Furthermore, the rotating disc method requires highly accurate positioning of the electrode with respect to the work piece. Unprecise positioning leads to a significant rise in process time by 300 - 500%.

These main disadvantages regarding serial production can be overcome by a combined μ-EDM process described briefly in the following. In a first step, tungsten carbide electrodes for the subsequent die sinking process are fabricated by wire erosion using a brass core wire-electrode 0.15 mm in diameter and coated with zinc. Using a special multi-clamping device, as much as five roughing and five finishing electrodes can be employed in one single process without need of any intervention by the operator. This process meanwhile has been optimized using thin wires with a diameter down to 30 μm which yield electrode structures with a width of less than 20 μm.

Fig. 5: Set of electrodes fabricated by thin wire erosion and used for the microstructurization of the reaction units by a μ-EDM die sinking process.

After electrode generation these are transferred to the die sinking machine using the same clamping mechanism and, thus, minimizing errors in adjustment between electrode and work piece. The use of roughing and finishing electrodes within the die sinking process provides homogeneous channel structures over the whole depth. The minimal channel width to be obtained by this technique using small comb-like electrode structures amounts to 30 – 40 μm at a surface roughness Ra of 0.3 - 0.6 μm.

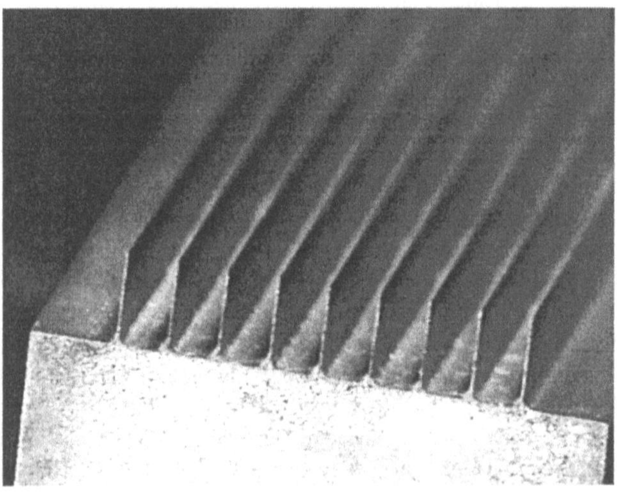

Fig. 6: Detailed view of a tungsten carbide electrode for μ-EDM die sinking of the cooling channels. Dimensions: width 50 μm, height 1.2 mm, length 50 mm.

This fabrication technique provides uniform microchannels with a high accuracy and improved control of channel geometry. Furthermore, the use of more complex electrode structures offers a flexible process for the generation of fluidic microstructures with complex geometries.

3 Conclusions

The integrated microreactor described here, serves as a flexible tool for process development as well as R&D purposes in microreaction technology. It is especially suited for liquid multiphase reactions employing immiscible liquids. The novel reaction conditions available, including efficient heat management, strongly influence the reaction process [7]. Moreover, the microstructuring process has been adapted to the specific needs of small lot production by developing a faster and more reliable μ-EDM structuring process suitable for metallic microreaction systems.

References

[1] Ehrfeld, W.; Golbig, K.; Hessel, V.; Konrad, R.; Löwe, H.; Richter, Th.: *Fabrication of components and systems for chemical and biological microreactors*, In: Microreaction Technology, Proceedings of the First International Conference on Microreaction Technology, Ed.: W. Ehrfeld, Springer, Berlin (1997) 72 - 90

[2] Ehrfeld, W.; Hessel, V.; Haverkamp, V.: *Microreactors*, Ullmann's Encyclopedia of Industrial Chemistry, Sixth Ed., 1999 Electronic Release, Wiley / VCH

[3] Stoldt, J.; Wurziger, H.; Krummradt, H.; Kopp, U.; Hohmann, M.; Schwesinger, N.: *Experiences with the use of microreactors in organic synthesis*, Published in this volume

[4] Bayer, T.; Pysall, D.; Wachsen, O.: *Micro mixing effects in continuous radical polymerization*, Published in this volume

[5] Ehrfeld, W.; Golbig, K.; Hessel, V.; Löwe, H.; Richter, Th.: *Characterization of mixing in micromixers by a test reaction: Single mixing units and mixer arrays*, Ind. Eng. Chem. Res. 38 (1999) 1075 - 1082

[6] Hessel, V.; Ehrfeld, W.; Haverkamp, V.; Löwe, H.; Schiewe, J.: *Generation of dispersions using multilamination of fluid layers in micromixers*, In: Dispersion Techniques for Laboratory and Industrial Production, Ed.: R.H. Müller, B. Böhm, Wissenschaftliche Verlagsgesellschaft Stuttgart, Accepted for publication (1999)

[7] Wörz, O.; Jäckel, K.P.; Richter, Th.; Wolf, A.: *Microreactors, a new efficient tool for optimum reactor design*, In: Process Miniaturization: 2nd International Conference on Microreaction Technology, March 9-12 1998, New Orleans, LA, Topical Conference Preprints, The American Institute of Chemical Engineers (1998) 183 - 185

[8] Ehrfeld, W.; Golbig, K.; Hessel, V.; Hofmann, Ch.; Löwe, H.; Storz, M.; Breysse, J.; Wörz, O.: *High temperature HCN generation in a complex integrated microreaction system*, Published in this volume

[9] Richter, Th.; Ehrfeld, W.; Erntner, D.; Gebauer, K.; Golbig, K.; Löwe, H.; Lange de Oliveira, A.; Schmidt, W.; Schüth, F.: *Microstructured reactor for consecutive heterogeneous / homogeneous gas phase reactions*, Published in this volume

[10] Hessel, V.; Ehrfeld, W.; Golbig, K.; Haverkamp, V.; Löwe, H.; Storz, M.; Wille, C.; Guber, A.; Jähnisch, K.; Baerns, M.: *Gas / liquid microreactors for direct fluorination of aromatics using elemental fluorine*, Published in this volume

[11] Kestenbaum, H.; Lange de Oliveira, A.; Schmidt, W.; Schüth, F.; Ehrfeld, W.; Gebauer, K.; Löwe, H.; Richter, Th.: *Synthesis of ethylene oxide in a microreaction system*, Published in this volume

[12] Hagendorf, U.; Janicke, M.; Schüth, F.; Schubert, K.; Fichtner, M.: *A Pt / Al_2O_3 coated microstructured reactor / heat exchanger for the controlled H_2 / O_2-reaction in the explosion regime*, In: Process Miniaturization: 2nd International Conference on Microreaction Technology, March 9-12 1998, New Orleans, LA, Topical Conference Preprints, The American Institute of Chemical Engineers (1998) 81 - 87

[13] Richter, Th.; Ehrfeld, W.; Wolf, A.; Gruber, H.-P.; Wörz, O.: *Fabrication of microreactor components by electro discharge machining*, In: Microreaction Technology, Proceedings of the First International Conference on Microreaction Technology, Ed.: W. Ehrfeld, Springer, Berlin (1997) 158 - 168

Fast Response Heating Module for Temperature Programmed GC Analysis in Microreaction Systems

J. Schiewe, W. Ehrfeld, T. Hang, H. Löwe, Th. Richter, X. L. Yan
Institut für Mikrotechnik Mainz GmbH, Carl-Zeiss-Str. 18-20, D-55129 Mainz, Germany,
Tel.: +49-6131-990-451, Fax: +49-6131-990-475, e-mail: schiewe@imm-mainz.de

A. A. Kurganov, K. K. Unger
Institut für Anorganische und Analytische Chemie der Universität Mainz,
Joh.-Becher-Weg 24,
D-55128 Mainz, Germany, Tel.: +49-6131-173361, Fax: +49-6131-173364

Abstract

For monitoring the reactant and product concentrations of a gas phase microreactor for ethylene oxide synthesis the development of a miniaturised device for fast GC analysis suitable for operation in conjunction with the microreactor is advantageous. Therefore, a method employing short (0.2 – 0.4 m) packed capillary columns was developed. Highly crosslinked polystyrene microspheres proved to be the most suitable adsorbent enabling a high separation efficiency. For fast temperature programming design concepts for an efficient heating module capable to be used with short capillary columns were developed. In principle, planar as well as cylindrical designs are suitable for integration of a resistive heating device, cooling microchannels and structures to accommodate a short GC capillary column. First experiments using a planar prototype module show that 10 to 15 times higher heating rates can be achieved compared to a standard GC oven, thus, leading to a 6-fold decrease in retention time. Further improvement employing a cylindrical design is deduced from simulation results.

1. Introduction

Typical state-of-the-art gas chromatography is usually performed using 15 - 30 m long capillary columns with 0.20 - 0.32 mm internal diameter allowing analysis times ranging between 10 and 60 minutes. The decrease in analysis time has long been the goal in separation sciences in order to increase the sample throughput on the one hand and to enable a fast process control for chemical reactors on the other hand. This can be achieved employing small diameter capillaries with a length of only 1 or 2 m [1]. With the use of short capillaries a miniaturisation of the gas chromatograph suggests itself. Terry et al. at Stanford University [2] realised a whole device comprising a 1.5 m capillary column, a micro thermal conductivity detector as well as a miniaturised injection valve on a silicon wafer. Although it was a big step forward towards fast, miniaturised GC analysis this first prototype lacked from inferior separation performance compared to standard technology. Nevertheless, parts of the overall system, such as the microstructured injector and detector components developed, have been implemented in portable GC systems on a commercial scale [3].

Besides the aspect of using narrow bore, short capillary columns a further decrease in analysis time can be achieved by increasing the heating and cooling rates of temperature programmed GC analysis. Therefore, replacing the conventional hot air column thermostat of a gas chromatograph was the focus of work recently reported. For instance, resistively heated metal-clad capillary columns were developed [4] enabling heating rates of up to 100 K/min. However, in order to decrease overall analysis cycle times, a combined improvement of heating, cooling and separation performance is needed.

2. Aim and motivation

The aim of the work presented here was the development of a device and a method for fast GC analysis of the reactor off-gas components of an ethylene oxide microreactor [5], [6]. This device, therefore, has to be miniaturised in order to be capable of working as a supplemental part of microreactors. In addition to these system integration issues the method should be fast and reliable to enable efficient process control based on the analytical results.

The analytes in the reactor off-gas are the reactant ethylene and the product ethylene oxide as well as side products like carbon dioxide and acetaldehyde. Capillary columns coated with liquid stationary phases which are routinely used in high speed GC analysis do not provide sufficient selectivity for the separation of carbon dioxide from oxygen and nitrogen, ethylene from other light hydrocarbons and ethylene oxide and acetaldehyde from alcohols and other polar oxidation products. Therefore, short packed capillary columns were utilised having small diameter solid particle packings.

Since the polarity of the analytes varies significantly, temperature programmed GC analysis is needed. Therefore, a miniaturised fast response heating module utilising resistive heating and fast cooling after each analysis run has been developed.

3. Results and Discussion

3.1. GC analysis using short packed capillary columns

A screening of stationary phases suitable for separation of all compounds listed above led to the use of a polystyrene polymer packing with a high degree of crosslinking as the best choice. Table 1 gives an overview of different packing materials tested.

Table1: Suitability of different packing materials for simultaneous separation of polar and nonpolar compounds.

Adsorbate	Chemical structure	Non polar compounds	Polar compounds
Porapak® Q	Copolymer of styrene, ethyl benzene and divinyl benzene	Unresolved peaks	Well separated using temperature gradient
LiChro-spher® C18	Silica gel with n-octadecyl groups	Unresolved peaks	Well separated using temperature gradient
HayeSep® A	Copolymer of styrene and ethyl glycol dimethacrylat	Efficient separation	Well separated using temperature gradient
MzPS	Strong crosslinked polystyrene	Excellent separation	Well separated using temperature gradient

The high separation efficiency of capillary columns packed with MzPS polymer particles (purchased from MZ-Analysentechnik GmbH, Mainz, Germany) allowed to diminish the capillary length necessary to achieve complete separation of all compounds to only 0.2 - 0.4 m. In figure 1A and 1B two chromatograms are compared illustrating the decrease in analysis time that can be achieved when employing short, packed columns.

The retention time of ethylene oxide decreases by almost a factor of four (from 330 seconds to 90 seconds) when replacing a standard capillary column by a 0.3 m short packed capillary column. (Note that in figure 1A in addition to the analytes mentioned also carbon dioxide, water and propylene oxide were separated with the latter being the peak with the longest retention time.) A particle size of 5 µm was found to be optimal with respect to pressure drop and volume flow rate of the carrier gas.

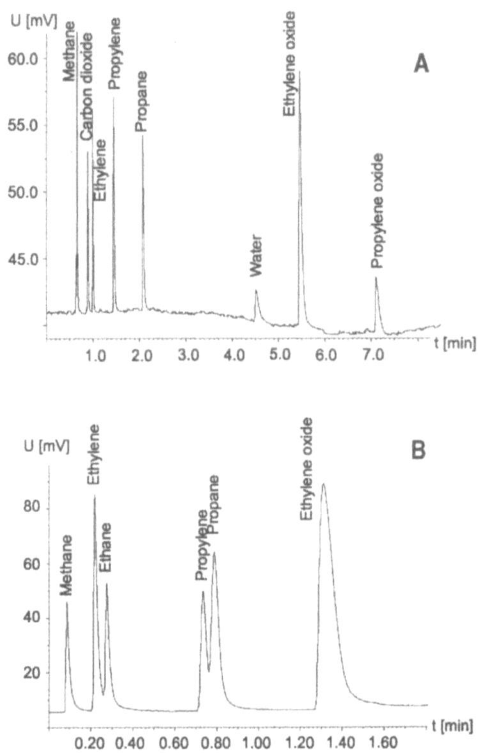

Figure 1: Comparison of the separation of a test sample of reactants and products of the gas phase oxidation of hydrocarbons: **A**) employing a standard 15 m gas-pro column (Advanced Separation Technologies, NJ, USA), and **B**) employing a 0.3 m long capillary with 5 µm MzPS packing, tempearture program: 40 °C to 180 °C with a heating rate of 0,3 K/s.

Experiments with even smaller particles of 1.5 µm and 3 µm, respectively, although having a higher specific separation efficiency (per unit length), showed no advantage. Since smaller particle packings lead to higher pressure drops shorter capillaries had to be used in order to keep the inlet gas pressure within the practical limit of approximately 65 bar. Therefore the integral separation efficiency over the length of the capillary could not be improved.

3.2. Development of a fast response heating module

In order to develop a fast heating module capable for use with short capillary columns different design concepts were evaluated. In principle, planar as well as cylindrical designs are possible consisting of a column support, a heating and a cooling device.

First experiments were performed employing a planar design (see figure 2). The module consists of a capillary column support (50 x 50 x 30 mm^3) made of aluminium located between resistive heaters on one side and a cooling fan on the other side. All components are held together by a frame made of PEEK. Thus, a compact device with overall dimensions of 80 x 80 x 75 mm^3 was realised. The two piece column support contained grooves on either part to accommodate capillary columns with outer diameters of 360 μm to 460 μm (standard capillary column dimensions).

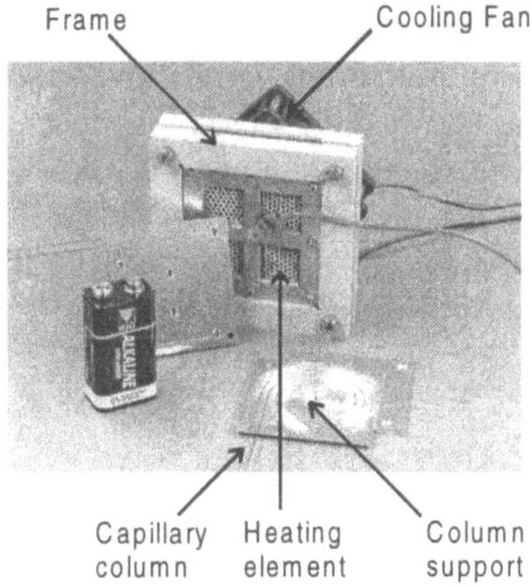

Figure 2: Planar heating device comprising resistive heaters, cooling fan and column support to accommodate 0.12 - 0.40 m long capillary columns.

Capillary material of variable length can be inserted into the grooves of the bottom part of the support. It is fixed between bottom and top part by screws. The support was structured in order to enable insertion of three different capillary lengths, as shown schematically in figure 3 (black line). This way, experiments with different particle size packings requiring different column lengths were possible.

Figure 3: Schematic drawing of different lenghts of column that can be fixed between the two parts of the support (top view). Black lines symbolise the capillary column.

Column and column support are heated by means of four resistive heaters each having a maximum heating power of 50 W. Cooling is achieved by operating an axial-flow fan known from conventional computer equipment. Attention was drawn to realise a uniform temperature distribution along the entire length of the capillary. The maximum temperature gradient measured on seven different spots on the aluminium support from center to edge was < 3 K. The mean heating rate of the device in the temperature range of 25 to 180 °C was found to be 2 K/s (120 K/min).
Figure 4 shows a typical experimental temperature-time curve. Heating was performed at constant voltage and since the resistance increases with rising temperature the power drops significantly above 180 °C leading to the inclined shape of the heating curve. However, since the maximum temperature required for analysis amounts to 180 °C, the linearity of the beginning part of the heating curve is sufficient. The cooling rate is a function of the temperature with an overall value of -0.5 K/s (-30 K/min) in the temperature range from 180 to 40 °C .

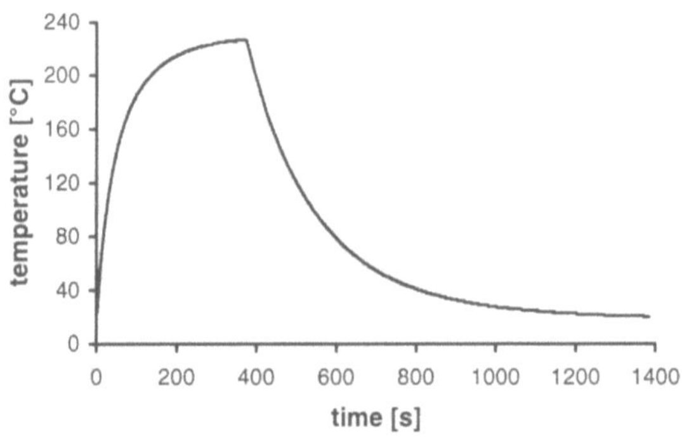

Figure 4: Experimental evaluation of heating and cooling behaviour of the planar device.

In figure 5 a typical chromatogram obtained using the planar heating module is shown. A further decrease of the ethylene oxide retention time to only 55 seconds could be achieved. Thus, an increase of the analysis speed by nearly a factor of two compared to chromatogram B in figure 1 was found.
The temperature program was set to run from 30 °C to 150 °C within 60 seconds. One total analysis took up 8 minutes including cooling to start again from the initial temperature of 30 °C. This is a clear gain of total analysis time compared to a conventional ethylene oxide analysis of about 25 to 35 minutes.

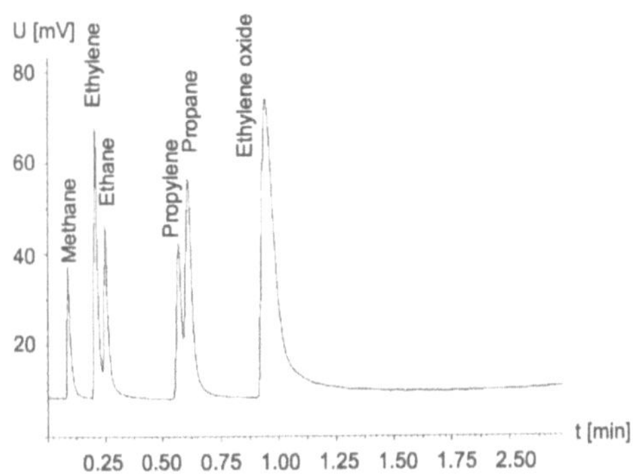

Figure 5: Separation of a test sample of reactants and products of the gas phase oxidation of hydrocarbons employing a 0.3 m long capillary with 5 µm MzPS packing. Linear temperature programming using the planar heating module from 30 to 150 °C and a heating rate of 2 K/s.

Further improvement should mainly focus on increasing the efficiency of cooling. Therefore, a cylindrical design concept was developed which accommodates a standard heating cartridge, axial microchannels for cooling as well as a column support. Figure 6 shows a schematic view of the device. Fast and efficient cooling can be achieved by allowing compressed air to pass axial microchannels.

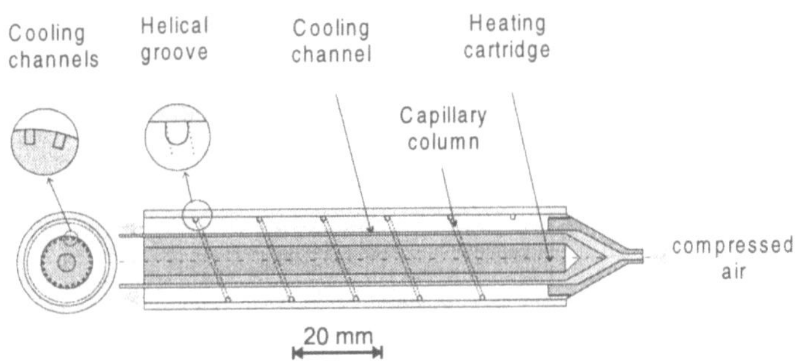

Figure 6: Cylindrical heating device comprising a heating cartridge, cooling microchannels and structures to accommodate short packed capillary columns.

So far simulations of the heat flow resulted in a proposed heating rate of 4 K/s (240 K/min) and a cooling rate of approximately –3 K/s (-180 K/min) when operating with pressurised air at 4 bar. Hence, a further decrease of the total analysis time (including cooling to initial temperature) of only one minute appears to be feasible, qualifying the device to be suitable for a quasi online reactor control. Table 2 gives an overview of heating and cooling rates of different devices.

Table 2: Heating and cooling rates for different devices.

	GC-oven *	planar modul*	cylindrical modul +
heating rate	0.4 K/s	2.0 K/s	~ 4.0 K/s
cooling rate	- 0.2 K/s	- 0.5 K/s	< - 3.0 K/s

* measured
+ simulated

4. Conclusions

A twofold approach to perform temperature programmed fast GC-analysis using a miniaturised module was successfully applied to analyse the off-gas components of ethylene oxide synthesis. On the one hand the use of short capillaries packed with highly crosslinked polystyrene particles led to four times shorter retention times. On the other hand the construction of a fast response heating module replacing the conventional hot air thermostat further improved the analysis speed by a factor of two.

The performance test of method and equipment developed showed that the separation of light hydrocarbons and their oxidation products can be completed within 55 seconds and a total analysis time (including cooling to initial temperature) of 8 minutes was achieved, proving clear advantages over conventional technology. The development of a cylindrical module, currently under way, will lead to a further decrease in analysis cycle time (including column cooling) to under 100 seconds.

5. References

[1] W. Engewald, GIT Labor-Fachzeitschrift 1/99 (1999), 1-4.
[2] S. C. Terry, J. H. Jerman, J. B. Angell, IEEE Transactions on Electron. Devices, Vol. ED-26, No. 12 (1979), 1880-1886.
[3] HP Mikro GC, Hewlett Packard, Waldbronn, Germany, 1999.
[4] Th. Theuerl, R. Blanke, W. Kleinböhmer, GIT Spezial Separation 1/99 (1999), 40-43.
[5] H. Löwe, W. Ehrfeld, K. Gebauer, K. Golbig, O. Hausner, V. Haverkamp, V. Hessel, Th. Richter, Process Miniaturization: IMRET 2, March 9-12 1998, New Orleans, La., Topical Conference Preprints, AIChE (1998), 63-74.
[6] Th. Richter, W. Ehrfeld, K. Gebauer, K Golbig, V. Hessel, H. Löwe, A. Wolf, Process Miniaturization: IMRET 2, March 9-12 1998, New Orleans, La., Topical Conference Preprints, AIChE (1998), 146-151.

A Silicon-based Microreaction System for Analytical Applications

Peter Woias [1], Karin Hauser [1], Erwin Yacoub-George [2], Bernd Hillerich [1]

1 Fraunhofer-Institute for Integrated Circuits and Systems (IMS-3), Hansastr. 27 d, D-80686 Munich, Germany

2 University of Regensburg, Institute for Medical Microbiology and Hygiene, Franz-Josef-Strauß-Allee 11, D-93042 Regensburg, Germany

Abstract

A silicon microreaction system is presented, which forms a microfluidic concept for well-established detection procedures of indicator-based colorimetric analysis. The system comprises capillary dosing systems for the precise delivery of sample and indicator solution, respectively, a novel active micromixer for a rapid and complete mixing of both components and a microcuvette for the detection of the resulting colorimetric signal. The microfluidic set-up is realized as compact modules to be integrated into a conventional flow injection analysis system for sample preparation and analysis control. The performance of the micromixer and the microreaction system as a whole is demonstrated with a well-established pH-detection method.

Keywords: capillary microdosing, micromixer, optical transmission cell, colorimetric analysis

1 Introduction

Today, a wide number of microfluidic systems for chemical analysis and separation technology is pursued under the topic „lab on a chip". Especially in the field of chemical analysis a trend is observable to mimick macrofluidic concepts, like flow injection analysis (FIA), with a microminiaturized set-up of pumps, valves, channel systems and solid state chemo- or biosensors for a selective substance detection [1,2]. Depending on the application such a system approach can solve a number of problems associated with bare chemical sensors [3]. However, as any sensor-based analytical system is tailored to the application mainly by choosing the appropriate chemo- or biosensors, a number of well-known problems often remains, like limited sensor lifetime or unsatisfying chemical sensitivity.

Chemical reaction techniques can be seen as an alternative to circumvent these sensor-related disadvantages. In this study, a prototype microchemical reaction system for chemical analysis was realized to combine well-established reaction procedures with a microfluidic set-up.

2 System Concept

To investigate the concept described above the microreaction system shown in Fig. 1 was designed. The system performs a standard colorimetric analysis technique, which is typically applied as a cuvette test e.g. in on-site water analysis. These assays usually require the manual pipetting of a defined sample volume to an indicator solution, which is pre-packed in a cuvette. After thorough mixing a chemical reaction of the indicator with the compound of interest is producing a color change, which is usually determined by inserting the cuvette manually into a photometer.

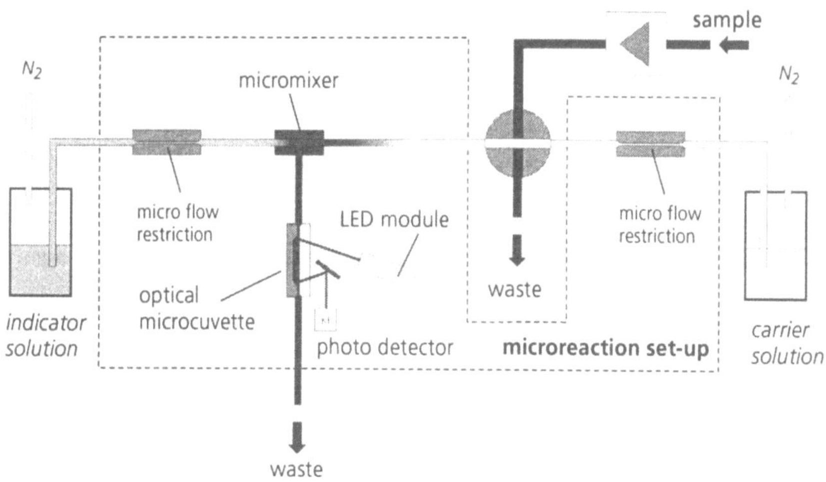

Fig. 1 Schematic set-up of the microreaction system

Aside from hand-held test kits, which are available from a number of commercial suppliers, this detection method is also found in automatic on-line-analyzers, which perform virtually the same reaction process in a flow-through system. For this as for other applications, microminiaturization is desirable from various reasons:

- reduction of analysis time due to reduced internal volumes and transport times,
- reduction of waste production,
- reduction of reagent consumption due to the lower flow rates required, and
- system miniaturization to allow a fully automated portable or mobile application and a reduction of manual processes.

In the set-up shown in Fig. 1 the sample is provided via a conventional FIA system. The carrier buffer of the FIA system as well as the indicator reagent are fed into a micromixer with two capillary microdosing systems, which utilize silicon microflow restrictors at the outlet of two Nitrogen-pressurized siphon bottles. The mixture of sample and indicator is transported into a micro transmission cell to perform a photometric analysis. The individual components and the system construction will be described in the following.

3 Microfluidic Components

3.1 Active Micromixer

Effective micromixing is a key function in miniaturized flow systems for chemical analysis as well as for microreaction technology. However, in contrary to a macroscopic set-up, mixing in microfluidic structures with Reynolds numbers around or below 1 is generally not assisted by turbulence but mainly dominated by relatively slow molecular diffusion processes. To solve this problem, most of the micromixers known today are following the principle of static mixing by multilamination to decrease the effective diffusion length and, thus, to speed the diffusion process. This is done by splitting the input streams into a multi-sandwich structure of alternating fluid laminates [4,5] or by injecting one input stream into the other via a micro nozzle array [6,7].

Static mixing concepts are sharing the common disadvantage, that multilamination requires relatively complicated microstructures and works best at channel or nozzle dimensions of several microns, which are always prone to bubble entrapment, clogging with particles and sedimentation. Moreover, it has been demonstrated, that the mixing efficiency of static mixers is depending on a sufficient throughput and decreases with a decrease of the flowrate [6]. However, in contrary to high-throughput devices desirable for chemical synthesis, a micromixer for chemical analysis has to be tailored for high efficiency particular at low throughput to conserve reagent and waste. For this purpose the active micromixer shown in Fig. 2 was realized as an alternative to static micromixing.

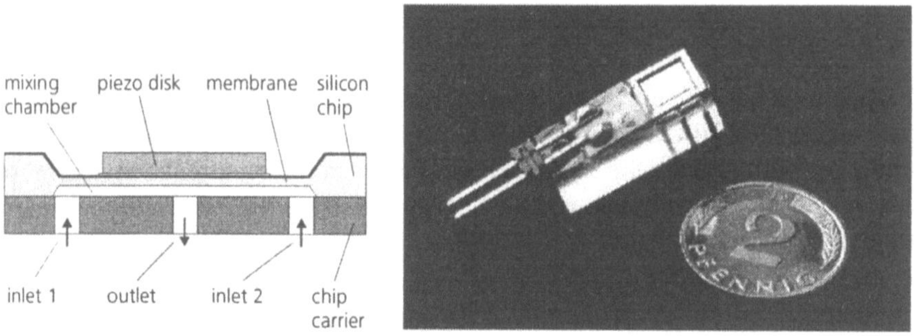

Fig. 2 Schematic set-up of the silicon micromixer (left) and photograph of a test structure on a PMMA chip carrier (right)

The micromixer is made from a piezoelectric membrane actuator, which is also used in silicon micropumps [8]. It consists of a 7x7 mm^2 silicon chip with a 5x5 mm^2 wide and 40 µm thick silicon membrane. Fabrication is done on 4 inch <100> wafers by anisotropic etching in KOH. During the etching process also a 15 µm deep cavity is formed under the membrane, which acts as a mixing chamber in the completed device. After wet chemical processing of the membrane, the wafer top side is metallized with alumina, individual chips are diced and, finally, a thin piezo disk is glued onto the metallization of each chip. The mixer is completed by mounting the silicon chip onto a chip carrier made from perspex or PEEK. The chip carrier employs electrical contacts for wire bonding of the piezo disk, two inlet channels located at opposite corners of the mixing chamber and an outlet channel, which is placed below the chamber center.

Fig. 3 shows various photographs taken from the outlet channel of the micromixer under various operation conditions. For test purposes the micromixer was loaded with water and a blue dye. Without excitation of the piezoactuator the fluids supplied via the inlet channels do not mix in the mixing chamber and appear as a clearly separated laminar flow pattern at the mixer outlet (Fig. 3a).

During operation the piezoactuator is driven with a square wave signal, which causes the silicon membrane to bend upward and downward around its center position. As a result, the mixing chamber volume is periodically increased and decreased, which causes a periodic inflow and outflow through all fluid ports and, therefore, a mixing effect inside the mixing chamber. As a second effect, longitudinal shock waves are generated in the outlet during the rapid downstroke of the membrane (Fig. 3b). These shock waves generate an effective three-dimensional perturbation of the previously straight laminar flow pattern. An increase of the actuation frequency decreases the longitudinal distance between the perturbation zones, until the generated microlaminates reach the dimension of the diffusion length, where a rapid and complete mixing by molecular interdiffusion occurs (Fig. 3c). In the experiment shown in Fig. 3, the complete conversion from laminated to mixed flow took place in less than one second.

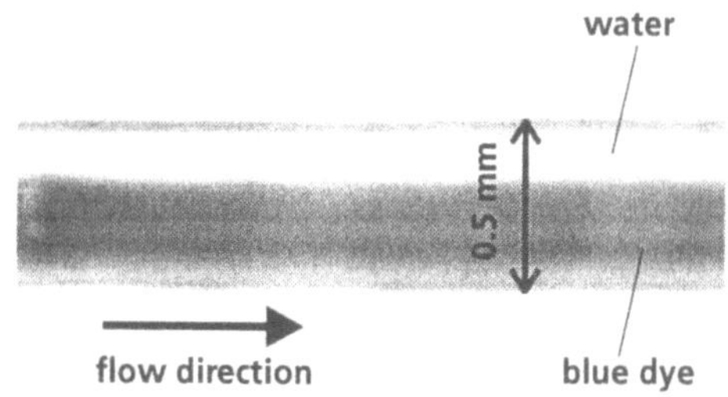

Fig. 3a mixing disabled: laminar flow at the micromixer outlet

Fig. 3b shock wave due to a single membrane downstroke

Fig. 3c mixing enabled: mixed flow at the micromixer outlet

3.2 Microflow Restrictors

The performance of any microreaction system is severely deteriorated by an insufficient fluid dosing accuracy. First investigations with the micromixer described above showed, that pulse-damped roller pumps and even piston pumps were not able to generate a sufficient flow stability. Therefore, a capillary microdosing concept based on silicon microflow restrictors was applied. For this purpose, all input reagents to the micromixer were supplied from nitrogen-pressurized siphon bottles. Silicon microflow restrictors were mounted at the outlet of the siphon bottles to set the desired flow in conjunction with an appropriate adjustment of the nitrogen gas pressure.

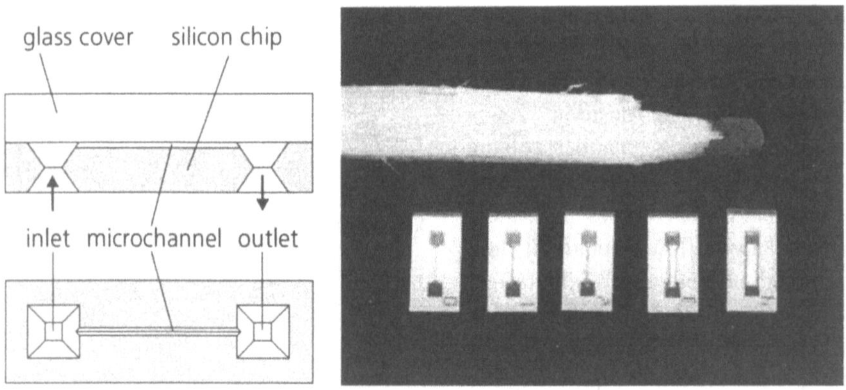

Fig. 4 Schematic set-up of the silicon microflow restrictor (left, top view and cross section) and photograph of devices with different channel geometries (right)

The microflow restrictors used in this study consist of two stacked chips made from silicon and glass, respectively. The silicon chip exhibits a straight v-shaped microchannel at the topside and two fluid ports on the bottom and is fabricated with the same technology as described for the micromixer. After wet chemical etching the silicon wafer is covered with a pyrex wafer by anodic bonding to seal the individual microchannels. For test purposes a number of different microchannel designs was realized with a common channel length of 2 mm and various upper widths of the v-shaped or trapezoid cross section, ranging from 28 μm to 1mm. For an easy exchange these chips were designed with a common chip size of 3.5x7.0 mm² and a common location of the fluid ports with a lateral center distance of 3.5 mm (see Fig. 4).

A detailled description of the performance achievable with this microdosing concept is given in [9]. The main advantage for this particular application is a constant and virtually pulsefree flow. Moreover, with the modular concept shown in Fig. 4, the dosing rate can easily be adjusted by selecting an appropriate microchannel.

3.3 Microcuvette

For the optical transmission measurement a silicon microcuvette was applied, which has already successfully been tested in a previous study [10]. This device uses the chip design of the microflow restrictors described above and again comprises of a silicon microchannel with inclined walls at an angle of 54.7° and fluid connections from the backside. To convert the silicon microchannel into a microcuvette, two opposite side walls of the microreactor are used as mirrors to reflect a light beam into and out of the microchannel axis. Thus, any change of optical absorbance inside the microchannel is immediately detected with the photodetector. For this particular study a microcuvette with an internal volume of 0.9 µl and an effective optical pathway of about 4 mm was applied.

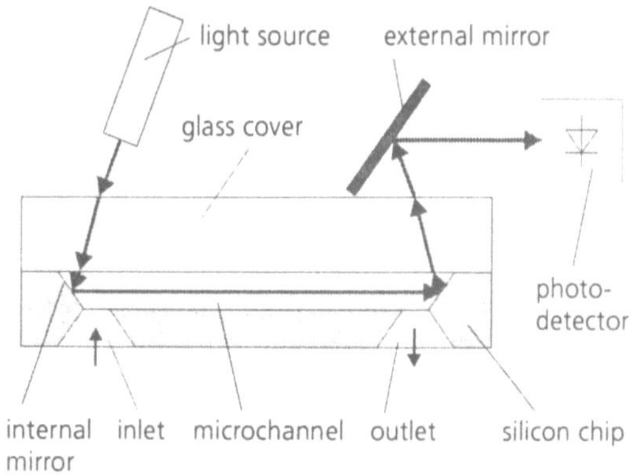

Fig. 5 Schematic set-up of the silicon microcuvette with optical read-out

4 System Integration

As shown in Fig. 6 the microfluidic components described above were integrated into several PEEK holders to build the microreaction set-up. The micromixer, the microcuvette chip, all interconnecting channels and three push-on tubing connectors were integrated into a fluidic module. To achieve a low dead volume the micromixer and the microcuvette were connected with a short channel with a diameter of 0.5 mm and a length of 4 mm. The optical set-up for the transmission measurement, comprising a LED or laser diode module, a silicon

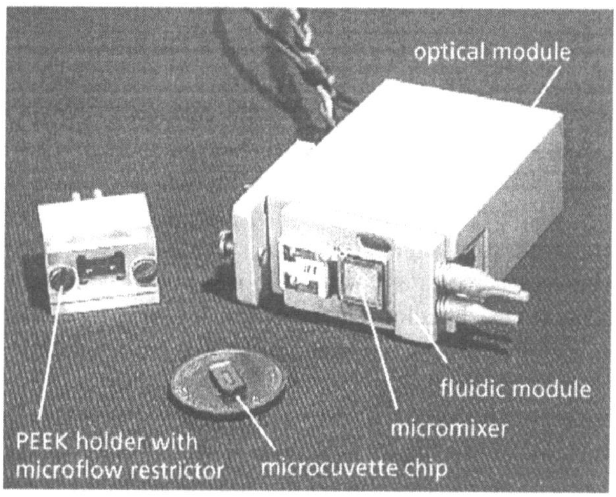

Fig. 6 Photograph of the microreaction set-up

photodetector with integrated transimpedance amplifier and a silicon mirror, were built into an optical module. Mounting flanges and two one-axis translatorial shift stages were provided on the optical module for an easy interconnection and an independent optical alignment of the microcuvette and the external mirror (see Fig. 5). The combined fluidic/optical module measures approximately 45x33x16 mm^3, which would leave enough space to add the microflow restrictors for a fully integrated version. Nethertheless, separate holders with push-on tubing connectors were realized for these components to allow an easier handling.

5 Experimental

5.1 Materials and Methods

pH-detection with bromethymol blue (BTB) was chosen as model reaction for all experiments. At the beginning of the experimental work an ethanol-soluble BTB salt was dissolved at a concentration of 3 mg/ml first and further diluted with water to achieve a BTB concentration of 40 µg/ml. For an easier preparation, a water-soluble BTB-Na salt was applied later on in the same concentration. Within its detection range the chosen indicator changes color from pale yellow at pH 5 to a dark blue at pH 8. To achieve a high measurement signal a red LED with a wavelength of 610 nm was chosen for the optical detection system, which proved to be close enough to the indicators absorption maximum at 617 nm.

As sample phosphate buffer in the pH-detection range of BTB was prepared by mixing the appropriate amounts of 1/15 M Na_2HPO_4 and KH_2HPO_4. The mixing ratio of sample and indicator set to 1 throughout all experiments by chosing identical microflow restrictors at the siphon bottles, which were also driven from the same nitrogen supply. The flow rate at the mixer outlet was measured frequently with a scale and the nitrogen pressure was corrected accordingly to compensate disturbing effects of temperature and viscosity inherent to the dosing concept [9].

As shown in Fig. 5 the optical detection system is integrating the optical absorbance over the microcuvette cross section. Therefore, the photodetector signal is also a direct criterion for the micromixer efficiency, as partial mixing will only produce a small colorized reaction area between undisturbed fluid zones of clear sample and indicator solution, which in total will not reach the absorbance of a homogeneously colorized cross section. Using this effect the micromixer characterization was done in a continuous flow mode with pH 8 buffer as a sample. To produce an uncolorized mixture for a photodetector baseline signal I_0 the micromixer was run first with buffer only. Afterwards BTB indicator was fed into one inlet and the photodetector signal I_S was recorded. The resulting optical absorbance E was calculated according to the Lambert-Beer law ($E = -\ln[I_S/I_0]$).

The colorimetric pH-detection was done in the FIA set-up shown in Fig. 1. To obtain a baseline signal the BTB indicator and water as a carrier were continuously fed into the micromixer. Buffer samples (0.5 ml) with varying pH were periodically injected into the carrier. The resulting changes of the photodetector signal from the baseline level were calculated as output signal.

5.2 Micromixer Characterization

For a reliable operation the operation set-point of the active micromixer had to be determined first. For a simple electrical operation it was decided to drive the piezoelectric actuator with a symmetric square wave signal at a duty cycle of 0.5, which also provides a desirable high slew rate of the membrane actuation. Fig. 9 shows the influence of the signal amplitude onto the mixing efficiency for an operation frequency of 100 Hz and a total throughput between 95 µl/min and 492 µl/min. As to be expected the mixing efficiency (i.e. the optical absorbance) is increasing with an increasing amplitude of the electric driving signal. Insufficient mixing is indicated by a decrease of the optical absorbance due to a lamination of buffer, indicator solution and reaction product inside the microcuvette. Depending on the individual membrane actuators, an optimum signal amplitude between 30 V and 40 V was found, which corresponds to a total membrane stroke between 6 µm and 7 µm. Higher signal amplitudes or unsymmetric signals as used for micropump operation [6] did not result in a further improvement, but in some cases even in a deterioration of the mixing efficiency. This can be attributed to the fact, that the downstriking membrane center will touch the chip carrier at high signal amplitudes and thus block the mixer outlet. This is resulting in a periodic modulation of the inlet flow, which can alter the average flow rate and thus the mixing behaviour.

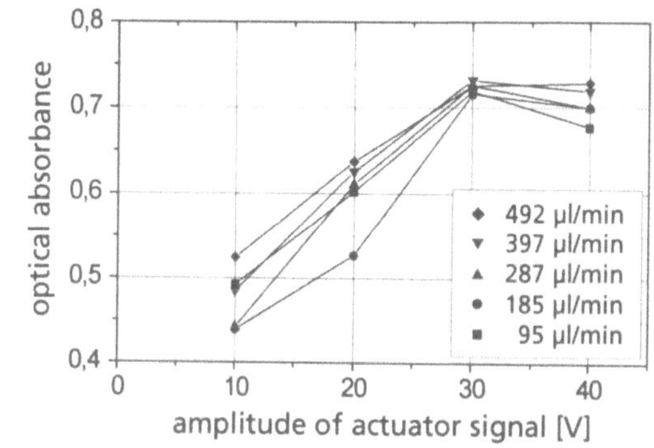

Fig. 8 Influence of the actuator signal amplitude on the mixing efficiency (frequency 100 Hz)

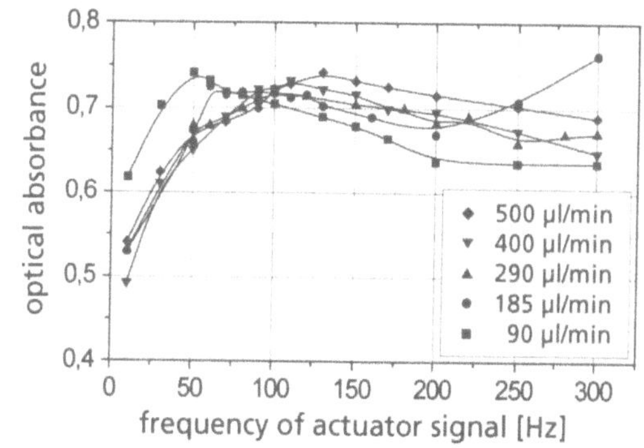

Fig. 9 Influence of the actuator signal frequency on the mixing efficiency (amplitude 30 V)

The influence of the actuation frequency on the micromixer performance is shown in Fig. 9 for a signal amplitude of 30 V and various flowrates. In the low frequency range under investigation all curves show an efficiency maximum at frequencies between 50 Hz and 125 Hz, which approximately is increasing with the flowrate. This effect can be explained by a flow-dependent variation of the perturbation zone distance in the mixer outlet. At higher frequencies the mixing efficiency usually goes down, which presumably is caused by an increased damping of the membrane vibration. For this study a value of 100 Hz was finally chosen as actuation frequency. Around this set-point all efficiency curves intersect, which reduces the influence of flowrate variations to a minimum.

5.3 Colorimetric pH-detection

Fig. 10 shows an example curve of the micromixer performance with the previously defined operation set-point. In this experiment, the photodetector signal of the silicon microcuvette is recorded during an injection of buffer via the FIA system. The signal baseline is obtained by an active mixing of the BTB-Na indicator with the carrier (water). After sample injection the mixromixer is disabled, which causes a switchover to laminar flow in the microcuvette with only a partial absorbance as described in 5.1. After micromixer reactivation the photodetector signal is rapidly decreasing to a stable absorbance value and is also instantly returning to the signal level of the laminar state after a second deactivation. During the active mixing of sample and indicator the outflow of the microcuvette was collected to be reinjected backward into the microcuvette at the end of the experiment. The resulting signal is equal with the signal of the active mixing phase, which indicates, that the micromixer is providing a complete mixing during the injection. A full pH-calibration curve is shown in Fig. 11. As to be seen the typical absorption behaviour of the indicator is accurately reproduced over the whole detectable pH-range.

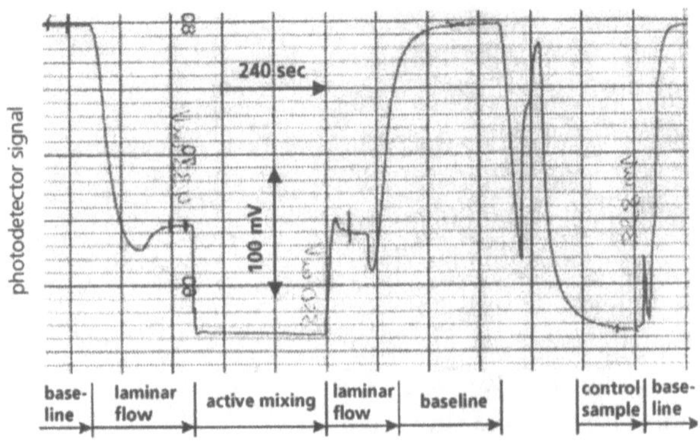

Fig. 10 FIA experiment with enabled and disabled mixing (sample: phosphate buffer pH 8, indicator: BTB-Na 40 µg/ml, flow rate: 90 µl/min each)

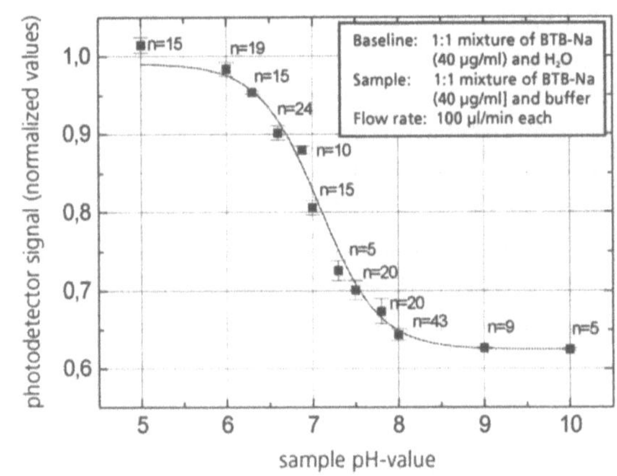

Fig. 11 Calibration curve for photometric pH-detection

6 Conclusions

In this publication a silicon-based microreaction system is described, which was developed to combine well-established reaction procedures of chemical analysis with a microfliudic set-up. The system is configurated around a novel micromixer, which performs the mixing of reagents by means of a vibrating silicon membrane. A characterization of the micromixer has shown a rapid and reproducible mixing in a flow range, which is typical for microchemical analysis systems [1,2]. Aside from the pH-detection chosen as test reaction a large number of colorimetric cuvette assays are commercially available. Thus, a microreaction-based concept of an „intelligent microcuvette" could be envisaged, which allows to integrate microreaction technology into commercial photometric detection systems with drastically reduced reagent consumption and waste production.

References

[1] M. Richter, A. Prak, J. Naundorf, M. Eberl, H. Leeuwis, P. Woias, A. Steckenborn, A Chemical Microanalysis System as a Microfluid System Demonstrator, Proc. Transducers 97, Chicago, USA, June 16-19, 1997, Vol. 1, pp. 303-306.

[2] S. Drost, W. Wörmann, P. Woias, B. Ross, O. Köster, W. Konz, B. Edler, W. Schuhmann, L. Meixner, R. Ferretti, Microanalytical System for Environmental Control, Proc. Transducers 97, Chicago/USA, June 16-19, 1997, Vol. 2, pp. 931-934.

[3] P. Woias, S. Koch, E. Müller, D. Barrow, J. Cefai, G. Curtis, H. Hughes, An ISFET-FIA system for high precision pH-recording, *Sensors & Actuators B*, 15-16, (1993), pp. 68-74.

[4] K.-P. Kämper et *al.*, Microfluidic Components for Biological and Chemical Microreactors, Proc. MEMS-97, Nagoya, Japan, Jan. 26-30, 1997, pp. 338-343.

[5] J. Branebjerg *et al.*, Fast mixing by lamination, Proc. IEEE MEMS-96, San Diego, USA, February 11-15, 1996, pp. 441-446.

[6] R. Miyake *et al.*, Micromixer with Fast Diffusion, Proc. MEMS-93, Fort Lauderdale, USA, February 7-10, 1993, pp. 248-253.

[7] A. Yotsumoto et *al.*, Fabrication of an integrated mixing/reaction micro flow cell for μTAS, Proc. μTAS '98, Banff, Canada, Oct. 13-16, 1998, pp. 185-188.

[8] P. Woias, R. Linnemann, M. Richter, A. Leistner, B. Hillerich, A silicon micropump with a high bubble tolerance and self-priming capability, Proc. μTAS '98, Banff, Canada, October 13-16, 1998, p. 383-386.

[9] M. Richter, P. Woias, D. Weiß, Microchannels for Applications in Liquid Dosing and Flow-Rate Measurement, *Sensors & Actuators A 62,* (1997), pp. 480-483.

[10] P.Woias, M.Richter, K.Hauser, E.Yacoub-George, H.Wolf, S.Koch, Silicon microreactors with photometric and chemiluminescent on-column detection, Proc. IMRET 2, New Orleans, USA, March 8-11, 1998, pp. 235-239.

Designing and Constructing Microplants

Dipl.-Ing. Henning Fink and Prof. Dr.-Ing. Manfred J. Hampe

FG Thermische Verfahrenstechnik, TU Darmstadt
Petersenstr. 30, 64287 Darmstadt, Germany
Tel.: +49 61 51/16-21 64, Fax: +49 61 51/16-45 16
e-Mail: hampe@tvt.tu-darmstadt.de

1. Why Microplant-Technology?

1.1 Advantages of Microplants

Economic reasons necessiate the reduction of time and costs required to develope new products or to design new processes. Nowadays Miniplants are useful for the design of new processes by supporting and validating the results of computer-simulation. The size of a Miniplant can easily reach the height of some meters covering the area of some square-meters. But the thermodynamics of the seperation processes are independent of the size of a plant. The new Microplant-technology downsizes the Miniplant with a factor 10^2, so that a Microplant can be constructed on a table. It reduces the feed of a laboratory-scale plant down to 10-50 ml/h. So the capital expenditure and volumes of the experimental setup are reduced, while the minimal hold-up reduces the safety risks if hazardeous substances are envolved.
A tool will be used as more frequently as the build up of an experiment is easier and cheaper. So a strictly modular building set guarantees the quick construction of the Microplants, quick dismanteling and the reusibility of units, pumps, control devices etc. Microplants are designed to operate automatically. The utilization of other materials than glass allows the operation of Microplants with alkaline media or under high pressure.

1.2 Field of application

A Microplant can't substitute a Miniplant in any case, but it completes the laboratory-technology. The main application of a Microplant will be during the basic-engineering of processes, while experiments should clarify the existence or behavior of substances in a process. Often, it is impossible to determine all secondary reactions and the enrichement of their unknown reaction products in recycle loops by using only computer-simulations. Microplants mirror chemical production plants or the processes of interest and contain the complete seperation sequences, all units, and all recycle loops. With conventional methods like gas-chromatography or HPLC the products of a Microplant can be analysed and all substances be identified.

The sequently modular construction of conventional software for process-simulation makes it difficult to simulate recycle loops. In case that the simulation of process-variants doesn't converge, unexperienced engineers tend to abandon these variants, even if, for other reasons, the process seems to be a good solution for the problem. In such a case, the experiment with a Microplant could help to check the results of the simulation.
The factor 10000 is between the size of a Microplant and a real plant. The apparatus of Microplants are not from the same type and geometry as the apparatus of the real plant. For example the distillation column in a Microplant is designed as a colimmator-column, but most of the real columns for production are plate or packing columns. The thermodynamics of the seperation process are the same for all types of columns, only their geometry differs. So the scale-up from the Microplant to the real plant is hardly possible and not aspired. The scale-up from an experimental plant to the realized solution will still remain the field for Miniplants. In a later stage of the developement process, after the reduction of a wide variety of process-variants by the use of Microplants, a Miniplant can validate the chosen type and geometry of the apparatus.

2. The design of Microplants

Traditional piping systems are no longer appropriate for Microplants. In designing a Microplant, not only the seperation units, e.g. distillation columns, and heat exchangers etc. have to be miniaturized, but also special considerations have to be given to the question of how to connect the different units and how to establish recycle loops.
The first of two proposals is a manifold, the other is a modular built piping system.

2.1 The manifold

The structure of a manifold ressembles a sandwich with different layers (see fig. 2.1):

- two cap-layers
- two sealing-layers
- a channel-layer

The channel-layer contains all the channels conducting product-streams, throttles, valves, measurement techniques. At its front there are junctions for pumps and other equipment which is installed outside of the manifold.
Channels connected by drill holes are milled in the top- and the bottom-side of the channel-layer, so streams can cross each other on different levels. Due

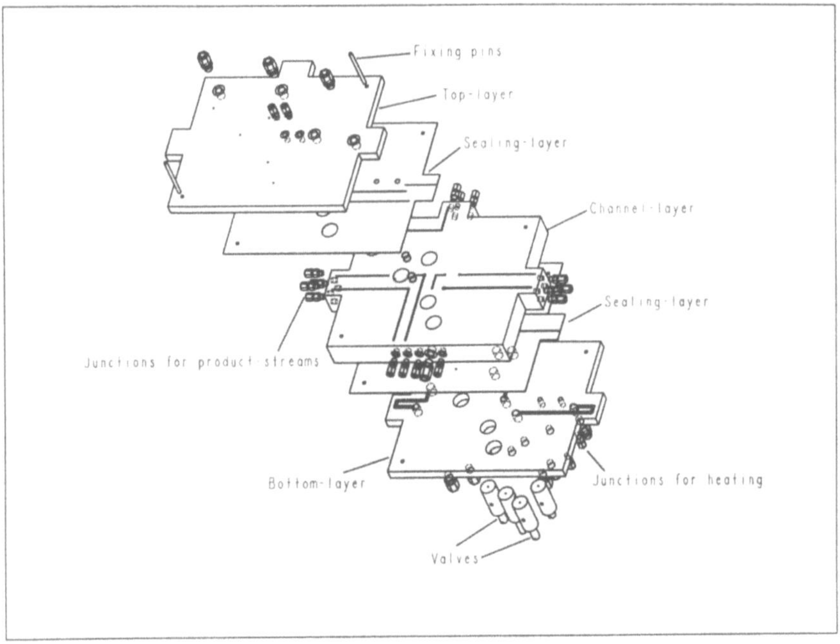

Fig. 2.1. The manifold

to the small flow-rates the velocities are in the order of magnitude of a few millimeters per second.
Elements of sintered metal tightly fit into drill holes throttle the product-streams. The hydrodynamic characteristics of the sintered elements are described by their pressure-drop as function of the flow-rate.

In the top- and bottom-layer channels, which accompany the product-streams, pipeline the heating fluid. The sealing-layers seperate the product and the heating fluid. Hence they are made from cupper, which has a high coefficient of thermal conduction in order to guarantee a good heat transfer. The manifold offers the possibility to realize channels with different cross-sectional-dimensions. So the channels can be adapted to the requirements of different flow rates. Also, in case of substances in the product streams running the risk of crystallizing out, the streams can be heated continuously remaining above the cristallization temperature.

2.2 The piping-system

In a system of parallel plates fixed at a wall the pipelines, fittings and apparatus are mounted. A special fixing system allows the mounting and dismanteling of the devices. All of them have a standardized size and fittings in

order to guarantee their interchangeability.
The piping-system is appropriate for the development of processes, whose structure changes frequently. The flexibility of the system allows an exchange of single elements and a quick and easy rebuilding of the process structure. The diameter of the pipelines is 1/8", the wall thickness is 0.6 mm. The pipes and apparatus are all connected by screwed-fittings.
The product-streams are heated by heat exchangers working with Peltier cells. A continuous heating as realized in the manifold requires an enormous technical effort.

3. The distillation-column

A distillation-column designed as a collimator-column for a feed of 10 ml/h of a mixture of water (68%) and methanole (32 %) has been constructed. The column works at atmospheric pressure. The collimator column is a type of distillation-column especially adapted to small flow-rates. In the aperture between an inner and an outer pipe the vapour streams up to the top of the column while the liquid flows down the walls.

Circular elements of sintered metal distribute feed and reflux over the surface of the inner and outer tube. Both capacity to distribute the liquid and the relation between pressure drop and flow-rate are influenced by the porosity and thickness of the sintered element. A small porosity assumed to be most convenient, because the average radius of the pores is small then. This implicates small drops on the surface of and small flow-rates through the element. As a disadvantage follows a higher pressure-drop. The statistical distribution of the pores in the sintered material determines its permeability for the liquid. Smaller porosity leads to less cavities in the material and a smaller probability that a multitude of pores forms a channel from the inner to the outer surface of the element. The actually utilized type has a porosity of 31%, an average CCE pore diameter of $11\mu m$ and a pore size distribution from $7\mu m$ to $14\mu m$. Due to the small flow-rate, the pressure drop is less than 1 mbar.

3.1 Calculation of the column's geometry

A procedure based on the model of Kuhn [Kuh42] and Jantzen [Jea54] was developed to calculate as well the steam- and liquid-flows in the column as the length and diameter of the collimator. Jantzen gives a formulare for the length l_{st} of a theoretical stage:

$$l_{st} = \frac{a \cdot m_{vap}}{4 \cdot b \cdot \eta_{vap}} \tag{3.1}$$

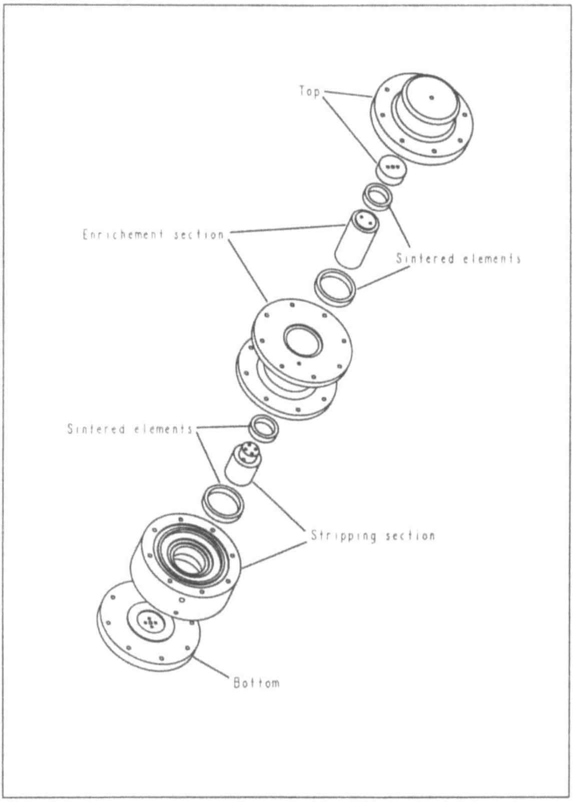

Fig. 3.1. The distillation-column

First, by the help of the process-simulation system ASPEN+, the streams running in the column and their number of stages were calculated. A stage of the model implemented in ASPEN is equal to a platform. In our column the seperation takes place continuously, hence by application of Kuhn's model the results of the ASPEN-simulation have been transformed into the length and diameter of a collimator.
The results of the calculations are 10 theoretical stages, a length of the still of 40 mm, an inner diameter of the collimator of 16 mm and an outer diameter of the collimator of 19 mm.

3.2 The slope of the temperature profile

The following facts are considered:

- in radial and tangential direction the walls of the column are adiabatic,
- no axial heat-flow in the vapour and liquid,

– boiling liquid,
– condensing vapour.

An ideale temperature profile, which only depends on the thermodynamics of the seperation, will be established in the aperture along the column (see figure 3.2, l=point at the z-axis, z=0 at the bottom ; L=length of the column).

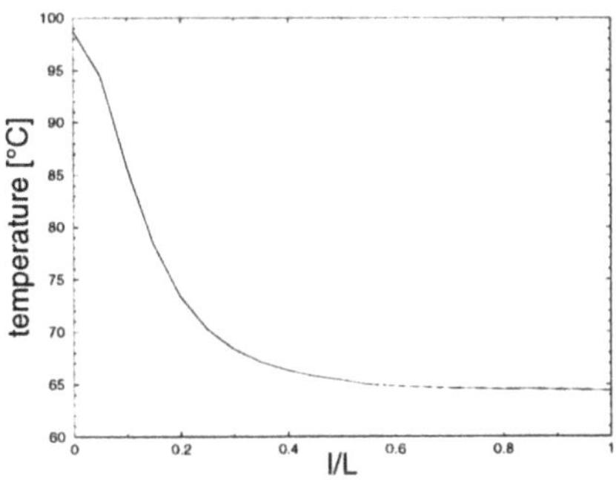

Fig. 3.2. The ideale slope of the temperature profile of the column

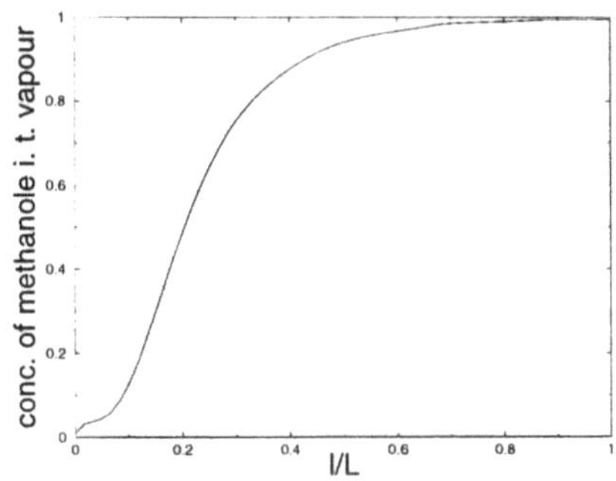

Fig. 3.3. The ideale concentration profile of the column

A concentration profile is directly related to the temperature profile (see figure 3.3).
Considering z=0 at the bottom of the column, it can be seen in figure 3.2, that over the first third of the length the temperature decreases from 98°C to 68°C. In an ideale case the temperature profile in the aperture will be the same as in the walls of the column. In combination with a well conducting material this steep gradient provokes a large axial heat flow through the walls. In reality the walls are not adiabatic, so this additional heat flow influences the ideal temperature profile in the aperture. Every disturbance of the ideal temperature profile impacts a diminution of the efficiency of the column.
In order to reduce the axial heat flow the column is constructed from polyvinylidene fluoride (PVDF), which has a coefficient of thermal conduction that is only $0.23\frac{W}{K \cdot m}$.The thermal, mechanical and chemical stability of PVDF are sufficient under the examined conditions.
In order to improve the isolation of the column, the temperature profile is examined by the use of a finite element analyse.

3.3 Wettability of the walls

The efficiency of the column is directly related to the establishement of a liquid film of constant thickness. PVDF is a material of low surface energy, so the wettability with liquids should be difficult. The contact angle of a system liquid-solid is an index to predict the capability of the liquid to spread over the surface. The wetting properties of different materials are compared by their measured contact angles. Doniec [Don88] gives a formular to calculate the minimal possible thickness δ of a continuous film:

$$\delta(\Theta) = 1,45085 \cdot \left(\frac{(1 - \cos\Theta) \cdot \eta_l^2 \cdot \sigma}{\rho_l^3 \cdot g^2} \right) \tag{3.2}$$

With his theory the critical minimal thickness of the continuous film is calculated to the range of 0.025 mm to 0.14 mm, depending on the position in the column and so the composition of the film. The resulting real film thickness s calculated with the model of Kuhn [Kuh42]

$$s = \left(\frac{3 \cdot \eta_l \cdot \rho_{vap} \cdot u \cdot a}{\rho_l \cdot g} \right)^{\frac{1}{3}} \tag{3.3}$$

is 0.02 mm.
Kuhn [Kuh42] considers a negligible thickness of the liquid film in relation to the width of the aperture. An increasing hold-up thickens the film, but for a width of the aperture of 1.5 mm, the condition of a negligible thickness of the liquid film [Kuh42] is only valid for thin films. A film of 0.15 mm would reduce the width of the aperture by 20%. So films with a thickness as thin as possible are aspired.
To improve the wetting properties of the column even for very thin films, special treatments and patterns of its inner surface are examined.
A sulphonation of the surface showed no amelioration whereas after a roughening of the surface with abrasive paper (grain size 60) the contact angle diminished at about 10%.
Actually a texturing of the surface with different profiles like rifles, knurls or meshes is examined and shows good results so far.

3.4 Construction of the column

Figure 3.4 shows a sectional drawing of the column. The elements are:

- outer part of the top (PVDF) ①
- inner part of the top (PVDF) ②
- sintered element (special steel) ③
- sintered element (special steel) ④
- inner part of the enrichement section (PVDF) ⑤
- outer part of the enrichement section (PVDF) ⑥
- inner part of the stripping section (PVCF) ⑦
- outer part of the stripping section PVDF) ⑧
- bottom (heating plate, special steel) ⑨

The outer parts of the top and enrichement section and the inner parts of the top, the enrichement section, the stripping section and the bottom are bolt togehter just as the outer parts of the enrichement section, the stripping

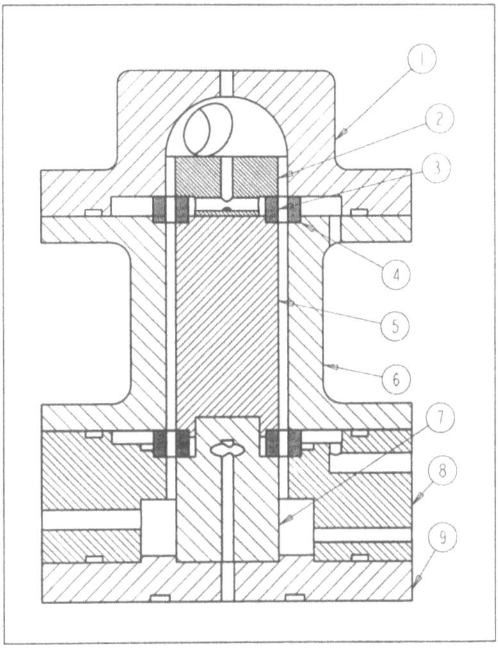

Fig. 3.4. The distillation-column

section and the bottom.
Due to the small dimensions of the column, it is not possible to screw every fittin so that some must be pasted. A series of experiments showed, that cyanogenacrylate is best adapted to connect the metallic pipes to the column.

3.5 Control technology of the column

The control of a micro-distillation column requires other features than a big column does. The hold-up of the micro-size column is 3 ml, so the column is sensitive to external heat flows. Additionally every measuring point acts as an interference on the column. This effect will be intensified by the unfavorable surface to volume ratio of the still and the big size of the sensors in relation to the size of the column and the liquid film. However, on the one hand the adjustment must be very fast on the other hand only a small derivation and overswing are acceptable.
The pressure in the still and so the boiling point of the liquid changes with the vapour flow. A precision resistor, dipping into the liquid at the base, and the heating system are integrated into a control loop to regulate the reboil rate.
The upper level of the liquid is controled by an adjustable siphon. A temper-

ature control in the bottom stops the heating in case of the column running dry.
The vapour at the top will be totally condensed and the stream will be devided into continuous product and reflux streams by use of a valve.
The streams are tempered by heat-exchangers with Peltier cells. The Peltier cells can heat or cool only by changing their electric polarity.

4. Conclusion

First experiments showed that the Microplant technology can be a complement to the Miniplant. To develope the Microplant to an industrial used technology, some further work will be necessary.
Based on the presented concepts, a Microplant will be constructed and experimentally analysed. As far as the column is concerned special research on the establishement of a liquid film should be done. The parameters of the column's control have to be determined and the concept should be tested at a realized Microplant.
The most important progress in the development of the manifold will be the integration of valves, pumps and the measuring system in the manifold. If the top and bottom layers exist in a general form and only the channel layer has to be adapted at the actually examined process, the use of a manifold would offer a great potential of work-time savings.

Notation

Latin letters
a=semiwidth of the aperture [m]
b=medium diameter of the aperture [m]
g=acceleration due to gravity $[\frac{m}{s^2}]$
l=length [m]
m=flow rate $[\frac{kg}{s}]$
s=film-thickness [m]
u=average velocity of the vapour $[\frac{m}{s}]$

Greek letters
δ=minimum thickness [m]
η=dynamical viscosity of liquid [Pa s]
Θ=contact angle [°]
ρ=density $[\frac{kg}{m^3}]$
σ=surface tension $[\frac{N}{m}]$

Subscripts
l=liquid
st=stage
vap=vapour

References

[Don88] A. Doniec. Flow of a laminar liquid film down a vertical surface. *Chemical Engineering Science, 43(4) pp. 847-854*, 1988.
[Jea54] E Jantzen and et al. Ringspaltsäulen zur hochwirksamen Destillation bei niedrigen Drücken. *Chem.Ing.Techn. 26, Nr.7*, 1954.
[Kuh42] Werner Kuhn. Die Destillation. *Helv. chim. Acta 25*, 1942.

A Modular Microreactor Design for High–Temperature Catalytic Oxidation Reactions

G. Veser* **and G. Friedrich**

Institut für Chemische Verfahrenstechnik,
Universität Stuttgart, D–70199 Stuttgart, Germany;

and

M. Freygang and R. Zengerle

Institute of Micromachining and Information Technology,
Hahn–Schickard–Gesellschaft, D–78052 Villingen–Schwenningen; Germany.

Introduction

High temperature catalytic oxidation reactions of hydrocarbons have again become the focus of intense research during the past decade, both due to their importance for energy production such as in catalytically stabilised burners as well as due to a renewed interest in alkane functionalisation through catalytic partial oxidation. In the first case, the catalyst serves to widen the range of operability of conventional gas burners and to render the burner less susceptible to fluctuations in inlet conditions. In the second case, the catalyst opens up new reaction pathways for the selective oxidation of saturated hydrocarbons towards more valuable chemicals, such as olefins and oxygenates. In both cases, the reaction systems are characterised by very short residence times due to the extremely fast reaction rates under high temperature conditions. This allows for very compact reactors with high throughput, which could make them an interesting alternative to existing technologies.

While high temperature catalytic partial oxidation reactions have been shown to proceed very selectively over noble metal catalysts (Schmidt, 1995), the high temperature conditions pose several problems both for the practical application of these reactions as well as for a scientifically sound understanding of the reaction mechanisms. From a practical point of view, handling hydrocarbon–air mixtures at high temperatures poses a safety problem due to the imminent danger

* *to whom all correspondance should be addressed (veser@icvt.uni–stuttgart.de).*
Financial assistance by the Deutsche Forschungsgemeinschaft (DFG) is gratefully acknowledged (G.V.).

of flames or explosions. From a scientific perspective, the common assumption that heterogeneously catalysed reaction pathways are so much faster than homogeneous reactions that the latter can be safely neglected is no longer necessarily true at very high temperatures. While it has been shown in detailed modeling studies that at least for some high temperature catalytic oxidations and in certain experimental parameter ranges homogeneous contributions should be negligible (Veser, 1999), experimental investigations of heterogeneous–homogeneous processes (typically using spectroscopic techniques such as laser–induced fluorescence spectrocopy) are rather difficult and therefore rare (Driscoll, 1987).

Here, microreactors, i.e. chemical reactors with characteristic dimensions in the sub–millimeter range, have a great potential to become valuable tools for the investigation and possible application of high temperature catalytic processes. Among their advantages for chemical processes are: the very small thermal inertia, which allows for a very direct control of the reaction temperature; their inherent safety due to both the very small reactor volume and the well controllable reaction conditions; and their small overall dimensions, allowing for an easy integration into existing processes and for their use where space require–ments are critical.

Furthermore, microreactors offer the advantage of a very large surface to volume ratio. This could allow for an effective suppression of homogeneous gas phase reactions, since free surfaces typically are strong sinks for radical species which are required to sustain a homogeneous reaction. Therefore, it should be possible to conduct a heterogeneously catalysed reaction involving a mixture of potentially flammable (if not explosive) gases in a microreactor without any danger of open flames and explosions. Thus, the process not only becomes intrinsically safe, but it should also become possible to study heterogeneously catalysed high–temperature reactions undisturbed by parallel homogeneous reaction pathways, i.e. it should be possible to clearly separate homogeneous and heterogeneous contributions to the overall reaction behavior.

In the following, we will describe the design and realisation of a simple, modular microreactor for the study of high temperature catalytic reactions. Reactor design and fabrication as well as first experimental experiences with this reactor will be reported, in which the H_2+O_2 reaction was used as a test case for a high temperature catalytic reaction. The current design will be critically assessed and next steps in the further development will be briefly discussed.

Reactor Design Considerations

The development of a microreactor as a research tool should take into consideration an easy handling of the reactor set–up, suitability of the micro–reactor for a range of reactions and reaction conditions as wide as possible, reusability of all parts of the reactor, and an easy and cheap fabrication techno–logy. Therefore, the decision was made for a *modular reactor design* in which the catalyst, the microreactor and the reactor housing are completely separate, exchangeable units.

While very advanced micro-manufacturing methods have been developed over the past 10 years, such as laser ablation methods and X-ray based lithographic methods (LIGA and similar methods), these techniques are at this point still expensive and require rather specific equipment (Wise 1991, Bacher 1994, Wengeng 1996). In contrast to this, silicon based micromachining is not only readily available, but also very cheap. It seems therefore well suited for exploratory studies at this still rather early stage of development for chemical microreactors, allowing for a comparatively quick and cheap prototyping of such reactors. Furthermore, good mechanical strength, high thermal conductivity, and the relative chemical inertness of silicon make it an ideal candidate particularly for high temperature studies. It is for these reasons that we choose silicon as the base material for our current microreactor design.

The reactor chamber is defined through two stacked silicon chips, into which channel and hole structures were etched (fig. 1). The catalyst is added in the form of a wire which is chucked into the channel structure. The stacked Si chips are then pressed together by the reactor housing, assuring a gas tight closure due to the flatness of the silicon surfaces down to the atomic scale.

The reactor design thus incorporates the following main considerations:

- easy exchange of the catalyst material by changing the wire material
- easy adjustment of the reactor wall properties by different pretreatments of the silicon chips
- easy handling of the housing and reuse of all components
- cheap batch processing of the silicon wafers.

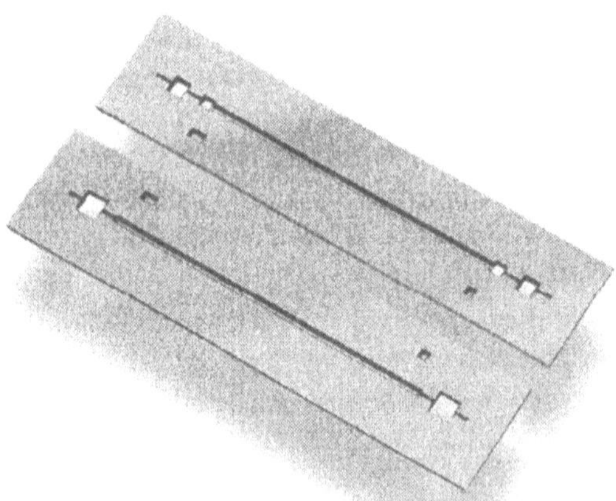

Fig. 1: *The etched Si-chips with the micro-reaction channel at the center. Holes for the electrical contacting of the catalytic wire as well as for the gas inlet and outlet can be seen on both ends of the channel.*

microreactor chips

For manufacturing the reactor chips in a way as simple as possible and at least cost, we chose a process with only two masks for photolithography and only one step of wet etching with KOH. The wafers were etched from both sides simultaneously to a depth of a little more than half the wafer thickness of about 525 μm. This resulted in the formation of a reaction chamber of some 0.167 mm^2 cross–sectional area and a length of about 20 mm. The cross sectional area of the inlet bore is designed to equal that of the gas channel in the rector chamber while the area of the gas outlet bore is 1.5 times larger. The exact dimensions of the channel bottle–necks in which the wire is fixed are defined by means of the common self stopping mechanism of the KOH wet etch process.

After etching the wafer in a fast and cheap batch process, several wafers were coated at 950°C with a silicon oxide thin film of 400 nm thickness for testing purposes. As a further modification, on some of these oxidized wafers LPCVD silicon nitride was deposited at a thickness of about 110 nm. The nitride films were then tempered for one hour at 1100°C to reduce intrinsic layer stresses. Thus, three different microchannel surfaces were available and tested with respect to their suitability to high temperature reactive conditions.

Catalyst

The catalyst is introduced into the micro–reaction channel in form of a wire. This has several advantages over a catalytic coating of the channel walls: the catalyst can be exchanged very easily between experiments, and can easily be examined ex–situ after the reaction; the same micro–chips can be re–used many times for different catalysts; and most metals are readily available in the form of wires with different diameters, and can also be handled easily in this form. By varying the wire diameter, the total catalytic surface area as well as the surface–to–volume ratio can readily be changed without the necessity for any changes in the reactor setup itself. Furthermore, this configuration allows for a very easy resistive (pre–)heating of the catalyst through electrical contacts from the out–side. Finally, no thermal degradation problems should be expected for this set–up, while such problems can become critical under high temperatures conditions due to thermal stresses when thin film coatings of the catalytic material on the reactor walls are used (Srinivasan 1997).

However, the use of catalytic wires also has some disadvantages compared to catalytically coated walls. First of all, the resistive heating of the wire – while very simple – does not allow for a precise zone–wise heating of the catalyst, but rather leads to a global heating over the whole reactor length. Furthermore, the positioning of the wire in the microchannel is obviously less well defined than for a thin film coating of the channel walls. And finally, the catalytic wire in the center of the reaction channel leaves the surface area of the channel walls exposed to the reactants, which could catalyse undesirable side reactions in some processes. While this should not be a problem for the type of reactions discussed in this paper, it cannot be excluded for other reaction systems. Nevertheless, the use of catalytic wires is a very simple and particularly flexible configuration for catalytic microreactor studies.

Reactor housing

The design goal for the microreactor housing was to ensure an easy and flexible handling of the microreactor itself, i.e. quick and easy access to the microreactor as well as an easy connection to standard (macroscopic) laboratory equipment. Furthermore, the equipment (i.e. the microreactor as well as the housing) had to be re–usable many times to keep development cost low.

The reactor housing is built completely from stainless steel (fig. 2). To assure an easy handling and connecting of the microreactor, the housing is built comparatively large with a diameter of about 50 mm and a thickness of about 7.5 mm. The inner surfaces of the housing were manufactured by spark–erosion to produce a flat, smooth surface contacting the silicon micro–chips (fig. 3). This not only helps to ensure a well defined positioning of the chips in the housing, but also allows for a gas–tight connection between the micro–chips and the housing at the gas inlet and outlet without the necessity for additional sealings.

Fig. 2: *The microreactor housing with the two gas inlets (lower right) and the gas outlet (upper left).*

Fig. 3: *The open reactor housing with the top and bottom parts of the microreactor. The catalytic Pt wire is visible in the microchannel (left).*

The catalytic wire in the microchannel is contacted electrically through small metal screws, allowing for a direct resistive heating of the catalyst for the purpose of igniting or thermostating the reaction. The reaction gases can be fed separately through two steel tubings (inner diameter= 0.5 mm, outer diameter= 1.5 mm). Connections to the peripherals can be made through standard Swage–lock–type connectors, allowing for an very easy connection to standard labora–tory equipment. The feed gases are mixed by perpendicular impingement at the reaction channel entrance. This mixing right at the beginning of the catalytic reaction zone helps to avoid possible homogeneous reactions between the reactants in the inlet tubing which can pose a problem for the type of high tem–perature oxidation reactions studied here.

EXPERIMENTAL MEASUREMENTS

The H_2+O_2 reaction

We tested the applicability of this simple microreactor set–up using the Pt–catalyzed H_2+O_2–reaction ($H_2 + 0.5\ O_2 = H_2O$) as a sample reaction. This reaction constitutes a particularly interesting test case for the current system for several reasons: The reaction is known to proceed very fast both homogeneously in the gas phase and heterogeneously catalyzed by platinum. It is characterized by particularly wide flammability limits (about 3 vol% – 75 vol% in air) and very high flame velocities, i.e. strong explosions (Lewis 1987). While the reaction is strongly exothermic and therefore interesting as a possible energy (heat) source, this danger of strong explosions restricts its use for practical applications severely since the reaction typically has to be conducted with mixtures well below the lower flammability limit to ensure safe operation under all conditions. Therefore, if it would be possible to effectively suppress homogeneous reactions (and thus open flames and explosions), the potential of this reaction could be much more efficiently exploited. (This can be shown with a simple estimate of the adiabatic temperature rise of the H_2+O_2–reaction: while this temperature rise at the lower flammability limit in air (3 vol% H_2 in air) is only about ΔT_{ad}= 250K, under stoichiometric conditions it would increase to about 1600K. This value could be further increased drastically by reducing the nitrogen content, i.e. replacing air by oxygen (see fig. 4). Although this calculated temperature rise constitutes a theoretical value which obviously cannot be achieved in a practical set–up due to heat losses and temperature limitations of the reactor materials, it nevertheless serves here to show the potential of this reaction as a heat source).

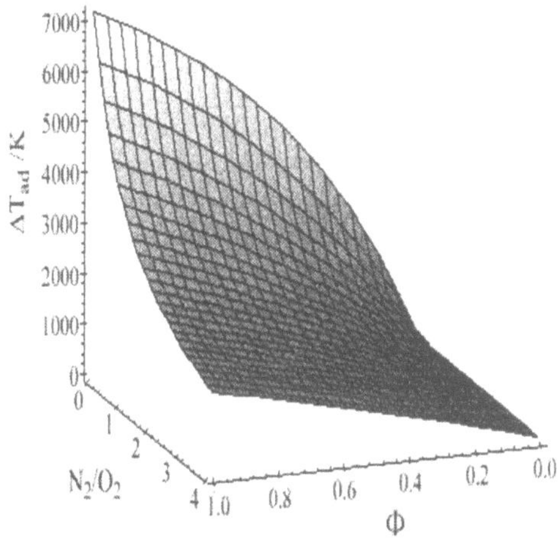

Fig. 4: *Adiabatic temperature rise vs N_2/O_2–ratio and H_2/O_2 –equivalence ratio (Φ) for a $N_2/O_2/H_2$–mixture with a fixed oxygen inlet flow of 0.2 slpm. (Φ=$[H_2/O_2]$/ $[H_2/O_2]_{stoich}$ runs from 0 for hydrogen–free gas streams to 1 at the stoichiometric point for H_2O formation).*

To test our microreactor set–up with this reaction system, a catalytic Pt–wire (diameter 0.15 mm, purity 99.99%) was chucked into the reaction channel. The reaction gases H_2, O_2, and N_2 were fed through standard mass flow controllers. A thermocouple (K–type) was positioned right at the microchannel exit of the microreactor to measure the temperature of the effluent reaction gases. The effluent gas stream was then fed to a cryostat to remove the water vapor formed in the reaction, and the remaining gases were fed to a flow meter to measure the total flow rate, and to a O_2 analyzer to measure the remaining oxygen content. Since the investigated reaction follows a very simple reaction stoichiometry, the extent of the reaction (i.e. hydrogen and oxygen conversions) could be calculated from both measures and were found to agree within experimental error (< 5%).

For ignition of the reaction, synthetic air was introduced at a flow rate of about 0.1 slpm (standard liters per minute), and then H_2 was added to the feed gas stream. In contrast to our experience in 'macroscopic' fixed–bed catalytic reactors, the reaction did not ignite at room temperature at any H_2/O_2 ratio. At this point, we do not yet know whether this is due to finite size effects in the microchannel, or whether the heat losses in the microreactor are simply too high for ignition of the reaction at room temperature conditions. More experiments and simulations will be needed to clearify this point. In order to ignite the reaction, the catalytic wire was therefore preheated resistively in about 5% H_2 in air. At an exit gas temperature of about 100°C, ignition of the catalytic reaction was usually observed. The temperature measured at the reactor exit jumps to about 300°C (i.e. close to the adiabatic reaction temperature), indicating ignition of the reaction. The whole ignition process takes place within a few seconds.

Qualitative tests

As a first test for the thermal stability of the silicon micro–chips under reaction conditions, some test runs were conducted with only the silicon micro–chips, i.e. without a reactor housing (fig. 5). In these experiments, the hydrogen content in the feed gas was adjusted to stoichiometric conditions after ignition of the reaction, and then the N_2 content was slowly reduced. Although no quantitative temperature measurements were possible with this setup, simple visual inspection showed that the chip was glowing red–hot near the entrance zone of the catalyst. We estimate the local reactor temperature therefore to be around 1000°C. It seems interesting, that by further increasing the total gas flow rate, this hot spot could be smeared out over the whole length of the catalyst, until it finally started to shift towards the exit zone of the microchannel. However, under no experimental conditions any indication for open flames or explosions in the micro–reaction channel was observed.

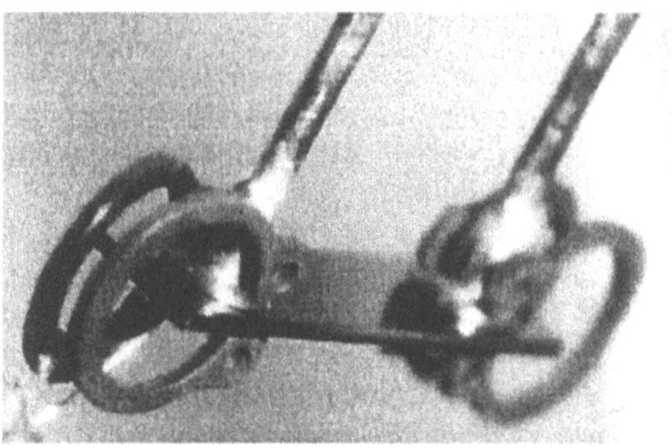

Fig. 5: *Simple microreactor set–up without reactor housing. Gas inlet and outlet can be seen on top the the Si chips, while in the right lower corner of the picture the catalytic wire is visible.*

Following these experiments, the microreactor chips were inspected for signs of degradation due to the exposure to these very high temperatures and reactive atmospheres. For none of the tested reactor materials (pure Si chips, as well as oxide and nitrite surfaces) any degradation could be observed. Furthermore, no sintering of the top and the bottom parts of the microreactor was observed either, making the exchange of the catalyst between runs a rather easy and quick task. Finally, no differences in the reactor behavior or the reaction conversions were found between the different reactor materials. This supports our assumption that at least for this reaction the Si–based microchannel walls are completely inert.

In contrast to the reactor walls, the Pt wire roughened visibly over the course of several experiments which is a well known phenomenon for Pt catalyzed oxidation reactions (Satterfield 1991). No evidence for a net loss of Pt was found in our experiments, however, in some runs, a breaking of the catalytic wire occurred in the microchannel. We do not attribute this to the roughening process of the catalyst but rather to localised thermal stresses at points where the hot catalytic wire touched the microchannel walls due to the considerable thermal expansion and bending of the wire at these extremely high reaction temperatures. While this does not pose a problem during the reaction, the broken catalytic wire had to be exchanged at the end of that particular experiment since no resistive heating was possible any more to ignite the reaction in subsequent runs. A well defined positioning of the catalytic wire and/or an additional means of pre–heating the microchannel should therefore be an important consideration in further developments of this microreactor.

In summarizing, these first tests showed quite convincingly that all three Si–based materials used for the microreactor are well suited for the application in these high temperature reactions, with respect to their thermal stability as well as their reactive inertness. Furthermore, the lack of flames and explosions over this very wide range of reaction conditions indicates, that homogeneous reactions seem indeed to be effectively suppressed in the fabricated microchannels.

Quantitative measurements

Following these qualitative tests, the microreactor was put into the reactor housing for quantitative measurements of reaction temperatures and conversions. The basic experimental procedure was the same as before: the reaction was ignited by heating the catalytic wire resistively in a lean $N_2/O_2/H_2$–atmosphere. After ignition, the hydrogen feed gas stream was varied and the exit gas temperature, total gas flow at the reactor exit, and oxygen concentration of the exit gas stream were measured.

Figure 6 shows a typical curve for the temperature of the effluent gases for a feed gas stream of 1 slpm synthetic air when adding increasing amounts of hydrogen (0 slpm to 0.4 slpm, i.e. stoichiometric composition), along with the corresponding H_2 and O_2 conversions. One observes that with increasing hydrogen content the exit gas temperature initially increases almost linearly up to an H_2 feed stream of about 0.17 slpm. Above this ratio, the temperature curve flattens, and above an H_2 flow rate of about 0.25 slpm it continues to rise linearly with a somewhat reduced slope. At the stoichiometric feed of H_2 and O_2 an exit gas temperature of close to 700°C is finally reached. To avoid possible problems due to overheating of the steel housing, we did not attempt to decrease the N_2 content in the feed gas in these experiments.

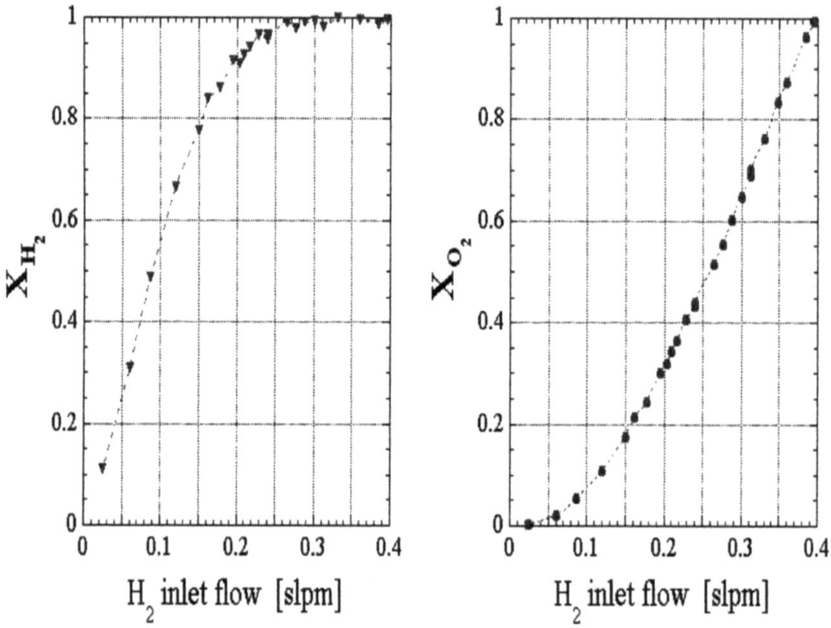

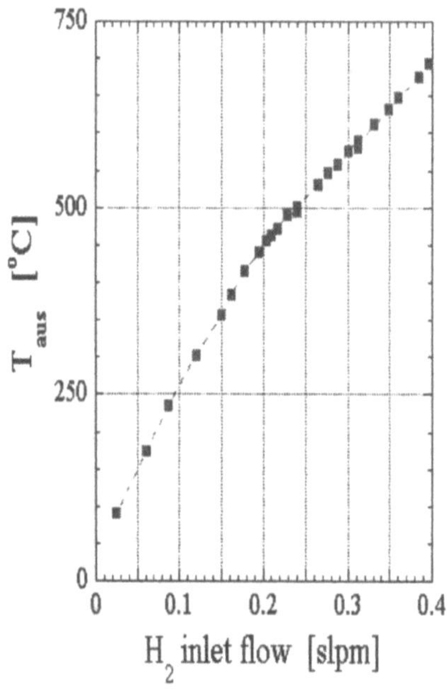

***Fig. 6: H**ydrogen and oxygen conversion and gas exit temperature (clockwise from left top graph) in the microreactor for increasing H_2 content in 1 slpm air.*

Similar to the exit gas temperature, the H_2 conversion initially increases almost linearly up to an H_2 feed gas stream of about 0.17 slpm. Then, the conversion curve flattens, and above flow rates of about 0.25 total conversion of H_2 is reached. The oxygen conversion also follows these trends very closely: between an H_2 inlet flow of 0 slpm and about 0.17 slpm, the oxygen conversion increases about parabolically, reflecting the increasing O_2 consumption due to a superposition of the (linearly) increasing H_2 content of the reactant gas stream and the (linearly) increasing H_2 conversion in this range. Above H_2 flow rates of about 0.25, the O_2 consumption increases linearly, since now complete H_2 conversion has been reached and any further increase in O_2 conversion can only be due to the increasing H_2 content of the gas inlet. Finally, at the stoichiometric point, complete O_2 conversion is reached. Thus, all three measurements show very consistent results.

In general, these experiments show a reactor behavior for this experimental set–up which shows no obvious differences from the expected behavior in 'macroscopic' catalytic fixed–bed reactors – except for the remarkable absence of flames and explosions even under stoichiometric conditions. Although the studied reaction system is a mechanistically very simple one, this can be taken as an indication that the current microreactor set–up should be well suited for investigations of other high temperature catalytic reactions, including more complex catalytic reaction systems such as catalytic partial oxidation reactions.

DISCUSSION

Possible applications

It seems remarkable that in the above experiments, full conversion could be achieved at flow rates in excess of 1 slpm. This total through–put is surprisingly high for such a small reaction channel, indicating that actually quite reasonable amounts of chemicals can be converted in such tiny reactors given fast enough reactions, such as high–temperature catalytic oxidations.

Under our experimental conditions, a power of about 72 W could be produced in a single microchannel at the stoichiometric point for the H_2+O_2–reaction. This output could be further increased by at least a factor of 2 by further increasing the flow rate and/or reducing the nitrogen content of the gas stream. A practical application for this potential could be seen in the use as a preheating system for an automotive exhaust catalyst (Veser, 1998). This preheating is currently the focus of intensive research (Kollmann, 1994; Kirchner, 1996), since 80–90% of the emissions during typical test cycles are produced during the cold start–up phase of the catalyst. Increasingly stringent legislative requirements can only be met if a substantial improvement in this start–up phase can be achieved.

One possibility to achieve this goal could be by using the H_2+O_2 reaction in a microreactor as a very quick and efficient – yet safe – way to chemically preheat the main automotive catalyst. Reacting a stoichiometric H_2/air mixture in a microreactor would allow to safely produce an effluent gas stream at temperatures around 700°C as seen in the above experiments. Such 'micro–

burners' could thus be used to quickly preheat the main exhaust catalyst without producing any additional emissions (other than water vapor). One can estimate that less than 100 microchannels (which could readily be parallelised onto just a few micro–chips) would suffice to feed enough heat to the catalyst to pre–heat it within about 10s to the required operation temperatures. Due to their small space requirements – which is a main concern in automotive applications – these microreactors could easily be positioned on the perimeter of the car exhaust pipes – or maybe even be implanted in the monolithic catalyst itself.

Another area in which the presented microreactor set–up could have interesting applications is the study of high–temperature catalytic partial oxidation of hydrocarbons as already pointed out in the introduction. These reactions have been shown to proceed very selectively to syngas and olefins over noble metal catalysts at high temperatures (about 800–1200°C) and very short contact times (about 1–50 ms) (Schmidt, 1995). From our current channel geometry and the total flow rate in the experiments we calculate an extremely short contact time of about 0.1 msec for our experiments. This demonstrates that contact time ranges even well below those of previous investigations can easily be reached in a microreactor. This range of extremely short contact times holds great potential for new reaction pathways in the complex network of parallel and consecutive reactions typical for hydrocarbon oxidation reactions (see for example Goetsch, 1996 and Witt, 1997). However, due to the extreme reaction conditions of these process routes and therefore the inherent danger of homogeneous reaction (and thus flames or explosions), they have barely been studied so far. As we could show in our experiments, homogeneous reactions can be quite effectively suppressed in a microreactor, indicating that this set–up should be an ideal candidate for the study of this class of reactions, in which the microchannel could be seen as an isolated, idealized single pore of a monolithic catalyst.

Further developments

The first tests of our Si–based catalytic microreactor demonstrated that the current set–up is in principle well suited for the intended goal of studying high–temperature catalytic oxidation reactions. Therefore, further developments will now focus on refining the current, very simple design without abandoning its simplicity and ease of use.

The most important next step in the reactor design is a high temperature resistant reactor housing. Obviously, stainless steel – while easy in handling and manufacturing – is typically limited to reaction temperatures below about 700 – 800°C. Therefore, the next housing should be made from a high temperature resistant material such as a high temperature ceramics.

Another important step will be the implementation of local heating elements on the micro–chips. This would allow to pre–heat the catalyst to ignition temperatures without the necessity of a resistive heating, thus circumventing the problems that occur when the catalytic wire breaks during an experiment. Furthermore, this would enable a zone–wise heating of the reaction channel, thereby allowing a more detailed 'engineering' of the reaction conditions in the microreactor by imposing temperature profiles along the reactor length. When not being used as heaters, such elements could also allow to measure temperature profiles along the reactor axis yielding additional important information about the local reaction progress. Such a configuration would make the catalytic microreactor a very flexible and well defined tool, for example for catalytic ignition–extinction studies which can be used to get very sensitive mechanistic information about essential reaction steps in catalytic oxidation reactions (Veser 1996, Ziauddin 1997).

Such local heaters could relatively easily be implemented in the current design by thin film design on the Si chips as already well established in the design of chemical sensors (Semancik 1993). Since these elements could be applied to the outside of the chips and thus would not be exposed to a reactive atmosphere, degradation of these films should be much less a problem than for thin film catalysts inside the reaction chamber. However, implementation of such heating and sensing elements would be facilitated by a more open construction of the reactor housing, so that the Si chips could be contacted directly from the outside.

These considerations have been taken into account in designing a new reactor housing (fig. 7), which incorporates a more open design in a high temperature–resistant ceramic housing. This new set–up is currently being tested.

Fig.7: *Ceramic microreactor housing (dimensions 50 mm * 20 mm * 20 mm) with gas inlet and outlet (top), electrical contacts (bottom) and the micro–reaction chips (center).*

SUMMARY

We presented the design for a very simple microreactor setup, based on a modular concept in which reactor housing, micro–reaction channel, and catalyst are three completely separate units, each of which can be (re–)used and exchanged independently. The overall emphasis in the realised system has been on flexibility and ease of handling of the microreactor. Silicon was used as a base material, into which micro–reaction channels were etched, forming a cheap and readily available technology for the manufacturing of high temperature–

resistant and chemically inert reactor walls. A catalytic wire can be chucked into this reaction channel, which is then set into a relatively large reactor housing which allows for an easy connection to standard macroscopic laboratory equipment.

This reactor set–up was tested with the Pt–catalyzed H_2+O_2 reaction. We could demonstrate that the reactor design is well suited for the intended task, with–standing very high temperatures without any degradation and very effectively suppressing homogeneous flames and explosions in this reaction system. Very high flow rates in excess of 1 slpm, and exit gas temperature of about 700°C, were achieved at complete conversion of stoichiometric H_2 feeds in air.

We see a promising potential application of the H_2+O_2 reaction in a micro–reactor as a 'micro–burner' for the preheating of an automotive exhaust catalyst. Due to the effective suppression of flames in a microchannel, the strong exothermicity of this reaction near stoichiometric conditions could be put to use in such a set–up. A further use of the current reactor set–up can be seen in the study of high–temperature catalytic partial oxidation of hydrocarbons.

References

W. Bacher et al. (1994), Naturwissenschaften 81, 536–545

D.J. Driscoll, K.D. Campbell, and J.H. Lunsford (1987) Adv. Catal. 35, 139–186

D. Goetsch and L.D. Schmidt (1996) Science 271, 1560–1562

T. Kirchner and G. Eigenberger (1996), Chem. Eng. Sci. 51, 2409–2418

B. Lewis and G. von Elbe (1981), 'Combustion and Flames and Explosions of Gases', Academic Press, New York.

K. Kollmann, J. Abthoff, and W. Zahn (1994), SAE Papers 940469.

C. Satterfield (1991), 'Heterogeneous Catalysis in Industrial Practice', McGraw–Hill, New York.

L. Schmidt, M. Huff and S. Bharadwaj (1995), Chem. Eng. Sci. 49, 3981–3994

S. Semancik and R.E. Cavicchi (1993), Appl. Surf. Sci. 70/71, 337–346

R. Srinivasan et al. (1997), AIChE J. 43, 3059–3069

G. Veser and L.D. Schmidt (1996), AIChE J. 47, 1077–1086

G. Veser, G. Friedrich, M. Freygang, and R. Zengerle, (1998) DSC–Vol. 66, ASME, 199–206

G. Veser and J. Frauhammer, Chem. Eng. Sci. (1999) in print.

R.S. Wengeng, C.J. Call, and M.K. Drost (1996), presentation at 1996 spring national meeting of AIChE.

K.D. Wise and K. Najafi (1991), Science 254, 1335–1342

P. Witt and L.D. Schmidt (1998), J. Catal., in print

M. Ziauddin, G. Veser, and L.D. Schmidt (1997), Catal. Lett. 46, 159–167

Microstructured Reactor for Consecutive Heterogeneous / Homogeneous Gas Phase Reactions

Th. Richter, W. Ehrfeld, D. Erntner, K. Gebauer, K. Golbig, O. Hausner, H. Löwe, A. Lange de Oliveira*, W. Schmidt*, F. Schüth*

Institut für Mikrotechnik Mainz GmbH, Carl-Zeiss-Str. 18-20, D-55129 Mainz, Germany, e-mail: richter@imm-mainz.de

*Max-Planck-Institut für Kohlenforschung, Kaiser-Wilhelm-Platz 1, D-45470 Mülheim / Ruhr, Germany

Abstract

Short reaction paths, small amounts of reactants as well as fast and effective changes in reaction conditions are outstanding features of microreaction systems. These characteristics are of interest especially with respect to very fast consecutive reaction sequences utilizing different thermal conditions or reactants. Here, a microreactor designed for combined heterogeneously catalyzed / homogeneous gas phase reactions is presented. Main features of the reactor are microstructured components located directly at the reaction zone. Focus of the work reported here is the feasibility of combined heterogeneous / homogeneous gas phase processes in microreaction systems by combining short distances and steep temperature gradients in order to assure fast processing of unstable reaction intermediates.

1 Introduction

Much attention in microreaction technology so far has been drawn to gas phase reactions [1-3]. Several heterogeneously catalyzed gas phase reactions have been investigated regarding the feasibility of new reaction regimes in microreactors, reaction mechanisms and industrial applicability. Within these different experimental approaches, several types of catalysts as well as different temperature regimes have been studied as briefly discussed in the following.

Industrial investigations on gas phase processes in microreactors showed the importance of effective heat transfer and isothermal reaction conditions for improved yields and selectivities using the oxidation of an unsaturated alcohol to an unsaturated aldehyde as an example [4]. More recently, investigations were expanded to high temperature gas phase reactions as well. Their feasibility in microreactors was verified recently for HCN-generation by the Andrussow process [5].

Basic investigations on the influence of uniform flow conditions and equal distribution of the catalytic components in microchannels indicate the formation of thermodynamically unstable products at increased yields [6].

The feasibility of non-equilibrium chemistry in microreactors was demonstrated for the catalytic partial oxidation of methane [7] using packed Rh / SiO_2 catalyst powders. Latest results confirmed the applicability and advantages of microstructured rhodium catalysts for this process as well [8]. The importance of temperature control and feedback as well as its influence on yields and selectivities were analyzed in silicon based membrane microreactors using the catalytic oxidation of ammonia over platinum [9].

Another important issue of the use of gas phase microreactors is their inherent safety in connection with the performance of processes within the explosive regime of the reactants. This, for example, was shown for the controlled oxidation of hydrogen / oxygen mixtures [10] as well as for the synthesis of ethylene oxide [11, 12].

As can be seen easily, a main advantage of microreaction technology for heterogeneously catalyzed gas phase reactions is given by the effective control of temperature profiles and residence times. Especially this should be the case for consecutive heterogeneous / homogeneous processes and their special requirements regarding reaction conditions in order to prevent radical species from decay or undesired side reactions, e.g. by combination, chain transfer or disproportionation. Moreover, fast and efficient mixing processes are crucial for immediate changes in reactant composition.

Major aim of the work reported here is to show the experimental feasibility of microreaction systems with respect to possible influences on reactivity of intermediates by fast changes in reaction conditions. For this purpose, a special microstructured reaction system was developed. Additionally, this work is focused on mechanistic investigations with respect to the direct oxidation route of propylene to propylene oxide which has not been established as an industrial process so far. However, this work still is in progress and will be reported on in the near future.

2 Proposed Mechanism

The example chosen here, the direct oxidation of propylene to propylene oxide, usually is achieved by wet chemical processes, e.g. the chlorohydrine route or the oxidation of organic peroxides. For environmental reasons, the direct oxidation process using oxygen would present a favorable process route but could not be established yet [13]. Main obstacles can be seen in significant changes in temperature along the reaction pathway when using conventional equipment, thus leading to notable formation of side products not easy to remove from the reactant mixture. Furthermore, no suitable catalysts for the direct oxidation route, providing low amount of side products and favorable reaction conditions, have been identified so far.

Reactor design and realization especially in microreactors strongly depend on the proposed mechanism. Here, a consecutive heterogeneous / homogeneous mechanism for the direct oxidation of propylene to propylene oxide was taken as a basis [14-16]. Its intermediate steps as well as the appropriate thermal conditions are given in figure 1.

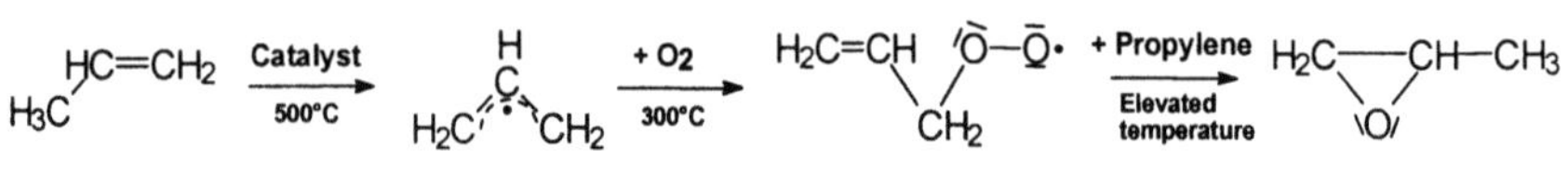

Fig. 1: Proposed mechanism for the formation of propylene oxide [14-16].

As a first step, activation of propylene is achieved by catalytic hydrogen abstraction in a temperature regime up to 500°C to yield allyl radicals. To prevent decomposition of these highly reactive radical species, fast mixing with oxygen is crucial in order to yield allylperoxy radicals within a subsequent homogeneous reaction step. Additionally, fast and efficient cooling to a temperature level of about 300°C is needed for increased stability of the allylperoxy radicals. These are assumed to work as the epoxidizing species to yield propylene oxide when reacting with excess propylene to propylene oxide at elevated temperatures.

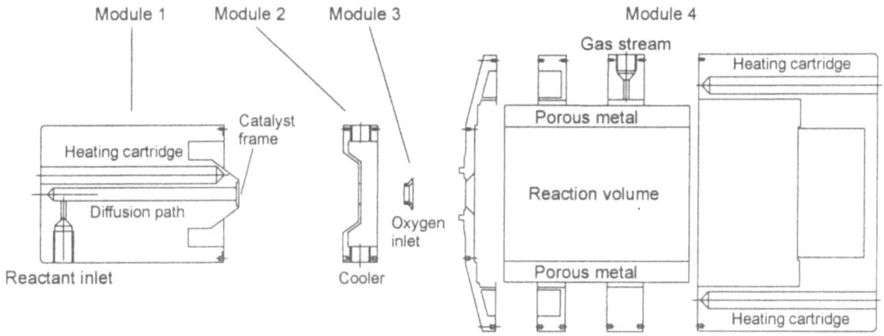

Fig. 2: Schematic view of the modular structure of the microstructured reactor for consecutive heterogeneous / homogeneous gas phase reactions.

3 Reactor Concept and Realization

The reactor consists of several independent modules, each designed specifically to ensure appropriate reaction conditions for each of the above mentioned reaction steps. This modular concept, shown in figure 2, allows an easy exchange or modification of reactor parts in order to change specific reaction conditions for research purposes. The reactor modules and their specific tasks are briefly described in the following.

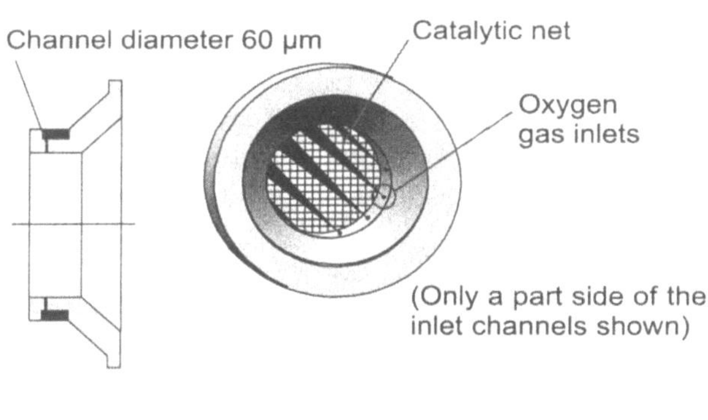

Fig. 3: Schematic view of the oxygen inlet showing the catalytic net and the oxygen inlets.

Inlet module (module 1)

Within this module the inlet stream (e.g. propylene diluted in nitrogen) is heated up to 500°C and mixed by diffusion. Heating is performed by three heating cartridges, located around the axis of the mixing channel in a triangular arrangement in order to ensure equal temperature profiles along the flow axis.

Catalyst support (module 2)

The catalyst support basically consists of a cylindrical tube. The catalyst carrier, in this case a net coated with the catalytic material, is placed inside the tube.

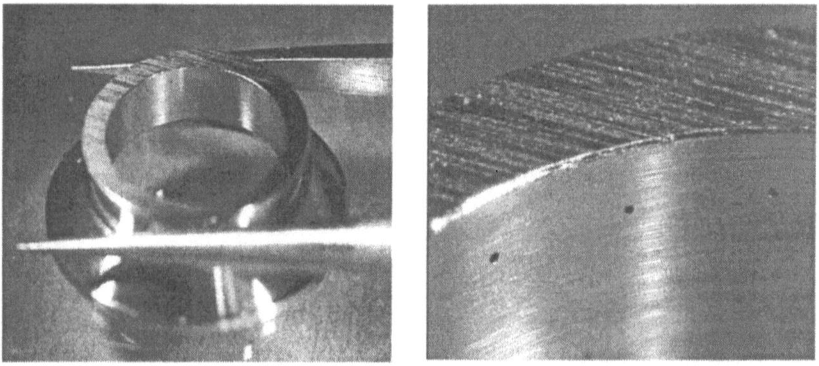

Fig. 4: Oxygen inlet. The small inlet orifices with a diameter of 60 μm are fabricated by μ-EDM drilling.

Oxygen inlet (module 3)

The systems main component features a microstructured oxygen inlet unit for the homogeneous gas phase process. Here, the reaction gas stream, already containing a certain amount of heterogeneously formed radicals, is guided through an "oxygen curtain" as indicated in figure 3. This stream of oxygen, essentially a thin sheet which is introduced through 60 μm wide inlet orifices within a changeable ring structure (see figure 4), assures mixing with the reactant stream as well as fast cooling down below 300°C in order to enhance stability of the generated allyl-peroxy radicals. Thus, within a distance of 500 μm between the catalytic surface and the oxygen inlet a residence time of less than two milliseconds can be achieved at a volume flow of 2.1 l / h at ambient conditions.

Homogeneous reaction volume (module 4)

The homogeneous reaction takes place within the so-called postcatalytic volume directly following the microstructured reactor parts. It is designed to prevent the radical species from decomposition by wall collision. For this purpose, a cartridge consisting of porous metallic material is inserted. A circular gas stream is directed from outside the cartridge towards the reactant stream in its inner core in order to prevent wall collision to a certain amount. Calculations showed that using a minimum diameter of 10 mm for the cartridge, a radical concentration of 50 % can still be expected within the reactant stream at the end of the postcatalytic volume. Therefore, nearly complete consumption and, consequently, enhanced conversion rates of the generated radical species can be assumed.

Fig. 5: Completely mounted reactor. The microstructured components are located in between the reactant inlets (located in front) and the postcatalytic volume.

4 Conclusions

The microstructured reactor described here is designed specifically for the investigation of combined heterogeneous / homogeneous reaction sequences employing radical species. Its modular construction serves research purposes and mechanistic investigations with respect to the direct oxidation of propylene to propylene oxide. The systems main component, a microstructured oxygen inlet unit for the homogeneous process step, is proposed to strongly influence the reaction process and to enhance performance by assuring a steep temperature gradient as well as fast reactant mixing within milliseconds. However, experimental characterization is still in progress and will be completed in the near future.

Acknowledgements

Funding by the German Federal Ministry of Education and Research within the scientific research project "Grundlegende Untersuchungen zur Entwicklung, Realisierbarkeit und Charakterisierung von Mikroreaktoren für industrierelevante Reaktionen in der Gasphase (Mikreak)" is gratefully acknowledged. Additionally, the authors would like to thank Mr. Andrej Wolf at IMM for valuable support in the field of μ-EDM microstructuring techniques.

References

[1] Ehrfeld, W.; Hessel, V.; Haverkamp, V.: *Microreactors*, In: Ullmann's Encyclopedia of Industrial Chemistry, Sixth Ed., 1999 Electronic Release, Wiley / VCH

[2] Ehrfeld, W.; Golbig, K.; Hessel, V.; Konrad, R.; Löwe, H.; Richter, Th.: *Fabrication of components and systems for chemical and biological microreactors*, In: Microreaction Technology, Proceedings of the First International Conference on Microreaction Technology, Ed.: W. Ehrfeld, Springer, Berlin (1997) 72 - 90

[3] Löwe, H.; Ehrfeld, W.; Gebauer, K.; Golbig, K.; Hausner, O.; Haverkamp, V.; Hessel, V.; Richter, Th.: *Microreactor concepts for heterogeneous gas phase reactions*, In: Process Miniaturization: 2nd International Conference on Microreaction Technology, March 9-12 1998, New Orleans, LA, Topical Conference Preprints, The American Institute of Chemical Engineers (1998) 63 - 74

[4] Wörz, O.; Jäckel, K.P.; Richter, Th.; Wolf, A.: *Microreactors, a new efficient tool for optimum reactor design*, In: Process Miniaturization: 2nd International Conference on Microreaction Technology, March 9-12 1998, New Orleans, LA, Topical Conference Preprints, The American Institute of Chemical Engineers (1998) 183 - 185

[5] Ehrfeld, W.; Golbig, K.; Hessel, V.; Hofmann, Ch.; Löwe, H.; Storz, M.; Breysse, J.; Wörz, O.: *High temperature HCN generation in a complex integrated microreaction system*, Published in this volume

[6] Wießmeier, G.; Hönicke, D.: *Strategy for the development of micro channel reactors for heterogeneously catalyzed reactions*, In: Process Miniaturization: 2nd International Conference on Microreaction Technology, March 9-12 1998, New Orleans, LA, Topical Conference Preprints, The American Institute of Chemical Engineers (1998) 24 - 32

[7] Tonkovich, A.L.Y.; Zilka, J.L.; Powell, M.R.; Call, C.J.: *The catalytic partial oxidation of methane in a microchannel chemical reactor*, In: Process Miniaturization: 2nd International Conference on Microreaction Technology, March 9-12 1998, New Orleans, LA, Topical Conference Preprints, The American Institute of Chemical Engineers (1998) 45 - 53

[8] Mayer, J.; Fichtner, M.; Schubert, K.; Wolf, D.: *A microstructured reactor for the catalytic partial oxidation of methane to syngas*, Published in this volume

[9] Franz, A.J.; Quiram, D.; Srinivasan, R.; Hsing, I-M.; Firebaugh, S.L.; Jensen, K.F.; Schmidt, M.A.: *New operating regimes and applications feasible with microreactors*, In: Process Miniaturization: 2nd International Conference on Microreaction Technology, March 9-12 1998, New Orleans, LA, Topical Conference Preprints, The American Institute of Chemical Engineers (1998) 33 - 38

[10] Hagendorf, U.; Janicke, M.; Schüth, F.; Schubert, K.; Fichtner, M.: *A Pt / Al_2O_3 coated microstructured reactor / heat exchanger for the controlled H_2 / O_2-reaction in the explosion regime*, In: Process Miniaturization: 2nd International Conference on Microreaction Technology, March 9-12 1998, New Orleans, LA, Topical Conference Preprints, The American Institute of Chemical Engineers (1998) 81 - 87

[11] Kestenbaum, H.; Lange de Oliveira, A.; Schmidt, W.; Schüth, F.; Ehrfeld, W.; Gebauer, K.; Löwe, H.; Richter, Th.: *Synthesis of ethylene oxide in a microreaction system*, Published in this volume

[12] Richter, Th.; Ehrfeld, W.; Gebauer, K.; Golbig, K.; Hessel, V.; Löwe, H.; Wolf, A.: *Metallic microreactors: Components and integrated systems*, In: Process Miniaturization: 2nd International Conference on Microreaction Technology, March 9-12 1998, New Orleans, LA, Topical Conference Preprints, The American Institute of Chemical Engineers (1998) 146 - 151

[13] Kahlich, D.; Wiechern, U.; Lindner, J.: *Propylene oxide*, In: Ullmann's Encyclopedia of Industrial Chemistry, A22 (1993) 239 - 258

[14] Daniel, C.; Keulks, G.W.: *The catalytic oxidation of propylene: I. Evidence for surface initiated homogeneous reactions*, J. Cat. 24 (1971) 529 - 535

[15] Daniel, C.; Monnier, J.R.; Keulks, G.W.: *The catalytic oxidation of propylene: III. Additional evidence for surface initiated homogeneous reactions*, J. Cat. 31 (1973) 360 - 368

[16] Keulks, G.W.; Krenzke, L.D.; Notermann, T.M.: *Selective oxidation of propylene*, Adv. Cat. 27 (1978) 183 - 225

MIX
Papier aus verantwortungsvollen Quellen
Paper from responsible sources
FSC® C105338

If you have any concerns about our products,
you can contact us on
ProductSafety@springernature.com

In case Publisher is established outside the EU,
the EU authorized representative is:
Springer Nature Customer Service Center GmbH
Europaplatz 3, 69115 Heidelberg, Germany

Printed by Libri Plureos GmbH
in Hamburg, Germany